Progress in Botany/Fortschritte der Botanik 54

# 54 Progress in Botany

Structural Botany Physiology
Genetics Taxonomy Geobotany

Fortschritte der Botanik
Struktur Physiologie Genetik
Systematik Geobotanik

Editors
H.-Dietmar Behnke, Heidelberg
Ulrich Lüttge, Darmstadt
Karl Esser, Bochum
Joachim W. Kadereit, Mainz
Michael Runge, Göttingen

Springer-Verlag
Berlin Heidelberg New York
London Paris Tokyo Hong Kong
Barcelona Budapest

With 56 Figures

ISBN 3-540-56358-X Springer-Verlag Berlin Heidelberg New York
ISBN 0-387-56358-X Springer-Verlag New York Berlin Heidelberg


Printed in Germany

Library of Congress Catalog Card Number 33-15850



Typesetting: M. Masson-Scheurer, 6654 Kirkel 2
31/3145-5 4 3 2 1 0 – Printed on acid-free paper

# Contents

## Review



## A. Structural Botany

## C. Genetics

## D. Taxonomy

## E. Geobotany

# List of Editors

## Section A:

Professor Dr. H.-D. BEHNKE, Zellenlehre, Universität Heidelberg, Im Neuenheimer Feld 230, W-6900 Heidelberg, FRG

## Section B:

Professor Dr. U. LÜTTGE, TH Darmstadt, Fakultät für Botanik, FB Biologie (10), Schnittspahnstraße 3–5, W-6100 Darmstadt, FRG

## Section C:

Professor Dr. h.c. K. ESSER, Lehrstuhl für Allgemeine Botanik, Ruhr Universität, Postfach 10 21 48, W-4630 Bochum 1, FRG

## Section D:

Professor Dr. J. W. KADEREIT, Institut für Spezielle Botanik und Botanischer Garten, Universität Mainz, Saarstraße 21, W-6500 Mainz, FRG

## Section E:

Professor Dr. M. RUNGE, Lehrstuhl für Geobotanik, Systematisch-Geobotanisches Institut der Universität, Untere Karspüle 2, W-3400 Göttingen, FRG

## Section F:

Professor Dr. J. W.KADEREIT, Institut für Spezielle Botanik und Botanischer Garten, Universität Mainz, Saarstraße 21, W-6500 Mainz, FRG

Professor Dr. M. RUNGE, Lehrstuhl für Geobotanik, Systematisch-Geobotanisches Institut der Universität, Untere Karspüle 2, W-3400 Göttingen, FRG

Professor Dr. Otto Kandler was born in Deggendorf, a small town in Bavaria, Germany, on October 23, 1920.

From 1946 to 1949 he was studying Biology and Chemistry at the University of Munich, where he also received his doctor's degree in 1949.

From then on he was teaching as Assistant Professor at the Botanical Institute until 1960. Within this period (in 1956/57) he worked as guest professor at Brookhaven Laboratories thanks to a Rockefeller Fellowship and at Radiation Laboratories in Berkeley.

From 1960 to 1968 he was Full Professor of Applied Botany at the Technical University of Munich, and from 1968 to 1986 Full Professor of Botany at the University of Munich, where he, since 1986, is Professor Emeritus.

Otto Kandler took up a variety of additional activities and responsibilities. They include, for instance, research in the field of dairy bacteriology, at the time when he was Head of the Institute of Dairy Bacteriology at the Dairy Research Station of the Technical University of Munich from 1957 until 1966, the function of Dean of the Faculty of Science and the membership in the Senate of the Technical University of Munich in 1964/65, the membership in the Senate and the Executive Board of the Deutsche Forschungsgemeinschaft (1970–76) and the Chief Editorship of various journals, such as "Zeitschrift für Pflanzenphysiologie" (1963–68) and "Systematic and Applied Microbiology" (1980–1992).

In 1972 he was elected as member of the Deutsche Akademie der Naturforscher LEOPOLDINA, and of the Bavarian Academy of Sciences in 1985. In 1989 he received the Ferdinand Cohn Medal for his Merits in Microbiology and in 1992 he was awarded the Bundesverdienstkreuz 1. Klasse (First Class Order of Merit of the Federal Republic of Germany). Otto Kandler is honorary member of several botanical and microbiological societies and received an honorary doctor's degree from the Technical University of Munich and the University of Gent (Belgium).

His work includes approximately 380 scientific publications.

**Review**

# Archaea Archaebacteria

By O. Kandler

## 1. Introduction

The recognition and establishment of a new distinctive phylogenetically coherent group of organisms, the archaebacteria, have changed our perception of the phylogenetic structure of the living world. This review will deal largely with the "discovery" of archaebacteria, their establishment as the domain Archaea, a newly defined taxon of highest rank, the description of their characteristics, and the biotechnological and evolutionary implications of archaeal research.

## 2. First Recognition and Establishment of Archaebacteria as the Domain Archaea

Following Chatton's (1937) seminal realization that all living beings fall into one of two cytologically defined classes, the prokaryotes and the eukaryotes, this so-called prokaryotic-eukaryotic dichotomy was generally regarded also as a phylogenetic distinction. On this basis, all extant life was commonly divided into the kingdoms Eukaryota and Prokaryota. However, recent phylogenetic analyses based on sequence comparison of various semantophoretic molecules (sensu Zuckerkandl and Pauling 1965) of numerous organisms have revealed deep phylogenetic clefts, not only between the eukaryotes and the prokaryotes, but also within the prokaryotes (Woese et al. 1976; Balch et al. 1977). Therefore, Woese and Fox (1977) proposed: "... that living systems represent one of three aboriginal lines of descent: (i) the eubacteria, comprising all typical bacteria; (ii) the archaebacteria, containing methanogenic bacteria; and (iii) the urkaryotes, now represented in the cytoplasmic component of eukaryotic-cells." Continued phylogenetic analyses based on 16/18S rRNA oligonucleotide catalogues have revealed that the group originally designated archaebacteria also encompasses the extreme halophiles and various thermoacidophilic microorganisms in addition to methanogens. They all exhibit unique biochemical and molecular traits. Thus, archaebacteria have been redefined by the following characteristics (Woese et al. 1978): "(i) the presence of characteristic tRNAs and ribosomal RNAs; (ii) the absence of peptidoglycan cell walls, and, in many cases, replacement by a largely proteinaceous coat; (iii) the occurrence of ether lipids built from pythanyl chains; and (iiii) in all cases known so far, their occurrence only in unusual habitats."

Progress in Botany, Vol. 54

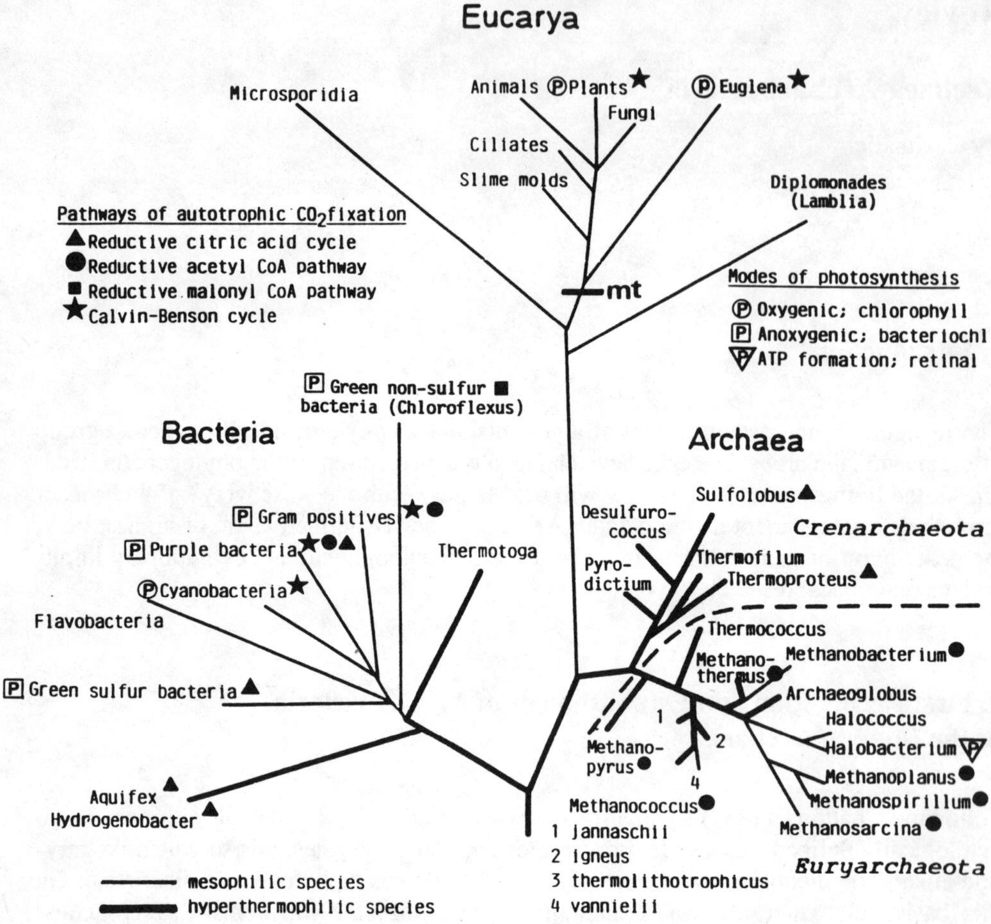


**Fig. 1.** Global phylogenetic tree of organisms and distribution of autotrophic pathways (after Woese et al. 1990; modified by Stetter 1992 and Kandler 1993). Branching order and branch lengths are based upon 16/18S rRNA sequence comparisons. The position of the root was determined by comparing known sequences of paralogous genes (ATPases; protein elongation factor) that diverged from each other before the three primary lineages emerged from their ancestral condition. (Gogarten et al. 1989; Iwabe et al. 1989)

When, thanks to improved techniques, a great number of completely analyzed RNA sequences became available, phylogenetic distances could be determined more precisely. The comparison of such complete 16/18S rRNA sequences confirmed the conclusion drawn from 16/18S RNA catalogues that the phylogenetic distance between the eukaryotic group and each of the two prokaryotic groups is of the same order of magnitude (Woese 1987). The large phylogenetic distances that separate the three groups from each other are also reflected in the degree of phenotypic differences between them, when molecular and biochemical characteristics are considered. In order to bring formal taxonomy into line with the phylogenetic tree (Fig. 1) emerging from sequence comparisons of ribosomal 16/18S RNA, a new global system of organisms has been proposed in which a new taxon, called a "domain" (Woese et al. 1990), exists above the level of kingdom.

Life would then comprise three domains, the Bacteria, the Archaea, and the Eucarya, each containing several kingdoms. Some of the traditional eukaryotic kingdoms, for instance those of animals and plants, fit the phylogenetic structure of the Eucarya and will thus be retained. Others, for instance those of fungi and protists, have to be redefined or newly created. The two prokaryotic domains require a completely new taxonomic structure at the level of kingdoms and classes. So far, only the Archaea have been formally subdivided into the kingdoms Crenarchaeota and Euryarchaeota, thus reflecting the obvious bifurcation of the archaeal subtree (Woese et al. 1990).

Presently, the concept of a tripartite world of life, originally derived from a relatively few 16/18S rRNA sequences and the unique biochemical traits of only one archaeal phenotype, the methanogens (Woese and Fox 1977), is well established. In addition to the numerous newly discovered representatives of ecologically and biochemically diverse phenotypes of Archaea, the tripartition concept rests today on signature analysis of a large collection of 16/18S rRNA sequences from more than 500 species of all major groups of prokaryotic and eukaryotic organisms (Winker and Woese 1991), and has been corroborated by sequence comparison of additional but different semantophoretic molecules, for instance 23S rRNA (Schleifer and Ludwig 1989), protein elongation factors (Iwabe et al. 1989; Schleifer and Ludwig 1989; Auer et al. 1991; Miyata et al. 1991) ATPases (Gogarten et al. 1989; Iwabe et al. 1989; Schleifer and Ludwig 1989; Miyata et al. 1991; Gogarten and Taiz 1992), and ribosomal proteins (Auer et al. 1990). The new concept is also supported by the distribution of a large number of important molecular, biochemical, and physiological features among the three groupings (examples are given in Table 3).

For retrospective reading on archaebacteria at the discovery stage, the proceedings of the first "International Workshop on Archaebacteria" held in Munich 27 June–1 July 1981 (Kandler 1982a)[1] are recommended, while the state of knowledge during the consolidation stage is reflected by the proceedings of the two complementary workshops on "Molecular Genetics of Archaebacteria", and on "Biology and Biochemistry of Archaebacteria", both held in Munich 1985 (Kandler and Zillig 1986)[1], and by the comprehensive overview on the various aspects of Archaea given in Vol. VIII "Archaebacteria" of the series, *The Bacteria* (Woese and Wolfe 1985). More recent developments have been discussed at the symposium on "Molecular Biology of Archaebacteria" held in Victoria, British Columbia, 31 July–5 August 1988 (lectures published 1989 in Can. J. Microbiol. 35:1–244). Current ideas on biochemical and biotechnological aspects have been presented at a symposium of The Biochemical Society on "Biochemistry and Biotechnology of Archaebacteria", Edinburgh, 4–5, September 1991 (Danson et al. 1992). The role of Archaea in the origin and early diversification of life has been discussed at the 3rd Recontres de Blois/France "Frontiers of Life" held in Blois/France June 1991 (Thran 1993),

---

[1] In addition to the proceedings cited above, which contain both the lectures and the poster abstracts, the full papers given at these workshops held in 1981 and 1985 are also published in Zentralbl. Bakteriol. Hyg. I, Abt. Orig. C 3:1–345 (1982), and in Syst. Appl. Microbiol. 7:1–397 (1986), respectively.

and at the Nobel symposium "Early Life on Earth" held in Karlskoga/Sweden May 1992 (Bengtson 1993).

## 3. Taxonomy and Ecology of Archaea

The present taxonomy of archaea is based on the branching order and phylogenetic distances delineated from molecular chronometers, mainly 16S rRNA, and on phenetic (chemical, physiological, etc.) traits. Because of an almost exponential increase in newly discovered archaea, especially hyperthermophiles growing optimally above 80 °C (Fig. 2) that exhibit unique molecular and physiological traits, the taxonomy of archaea is still in a development stage. At present, the two kingdoms, Crenarchaeota and Euryarchaeota, which represent the two primary branches of the domain Archaea, comprise about 100 species. They are affiliated to 40 genera and grouped into 3 crenarchaeal and 7 euryarchaeal orders (Table 1). The species described before 1988 are listed in *Bergey's Manual of Systematic Bacteriology* (König and Stetter 1989), whereas the more recently described genera and species are compiled in Table 2. The rank "family" is not considered in Tables 1 and 2 since, in many cases, the affiliation of genera to families is still elusive.

Physiologically and ecologically, the kingdom Crenarchaeota is a relatively homogeneous group and comprises exclusively hyperthermophilic sulfur-dependent Archaea which derive energy either from sulfur reduction (Thermoproteales, Pyrodictiales) or sulfur oxidation (Sulfolobales). The Crenarchaeotes are confined to submarine or terrestrial volcanic niches and may thus be considered to represent an archaic ancestral ecotype, an assumption also expressed by their name (κρηνη: spring, fount; αρχαιος: ancient,

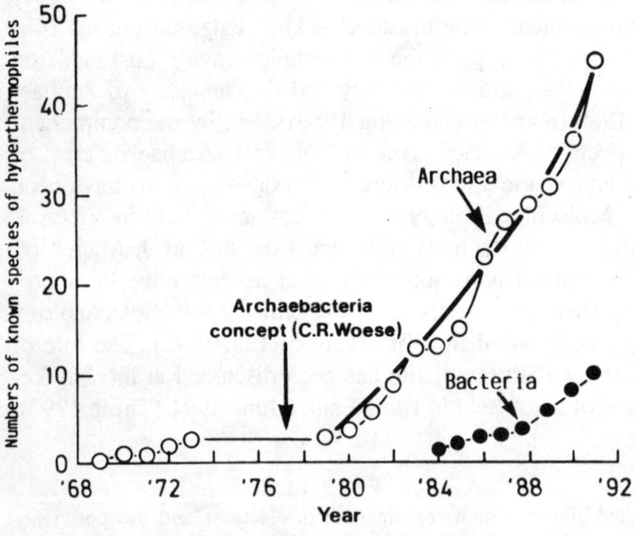


**Fig. 2.** Increase in the total number of known hyperthermophilic species after Woese's publication of the archaebacteria concept (Woese and Fox 1977)

**Table 1.** Affiliation of genera to the orders of the domain Archaea

| Kingdom | Order | Genus | Number of strains |
|---|---|---|---|
| Crenarchaeota | Thermoproteales | *Thermoproteus* | *3* |
| | | *Pyrobaculum* | *2* |
| | | *Thermophilum* | *2* |
| | | *Desulfurococcus* | *4* |
| | | *Staphylothermus* | *1* |
| | "Pyrodictiales" | *Pyrodictium* | *3* |
| | | *Hyperthermus* | *1* |
| | | *Thermodiscus* | *1* |
| | Sulfolobales | *Sulfolobus* | *5* |
| | | *Acidianus* | *2* |
| | | *Metallosphaera* | *1* |
| | | *Desulfurolobus* | *1* |
| | | *Styigolobus* | *1* |
| Euryarchaeota | Thermococcales | *Thermococcus* | *5* |
| | | *Pyrococcus* | *3* |
| | "Archaeoglobales" | *Archaeoglobus* | *3* |
| | Thermoplasmales | *Thermoplasma* | *2* |
| | Methanococcales | *Methanococcus* | *9* |
| | Methanobacteriales | *Methanobacterium* | *18* |
| | | *Methanobrevibacter* | *3* |
| | | *Methanothermus* | *2* |
| | Methanomicrobiales | *Methanomicrobium* | *2* |
| | | *Methanococcoides* | *2* |
| | | *Methanogenium* | *9* |
| | | *Methanocorpusculum* | *4* |
| | | *Methanohalobium* | *1* |
| | | *Methanohalophilus* | *3* |
| | | *Methanolobus* | *2* |
| | | *Methanolacina* | *1* |
| | | *Methanoplanus* | *2* |
| | | *Methanosaeta* (formerly *Methanothrix*) | *2* |
| | | *Methanosarcina* | *6* |
| | | *Methanosphaera* | *2* |
| | | *Methanospirillum* | *1* |
| | Halobacteriales | *Halobacterium* | *12* |
| | | *Haloarcula* | *5* |
| | | *Haloferax* | *3* |
| | | *Halobacterioides* | *1* |
| | | *Halococcus* | *4* |
| | | *Natronococcus* | *1* |
| | | *Natronobacterium* | *4* |

**Table 2.** Recently described new taxa of Archaea, not listed in *Bergey's Manual of Systematic Bacteriology* (1989)

| Kingdom | Order | Newly described genera and species | References |
|---|---|---|---|
| *Crenarchaeota* | Sulfolobales | *Sulfolobus metallicus* sp. nov. | Huber G. et al. (1991) Syst. Appl. Microbiol. 14:372–378 |
| | | *Sulfolobus shibatae* sp. nov. | Grogan D. et al. (1990) Arch. Microbiol. 154:594–599 |
| | | *Hyperthermus* gen. nov. | Zillig W. et al. (1991) Int. J. System. Bacteriol. 41:169–170 |
| | | *Hyperthermus butylicus* sp. nov. | Zillig W. et al. (1991) Int. J. System.Bacteriol.41:169–170 |
| | | *Stygiolobus* gen. nov. | Segerer A. H. et al. (1991) Int. J. System. Bacteriol. 41:495–501 |
| | | *Stygiolobus azoricus* sp. nov. | Segerer A. H. et al. (1991) Int. J. System. Bacteriol. 41:495–501 |
| | | *Metallosphaera* gen. nov. | Huber G. et al. (1989) Syst. Appl. Microbiol. 12:38–47 |
| | | *Metallospaera sedula* sp. nov. | Huber G. et al. (1989) Syst. Appl. Microbiol. 12:38–47 |
| | Pyrodictiales | *Pyrodictium abyssi* sp. nov. | Pley U. et al. (1991) Syst. Appl. Microbiol. 14:245–253 |
| | | *Hyperthermus* gen. nov. | Zillig W. et al. (1991) Int. J. System. Bacteriol. 41:169–170 |
| | | *Hyperthermus butylicus* sp. nov. | Zillig W. et al. (1991) Int. J. System. Bacteriol. 41:169–170 |
| | Thermoproteales | *Thermoproteus uzonensis* sp. nov. | Bonch-Osmolovskaya E. A. et al. (1990) Arch. Microbiol. 154:556–559 |
| | | *Pyrobaculum* gen. nov. | Huber R. et al. (1987) Arch. Microbiol. 149:95–101 |
| | | *Pyrobaculum islandicum* sp. nov. | Huber R. et al. (1987) Arch. Microbiol. 149:95–101 |
| | | *Pyrobaculum organotrophum* sp. nov. | Huber R. et al. (1987) Arch. Microbiol. 149:95–101 |
| *Euryarcheota* | Thermococcales | *Thermococcus litoralis* sp. nov. | Neuner A. et al. (1990) Arch. Microbiol. 153:205–207 |
| | | *Thermococcus stetteri* sp. nov. | Miroshnichenko M. L. et al. (1989) Syst. App. Microbiol. 12:257–262 |
| | Archaeoglobales | *Archaeoglobus profundus* sp. nov. | Burggraf S. et al. (1990) Syst. Appl. Microbiol. 13:24–28 |

**Table 2** (continued)

| Kingdom | Order | Newly described genera and species | References |
|---|---|---|---|
| | Thermoplasmales | *Thermoplasma volcanium* sp. nov. | Segerer A. et al. (1988) Syst. Appl. Microbiol. 10:161–171 |
| | Methano-bacteriales | *Methanobacterium espanolae* sp. nov. | Patel G. B. et al. (1990) Int. J. System. Bacteriol.40:12–18 |
| | | *Methanobacterium ivanovii* sp. nov. | Jain M. K. et al. (1987) Syst. Appl. Microbiol. 9:77–82 |
| | | *Methanobacterium palustre* sp. nov. | Zellner G. et al. (1989) Arch. Microbiol. 151:1–9 |
| | | *Methanobacterium thermophilum* sp. nov. | Laurinavichus K. S. et al. (1987) Mikrobiologiya 57:1035–1041 |
| | Methanococcales | *Methanococcus deltae* sp. nov. | Corder R. E. et al. (1983) Arch. Microbiol. 134:28–32 |
| | | *Methanococcus igneus* sp. nov. | Burggraf S. et al. (1990) Syst. Appl. Microbiol. 13:263–269 |
| | Methano-microbiales | *Methanocorpusculum* gen. nov. | Zellner G. et al. (1987) Arch. Microbiol. 147:13–20 |
| | | *Metanocorpusculum bavaricum* sp. nov. | Zellner G. et al. (1989) Arch. Microbiol. 151:381–390 |
| | | *Methanocorpusculum labreanum* sp. nov. | Zhao Y. et al. (1989) Int. J. System. Bacteriol. 39:10–13 |
| | | *Methanocorpusculum parvum* sp. nov. | Zellner G. et al. (1987) Arch. Microbiol. 147:13–20 |
| | | *Methanocorpusculum sinense* sp. nov. | Zellner G. et al. (1989) Arch. Microbiol. 151:381–390 |
| | | *Methanogenium liminatans* sp. nov | Zellner G. et al. (1990) Arch. Microbiol. 153:287–293 |
| | | *Methanogenium organophilum* sp. nov. | Widdel F. et al. (1988) Arch. Microbiol. 150:477–481 |
| | | *Methanohalobium evestigatus* sp. nov. | Zhilina T. N. et al. (1987) Dokl. Akad. Nauk SSSR 293:464–468 |
| | | *Methanohalophilus* gen.nov. | Paterek J.R. et al. (1988) Int. J. System. Bacteriol. 38:122–123 |
| | | *Methanohalophilus mahii* sp. nov. | Paterek J.R. et al. (1988) Int. J. System. Bacteriol. 38:122–123 |
| | | *Methanohalophilus oregonense* sp. nov. | Liu Y. et al. (1990) Int. J. System. Bacteriol. 40:111–116 |
| | | *Methanohalophilus zhilinae* sp. nov. | Mathrani I. M. et al. (1988) Int. J. System. Bacteriol.38:139–142 |
| | | *Methanolacinia* gen. nov. | G. Zellner et al. (1989) J. Gen. Appl. Microbiol. 35:185–202 |

**Table 2** (continued)

| Kingdom | Order | Newly described genera and species | References |
|---|---|---|---|
| | | *Methanolacinia paynteri* comb. nov. | G. Zellner et al. (1989) J. Gen. Appl. Microbiol. 35:185–202 |
| | | *Methanosaeta* gen. nov. | Patel G. B. et al. (1990) Int. J. System. Bacteriol. 40:79–82 |
| | | *Methanosaeta concilii* sp. nov. | Patel G. B. et al. (1990) Int. J. System. Bacteriol. 40:79–82 |
| | | *Methanosphaera cuniculi* sp. nov. | Biavati B. et al. (1988) Appl. Environ. Microbiol. 54:768–771 |
| | Methanopyrales | *Methanopyrus* gen. nov. | Kurr M. et al. (1991) Arch.Microbiol. 156:239–247 |
| | | *Methanopyrus kandleri* sp. nov. | Kurr M. et al. (1991) Arch.Microbiol. 156:239–247 |
| | Halobacteriales | *Haloarcula japonica* sp.nov. | Takashina T. et al. (1990) Syst. Appl. Microbiol. 13:177–181 |
| | | *Halobacterium distributus* sp. nov. | Zvyagintseva I. S. et al. (1987) Mikrobiologiya 56:839–844 |
| | | *Halobacterium lacusprofundi* sp. nov. | Franzmann P. D. et al. (1988) Syst. Appl. Microbiol. 11:20–27 |
| | | *Halococcus saccharolyticus* sp. nov. | Montero C. G. et al. (1989) Syst. Appl. Microbiol. 12:167–171 |
| | | *Halococcus turkmenicus* sp. nov. | Zvyagintseva I. S. et al. (1987) Mikrobiologiya 56:839–844 |
| | | *Natronobacterium magadii* sp. nov. | Lodwick D. et al. (1991) Syst. Appl. Microbiol. 14:352–357 |

primitive; Woese et al. 1990). On the other hand, the kingdom Euryarchaeota comprises several diverse ecotypes, partly adapted to more "modern" habitats. Thus, it shows a much wider ecological amplitude, a characteristic also expressed by its name (ευρυς: broad, wide, spacious; Woese et al. 1990). One of the two deepest branches within the euryarchaeal subtree, the hyperthermophilic, organotrophic order Thermococcales, resembles the crenarchaeal phenotype, especially that of the organotrophic representatives of the Thermoproteales.

The dominating ecotype, at least with respect to the number of taxa and the range of ecological diversity, is represented by the strictly anaerobic, chemolithotrophic methanogens. This ecotype comprises the extremely hyperthermophilic (growth up to 110 °C) Methanopyrales (Kurr et al. 1991; Rospert et al. 1991), a branch closest to the common root of the two kingdoms of Archaea (Burggraf et al. 1991); the Methanococcales, a later branch which comprises both hyperthermophilic and mesophilic species; the Methanobacteriales, comprising predominantly mesophilic genera; and the large order Methanomicrobiales, which does not contain hyperthermophilic but mainly mesophilic and only a few moderate thermophilic species (growth up to 75 °C). Methanogens represent the only archaeal phenotype with global distribution. They live in all anaerobic thermo- and mesophilic habitats (water, soil, intestine of animals) where methanogenic substrates, e.g., $CO_2$, $H_2$ or simple breakdown products of organic matter (acetate, formate, methylamines), are available.

From an evolutionary perspective, the Methanomicrobiales are the most interesting methanogenic grouping, for only within this group there are methanogens that produce methane from carbon sources other than $CO_2$, for instance from methanol, actate, various methylamines, etc., and may live in niches unusual for methanogens, for instance, in water with elevated salinity or alkalinity. Several taxa of the Methanomicrobiales also show distinctly different morphologies, ranging from regular and irregular cocci to sarcina-like clusters, rods, and trichomes. The grouping of the species of Methanomicrobiales resulting from a detailed phylogenetic analysis (Rouvière et al. 1992) does not fully agree with the present taxonomic subdivision of this order and thus calls for future taxonomic revision. Interestingly, Methanomicrobiales are a sister group of the Halobacteriales (Lodwick et al. 1991; Kjems et al. 1992; Rouvière et al. 1992), the most derived group of Euryarchaeota that does not comprise any thermophiles or anaerobes but is fully adapted to aerobic life in extremely high saline or alkaline waters.

Methanomicrobiales are also specifically related to *Archaeoglobus* (Woese et al. 1991), a marine hyperthermophilic, chemolithotrophic sulfate reducer, and show some relatedness to *Thermoplasma*, an aerobic acidophilic archaeote living in heated solfatara fields (Segerer et al 1988) and smoldering coal piles (Brock 1978).

Thus, it has been suggested that the presumably methanogenic ancestral lineage of the Methanomicrobiales may have spawned several derived non-methanogenic phenotypes, adapted to oxygen or oxygen-rich electron acceptors: the Halobacteriales, the Archaeoglobales, and the Thermoplasmales (Woese et al. 1991). The radiation within the kingdom Euryarchaeota may thus be understood as the result of a parallel evolution of two hyperthermophilic lineages: the Thermococcales, resembling the sulfur metabolizing, basically chemolithotrophic crenarchaeotes, and the strictly chemolithotrophic, hyperthermophilic, methanogenic Methanopyrales. During this evolution towards life in a "modern" world, the two lineages may have exchanged genetic information which then led to the phenetic variety within the final radiations.

## 4. Characteristics of Archaea

Of the numerous characteristics which distinguish Archaea from Bacteria and Eucarya, here emphasis is put on those which played an important role in recognizing archaebacteria and establishing these organisms as the domain Archaea (Table 3).

### a) Morphology and Cytology

Size and morphology of archaea cells are basically the same as in bacteria. The cells may be motile by flagella, they may form chains and clusters, or, in a few cases, fila-

**Table 3.** Distribution of discrete characters between the domains Bacteria and Eucarya, and the two kingdoms of the domain Archaea

| Characters | | Bacteria[a] | Archaea | | | Eucarya | References |
|---|---|---|---|---|---|---|---|
| | | | Euryarchaeota | | Crenarchaeota | | |
| | | | Methanogens | Halophiles | | | |
| Embden-Meyerhof pathway | catabolic | + | – | – | – | + | |
| | anabolic | + | + | + | ? | + | Danson (1989) |
| Entner-Doudoroff pathway | | + | Modified | Modified | Modified | – | Danson and Hough (1992) |
| Photosynthetic $CO_2$ fixation | | + | – | – | – | –[d] | |
| Chemoautotrophy based on the redox couples | $H_2/S^0$ | – | – | – | + | – | |
| | $H_2/CO_2$ | +[b] | +[c] | – | – | – | Fuchs et al. (1992) |
| | $H_2/O_2$ | + | – | – | + | – | |
| Growth at ca. 100 °C | | + | + | – | + | – | Stetter et al. (1990) |
| Acylester lipids | | + | – | – | – | + | |
| Isopranyl ether lipids | | – | + | + | + | – | Langworthy (1985) |
| Glycerolipids | dextrorotatory (L or sn- 2, 3 glycerol) | – | + | + | + | – | Kates (1992) |
| | levorotatory (D or sn- 1, 2 glycerol) | + | – | – | – | + | |
| Rigid cell wall sacculus | murein | +(–) | – | – | – | – | Kandler and König (1985) |
| | non-murein polymers | – | +(–) | –(+) | – | –(+) | Kandler and König (1993) |

| | | | | | | |
|---|---|---|---|---|---|---|
| Cytoskeleton | – | – | – | – | + | |
| Size of ribosomes | 70S | 70S | 70S | 70S | 80S | Matheson (1992) |
| RNA polymerase (large subunits) | ßß' | B'B"A'A" | B'B"A'A" | BA'A" | BA | Zillig et al. (1993a, 1993b) |
| Unique modified nucleotides in tRNA | – | + | + | + | – | McCloskey (1986) Edmonds et al. (1991) |
| ATP synthase (ATPase) | F | V | V | V | V | Kibak et al. (1992) |
| Transcription units | + | + | + | + | – | |
| Gene order in RNA operon | b | b | b | c | e | Zillig et al. (1988) Forterre et al. (1992) |
| Shine-Dalgarno sequences | + | + | + | + | – | |

[a] (), Minority of the species.
[b] Reduced end product of energy-yielding reaction: acetate.
[c] Reduced and product of energy-yielding reaction: methane.
[d] Photosynthesis by chloroplasts, derived from bacteria via endosymbiosis.

ments possessing rigid proteinaceous sheaths. Multiplication occurs by binary fission, constriction, budding, fragmentation, or by still unknown mechanisms. Illustrations of the various forms are compiled in *Bergey's Manual of Systematic Bacteriology* (König and Stetter 1989). Cytologically, archaea also resemble bacteria in possessing 70S ribosomes and circular DNA, but they lack a true nucleus and organelles surrounded by a double membrane.

### b) Cell Walls and Cell Envelopes

Archaea do not contain murein, the typical cell wall polymer of bacteria (Schleifer and Kandler 1972). Instead, some of the euryarchaeotes possess rigid cell wall sacculi consisting of different polymers such as pseudomurein (Kandler 1982b) and methanochondroitin (Kreisl and Kandler 1986) in some of the methanogens, while acid heteropolysaccarides are found in halococci (Schleifer et al. 1982) and glucosaminoglycan in basophilic cocci (Kandler and König 1993). Since these cell walls form septa during cell division and exhibit about the same width as that of murein cell walls, morphologically, they cannot be distinguished from murein cell walls in electron micrographs of thin sections of the cells. Some members of the Euryarchaeota and none of the Crenarchaeota possess rigid cell wall sacculi but cell envelopes formed by a so-called surface layer (S-layer) consisting of regular arrays of protein or glycoprotein subunits attached to the cytoplasmic membrane (Baumeister et al. 1989, 1990). The distribution of the chemically different types of cell walls and cell envelope polymers among the various lineages of Archaea (Fig. 3) show that even pairs of late branches have developed basically different cell wall polymers. The uniqueness of the chemical structure of the different polymers is also reflected in the unusual features found in the biosynthetic pathways of pseudomurein (Hartmann and König 1990), methanochondroitin (Hartmann and König 1991), and the S-layer glycoprotein of *Halobacterium* (Lechner and Wieland 1989). Each of these biosynthetic pathways involve unique steps typical of the respective taxonomic grouping of Archaea. These findings corroborate the assumption (Kandler 1982b) that bacteria radiated only after the invention of the murein sacculus by the common ancestor of bacteria, while radiation within the Archaea occurred before the invention of cell walls, or cell envelope polymers. Only after this radiation, have envelope polymers evolved independently from each other in some of the lineages. The results of the chemical and fine structural studies on cell wall polymers and S-layers have been reviewed by Kandler and König (1985, 1993).

### c) Metabolism

#### α) Central Metabolism

The pathways of central metabolism, which provide the link between the degradative and biosynthetic processes, are basically the same in the three domains, thus confirming the unity of life. However, distinct modifications are observed in archaea. For instance, the

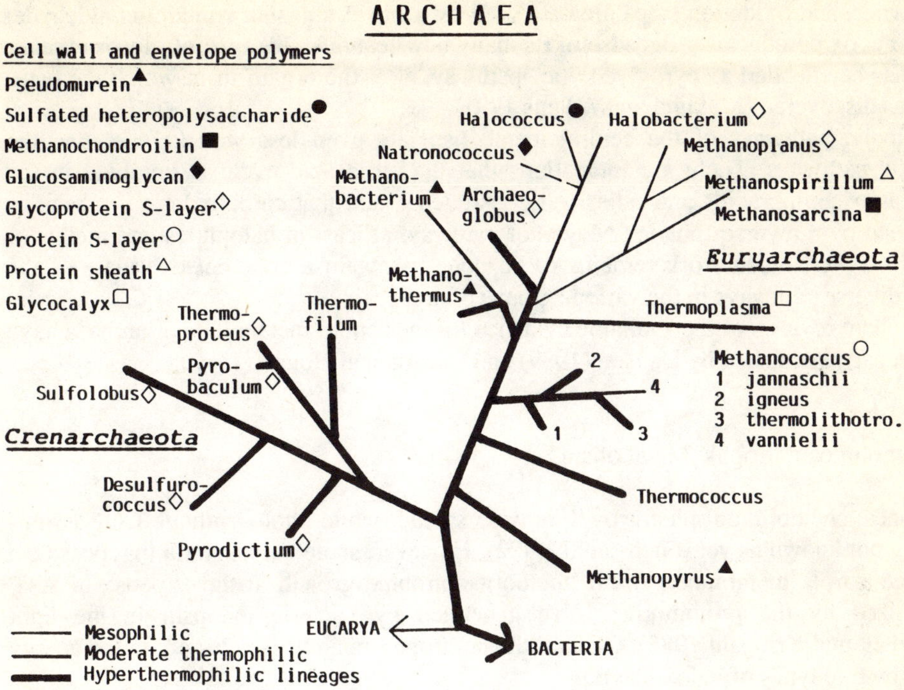


**Fig. 3.** Distribution of cell wall and cell envelope polymers among Archaea. The depicted portion of the phylogenetic tree is taken from Fig. 1

Embden-Meyerhof glycolytic pathway, common in eucaryotes and anaerobic bacteria, does not operate in the degradative (catabolic) direction in archaea because ATP phosphofructokinase, the key enzyme at the entrance of this pathway, is missing. However, all other enzymes are present. Thus, depending on the group of archaea, sugar breakdown is initiated by one of the various versions of the "phosphorylated" or "non-phosphorylated" Entner-Doudoroff pathways which start with the oxidation of glucose-6-phosphate or glucose and lead to pyruvate and glyceraldehyde or glyceraldehyde phosphate. These intermediates are decarboxylated to acetyl-CoA which is either converted to acetate and ATP by an ADP-dependent acetyl-CoA synthase, as in the case of glucose fermentation in *Pyrococcus furiosus* (Mukund and Adams 1991; Schäfer and Schönheit 1992) where acetate is an end product of fermentation; or acetyl-CoA is oxidized to $CO_2$ by a complete or partial citric acid cycle with $O_2$ (halophiles) or elemental sulfur (most heterotrophic Crenarchaeota) serving as electron acceptor. Evidence for the oxidation of acetate in *Thermoproteus tenax* and *Pyrobaculum islandicum* via the citric acid cycle with sulfur or thiosulfate as electron acceptor has been reported (Schäfer et al. 1992).

Knowledge of enzymes and cofactors involved in the various versions of catabolic pathways in the different groups of archaea is still very fragmentary. Only the fermentation of glucose to $CO_2$, acetate, and $H_2$ by the hyperthermophilic *Pyrococcus furiosus* has been fully elucidated (Mukund and Adams 1991: Schäfer and Schönheit 1992). Here, ferredoxin instead of the common pyridine nucleotides (NAD or NADP) serves as electron carrier, and oxidation steps are catalyzed by a novel tungsten containing aldehyde: ferredoxin oxidoreductase. Because of its many new features, this type of glucose breakdown has been called a "pyroglycolytic" pathway, after the organism in which the pathway was discovered (Mukund and Adams 1991).

Anabolic pathways of the central metabolism are even less well known than the catabolic pathways. There are indications that the citric acid cycle is involved in the synthesis of amino acids and other cell constituents, and that carbohydrate synthesis is performed by a reverse Embden-Meyerhof pathway, at least in halophiles and methanogens. However, much work remains to be done to obtain a coherent overview of the highly diverse pathways in the various groups of archaea.

Excellent reviews of the available evidence for the central metabolism of archaea have been recently published by Danson (1989) and Danson and Hough (1992).

### β) Chemolithoautotrophic Metabolism

In archaea, chemolithotrophic growth is wide spread, while photosynthetic $CO_2$ assimilation is not known as yet (Fuchs et al. 1992). However, some halobacteria that possess a so-called purple membrane exhibit photoorganotrophic growth at the expense of ATP synthesized by the protomotive force generated by bacteriorhodopsin in the light (Oesterhelt and Krippahl 1983). Chemolithoautotrophy in archaea is based on the energy supply by five types of redox reactions.

1. The reduction of $CO_2$ or $C_1$ compounds with $H_2$ to methane or the disproportionation of acetate to $CO_2$ and methane.
2. Reduction of elemental sulfur or thiosulfate with $H_2$ to $H_2S$.
3. Reduction of sulfate or other oxosulfur compounds with $H_2$ to $H_2S$.
4. Oxidation of elemental sulfur or $H_2S$ with $O_2$ to $SO_4^{2-}$.
5. Oxidation of $H_2$ with $O_2$ to $H_2O$.

Methane formation is restricted to the four orders of methanogens within the Euryarchaeota. $CO_2$ fixation proceeds via the acetyl-CoA pathway (Wolfe 1992) which involves several unique co factors occurring only in methanogens (for details, see review DiMarco et al. 1990). Although chemolithoautotrophy based on methane formation is not found in any organism outside the Euryarchaeota, the formation of acetate from $CO_2$ and $H_2$ as a final product (Wood 1990), observed in a small group of acetogenic chemolithoautotrophic bacteria, has much in common with the pathway of methane formation. In the near future, alignments of sequences of the respective enzymes of both groups of organisms will show whether the two pathways share a common phylogenetic root.

Chemolithoautotrophy based on energy obtained from the reduction of elemental sulfur with $H_2$ is only observed in the orders Thermoproteales and Pyrodictiales. Here, $CO_2$ assimilation proceeds not via the Calvin-Benson cycle, but most likely via a reductive citric acid cycle (Strauss et al. 1992). Also, the sulfur-respiring organotrophic Gram-negative bacterium *Wolinella succinogenes* is able to gain energy from the reduction of elemental sulfur with $H_2$ (Fuchs et al. 1992); however, this organism requires acetate for growth. At present, it is not known whether the mechanisms and enzymes involved in the energy conversion during sulfur reduction in archaea and bacteria are phylogenetically related.

Sulfate reduction with $H_2$ and oxidation of elemental sulfur with $O_2$ are used as energy sources by several groups of bacteria and archaea. Among archaea, sulfate reduction and oxidation of sulfur are used for chemolithoautotrophic growth by *Archaeoglobus* and by members of the Sulfolobales, respectively. Isotope studies indicate that *Sulfolobus* assimilates $CO_2$ via a reductive citric acid cycle (Kandler and Stetter 1981). However, no studies on the path of carbon in $CO_2$ assimilation by *Archaeoglobus* are available so far. Until recently, only bacteria have been known to grow chemolithoautotrophically at the expense of $H_2$ oxidation with $O_2$. However, recently, Stetter (1993) reported that members of the Sulfolobales and some newly isolated hyperthermophilic, microaerophilic strains, which may belong to the order Pyrodictiales, are also able to utilize this energy source.

Much work remains to be done to elucidate the various chemolithoautotrophic pathways in archaea to the same extent as in methanogens.

### d) Lipids

The nature of membrane lipids has been one of the early distinctions between Bacteria and Eucarya, on the one hand, and Archaea on the other (Langworthy 1985). Whereas Eucarya and Bacteria contain straight-chain fatty acids and levorotatory fatty acid ester glycerol lipids exhibiting the D- or sn-1,2-glycerol configuration, Archaea contain isoprenoid and hydroisoprenoid hydrocarbons and dextrorotatory isopranyl glycero ether lipids exhibiting the L- or sn-2-3-glycerol configuration. Acyl-ester lipids are missing. In Archaea, the main structures are $C_{20}$–$C_{20}$ diacyl glycerol diether, mainly sn-2,3-diphytanylglycerol and its dimer, dibiphytanyldiglycerol tetraether, in which two $C_{20}$–$C_{20}$ diether moieties are linked head to head. Several variants of the diphytanylglycerol diether and the dibiphytanyl tetraether core lipid structures are found in the different groups of archaea. For example, both the $C_{20}$–$C_{25}$ diether and $C_{25}$–$C_{25}$ diether core lipids occur in the basophilic and extreme halophilic genera *Natronococcus* and *Natronobacter* (De Rosa et al. 1982, 1983), and in the extremely halophilic genus *Halococcus* (Moldoveanu et al. 1990). A macrocyclic $C_{40}$ diether is the major core lipid of the thermophilic methanogenic methanococci (Comita and Gagosian 1983; Tricone et al. 1992), and lipids derived from dibiphytanylglycerol nonitoltetraether containing one to four cyclopentane rings are common in hyperthermophiles (De Rosa et al. 1983; Trincone et al. 1992). In most cases, the lipids are glycosylated or phosphorylated. Detailed reviews on the multitude of variations and on the known evidence regarding lipid biosynthesis via the mevalonic acid pathway are available (Langworthy 1985; Kates 1988, 1992).

e) Modified Nucleosides in Transfer RNA

Early comparative studies on the patterns of modified nucleosides of tRNA within the three domains revealed that ribothymidine found in Eucarya and Bacteria is replaced by pseudouridine or 1-methylpseudouridine in Archaea (McCloskey 1986). This finding was considered important evidence in support of the archaebacteria concept (McCloskey 1986). Further studies (Edmonds et al. 1991) showed that out of a total of 75 modified nucleosides known to occur in organisms, 33 are found in Archaea. Nine of these occur only in Archaea, 7 are common in Archaea and Eucarya, 5 in Archaea and Bacteria, and 12 in members of all three domains. The overall pattern of modification of nucleosides in Archaea is more typical of Eucarya than of Bacteria. Among the archaea, modified nucleosides are more common in hyperthermophiles, which contain up to eight types of modified nucleosides, than in mesophiles, or in the moderate thermophilic *Thermoplasma* which contains only one or two types of modified nucleosides (Edmonds et al. 1991).

f) DNA-Dependent RNA Polymerases

Archaeal RNA polymerases show a complex structure comprising 7 to 12 archaeal subunits of different sizes (Zillig et al. 1982, 1993a). Thus, they resemble more closely the complex RNA polymerases of eukaryotes, also consisting of up to 13 subunits, than those of bacteria which contain only 5 subunits and show the uniform structure $\beta'\beta\alpha_2\sigma$. Detailed immunological and sequence studies revealed that the large subunits labeled B and $\beta$ in Archea and Bacteria, respectively, are homologous to the large subunit B of the eukaryotic RNA polymerase 2 located at chromosome II of *Saccharomyces cerevisiae* and that the large subunits labeled A and A" or $\beta'$ in Archaea and Bacteria, respectively, are homologous to the large subunit A of the eukaryotic RNA polymerase 2, located at chromosome IV of *Saccharomyces cerevisiae* (Zillig et al. 1993a, b, 1993b). In spite of the overall homology between the large subunits in representatives of the three domains, the enzymes differ in a number of details. For example, eukaryotes possess three specialized RNA polymerases producing different types of RNAs, whereas in Archaea and Bacteria one and the same RNA polymerase synthesizes all types of RNA; eukaryotic and archaeal polymerases contain many more additional small subunits than bacterial polymerases are, insensitive to rifampicin which inhibits initiation in bacteria and are also insensitive to $\alpha$-amanitin which inhibits nuclear RNA polymerase I and II of eukaryotes. In summary, the structure and properties of RNA polymerase indicate that these enzymes certainly belong to the common heritage of the three domains, but developed many autapomorphic characteristics after the three lineages diverged from each other.

g) ATP Synthetase (ATPase)

Proton-pumping ATPases are present in all organisms and show highly conserved sequences. The F-ATPases of bacteria, mitochondria and chloroplasts were found to be homologous to V-ATPase found in endomembranes of eukaryotic cells and in plasma membranes of archaea. The two V-ATPases are structurally more similar and show higher sequence homology between each other than toward F-ATPase. This finding indicates a closer phylogenetic relationship between archaea and Eukarya than between Archaea and Bacteria. The unique feature of both ATPases is the presence of noncatalytic and catalytic subunits which evolved from a common ancestral gene by gene duplication in the common ancestor of all living organisms (Kibak et al. 1992). Such so-called paralogous genes can be used as an outgroup to determine the root of the global tree of life (Gogarten et al. 1989; Iwabe et al. 1989) which shows Archaea and Eucarya to be sister groups (Fig. 1). The same result is obtained when another pair of paralogous genes, the protein elongation factors Ef-Tu/1 $\alpha$ and Ef-G/2, are employed (Auer et al. 1989; Iwabe et al. 1989; Miyata et al. 1991).

### h) Genome Organization

Archaeal DNA has about the same size and the same sequence complexity and forms a single circular chromosomes as does bacterial DNA. Also, the same genetic elements and viruses (Zillig et al. 1986; Palm et al. 1991; Schleper et al. 1992) are common. Transcription units including initiation codons and terminators, flanked by Shine-Dalgarno sequences (ribosome-binding sites complementary to the 16/18S rRNA terminal sequences), have been found (Zillig et al. 1988). Several putative histones have been isolated from archaea (for review, see Laine et al. 1990). Only two of them have been studied in more detail. One of them, HTa, exhibits structural and sequence similarity with the bacterial DNA-binding protein HU (De Lange and Williams 1981), while the other, HMf, exhibits much better sequence similarity with the eukaryotic histones involved in nucleosome formation, and forms nucleoprotein particles reminiscent of eukaryotic nucleosomes (Sandmann et al. 1990). Although there are distinct structural and organizational variations among the genomes of the various groupings of archaea, the overall structure and organization of the archaeal genome are distinctly reminiscent of the bacterial genome. However, some features of eukaryotic chromosome organizations are also found (for review, see Zillig et al. 1988; Forterre et al. 1992).

## 5. Biotechnological Potential of Archaea

Although methanogens have been used for sewage sludge stabilization and methane production since the beginning of this century, their wide biotechnological potential and that of their fellows, the "archaebacteria", became apparent only during the 1980s, when, triggered by Woese's archaebacteria concept, more and more new phenotypes of Archaea became known. Thus, for the workshop "Futuristic aspects in biotechnology" held during the Third European Congress on Biotechnology in Munich, September 1984, a lecture on "Archaebacteria – biotechnological implications" was scheduled. Although only a few examples of new biotechnological applications of archaea were reported or suggested, the lecture closed with the appeal: Biotechnologists should neither be the last nor the least among the explorers of this new continent of living beings" (Kandler 1985).

In the meantime, research on the biotechnological aspects of archaea and other "extremophiles" has increased fueled by the discovery of ever more archaea and bacteria living under extreme environmental conditions (Fig. 2) and by consecutive "Biotechnology Action Programmes (BAP) 1981–1989" (Vassarotti and Magnien 1990) of the Commission of the European Community, who supported joint research projects of European laboratories on biotechnologically oriented work on extremophiles with several million ECU. Whereas the first European "Biomolecular Engineering Programme BEP 1982–1986" (Magnien 1986) focused on studies of the biotechnological potential of *Sulfolobus solfataricus*, support has been extended to extremophiles in general by the consecutive action programs under the heading "European laboratories without walls" (Aguilar 1989, 1992; Magnien et al. 1989; Vassarotti and Magnien 1990).

Reports on current, biotechnologically oriented research can be found in the abstract book of the International Conference, Thermophiles: Science and Technology, Reykjavik, Iceland 2, 23rd–26th August 1992, Ice Tec Keldnaholt, 112 Reykjavik, Iceland. This book contains extended abstracts including short reference lists.

### a) Methanogens

In addition fo further developments concerning the traditional use of methanogens in methane production and sewage stabilization, which will not be treated in this report, the following aspects may gain biotechnological interest in the future:

Ethane production from ethanol (Belay and Daniels 1988); conversion of halogenated hydrocarbons into harmless gases (Belay and Daniels 1987; Genthner et al. 1989); production of industrial quantities of corrinoids (Mazumder et al. 1987; Silveira et al. 1991); suppression of methanogens in rumen in order to minimize the energy drain and to increase feed efficiency in ruminants (Russel and Strobel 1989); the production of 5-deazaflavin $F_{420}$ and other unique coenzymes as speciality products from methanogens (recently discussed by Daniels 1992).

### b) Halobacteria

Traditionally, halobacteria play a role in food technology because of the spoilage of salted, stored food by proteinase production. Thus, methods for controlling the growth of halophiles in such food is of interest (Kamekura and Seno 1991). Recently, Rodriguez-Valera (1992) discussed the following possible biotechnological applications of halobacteria: screening for anticancer agents due to the similarities between halobacterial and eukaryotic cells at the molecular level; replacement of some presently used conventional enzymes by those from halobacteria because of the supposed higher stress resistance of the latter; substitution of acid polysaccharides from plants (algae) by those from halobacteria; and mass production of polyhydroxy acid esters for the production of bioplastics. However, none of these proposals can be presently realized.

The most sophisticated and substantiated biotechnological use of a unique product of halobacteria is the exploitation of the light-driven proton pump operating in the purple membranes of *Halobacterium halobium.* Purple membranes consist of bacteriorhodopsin with a retinal chromophore as light acceptor. After genetechnological modification it can be used for holographic applications (Hampp et al. 1992), optical signal processing, optical pattern recognition, optical associative memories and neural networks, optical memories and photoelectric devices (Bräuchle et al. 1991). "Biochips" for an optical computer may be possible in the not too distant future.

### c) Thermoacidophilic Archaea

Biotechnological interest in thermoacidophiles (*Sulfolobus, Acidianus, Metallosphaera, Thermoplasma*) is primarily restricted to ore leaching (Brierley and Brierley 1986; Le Roux and Wakerley 1988; Norris 1990, 1992) and desulfurization of coal (Bos and Kuenen 1990; Lucchese et al. 1992; Holst et al. 1992a, b). A broad temperature range and high tolerance to dissolved heavy metals are prerequisites for an organism to be useful for this purpose (for review, see Norris 1992).

### d) Archaea as a Source of Thermophilic Enzymes

The discovery of hyperthermophiles initiated the still ongoing activities in the search for "hyperthermophilic" enzymes for industrial applications, especially in the food industry, where limited heat-stable enzymes obtained from moderately thermophilic bacteria, mostly bacilli, have been in use for some time. As yet, detailed studies on about 50 hyperthermophilic enzymes are available (for review, see Cowan 1992). Some of these enzymes are already strongly competing with "traditional" thermophilic enzymes. For instance, Pfu DNA-polymerase from *Pyrococcus furiosus* (Lundberg et al. 1991) is aggressively advertised by STRATGENE Ltd. not only because of its higher heat stability but also because of its improved translation fidelity compared to Taq DNA-polymerase obtained from *Thermus aquaticus* (Mullis et al. 1986), so far widely used in PCR techniques. Other enzymes have found unforeseen applications, for instance, heat-stable amylopullanases and β amylases from *Pyrococcus furiosus* have been found to retard staling of bread and to prolong the shelf life of freshly baked bread for days when added to the dough before baking (United States Patent No. 5,023,094, June 11, 1991; Jones 1992). Still others have added to the set of genetechnological tools, for instance, restriction enzymes of unusual specificity from *Methanococcus aeolicus* (Schmid et al. 1984). Thermophilic enzymes may also be employed to carry out reactions in the gas phase, for instance, the oxidation of alcohols to the respective aldehydes (Legoy 1992). Other examples and biotechnological aspects of thermophilic enzymes from archaea have been comprehensively treated by Cowan (1992).

## 6. Archaea and the Origin of Life

Until recently, the first organisms on earth were assumed to have been mesophilic heterotrophs, thriving in a prebiotic broth, whereas a thermophilic origin of life was thought to be most unlikely, as pointed out by Walker et al. (1983): "If the earliest forms of life evolved in a high-temperature regime, one might expect to see a phylogenetic correlation between degree of 'primitiveness' and thermophily among extant microbes, but this is not the case."

However, the scene has been changing radically since the discovery of a large number of hyperthermophilic archaea and bacteria (Fig. 2). They exhibit the expected phylogenetic correlation between "primitiveness" and thermophily, since their lineages represent the deepest branchings close to the common root of the phylogenetic tree (Fig. 1). Most of the hyperthermophilic archaea and bacteria are obligately or facultatively $H_2/SO_4/S^0$, $H_2/CO_2$, $H_2/SO_4$, or $H_2/O_2$ chemolithoautotrophic, but none is photoautotrophic. This suggests a priority of chemoautotrophy over photoautotrophy, as recently pointed out by Stetter (1993a, 1993b) and Kandler (1992, 1993). Such ideas are compatible with Wächtershäuser's proposal that life was incited in the heated hydrosphere of a primeval inorganic world by $CO_2$ reduction and organic synthesis driven by inorganic redox energy and electron flow, presumably originating from pyrite formation ($H_2S + FeS = FeS_2 + H_2$; $G^0 = -41.9$ kJ/mol) (Wächtershäuser 1988a, b, 1990, 1993a, 1993b). This common geochemical process was shown to occur spontaneously at elevated temperatures (Drobner et al. 1990) and to lead also to the reduction of a variety of organic compounds (Blöchl et al. 1992). At a very early stage of biochemical evolution, other redox energy sources based on hydrogen oxidation, which are still employed by extant chemolithoautotrophic hyperthermophiles, may have supplemented and finally replaced the aboriginal redox energy source, when the first primitive hydrogenase activity had evolved. Thus, a wider range of substrates became available and the evolving life could venture into a larger variety of habitats (Kandler 1992, 1993).

Therefore, archaea and the hyperthermophilic bacteria may be considered as archaic witnesses of an early stage of diversification of life still based on inorganic redox energy sources (Kandler 1992, 1993; Stetter 1993a, 1993b; Wächtershäuser 1993a, 1993b) such as $H_2/S^0$, $H_2/SO_4$, $H_2/CO_2$, and $H_2/O_2$. Heterotrophic descendants utilizing the remains of the chemolithoautotrophs may have initiated an early carbon cycle.

Obviously, such scenarios contrast to the traditional "broth" hypothesis based on the chaos of prebiotic organic matter originating by chance from the interaction of radiation and electrical discharge with inorganic matter (Oparin 1938; Miller and Urey 1959; Chang et al. 1983). Thus, the discovery of archaea and bacteria exhibiting "archaic" metabolic and ecological traits such as chemolithoautotrophy by a non-Calvin-Benson cycle $CO_2$ fixation (Kandler and Stetter 1981, Fuchs et al. 1992), and hyperthermophily (Stetter et al. 1990; Stetter 1993b), together with Wächtershäuser's proposal for an origin of life pulled by geochemical (pyrite) energy (Wächtershäuser 1988a, b, 1993a, 1993b), opened a new vista of the origin of life.

## References

**Aguilar A** (1989) In: da Costa MS et al. (eds) Microbiology of extreme environments and its potential for biotechnology: European laboratories without walls on extremophiles. Elsevier, Lindon, pp 1–5. **Aguilar A** (1992) In: Geirsdóttir AM, Brown HP, Sjenstad T (eds) Abstract book. International conference thermophiles: science and technolgy. IceTec Reykjavik, Iceland, p 2. **Auer J**, Lechner K, Böck A (1989) Can J Microbiol 35:200–204. **Auer J**, Spicker G, Böck A (1990) Syst Appl Microbiol 13:354–360. **Auer J,** Spicker G, Mayerhofer L et al. (1991) Syst Appl Microbiol 14:14–22.

**Balch WE**, Magrum LJ, Fox GE et al. (1977) J Mol Evol 9:305–311. **Baumeister W,** Wildhaber I, Phipps BM (1989) Can J Microbiol 35:215–227. **Baumeister W,** Santarius U, Volker S et al. (1990) Syst Appl Microbiol 13:105–111. **Belay N,** Daniels L (1987) Appl Environ Microbiol 53:1604–1610. **Belay N,** Daniels L (1988) Antonie van Leeuwenhoek J Microbiol 54:113–125. **Bengtson S** (ed) (1993) Early life on earth. Proceedings of Nobel Symposium '84, Columbia Univ Press, New York. **Blöchl E,** Keller M, Wächtershäuser G et al. (1992) Proc Natl Acad Sci USA 89: 8117–8120. **Bos P**, Kuenen JG (1990) In: Brock TD (ed) Thermophiles: general, molecular, and applied microbiology. Wiley New York, pp 279–305. **Bräuchle C,** Hampp N, Oesterhelt D (1991) Adv Mater 3:420–428. **Brierley JA,** Brierley CL (1986) In: Brock TD (ed) Thermophiles: general molecular and applied microbiology. Wiley, New York, pp 279–305. **Brock TD** (1978) Starr MP (ed) Thermophilic microorganisms and life at high temperatures. Springer, New York Heidelberg Berlin, pp 1–465. **Burggraf S,** Stetter KO, Rouviere P et al. (1991) Syst Appl Microbiol 14:346–351.

**Chang S,** Des Marais D, Mack R et al. (1983) In : Schopf JW (ed) Earth' earliest biosphere – its origin and evolution. Princeton Univ Press, Princeton. **Chatton E** (1937) Titre et travaux scientifiques. Sottana Italie, Sète. **Comita PB,** Gagosian RB (1983) Science 222:1329–1331. **Cowan DA** (1992) In: Danson MJ, Hough DW, Lunt GG (eds) The archaebacteria: biochemistry and biotechnology. Portland Press, London, pp 149–169.

**Daniels L** (1992) In: Danson MJ, Hough DW, Lunt GG (eds) The archaebacteria: biochemistry and biotechnology. Portland Press, London, pp 181–193. **Danson MJ** (1989) Can J Microbiol 35:58–64. **Danson MJ,** Hough DW (1992) In: Danson MJ, Hough DW, Lunt GG (eds) The archaebacteria: biochemistry and biotechnology. Portland Press, London, pp 7–22. **Danson MJ,** Hough DW, Lunt GG (eds) (1992) The archaebacteria: biochemistry and biotechnology. Portland Press, London, pp 1–212. **De Lange R,** Williams L (1981) J Biol Chem 256:905–911. **De Rosa M,** Gambacorta A, Nicolaus B et al. (1982) J Gen Microbiol 128:343–348. **De Rosa M,** Gambacorta A, Nicolaus B et al. (1983) J Gen Microbiol 129:2333–2337. **DiMarco AA,** Bobi TA, Wolfe RS (1990) Annu Rev Biochem 59:355–394. **Drobner E,** Huber H, Wächtershäuser G et al. (1990) Nature 346:742–744.

**Edmonds ChG,** Crain PF, Gupta R et al. (1991) J Bacteriol 173:3138–3148.

**Forterre P,** Charbonnier F, Marguet E et al. (1992) In: Danson MJ, Hough DW, Lunt GG (eds) The archaebacteria: biochemistry and biotechnology. Portland Press, London, pp 99–112. **Fuchs G,** Ecker A, Strauss G (1992) In: Danson MJ, Hough DW, Lunt GG (eds) The archaebacteria: biochemistry and biotechnology. Portland Press, London, pp 29–39.

**Genthner BRS,** Price WA II, Pritchard PH (1989) Appl Environ Microbiol 55:1466–1471. **Gogarten JP**, Taiz L (1992) In: Hartmann H (ed) Origin of life. Neil Patterson Publishers. **Gogarten JP,** Kibak H, Dittrich P et al. (1989) Proc Natl Acad Sci USA 86:6661–6665.

**Hampp N,** Thoma R, Bräuchle C et al. (1992) AIP Conf Proc 262 (Mol Electron Sci Technol): 181–190. **Hartmann E,** König H (1990) Naturwissenschaften 77:472–475. **Hartmann E,** König H

(1991) Biol Chem Hoppe-Seyler 372:971–974. **Holst O,** Larsson L, Olsson G et al. (1992a) In: Geirsdóttir AM, Brown HP, Sjenstad T (eds) Abstract book. International conference thermophiles: science and technology. IceTec Reykjavik, Iceland, pp 143–146. **Holst O,** Larson L, Olsson G et al. (1992b) In: Geirsdóttir AM, Brown HP, Sjenstad T (eds) Abstract book. International conference thermophiles: science and technology. IceTec Reykjavik, Iceland, p 113.

**Iwabe N,** Kuma K, Hasegawa M et al. (1989) Proc Natl Acad Sci USA 86:9355–9359.

**Jones BE** (1992) In: Geirsdóttir AM, Brown HP, Sjenstad T (eds) Abstract book International Conference Thermophiles: Science and Technology. IceTec Reykjavik Iceland, p 114.

**Kamekura M,** Seno Y (1991) In: Rodriguez-Valera F (ed) General and Applied Aspects of Halophilic Microorganisms. Plenum Press New York, pp 305–365. **Kandler O** (ed) (1982a) Archaebacteria. Gustav Fischer, Stuttgart, pp 1–366. **Kandler O** (1982b) Zentralbl Bakteriol Hyg I Abt Orig C 3:149–160. **Kandler O** (1985) In: Proceedings Third European Congress on Biotechnology, vol IV.VCH, Weinheim, pp 551–560. **Kandler O** (1992) In: Danson MJ, Hough DW, Lunt GG (eds) The archaebacteria: biochemistry and biotechnology. Portland Press, London, pp 195–207. **Kandler O** (1993) In: Bengtson S (ed) Early life on earth. Proceedings of Nobel Symposium '84. Columbia Univ Press (in press). **Kandler O,** König H (1985) In: Woese CR, Wolfe RS (eds) The bacteria, vol VIII. Archaebacteria. Academic Press, Orlando, pp 413–452. **Kandler O,** König H (1993) In: Kates M, Kushner DJ, Matheson AT (eds) The biochemistry of Archaea (archaebacteria). Elsevier, Amsterdam (in press). **Kandler O,** Stetter KO (1981) Zentralbl Bakteriol Hyg I Abt Orig C 2:111–121. **Kandler O,** Zillig W (eds) (1986) Archaebacteria '85. Gustav Fischer, Stuttgart, pp 432. **Kates M** (1988) In: Karnovsky ML, Leaf A, Bolis LC (eds) Biological membranes: aberrations in membrane structure and function; structure, physical properties and function of archaebacterial lipids. Alan Liss, New York, pp 357–384. **Kates M** (1992) In: Danson MJ, Hough DW, Lunt GG (eds) The archaebacteria: biochemistry and biotechnology. Portland Press, London, pp 51–72. **Kibak H,** Taiz L, Stoke T, Bernasconi P, Gogarten JP (1992) J Bioenerg Biomembr 24:415–424. **Kjems J,** Larsen N, Dalgaard JZ et al. (1992) Syst Appl Microbiol 15:203–208. **König H,** Stetter KO (1989): In: Staley JT, Bryant MP, Pfennig N (eds) Bergey's manual of systematic bacteriology, vol 13. William & Wilkins, Baltimore, pp 2171–2253. **Kreisl P,** Kandler O (1986) Syst Appl Microbiol 7:293–299. **Kurr M,** Huber R, König H et al. (1991) Arch Microbiol 156:239–247.

**Laine B,** Chartier F, Imbert B et al. (1990) In: Belaich JP, Bruschi M, Garcia JL (eds) FEMS Symp. Plenum Press, New York, pp 291–301. **Langworthy TA,** (1985) In: Woese CR, Wolfe RS (eds) The bacteria vol VIII. Archaebacteria. Academic Press, Orlando, pp 459–497. **Lechner J,** Wieland F (1989) Annu Rev Biochem 48:173–194. **Legoy MD** (1992) In: Geirsdóttir AM, Brown HP, Sjenstad T (eds) Abstract book. International conference thermophiles: science and technology. IceTec Reykjavik, Iceland, p 116A. **Le Roux NW,** Wakerley DS (1988) In: Norris PR, Kelle DP (eds) Biohydrometall. Proc Int Symp, Science and Technology Letter Kew, pp 305–317. **Lodwick D,** Ross HNM, Walker JA et al. (1991) Syst Appl Microbiol 14:352–357. **Lucchese G,** Scolla G, Morana A (1992) In: Geirsdóttir AM, Brown HP, Sjenstad T (eds) Abstract book. International conference thermophiles: science and technology. IceTec Rejkjavik, Iceland, p 68. **Lundberg KS,** Shoemaker DD, Adams MWW et al. (1991) Gene 108:1–6.

**Magnien E** (ed)(1986) Biomolecular engineering in the European Community. Martinus Nijhoff Publishers for the CEC Brussels, The Hague. **Magnien E,** Aguilar A, Wragg P et al. (1989) Biofuture 84:17–30. **Matheson AT** (1992) In: Danson MJ, Hough DW, Lunt GG (eds) The archaebacteria. biochemistry and biotechnology. Portland Press, London, pp 89–98. **Mazumder TK,** Nishio TK, Fukuzaki N et al. (1987) Appl Microbiol Biotechnol 26:511–516. **McCloskey JA** (1986) Syst Appl Microbiol 7:246–252. **Miller SL,** Urey HC (1959) Science 130:245–251. **Miyata T,** Iwabe N, Kuma K et al. (1991) In: Osawa S, Honjo T (eds) Evolution of life. Springer, Berlin Heidelberg New York, pp 337–351. **Moldoveanu N,** Kates M, Monter CG et al. (1990) Biochem Biophys Acta

1046:127–135. **Mukund S**, Adams MWW (1991) J Biol Chem 266:14208–14216. **Mullis K**, Falcona F, Saiki R et al. (1986) Quant Biol 51:263–273.

**Norris PR** (1990) In: Ehrlich HL, Brierley CI (eds) Microbial mineral recovery. McGraw-Hill, New York, pp 3–27. **Norris PR** (1992) In: Danson MJ, Hough DW, Lunt GG (eds) The archaebacteria: biochemistry and biotechnology. Portland Press, London, pp 171–180.

**Oesterhelt D,** Krippahl G (1983) Ann Microbiol (Paris) 134B:137–150. **Oparin AI** (1938) The origin of life. MacMillan, New York, pp 270.

**Palm P,** Schleper C, Grampp B et al. (1991) Virology 185:242–250.

**Rodriguez-Valera F** (1992) In: Danson MJ, Hough DW, Lunt GG (eds) The archaebacteria: biochemistry and biotechnology. Portland Press, London, pp 135–148. **Rospert S,** Breitung J, Ma K et al. (1991) Arch Microbiol 156:49–55. **Rouvière P,** Mandelco L, Winker S et al. (1992) Syst Appl Microbiol 15:363–371. **Russell JB,** Strobel HJ (1989) Appl Environ Microbiol 55:1–6.

**Sandman K,** Krzycki JA, Dobrinski B et al. (1990) Proc Natl Acad Sci USA 87:5788–5791. **Schäfer Th,** Schönheit P (1992) Arch Microbiol 158:188–202. **Schäfer Th,** Selig M, Schönheit P (1992) In: Geirsdóttir AM, Brown HP, Sjenstad T (eds) Abstract book. International conference thermophiles: science and technology. IceTec Reykjavik, Iceland, pp 154–155. **Schleifer KH**, Kandler O (1972) Bacteriol Rev 36:407–477. **Schleifer KH**, Ludwig W (1989) In: Fernholm B, Bremer K, Jörnwall H (eds) The hierarchy of life. Exacta Medica, Amsterdam, pp 103–117. **Schleifer KH**, Steber J, Mayer H (1982) Zentralbl Bakteriol Hyg I Abt Orig C 3:171–178. **Schleper C,** Kubo K, Zillig W (1992) Proc Natl Acad Sci USA (in press). **Schmid K,** Thomm M, Laminet A et al. (1984) Nucleic Acid Res 12:2619–2628. **Segerer A**, Langworthy TA, Stetter KO (1988) Syst Appl Microbiol 10:161–171. **Silveira RG,** Nihsio N, Nagai S (1991) J Ferm Bioeng 71:28–34. **Stetter KO** (1991) Arch Microbiol 156:239–247. **Stetter KO** (1993) In: Thran TV (ed) Frontiers of life. Editions Frontiers, Gif sur Yvette France (in press). **Stetter KO** (1993b) In: Bengtson S (ed) Early life on earth. Proceedings of Nobel Symposium '84. Columbia Univ Press (in press). **Stetter KO**, Fiala G, Huber G et al. (1990) FEMS Microbiol Rev 75:117–124. **Strauss G,** Eisenreich W, Bacher A et al. (1992) Eur J Biochem 205:853–866.

**Thran TV** (ed) (1993) In: Frontiers of life. Editions Frontiers, Gif sur Yvette France (in press). **Trincone A,** Nicolaus B, Palmieri G et al. (1992) Syst Appl Microbiol 15:11–17.

**Vassarotti A,** Magnien E (eds) (1990) Biotechnology R & D in the EC: Biotechnology Action Programme (BAP) 1981–1989, vols I and II. Elsevier, Brussels.

**Wächtershäuser G** (1988a) System Appl Microbiol 10:207–210. **Wächtershäuser G** (1988b) Microbiol Rev 52:452–480. **Wächtershäuser G** (1990) Proc Natl Acad Sci USA 87:200–204. **Wächtershäuser G** (1993a) In: Thran TV (ed) Frontiers of life. Editions Frontiers, Gif sur Yvette France (in press). **Wächtershäuser G** (1993b) In: Bengtson S (ed) Early life on earth. Proceedings of Nobel Symposium '84. Columbia Univ Press (in press). **Walker JCG,** Klein C, Schidlowski M et al. (1983) In: Schopf JW (ed) Earth's earliest biosphere – its origin and evolution. Princeton University Press, Princeton. **Winker S,** Woese CR (1991) Syst Appl Microbiol 14:305–310. **Woese CR** (1987) Microbiol Rev 51:221–271. **Woese CR**, Fox GE (1977) Proc Natl Acad Sci USA 74:5088–5090. **Woese CR,** Wolfe RS (eds) (1985) The bacteria, vol VIII. Academic Press, Orlando, pp 1–582. **Woese CR,** Sogin M, Stahl D et al. (1976) J Mol Evol 7:197–213. **Woese CR,** Magrum LJ, Fox GE (1978) J Mol Evol 11:245–252. **Woese CR**, Kandler O, Wheelis ML (1990) Proc Natl Acad Sci USA 87:4576–4579. **Woese CR,** Achenbach L, Rouviere P et al. (1991) Syst Appl Microbiol 14:364–371. **Wolfe RS** (1992) In: Danson MJ, Hough DW, Lunt GG (eds) The archaebacteria: biochemistry and biotechnology. Portland Press, London, pp 41–49. **Wood HG** (1990) In: Schlegel HG, Bovien B (eds) Autotrophic bacteria. Science Publishers, Madison, pp 33–52.

**Zillig W,** Stetter KO, Schnabel R et al. (1982) Zentralbl Bacteriol Hyg I Abt Orig C 3:218–227.
**Zillig W,** Gropp F, Henschen A et al. (1986) Syst Appl Microbiol 7:58–66.
**Zillig W,** Palm P, Reiter W-D et al. (1988) Eur J Biochem 173:437–482. **Zillig W,** Palm P, Klenk HP (1993a) In: Thran TV (ed) Frontiers of life. Editions Frontiers, Gif sur Yvette France (in press). **Zillig W,** Palm P, Langer D et al. (1992) In: Danson MJ, Hough DW, Lunt GG (eds) The archaebacteria: biochemistry and biotechnology. Portland Press, London, pp 79.88. **Zillig W,** Palm P, Klenk HP et al. (1993b) In: Kates M, Kushner DJ, Matheson AT (eds) The biochemistry of Archaea (archaebacteria). Elsevier, Amsterdam (in press). **Zuckerkandl E,** Pauling L (1965) J Theor Biol 8:357–366.

Prof. Dr. Otto Kandler
Botanisches Institut der
Ludwig-Maximilians-Universität
Menziger Straße 67
8000 München 19, Germany

# A. Structural Botany

## I. Cytology and Morphogenesis of the Prokaryotic Cell

By Frank Mayer and Holger Gerberding

### 1. Eubacterial Cell Wall Structure and Assembly Processes

#### a) Cell Wall Morphology and Assembly in Gram-Positive Bacteria

The peptidoglycan sacculus of a bacterial cell which is located at the outside of the cytoplasmic membrane is considered to be the load-bearing structure of Gram-positive and Gram-negative eubacteria. Numerous reports over the last few years based on electron microscopic analysis of freeze-substituted thin-sectioned cells and pulsed addition of radiolabeled diaminopimelic acid to growing cells have improved our knowledge concerning fine structural aspects and cell wall assembly processes.

Freeze-substituted Gram-positive eubacteria reveal a cell morphology resembling more a fibrous matrix frayed at the outer face and extending up to 30 nm beyond the cytoplasmic membrane rather than an amorphous condensed aggregation as seen by older conventional electron microscopic embedding techniques (Umeda et al. 1987; Cook et al. 1988; Reid et al. 1988; Graham and Beveridge 1990a). Optical densitometry scanning data of freeze-substituted *Bacillus subtilis* cell walls indicate a significant decreasing intensity of the cell wall density with increasing distance from the cytoplasmic membrane (Graham and Beveridge 1990b).

Data obtained from pulse labeling experiments suggest a monomer addition process without a secondary cross-linking activity during the assembly of glycan monomer peptide chains into the cylindrical wall of *Bacillus megaterium* (Gally et al. 1991a, b). The process of incorporation is characterized by incoming glycan monomer peptides which form cross-linkages with acceptor peptides at the inner face of the preexisting wall only when they are in appropriate steric disposition resulting in an initial distribution of monomers, dimers, trimers, and higher oligomers which is constant throughout the continuously from inside-to-outside growing multilayered cell wall. As a consequence of an exponentially increasing cell volume during the division cycle, lateral tension of wall polymers increases with increasing distance from the innermost wall surface, generating a maximal stretching in the middle region of the cell wall. Due to the activity of autolysins (Koch and Doyle 1986) and partial physical breakage, the outermost cell wall layer loses integrity concerning structural dissection followed by subsequent release of soluble wall polymer products.

Progress in Botany, Vol. 54

b) Heteromorphism of Cell Envelope in Gram-Negative Bacteria

Recently, a variety of freeze-substituted Gram-negative eubacteria were analyzed by Graham et al. (1991) revealing a heteromorphism of the cell envelope. Asymmetrically structured outer membrane portions enclose at least three distinguishable types of periplasm. The more common type, represented by *Escherichia coli*, consists of a highly condensed thick periplasmic gel in which the peptidoglycan layer could not be detected. However, a neutron small-angle scattering study revealed a "more than single-layered" peptidoglycan array (Labischinski et al. 1991). Thin peptidoglycan layers were found in *Leptothrix discophora* and *Caulobacter crescentus*, exhibiting an electron-translucent periplasmic space. An intermediate periplasmic density was found in *Pseudomonas aeruginosa* and *Vibrio cholerae* with nodes of peptidoglycan connecting inner and outer membrane.

c) Cell Wall Synthesis and Turnover in Gram-Negative Bacteria

As compared to the cell wall synthesis and turnover of wall polymers observed in Gram-positive eubacteria, Gram-negative eubacteria analyzed so far exhibit a different kind of mechanism of wall synthesis during the division cycle. Since no turnover of peptidoglycan was demonstrated to occur in *Salmonella typhimurium* (Cooper 1988) or *Escherichia coli* (Cooper and Hsieh 1988) a comprehensive description of wall synthesis of rod-shaped Gram-negative eubacteria was presented by Cooper (1991a, b) specifying in more detail the surface stress model proposed earlier by Koch (1983). The thin peptidoglycan multilayer is composed of a single taut-bonded load-bearing monolayer and several non-load-bearing layers beneath. The load-bearing monolayer is connected by cross-linkages with the nonload-bearing layers which are loosely connected with one another. The integrity of the load-bearing structure is maintained during the division cycle. Stressed bonds are cut successively within the load-bearing monolayer, generating an enlargement of the cell surface. New glycan monomer peptides are incorporated prior to bond cutting by diffuse intercalation. A "smart autolysin" was proposed by Koch (1990) that would not hydrolyze a stressed bond before there is another bond in place. The increase in cell mass occurs exponentially concerning the amount of the cytoplasm and the internal volume of the cell. However, the cell density remains constant during the division cycle (Kubitschek 1987). Therefore, the rate of cell surface synthesis was considered to be close to exponential.

## 2. Archebacterial Crystalline Surface Layers

a) Structure and Assembly of a Coccoid Archebacterium

Periodic macromolecular monolayers are known for almost four decades (Houwink 1953) as the outermost part of a bacterial cell envelope and are characterized in a continuously increasing number of prokaryotes (Sleytr et al. 1988). Previously defined as porous crystalline surface layers (S-layers) (Sleytr and Messner 1983), the planar array is composed of identical proteinaceous or glycoproteinaceous particles arranged symmetrically in unit cells. In a recent report, Pum et al. (1991) described new insights into the molecular arrangement and the process of assembly of the crystalline surface layer which is the only component of the cell wall of the lobed coccoid archaebacterial methanogen *Methanocorpusculum sinense*. Based on data which were achieved by freeze-etched preparations of whole cells and negatively stained cell envelope fragments subsequently analyzed by standard digital image processing, a hexagonally arranged symmetry of the morphological S-layer unit cell was reported consisting of six monomers arranged around the threefold axes. Lattice faults within the closed protein crystal determining the cell shape of this lobed bacterium were identified as pentagons and heptagons in the hexagonally composed network. These sites of edge dislocations and wedge disclinations are seen as positions for the incorporation and self-assembly of new morphological units into the lattice during the process of morphogenesis related to the increasing volume of a growing cell and as initiation points of cell division.

b) Structural Aspects in a Filamentous Archebacterium

Beveridge et al. (1990, 1991) investigated aspects of the morphogenesis of the filamentous archaebacterium *Methanospirillum hungatei*. In these studies energy-dispersed X-ray spectroscopy, electron and scanning tunneling microscopy were combined to characterize ultrastructural features of the S-layer sheath and of the multilaminar cell spacer plugs by which rod-shaped cells are separated in multicellular filaments, elucidating a unique cell permeability phenomenon for this bacterium. The spacer plugs are composed of three layers, two particulate layers and a holey layer in between. The layers are connected with the S-layer sheath and they are located at each cell pole. The spacer plugs are separated from one another by an amorphous central zone (Shaw et al. 1985). The particulate layers show 12.5 by 7 nm holes permeating them, the holey layer contains circular 12.5 nm holes. The layers exhibit a hexagonally arranged symmetry of the morphological unit cell. However, the pore morphology of the S-layer sheath is defined by small holes 2 nm in diameter. This sharp exclusion cut-off allows only molecules with an $M_r$ less than 50 to permeate the S-layer sheath (subtrates and end products). Larger molecules such as cofactors and vitamins enter through the terminal plugs and diffuse along the filament-cell axis. Since flagella are located at the ends of the filaments (Southam et al. 1990), it was suggested that some mode of cell-cell communication occurs to stimulate and to coordinate flagellar rotation in *M. hungatei*.

## 3. Periplasmic Space

Textbooks on bacterial cytology usually describe the periplasm as a space, between cytoplasmic and outer membrane, about 7 nm wide, and a value of 20% of the total cell volume is given for its size. Application of advanced techniques of electron microscopic sample preparation (Dubochet et al. 1983; Hobot et al. 1984; Leduc et al. 1985) had resulted in much higher values for the width, i.e., 15 to 33 nm.

Anthony (1988) reported that in certain Gram-negative bacteria there are large amounts (up to 15% of the soluble bacterial proteins) of redox carriers in the periplasm. Thus assumptions on the width of the periplasmic layer have implications for the view on the interaction of the electron carriers and the mechanism of electron transport. Van Wielink and Duine (1990) communicated unpublished results indicating that the periplasmic layer is indeed much wider than assumed in the textbooks, and that the components of the periplasmic-localized electron transport chain are distributed randomly in the periplasm.

## 4. New Features of the Flagellar Base

The flagellar base harbors the machinery for filament rotation, chemotactic signal reception, and flagellar protein export. The structural modules coupling the energy derived from "down-hill" ion (proton, cation) transport to force generation must form part of the membrane-associated base. Khan et al. (1991) studied, by application of rapid-freeze techniques, the structure of the flagellar base in *Salmonella typhimurium*. They detected that the flagellar base has a bell-shaped extension reaching 50 nm into the bacterial cytoplasm. Thus the flagellar apparatus is more extensive than previously supposed. Functional implications may be expected.

## 5. Cell-Cell Interactions in Myxobacteria

Social and developmental biology of myxobacteria is a field which has attracted attention for many years. Shimkets (1990) has summarized classical and recent findings regarding this unique group of soil bacteria. Besides locomotion by gliding, contact-mediated cell-cell interactions have been a topic for research because they play an important role in the biology of myxobacteria. Behmlander and Dworkin (1991) now observed, by high-resolution scanning electron microscopy, that specific fibrils (30 nm in diameter, probably consisting of polysaccharides; Shimkets 1990) are produced most extensively by cells that are in close contact with each other. Expression of these fibrils was shown to be differentially regulated between adventuour (individual) and socially (group) motile cells. The authors postulate that the differential expression suggests the existence of a mechanism for the regulation of fibril biosynthesis that functions within the overall system governing social interactions in myxobacteria.

## References

**Anthony C** (1988) In: Anthony C (ed) Bacterial energy transduction. Academic Press, London, pp 293–316.

**Behmlander RM**, Dworkin M (1991) J Bacteriol 173:7810–7821. **Beveridge TJ**, Southam G, Jericho MH, Blackford BL (1990) J Bacteriol 172:6589–6595. **Beveridge TJ**, Sprott GD, Whippey P (1991) J Bacteriol 173:130–140.

**Cook RL**, Harris RJ, Reid G (1988) Curr Microbiol 17:159–166. **Cooper S** (1988) J Bacteriol 170:422–430. **Cooper S** (1991a) Bacterial growth and division. Academic Press, San Diego. **Cooper S** (1991b) Microbiol Rev 55:649–674. **Cooper S**, Hsieh ML (1988) J Gen Microbiol 134:1717–1721.

**Dubochet J**, McDowall AW, Menge B, Schmidt EN, Lickfeld KG (1983) J Bacteriol 155:381–390.

**Gally DL**, Hancock IC, Harwood CR, Archibald AR (1991a) J Bacteriol 173:2548–2555. **Gally DL**, Harwood CR, Archibald AR (1991b) Lett Appl Microbiol 12:54–58. **Graham LL**, Beveridge TJ (1990a) J Bacteriol 172:2150–2159. **Graham LL**, Beveridge TJ (1990b) J Bacteriol 172:2141–2149. **Graham LL**, Harris R, Villiger W, Beveridge TJ (1991) J Bacteriol 173:1623–1633.

**Hobot JA**, Carlemalm E, Villiger W, Kellenberger E (1984) J Bacteriol 160:143–152. **Houwink AL** (1953) Biochim Biophys Acta 10:360–366.

**Khan S**, Khan IH, Reese TS (1991) J Bacteriol 173:2888–2896. **Koch AL** (1983) Adv Microbiol Physiol 24:301–366. **Koch AL** (1990) Res Microbiol 141:529–541. **Koch AL**, Doyle RJ (1986) FEMS Microbiol Rev 32:247–254. **Kubitschek HE** (1987) Crit Rev Microbiol 14:73–97.

**Labischinski H**, Goodell EW, Goodell A, Hochberg ML (1991) J Bacteriol 173:751–756. **Leduc M**, Frehel C, Van Heyenoort J (1985) J Bacteriol 161:627–635.

**Pum D**, Messner P, Sleytr UB (1991) J Bacteriol 173:6865–6873.

**Reid G**, Cook RL, Harris RJ, Rousseau D, Laxford H (1988) Curr Microbiol 17:151–158.

**Shaw PJ**, Hills GJ, Henwwod JA, Harris JE, Archer DB (1985) J Bacteriol 161:750–757. **Shimkets LJ** (1990) Microbiol Rev 54:473–501. **Sleytr UB**, Messner P (1983) Annu Rev Microbiol 37:311–339. **Sleytr UB**, Messner P, Pum D, Sara M (1988) Crystalline bacterial cell surface layers. Springer, Berlin Heidelberg New York. **Southam G**, Kalmokoff ML, Jarrell KF, Koval SF, Beveridge TJ (1990) J Bacteriol 172:3221–3228.

**Umeda A**, Ueki Y, Amako K (1987) J Bacteriol 169:2482–2487.

**Van Wielink JE**, Duine JA (1990) TIBS 15:136–137.

Prof. Dr. Frank Mayer, Dr. Holger Gerberding
Institut für Mikrobiologie
Universität Göttingen
Grisebachstraße 8
W-3400 Göttingen, Germany

# II. Cell Architecture and Cellular Morphogenesis of Eukaryotic Algae (Rhodophyta and Phaeophyta)

By Diedrik Menzel

## Introduction

The architecture of algal cells and the ways in which these cells change shape and internal organization have attracted the curiosity of researchers ever since the first alga was placed under a microscope. It is not surprising, therefore, that publishing in this area has been substantial. In the past decade the literature has continued to grow exponentially. This enormous mass of information makes it necessary to narrow down the scope of the present chapter by focusing on the cell rather than on tissue features and thallus development. Morphogenesis is thus defined as the change of shape and function as the cell moves towards its developmental destination.

References before 1981 are covered by Brawley and Whetherbee (1981) and Leedale (1982). Additional and supplementary information on cell structure and morphogenesis of the algae may be found in the following recent review articles and monographs: Darley and Marshal (1982), Round and Chapman (1982–1990), Wiessner et al. (1983, 1986, 1990), Bold and Wynne (1985), South and Whittick (1987), Lobban et al. (1988), Coleman et al. (1989), Gibor (1989), Kumar (1989, 1990), Harold (1990), van den Hoek (1992).

Because of their diversity in cellular architecture, algae are particularly rich in suitable model systems for the study of structure-function relationships. On the subcellular level, the cytoskeleton is most likely involved in such vital processes as cellular compartmentation, cell division, locomotion, intracellular transport, polarity, and cellular morphogenesis. It is, therefore, not a mere coincidence that studying cytoskeletal structure and function is becoming a major new research area in phycology (Melkonian et al. 1991; Melkonian 1992; Menzel 1992).

This chapter is the first part of a review covering Rhodophyta, Phaeophyta, and Chlorophyta. The second part is scheduled to appear in the next issue of Progress in Botany. Priority has been given to these three major algal divisions not only due to space limitations, but also because recent progress in research on cell structure and morphogenesis in the remainder of the algae, notably chromophytan taxa, have been duly covered in other publications:

Margulis et al. (1990) have compiled essential information on all known protoctistan heterotrophic and phototrophic taxa in a large monograph, which is superbly suited for quick overviews. Two collections of articles on the biology and phylogeny of chromophyte algae can be found in Green et al. (1989) and Kristiansen et al. (1989). Cell structure and cytoskeletal features of

Progress in Botany, Vol. 54

phytoflagellates (Chlorophytes and Chromophytes) have been addressed by Melkonian, Andersen, Schnepf and their coauthors (1991). Another recent monograph is dedicated to motility phenomena and underlying cytoskeletal mechanisms in unicellular algae (Melkonian 1992). Diatom locomotion, cell structure, morphogenesis, mitosis and cytokinesis have been examined in several excellent reviews (Edgar and Pickett-Heaps 1984; Pickett-Heaps et al. 1990; Pickett-Heaps 1991), and a comprehensive treatise of the entire class has also appeared recently (Round et al. 1990). Chapters on diatom mitosis, the cytoskeleton in the Chrysophyceae and in the xanthophyte *Vaucheria* are included in Menzel (1992). Photosynthetic protoctists such as cryptophytes, euglenoids and dinoflagellates, which traditionally have been treated as algae, but most likely share common ancestors with protozoan taxa (Melkonian 1991), have also been covered elsewhere in sufficient detail (Spector 1984; Corliss 1987; Willey et al. 1989; Margulis et al. 1990; Melkonian 1991; Triemer and Farmer 1991; Menzel 1992).

## 1. Rhodophyta

Research on the red algae has traditionally placed strong emphasis on taxonomy, morphology, and phytogeography. Observations on cell structure and cellular morphogenesis have been primarily addressed in the context of clarifying phylogenetic relationships between red algal taxa (i.e., Gabrielson et al. 1985). This trend has gradually changed over the past decade in favor of genuine cytology studies, as documented in many review articles and monographs (Waaland 1980, 1986; Brawley and Wetherbee 1981; Johansen 1981; Kraft 1981; Goff 1982; Leedale 1982; Sheath 1984; Gabrielson and Garbary 1986; Woelkerling 1988). A new standard has recently been set by Cole and Sheath (1990) and their team of authors, who have given an exhaustive account of our present knowledge on the biology of the red algae.

### a) General Aspects of Cell Structure

#### $\alpha$) Organellar Associations

Conspicuous structural associations between organelles may occur in red algal cells, depending on the cell type and the developmental stage. The most typical is that between dictyosomes and mitochondria (Scott 1986; Pueschel and Magne 1987; Pueschel 1990).

In trichoblasts of the genus *Polysiphonia* (Ceramiales) almost all components of the endomembrane system become involved in the formation of a structural unit (Oates and Cole 1989). In this case an endoplasmic reticulum (ER) cisterna is sandwiched between the dictyosome and the mitochondrion, which at the side of association becomes stretched out and flattened, so that vis-a-vis from the dictyosome a proplastid or a large osmiophilic body can also become associated.

The functional significance or organelle associations like these is not known, but Oates and Cole (1989) discuss the possibility that particularly in large cells such as red algal hair cells a close structural apposition between the membranous organelles could be advantageous for cell function. Similar observations have been made in the large diatom

*Coscinodiscus* (Schmid 1988) and in the xanthophycean coenocyte *Vaucheria* (Ott 1992). At least in the latter case, it is conceivable that a structural linkage between the two organelles could be of functional advantage for the cell repair mechanism, which is designed to rapidly close mechanical wounds in the coenocytic filament.

An unusually strong association of dictyosomes with the surface of interphase nuclei has been described in *Rhodella reticulata* (Bangiophycidae, Pueschel 1990). Although this is apparently a rare type of association, it is of some interest because it involves a fibrous band interposed between the nuclear envelope (NE) and the dictyosome. It should be interesting to see, whether this cytoskeletal material is identical with actin (see Sect. 1c).

In *Gracilaria verrucosa* (Gigartinales, Delivopoulos and Tsekos 1986), pieces of the annulate lamella (AL, see Sect. 1b, below), are tightly sandwiched between the nuclear surface and mitochondria. Unfortunately, this quite striking variant of organellar association has also not yet found an explanation.

The nucleus associated organelle (NAO), a further item on this list, will be discussed below (see Sect. 1c).

### β) Dictyosomes

Red algal dictyosomes are known to produce many types of vesicles varying in dimension, content, and stage of appearance. This is particularly striking during successive stages of spore and spermatia formation (for summary see Brawley and Wetherbee 1981; Pueschel 1990). Perhaps the most distinctive population is that of the cored vesicles appearing in late carpospore development. These vesicles apparently form by the detachment of entire distal cisternae from the Golgi complex (see, for instance, Tsekos 1985) and may contain a core of granular or striated electron-dense material (Kugrens and Delivopoulos 1986), which is believed to confer adhesiveness to the spore surface upon extrusion. The core presumably originates from the region of the laminated central part of the cisterna, whereas the loose, fibrous matrix comes from the peripheral part of the cisterna. Cisternal progression is apparently also the major route for Golgi products in the scale-forming phytoflagellates (see discussion in Melkonian et al. 1988). The apparent contradiction that arises by this mode of Golgi vesicle formation to the presently held concept of polysaccharide and glycoprotein processing by shuttling of vesicles between stable cisternal compartments (Farquhar 1985) has not yet been resolved.

### γ) Mitochondria

Mitochondrial shape can be as variable in red algal cells as in any other plant cell. In the extreme, individual mitochondrial profiles seen in ultrathin sections may all be part of a single reticulate chondriome as has recently been demonstrated by serial sectioning analysis of *Rhodella reticulata* (Porphyridiales, Broadwater and Scott 1986).

Another item on the long list of unidentified mitochondrial inclusions (see summary in Brawley and Wetherbee 1981) has recently been described by Tsekos and Diannelidis (1990), who observed very fine, perhaps helically wound filaments in the mitochondrial matrix of developing carpospores of *Gigartina teedii*. Their suggestion, that these structures could be aspects of the replicating mitochondrial genome, requires further experimental clarification.

### δ) Endoplasmic Reticulum (ER)

In certain stages of carpospore development the ER can dramatically proliferate (Tsekos and Schnepf 1985, see Sect. 1f). It can also form a continuous layer around dividing nuclei (see Sect. 1c). During tetraspore formation, the ER occurs in several specialized subtypes associated with nuclear function, cytokinesis, and mucilage production.

An unusual elaboration of the ER has been known from earlier studies to occur in the unicells of the Porphyridiales (Deason et al. 1983). This has recently been re-examined and described in further detail in *Rhodella violacea* (Patrone et al. 1991). In this species, the ER consists of a conspicuous peripheral sheath, which is separated from the plasmalemma by a thin fibrous layer. Tubular connections of 18 nm diameter run through this layer between the ER sheath and the cell membrane procuring a continuation between the exterior and the ER lumen. It is tempting to suggest that this peculiar structural arrangement is designed for extrusion of mucous material, and that this could be responsible for the gliding motility common in these unicells, but further work is required to support this hypothesis.

Open connections between the ER and the exterior have also been observed in rhizoidal cells of parasitic red algae (Goff 1982) and in developing tetraspores (Vesk and Borowitzka 1984; see Sect. 1gα).

ER is thought to be responsible for the formation of mucilage sacs, which are prominent in all stages of gonimoblast development including the final stage of carpospore formation and may also occur in tip cells (Delivopoulos et al. 1989). They are large irregularly outlined vacuoles filled with a loosely spread, fibrous material. Aspects of local swelling of the intracisternal ER lumen have been interpreted as early stages in the formation of these structures (Tsekos 1985; Delivopoulos and Tsekos 1986). In gonimoblast cells of *Gracilaria verrucosa* (Gigartinales), similar mucilage sacs appear to be formed by swelling and proliferation of the nuclear envelope (Delivopoulos and Tsekos 1985a). In advanced stages they may become surrounded by additional membranes collapsing onto each other, sometimes building up to conspicuous concentric lamellae (Delivopoulos and Kugrens 1984b; Delivopoulos and Tsekos 1985a; Tsekos 1985; Tsekos and Schnepf 1985). It may be noted that myelinic conglomerations of membrane material occur frequently in red algal cells, and it has recently been suggested that this material acts as precursor or transient store in processes of rapid membranes build up (Oates and Cole 1992).

Another unusual type of vacuole which has an electron-dense canal associated with it has been termed the fibrous vacuole associated organelle (FVAO, Broadwater and Scott 1983). FVAOs have been found in spermatangia as well as in tetrasporangia of ceramialean red algae, and probably also in members of other orders such as the Bonnemaisoniales.

The genesis of this vacuolar compartment is rather complex and may not yet be completely understood. It seems to start out from a curling up of a smooth cup-shaped ER cisterna. In its mature form, the FVAO consists of a slightly curved flat membrane cisterna, filled with an electron-dense, fuzzy material and openings on both sides into large vesicular protuberances, which become juxtaposed to each other as they enlarge. One of the openings is distinct from the other by the presence of a membrane coat on the vesicle inside in the neck region. In the course of spermatial development, these protuberances enlarge further to become the spermatial vesicles (Broadwater and Scott 1983). The long electron-dense ER canal was originally mistaken for a cryptic flagellum (Tripodi and de Masi 1983). Broadwater and Scott (1983) proposed that the electron-dense inner membrane coat must be functionally involved in the processing of polysaccharide precursor material that eventually contributes to the spermatial surface covering.

The mechanism of sugar polymer formation in any of the vacuole and vesicle types in red algae is not known. Tsekos et al. (1985) have found clearly delineated particle tetrads on the inner face of vacuole membranes in *Gigartina* and *Ceramium*, and speculated whether these structures could be involved in linear polysaccharide synthesis in analogy to the cellulose synthetase rosette structures occurring in green plant cells on the plasmalemma face and within dictyosome vesicles (Emons 1991).

### ε) Rhodoplasts

General aspects of the rhodoplast fine structure and development have been treated elsewhere (Gantt 1986; Pueschel 1990). An intimate relationship between the nucleus and the pyrenoid has been described in two *Rhodella* species (Patrone et al. 1991), where a finger-like projection from the nucleus extends deeply into the pyrenoid. The function of this association has not been elucidated.

The mechanism of plastid division in algae and land plants has received much attention lately (Oross and Possingham 1989). An important contribution to this field of research has been made by Mita et al. (1986), who discovered a ring-like structure in the istmus of the dividing plastid in *Cyanidium caldarium* (an acidothermophilic red alga of unresolved systematic affinities, see Garbary and Gabrielson (1990). Fine structural preservation of this structure has become possible by the use of freeze substitution methodology. On the basis of specific staining with rhodamin-phalloidin, Mita and Kuroiwa (1988) concluded that the fine fibers constituting the ring should be actin filaments. They further showed that actin filaments must be functionally involved, since chloroplast division is inhibited by cytochalasin B whereas nuclear division is not.

A distinct feature of rhodoplasts is that they do not contain starch grains. Instead, floridean starch, an alpha-1,4-glucan with alpha-1,6-branch points, is formed in the cytoplasm in close association with mitochondria or dictyosomes. This is another common case of organelle association (see Sect. 1aα, above). Both of these organelles could be potential metabolic sinks for carbohydrates. Starch build-up also occurs either in the cytoplasm next to a pyrenoid, if it is situated close to the plastidal surface, or in a cytoplasmic pocket between a sheath of ER and the nuclear envelope. This has been observed for instance in *Faucheocolax* (Rhodymeniales, Delivopoulos and Kugrens 1984b). Rapid degradation of starch is suggested to involve the closure of a complete ER cisterna around a starch grain. In developing carpospores of *Chondria tenuissima* (Rhodomelaceae, Ceramiales, Tsekos 1985) these ER cisterna appear to undergo conspicuous swelling when the starch grain is beginning to shrink. However, there does not seem to be a general rule as to how starch synthesizing and degrading pathways are compartmentalized in red algae. Perhaps a radially arranged fibrous matrix within the grains (Sheath et al. 1981) may provide the necessary microcompartmentation for mutually exclusive metabolic pathways.

### ζ) Cell Walls

Composition and structure of the red algal cell walls are rather heterogeneous. Chemical analysis shows that cellulose is undoubtedly present in the majority of red algae, although it is often only a minor component, and occurs in a less crystalline packing than in other algae. It is even missing entirely in the generic phase of the Bangiophycidae[1], in which case it has become replaced by a crystalline mannan as the major structural polysaccharide (Gretz et al. 1986).

Structurally complex heteropolysaccharides constitute the bulk of the cell wall, among them xylans and especially sulfated galactans, of which agars and carragenans are best known, and of growing commercial interest (Whyte et al. 1985).

Despite the fact that production of polysaccharides and mucilages represents a significant fraction of the cell's synthetic machinery, and vesicles transporting all these materials to the cell surface comprise a considerable portion of the endomembrane system, we know surprisingly little about the role of all these polysaccharide fractions in morphogenesis and cell function.

Studies employing conventional polysaccharide histochemistry revealed that chemically different polysaccharide fractions occur in the three main layers: outer coating, exterior wall and interior wall of *Bangia atropurpurea* (Cole et al. 1985) with the most strongly sulfated type in the exterior wall. In addition, female thalli appeared to contain a carboxylated polysaccharide type missing from all other stages. The employment of recently developed carbohydrate-carbohydrate hybridization probes demonstrated that different carrageenan subtypes can be localized to different wall layers as well. This study also localized the stronger sulfated types to the outer wall layers and the less strongly sulfated ones to the inner layers (Zablackis et al. 1991).

Furthermore, fluorochrome-conjugated lectins have been used to differentially stain cell wall layers, intercellular mucilage, and intracellular starch grains (Diannelidis and Kristen 1988). When used on Western blots, biotinylated lectins appear to bind to tissue specific sets of glycoproteins isolated from *Porphyra perphorata* blades (Kaska et al. 1988). Some of these could be constituents of the cell wall, but further analysis is necessary before we know more about the role of these proteins.

### ε) Pit Plugs

Pit connections are formed in red algae as a result of an incomplete cytokinesis leaving a septal pore in the separating wall between recently divided daughter cells. The pore is then blocked by a plug core which may, depending on the taxonomic group, be covered by additional cap structures and a membrane-like layer (Pueschel 1980, 1987, 1989; Pueschel and Cole 1982).

Originally, pit plugs were thought to be restricted to the evolutionary advanced red algae (Florideophycidae, see Footnote 1) and some advanced members of the Bangiophycidae possessing

---

[1] In this chapter the recognition of the Florideophycidae and Bangiophycidae as subclasses in the class Rhodophyceae, division Rhodophyta, is followed sensu Bold and Wynne (1985). The reader may, however, notice that justification for a retention of the two subclasses is dwindling.

filament elongation by apical growth (Pueschel and Cole 1982), however, subsequent work has proven the existence of pit plugs in *Porphyra, Bangia,* and *Rhodochaete* (Pueschel 1987; Pueschel and Magne 1987), and another recent study has re-examined the situation in *Compsopogon coeruleus* (Scott et al. 1988), demonstrating a rather simple version of pit plugs lacking a cap structure and cap membrane in this freshwater bangiophyte.

The cap stains with periodic acid Schiff (PAS) and other stains of similar specificity, indicating the presence of polysaccharides, and with mercuric bromphenol blue, indicating the presence of proteins in all major groups of Florideophycidae (Trick and Pueschel 1991). However, the employment of FITC-labeled lectin probes has not given any positive result on the plugs in *Bossiella californica* (Corallinales), although cell wall components are readily stained. Of the 23 hydrolytic enzymes tested on this species only three, i.e., proteinase K, pronase E, and pancreatin, were effective, all of them on the cap layer but not the core (Trick and Pueschel 1990). This could be taken as indication that the proteins in the plug core are made inaccessible for the enzymes by masking with additional yet unknown plug components.

Despite intense work, the function of red algal pit plugs is still a matter of debate (Pueschel 1990). It is obvious that they constitute a direct structural linkage between cells and thus serve to stabilize the mechanical integrity of the thallus. When considering a role in cell communication, one may take into account that the plasma membrane between adjacent cells is continuous, and electrical potential differences can travel between cells, even though the septal pores are obstructed by plugs (Bauman and Jones 1986). In addition, however, it is likely that communication is accomplished by exchange of molecules through the plug. Some chemical factors indeed seem to travel from one cell to the next via the pit connection (Koslowsky and Waaland 1987). On the other hand, dye coupling does not exist in *Griffithsia*, i.e., lucifer yellow injected in one cell is prevented from diffusing to the adjacent cells (Bauman and Jones 1986). Although this renders a general transport function for pit connections unlikely, it does not rule out the nutrient transport function in specialized cells, as originally proposed by Wetherbee (Wetherbee 1979; Wetherbee and Kraft 1981). It is probably correct to assume that pit plugs can do more than one thing, but at present there is simply too little experimental evidence in favor of any general hypothesis.

### b) Structure and Behavior of the Interphase Nucleus

Nuclear shape can vary in the red algae, ranging from highly condensed forms such as those in the planetic nuclei of parasitic species during infection of host cells (Goff and Coleman 1984) to rather lobed and irregular forms found in early stages of carposporophyte development. Recently, most striking images of extremely irregular shapes have been obtained by using fluorescent DNA-stains in the endopolyploidic nuclei of *Scagelia, Wrangelia* and other Ceramiales (Goff and Coleman 1990b).

Pueschel (1990) speculated that rhodophytan nuclei may not contain a lamina, the fibrous layer lining the inside of animal (Bourgeois and Hubert 1988) and perhaps also higher plant nuclei (Beven et al. 1991). Indeed, such degenerations of shape as mentioned above are reminiscent to hereditary defects on nuclear lamina structure in some animal cell lines (Willingale-Theune et al. 1989).

Other unusual features are accumulations of vesicles, membrane stacks, and paracrystalline inclusions in the nucleus often associated with the inner face of the nuclear envelope (NE) and annulate lamellae (AL) outside and sometimes even inside the nucleus. Such "oddities" were observed throughout the 1970s (Brawley and Wetherbee 1981) and kept turning up in fine structural studies through the 1980s as well (Pueschel 1990) without any clarification of their functions. An interesting structural detail that has not been observed previously was the occurrence of microfilament-like bundles inside the nuclei and in the cytoplasm of fusion cells of *Faucheocolax* (Rhodymeniales) at the onset of their degeneration (Delivopoulos and Kugrens 1984a). Similar bundles of microfilaments have been seen in members of the Porphyridiales (Deason et al. 1983 and references cited there). It remains to be seen whether a function can be assigned to these structures.

#### α) Polygenomy

Polygenomy has interesting consequences for cellular morphogenesis. Nuclei in the Florideophycidae exhibit a large variability in their per-cell number, shape, and ploidy levels. This is particularly evident in the order Ceramiales. In these algae the nuclei are regularly, almost hexagonally, distributed throughout the parietal cytoplasm of both phases of the life cycle. Goff and Coleman (1987) have found that in the diploid phase (tetrasporophyte) nuclei were spaced twice as widely from each other than in the haploid phase (gametophyte). Consequently, a diploid nucleus controls a larger volume of cytoplasm with a greater number of chloroplasts than a haploid one.

Since intracellular transport of metabolites is not aided by organelle transport in red algae, this spatial arrangement may reflect the maximum distance over which gene products can translocate by diffusion. Goff and Coleman (1987) speculated that the cytoplasmic domains, each of which contains a single nucleus, may be considered as cell units without delimiting membranes and walls. In essence, a *Griffithsia* cell would then represent a "functionally multicellular" structure.

Polyploidy may have other benefits with respect to cellular fitness besides consequences for cell size and morphogenetic capacities. Goff and Coleman (1990b) speculate that it could serve as a genetic buffer minimizing the chances for deleterious mutations to take effect in the cell that carries the mutation, but still allowing the propagation of mutations to the following generation. This issue, however, is rather complicated by the fact that we do not know which and how many copies of the multiple genome are really utilized for gene expression in polygenomic cells.

#### β) Nuclear Transfer and Cell Fusion

The ability to fuse cells with one another has a profound impact on cell morphogenesis and hence has had remarkable consequences for the evolution of developmental strategies in red algae. For instance, cell fusion occurs as integral part in cellular differentiation following the fertilization of the carpogone (see Sect. 1f, below). In most red algae, the carpogone fuses with other supporting cells soon after karyogamy, and transfers its freshly formed zygotic nucleus or its derivative nuclei into the auxiliary cell, from which carposporophyte development eventually may be initiated. One feature that most species have in common despite many variations in postfertilization development, is that, due to the nuclear transfer, zygotic nuclei occur simultaneously in the same cell with vegetative haploid nuclei (Pueschel 1990).

Although the biological consequences of nuclear transfer are still very little understood, it may be expected that the extra nuclei will have an impact on the physiological activity and developmental capacity of the recipient cell. Apparently, problems arising from interferences between genomes are resolved at least in the order Ceramiales by separating the two types of nuclei either by segregating the haploid nuclei into so-called disposal cells (Huisman and Kraft 1992), or into the foot cell, one of the two daughter cells, resulting from the division of the auxiliary cell (Delivopoulos 1990). In the foot cell both nuclear generations are mixed, whereas the other daughter cell (the gonimoblast) receives only one diploid nucleus and is, therefore, the founding cell for the diploid carposporophyte generation.

In the foot cell, the diploid nucleus takes over by an as yet undiscovered mechanism, which overrides whatever control the haploid nuclei were exerting on cell function. Only rarely do haploid nuclei coexist with diploid ones without being degraded (for instance in members of the Gigartinaceae complex, see Hommersand and Fredericq 1992, and in artificial parasexual hybrids, see below).

### γ) Filament Repair

Fusion processes also play an important role in filament repair following wounding. Waaland and coworkers (summarized in Waaland 1990) have studied this phenomenon in cell filaments of several species in the order Ceramiales. They found that upon destruction of the cell content of an intercalary cell, repair cells are cut off from the immediately adjacent cells. The repair cells grow towards each other and fuse so that the original status of the filament is restored. Principally, intercalary cells in the Ceramiales do not divide, with the exception of those that give rise to side branches. The question, therefore, is how these adjacent cells become re-programmed to enter the division cycle.

In *Callithamnion* and *Platythamnion*, both uninucleate genera of the Ceramiales, the repair mechanism has become somewhat modified in that repair cells do not fuse but establish contact by secondary pit connections (*Callithamnion*) or repair cells do not form at all; instead the cells above and below the destroyed cell begin to elongate, grow towards each other, and connect by secondary pit plugs (*Platythamnion*, Kim et al. 1988).

Waaland and coworkers have succeeded in purifying and chemically characterizing a peptide hormone, rhodomorphin, which species-specifically elicits repair cell formation (Watson and Waaland 1986). It is apparently produced by the repair root cell and guides the direction of growth of both cells towards each other. One can therefore speculate that re-programming involves gene regulation by rhodomorphin, but initially other factors should be required to trigger hormone production.

The two cells that are formed in response to wounding and destined to fuse are structurally and functionally different from each other and also from the intercalary cells. Apparently, the route of differentiation that follows division is dictated by the cellular environment. A cell apical to the wound will always form a root-like repair cell and the one issued from the cell basal to the wound will become a shoot-like repair cell (Waaland 1990).

Exogenously applied rhodomorphin can induce repair shoot cell formation and eventually allows fusion to occur between somatic cells of two separate plants, even if one is haploid and the other diploid, or one is male and the other female (Waaland 1978).

In a recent study, Hwang and coworkers (1991) succeeded in producing hybrids between male and females and hybrids between cells of either sex with cells of the tetrasporophytic stage. The resulting fusion cells developed three types of reproductive structures (spermatangia, carpogones, and tetrasporangia) when they were hybrids between male and female, and two types (one gametic stage and tetrasporangia) when they were hybrids between either sex and tetrasporophytes. In the case when all three reproductive structures were formed on one plant, tetrasporangia and spermatangia were always borne on the same cell, whereas carpogones were always separated from them on different cells. Since some of the tetraspores were viable, Hwang et al. hypothesized that a certain number of nuclei had actually fused to produce zygotic nuclei and that these are the ones that initiate formation of the tetraspore mother cell. They further speculate that the same cell could then cut off spermatangial mother cells due to the presence of active male nuclei, whereas the female nuclei remained quiescent. By unequal segregation of nuclei in the course of cell division, some daughter cells could have received more female than male nuclei alongside the diploid nuclei. It is, therefore, possible, that under these circumstances carpogones develop. This is consistent with the finding that female/tetrasporophyte hybrids can form carpogones from one cell and tetraspores from another. Clearly, at this point, more cytological and fine structural data are needed to understand how mixed reproductive phases can be produced from one cell.

Hybrids can also be formed between geographically distant isolates of the same species. Surprisingly, such hybrids revealed incompatibilities in their plastomes, resulting in the selective destruction of one of the two rhodoplast populations (Koslowsky and Waaland 1987). In conclusion, these advances in cell fusion work justify the expectation that parasexual hybrids in red algae could soon provide the means in future work to study control mechanisms on the gene level between plastomes in the same cytoplasmic environment and between a foreign plastome and the nuclear genome.

### δ) Host-Parasite Interactions

Cell fusion constitutes the major cytological event associated with red algal host-parasite interactions (Goff 1982). In each case of fusion, at least one nucleus and sometimes numerous nuclei are transferred from the donor to the recipient cell (Goff and Coleman 1990b). Although the experimental evidence is still scarce, it is very likely that the host cell metabolism is at least partially placed under the control of the transferred parasite nuclei, as discussed above for the situation during carposporophyte development.

In the case of adelphoparasites the infecting nuclei continue to divide and eventually outnumber the host nuclei (Goff and Coleman 1990b). In alloparasitic interaction, the parasite nuclei remain small with highly condensed chromatin suggesting reduced synthetic activity, but even then changes in host cell morphology indicate some parasite control over the host metabolism (Goff and Coleman 1984).

Genome transfer in red algae is a fascinating new area of research, and it should soon become possible to select suitable experimental model systems for a rigorous study of the control of cell development involving manipulations of the gene dosages and/or the introduction of alien genetic information via heterokaryosis into compatible cellular environments. This possibility has recently been extended even further by the discovery of red algal plasmids, and expectations might indeed be justified that these plasmids may be suitable for gene engineering and transformation in analogy to bacterial plasmids (Goff and Coleman 1990a).

### c) Cytoskeleton, Cell Division, and Mitosis

Patterns of mitosis and cell division relate to the mode of growth and developmental differentiation. How these may eventually link up with the formation of a species-specific thallus organization has recently been discussed by Coomans and Hommersand (1990). One of the major unresolved questions currently is, what actually happens on the level of the cell? Which factors control the plane of cell division and dictate the development of the progeny cells? Much of the regulation of these processes will most likely involve the cytoskeleton. However, detailed information on the red algal cytoskeleton is just beginning to emerge, so that we can only now start to fit together the first available pieces of information.

Scott and Broadwater (1990) present a compilation of all currently available fine structural informations on cell division in the red algae and come to the tentative recognition of five patterns of mitosis which may be grouped with the type IIa/IIb spindles sensu Stewart and Mattox (1980). This type of spindle is either completely closed or may be fenestrated at the poles.

Characteristic for red algae is the presence of the nucleus-associated organelle (NAO), which in the majority of species is ring-like, duplicates some time between telophase and prometaphase, and behaves during mitosis much like the centrioles in protists and animals.

Depending on the species, the NAO can form intimate associations with the nuclear envelope during prophase. The most intriguing one is the formation of a small tubular extension of the outer nuclear envelope that runs through the center hole of the NAO and then possibly links up with the ER system. The formations of depressions in the NE, which in the extreme may transform to transnuclear channels preceding polar fenestration, takes place in the immediate vicinity of the NAO and may therefore be caused by it in an as yet unknown way.

Cytokinesis involves a fibrous ring at least in the unicellular bangiophytes (Scott 1986) in analogy to the contractile ring found in many other protists, algae and animal cells, whereas it is not yet clear by which mechanism the edge of the incipient cleavage furrow in the higher red algae is driven.

A variant of cytokinesis, termed plastokinesis (Oates and Cole 1992), has recently been reported in *Bangia vermicularis*. This type involves multiple narrow cleavage channels which are sequentially initiated from the periphery and progress towards the center of the cell. The rhodoplast is cleaved by this process long before karyokinesis occurs. The cleavage channel is a meandering, ill defined membrane invagination. It partitions the cytoplasm, but at the same time the two newly created membrane surfaces remain appressed to each other until after karyokinesis, when polysaccharide material begins to become extruded by lomasome-like membrane specializations. At present, it is difficult to envision the mechanism maintaining spatial order and driving the formation of the cleavage channels in this type of cleavage.

The current status of our knowledge on the cytoskeleton in the red algae has been discussed recently by Scott and Broadwater (1990), who state that immunofluorescence work is just starting to be applied to selected species. Meanwhile, Garbary et al. (1992) have succeeded in visualizing, for the first time, actin filaments in several species of red algae. Fluorochrome-labeled actin filaments were seen as a fine longitudinally oriented network of fibers, rope-like structures, and spikes. They also showed actin staining in the cleavage furrow, thus confirming the contractile ring hypothesis for the higher multicellular red algae. Surprisingly, actin staining occurred also in the areas of pit plugs, but clearly, a final evaluation of the exact conformation of the actin cytoskeleton will have to wait until the staining protocol has been refined further and more species have been examined.

The same may be said about MT staining. Garbary et al. (1992) managed to visualize MTs by indirect immunofluorescence microscopy in only one species (*Griffithsia pacifica*). MTs were localized around the nuclei and in extensive arrays in the cytoplasm so as to hold the nuclei in place.

### d) Vegetative Morphogenesis

Thallus organization in the red algae is governed to a large extent by growth and division of the apical cells. In addition, changes in cell shape, particularly cell enlargement, play an important role in cellular morphogenesis and hence for the final shape of the thallus.

#### α) Cell Expansion

Intercalary cells in the filament axis of ceramialean species may increase in length by a factor of 100 and more (Waaland 1990). It has long been thought that this type of growth should be accomplished by the localized extrusion of wall loosening enzymes in combination with deposition of new wall material. The use of so-called fluorescent brighteners has made it possible to actually visualize these areas on the light microscopical level.

Using this technique as long as two decades ago, Waaland et al. (1972) discovered a band-like expansion of intercalary cells in the Ceramiaceae. Since then, further work has led to the recognition of three different types of cell elongation in the red algae (Garbary and Belliveau 1990), elongation of the tip cell by apical deposition of wall material, band-like expansion of intercalary cells (Ceramiales), and expansion of intercalary cells by diffuse deposition of wall material (*Audouinella*, Acrochaetiales).

Garbary et al. (1988) showed that elongation and specifically the position of cell elongation bands are controlled by the apical cell. After removal of this cell, all axial cells located further down the filament exhibited a perturbed pattern of cell wall growth either by formation of additional elongation bands between the major apical and basal bands or by a general diffusion of the banding pattern.

Recently, Sylvester and Waaland (1992) provided evidence for the operation of a second mechanism of cell wall formation, which is restricted to the middle region of the cell and consists of the massive extrusion of large polysaccharide filled vesicles. This process occurs during daytime when the chloroplasts accumulate as a dark band coinciding with the zone of vesicle extrusion.

### β) Side Branch Initiation

The pattern of branching in the Ceramiales, i.e., whether a cell bears one or two side branches, and the morphology of the carpogonial branch are thought to be genetically determined. However, the decision whether a side branch is formed or not and hence the exact distance between branches along the filament, including vegetative and carpogonial branches, seems to be more flexible and could depend on growth conditions and light intensity (Waaland 1990). This also seems to apply for members of other red algal orders such as the Gelidiales (D'Antonio and Gibor 1985).

The morphogenetic capacity of axial cells leading to the formation of the cladomothallic architecture in the Ceramiaceae is determined by the growth characteristics of the tip cell and a number of subapical cells. Lambert et al. (1990) addressed the question in *Antithamnion* as to what may distinguish a subapical axial cell that gives rise to a cladome branch from another one that give rise to a pleuridium. They find indeed spatiotemporal parameters that differ, so that one can predict the fate of a cell at a very early stage in development. However, this does not tell us yet what makes two cells develop differently.

### γ) Protoplasts and Cell Culture

Due to the complexity of cell wall chemistry and the lack of suitable cell wall-degrading enzymes, studies on algal protoplasts have been lagging behind until recently. This situation has changed since enzyme preparations have been isolated from marine bacteria (*Pseudomonas carrageenovora*, Gross 1990) or from the gut extracts of marine snails and mollusks (Polne-Fuller and Gibor 1990). This development has been promoted by the expectation that the protoplast technology could be exploited for the selection of new strains and for improving crop yields of commercially important agarophytes.

Only few researchers have recognized the value of protoplasts for the study of cell development and tissue differentiation. Polne-Fuller and Gibor (1984) have shown that protoplasts, derived from the five principle cell types present in the *Porphyra* blade, regenerate differently, depending on the developmental program of the original cell type. Actively dividing cells from the center of the blade apparently reproduced their in situ morphology, whereas vegetative cells from the areas between the patches of generative tissue at the apical margins of the blade adopted a more filamentous morphology (see also Wang et al. 1987). The presence of sorbitol or the condition of nitrogen starvation caused blade cell protoplasts to reform a highly aberrant cell wall with large gelluous blisters attached to it, suggesting that a metabolic shift towards higher synthesis and structurally uncontrolled extrusion of polysaccharides has occurred (Polne-Fuller and Gibor 1990).

In a somewhat different approach, Liu and Kloareg (1991) studied the effects of growth-stimulating hormones on tissue explants of *Porphyra umbilicalis* (Bangiales). Three-dimensional calli developed at the cut edges only in the presence of both naphthalene acetic acid and cytokinin. The authors suggest that this does not necessarily mean that both compounds exist and have a natural function in the alga, but rather that endogenous growth regulation became artificially overridden by the exogenous growth hormones. As yet there has been no success in inducing differentiation in such calli. However, it may soon be possible to apply more physiological conditions, since cytokinins and indole derivatives have recently been identified in red algae (Bernart and Gerwick 1990; Zhang et al. 1991), and there is an increasing awareness of the possible role of growth hormones for the control of development in algae (Bradley 1991; Evans and Trewavas 1991). Regeneration has not been studied on the fine structural level, and it is not known whether pit connections form or how else cells adhere to each other. The situation could be rather reminiscent of spore development, for instance, in a florideophycean species, *Pterocladia cappillacea* (Gelidiales, Liu et al. 1990); single cells become easily separated from the callus and start new calli, indicating that no stable connections form.

### e) Spermatangia and Spermatia

Information on the structure and development of spermatia is relatively scarce. One distinctive feature of the mucilage-covered, flagella-less cells is the formation of spermatangial vesicles (Pueschel 1990). Their amorphous or fibrous contents, which may undergo changes in appearance with maturity, is destined for extrusion upon release of the spermatia (i.e. see Fetter and Neushul 1981). Spermatangial vesicles are the product of the FVAO, as discussed above (Sect. 1aδ).

Recently, a new type of surface sculpturing has been discovered in *Spyridia filamentosa* (Ceramiales), where the spermatia are decorated by six to eight stalked knobs (Broadwater et al. 1991). The knobs are preformed in the spermatangial vesicles, whereas the stalks begin to form interposed between the extruded knob and the cell surface and elongate to a length of 1 μm after release of the spermatia from the spermatangia. This causes the appendage to remain anchored on the cell surface.

The electron micrographs presented by Broadwater and coworkers show concave depressions in the cell surface underlying each of the forming stalks and the contour of the membrane in this area exhibits a type of small indentation reminiscent of those seen in specialized membrane depressions responsible for cellulose spike formation in the green alga *Acanthosphaera zachariasi* (Chlorococcales) by Schnepf et al. (1982). One could therefore speculate that this particular membrane specialization may be involved in the rapid synthesis of cellulose or another linear polysaccharide.

### f) Carpogone and Postfertilization Processes

Cytological features of the female reproductive tissue has received immensely broad coverage in the red algal literature. The basic principles of the cytoarchitecture of its major cell types have been identified, and many of the cytological details such as types of vesicles, organellar associations, pit plug degradation, and membraneous structures have been described for various species over and over again (Brawley and Wetherbee 1981;

Pueschel 1990). It has also been found that post-fertilization events are precisely timed (O'Kelly and Baca 1984). Yet the role of a particular cell for the developmental fate of the carposporophyte and the functional meaning of many of the observed intracellular components for a given cell are still widely debated.

The peculiar feature of postfertilization development in the red algae is that the diploid phase of the life cycle remains physically connected to the female gametophyte by establishing a parasite-like relationship with it. This is characterized by various patterns of cellular fusion and nuclear transfers (see Sect. 1bβ, above) between the zygotic tissue (carposporophyte) and the haploid tissue (gametophyte) followed by special adaptations in morphology and functions of the vegetative cells of the gametophyte in the immediate contact area.

On the side of the carposporophyte, a sequence of cell differentiations takes place, leading from the formation of gonimoblast initials to gonimoblasts and finally to carpospores. This principle sequence has undergone numerous modifications, bypasses, shortcuts, and elaborations. A detailed description of these processes on the level of the light microscope has recently been published by Hommersand and Fredericq (1990).

In the Calosiphoniaceae (Gigartinales), the fertilized carpogone gives rise to multiple connecting filaments which make contact with distantly located auxiliary cells. This developmental pattern produces multiple carposporophytes from a single fertilization event (Wilce and Sears 1991). In the Gracilariales an even more profoundly proliferating gonimoblast tissue has evolved that differentiates into functionally distinct subtypes. One is destined to form carpospores, whereas the other is vegetalized and establishes numerous nutritive channels by fusion with each other and with underlying cells of the cystocarpic floor (Fredericq and Hommersand 1989).

### α) Interaction of Spermatia with the Carpogone

Contact between the spermatia and the carpogone, the receptive cell of the female gametophyte, usually takes place at the trichogyne, a hair-like extension of the carpogone. The adhesive substance deposited either on the trichogyne or spermatia surface is not yet known, but is has been observed that it does not confer strict species specifity, since spermatia from several related species may adhere to the same trichogyne, and their respective nuclei may actually enter the trichogyne (Magruder 1984). Since the small trichogyne diameter allows the passage of only one nucleus at a time, the possibility for polyspermy is rather low; nevertheless, contraction of the trichogyne cytoplasm away from the freshly fused zygote nucleus indicates that mechanisms exist to prevent polyspermy.

The question of which aspect of the cell wall chemistry in a bangiophyte changes in order for the trichogyne or the carpogone surface to become receptive has been examined by Cole et al. (1985). It turned out that the number of sulfate groups on the wall polysaccharides decreased, whereas the content of carboxylated carbohydrates in the walls increased. This does not yet explain the mechanism of adherence, but it is consistent with the view that sulfate-modified polysaccharides confer a stronger resistance to the cell wall and hence their removal should be a prerequisite to facilitate the formation of the entrance canal in the trichogyne wall by hydrolytic enzymes and subsequent cell-cell interaction during fertilization.

### β) Recognition Events Associated with Nuclear Transfer

The advantage of nuclear transfer for carposporophyte development is still an enigma, but it has been speculated that cell contact preceding fusion of the carpogone or a connecting cell issued from the carpogone with neighboring cells requires a recognition step which could provide a more specific selection barrier at which zygote nuclei that are a product of crossing between incompatible species can be rejected (Boo and Lee 1983).

Even when the carpogone has been fertilized by a compatible spermatial nucleus, recognition processes in post-fertilization development are still very important and appear exceedingly complicated, as in genera of red algae that initiate connecting filaments from the carpogone in order to reach distantly located auxiliary cells. This behavior requires directional guidance and communication between cells (Knauss and Hommersand 1989). In analogy to the repair mechanisms discussed above, the involvement of diffusible hormones generated by the auxiliary cell may be suspected.

### γ) Fine Structure of Carpogone and Carposporophyte

Details of the migration of sperm nuclei through the trichogyne and fusion with the carpogonial nucleus have not been observed on the fine structural level. The exact mode of nuclear transfer from the carpogone to the auxiliary or any other cell is also still undetermined.

Some of the complexity on the fine structural level that carposporophyte development involves may be derived from the work on *Polysiphonia* (Ceramiales, Wetherbee 1980; Broadwater and Scott 1982) representative of many other studies (Pueschel 1990).

Prior to fertilization in *Polysiphonia*, densely packed smooth ER extends like a channel from the carpogonial cell through the three cells of the carpogonial branch to the support cell, interrupted only by pit connections. Broadwater and Scott (1982) suggested that this arrangement indicates an increased transport/communication between the carpogone and the support cell through this file of carpogonial branch cells. This arrangement is consistent with the idea that at the moment of fertilization a chemical message is released from within the carpogone, which is then transported through the SER/pit connection channel to the support cell in order to initiate the formation of the auxiliary cell and synchronize the events of subsequent fusion of the auxiliary cell with the carpogone. The same cells later dissolve their pit connections, to establish continuity. In this stage the SER channels are no longer obvious.

The factors controlling the selectivity, direction, and extent of pit plug degradation are not known. Delivopoulos and Kugrens (1984a) noted that the successive inclusion of neighboring cells into the fusion cell system in *Faucheocolax* (Rhodymeniales) involves degeneration of pit plugs progressing from the fusion cell side, suggesting that degrading enzymes may be produced in the fusion cells and then spread centrifugally to the neighboring cells. The same situation is seen in other species as well (Wetherbee 1980), but it is also possible that local control is exerted by a proliferating network of ER tubules, often seen next to the degenerating plugs. In the case of *Polysiphonia* (Broadwater and Scott 1982), only those pit connections are bound to dissolve whose cones are not covered by an electron-dense membrane, indicating that it is the composition and structure of the pit plugs that determine its fate.

In the Bangiophycidae, the cytology of carpogone development and postfertilization processes is much simpler. In most species, carpogones become distinguishable from vegetative blade cells by the appearance of a shallow surface protuberance representing the trichogyne to which spermatia attach. Carpospores form directly from the fertilized carpogone. In *Porphyra abbottae* a somewhat more complex pattern of carpogone development has been reported (Cannon 1989), where unusually large procarpogonial mother cells are formed on the blade at distances of several 100 μm apart from each other. Clusters of carpogones are cut off from these cells in a burst of mitotic activity. During this time, the cytoplasmic content of the procarpogonial mother cells is said to appear "pulsating" (Cannon 1989), which probably refers to some kind of hitherto unknown organelle motility.

## g) Sporangia and Spores

Due to the complex, often tripartite, life histories of red algae, several types of spores may be formed either to repeat the identical phase of the life cycle or to initiate the next stage. Four major spore types (monospores, carpospores, tetraspores and polyspores) may be distinguished and these may be grouped in as many as 11 variant forms (Guiry 1990). Despite this diversity, the general fine structural character is very similar in all types of spores and has been amply documented in earlier reviews (Brawley and Wetherbee 1981; Guiry 1990; Pueschel 1990). Formation of spores also shares many fine structural characteristics, although there might be differences in details. These will be discussed in the context of tetraspore formation, for monospore and polyspore formation see Hymes and Cole (1983) and Sheath et al. (1987).

### α) Tetraspore Formation

A few detailed fine structural accounts of tetrasporogenesis have been published during the past decade. Vesk and Borowitzka (1984) have reviewed the older literature and compared their data of the zonately dividing tetrasporangia of the coralline species *Haliptilon cuvieri* with the situation in other species possessing tetrahedally or cruciately dividing sporangia. Another more recent study has provided additional information on meiosis preceding tetraspore formation in *Dasya baillouviana* (Ceramiales, Broadwater and Scott 1986). Both studies place much emphasis on the various elaborations of the ER (see also Sect. 1aδ). It is accumulated in massive polar complexes at meiosis I in *Dasya*. In *Haliptilon*, the ER exists in the form of several layers of circumferential cisterna around the nucleus, and is embedded in an unusually dense cytoplasmic matrix (PNER).

This extensive PNER undergoes a remarkable change in orientation from circumferential to radial relative to the surface of the nucleus at the termination of meiosis. Since the degree of elaboration of the PNER varies among species and also the exact stage of sporogenesis at which it appears, it is questionable whether it serves a function vital for the mechanism of sporogenesis in general. Vesk and Borowitzka (1984) have suggested that it may rather reflect a general principle in the red algae to compartmentalize nuclear domains in multinuclear cells (see also Phillips and Scott 1981). Such a principle may also be realized in other multinucleate algae (Liddle et al. 1982).

With respect to the electron-dense appearance of the cytoplasmic matrix around the nucleus, the authors have drawn a parallel to the perinuclear dense bodies in *Acetabularia* (Dasycladaceae,

Chlorophyta), which are thought to contain extranuclear DNA and RNA and could represent organelles of supplementary extranuclear RNA processing and storage (Berger and Schweiger 1986). With this in mind, Vesk and Borowitzka (1984) have offered the speculation that the particular arrangement of endomembranes seen in *Haliptilon* could be part of an extremely optimized synthetic machinery involved in vital aspects of sporogenesis, i.e., spore wall formation.

Another interesting feature described for the first time in *Haliptilon* sporogenesis is the appearance of circumferential inner wall ridges in the cell division plane at the onset of cytokinesis. These ridges are composed of a conspicuous paracrystalline material of unknown composition. They appear to act as initiation sites, from where the cleavage furrows begin to grow inwards. Nothing is yet known on the origin of this material and the cellular structures which aid in its positioning. The formation of a pre-cytokinetic ring-like domain in the future division plane is known from bacteria to higher plants (Bi and Lutkenhaus 1991; Loyd 1991), but in all cases the mechanism that spatially controls cytokinesis has not yet been discovered.

Invaginating membrane tubules have been described in tetraspores of *Haliptilon* (Vesk and Borowitzka 1984) and in other red algal spores (Avazini and Honsell 1984; Mandura et al. 1986). In *Haliptilon* they appear at the plasmalemma surface concomitant with the formation of an electron-dense wall layer around the spore. The membrane tubules are open to the exterior and they are implicated in the direct extrusion of wall material from the ER compartment just as discussed above (see Sect. 1aδ, above) for the peripheral ER system in *Rhodella* (Patrone et al. 1991).

Tubular structures become also associated with plugs between carpospores and supporting cells, particularly in the stage of pit plug degradation that precedes spore release. Delivopoulos and Kugrens (1985) tend to believe that in *Gloiosiphonia* (Cryptonemiales) these tubules represent membrane channels directing hydrolytic enzymes towards the plugs. However, Delivopoulos and coworkers (Delivopoulos and Kugrens 1984b; Delivopoulos and Diannelidis 1991) have found similar structures in *Ceramium strictum* and *Faucheocolax* associated with plugs between mature carpospores, and this time they referred to them as contorted microtubules.

#### β) Spore Germination

Spore germination in many red algae does not proceed by expression of growth patterns commonly found in the adult thallus. It rather builds up stepwise by successively invoking intermediate morphogenetic programs. These appear to produce a base from which the adult thallus develops.

Well-known examples are the Corallinaceae, where spore germination involves the formation of a disk stage containing eight or more cells which are arranged in a species-specific geometrical pattern, and are produced by division of the spore within the confinement of the spore wall. Formation of the mature thallus proceeds from this germination disk (Chamberlain 1984; Jones and Woelkerling 1984).

Similar germination discs develop in the Gigartinales where, however, the sporeling morphology seems to depend on the density of spores per surface area. Beyond a threshold density, the morphology tends to become more filamentous, reducing the contact area of the individual spore with the support, so that these spores can be removed by water currents more easily (Sylvester and Waaland 1984). This behavior speaks again for some kind of chemical signal released from and processed by the spores as discussed above in conjunction with repair mechanisms (Sect. 1bγ).

Spore germination in the Bangiales leads directly to the filamentous thallus structure of the conchocelis phase (sporophyte), whereby the first septum is positioned either after germination in the neck of the outgrowth or it partitions the spore before the germination tube is formed (Pueschel and Cole 1985). This implies a lack of spatial control over cytokinesis as opposed to the situation in *Haliptilon* tetrasporangia discussed above.

## 2. Phaeophyta

A significant part of macrothallic benthic algae of the worlds marine flora is represented by the brown algae, some of which reach tens of meters in length. There is not a single unicellular taxon in this algal division. This has fueled discussions on the nature of the direct ancestors (Clayton 1984). However, considering the architecture of the cell, and particularly of the unicellular reproductive zoids (see below) one is impressed by the many features they share with a sister class of algae, the Chrysophyceae (Clayton 1989). Even more astounding is the fact that zoid ultrastructure is remarkably constant in all phaeophytan taxa, whereas vegetative thallus morphology can vary widely from inconspicuous, microscopically small filaments to giant kelps.

### a) Vegetative Cell Structure

The general features of cell ultrastructure in the Phaeophyta are well known and have been the subject of numerous studies (Brawley and Wetherbee 1981).

Stalked pyrenoids protrude from the chloroplasts into the vacuole and are surrounded by an electron-translucent sheath of starch (chrysolaminarin). An ER cisterna (CER) in connection with the nuclear envelope (NE) wraps around the chloroplast, a circular plastid genophor runs along the chloroplast margin, and cells are connected through the cell walls by numerous pit fields. Furthermore, MTs are frequently seen between the CER and the chloroplast envelope. Dictyosomes occur close to the nucleus directly at one side of the centrioles. This association is apparently typical of many brown algae (see also Pueschel and Stein 1983) but not all. In some orders (Fucales, Dictyotales, Sphacelariales), dictyosomes are lined with their forming faces parallel to the nuclear surface (Katsaros and Galatis 1985, 1988) but they do not keep a particularly close association with the centrioles. As in the Rhodophyta, numerous inclusion bodies occur in brown algal cells, the most prominent ones are physodes and iridescent bodies (Brawley and Wetherbee 1981; Pellegrini et al. 1989).

Fine structural preservation has been notoriously difficult in the Phaeophyta (Hallam and Luff 1988). Damage to the cell structure by routine fixation methods is thought to be caused by the highly reactive phenolic contents of the abundant intracellular physodes. Clayton and Beakes (1983) systematically tested fixation protocols and came to the conclusion that addition of caffeine in combination with glutaraldehyde/osmium fixation gave the best results, possibly because caffeine improved retention of the phenolic material within the physodes.

Phaeophyta grow either by apical growth or by intercalary meristems. Apical growth has some unusual features not seen, for instance, in the Rhodophyta. Usually, the apical cell actively divides in one or consecutively in varying planes thus contributing to the growth of the thallus. However, Clayton and Shankly (1987) established that this is not so in *Splachnidium rugosum*, a species with uncertain systematic affinities. Here, the apical cell is embedded in a depression of the apical meristem, but does not divide itself; it merely induces its neighboring cells to continually divide and differentiate cortical cells, hair cell initials, and conceptacle initials.

The apical cell in *Splachnidium* has a pronounced polar organization, with the pear-shaped cell body protruding outward and a long basal appendage (tail) penetrating between the neighboring cells into the central, slime-filled cavity of the thallus. The nucleus in the basal part of the cells is surrounded by numerous dictyosomes, and masses of ER fill the tapering part of the tail. This is taken as an indication of synthetic activity possibly in conjunction with growth hormone production. Interestingly, the conceptacle initials, which begin to differentiate subapically from the meristem, repeat the developmental pattern seen in the apical cell. They build up the same cell shape: pear-shaped cell body and tapering basal portion extending between the cells of the cortical tissue into the thallus cavity, and they promote mitotic activities in their immediate surrounding. This, in turn, leads again to cellular proliferation and eventually to the formation of the conceptacle. Why only these cells form numerous plasmodesmata with their neighboring cells, whereas the tip cells do not, has not been clarified.

Members of the Fucales also possess a quiescent apical cell, which is thought to be the source of growth stimulation for the surrounding meristem. Because of the similar growth pattern, Clayton and Shankly (1987) propose that the fucoid taxa and *Splachnidium* share a common ancestor.

The developmental repertoire of a tip cell may even include both growth stimulation and active participation by cutting off daughter cells. In *Hormosira banksii* (Fucales) it is a group of four tip cells that control growth of the apical meristem in alternating phases causing the development of a segmented, cylindrical thallus (Clayton et al. 1985).

Examples for actively dividing tip cells are seen in *Dictyota* species. Katsaros and Galatis (1985) studied the division pattern of the tip cell and the adjacent files of subapical cells leading to the typical dichotomously branched, trilaminate thallus. The division of the tip cell is asymmetric, producing a large, apical and a flat discoidal, subapical cell. It is not clear in this and in any phaeophytan cytokinesis, how the cleavage membrane is positioned, and there are no obvious fine structural indicators in *Dictyota* that could explain the unequal size of the two daughter cells except for a slightly asymmetric position of the division nucleus.

Division of the subapical cell (termed discoid segment) is again asymmetric, leading to a large and a smaller epidermal cell (Katsaros and Galatis 1985). The larger cell divides again, producing the medullary initial and another epidermal cell. Hence, the most recently formed epidermal cell and the medullary cell are sibling clones. However, it is evident that they follow different developmental programs, whereas the cells of upper and lower epidermis, which are not direct siblings, develop into structurally and functionally indistinguishable cell populations. At present, concepts are much desired that could help to identify the morphogenetic switches controlling the fate of these cells.

Numerous fine structural differences are becoming manifest in almost any organelle between medullary and epidermal cells of *Dictyota* starting only four cells behind the tip cell (Katsaros and Galatis 1985). Most dramatically, the medullary cell continues to increase in size simply by enlarging its vacuolar compartment, whereas the epidermal cell will have to undergo two or three more divisions to match this amount of expansion. In the anticlinal walls of the epidermal cells, cell wall thickenings form by an as yet undetermined mechanism, and physodes become almost exclusively positioned against the external cell walls.

In *Dictyopteris membranaceae*, another species of the Dictyotales, cortical cells underneath the epidermal layer resemble the medullary cells of *Dictyota* except for the increased thickness of their walls. In addition, a third cell type emerges in the center of the blade causing the development of prominent axial midribs in the thallus. Because these cells are connected among each other by extended primary pit fields, Katsaros and Galatis (1988) speculate that they have a function in photoassimilate transport as has been known for the trumpet cells (sieve element-like cells) in the Laminariales, although the pore diameter is smaller in the midrib cells.

In filamentous thalli like those of *Sphacelaria*, tip cells are capable of regenerating an entire new thallus when separated from their subapical cells (Ducreux 1983). With regard to the regeneration capacity, the subapical cell is almost as competent as the tip cell. Ducreux (1984) has studied the morphogenetic reprogramming of the subapical cell during apex regeneration in *Sphacelaria cirrosa* (Sphacelariales). He demonstrated that after removal of the apical cell, the subapical cell will either transform directly into a new tip cell, or will, depending on its age, divide first into two smaller nodal cells of which the top one becomes the new apical cell. This indicates that the capacity to invoke a new morphogenetic program is cell cycle controlled.

In addition, Ducreux found evidence that there must be control over the subapical cell by the subtending cells as well. If, for instance, all cells subjacent to the subapical cell are also removed when the tip cell is removed, a different pattern of regeneration occurs. The subapical cell runs first through at least four successive transversal and a number of more longitudinal divisions, creating several tiers of small cells without increasing the diameter of the filament. Only after this nodal morphology has been established, can one or two new axes differentiate at the apical region of the regenerate and a rhizoid from the basal region.

However, if the subjacent cells had been removed more than 12 h after removal of the tip cell, a normal apical cell was regenerated. It is very likely, but not proven, that specific growth hormones are exchanged between cells, creating polar gradients within the filament axis, which, in turn, affect the polarization and mitotic activity of the cells. In this context it may be interesting to note that cytokinins have indeed been isolated from brown algae (Zhang et al. 1991).

On the cytological level, the only perceptible character that forecasts the beginning of reprogramming was a slight relocation of physodes and the nucleus towards the apical cell pole. The mechanism by which the new polar distribution of organelles was affected is not known.

Katsaros and Galatis (1990) have recently addressed apical regeneration in *Halopteris filicina*, another species of the Sphacelariales, that exhibits a somewhat more complex morphology. The authors report that after the destruction of the apical cell, the next one to three cells subjacent to the apical cell degenerate, before the next cells further down begin to regenerate often several new apical cells. Unfortunately, this interesting behavior has not been followed up further. The authors have speculated, however, that apical regeneration requires the cells to dedifferentiate, before tip growth activity commences, but they have presented little evidence in support of this idea.

Increase in thallus diameter or development of specialized tissue may take place independently of apical growth or intercalary meristems. This activity is due to the capacity of the epidermis (meristoderm) to remain mitotically active. One of many examples is the formation and differentiation of a mucilage duct system in the blades of the large kelps (Laminariales). Grenville et al. (1982) showed that rows of secretory cells are cut off from the meristoderm toward the thallus interior. These cells proliferate further and separate from each other along their anticlinal walls so that a continuous cavity is formed, which runs as a branched network through the cortex of the entire blade. Histochemical evidence has indicated that the ducts are filled with acidic and sulfated polysaccharides and some lipid. The duct system is apparently not involved with sieve element formation or function. It is already apparent in young plants before the photosynthate system has developed. There has also been no unequivocal evidence for the existence of openings to the blade surface. Hypotheses for the function of the duct system range from a role in wound response and grazing deterrent, resistance to desiccation, and osmoregulation.

### b) Mitosis, Cytokinesis, and the Cytoskeleton

Taking up again on the issue of unequal cell division, two types of division have been observed in the highly polarized tip cells of *Sphacelaria tribuloides* (Katsaros et al. 1983), both being asymmetric. One separates the apical part that is rich in cytoplasm from the vacuolated lower part, thus giving rise to a new tip cell and a subapical, axial cell. The other type of division cuts off a small subapical lateral protuberance from the rest of the cell, which gives rise to a hair cell. While the hair cell grows out, the tip cell continues to elongate. In the latter case the nucleus is positioned closer to the tip prior to division, the polar axis is tilted towards the newly formed subapical protuberance and the cleavage plane is positioned in the neck of the protuberance, whereas in the former case the nucleus remains at its position in the apical one-third of the cell, and the cleavage plane cuts at right angles through this cell. All other fine structural features of cell division are very similar in both types of division. The nucleus is associated neither with dictyosomes nor with ER at any stage of mitosis.

Katsaros et al. (1983) show that nuclear fenestrations are formed at the poles some time before prophase and the rest of the nuclear envelope breaks down in late anaphase to reassemble at telophase. Electron dense kinetochor-like structures are visible throughout metaphase and anaphase in *Sphacelaria*. The daughter nuclei separate widely and regain interphase configuration before the septum is formed. Again, it has not been established by which mechanism the septum forms. Katsaros et al. (1983) report that it grows centripetally, but unlike a cleavage furrow, it appears interrupted and wall deposition proceeds nonuniformly. The process of plasmodesmata formation in this wall is also unclear.

In *Carpomitra cabreae* (Sporochnales, Motomura and Sakai 1985) polar depressions are very pronounced and an unusually dense array of microtubules radiates from the centrioles into the depression. The MTs seem to anchor in an electron-dense lining on the cytoplasmic face of the nuclear depression. Similar electron-dense layers have been observed in phaeophytan taxa, where MTs coming from the pericentriolar area towards the nuclear depression are more disperse

(*Cutleria*, La Claire 1982). In *Carpomitra cabreae* the spindle is closed with the exception of the polar gaps. Kinetochors are absent, whereas they were present in *Sphacelaria*.

In *Halopteris* (Sphacelariales), unequal cell division, anticipated by asymmetric positioning of the nucleus, is also the primary mode of hair cell differentiation. Katsaros and Galatis (1990) find numerous dictyosomes, vesicles and ER membranes in the diaphragm, which divides the hair protuberance from the main cell, but no other structures that could help explain its origin. Hence, it must be concluded that the structural basis of cytokinesis in the brown algae is currently unknown. The lack of relevant fine structural data is probably not due to a failure of visualizing cytoskeletal structures. At least MTs are clearly seen in the thin cytoplasmic strands of medullary cells in *Halopteris* (Katsaros and Galatis 1990).

These problems may become resolved in the near future by more stringent application of new visualization techniques for cytoskeletal elements. In fact, the first immunofluorescent images produced by Katsaros et al. (1991) have meant a great step forward in our understanding of subcellular organization in these algae. In *Ectocarpus siliculosus* a single MT-center seems to exist at the interphase nucleus from where MTs radiate along the nuclear surface, further into the cytoplasm, and partly along the cell wall. The MT-center is identical with the centrioles and contains, perhaps not surprisingly, a centrin homolog. Centrin is a contractile, Ca-binding phosphoprotein, originally discovered in the rhizoplasts of a prasinophyte flagellate, but it is now found almost ubiquitously colocalized with animal, protist, and algal centrosomes or basal bodies respectively (Melkonian 1991; Salisbury 1992). In *Ectocarpus* the centrioles are seen to separate at an yet not clearly defined premitotic stage and establish opposite mitotic poles.

In *Dictyota dichotoma* the situation is quite similar (Katsaros and Galatis 1992), except that the majority of cells contain two oppositely oriented MT-centers, indicating that the centrioles have divided and separated very early in the cell cycle. It is conceivable that these two oppositely oriented MT-arrays, present in each cell, could somehow be responsible for the asymmetric positioning of the nucleus preparing for an unequal division, but it will take further detailed examination to test this hypothesis.

Immunofluorescence of dividing nuclei in *Dictyota dichotoma* has largely confirmed the concept of mitosis in Phaeophyta based on earlier fine structural data (Katsaros and Galatis 1992). In addition, these data reveal a hitherto unknown MT array, the perinuclear MT cage, which encloses both daughter nuclei at the onset of cytokinesis. This cage is somehow bisected by the cytokinetic partition membrane. It may be possible that the partition membrane could be directed by actin filaments, and indeed, putative actin filaments have been seen associated with the cleavage furrow at least in one case (Brawley et al. 1977). However, this observation requires further confirmation. At present, we have to admit that we know little more about cytokinesis in Phaeophyta than 15 years ago.

### c) Cell Wall

The cell wall in the Phaeophyta is primarily composed of cellulose as structural component and alginic acid and fucoidan as outer layer components. The external wall layer can be rich in surface features. Oates and Cole (1990) examined the cell filaments of two *Desmarestia* species by a number of microscopical techniques showing surface ornamentations that differed in extent and pattern between the two species. The projecting ridges of the surface ornamentation originated from the outermost wall layer. Around the apex of the tip cell, a circular demarcation line divided the richer subapical surface pattern from the smoother dome surface.

Although the mode of extrusion and the spatial and temporal controls of removal and deposition of wall material have not been clarified, it is obvious from fine structural observations that the outer wall layers frequently break away and become replaced by new ones. This is seen in *Compsonema* (Scytosiphonales, Pedersen and Sokhi 1990) or, for instance, in *Hormosira* (Clayton et al. 1985) where the term "exfoliation" has been used to describe this phenomenon. In all cases, it may be justified to assume that this is an effective mechanism to periodically remove epiphytic organisms from the thallus surface in addition to other defense mechanisms such as the extrusion of phenolic, antimicrobial growth inhibitors.

### d) Motile Reproductive Cells

Previous work on zoid structure in the Phaeophyta has been summarized by Moestrup (1982). In the past decade several more species have been studied, confirming that the zoid structure in the Phaeophyta is indeed very uniform, even when considering the more specialized eyespot-less spermatozoids in the Laminariales. This may be taken as an indication that the division Phaeophyta has a monophyletic origin. A systematic study involving a representative cross-section through the genera in this division, however, has not yet been conducted.

Henry and Cole (1982b) described gametogenesis and gamete structure in several species of the order Laminariales. Although some details of their observations might have to be re-evaluated with respect to the recognition of basal body root structures (see below), the main ductus of their interpretation is still valid.

In the small, filamentous gametophyte, one sperm is produced per antheridium. The gamete is folded within the antheridium with both flagella placed parallel to each other. From the anterior basal body an MT-band runs into the anterior lobe, which is folded forward and bends down along the ventral side of the gamete. Upon release this part swings upward, the MT-band loops around at the then most anterior end of the cell body. Concomitantly, the basal bodies flip apart forming a large angle between each other. This rearrangement brings the insertion point of the flagella to a lateral position. Two opposite rows of mastigonemes are present in the proximal half of the anterior flagellum. Like the zoids in other species, the anterior flagellum bears an acronemal extension which is made of the two central MTs. Circumstantial evidence points to a role of the acronema in the initial adhesion of the brown algal spermatozoid to the egg surface (Moestrup 1982)

The posterior naked fagellum is unusually long in the Laminariales, but as in other brown algal genera, it tapers out by a reduction in the flagellar doublets to singlet MTs, which become less regularly spaced. This feature is absent from the zoospores of Laminaria. Zoospores differ also in their possession of cored Golgi vesicles believed to contain adhesive material, but in all other aspects gametes and spores are very similar (Henry and Cole 1982a).

As in oogamous species of the Fucales, Desmarestiales, Sporochnales and Dictyotales, the male gamete of the Laminariales lacks an eyespot. This feature is interpreted as an adaptation to the chemotaxis of brown algal spermatozoids (Maier and Müller 1986), whereas eyespots are always present in isogamous and anisogamous species.

It has been generally accepted that spermatozoids of the Dictyotales lack the posterior flagellum, but in fact only a rather small number of taxa has been specifically examined yet and all of them contained a vestigial basal body. In one case the posterior basal body was reported as producing a short flagellar stump (*Zonaria angusta*, Phillips and Clayton 1991).

Clayton (1984) studied gamete structure and release from the antheridia in *Scytosiphon* (Scytosiphonales). Gametes are formed in plurilocular gametangia, which upon maturity dissolve their inner walls so that gametes swim around in the central cavity formed by wall digestion, before they are released through the outer (more resistant wall). Hydrolytic enzymes are assumed to be discharged from the major Golgi vesicle fraction accumulating in the sperm during spermatogenesis just prior to wall disintegration. Gametes in *Scytosiphon* are negatively phototactic and slightly anisogamous. They have also have a prominent eyespot and the posterior flagellum possesses a parabasal body closely appressed to the cell surface over the eyespot.

The question of eyespot function in the sperm of *Ectocarpus siliculosus* has recently been addressed by Kreimer et al. (1991) using confocal laser reflection microscopy. The authors present evidence for the assumption that the eyespot could be capable of channeling light onto the posterior flagellum which carries the basal swelling and which has recently been identified as the location of an autofluorescing flavin-like photoreceptor (Kawai 1988).

According to Clayton (1984), the flagellar root system in *Scytosiphon* gametes is built by a seven-membered MT-band originating from the basal body of the anterior flagellum. The band runs towards the anterior end, loops back, returns along the plasmalemma, passes the basal bodies, and terminates over the eyespot. A band of electron-dense material is also present at each basal body. The latter structures are possible candidates for the centrin-positive material localized near the basal bodies by Katsaros et al. (1991).

A similarly looping anterior MT-root had been described in the spermatozoids of the Laminariales (Henry and Cole 1982b). However, from more recent studies employing serial sectioning (O'Kelly and Floyd 1984) it appears that this apparently looping rootlet splits up into two separate seven-membered entities now termed MAR (major anterior rootlet) and BR (bypassing rootlet). The MAR starts from the anterior basal body and proceeds in an arc towards the anterior end of the cell, and the BR comes down from the anterior end, passes by the basal bodies and runs further into posterior direction.

O'Kelly and Floyd (1985) have compared the flagellar rootlet organization published at that time in several phaeophytan genera with other chrysophycean taxa and came to the conclusion that four of the five types of MT-rootlets present in the Chrysophyta have perfectly matching counterparts in the phaeophytan rootlets. Only the MPR (major posterior rootlet) was missing in the phaeophyte zoids. This rootlet originates from one side of the posterior flagellum and leads into posterior direction.

Another, more recent, study on *Laminaria angustata* sperms (Motomura 1989) has confirmed the configuration favored by O'Kelly and Floyd, and this view has also been adopted by Anderson (1991) in his recent assessment of the chromophytan cytoskeleton.

In addition, Motomura (1989) has drawn attention to a subtle difference in zoospore and gamete structure as seen in the scanning electron microscope (SEM) in that the flagella in the zoospores of *Laminaria angustata* emerge from the cell body at the laterally located basal body/flagellum transition and run along shallow grooves in the surface, whereas in the sperm they also insert laterally, but are hidden within gullets over the entire length of the cell body. This feature may have been overlooked in other studies.

Composition and structure of the gamete surface may be considered an integral part in the process of recognition and fertilization. This aspect has recently been addressed by Jones et al. (1988), who have raised monoclonal antibodies (MABs) against *Fucus* sperm surface antigens. Twelve of them recognized different regions of the cell surface, the flagellar tip, and the mastigonemes. Some of them stained the egg surface as well, and one exhibited species specificity. These findings indicate that macromolecules must be targeted to specific areas of the gamete surface and that mechanisms must exist to maintain spatial organization on the level of the plasmalemma. The authors have recently extended their studies by identifying three groups of glycoproteins that are recognized by these antibodies (Jones et al. 1990). Unfortunately, the same carbohydrate side chains present on more than one glycoprotein turn out to be the most immunogenic determinants, so that overlapping staining patterns were obtained on Western blots. Obviously, this approach needs further refinement in order to improve the antibody specifity, before spatial organization of the cell surface can be correlated with a specific molecular component.

e) Oogonia

*Laminaria* species have been thought to fertilize oogamously. This definition may now need some careful re-examination after Motomura and Sakai (1988) discovered flagella in the eggs of *Laminaria angustata*.

The flagella appear to be nonfunctional, since they are shed at the moment when the egg is liberated, and their axonemal organization appears somewhat aberrant. Classical gametic microtubule rootlet are absent, but in a recent immunofluorescent study, Motomura (1991) was able to show that MTs converge on this basal body pair in a radial fashion and that the nucleus is pulled out into a pointed shape towards these basal bodies. This seems to indicate that the flagellar apparatus may not be entirely functionless. Motomura and Sakai speculate that *Laminaria angustata* could represent an intermediate evolutionary state between anisogamy and oogamy.

### f) Fertilization and Early Postfertilization Processes

Postfertilization processes have been studied in the past mostly on species of the Fucales (Brawley and Wetherbee 1981). Until recently, there has been no other detailed cytological study on the fate of the sperm organelles after plasmogamy in any other brown algal species. This gap has now been filled by observations on postfertilization fine structure in *Laminaria angustata* (Motomura 1990). According to Motomura, mitochondria and chloroplasts are maternally inherited. This is a consequence of digestion of the sperm mitochondria in lysosome-like vacuoles in the zygote. The chloroplasts become incapacitated by what appears to be a suppression of division so that they simply do not survive beyond the first couple of cell divisions.

On the other hand, the centrioles become paternally inherited, while the maternal ones simply disappear. These conclusions are at variance with those made previously by Brawley et al. (1976) on fertilized *Fucus* eggs, who presented evidence that the maternal centrioles are important in initiating spindle assembly.

Motomura (1991) has presented further evidence by MT-immunofluorescence of fertilized eggs in support of his hypothesis of paternal inheritance of the centrioles. He showed that parthenogenetically germinating eggs eventually aborted their development because of aberrant spindle formation, which he believes is caused by a gradual loss of maternal centriole material. Still, the discrepancy to the *Fucus* data remains unresolved and will require the examination of a wider range of species before the dispute can be settled.

### g) Polyspermy Block

To prevent polyspermy, *Fucus* eggs depolarize their membrane potential from about –60 mV to above 45 mV. By experimentally manipulating the membrane potentials of *Fucus* eggs, Brawley (1991) was able to demonstrate that sperms cannot adhere to eggs maintained at these higher potentials, and swim away from the egg surface after 1–3 min. This behavior is consistent with the idea that blocking polyspermy is really a three-step response: the first action is the rapid depolarization of the membrane potential triggered by the successful sperm; this is followed by an intermediate, yet unknown process in the range of minutes to further discourage other sperms, and completed soon after that by cell wall synthesis, which permanently blocks access to the egg membrane.

The author hypothesizes that the intermediate block could be accomplished by selective destruction of glycoprotein receptors on the egg surface. Candidates for such receptors have been discovered by Bolwell et al. (1980), who showed that a glycoprotein containing alpha-fucoid and alpha-mannoside residues isolated from the egg plasma membrane was able to block sperm attachment to the egg, when the sperm was preabsorbed with it.

Calcium influx could be responsible for the start of cell wall synthesis. In contrast to the situation during cell wall formation triggered naturally by a sperm, where cell wall formation starts at the point of sperm entry, treatment with Ca-ionophore causes a patchy appearance of wall material all over the egg surface (Brawley and Bell 1987). However, a rather high concentration of ionophore had to be used. Low sodium artificial seawater media, using N-methylglucamine as substitution for sodium, cause a similar patchy secretion of cellulose, prompting the authors to suggest the operation of a $Na^+/Ca^{2+}$ antiporter. This also explains why Ca-ionophores had relatively little effect as long as $Na^+$ was present.

It is not yet clear whether alginic acid is produced first and then mixed with cellulose or whether both are deposited simultaneously. In any case, it is interesting to note that wall synthesis forced by a disturbance of Ca-homeostasis leads to a change in cell shape, as in normal fertilization, although this is not followed by division as in fertilized eggs.

The importance of $Ca^{2+}$ for cell wall formation and morphogenesis prompted Brawley and Roberts (1989) to search for Ca-binding proteins in the *Fucus* zygote by making use of the $^{45}Ca^{2+}$ overlay technique. Gametes, eggs, and zygotes had a 17-kDa and a 30-kDa polypeptide in common, the 17-kDa band was identified as calmodulin. $^{125}I$-calmodulin overlay assay was used to identify a 45-kDa calmodulin-binding band in sperm and a 72-kDa band in zygotes. These data suggest that the expression of these proteins is under developmental control. It is the first rigorous approach on the biochemical level towards an understanding of $Ca^{2+}$ signaling during fertilization and embryogenesis in Phaeophyta.

### h) Cell Polarity and Germination

The fertilized *Fucus* egg is a traditionally attractive model for the study of polarization of an originally unpolarized cell.

Work with this cell system has revealed essentially three phases in response to a unilateral light stimulus: (1) formation of a polar axis, (2) fixation of the axis, and (3) the actual response on the structural level (Brownlee 1990). It is generally accepted that in the *Fucus* egg the direction of light is perceived by localized shading of a blue light photoreceptor uniformly distributed in the plasma membrane (Dring 1988), and that one of the first measurable parameters that changes upon stimulation is an influx of cations in the shaded area. $Ca^{2+}$ has been favored for a long time as the most important ion species in the polarized influx, but Kropf (1992b) has recently re-examined the experimental evidence and has come to the conclusion that this issue is still controversial. The most convincing evidence against a role of calcium was that axis fixation works perfectly well even in the complete absence of external calcium, so that Ca-influx does not appear to be a relevant component in the mechanism of axis formation. Kropf (1992b) argues, however, that the data support a function of $Ca^{2+}$ in cell wall formation triggered by the sperm and in growth stimulation of the germinating rhizoid pole once the axis has been established.

Considering the role of the cytoskeleton, it has been established that photopolarization and axis fixation requires a structurally undisturbed actin network and an intact cell wall, whereas microtubules are dispensable (Kropf 1992a). Actin becomes predominantly localized underneath the plasma membrane at the presumptive rhizoid pole, but it is not at all clear what its function is. Brawley and Robinson (1985) propose that the actin submembrane cytoskeleton interferes with the ion currents. In another model (the axis fixation model), the assumption is made that the actin network binds to the wall polysaccharides via transmembrane bridges (Kropf 1992a). Possible candidates for such transmembrane links could be polypeptides such as vitronectin, vinculin, and beta-integrin homologs. Following this proposition one step further, Quatrano et al. (1991) have detected bands in Western blots of *Fucus* protein extracts that cross-reacted with antibodies against these transmembrane proteins. If these promising results can be substantiated further, it would mean that significant progress can be made in the near future towards understanding cellular morphogenesis.

## References

**Andersen RA** (1991) Protoplasma 164:143–159. **Avazini A**, Honsell P (1984) Protoplasma 119:156–158.

**Baumann RW**, Jones BR (1986) J Phycol 22:49–56. **Berger S**, Schweiger H-G (1986) J Cell Sci 80:1–11. **Bernart M**, Gerwick WH (1990) Phytochemistry 29:3697–3698. **Beven A**, Guan YH, Peart J, Cooper C, Shaw P (1991) J Cell Sci 98:293–302. **Bi E**, Lutkenhaus J (1991) Nature 354:161–164. **Bold HC**, Wynne MJ (1985) Introduction to the algae. Structure and reproduction, 2nd edn. Prentice-Hall, Englewood Cliffs, New Jersey. **Boo SM**, Lee IK (1983) Korean J Bot 26:141–150. **Bourgeois CA**, Hubert J (1988) Int Rev Cytol 111:1–52. **Bolwell GP**, Callow JA, Evans LV (1980) J Cell Sci 43:209–224. **Bradley PM** (1991) J Phycol 27:317–321. **Brawley SH** (1991) Dev Biol 144:94–106. **Brawley SH**, Bell E (1987) Dev Biol 122:217–226. **Brawley SH**, Robinson KR (1985) J Cell Biol 100:1173–1184. **Brawley SH**, Roberts DM (1989) Dev Biol 131:313–320. **Brawley SH**, Quatrano RS, Wetherbee R (1977) J Cell Sci 24:275–294. **Brawley SH**, Wetherbee R (1981) Cytology and ultrastructure. In: Lobban CS, Wynne MJ (eds) The biology of the seaweeds. Univ California Press, Berkeley, pp 248–299. **Brawley SH**, Wetherbee R, Quatrano RS (1976) J Cell Sci 20:233–254. **Broadwater ST**, Scott J (1982) J Phycol 18:427–441. **Broadwater S**, Scott J (1983) Phycologia 23:225–233. **Broadwater S**, Scott J (1986) J Cell Sci 84:213–219. **Broadwater ST**, Scott JL, West JA (1991) Phycologia and molecular aspects of photomorphogenesis in brown algae. In: Herring PJ, Whitfield M, Maddock L (eds) Light and life in the sea. Univ Press, Cambridge, pp 115–126.

**Cannon MI** (1989) J Phycol 25:612–615. **Chamberlain YM** (1984) Phycologia 23:433–442. **Clayton MN** (1984) Prog Phycol Res 3:11–46. **Clayton MN** (1989) Brown algae and chromophyte phylogeny. In: Green JC, Leadbeater BSC, Diver W (eds) The chromophyte algae: problems and perspectives. Claredon, Oxford, pp 229–253. **Clayton MN**, Beakes GW (1983) J Phycol 19:4–16. **Clayton MN**, Shankly CM (1987) J Phycol 23:296–307. **Clayton MN**, Hallam ND, Luff SE, Diggins T (1985) Phycologia 24:181–190. **Cole KM**, Sheath G (eds) (1990) Biology of the red algae. Univ Press, Cambridge. **Cole KM**, Park CM, Reid PE (1985) J Phycol 21:585–592. **Coleman AW**, Goff LJ, Stein-Taylor JR (eds) (1989) Algae as experimental systems in cell biology. Liss, New York. **Coomans RJ**, Hommersand MH (1990) Vegetative growth and organization. In: Cole KM, Sheath RG (eds) Biology of the red algae. Univ Press, Cambridge. **Corliss JO** (1987) Int Rev Cytol 100:319–370. **Craigie JS** (1990) Cell walls. In: Cole KM, Sheath RG (eds) Biology of the red algae. Univ Press, Cambridge, pp 221–257.

**D'Antonio CM**, Gibor A (1985) Bot Mar 28:313–316. **Darley WM** (1982) Algal biology: a physiological approach. Basic microbiology. Blackwell, Oxford. **Deason TR**, Butler GL, Rhyne C (1983) J Phycol 19:104–111. **Delivopoulos SG** (1990) Cytobios 63:177–184. **Delivopoulos G**, Diannelidis BE (1991) Ann Sci Nat Bot Paris 13:9–21. **Delivopoulos SG**, Kugrens P (1984a) J Cell Sci 72:307–319. **Delivopoulos SG**, Kugrens P (1984b) Am J Bot 71:1245–1259. **Delivopoulos SG**, Kugrens P (1985) Am J Bot 72:1926–1938. **Delivopoulos SG**, Tsekos I (1985a) Flora

177:309–315. **Delivopoulos SG**, Tsekos I (1985b) New Phytol 101:605–612. **Delivopoulos SG**, Tsekos I (1986) Bot Mar 29:27–35. **Delivopoulos SG**, Pellegrini M, Pellegrini L (1989) Cytobios 57:177–184. **Diannelidis BE**, Kristen U (1988) Bot Mar 31:163–170. **Dring MJ** (1988) Annu Rev Plant Physiol Plant Mol Biol 39:157–174. **Ducreux G** (1983) Phycologia 22:415–429. **Ducreux G** (1984) J Phycol 20:447–454.

**Edgar LA**, Pickett-Heaps JD (1984) Prog Phycol Res 3:47–88. **Emons AMC** (1991) Role of particle rosettes and terminal globuls in cellulose synthesis. In: Haigler CH, Weimer PJ (eds) Biosynthesis and biodegradation of cellulose. Dekker, New York, pp 71–97. **Evans LV**, Trewavas AJ (1991) J Phycol 27:322–326.

**Farquhar MG** (1985) Annu Rev Cell Biol 1:447–488. **Fetter R**, Neushul M (1981) J Phycol 17:141–159. **Fredericq S**, Hommersand MH (1989) N Z J Bot 27:521–530.

**Gabrielson PW**, Garbary D (1986) CRC Crit Rev Plant Sci 3:325–366. **Gabrielson PW**, Garbary DJ, Scagel RF (1985) BioSystems 18:336–346. **Gantt E** (1986) Phycobilisomes. In Staehelin LA, Arnzten CJ (eds) Encyclopedia of plant physiology, Photosynthetic membranes and light harvesting systems. Springer, Berlin Heidelberg New York. **Garbary DJ**, Gabrielson PW (1990) Taxonomy and evolution. In: Cole KM, Sheath RG (eds) Biology of the red algae. Univ Press, Cambridge, pp 477–498. **Garbary DJ**, Belliveau DJ (1990) Phycologia 29:98–102. **Garbary D**, Belliveau D, Irwin R (1988) Can J Bot 66:1308–1315. **Garbary DJ**, McDonald AR, Duckett JG (1992) New Phytol 120:435–444. **Gibor A** (1989) Int Rev Cytol 118:93–114. **Goff LJ** (1982) Prog Phycol Res 1:289–369. **Goff LJ**, Coleman AW (1984) Proc Natl Acad Sci USA 81:5420–5424. **Goff LJ**, Coleman AW (1987) J Cell Biol 104:739–748. **Goff LJ**, Coleman AW (1990a) Curr Genet 18:557–565. **Goff LJ**, Coleman AW (1990b) DNA: microspectrofluorometric studies. In: Cole KM, Sheath GR (eds) Biology of the red algae. Univ Press, Cambridge, pp 43–71. **Green JC**, Leadbeater BSC, Diver WI (eds) (1989) The chromophyte algae: problems and perspectives. Clarendon, Oxford. **Grenville DJ**, Peterson RL, Barrales HL, Gerrath JF (1982) J Phycol 18:232–240. **Gretz MR**, Aronson JM, Sommerfeld MR (1986) Bot Mar 29:91–96. **Gross W** (1990) J Microbiol Methods 12:217–223. **Guiry MD** (1990) Sporangia and spores. In: Cole KM, Sheath RG (eds) Biology of the red algae. Univ Press, Cambridge, pp 347–376.

**Hallam ND**, Luff SE (1988) Br J Phycol 23:337–346. **Harold FM** (1990) Microbiol Rev 54:381–431. **Henry EC**, Cole KM (1982a) J Phycol 18:550–569. **Henry EC**, Cole KM (1982b) J Phycol 18:570–579. **Hommersand MH**, Fredericq S (1990) Sexual reproduction and cystocarp development. In: Cole KM, Sheath RG (eds) Biology of the red algae. Univ Press, Cambridge, pp 305–345. **Hommersand M**, Fredericq S (1992) Phycologia 31:300–325. **Huisman JM**, Kraft GT (1992) Phycologia 31:127–137. **Hwang MS**, Kim H-S, Lee IK (1991) J Phycol 27:441–447. **Hymes BJ**, Cole KM (1983) Can J Bot 61:3377–3385.

**Johansen HW** (1981) Corraline algae, a first synthesis. CRC, Boca Raton. **Jones JL**, Callow JA, Green JR (1988) Planta 176:298–306. **Jones JL**, Callow JA, Green JR (1990) Planta 182:64–71. **Jones PL**, Woelkerling WmJ (1984) Phycologia 23:183–194.

**Kaska DD**, Polne-Fuller M, Gibor A (1988) J Phycol 24:102–107. **Katsaros C**, Galatis B (1985) Br Phycol J 20:263–276. **Katsaros C**, Galatis B (1988) Br Phycol J 23:71–88. **Katsaros C**, Galatis B (1990) Br Phycol J 25:63–74. **Katsaros C**, Galatis B (1992) Protoplasma (in press). **Katsaros C**, Galatis B, Mitrakos K (1983) J Phycol 19:16–30. **Katsaros C**, Kreimer G, Melkonian M (1991) Bot Acta 104:87–92. **Kawai H** (1988) J Phycol 24:114–117. **Kim H-S**, Kim GH, Lee IK (1988) Korean J Phycol 3 (1):15–27. **Knauss ME**, Hommersand MH (1989) Br phycol J 24:39–52. **Koslowsky DJ**, Waaland SD (1987) J Phycol 23:638–648. **Kraft GT** (1981) The Rhodophyta: morphology and classification. In: Lobban CS, Wynne MJ (eds). The Biology of seaweeds. Univ California Press, Berkeley, pp 6–51. **Kreimer G**, Kawai H, Müller DG, Melkonian M (1991) J Phycol 27:268–276. **Kristiansen J**, Cronberg G, Geissler U (eds) (1989) Chrysophytes, development and perspectives. Proc 2nd Int Chrysophyte Symp Berlin 1987. Nova Hedwigia Beih 95. **Kropf DL** (1992a) Role of the cytoskeleton in cellular morphogenesis of zygotes of fucoid algae. In: Menzel D (ed) The cytoskeleton of the algae. CRC, Boca Raton, pp 79–92. **Kropf DL** (1992b) Microbiol Rev (in press). **Kugrens P**, Delivopoulos SG (1986) J Phycol 22:8–21. **Kumar HD** (1989) Algal cell biology, 2nd edn. Affiliated East-West Press, New Delhi. **Kumar HD** (1990) Introductory phycology. Affiliated East-West Press, New Delhi.

**La Claire II JW** (1982) Phycologia 21:273–287. **Lambert C**, Buis R, Lhardyhalos MT (1990) Can J Bot 68:1569–1573. Leedale GF (1982) Prog Bot 44:32–42. **Liddle L**, Thomas J, Scott J (1982) J Phycol 18:257–264. **Liu XW**, Kloareg B (1991) C R Acad Sci Ser III 312:517–522. **Liu XW**, Rochas C, Kloareg B (1990) J Appl Phycol 2:297–303. **Lloyd CW** (1991) Development Suppl1:55–65. **Lobban CS**, Chapman DJ, Kremer BP (1988) Experimental phycology, a laboratory manual. Cambridge Univ Press, Cambridge.

**Magruder WH** (1984) J Phycol 20:436–440. **Maier I**, Müller DG (1986) Biol Bull 170:145–175. **Mandura A**, Boney AD, Bowes BG (1986) Nova Hedwigia 42:277–282. **Margulis L**, Corliss JO, Melkonian M, Chapman DJ (1990) Handbook of Protoctista. The structure, cultivation, habitats, and life cycles of the eukaryotic microorganisms and their descendants exclusive of animals, plant, and fungi. Jones & Barlett, Boston. **Melkonian M** (1991) Prog Bot 52:271–307. **Melkonian M** (ed) (1992) Algal cell motility. Chapman and Hall, New York. **Melkonian M**, Becker D, McFadden GI, Reize IB (1988) Experimental analysis of membrane traffic during secretion of scales in green algae. In: Cell-free analysis of membrane traffic, Liss, New York, pp 297–301. **Melkonian M**, Andersen RA, Schnepf E (1991) Protoplasma 164:1–167. **Menzel D** (ed) (1992) The Cytoskeleton of the Algae. CRC, Boca Raton. **Mita T**, Kuroiwa T (1988) Protoplasma Suppl 1:133–152. **Mita T**, Kanbe T, Tanaka K, Kuroiwa T (1986) Protoplasma 130:211–214. **Moestrup O** (1982) Phycologia 21:427–528. **Motomura T** (1989) Jpn J Phycol 37:105–116. **Motomura T** (1990) J Phycol 26:80–89. **Motomura T** (1991) J Phycol 27:248–257. **Motomura T**, Sakai Y (1985) Jpn J Phycol 33:199–209. **Motomura T**, Sakai Y (1988) J Phycol 24:282–285.

**Oates BR**, Cole KM (1989) Br phycol J 24:339–345. **Oates BR**, Cole KM (1990) Phycologia 29:529–532. **Oates BR**, Cole KM (1992) Protoplasma (in press). **O'Kelly CJ**, Baca BJ (1984) Phycologia 23:407–417. **O'Kelly CJ**, Floyd GL (1984) Protoplasma 123:18–25. **O'Kelly CJ**, Floyd GL (1985) Phycologia 24:263–274. **Oross JW**, Possingham JV (1989) Protoplasma 150:131–138. **Ott D** (1992) The role of the cytoskeleton in organelle translocation in *Vaucheria longicaulis*. In: Menzel D (ed) The cytoskeleton in the algae. CRC, Boca Raton, pp 255–272.

**Patrone LM**, Broadwater ST, Scott JT (1991) J Phycol 27:742–753. **Pedersen PM**, Sokhi G (1990) Nord J Bot 10:547–556. **Pellegrini L**, Gaillard J, L'Hardy-Halos MT (1989) Iridescent bodies and physodes in eukaryotic algae. In: Kumar HD (ed) Phycotalk, vol 1. Rastogi, Subhash Bazar, Meermut, pp 107–124. **Phillips JA**, Clayton MN (1991) Phycologia 30:205–214. **Phillips D**, Scott J (1981) Protoplasma 106:329–341. **Pickett-Heaps J** (1991) Int Rev Cytol 128:63–108. **Pickett-Heaps J**, Schmid A-MM, Edgar LA (1990) Prog Phycol Res 7:1–168. **Polne-Fuller M**, Gibor A (1984) J Phycol 20:609–616. **Polne-Fuller M**, Gibor A (1990) J Phycol 26:274–282. **Pueschel CM** (1980) Phycologia 19:210–217. **Pueschel CM** (1982) Br Phycol J 17:333–341. **Pueschel CM** (1987) J Phycol 23:150–156. **Pueschel CM** (1989) J Phycol 25:625–636. **Pueschel CM** (1990) Cell structure. In: Cole KM, Sheath (eds) Biology of the red algae. Univ Press, Cambridge, pp 7–41. **Pueschel CM**, Cole KM (1982) Am J Bot 69:703–720. **Pueschel CM**, Cole KM (1985) J Phycol 21:146–154. **Pueschel CM**, Magne F (1987) Cryptogram Algol 8:201–209. **Pueschel CM**, Stein JR (1983) J Phycol 19:209–215.

**Quatrano RS**, Brian L, Aldridge J, Schultz T (1991) Development 1991 (Suppl 1):11–16.

**Round FE**, Chapman DJ (1990) Progress in phycologial research, vols 1–7. BioPress, Bristol. **Round FE**, Crawford RM, Mann DG (1990) The diatoms. Biology and morphology of the genera. Univ Press, Cambridge.

**Salisbury JL** (1992) Centrinbased calcium-sensitive contractile fibres of the green algae. In: Menzel D (ed) The cytoskeleton of the algae. CRC, Boca Raton, pp 292–410. **Schmid A-M** (1988) Plant Syst Evol 158:211–223.

**Schnepf E**, Deichgräber G, Herth W (1982) Protoplasma 110:203–214. **Scott J** (1986) Can J Bot 64:516–524. **Scott J**, Broadwater S (1990) Cell division. In: Cole KM, Sheath RG (eds) Biology of the red algae. Univ Press, Cambridge, pp 123–145. **Scott J**, Thomas J, Saunders B (1988) Phycologia 27:327–333. **Sheath RG** (1984) Prog Phycol Res 3:89–157. **Sheath RG**, Cole KM, Hymes BJ (1987) Phycologia 26:1–8. **Sheath RG**, Hellebust JA, Sawa T (1981) Phycologia 20:292–297. **South GR**, Whittick A (1987) Introduction to phycology. Blackwell, Oxford. **Spector DL** (1984) Dinoflagellates. Academic Press, New York. **Stewart KD**, Mattox K (1980) Phylogeny of phytoflagellates. In: Cox ER (ed) Phytoflagellates. Elsevier, New York, pp 433–462. **Sylvester AW**, Waaland JR (1984) Phycologia 23:427–432. **Sylvester AW**, Waaland JR (1992) Planta 186:273–281.

**Trick HN**, Pueschel CM (1990) Phycologia 29:403–409. **Trick HN**, Pueschel CM (1991) Phycologia 30:196–204. **Triemer RE**, Farmer MA (1991) Protoplasma 164:91–104. **Tripodi G**, de Masi F (1983) Plant Syst Evol 143:197–206. **Tsekos I** (1985) Protoplasma 129:127–136. **Tsekos I**, Diannelidis BE (1990) Flora 184:78–80. **Tsekos I**, Schnepf E (1985) Plant Syst Evol 151:1–18. **Tsekos I**, Reiss H-D, Schnepf E (1985) Naturwissenschaften 72:489–490. **van den Hoek C** (1992) Algen: Einführung in die Phykologie, 2te Aufl. Thieme, Stuttgart (in press).

**Vesk M**, Borowitzka MA (1984) J Phycol 20:501–515.

**Waaland SD** (1978) Planta 138:65–68. **Waaland SD** (1980) Development in red algae: elongation and cell fusion. In: Gantt E (ed) Handbook of phycological methods. Developmental and cytological methods. Univ Press, Cambridge, pp 85–94. **Waaland SD** (1986) Cell elongation in the red alga *Griffithsia*, control by light, ions, and an endogenous glycoprotein hormone. In: Wiessner W, Robinson DG, Starr RC (eds) Algal development, molecular and cellular aspects. Springer, Berlin Heidelberg New York, pp 42–49. **Waaland SD** (1990) Development. In: Cole KM, Sheath RG (eds) Biology of the red algae. Univ Press, Cambridge, pp 259–273. **Waaland SD**, Waaland JR, Cleland R (1972) J Cell Biol 54:184–190. **Wang S**, Sun Y, Lu A, Wang G (1987) Chin J Oceanol Limnol 5:217–221. **Watson BA**, Waaland SD (1986) Plant Cell Physiol 27:1043–1050. Wetherbee R (1979) Science 204:858–859. **Wetherbee R** (1980) J Ultrastruct Res 70:259–274. **Wetherbee R**, Kraft GT (1981) Protoplasma 106:167–172. **Whyte JNC**, Hosford SPC, Englar JR (1985) Carbohydr Res 140:336–341. **Wiessner W**, Robinson DG, Starr RC (1983) Compartments in algal cells. Springer, Berlin Heidelberg New York. **Wiessner W**, Robinson DG, Starr RC (eds) (1986) Algal development. Molecular and cellular aspects. Springer, Berlin Heidelberg New York. **Wiessner W**, Robinson DG, Starr RC (1990) Experimental phycology, vol 1. Cell walls and surfaces, reproduction, photosynthesis. Springer, Berlin Heidelberg New York. **Wilce RT**, Sears JR (1991) Phycologia 30:151–169. **Willey RL**, Walne PL, Kivic P (1989) CRC Crit Rev Plant Sci 7:303–340. **Willingale-Theune J**, Schweiger M, Hirsch-Kaufmann M, Meek AS, Paulin-Levasseur M, Traub P (1989) J Cell Sci 93:651–665. **Woelkerling WJ** (1988) The corraline red algae: an analysis of the genera and subfamilies of nongeniculate Corralinaceae (Rhodophyta). Oxford Univ Press, London.

**Zablackis E**, Vreeland V, Doboszewski B, Laetsch WM (1991) J Phycol 27:241–248. **Zhang W**, Chapman DJ, Phinney BO, Spray CR (1991) J Phycol 27:87–91.

PD Dr. Diedrik Menzel
Max-Planck-Institut für Zellbiologie
Rosenhof
W-6802 Ladenburg, Germany

# III. Phloem Regeneration

By Rainer Kollmann and Alexander Schulz

## 1. Introduction

In higher plants development of the various organs is closely correlated with primary vascularization of the plant body. During regular growth and replacement of the organs, the procambium and cambium continuously produce new vascular elements for the transport of water and nutrients, thereby always maintaining contact to the pre-existing vascular system.

For loading and unloading of organic nutrients above all a close interaction between phloem elements and the assimilate producing respectively consuming tissues is important. The highly specialized sieve elements with their companion cells approach the photosynthetic active mesophyll cells as well as any growth and storage tissue of shoots and roots up to less than a hundred micrometers; and the various source and sink regions are completely interconnected by the vascular system.

Any interruption of the vascular continuity interferes with vital functions of individual organs and the whole plant, respectively. Wounding and vascular disruption may be encountered by various environmental events in the life of a plant. Thus, plants must be provided with a rapid wound-response system for re-establishing the continuity of the transport system.

This review will cover the research on phloem regeneration with special regard to (1) healing of experimentally set interruptions, i.e., the induction, decisive steps in cellular reprogramming, and differentiation of wound-bypassing phloem, and (2) grafting processes, dealing with the question of establishing secondary symplastic connections with regard to compatibility between the graft partners. The term "wound phloem" and its derivatives – wound-sieve elements, wound-sieve tubes – will be used for both types of wound-*induced* phloem.

For reviews of the special cytology of regular phloem see Behnke 1983, 1990; Behnke and Sjolund 1990; for phloem transport Baker and Milburn 1989; Eschrich and Eschrich 1989; for structure and function of xylem elements Sauter 1986.

Progress in Botany, Vol. 54

## 2. Phloem Regeneration After Wounding

Many herbaceous dicotyledons are capable of phloem regeneration around a wound that is experimentally set by the severance of vascular bundles. The degree of phloem interruption influences speed and intensity of regeneration (Fig. 1). If single vascular bundles are severed in *Impatiens* or *Coleus* stem internodes, a few wound-sieve tubes develop within 6 days and reconnect the cut bundle or contact intact bundles (Fig. 1A; von Kaan-Albest 1934). If all but one or two bundles are interrupted in *Cucurbita*, regeneration proceeds faster so that all severed bundles are bridged 3 days after wounding (Fig. 1B; Schulz 1990). In the primary root of *Pisum*, severance of all phloem connections can cause a reconnection of the stele by wound-sieve tubes within less than 60 h (Fig. 1C; Schulz 1987). Crossing pericycle and endodermis (Schulz 1986a), they bypass the wound gap in formerly quiescent cortical tissue that otherwise would never give rise to vascular elements. Wound-xylem elements appear later than phloem elements and are always found closer to the wound (Robbertse and McCully 1979; Schulz 1987). Following this short-term process regeneration in pea roots continues, and a cambium develops in between the parallel aligned phloem and xylem strands (Hardham and McCully 1982a; Schulz 1987). Quite similar to regular bundle development, this cambium gives rise to *secondary* wound-sieve elements at the phloem side, and *secondary* tracheary elements at the xylem side (Schulz 1987). In contrast to this, wound phloem does not develop in plants possessing a high number of regular phloem anastomoses (Behnke and Sukkri 1971). These anastomoses are apparently sufficient to shunt assimilates from cut to intact bundles (Aloni and Peterson 1990).

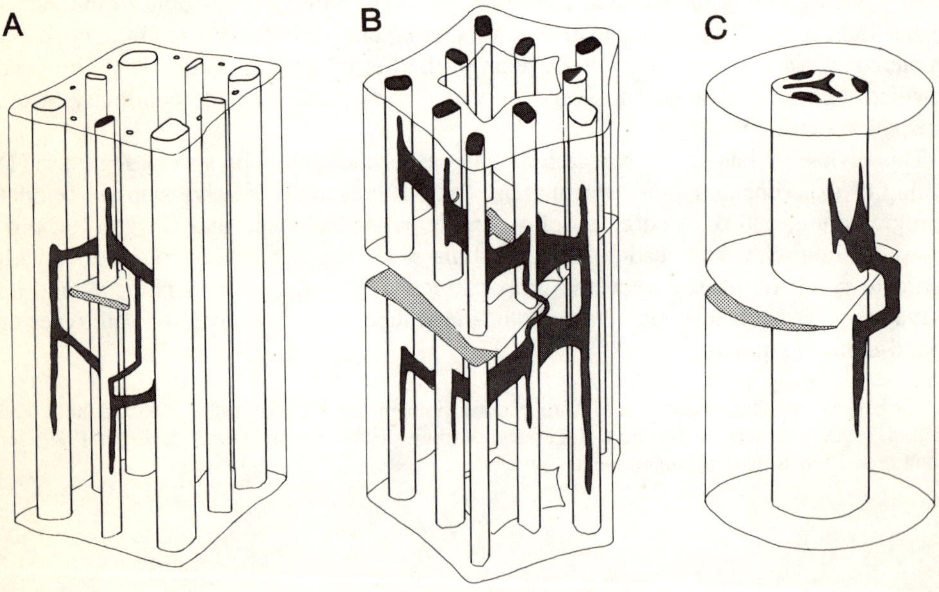


**Fig. 1 A–C.** The course of wound phloem in **A** *Coleus*, **B** *Cucurbita* stem internodes, and **C** *Pisum* roots where single bundles and the entire stele, respectively, were severed. (Schulz 1990)

## a) Tissue Changes Preceding Wound-Phloem Development

Following wounding, formerly quiescent cells enter mitosis. While those at the wound surface establish a wound callus (Kahl 1985), others at some distance from the cut are reprogrammed for vascular regeneration. In the pea-root system, new walls appearing some 26 h post-wounding (Robbertse and McCully 1979) indicate a change in tissue polarity, recognizable much earlier by a change in orientation of cortical microtubules, starting already 5 h post-wounding (Hush et al. 1990). The re-orientation of the cortical microtubules is followed by a centripetal movement of the nucleus within newly developed, transvacuolar cytoplasmic strands that are guided by cytoplasmic microtubules (Schulz 1988) and follow the new polarity (Kirschner and Sachs 1978). This takes place in the cortex of pea roots in a region with a raised esterase activity (Rana and Gahan 1983). Still preceding any divisions, an arc of cells in the cortex is predestined for the differentiation of xylem elements, occurring 2 days later, 20 h post-wounding (Rana and Gahan 1983 – phloem was not investigated in this study), exactly at the time when auxin starts to pass experimental wounds (Gersani 1985).

Within a single cell, the transvacuolar strands transform into a phragmosome. After caryokinesis, the developing cell plate follows the cytoplasm of the phragmosome centrifugally and fuses with the original cell wall (Schulz 1988; see also Venverloo 1990). Preprophase bands of microtubules occur at these locations prior to mitosis (Hardham and McCully 1982a). Following the first division, the mitotic activity continues and subsequent divisions cut the original cells more or less parallel to the first one. Some of the small cells thus formed are precursors for the now starting phloem development and determine size and shape of sieve elements (e.g., Fig. 2C). The members of wound-sieve tubes are linked by sieve plates that always arise in primary pit fields and other wall portions of the original cells (Schulz 1987).

In all systems described to date, the parenchyma cells were shown to divide prior to wound-sieve element development. This contrasts with the differentiation of xylem elements after wounding that may result from a direct conversion of a parenchyma cell (Hammersley and McCully 1980). Underlining the rule that a companion cell is an obligatory sister cell of wound-sieve elements (Behnke and Schulz 1980; Schulz 1986a), the minimal number of divisions found to precede this differentiation was one (Schulz 1990). Cortex parenchyma cells converted directly into sieve elements only when colchicine was applied in such a high concentration that also root growth was inhibited (Hardham and McCully 1982b). Usually, the division pattern valid for the development of a wound-sieve element reflects that of regular phloem, i.e., exhibits a few divisions of the original cell prior to the last, unequal one (Behnke and Schulz 1980; Schulz 1986a).

## b) Ultrastructure

If the differentiation of wound-sieve elements is to be compared with that of regular phloem, the events are surprisingly similar (Behnke and Schulz 1980; Schulz 1986a, 1990). Once programmed by the unequal division of the prospective sieve-element/companion-cell complex, the differentiation follows the regular pattern (see Evert 1990).

There are, however, indications that differentiation speeds up when regeneration is pressing, as in the pea root system (Fig. 1C). Here, only small amounts of P-protein appear in the first wound-sieve elements (Schulz 1990). In addition, the elongating *Pisum*-wound sieve tubes have only a small number of differentiating members (compared to protophloem) in the nucleate stages (see Fig. 2; Schulz 1986a).

Most differences between wound- and bundle-sieve elements result from the different tissues of origin. Instead developing from proplastids, in wound-sieve elements plastids differentiate from chloro- or amyloplasts. At maturity remnants of the former plastidal type – amyloplast-starch grains and grana thylakoids – have disappeared and the typical inclusions of sieve-element plastids are formed, i.e., spherical starch grains in the S-type plastids of *Coleus* and additionally protein crystals in the P-type plastids of *Pisum* (Behnke 1981). The starch grains are histochemically of the same composition as regular sieve-tube starch (Behnke and Schulz 1983).

The sieve-plate pores arise from single plasmodesmata or those in primary pit-fields and show the same course of differentiation as regular sieve pores, even though they originate in the rather thick walls of the original cells (Behnke and Schulz 1980; Schulz 1987). Median cavities of the primary pit fields or their remnants were encountered during pore development, but had no part in the final channel of the mature pores (Schulz 1987, 1990). If two or more plasmodesmata are closely aligned to each other, they seem to fuse into one future pore (Schulz 1990).

### c) The Induction of Wound-Sieve Tubes and Their Contact to Bundle Phloem

The control of vascular regeneration has been the subject of numerous investigations. Since von Kaan-Albest (1934), the physiological factors of phloem regeneration were often demonstrated to originate in leaves and buds above the wound. These factors were most intensively studied in the *Coleus* stem system. Because of the difficulty to detect sieve tubes, xylem regeneration was preferentially investigated (for reviews see Jacobs 1984; Warren Wilson and Warren Wilson 1984; Roberts et al. 1988). The effects of excised organs above the wound incision can be compensated for by an appropriate concentration of auxin (Jacobs 1970; Stoddard and McCully 1980). In addition to auxin – the main regulatory factor in vascular differentiation (Sachs 1984; Aloni 1987) – also cytokinins were shown to be effective in phloem regeneration when applied below the wound (Aloni et al. 1990).

Microscopy revealed that the course of wound-sieve tubes coincides with the path of xylem elements (Jacobs 1970; Robbertse and McCully 1979). Since wound xylem appears later than wound phloem, the general course in the tissue might be coordinated by the phloem. The balance between xylem and phloem differentiation was discussed to depend upon the amount of sucrose applied in addition to a given concentration of auxin (Wetmore and Rier 1963). This concept was disputed by Aloni (1980, 1987; see also Roberts 1988), who could show that the only effect of sucrose was an enhanced callose production at sieve plates, thus facilitating phloem detection. In contrast, the controlling factor for the quality of the vascular system formed seems to be the amo levels led to the development of phloem alone, while at higher concentr was found in tissue cultures (Aloni 1980). Considering this together wi course of phloem and xylem strands, Aloni (1980) discussed the need phloem-localized factor that induces xylem differentiation.

In conclusion, auxin seems to be involved in both the change of tissu ceding mitotic activity (see Chap. A.III.p.65; Sachs 1984), and the quantity and quality of the newly formed vascular elements (see also Sachs and Cohen 1982; Sachs 1984). Cytokinin, in contrast, could be involved in the induction and maintenance of meristematic activity around the wound (Aloni 1987; Roberts et al. 1988).

Although the general importance of both phytohormones for vascular regeneration is evident, their presence or flow in the wound region is not sufficient to explain how a given cell is reprogrammed for sieve-element differentiation. This is illustrated by the fact that wound-sieve tubes in *Pisum* roots elongate from both, the basal and apical stelar stump, into the cortex (Robbertse and McCully 1979; Schulz 1986a), i.e., below the wound opposite to the acropetal auxin transport in roots (Feldman 1981).

Another problem of phloem regeneration concerns the question of in which way pre-existing and newly differentiating phloem are interlinked. Since Eschrich's article (1953) it has been consistently observed that wound-sieve elements do no connect directly to pre-existing mature bundle sieve tubes, but open a new longitudinal tier of sieve elements of cambial origin (Eschrich 1953; Benayoun et al. 1975; Behnke and Schulz 1980). From the cytological point of view, the differentiation of a sieve pore between an enucleate sieve element and a nucleate wound-sieve element is hardly conceivable, since this requires participation of both cells. The importance of the mode of contacts between regular, new-developing phloem (e.g., from leaves) and pre-existing phloem was recently emphasized by Eschrich (1989).

Utilizing the clear conditions given by the pea-root system, Schulz reconstructed one severed phloem bundle with serial ultrathin sections 48 h post-wounding, so that incomplete wound-sieve tubes were expected (Schulz 1986a, b). According to this investigation, the mode of secondary contact formation with existing phloem is easier to understand when the question is put the other way round: How are wound-sieve tubes induced? Figure 2 summarizes the probable succession of events in a time series and simplifies the exact reconstruction given by Schulz (1986b).

According to this series, a bundle-sieve tube is blocked after severance. The first sieve plates above and below the incision are instantly plugged by P-protein and constricted by callose synthesized in a rapid wound response (Evert 1982; Schulz 1986b). Adjacent to this sieve tube, which was fully mature at the moment of wounding, another sieve tube was already in the stage 2 or 3 (see legend of Fig. 2) of differentiation. Lateral sieve areas and single pores could be demonstrated between both, blocked and new bundle sieve tubes (Fig. 2A; Schulz 1986b). The new sieve tube is fully mature 48 h post-wounding, but ends blindly, i.e., without a sieve plate at its wound-opposed end. Its sieve plates do not show a wound-response besides the known artifacts introduced by EM-fixations. A neighboring cambial cell, early activated by wounding, enters sieve element differentiation (Fig. 2A) as well as cells in the same longitudinal series (a lateral extension of bundle phloem following wounding was also described by Benayoun et al. 1975; Aloni and Jacobs 1977; Behnke and Schulz 1980). Symplastic contacts between this series and the

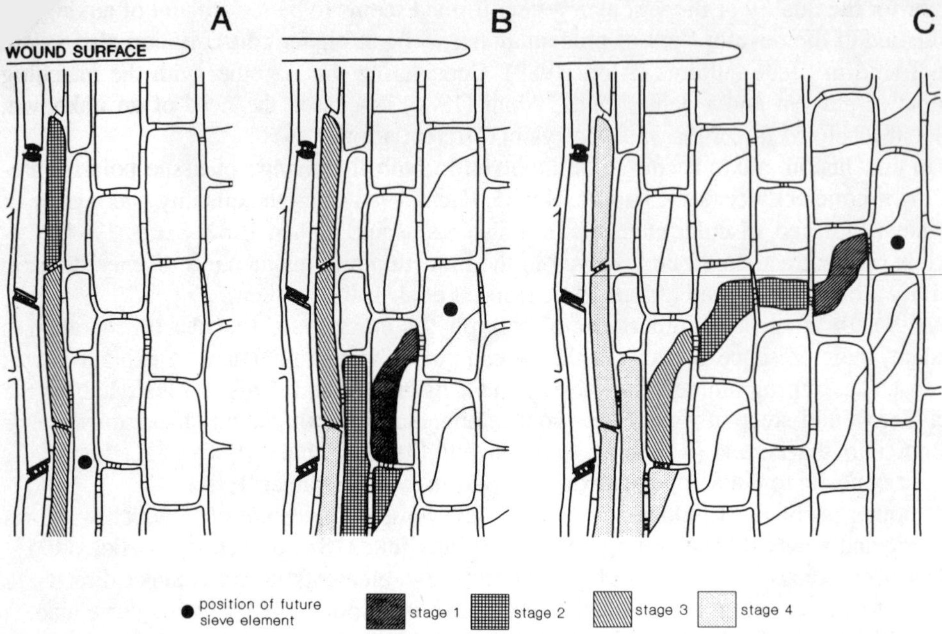


**Fig. 2 A–C.** Development of a wound-sieve tube arising from stelar phloem, shown in a time series. New sieve elements are shaded corresponding to their developmental stage. (After Schulz 1986b). The stages exhibit the cells with **1** dense cytoplasm without specific characters; **2** appearance of specific characters like P-protein, plastidal inclusions, and callose; **3** disintegration of nucleus, vacuole, Golgi apparatus, and ribosomes, and start of sieve-pore widening; **4** a fully mature, enucleate protoplast and pores widened throughout the wall

former sieve tube consisting of lateral sieve areas and single sieve pores exist already 48 h post-wounding (Schulz 1986b). Figure 2B depicts that the changed tissue polarity in the neighboring original parenchyma cell, equally connected by plasmodesmata/future pores, offers the chance for a deviative phloem course. Here, a wound-sieve tube is induced which elongates towards the cortex, showing the described sequence of differentiating sieve elements (Fig. 2B, C). Wound-sieve tubes departing from the basal and apical stelar stump can develop a common sieve plate when making a contact, because the "tip" members of both sieve tubes in the cortex are still nucleate. Thus, completion of a phloem bridge is possible where these sieve tubes happen to reach neighboring cortex cells that are linked by plasmodesmata. This may occur at any location of the dedifferentiated cortex tissue. At vascular bundles it is necessary that, after wounding, the phloem can be supplemented by sieve tubes and single elements that mediate between bundle and wound phloem. Accordingly, a prerequisite for wound phloem contacts is a reservoir of immature phloem or cambial cells in the bundle. The lack of immature cells in older bundles might be the reason that in monocotyledons wound-vascular elements had formed in very young internodes only (Aloni and Plotkin 1985).

*Homoiogenetic induction* (see von Kaan-Albest 1934) of sieve elements (reflected ultrastructurally by a sequence of differentiating elements) is the explanation for both the continuity of phloem development – pre-existing phloem is always the basis for new differentiation – and for the establishment of secondary contacts during regular and disturbed plant growth. The path of the inductive stimulus was discussed to be symplastic, since the distribution of plasmodesmata in the early stage of regeneration seems to direct the course of wound-sieve tubes (Schulz 1986a, 1987). The branching of wound-sieve tubes at 3- or more-poled members (Eschrich 1953; Schulz 1986a, b) is also indicative for the key role of the presence and distribution of plasmodesmata in sieve-element induction. The described sequence with gradual transitions from meristematic cells to fully mature sieve elements accords well with that in protophloem-sieve tubes, (see e.g., Thorsch and Esau 1981), and seems to be a general means of phloem development. Further investigations are needed to analyze the unknown, symplastically spreading stimulus that induces sieve-element differentiation.

### d) Translocation in Wound Phloem

The function of newly developed wound phloem was tested in *Impatiens* stems and *Pisum* roots by the application either of the phloem-mobile dye fluorescein or radioactive sucrose to source leaves, or by exposing a source leaf to $^{14}CO_2$. Fluorescence microscopy and autoradiography consistently showed that wound-sieve tubes and their companion cells are marked/labeled when complete phloem bridges existed in the respective systems (Eschrich 1953; Jacobsen and Eschrich 1990: 6 days post-wounding; Schulz 1987: 3 days post-wounding). Phloem transport in the severed bundles ceases in immediate response to wounding and resumes slowly in adjacent cell tiers (Eschrich 1953; Gersani et al. 1980; Schulz 1986c, 1987). This corresponds to the described wound-phloem induction: transport seems to resume first in those bundle-sieve tubes which were nucleate at the moment of wounding and fully mature 2 days post-wounding (see Fig. 2; Schulz 1986c, 1987). These, as well as recent investigations on regular phloem (e.g., Grignon et al. 1989; Aloni and Peterson 1990), demonstrate that fluorescein (or carboxyfluorescein), though not naturally occurring in plants, is a valuable, easy to handle label for phloem transport. The claim that fluorescein is only binding to proteins in the sieve tubes and therefore not reflecting mass flow of assimilates (Eschrich and Eschrich 1989; Jacobsen and Eschrich 1990) is not corroborated by microscopy. Neither in wound-phloem of *Impatiens* (Eschrich 1953) nor in that of *Pisum* (Schulz 1987) was an inhomogeneous labeling of the sieve element protoplasts described. This should be expected at the sieve plates, if the dye bound to P-protein.

The comparison of the structural regeneration and the beginning of fluorescein translocation in wound phloem of pea roots indicated that an effective transport of assimilates around the wound is only possible after the wound-sieve elements have matured (Schulz 1987). The timing was confirmed with [$^{14}C$] sucrose as label (Schulz and Gersani 1990). This contrasts to the observed spread of fluorescein into the starch sheath in the *Impatiens*-stem system preceding wound-sieve tube staining (Eschrich 1953). Radioactive assimilates in this system, however, were obviously restricted to wound-sieve tubes (Jacobsen and Eschrich 1990).

A more dynamic picture of the functional regeneration of phloem in wounded pea roots is formed when the transport *rate* of radioactive sucrose is analyzed (Schulz and

Gersani 1990). Radioactivity was recovered from the above-wound root/shoot transition and the entire primary root below the incision at four different transport periods (30 to 120 min), in order to calculate the translocation rate. By comparison of the transport rate in both parts at increasing regeneration periods, the process of the functional regeneration of the phloem was summarized as consisting of five steps: (1) phloem transport is blocked in response to wounding, (2) the meristematic activity of the around-wound area is locally supplied by sucrose, (3) sucrose accumulates above the wound, (4) activity is drained into the below-wound region and, eventually, (5) phloem transport into the root returns to normal within day 4 after stele severance (Schulz and Gersani 1990).

## 3. Phloem Regeneration in Graft Unions

In successful grafting a reconnection of the cut vascular bundles of scion and rootstock is secured by regeneration of xylem and phloem elements from callus tissue. Grafts are functional units with respect to the exchange of water and nutrients between the partners. Apoplastic and symplastic pathways across the graft union are warranted by an intimate connection of conducting elements and parenchyma cells.

Xylem connections in the graft union have often been described (see, e.g., Monzer and Kollmann 1986); phloem continuity with special regard to symplastic cell junctions, however, was an open question until quite recently. Problems in the past mainly consisted of localizing the exact boundary between cells of scion and stock. The question as to any symplastic communication between living cells of the graft partners was significant in so far as plasmodesmata and sieve pores at the graft interface must have been formed secondarily in nondivision walls and, in the case of heterografts, between cells of different species. An answer to this problem was obtained by studies on heterografts with species-specific cell markers (Kollmann and Glockmann 1985). Progress in research on graft-union development has been achieved during the last 15 years; it was summarized by McCully (1983), Kollmann and Glockmann (1990), and Kollmann (1992).

As for the grafting technique, a most exact matching of the different tissues of scion and stock with superimposed cambia promotes the process of regeneration. Above all, short distances between the cut vascular bundles and compatibility of the graft partners are important.

### a) Graft-Union Development

In herbaceous plants, the development of a graft union follows a sequence of several distinct steps. Within a few days after transplantation the process starts with cell proliferation at the wound surfaces and the development of wound callus filling in the gap between stock and scion. Depending on the degree of compatibility of the graft partners, a more or less prominent isolation layer (necrotic layer) is formed at the graft interface. It consists of break-down products of cut cells and of various substances such as pectinaceous and phenolic compounds secreted by the living callus cells (Stoddard and McCully

1979; Moore and Walker 1981; Deloire and Hébant 1982; Moore 1982; Tiedemann 1989; Kollmann and Glockmann 1990). In compatible grafts, the isolation layer eventually becomes disrupted and dissolved, and callus cells of scion and stock interdigitate, thus forming a mixed callus. Within this tissue of heterogeneous origin, individual cells of the graft partners are interconnected symplastically by secondarily formed plasmodesmata (Kollmann and Glockmann 1985; Kollmann et al. 1985).

About 3 days after grafting, tracheids and tracheary elements start differentiation within the mixed callus, either directly from individual callus cells or after several subdivisions (Stoddard and McCully 1979, 1980; Jeffree and Yeoman 1983). The first developing xylem elements correspond to the wound xylem, which usually forms when vascular bundles are severed (Hardham and McCully 1982a). There is, however, one essential difference between both types of wound xylem: in the case of wounding, xylem elements originate from dedifferentiated parenchyma cells, whereas in graft unions the first emerging tracheids and tracheary elements regenerate from undifferentiated callus cells. In both cases, however, formation of wound xylem represents a short-term regeneration according to Behnke (1983) and Warren Wilson and Warren Wilson (1984). Continuous xylem strands reconnecting the cut vascular bundles of scion and stock are established this way within 4 to 6 days following grafting (Stoddard and McCully 1980; Kollmann and Glockmann 1990). In contrast to the situation in wound healing (Robbertse and McCully 1979), in graft unions wound xylem principally regenerates prior to wound phloem. This should not be surprising, since an immediate water supply is essential for the implanted scion (Weatherhead and Barnett 1986).

Phloem regeneration in the graft union of herbaceous plants occurs within 4–6 days after grafting; differences may exist with the different graft systems and techniques employed (Tiedemann 1989). As a rule, differentiation in compatible heterografts occurs 1 or 2 days later than in homografts. On principle, the way of differentiation conforms to the wound-phloem regeneration as shown in Section 2 (see also Behnke and Schulz 1980; Hardham and McCully 1982a, b; Schulz 1986a, b). Sieve elements, together with their companion cells and associated phloem parenchyma cells, are formed by one or several subdivisions of callus cells (Tiedemann 1989). A complete phloem bridge interconnecting the cut vascular bundles across the graft union in herbaceous plants is regenerated about 6–9 days following grafting (Stoddard and McCully 1979, 1980; Kollmann et al. 1983; Tiedemann 1989). As shown with wound xylem, a vascular cambium is not involved in the process of the early phloem development. Comparable to the situation in wound xylem, the first sieve elements formed in a graft union, therefore, serve as a rapid, short-term regeneration.

With further development of the graft union, however, vascular cambium develops between wound phloem and xylem in connection with the cambia of the vascular bundles of scion and stock. This cambium bridging the graft union gives rise to secondary phloem and xylem. In this way, in graftings of herbaceous plants, an elaborate vascular system interconnecting scion and stock is developed within 2–3 weeks after transplantation (Stoddard and McCully 1979; Moore and Walker 1981; Deloire and Hébant 1982; Moore 1984a, b; Tiedemann 1989).

Cambium-derived phloem elements and wound phloem can be discriminated structurally: while the former represent the typical prosenchymatic cell type, the small wound-sieve elements having irregular shapes, together with their associated companion and parenchyma cells, disclose their origin from a parenchymatic cell.

### b) Translocation Between the Graft Partners in Relation to Phloem Regeneration

The interrelationship between phloem regeneration in graft unions and translocation of assimilates from the scion to the stock has been studied by Rachow-Brandt and Kollmann (1992a). Using $^{14}C$-labeling techniques, it was found that translocation across the graft interface in compatible heterografts *(Lycopersicon/Solanum)* and in homografts of *Lycopersicon* and *Solanum*, respectively, started 5–7 days after grafting and reached a maximum at days 8–12. In 10-day-old *Lycopersicon/Solanum* grafts, maximal amounts of assimilates were translocated into the stock (more than 80% of the label exported from a $^{14}CO_2$-fed scion leaf), most likely due to the prominent sink of the *Solanum* rootstock. The start of translocation and the course of import obviously reflected the onset and further development of phloem regeneration in the graft union. Actually, in various heterografts, such as *Lycopersicon/Solanum* (Rachow-Brandt and Kollmann 1992a) or *Cucumis/Cucurbita* (Tiedemann 1989), and in corresponding homografts, it could be shown that increasing numbers of sieve tubes in the graft union correlated with an increasing rate of translocated photoassimilates. In *Lycopersicon/Solanum* grafts, which were older than 10 days, a further phloem differentiation had no effect on the rate of translocation. The results indicate that (1) in compatible heterografts and in homografts the assimilate transport across the graft union takes place via the phloem and (2) not all sieve elements, regenerated during development of the graft union altogether, were functional. Maybe that the early-formed wound phloem functions in translocation only during the short-term regeneration of vascular tissue. In the less compatible heterograft *Vicia/Helianthus*, a strong correlation between phloem regeneration and assimilate transport across the graft union could not be found. Here, phloem regeneration in the graft union began as early as 5 days after grafting and proceeded over graft-union development; $^{14}C$-assimilate import into the rootstock, however, did not start before day 10 and reached only 2% of the label exported from a fed scion leaf. In the corresponding homografts, transport always started earlier and reached rates of about 30% (*Helianthus*) and 40% (*Vicia)*, respectively (Rachow-Brandt and Kollmann 1992a). Most likely, discrepancies in the relationship between structure and function of the regenerating phloem reflect differences in compatibility between the partners in the various homo- and heterografts.

In a further study on the function of graft unions in dependence on their development, Rachow-Brandt and Kollmann (1992b) presented evidence that in the heterografts *Lycopersicon/Solanum* and *Vicia/Helianthus* as well as in the corresponding homografts assimilates on their way from scion to the stock are unloaded into the apoplast to some extent. Unloading (demonstrated by washing-out experiments, by measurements of the activity of acid invertase, and by determination of the disaccharid/monosaccharid ratio) was significantly higher in the graft union than in internodes of scion and stock above and below the graft connection. Furthermore, the rate of unloading was higher in young graft unions than in older connections. It is noteworthy that the less compatible *Vicia/Helianthus* heterograft exhibits higher unloading than the compatible *Lycopersi-*

*con/Solanum* heterograft. Whether this is due to the differences in compatibility remains open to question, all the more since reasons for an unloading of assimilates in the graft union are unknown so far. Any correlation to an apoplastic transfer between the graft partners as well as to an enhanced metabolism of the growing callus might be discussed. In this context the nature of cell communication with special regard to the existence of plasmodesmata and sieve pores at the graft interface above all needs further consideration.

### c) Symplastic Connections Between the Graft Partners

The until recently open question as to the existence of symplastic communications between the graft partners (see the review by McCully 1983) has now been answered unequivocally. By investigation of the *Vicia/Helianthus* heterograft, Kollmann and Glockmann (1985, 1990) were able to identify the graft interface at the cellular level by species-specific cell markers and thus disclose cell connections between the graft partners. It could be shown that in the interspecific fusion walls of *Vicia/Helianthus* grafts various types of plasmodesmata exist between parenchyma cells; likewise in interspecific sieve plates, sieve pores have been found interconnecting the *Vicia* and *Helianthus* sieve tubes symplastically. Both types of cell connections must have been formed secondarily in nondivision walls. A common characteristic of secondary plasmodesmata and, less frequently, of secondary sieve pores are median cavities with multiple branching strands. The results with heterografts could be verified with homografts and autografts. Lacking partner-specific cell markers, however, in these cases the graft interface could be discerned only indirectly by examining complete serial sections of the graft union (Monzer and Kollmann 1986; Tiedemann 1989; Kollmann and Glockmann 1990).

Symplastic continuity could thus be demonstrated at graft interfaces between living cells of the vascular tissue of all homografts and heterografts so far examined. The structural findings should always be taken into consideration when the mechanism of phloem transport between the graft partners is to be discussed. In the less compatible *Vicia/Helianthus* heterograft, the interspecific sieve pores were, due to heavy callose deposits, narrower than primarily formed intraspecific pores of the graft partners, and the pore content (P-Protein, ER) was condensed. Whether this phenomenon represented symptoms of incompatibility could not be decided. As a peculiarity of secondary plasmodesmata, half strands, branched or unbranched, have been found at the graft interface (Kollmann et al. 1985). These half plasmodesmata, traversing the wall part of only one partner cell without connecting the abutting cell, predominate in incompatible heterografts and in fusion walls between cells of different developmental stages. Occasionally, interspecific sieve element/companion cell complexes were identified at the graft interface; plasmodesmata interconnecting those heteroplasts were of the usual one-sided, branched type. These observations clearly demonstrate that (1) the ontogenetic interrelationship between a sieve element and its companion cell must not always be a prerequisite for their interaction and (2) cells of unrelated plants should be able to cooperate in fundamental processes as a functional unit (Kollmann and Glockmann 1990).

### d) Mechanism of Secondary Plasmodesmata Formation

The mechanism of formation of secondary plasmodesmata at graft interfaces has recently been investigated with the heterograft *Vicia/Helianthus* by Kollmann and Glockmann (1991). For their studies on cellular interactions the authors used the *in-vitro grafting technique* of explanted internodes introduced by Parkinson and Yeoman (1982). This method has proved to be a reliable system for studying the communication between the graft partners at various defined stages of graft-union development. Early steps of plasmodesmata formation between contacting callus cells could thus be described. The whole process, it was found, is correlated with a striking modification of the cell walls involved. By local thinning and loosening of adjacent wall parts, cell membranes and associated ER cisternae of joining cells come in close contact, thus being able to fuse. During subsequent thickening of the modified wall parts, by wall material secreted from Golgi vesicles, strands of cytoplasm with enclosed ER tubules are entrapped. The cytoplasmic strands represent, both in form and fine structure, half, branched, or unbranched plasmodesmata. When corresponding half plasmodesmata of adjoining cells fuse across the remnants of loosened wall material, continuous secondary plasmodesmata will be formed. The mechanism (summarized in Fig. 3) resembles to some extent the formation of primary plasmodesmata. There is, however, one essential difference concerning the origin of both types of cell connections: while primary plasmodesmata interconnect cells from the very beginning during cell plate formation, within secondary plasmodesmata a fusion of half plasmodesmata is the prerequisite for any symplastic connection. The formation of continuous secondary plasmodesmata requires a high coordination and cooperation between two adjacent cells. In those cases where synchronization in wall modification between cell partners is lacking, only half plasmodesmata or discontinuous strands will be established.

The mechanism of de-novo formation of plasmodesmata as described with callus cells in graft unions is in agreement with a model proposed by Monzer (1990, 1991) for the mode of development of half plasmodesmata in plant protoplasts during wall regeneration. Most likely one unifying mechanism of secondary plasmodesmata formation exists with far-reaching similarities to the development of primary cell connections. The uniform fine structure of primary and secondary plasmodesmata supports this concept.

The development of secondary sieve pores in nondivision sieve plates at graft interfaces has not yet been studied in detail. However, there can be no doubt about their development from secondary plasmodesmata following the same mode as described for

**Fig. 3 a–n.** Diagram of the secondary formation of plasmodesmata at graft interfaces. **a** Approaching callus cells of scion (*sc*) and stock (*st*) with pectic material in between. *Arrows* indicate location of details **b–g** and **h–n**, respectively. **b–g** Formation of continuous secondary plasmodesmata by plasmalemma and ER fusion of adjoining cells (**e, f**) within wall parts thinned synchroneously by both cell partners. Elongation of branched and single strands during rebuilding of the modified wall parts (**f, g**). **h–n** Formation of discontinuous secondary plasmodesmata within opposing wall parts with different degree of thinning. In the upper (completely thinned) wall part half plasmodesmata and in the lower (incompletely thinned) wall part strands shorter than half plasmodesmata are formed. *w* cell wall; *p* pectic substances; *GV* Golgi vesicle; * newly deposited wall material; *arrowheads* 5 nm particles linking ER and cell membrane. (Kollmann and Glockmann 1991)

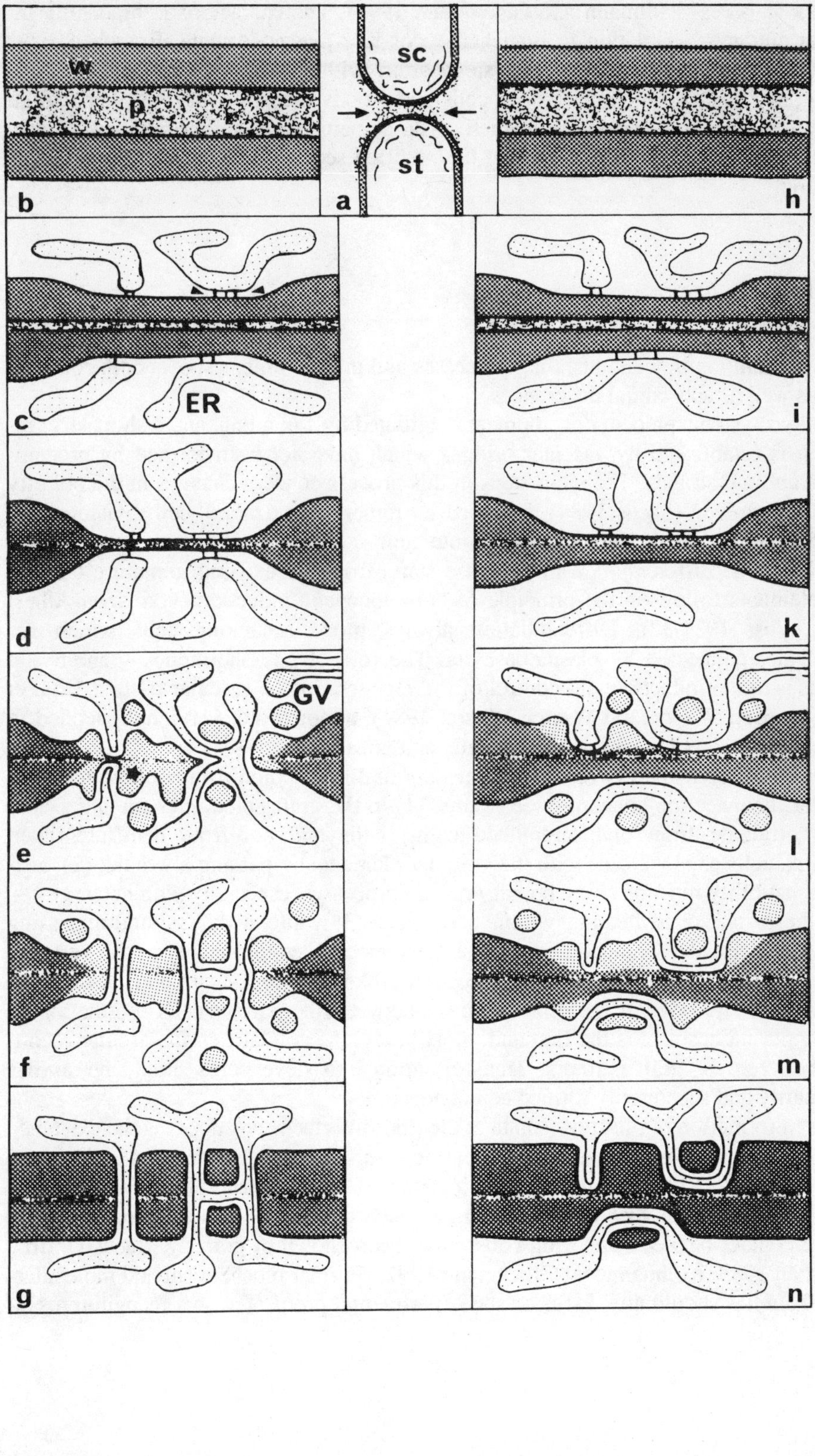
w
p
sc
st
b
a
h
c
ER
i
d
k
e
GV
l
f
m
g
n

primary sieve pores (Kollmann and Glockmann 1990). The characteristic branching indicates an ontogenetic relation to branched secondary plasmodesmata. Branched sieve pores to some extent resemble lateral sieve areas with median cavities and multiple branching, as known from conifer sieve cells and woody dicotyledons. Whether regular sieve areas of that kind, generally assumed so far as being of primary origin, might also be formed secondarily or, at least, might be modified secondarily, remains open to further research.

## 4. Conclusion

Phloem regeneration in wound healing processes and in graft unions shows both common features as well as substantial differences.

In the two systems phloem development is initiated by wounding, and a short-term regeneration is established by vascular bridges which have not been formed by procambium or cambium strands. The main steps in this process concern changes in the polarity of the cells involved (Sachs 1984), followed by remeristematization, cell division, and, finally, differentiation of the derivatives into new conducting elements. Induction of wound-sieve tube differentiation in any case starts from pre-existing immature bundle-phloem elements following the principle of homoiogenetic induction (von Kaan-Albest 1934; cf. Jost 1931/32). Differentiation always proceeds among cells which are symplastically connected by plasmodesmata. The role of phytohormones – auxin and cytokinins – in the induction and control of the regeneration of vascular tissue (Aloni et al. 1990; Warren Wilson and Warren Wilson 1984), although not yet comprehended in detail, should be identical with wound healing and grafting processes.

The main differences between wound healing and graft-union formation with regard to vascular reconnection concern three events: (1) in the graft union, phloem and xylem elements originate from undifferentiated callus cells and not from dedifferentiated parenchyma cells as is the case with the vascular elements by-passing a wound; (2) various incompatibility reactions may impair grafting processes, especially with heterografts, but not those of wound healing within one plant; (3) while within wound-bypassing phloem all cells of the vascular bridge are interconnected symplastically at the very beginning, within graftings symplastic connections, like plasmodesmata and sieve pores must be formed de novo at the graft interface between the partner cells. Formation of secondary plasmodesmata is the first and, most likely, the crucial step in vascular reconnection between the graft partners. Transformation into sieve pores should be similar with primarily and secondarily formed cell connections.

De-novo formation of plasmodesmata at the graft interfaces presupposes the existence of a recognition system between the cell partners as suggested by Yeoman and his coworkers (Yeoman et al. 1978; Jeffree and Yeoman 1983; Yeoman 1984). The prerequisites for the function of such a system, e.g., a synchronous cell wall thinning and plasmalemma contact between opposing cells, have been shown at graft interfaces (Jeffree and Yeoman 1983; Kollmann and Glockmann 1991). Further biochemical and molecular-biological studies should now focus on the experimental proof of such a recognition system.

*Acknowledgments.* We thank Mrs. C. Glockmann for critically reading the manuscript.

## References

**Aloni R** (1980) Planta 150:255–263. **Aloni R** (1987) Annu Rev Plant Physiol 38:179–204. **Aloni R**, Jacobs WP (1977) Am J Bot 64:615–621. **Aloni R**, Peterson CA (1990) Planta 182:583–590. **Aloni R**, Plotkin T (1985) Planta 163:126–132. **Aloni R**, Baum SF, Peterson CA (1990) Plant Physiol 93:982–989.

**Baker DA**, Milburn JA (eds) (1989) Transport of photoassimilates. Longman, Harlow, 384pp. **Behnke H-D** (1981) Nordic J Bot 1:381–400. **Behnke H-D** (1983) Progr Bot 45:18–35. **Behnke H-D** (1990) Naturwissenschaften 77:1–11. **Behnke H-D**, Schulz A (1980) Planta 150:357–365. **Behnke H-D**, Schulz A (1983) Protoplasma 144:125–132. **Behnke H-D**, Sjolund RD (eds) (1990) Sieve elements – comparative structure, induction and development. Springer, Berlin Heidelberg New York, 305pp. **Behnke H-D**, Sukkri B (1971) Z Pflanzenphysiol 66:82–92. **Benayoun J**, Aloni R, Sachs T (1975) Ann Bot 39:447–457.

**Deloire A**, Hébant C (1982) Ann Bot 49:887–891.

**Eschrich W** (1953) Planta 43:37–74. **Eschrich W** (1989) In: Baker DA, Milburn JA (eds) Transport of photoassimilates. Longman, Harrow, pp 206–263. **Eschrich W**, Eschrich B (1989) Progr Bot 51:80–92. **Evert RF** (1982) Bio Science 32:789–795. **Evert RF** (1990) In: Behnke H-D, Sjolund RD (eds) (1990) Sieve elements – comparative structure, induction and development. Springer, Berlin Heidelberg New York, pp 103–137.

**Feldman LJ** (1981) Plant Physiol 67:278–281.

**Gersani M** (1985) J Exp Bot 36:1809–1816. **Gersani M**, Lips SH, Sachs T (1980) J Exp Bot 31:783–789. **Grignon N**, Touraine B, Durand M (1989) Am J Bot 76:871–877.

**Hammersley DRH**, McCully ME (1980) Plant Sci Lett 19:151–156. **Hardham AR**, McCully ME (1982a) Protoplasma 112:143–151. **Hardham AR**, McCully ME (1982b) Protoplasma 112:152–166. **Hush JM**, Hawes CR, Overall RC (1990) J Cell Sci 96:47–61.

**Jacobs WP** (1970) Int Rev Cytol 28:239–273. **Jacobs WP** (1984) In: Scott TK (ed) Hormonal regulation of development II. Pirson A, Zimmermann MH (series eds) Encycl Plant Physiol NS vol 10. Springer, Berlin Heidelberg New York, pp 149–170. **Jacobsen KR**, Eschrich W (1990) Planta 181:335–342. **Jeffree CE**, Yeoman MM (1983) New Phytol 93:491–509. **Jost L** (1931/32) Z Bot 25:481–522.

**Kahl G** (1985) Bull Soc Bot Fr 132, Actual Bot:143–155. **Kirschner H**, Sachs T (1978) Isr J Bot 27:131–137. **Kollmann R** (1992) Biol Unserer Zeit 22:264–273. **Kollmann R**, Glockmann C (1985) Protoplasma 124:224–235. **Kollmann R**, Glockmann C (1990) In: Behnke H-D, Sjolund RD (eds) Sieve elements – comparative structure, induction and development. Springer, Berlin Heidelberg New York, pp 219–237. **Kollmann R**, Glockmann C (1991) Protoplasma 165:71–85. **Kollmann R**, Dörr I, Schulz A, Behnke H-D (1983) Ber Dtsch Bot Ges 96:117–132. **Kollmann R**, Yang S, Glockmann C (1985) Protoplasma 126:19–29.

**McCully ME** (1983) In: Moore R (ed) Vegetative compatibility responses in plants. Baylor Univ Press, Waco, pp 71–88. **Monzer J** (1990) In: Robards AW et al. (eds) Parallels in cell to cell junctions in plants and animals. NATO ASI Series H vol 46. Springer, Berlin Heidelberg New York, pp 185–197. **Monzer J** (1991) Protoplasma 165:86–95. **Monzer J**, Kollmann R (1986) J Plant Physiol 123:359–372. **Moore R** (1982) J Exp Bot 33:533–540. **Moore R** (1984a) Plant Cell Rep 3:172–175. **Moore R** (1984b) Ann Bot 54:127–133. **Moore R**, Walker DB (1981) Am J Bot 68:820–830.

**Parkinson M**, Yeoman MM (1982) New Phytol 91:711–719.

**Rachow-Brandt G**, Kollmann R (1992a) J Plant Physiol 139:579–583. **Rachow-Brandt G**, Kollmann R (1992b) J Plant Physiol 139:584–589. **Rana MA**, Gahan PB (1983) Planta 157:307–316. **Robbertse PJ**, McCully ME (1979) Planta 145:167–173. **Roberts LW** (1988) In: Roberts LW, Gahan PB, Aloni R (eds) Vascular differentiation and plant growth regulators. Springer, Berlin Heidelberg New York, pp 63–88. **Roberts LW**, Gahan PB, Aloni R (eds) (1988) Vascular differentiation and plant growth regulators. Springer, Berlin Heidelberg New York, 154 pp.

**Sachs T** (1984) In: Barlow PW, Carr DJ (eds) Positional control in plant development. Cambridge Univ Press, London, pp 193–224. **Sachs T**, Cohen D (1982) Differentiation 21:22–26. **Sauter JJ** (1986) Progr Bot 48:388–405. **Schulz A** (1986a) Protoplasma 130:12–26. **Schulz A** (1986b) Protoplasma 130:27–40. **Schulz A** (1986c) In: Cronshaw J, Lucas WJ, Giaquinta (eds) Phloem transport. Liss, New York, pp 183–186. **Schulz A** (1987) Planta 170:289–299. **Schulz A** (1988) Protoplasma 143:176–187. **Schulz A** (1990) In: Behnke H-D, Sjolund RD (eds) (1990) Sieve elements – comparative structure, induction and development. Springer, Berlin Heidelberg New York, pp 199–217. **Schulz A**, Gersani M (1990) J Plant Physiol 136:599–605. **Stoddard FL**, McCully ME (1979) Can J Bot 57:1486–1501. **Stoddard FL**, McCully ME (1980) Bot Gaz 141:401–412.

**Thorsch J**, Esau K (1981) J Ultrastruct Res 75:339–351. **Tiedemann R** (1989) J Plant Physiol 134:427–440.

**Venverloo CJ** (1990) Protoplasma 155:85-94. **von Kaan-Albest A** (1934) Z Bot 27:1–94.

**Warren Wilson J**, Warren Wilson PM (1984) In: Barlow PW, Carr DJ (eds) Positional control in plant development. Cambridge Univ Press, London, pp 225–280. **Weatherhead J**, Barnett JR (1986) Ann Bot 57:593–598. **Wetmore RH**, Rier JP (1963) Am J Bot 50:418–430.

**Yeoman MM** (1984) In: Linskens HF, Heslop-Harrison J (eds) Cellular interactions. Pirson A, Zimmermann MH (series eds) Encycl Plant Physiol. NS vol 17. Springer, Berlin Heidelberg New York, pp 453–472. **Yeoman MM**, Kilpartrick DC, Miedzybrodzka MB, Gould AR (1978) Symp Soc Exp Biol 32:139–160.

Professor Dr. Rainer Kollmann
Dr. Alexander Schulz
Botanisches Institut
der Christian-Albrechts-Universität
Olshausenstraße 40
W-2300 Kiel 1, Germany

# IV. Reproductive Development in Seed Plants: Research Activities at the Intersection of Molecular Genetics and Systematic Botany

By Rolf Rutishauser

## 1. Introduction

The following review summarizes new comparative ontogenetic studies on flowers and inflorescences in seed plants (including gymnosperms). With few exceptions, this discussion is restricted to literature published in the years 1989–1992. Emphasis is given on structural dynamics in ontogeny and phylogeny, on homeosis and heterochrony. Since the publication of my earlier reviews covering the years 1985–1988 (Prog. Bot. 49:51ff, 51:54ff), the genetic-molecular approach has been considered very successful, a "major breakthrough in the study of flowering in particular and plant development in general" (Lord 1991). The following review provides insights into the exciting world of flower development of *Antirrhinum*, *Arabidopsis*, and *Petunia* based on developmental mutants and new genetic-molecular methods. The models describing and explaining floral morphogenesis in these organisms seem to fit various other seed plants. Although the terms "genetic control" and "genetic program" are inappropriate to some degree (see criticism in Nijhout 1990), it was not always possible to avoid them in this review.

The following pages may be stimulatory for systematic and evolutionary botanists who try to reconstruct seed plant phylogeny focusing on reproductive developmental data.

The distinction of six subclasses within the dicotyledons (Magnoliidae, Hamamelidae, Caryophyllidae, Dilleniidae, Rosidae, and Asteridae) made in this review follows Cronquist's (1981) system (see Dahlgren 1989a, b; Martensen and Probst 1990; Huber 1991; Lidén 1992, for alternative angiosperm classifications).

## 2. From Vegetative to Floral Apices – a Developmental Continuum

It is beyond the scope of my review to summarize all new results on floral induction (see Prog. Bot. 51:54ff). News on the transition to flowering in angiosperms and its internal control were discussed by Bernier (1989), McDaniel (1989, 1991: "a blackbox with a bright future"), and Meeks-Wagner (1992). The role of phytohormones (cytokinins, gibberellins) in flower initiation was reviewed by Bernier et al. (1990), Metzger (1990), and Pharis et al. (1989). Negrutiu and Sala (1991) summarized these topics in a special issue on sex determination and differentiation in plants (for dioecious plants, see also Chattopadhyay and Sharma 1991). New insights into the genetic control of floral development in *Arabidopsis thaliana* (Brassicaceae), *Antirrhinum majus* (Scrophulariaceae), and

Progress in Botany, Vol. 54

other model species were provided by Bowman and Meyerowitz (1991), Carpenter and Coen (1990), Carpenter et al. (1991), Chasan (1991), Coen (1991), Coen and Carpenter (1992: "the power behind the flower"), Coen and Meyerowitz (1991: "the war of the whorls"), Coen et al. (1991), Drews et al. (1991), Gasser (1991), Lord (1991), Meyerowitz et al. (1989, 1991), Schwarz-Sommer et al. (1990), van Loon and Bruinsma (1992: "the new plant physiology"), van Tunen and Angenent (1991).

Shoot meristem activity can be summarized as a series of overlapping developmental phases: vegetative → inflorescence → floral. This progression is likely to occur by sequential changes in apical meristem activity that result in the modification of a genetically controlled meristem developmental program (Sussex 1989; Poethig 1990; Meeks-Wagner 1992). Regulatory genes involved in floral organogenesis have been isolated (see Sect. 5.d).

Flowering is normally a one-way process, although plants have adopted a modular strategy of development. Once a vegetative apex has begun to initiate flower buds or flower parts, it is usually committed to a determinate mode of growth. Battey and Lyndon (1990) and Donnison et al. (1991) described flower reversion with a switch back from determinacy to leaf production. Flower reversion in, e.g., *Anagallis, Pharbitis*, and *Impatiens* (triggered by defined environmental conditions), indicates that the flowering process must be regarded as a continuum. Thus, physiological stages, such as commitment to flower, and even morphological stages, such as different floral organ types, are artificial to varying extents.

The conservation of the floral organ patterns in several flowering plant taxa (e.g., Asteridae, see Sect. 4.d) demonstrates that they are primarily genetically determined, although biophysical phenomena and other epistatic modifiers may be involved in the manifestation of the "genetic program" (Gasser 1991; Rieseberg 1992). What is the mechanism by which organ position is established in reproductive shoot apices? Employing a dental impression technique on apices in *Anagallis* (Primulaceae), Green and collaborators have been able to follow the patterns of cell division and expansion in the surface layer of a single (living!) meristem during the floral transition (Green 1991; Green et al. 1991; Hernandez et al. 1991). They proposed that the biophysics of the surface layer establishes a feed-forward mechanism whereby lateral organs are positioned specifically on the surface of the meristem (see criticism against this and similar "relay models" in Bowman and Meyerowitz 1991; Meyerowitz et al. 1991; Negrutiu et al. 1991).

Cell lineage studies (clonal analysis) using mutagens or transposable elements for marking meristematic cells demonstrated that differentiation of floral organs can be traced to areas (layers) of the floral apex in, e.g., maize (Kelly Dawe and Freeling 1990; Dellaporta et al. 1991; Irish 1991). These studies show, however, that the fates of meristematic cells are not absolutely dependent on their lineage.

## 3. Developmental Pathways of Inflorescences

### a) Computer Simulation, Terminological Questions

Prusinkiewicz and Lindenmayer (1990, pp. 71–97) showed in a very aesthetic form how the various types of inflorescences in flowering plants can be stimulated using L-systems. The developmental branching models (of, e.g., *Mycelis*) assist in formulating plausible hypotheses related to the mechanisms that may control inflorescence development in nature.

In many flowering plants it is difficult or arbitrary to clearly distinguish a reproductive terminal region (inflorescence, frequently with simultaneous branching) and a basal vegetative region (often with retardive branching). For example, single flowers in the rheophyte *Indotristicha ramosissima* (Podostemaceae) may arise endogenously from the root and both endogenously and exogenously from the shoot axis (Rutishauser and Huber 1991). Weberling's (1989a) English introduction into the morphology of inflorescences and flowers (with a profitable glossary) may help botanists struggling with German plant typology. Another volume on comparative structure of inflorescences was published by Troll and Weberling (1989) covering families with determinate (monotelic) inflorescences (e.g., Caryophyllaceae, Saxifragaceae, Vitaceae). Unfortunately, the ontogenetic processes which are involved in deviating inflorescence types are not elucidated in many cases. Hagemann (1990) compared acrogenous branch systems of several woody angiosperms including phylogenetic considerations. He tried to overcome some incongruities between inflorescence typology sensu Troll (see Weberling 1989a) and whole plant architecture sensu Hallé et al. (1978). Kunze (1989) discussed evolutionary processes leading from the determinate (monotelic) to the indeterminate (polytelic) inflorescence type, e.g., loss of the terminal flower (truncation) and homogenization of the distal branches. According to Kuznetsova (1988, 1991: "pseudocyclic theory"), Troll's general inflorescence types are nonhomologous in different angiospermous taxa. Terminal flowers may also be observable in families with usually indeterminate inflorescences (Brassicaceae, Scrophulariaceae), due to developmental mutations, e.g., the terminal flower mutants in *Arabidopsis* and *Antirrhinum* (Meyerowitz et al. 1989; Coen 1991; Coen et al. 1991; Shannon and Meeks-Wagner 1991; Alvarez et al. 1992).

### b) Inflorescence Development in Grasses (Poaceae)

An overview on structural and functional diversity of grass spikelets was presented by Clayton (1990). With regard to grass spikelet development, Chapman (1992) focused on the fact that physiological studies have mainly centered on maize (*Zea*), wheat (*Triticum*), and rice (*Oryza*), although these are relatively atypical grasses. Less familiar genera (e.g., *Eragrostis, Phragmites*) may provide problems that are unusual and stimulating for physiologists and developmental biologists!

Reproductive development was studied in some cereals and/or their wild relatives: *Avena* (Landes and Porter 1990), *Lolium* (Perilleux et al. 1991), *Setaria* (Blaise et al. 1992), *Triticum* (Sibony and Pinthus 1988; Jedel and Hunt 1991), *Zea* (Sundberg and Orr

1990; Camara-Hernandez and Gambino 1991a, b; Dellaporta et al. 1991). Perilleux et al. (1991) and Blaise et al. (1992) quantified the developmental stages of inflorescence development in *Lolium temulentum* and *Setaria italica* in time and space. Inflorescence development in two annual teosintes (*Zea mays* subsp. *mexicana, Z. mays* subsp. *parviglumis*) was compared to perennial teosinte (*Zea diploperennis*) by Sundberg and Orr (1990). The precocious formation of tassel branches in annual teosintes suggests a possible mechanism for the transition from a distichous spike (typical of teosinte) to a polystichous spike which is typical of maize (*Zea mays* subsp. *mays*). Doebley and Stec (1991) added a genetic analysis of the morphological differences between maize and teosinte with some implications for the origin of maize. Veit et al. (1991) examined the combinatory effects of mutations on inflorescence development in maize.

### c) Inflorescence Development in Leguminosae and Some Other Dicotyledons

Weberling (1989b) presented an overview on inflorescences in the Leguminosae, including evolutionary tendencies. The basic inflorescence type is a double or multiple (indeterminate) raceme in which the main axis ends without a terminal flower. Only a few exceptions with terminal flowers are known: e.g., *Gleditsia triacanthos*. Weberling's paper may stimulate comparative studies on inflorescence development in various Leguminosae.

Some developmental work was already done within the last years, e.g., in Caesalpinioideae: *Gleditsia* (Tucker 1991); Mimosoideae: *Acacia bayleyana* (Derstine and Tucker 1991), *Mimosa* (Ramirez-Domenech and Tucker 1989); Papilionoideae: *Cajanus* (Turnbull 1991), *Lespedeza* (Nemoto and Ohashi 1990), *Pisum* (Tucker 1989a; Singer et al. 1990), *Psoralea* (Tucker and Stirton 1991; Tucker 1992c), *Trifolium* (Retallack and Willison 1990a, b).

Cymes are very unusual among legumes. Partial inflorescences somewhat similar to dichasial cymes were found in *Psoralea pinnata* by Tucker and Stirton (1991). An extra protective sheath ("cupulum", resulting from interprimordial growth of 3–4 sterile bracts) below each flower is unique to *Psoralea*. The *"determinate"* mutant of *Pisum* was thought to have determinate inflorescences. A developmental study by Singer et al. (1990) has shown that the apparent terminal flower arises in a lateral position, whereas the apical meristem senesces.

Some additional studies on inflorescence development in dicotyledons may be mentioned here. Steeves and Steeves (1990) described the development of the raceme-like determinate inflorescences in *Amelanchier alnifolia* (Rosaceae). Hufford (1989a) studied reproductive development in *Petalonyx*, the only genus within Loasaceae with racemes instead of cymose inflorescences. Okamoto (1989a, b) added new facts and ideas on cupule formation in Fagaceae (*Castanea, Fagus, Lithocarpus*). In order to explain the fagaceous cupules both the intercalary growth theory and the higher order dichasial branch theory are incomplete to some degree but complement each other.

### d) Inflorescence Development in Asteraceae and Other Angiosperms with Pseudanthia

Pseudanthia (inflorescences with aggregated flowers) have evolved in 40 angiosperm families as listed by Classen-Bockhoff (1990) who also discussed the various inflorescence types as a structural basis for pseudanthium formation.

Disk florets and receptacular bracts arise from common primordia in various Asteraceae: *Helianthus* (Steeves and Sussex 1989), *Helipterum* and *Helichrysum* (Sharman and Sedgley 1988), *Microseris* (Battjes et al. 1992). Generally reported for the capitula of most Asteraceae is an acropetal initiation order of the florets and their bracts. As a deviation from this acropetal pattern Harris et al. (1991) described a bidirectional order of floral initiation within the heads of *Erigeron philadelphicus*. Scanning electron micrographs indicate that the outer disk flowers appear first. The ray flowers then initiate basipetally following the parastichies established by the disk flowers, whereas the inner disk flowers arise acropetally. Bachmann and Chambers (1990) studied the heritable variation for heterocarpy in *Microseris bigelowii*. A numerical canalization is observed, whereby the number of the peripheral hairy achenes roughly depends on capitulum size, but absolute numbers related to the Fibonacci series are preferred. Battjes et al. (1992), Vlot and Bachmann (1991), and Vloth et al. (1992) studied the genetic and nongenetic variation in the number of pappus parts in *Microseris* that varies from 0–10, depending on the species and the strains. Comparing two strains in *Microseris pygmaea* (Asteraceae), Battjes et al. (1992) found that the larger number of florets in one strain is mainly a result of smaller floret primordia as compared to the other strain. Thus, differences in numbers of organs on a determinate apex is a function of the size of the apex and the size of the organ primordia.

Classen-Bockhoff (1992) studied the formation of the compound heads (syncephalia) in *Dyssodia decipiens* and *Eroeda capensis*, the only two known examples of Asteraceae with ligulate flowers restricted to the outer border of the peripheral secondary heads. In accordance with Bachmann's (1983) "field theory", it is supposed that the concentration of a morphogen determines the forming of ligulate and tubular flowers. As in Asteraceae with simple heads, the inflorescence apex starts with the initiation of series of ligulate flowers before it becomes fractioned into secondary heads. The complex head of *Gundelia tournefortii* is composed of several groups of flowers (Classen-Bockhoff et al. 1989). Each group terminates in a large flower which arises first. Additionally, each flower is surrounded by sterile bracts. This complex inflorescence may be derived from other Asteraceae by aggregation, reduction, centralization and ontogenetic abbreviation. In this phylogenetic line secondary monotely is postulated.

### e) Cone and Ovule Ontogeny in Gymnosperms (Mainly Conifers)

An overview on sex distribution and regulation of sex expression in gymnosperms was presented by Owens and Hardev (1990). Stevenson (1990) described the reproductive morphology of the Cycadales. Pharis et al. (1989) reviewed the promotion of flowering in conifers by gibberellins. Takaso and colleagues published an informative series of papers on female cone and ovule development in several conifers, especially Cupressaceae:

*Callitris* (Takaso and Tomlinson 1989b), Podocarpaceae: *Phyllocladus* (Tomlinson et al. 1989), Taxodiaceae: *Cryptomeria*, *Glyptostrobus*, *Taxodium*, *Sciadopitys* (Takaso and Tomlinson 1989a, 1990, 1991). Special emphasis is given to the developmental timing of ovule-supporting structures ("ovuliferous scale") in the axil of the fertile bracts. The results support a direct comparison with the extinct transition conifers *Pseudovoltzia* and *Aethophyllum* but do not fully support Florin's (1954) generalized model for the axillary complex ("branch") of female cones. In *Sciadopitys* the ovuliferous scale starts as a narrow ridge of tissue in the axil of each fertile bract. This ridge develops a series of 9(–12) apical lobes in centrifugal order, each of which is the primordium of a future tooth on the ovuliferous scale. There is a one-to-one ratio between lobes and ovules. In *Callitris* there is less evidence for a branch homologue because the ovules arise directly in the bract axil in basipetal sequence and without any "ovuliferous scales". Within the two trimerous whorls of bracts there are up to 63 ovules surrounding the shoot apex that becomes a sterile central columella. Again in *Phyllocladus* there is no ovuliferous scale. An aril develops late, comparable to *Taxus* (Tomlinson et al. 1989). Stoffberg (1991a, b) added ontogenetic studies on *Podocarpus* from southern Africa.

## 4. Developmental Pathways of Flowers

### a) How to Define Whorls and Cycles

Traditional morphologists defined the term "whorl" as a full cycle of organs that are simultaneously initiated or nearly so. In one flower there are normally 3–6 organs (e.g., petals, stamens) per whorl or cycle that can be delimited from preceding and subsequent whorls or cycles (Endress 1990, 1992). Depending on the evolutionary interpretation of the brassicaceous flower, the floral androecium of *Arabidopsis* (and most other Brassicaceae) consists either of one hexamerous whorl, or of two whorls: an inner one with four stamens, an outer one with two stamens (Endress 1992). Molecular geneticists (e.g., Coen 1991; Meyerowitz et al. 1991) used the term whorl in a much broader sense, i.e., for all organs of a particular type in both helical and cyclic flowers. According to Coen (1991), the domains should be a reflection of organ identity rather than the number or arrangement of organs (see Sect. 5. d). Thus, the six stamens of the *Arabidopsis* flower belong (by definition) to only one whorl (Meyerowitz et al. 1991). Moreover, the flower of *Helleborus niger* (Ranunculaceae) have only four whorls, with up to 100 stamens in the androecial whorl (Coen 1991). According to this terminology, it seems possible that the domains of expression of the various floral homeotic genes might correspond to similar whorls and regions in many flowering plants (Meyerowitz et al. 1989, 1991; Bowman and Meyerowitz 1991; but see Sect. 7).

The broader whorl definition (sensu Coen 1991) coincides with the definition for complex leaf whorls (Rutishauser and Sattler 1986), i.e., a ring-like arrangement of lateral appendages (leaves) which may be initiated simultaneously or successively. Each leaf whorl of, e.g., *Acacia longipedunculata* (Leguminosae) consists of 15–27 phyllodes and as many stipules, arranged in 2(–3) cycles with centrifugal initiation. What was labeled "whorl" by the traditional morphologists, should be called cycle, as will be done in this review. Thus, there are three primary patterns of floral phyllotaxis: cyclic, helical (spiral), and irregular (erratic). Helical and irregular patterns are called acyclic (cf. Weberling 1989a).

## b) Flexibility of Floral Construction, Meristic Variation, and Chaotic Flowers in Angiosperms

Among the flowering plants, flowers are relatively closed systems of considerable complexity. Their structural elements are arranged in certain patterns constrained by developmental conditions, including gene activity and developmental buffering ("numerical canalization", Endress 1990). Besides a few taxa with relatively stable merosity, e.g., Saururaceae (Liang and Tucker 1989, 1990), the primitive subclasses Magnoliidae and Hamamelididae show a floral construction that varies considerably with respect to the number of structural elements and their positioning (Endress 1989c, 1990). This "unstable" condition is contrasted by highly synorganized flowers with cyclic phyllotaxis in Asteridae and orchids where "fusion events" (e.g., interprimordial growth) are frequent (see Sect. 4.d, f).

In Magnoliidae flowers may possess no or up to 70 perianth members, and one to more than 1800 stamens and/or carpels. The floral appendages may arise in a helical, cyclic, or irregular (erratic) pattern, combined with centripetal primordial inception. *Eupomatia* (Eupomatiaceae) shows helical inception of 50–150 stamens and staminodes. In the monotypic Trochodendraceae (with reduced perianth) stamen inception may be helical or nearly cyclic (Endress 1990). In Annonaceae the very numerous stamens are initiated in a nearly cyclic pattern (*Monanthotaxis whytei*, Ronse Decraene 1990) or in an irregular pattern, i.e., neither helical nor cyclic (*Annona*, e.g., *A. cherimolia*; Moncur 1988; Endress 1990). In the annonaceous species *Rollinia deliciosa* irregular stamen inception starts first in the three antesepalous corners of the floral apex (Moncur 1988). Irregular floral phyllotaxis may be more frequent in flowers with reduced perianth, e.g., in *Achlys* (Berberidaceae, Endress 1989a).

Flowers with an unstable organization, i.e., a high degree of meristic variation and irregular primordial patterns, are not restricted to the primitive angiosperms. Irregular carpel initiation was found in female flowers of *Andruris* cf. *andajensis* (Triuridaceae, monocotyledons) by Ruebsamen-Weustenfeld (1991). According to Lehmann and Sattler (1989) and Ronse Decraene and Smets (1990), irregular stamen initiation is observable in *Begonia semperflorens* (Begoniaceae). Irregular stamen patterns combined with centrifugal primordial inception were found in polyandrous members of the Capparales, e.g., *Oceanopapaver*, *Polanisia* (see Karrer, cited in Endress 1992).

Meristic variation of monocotyledonous flowers was explored by Charlton and Posluszny (1991) in *Potamogeton* (Potamogetonaceae); and by Lehmann (1991), Lehmann and Sattler (1992), Scribailo and Tomlinson (1992) in *Calla palustris* (Araceae). The developing *Potamogeton* flower passes through two successive phases of organ initiation: one in which the (mostly four) perianth and superposed stamen primordia arise, and one in which the gynoecial primordia arise. The carpel number (from one to four or even up to 13) seems to depend on the relative size of the carpel primordia and floral apex. In *Calla palustris* the flowers develop no perianth but a series of 9–12 stamens, initiated irregularly in centripetal order. Flowers at the spadix base usually show a reduced stamen number.

Within the Leguminosae, *Ceratonia* and *Gleditsia* show an unusual degree of variability in floral characters that are normally canalized in development (Tucker 1991, 1992a, b, c). Both are atypical for legumes in having an acyclic ("helical" sensu Tucker) order of organogeny, radial symmetry, and a terminal, sometimes polymerous flower in the inflorescence. Tucker (1991) proposed that the absence of subtending floral bracts in *Gleditsia* influences development so as to favor radial symmetry and establishment of other "chaotic" characters. Patterns of floral development more typical for legumes (including simultaneous or unidirectional inception of perianth and stamens, with or without ring or common primordia) were described by Tucker and her collaborators in *Mimosa* (Ramirez-Domenech and Tucker 1989, 1990), *Pisum* (Tucker 1989a), *Psoralea* (Tucker and Stirton 1991), and *Ateleia* where four of the five petals are lost (Tucker 1990).

It is quite fashionable now to detect "chaotic" flowers in various angiosperms (Barabé 1991; Tucker 1991, 1992b). For example, the term chaotic was used for flowers with irregular phyllotaxis (e.g., in *Achlys*, Endress 1989a). According to Lehmann and Sattler (1992), stamen initiation in *Calla* could be mentioned as another example of chaos. As suggested by them, it is best to avoid the term "chaos" in floral morphogenesis because today that term often refers to the nonlinear mathematical phenomenon of deterministic chaos that has not been quantitatively demonstrated in patterns of floral development.

### c) Polyandrous Flowers in Dilleniidae and Rosidae ("Mid-level Dicotyledons")

Weberling (1989a) gave an introduction to the various types of polymerous androecia in angiosperms. Two different patterns of centrifugal stamen inception were found by Ronse Decraene and colleagues in various taxa of the subclass Dilleniidae (Moncur 1988, Ronse Decraene 1989, 1990; Ronse Decraene and Smets 1991a). It may start with antepetalous common primordia that develop into stamen fascicles (*Harungana, Reaumuria*), or it may start with a ring wall, i.e., a broad circular primordium (*Bixa, Cochlospermum, Rheedia*).

Once it was thought that centrifugal stamen initiation is characteristic for the subclass Dilleniidae, whereas all polyandrous members of the Rosidae show a centripetal initiation pattern. Centrifugal and centripetal androecial development, however, is not always restricted to the Dilleniidae and the Rosidae, respectively, as already discussed by Weberling (1989a). Hufford (1989b, 1990) described both patterns within the family Loasaceae (Dilleniidae). Polystaminate members of this family may possess either centripetal or centrifugal androecial development; however, the developmental patterns found by Hufford are too complex to be characterized as simply being centripetal among Mentzelioideae and centrifugal among Loasoideae.

In the Myrtales (Lythraceae: *Lagerstroemia*) and the Sapindales (Zygophyllaceae: *Nitraria*), both Rosidae, the centrifugal initiation pattern is present (Ronse Decraene and Smets 1991b, c). Polyandry in this order is tightly linked to corolla development and receptacular growth (Moncur 1988; Drinnan and Ladiges 1989a, b, c, 1991a, b; Ronse Decraene and Smets 1991b). In the Lythraceae (*Lagerstroemia*) receptacular growth takes place between stamens and petals, thus providing space for centrifugal stamen initiation. In the Punicaceae (*Punica*) and several genera in the Myrtaceae, growth of the receptacle occurs between the androecium and gynoecium, leading to centripetal development of the stamens.

In the mimosoid *Acacia bayleyana* (Rosidae) multistameny is due to sectorial proliferation of stamen primordia (Derstine and Tucker 1991). In *Theobroma cacao* (Sterculiaceae) the five fertile pairs of stamens are initiated as five "common primordia" in antepetalous position prior to the five antesepalous staminodes as is typical for Malvales. Later, all 15 androecial members develop a common tube at their base (Bayer and Hoppe 1990). Other malvalean flowers (*Durio, Matisia*) with androecial "common primordia" in antepetalous position (or nearly so) were studied by Moncur (1988).

### d) Development of Highly Synorganized Flowers with Cyclic Phyllotaxis (Asteridae, Orchids)

Asteridae and orchids provide beautiful examples of highly synorganized flowers. Synorganization is the intimate structural connection of two or several neighboring structural elements to form a functional system as shown by Endress (1990) and Kunze (1991) in flowers of various Asclepiadaceae. Flowers of both Asteridae and orchids show cyclic phyllotaxis and relatively stable merosity (organ number). Numbers of structural elements in flowers of many Asteridae were summarized by Endress (1990). With a few exceptions (e.g., Hydrostachyaceae without perianth), Asteridae flowers are characterized by four floral whorls. Calyx and corolla are most often pentamerous. There are often 2–5 stamens (exceptionally up to 16) and quite regularly 2–3 carpels (exceptionally up to 27). The Lentibulariaceae may be mentioned as an Asteridae family with a fixed number and position of structural elements in the flowers (2–5 sepals, 5 petals, 2 stamens, 2 carpels). This contrasts considerably with the high degree of developmental plasticity of the vegetative body in this family (especially *Utricularia*, see Brugger and Rutishauser 1989; Taylor 1989; Sattler and Rutishauser 1990).

Some of the floral development work done in *Antirrhinum* and its mutants is summarized in Section 5.d. Floral morphogenesis in other scrophulariaceous genera was studied by Armstrong and Douglas (1989: *Rhinanthus, Scrophularia*), Awasthi et al. (1991: *Lindenbergia*), Magin et al. (1989: *Craterostigma*), and Wunderlin (1992: *Alonsoa, Digitalis, Erinus, Nemesia, Wulfenia*).

Armstrong and Douglas (1989) elucidated the ontogenetic basis for corolla aestivation with two patterns that delimit two subfamilies of Scrophulariaceae. The early developmental stages of corolla formation appear to be systematically conservative, unambiguous characters that lead to characteristic patterns of corolla aestivation. Endress (1992) also presented the floral developmental patterns of genera from other Scrophulariales families (e.g., Acanthaceae: *Thunbergia*, Bignoniaceae: *Jacaranda*).

Apart from the Scrophulariales new developmental data were presented from flowers belonging to other Asteridae orders, e.g., Gentianales: *Asclepias* (Endress 1990), Solanales: *Nicotiana, Solanum* (Diggle 1991; Hill and Malmberg 1991), Lamiales: *Ceballosia* (Hilger 1989), Callitrichales: *Hydrostachys* (Leins and Erbar 1990), Dipsacales: *Dipsacus, Morina, Patrinia,*

*Triplostegia* (Hofmann and Göttmann 1990), Campanulales: *Campanula, Jasione, Lobelia, Selliera, Stylidium* (Erbar and Leins 1989; Leins and Erbar 1989; Erbar 1991), and Asterales: e.g., *Microseris* (see Sect. 3.d).

Especially frequent in Asteridae are flowers with compound organs resulting from "congenital fusion" between individual floral parts (see Sect. 4.f). Corolla (and androecium) formation often starts with a common annular primordium, e.g., in *Campanula, Morina* (Erbar and Leins 1989; Hofmann and Göttmann 1990; Erbar 1991).

Comparative developmental and functional research was done in various orchids (Moncur 1988; Dannenbaum et al. 1989; Kurzweil 1989, 1990, 1991; Schick 1989; Kurzweil et al. 1991; Kurzweil and Weber 1992). Kurzweil and colleagues surveyed floral development in many orchid genera paying special attention to the gynostemium and its compound structure. The stylar portion of the inferior ovary and the androecial members (1 stamen, 2 staminodes) contribute to a various degree to the gynostemium depending on the (sub)tribe within the Orchidoideae. The stigma is developmentally derived from the median carpel apex (which is very unusual in orchids) or from the median carpel region below the rostellum (Kurzweil 1990, 1991).

Similar to the orchid gynostemium there is a column with both stigma and anthers at its top in *Stylidium* (Stylidiaceae, Asteridae). Function, structure, and development of this irritable column in *Stylidium* was studied by Erbar (1992), and Findlay and Findlay (1989). According to Erbar (1992), this column should be interpreted as a receptacular tube because no distinct carpel primordia have been found. The inferior ovary of *Stylidium* results from intercalary growth in the peripheral parts of the receptacle below the calyx, corolla, and stamen primordia (see more on inferior ovaries in the next section).

### e) Some Comments on Gynoecium Development

In spite of the essays presented by Brückner (1991) and Cresens and Smets (1989), the gynoecium (including the carpel as its construction unit) remains a "problem child of floral morphology and evolution", at least for botanists who still try to draw imaginary borderlines between "fused" organs! Criticizing adherents of pseudomonomery in certain Araceae, Scribailo and Tomlinson (1992) argued that the *Calla* gynoecium shows no evidence of a pseudomonomerous condition either in development or vasculature (but see Lehmann and Sattler 1992, who found evidence of more than one gynoecial primordium in *Calla*). There are other aroid genera with obviously more than one locule (carpel) per ovary. Mayo (1989) found an extraordinarily variable locule number in *Philodendron*, particularly in the subgenus *Meconostigma*. Ovaries may have 2–47 locules. Even within a single spadix the locule number normally varies (e.g., from 8 to 13 in *Philodendron bipinnatum*).

Evidence from developmental mutants may even help to solve meristic problems, e.g., in carpel number of the gynoecium in Brassicaceae. As nicely described by Smyth et al. (1990) the gynoecium of *Arabidopsis* (wild-type) starts its development as a ring wall (common primordium), although it is thought to have evolved through the congenital fusion of two or four carpels (Endress 1992). Molecular geneticists, e.g., Kunst et al. (1989), provided additional arguments in favor of the "two-carpel" hypothesis for the

*Arabidopsis* gynoecium. When the gynoecia of "apetala-2" mutants are improperly fused, no more than two separate parts are observed, and the split is always in a median plane with a row of ovules on each unfused margin. In other capparidalean families it is easier to decide how many carpels are involved in the formation of the superior ovary, e.g. *Tovaria* with 5–8 carpels (Fisel and Weberling 1990).

Several botanists focused on developmental patterns of inferior ovaries (see also previous section). In the pentamerous monocotyledonous genus *Pentastemona* (Pentastemonaceae) the five stamen primordia develop on the rim of a concave apex that gives rise to the inferior ovary (van Heel 1992). Charpentier et al. (1989a, b) described female flower development in Begoniaceae comparing the inferior ovary of *Begonia* (*B. dregei, B. horticola*) with the semi-inferior ovary of *Hillebrandia*. According to them, the ovary wall in *Begonia* may be interpreted as an intermediate structure between axial and appendicular organs, difficult to compare directly with a set of fused carpels and perianth members (as postulated by the classical theory of the flower). The inferior ovary in *Amelanchier ovalis* (Rosaceae-Maloideae) was studied by Steeves et al. (1991) who concluded from their histogenetic studies that the ovarian wall and most of the septal tissue arise from intercalary growth and cannot be considered as derived from the gynoecial primordia. The appendicular-receptacular debate of the inferior ovary continues!

### f) Fusion Events During Floral Morphogenesis

Experimental and molecular biologists show an increasing interest in fusion events during flower development and evolution (see review by Verbeke 1992a). The distinction of postgenital fusion (also called ontogenetic fusion or surface fusion) and congenital fusion (also called phylogenetic fusion or zonal growth), however, was already done by traditional morphologists (as summarized by Weberling 1989a). Postgenital fusion is based on epidermal cell redifferentiation between contacting floral parts. According to Verbeke (1992b), the postgenital fusion of the two carpels in *Catharanthus* (Apocynaceae) seems to depend on diffusible factors that trigger epidermal cell differentiation.

Congenital fusion is a process that is not observable during floral development. It is based on the assumption that the ancestral condition was one of separate entities and that organ individuality was lost in the course of evolution. Sattler (1978) tried to replace the term "congenital fusion" by "interprimordial growth" or "shifting", i.e., processes that are observable during development. According to Verbeke (1992a), the mechanism of congenital fusion "lies in the establishment of some sort of basal or intercalary meristem that replaces the apical growth normally responsible for the formation of separate organs". Various examples of congenital fusion of carpels with other floral parts are discussed in Sections 4d, e.

Ring primordia or common primordia in early floral morphogenesis may or may not be accepted as evidence for congenital fusion of, e.g., stamens in polymerous androecia (see Sect. 4c). The Asteridae with their congenitally fused corollas mainly show two developmental patterns: early and late sympetaly (as surveyed by Erbar 1991). The corolla tube in Asteridae originates as a ring primordium either before or at the same time as the corolla lobe (early sympetaly) or by interprimordial growth of initially free corolla lobes (late sympetaly). There are orders in the Asteridae where only early sympetaly occurred in all families studied (e.g., Dipsacales, Campanulales-Asterales). Sympetaly can also be early, late, and transitory within a single family (e.g., Gentianaceae).

## 5. Homeosis, Heterochrony, and Developmental Mutants

### a) Fashionable Concepts

Since 1989 (Prog. Bot. 51:60–63) homeosis and heterochrony have become quite fashionable concepts in describing and explaining the diversity of reproductive developmental patterns in plants (Hill and Lord 1990; Lord 1991; Negrutiu and Sala 1991; Negrutiu et al. 1991; Sattler 1992). According to van Tunen and Angenent (1991), homeosis and heterochrony are even the two processes by which the formation of the flower and its organs is determined! The terms homeosis and heterochrony are used to describe processes that govern the placement and timing of parts in an organism or complex organ structure like the flower. The term homeosis (which is often but not always synonymous to heterotopy) describes the spatial transference of one part of an organism's body to a new position where it normally does not occur (Sattler 1988; Lord 1991). Heterochrony is defined as alterations in developmental timing. Molecular and developmental geneticists (e.g., Negrutiu et al. 1991) normally distinguish between homeotic and heterochronic genes, both being regulatory in the control of, e.g., floral morphogenesis. There is, however, a close relationship between homeosis and heterochrony in plants, which explains the additive and sequential nature of their development. Floral organs can switch identities by misplacement and wrong timing (Negrutiu and Sala 1991).

There is now the tendency to use the concepts of homeosis and heterochrony everywhere in evolutionary considerations. Depending on somewhat sloppy definitions of homeosis and heterochrony all evolutionary change can be categorized as some type of homeosis or heterochrony (see Raff and Wray 1989; Lord 1991; McKinney and McNamara 1991; Mabee 1991: "pan-heterochrony"). The future will show whether or not such a general usage of these concepts is justified. Hay and Mabberley (1991), in a paper on "transference of function" and the origin of the monocotyledonous family Araceae, assumed that homeotic saltatory transformation of both reproductive and vegetative shoots is involved. According to Kato (1990, 1991), the Ophioglossaceae may be regarded as living progymnosperms. Kato proposed a new archetype model for the angiosperm carpel starting with the three-dimensional construction of ophioglossoid fertile leaves (with epiphyllous sporophores). Ovular paedomorphosis (i.e., heterochrony) may have led to the angiosperm carpels with adaxial ovules. According to this hypothesis, the adaxial position of the ovules arose prior to the evolutionary process leading from gymnospermy to angiospermy. Alternative views on the origin of the angiosperm gynoecium are summarized in Brückner (1991), Friis (1992), Friis and Endress (1990).

### b) Naturally Occurring Homeosis in Flowering Plants

In this section some floral developmental studies are summarized where the concept of homeosis was or may be used to explain natural variation of floral construction.

Studying floral development in *Ranalisma humile* (Alismataceae) Charlton (1991) found six organogenetic sites above and alternating with the first six stamens. These sites may be indifferently occupied by stamen or carpel primordia. Also possible is an extra petal in a position where one of the first six stamens would be expected. Naturally occurring homeosis was also identified in Leguminosae. Tucker (1989b, 1992c) described transformation of petals to stamens in the caesalpinioid genus *Saraca*. In *Saraca declinata* five primordia are initiated in the usual petal sites, but none become petals. Four of them differentiate as stamens, the fifth remains rudimentary.

Homeosis has played an important role in the evolution of the flowers of the Zingiberales, especially those of the ginger group (Kirchoff 1991). In the Zingiberaceae, two members of the outer androecial whorl are replaced by a lip, and two members of the inner androecial whorl are replaced by petaloid structures. The flowers of the Cannaceae and Marantaceae have – besides 3–4 variously modified staminodes – one stamen-petalum mosaic organ (with one-half of a fertile anther and one-half of a petalum). The Zingiberales, therefore, show examples of total and partial stamen replacement, i.e., total and partial homeosis.

According to Lord (1991) and Negrutiu and Sala (1991), homeosis is only exceptionally complete in plants, and more often only partial, leading to mosaic or chimeric organs (see more in Sect. 5d). The petaloid stamens occurring in nymphaeaceous flowers (e.g., in *Euryale, Nymphaea*) may also be due to partial homeosis (Weberling 1989a; Chen et al. 1991; see also Hill and Lord 1989; Alvarez et al. 1992 on partial homeosis in flowers).

In some *Eucalyptus* groups studied by Drinnan and Ladiges (1989a, b, c, 1991a, b) the formation of the polymerous androecium starts with common primordia that arise on the basal corolla component (i.e., the staminophore of the mature flower). Stamen inception on the corolla (not on the floral apex) may be explained by homeosis (heterotopy) and heterochrony.

A special case of homeosis (heterotopy) was described by Marquez-Guzman et al. (1989), Martinez and Ramos (1989), and Stevens (1991a) in *Lacandonia schismatica* (Lacandoniaceae, monocotyledons). This plant is unique among angiosperms in having stamens that are in the center of the flower and that are surrounded by carpels, the reverse of the normal arrangement that occurs in the closely related family Triuridaceae (especially *Sciaphila* spp., see Ruebsamen-Weustenfeld 1991).

### c) Naturally Occurring Heterochrony in Flowering Plants

Methods assigning a metric age to ontogenesis are a prerequisite to any comparison and analysis of temporal patterns in terms of heterochrony and for the comprehension of form as a spatial and temporal process. Lord and collaborators progressed in kinematic analyses of surface growth patterns in *Lilium* flower organs based on sequential marking experiments of living material (Lord and Gould 1989; Crone and Lord 1991). Smyth et al. (1990) made a quantitative study on flower development in *Arabidopsis*. Under controlled growing conditions the flower buds took 13.25 days to progress through the 12 stages distinguished from initiation to opening. These quantitative studies are necessary if we want to learn how wild-type patterns are altered to create divergent floral forms in mutants as well as during evolution (Bowman et al. 1991a; Crone et al. 1992).

Ritterbusch (1989, 1990) proposed the normalized-age concept that allows a quantitative comparison of flower ontogenesis among related taxa with many-flowered inflorescences, e.g., species of *Calceolaria, Digitalis*, and *Pedicularis* (Scrophulariaceae). The various heterochronic changes needed to derive the floral developmental patterns of 7 genera (9 species) in Scrophulariaceae (*Alonsoa, Calceolaria, Digitalis, Erinus, Nemesia, Pedicularis, Wulfenia*) were presented by Wunderlin (1992). Guerrant (1989) compared size and allometric development in the flowers of two closely related species of Limnanthaceae. The small flowers of the autogamous *Limnanthes floccosa* are hypothesized to show the effects of paedomorphosis (or evolutionary juvenilization by the process of progenesis) relative to the large flowered, xenogamous *Limnanthes alba*. Kellogg (1990) studied allometric growth of florets in *Poa* species. The differences between them presumably reflect heterochronic changes in a developmental pattern that shows underlying similarity between the species.

The formation of floral gender can be based on heterochronic changes, especially when the floral forms in early development stages look alike, e.g., in *Asparagus officinalis*: male/female (Bracale et al. 1991), *Carica papaya*: male/female/bisexual (Moncur 1988), *Fraxinus pennsylvanica*: male/female (Remphrey 1989), *Mangifera indica*: male/bisexual (Moncur 1988), and *Silene alba*: male/female with and without *Ustilago* infection (Ye et al. 1991). The developmental processes that underlie the differences in floral sex expression of hermaphrodite and staminate flowers of *Solanum hirtum* (Solanaceae) were examined by Diggle (1991). Gynoecial sterility in staminate floral buds does not occur until 6–7 days before anthesis.

More or less late divergence of the ontogenetic pathways is found during the development of the 2–3 floral morphs in heterostylous plants (e.g., *Averrhoa carambola, Lythrum salicaria, Oxalis cernua, Pontederia cordata*) as described by Moncur 1988; Richards and Barrett 1992).

### d) Developmental Mutants and the *Arabidopsis-Antirrhinum* Model of Floral Morphogenesis

Mutations that cause abnormal pattern formation in the development of flowers have been known for centuries (see more in Meyerowitz et al. 1989; Penzig 1921–1922). Negrutiu et al. (1991) distinguish three types of developmental mutants in plants, especially floral formation (see also Sect. 2). Homeotic (heterotopic) mutants develop normal or wrong organs at a wrong place. Heterochronic mutants alter developmental timing and duration of events. Meristic mutants change the number of organs or suborgan components within the flower. A clear-cut distinction of these mutants, however, is not always possible. Some mutants are heterochronic, in addition to being homeotic, e.g., the "*floricaula*" mutants in *Antirrhinum* (Coen et al. 1990, 1991; Schwarz-Sommer et al. 1990; Huijser et al. 1992a, b). They reiterate early developmental pathways, thus leading to compound inflorescences instead of simple racemes. Similar mutants (e.g., "*leafy*"), regulating inflorescence branching, were also found in *Arabidopsis* (Schultz and Haughn 1991).

Many developmental floral mutants are known in the two model organisms *Arabidopsis* and *Antirrhinum.* The switch to determinacy is an important property of pattern formation in flowers (Bowman et al. 1991a; Chasan 1991). In *Arabidopsis* and *Antirrhinum,* the genes *AGAMOUS* (*AG*) and *PLENIFOLIA* (*PLENI*), respectively, suppress indeterminate growth of the floral meristem once it has formed four whorls of floral organs because *ag* and *pleni* mutations cause the floral meristem to proliferate indefinitively. Several groups studying *Arabidopsis, Antirrhinum,* and a few other model plants have used genetic and molecular data to formulate a model for the pattern-forming process (including organ specification) in floral development. This "*Arabidopsis-Antirrhinum* model" (including slightly modified versions) of Bowman, Coen, Lord, Meyerowitz, Schwarz-Sommer and their colleagues predicts the phenotypes of normal and double flowers in both *Arabidopsis* and *Antirrhinum* (see, e.g., Schwarz-Sommer et al. 1990; Bowman et al. 1991b; Coen 1991; Coen and Meyerowitz 1991; Lord 1991; Sommer et al. 1991). The distantly related flowering plants *Arabidopsis* and *Antirrhinum* seem to have homologous mechanisms in floral pattern formation. Although there are four whorls in these flowers (P = petals, S = sepals, A = androecium, G = gynoecium), only three sets of organ identity genes are postulated, each set controlling organ initiation and differentiation in two adjacent whorls. Thus, there are three overlapping domains of the organ identity genes: P-S/S-A/A-G.

Hill and Lord (1989) compared flower development in *Arabidopsis* wild-type (WT) and "*pistillata*" mutant in order to determine when and where developmental divergence takes place in the mutant compared to the WT. They found that early flower development in "*pistillata*" was normal until stamen initiation when obvious disruptions occurred with the formation of carpeloid structures. Petal initiation was normal but petal differentiation was disrupted by showing a divergence onto the sepal pathway. Rather than sepals replacing petals in a one-to-one organ replacement (typical for complete homeosis), the organs of whorl 2 form a developmental mosaic of petal and sepal features (typical for partial homeosis). Schwarz-Sommer et al. (1990) and Sommer et al. (1991) described a similar "*deficiens*" mutant in *Antirrhinum.* Their genetic data show that this gene acts early in stamen development (at initiation), and later in petal development (during differentiation). As summarized by Lord (1991) there are other homeotic mutants in *Arabidopsis,* e.g., "*agamous*" with a change in both stamen differentiation and carpel initiation. These regulatory genes (organ identity genes) – with overlapping domains – control late development (differentiation) in one floral whorl and early development (initiation) in the next inner one. The *Arabidopsis-Antirrhinum* model of Lord (1991) and others, therefore, seems to explain the formation of intermediate organs as developmental mosaics (Hill and Lord 1989; Meyerowitz et al. 1989; Poethig 1990; Schwarz-Sommer et al. 1990; Alvarez et al. 1992): The developmental fate of an organ ("organ identity") may gradually change from primordial initiation to later stages of differentiation.

Homeotic mutants involved in floral development were also analyzed in Solanaceae: in tomato (*Lycopersicon*) by Pnueli et al. (1991), Sawhney (1990, 1992), Sawhney and Rastogi (1990); in *Petunia* by Gerats (1991), Turlier and Alabouvette (1989), Turlier et al. (1991), van Tunen and Angenent (1991). The developmental pathways of a normal and a female sterile mutant of *Petunia hybrida* were compared by Turlier et al. (1991) showing the initiation sites for the partial masculinization of the gynoecium. According to van Tunen and Angenent (1991), the *Petunia* mutants indicate that the *Arabidopsis*-

*Antirrhinum* model for floral morphogenesis (see above) has to be modified or extended. Pnueli et al. (1991) described the MADS box genes and their expression during floral development in tomato. They are homologous to floral homeotic genes in *Antirrhinum* and *Arabidopsis* (cf. Sommer et al. 1991; Huijser et al. 1992a, b). In the "*solanifolia*" (*sf/sf*) mutant of tomato, the nonfusion of stamens and carpels is associated with an increase in the size of the apex and the number of organs. This mutant, therefore, may be classified as being heterochronic as well as meristic.

Comparing single (wild-type) and double flowers in *Potentilla fruticosa* (Rosaceae), Innes et al. (1989) found identical initiation patterns of petals and stamens, starting with a pentagonal ridge surrounding the floral apex. The five primordia at the vertices always develop into petals. In single flowers all other primordia on the ridge develop into stamens. In double flowers the primordia adjacent to those at the vertices develop petaloid characteristics. The extra petals in double flowers are thus found in positions otherwise occupied by stamens. Similar are the early developmental stages of single and double forms of male flowers in *Begonia* (Lehmann and Sattler 1989; Ronse Decraene and Smets 1990).

## 6. Cladistic Analysis Using Floral Developmental Characters

Both traditional systematists and cladists try to solve similar problems (see Stuessy 1990; Stevens 1991b): to select outgroups, to intuitively choose the qualitative or quantitative characters best suited for systematic or cladistic analysis, and to decide which character states may be the more primitive ones within a certain taxon. Some new cladistic analyses using ontogenetic characters are already summarized by Lidén (1992). A few additional studies combining floral ontogenetic data with cladistic analysis shall be mentioned here.

Linder and Kurzweil (1990) tried to combine developmental floral morphology and phylogeny in Disinae (Orchidaceae). Similar to Williams et al. (1990) they concluded that ontogenetic criteria are less useful than outgroup criteria in determining character polarity. Many systematists based their cladistic analysis also on floral developmental data (e.g., Hufford 1992: Rosidae s.l.; Hufford and Dickison 1992: Cunoniaceae). Drinnan and Ladiges (1989a, b, c, 1991a, b) used corolla and androecium development in various *Eucalyptus* groups and *Angophora* for cladistic analysis. One of the synapomorphies of some (not all) eucalypts is the stamen inception on the corolla base (see Sect. 5.b). Constructing cladistic relationships in Hamamelidaceae-Hamamelidoideae, Endress (1989b) used centrifugal polyandry as synapomorphy for the *Fothergilla-Parrotiopsis* clade, whereas centripetal polyandry is a derived character restricted to *Matudaea*.

Only few exponents of traditional systematics explained why they refused cladistics, e.g., Stützel (1990), who studied gynoecial development in the related families Eriocaulaceae and Xyridaceae (Commelinales). In both families there are genera with and without stylar appendages. If the Eriocaulaceae have evolved from Xyridaceae-like ancestors, the occurrence of stylar appendages may be accepted as a derived character within Xyridaceae which was lost again in 2 out of 12 genera in Eriocaulaceae (*Eriocaulon, Mesanthemum*). According to Stützel, such a situation is difficult to visualize by a cladogram due to other conflicting characters.

## 7. Evolution of Floral Ontogenies

It may be taken as a tentative framework that the patterns of early floral development are relatively conservative in evolution (Armstrong and Douglas 1989; Tucker 1989b, 1992c; Mione and Bogle 1990; Leins and Erbar 1991). Plant examples of von Baer's (1828) biogenetic rule in plants are discussed by Barabé (1990). As in zoology, limitations to this rule have been discussed for plants, especially flowers. Changes in early ontogeny in closely related groups are known as "*archallaxis*" (an extreme form of heterochrony protracted to the beginning of development, see Endress 1990). Hence, early developmental stages are not always conservative.

An example of archallaxis is the reversal of the positional relationship of the two innermost whorls in the flowers of *Lacandonia* (Lacandoniaceae, Triuridales) where carpel initiation precedes stamen inception, in contrast to all (?) other angiosperms (Stevens 1991a; see Sect. 5b).

The role of floral development in studies of legume evolution was explored by Tucker and colleagues (as summarized by Tucker 1992b, c; see Sects. 3c, 4b). They distinguished early, mid-stage, and late developmental events in order to elucidate phylogeny (Ramirez-Domenech and Tucker 1989, 1990; Tucker 1989b, 1992b). Tucker (1992c, p. 692) concluded with respect to legume systematics: "Floral development among species in a genus is generally very similar, with any differences usually occurring only terminally. In contrast, ontogenies among related genera show more differences, and those differences may be accelerated, appearing earlier in ontogeny than interspecific differences." Thus, floral development repeats evolution (at least to some degree). Tucker (1992c) also demonstrated convergence in legumes by comparing differing developmental pathways of specialized floral features such as dorsiventral heterostameny and eccentric corolla among species of *Cassia, Senna*, and *Chamaecrista*.

There is still considerable interest in developmental patterns of polymerous androecia in order to test phylogenetic hypotheses on the subclass level in angiosperms (see Weberling 1989a). Leins and Erbar (1991) combined early androecial development with the evolution of primitive and mid-level dicotyledons. The fascicled androecia found in *Paeonia* (Paeoniaceae) and other Dilleniidae can be regarded as having coexisted with "magnolioid" spiral androecia since early in flowering plant evolution. Simple (diplo- or haplostemonous) androecia of the dicotyledons of the middle evolutionary level can be derived from fascicled androecia (as typical for Dilleniidae, including *Paeonia*, whereas a derivation of simple androecia from spiral "magnolioid" androecia occurred probably only rarely. Leins and Erbar (1991) argue that "this conception is supported by the fact that in *Paeonia*, which is archaic in several floral features, the five fascicle primordia (at least by their position) fit in the basic floral spiral, which follows the ... 'golden' divergence angle." It is questionable whether this argument is strong enough to refute the opposite way of character polarization, i.e., the possibility to derive complex androecia from simple ones. This was done by Ronse Decraene and Smets (1991a), who concluded that the complex androecium in some Dilleniidae (e.g., in *Harungana* and other Clusiaceae) may have a diplostemonous origin. With respect to the Capparales Endress (1992) formulated a similar hypothesis: the hexamerous androecium (as typical for most Brassicaceae and many Capparaceae) is probably the evolutionarily primitive state and the polymerous androecia (as typical for various genera in Capparaceae) can be derived from it.

Geneticists and molecular biologists (e.g., Meyerowitz et al. 1989; Bowman and Meyerowitz 1991; Coen 1991; Pnueli et al. 1991; Sawhney 1992) also discussed (hypothetical) links between floral evolution and the floral ontogenies in *Arabidopsis, Antirrhinum,* and *Lycopersicon,* based on genetic-molecular analysis of floral mutants and on regulatory genes with highly conserved DNA sequences. Bowman and Meyerowitz (1991) suggested that much of the genetic machinery required for the development of flowers may have evolved from a small number of progenitor genes. Early duplication and divergence could have produced a small number of gene families whose members direct floral development in the various groups of angiosperms. When quite archaic angiosperms (e.g., the Chloranthaceae) are considered, the "*Arabidopsis-Antirrhinum* model" of floral morphogenesis (see Sect. 5d), however, needs to be strongly modified. For example, the flower of *Sarcandra* (Chloranthaceae) consists of only one stamen and one carpel, i.e., only two monomerous whorls according to the geneticists' whorl definition (see Sect. 4a; Endress 1992).

According to Coen (1991), radial symmetry was likely the ancestral form of the flowers in angiosperms and zygomorphy has evolved several times in independent evolutionary lines. This idea (consistent with taxonomic studies) is corroborated by the observation that mutations to radial symmetry were found in many zygomorphic species but that no well-defined mutations to zygomorphy have been described in actinomorphic species. In the case of terminal peloria in, e.g., *Antirrhinum,* there is a double atavism because both the flower shape (peloric, i.e., radial symmetric) and the inflorescence type (determinate) are ancestral conditions, at least in Scrophulariaceae. Based on developmental and experimental studies of the "*solanifolia*" mutant of *Lycopersicon* (Solanaceae; see Sect. 5d), Sawhney (1992) concluded that during the course of evolution plants with small floral apices were selected and that the reduction in apex size was achieved by mutations that affect endogenous plant growth regulators.

## 8. Outlook

This review emphasizes research activities at the intersection of molecular genetics and systematic botany. It is important to add three wishes for future research. The molecular geneticists should become more aware of the great diversity of floral construction in angiosperms which is far beyond the patterns observable in a few model organisms (*Antirrhinum, Arabidopsis, etc.*) and their mutants, especially when the primitive angiosperms are considered (cf. Endress 1992; see Sect. 4b). The systematic and structural botanists interested in comparative floral development should try to present their results in a way that is relevant to the research of developmental and molecular geneticists. Moreover, it is important to present comparative data in floral development using an approach as dynamic as possible (Posluszny and Gerrath 1992; Sattler 1992). The surveys on floral developmental patterns in many legumes and orchids by Tucker (e.g., 1992c), Kurzweil (e.g., 1991), and their colleagues may serve as examples how "progress in botany" will continue in order to elucidate ontogeny and phylogeny. The near future will show whether Coen (1991, p. 275) is right in predicting that the molecular and genetic analysis of homeotic genes and their interaction with other genes "will provide a new foundation for our understanding of floral development and evolution."

*Acknowledgements.* The author is grateful to Dr. H. H. Hilger, Berlin, and Dr. B. K. Kirchoff, Greensboro (NC), for sending their updated literature lists. Mr. R. Bolli, Dr. P. K. Endress, Mr. M. Reut, Mr. Ch. Wagner, Zürich, and Dr. R. Sattler, Montreal, provided valuable comments on the manuscript.

## References

**Alvarez J,** Guli CL, Yu X-H, Smyth DR (1992) Plant J 2:103–116. **Armstrong JE**, Douglas AW (1989) Bull Torrey Bot Club 116:378–389. **Awasthi DK**, Rawat R, Kumar A, Kumar V (1991) Bot Mag (Tokyo) 104:59-71.

**Bachmann K** (1983) Evolutionary genetics and the genetic control of morphogenesis in flowering plants. In: Hecht MK, Wallace B, Prance GT (eds) Evolutionary biology, vol 16. Plenum, New York, pp 157–207. **Bachmann K,** Chambers KL (1990) Beitr Biol Pflanz 65:123–146. **Barabé D** (1990) Année Biol 29:89–132. **Barabé D** (1991) Acta Biotheor 39:157–159. **Battey NH**, Lyndon RF (1990) Bot Rev 56:162–189. **Battjes J**, Bachmann K, Bouman F (1992) Bot Jahrb Syst 113:461–475. **Bayer C**, Hoppe JR (1990) Beitr Biol Pflanzen 65:301–312. **Bernier G** (1989) Events in the floral transition of meristems. In: Lord EM, Bernier G (eds) Plant reproduction: from floral induction to pollination. American Society of Plant Physiologists, Rockville, pp 42–50. **Bernier G,** Lejeune P, Jacqmard A, Kinet J-M (1990) Cytokinins in flower initiation. In: Pharis RP, Rood S (eds) Plant growth substances 1988. Springer, Berlin Heidelberg New York, pp 486–491. **Blaise MO,** Girardin P, Millet B (1992) Agronomie 12:141–156. **Bowman JL,** Meyerowitz EM (1991) Genetic control of pattern formation during flower development in *Arabidopsis*. In: Jenkins GI, Schuch W (eds) Molecular biology of plant development. Society for Experimental Biology, Cambridge, pp 89–115. **Bowman JL**, Drews GN, Meyerowitz EM (1991a) Plant Cell 3:749–758. **Bowman JL**, Smyth DR, Meyerowitz EM (1991b) Development 112:1–20. **Bracale M,** Caporali E, Galli MG, Longo C, Marziani-Longo G, Rossi G, Spada A, Soave C, Falavigna A, Raffaldi F, Maestri E, Restivo FM, Tassi F (1991) Plant Sci 80:67–77. **Brückner C** (1991) Gleditschia 19:3–14. **Brugger J,** Rutishauser R (1989) Bot Helv 99:91–146.

**Camara-Hernandez J**, Gambino S (1991a) Maydica 35:1–12. **Camara-Hernandez J**, Gambino S (1991b) Bot J Linn Soc 107:211–216. **Carpenter R,** Coen E (1990) Genes Dev 4:1483–1493. **Carpenter R,** Coen E, Huijser P, Schwarz-Sommer Z, Sommer H (1991) Flowering Newslett 12:38–41. **Chapman GP** (1992) Plant Growth Regul 11:33–36. **Charlton WA** Can J Bot 69:2790–2796. **Charlton WA,** Posluszny U (1991) Bot J Linn Soc 106:265–293. **Charpentier A,** Brouillet L, Barabé D (1989a) Can J Bot 67:559–572. **Charpentier A**, Brouillet L, Barabé D (1989b) Can J Bot 67:3625–3639. **Chasan R** (1991) Plant Cell 3:845–847. **Chattopadhyay D,** Sharma AK (1991) Feddes Rep 102:29–55. **Chen WP,** Zhang S-M, Xu X-S, Zhang W-M, Shi G-X (1991) Acta Bot Sin 33:104–109. **Classen-Bockhoff R** (1990) Plant Syst Evol 171:57–88. **Classen-Bockhoff R** (1992) Flora 186:1–22. **Classen-Bockhoff R**, Froebe HA, Langerbeins D (1989) Flora 182:463–479. **Clayton WD** (1990) The spikelet. In: Chapman GP (ed) Reproductive versatility in the grasses. Cambridge Univ Press, Cambridge, pp 32–51. **Coen ES** (1991) Annu Rev Plant Physiol Plant Mol Biol 42:241–279. **Coen ES,** Carpenter R (1992) New Sci 134:24–27. **Coen ES,** Meyerowitz EM (1991) Nature 353:31–37. **Coen ES**, Romero JM, Doyle S, Elliott R, Murphy G, Carpenter R (1990) Cell 63:1311–1322. **Coen ES**, Doyle S, Romero JM, Elliott R, Macrath R, Carpenter R (1991) Development (Suppl) 1:149–155. **Cresens EM,** Smets EF (1989) Bull Jard Bot Nat Belg 59:377–409. **Crone W,** Lord EM (1991) Dev Biol 143:408–417. **Crone W,** Eckard KJ,

Lord EM (1992) Am J Bot 79:28. **Cronquist A** (1981) An integrated system of classification of flowering plants. Columbia Univ Press, New York, 1280 pp.

**Dahlgren G** (1989a) The last dahlgrenogram. System of classification of the dicotyledons. In: Tan K (ed) The Davis and Hedge Festschrift. Edinburgh Univ Press, Edinburgh, pp 249–260. **Dahlgren G** (1989b) Bot J Linn Soc 100:197–203. **Dannenbaum C,** Wolter M, Schill R (1989) Bot Jahrb Syst 110:441–460. **Dellaporta SL,** Moreno MA, Delong A (1991) Development (Suppl) 1:141–147. **Derstine KS,** Tucker SC (1991) Am J Bot 78:816–832. **Diggle PK** (1991) Am J Bot 78:377–393. **Doebley J,** Stec A (1991) Genetics 129:285–295. **Donnison IS,** Francis D, Hughes SG (1991) Flowering Newslett 12:27–31. **Drews GN,** Weigel D, Meyerowitz EM (1991) Curr Opin Genet Dev 1:174–178. **Drinnan AN,** Ladiges PY (1989a) Plant Syst Evol 165:227–237. **Drinnan AN,** Ladiges PY (1989b) Plant Syst Evol 165:239–254. **Drinnan AN,** Ladiges PY (1989c) Plant Syst Evol 166:183–196. **Drinnan AN,** Ladiges PY (1991a) Aust Syst Bot 4:539–551. **Drinnan AN,** Ladiges PY (1991b) Aust Syst Bot 4:553–562.

**Endress PK** (1989a) Bot Acta 102:159–163. **Endress PK** (1989b) Phylogenetic relationships in the Hamamelidiodeae. In: Crane PR, Blackmore S (eds) Evolution, systematics, and fossil history of the Hamamelidae, vol 1. Clarendon Press, Oxford, pp 227–248. **Endress PK** (1989c) Mem N Y Bot Gard 55:5–34. **Endress PK** (1990) Biol J Linn Soc 39:153–175. **Endress PK** (1992) Int J Plant Sci 153:S106–S122. **Erbar C** (1991) Bot Jahrb Syst 112:417–451. **Erbar C** (1992) Can J Bot 70:258–271. **Erbar C,** Leins P (1989) Bot Jahrb Syst 111:29–55.

**Findlay N,** Findlay GP (1989) Aust J Bot 37:81–101. **Fisel KJ,** Weberling F (1990) Bot Jahrb Syst 111:365–387. **Florin R** (1954) Biol Rev Camb Philos Soc 29:367–389. **Friis EM** (1992) Prog Bot 53:318–340. **Friis EM,** Endress PK (1990) Adv Bot Res 17:99–162.

**Gasser CS** (1991) Annu Rev Plant Physiol Plant Mol Biol 42:621–649. **Gerats AGM** (1991) Plant Sci 80:19–25. **Green PB** (1991) Flowering Newslett 12:7–11. **Green PB,** Havelange A, Bernier G (1991) Planta 185:502–512. **Guerrant EO** (1989) Early maturity, small flowers and autogamy: a developmental connection? In: Bock JH, Linhart YB (eds) The evolutionary ecology of plants. Westview Press, Boulder, pp 61–84.

**Hagemann W** (1990) Acta Biotheor 38:207–242. **Hallé F,** Oldeman RAA, Tomlinson PB (1978) Tropical trees and forests. Springer, Berlin Heidelberg New York, 458 pp. **Harris EM,** Tucker SC, Urbatsch LE (1991) Am J Bot 78:108–121. **Hay A,** Mabberley DJ (1991) Bot Jahrb Syst 113:339–428. **Hernandez LF,** Havelange A, Bernier G, Green PB (1991) Planta 185:139–147. **Hilger HH** (1989) Plant Syst Evol 166:119–129. **Hill JP,** Lord EM (1989) Can J Bot 67:2922–2936. **Hill JP,** Lord EM (1990) The role of developmental timing in the evolution of floral form. In: Raff R (ed) Seminars in developmental biology. W B Saunders, London, pp 281–287. **Hill JP,** Malmberg RL (1991) Planta 185:472–478. **Hofmann U,** Göttmann J (1990) Bot Jahrb Syst 111:499–553. **Huber H** (1991) Angiospermen: Leitfaden durch die Ordnungen und Familien der Bedecktsamer. Fischer, Stuttgart, 170 pp. **Hufford LD** (1989a) Plant Syst Evol 163:211–226. **Hufford LD** (1989b) Nord J Bot 9:217–227. **Hufford LD** (1990) Can J Bot 68:402–419. **Hufford LD** (1992) Ann Mo Bot Gard 79:218–248. **Hufford LD,** Dickison WC (1992) Syst Bot 17:181–200. **Huijser P,** Klein J, Lönnig W-E, Meijer H, Saedler H, Sommer H (1992a) EMBO J 11:1239–1249. **Huijser P,** Klein J, Lönnig W-E, Meijer H, Saedler H, Sommer H (1992b) Flowering Newslett 13:38–43.

**Innes RL,** Remphrey WR, Lenz LM (1989) Can J Bot 67:1071–1079. **Irish VF** (1991) Curr Opin Genet Dev 1:169–173.

**Jedel PE,** Hunt LA (1991) Ann Bot 68:501–509.

**Kato M** (1990) Bot J Linn Soc 102:303–311. **Kato M** (1991) Taxon 40:189–194. **Kellogg EA** (1990) Evolution 44:1978–1989. **Kelly Dawe R,** Freeling M (1990) Dev Biol 142:233–245. **Kirchoff BK** (1991) Am J Bot 78:833–837. **Kunst L,** Klenz JE, Martinez-Zapater J, Haughn GW (1989) Plant Cell 1:1195–1208. **Kunze H** (1989) Plant Syst Evol 163:187–199. **Kunze H** (1991) Plant Syst Evol 176:227–253. **Kurzweil H** (1989) Lindleyana 4:1–5. **Kurzweil H** (1990) Bot J Linn Soc 102:61–83. **Kurzweil H** (1991) Bot Jahrb Syst 112:273–293. **Kurzweil H,** Weber A (1992) Nord J Bot 12:39–61. **Kurzweil H,** Linder HP, Chesselat P (1991) Plant Syst Evol 175:161–223. **Kuznetsova TV** (1988) Flora 181:1–17. **Kuznetsova TV** (1991) Itogi Nauki Techniki (Moscow) Ser Bot 12:51–174.

**Landes A,** Porter JR (1990) Ann Bot 66:41–50. **Lehmann N** (1991) Am J Bot 78(6,Suppl):30. **Lehmann N,** Sattler R (1989) Floral development and homeosis in *Begonia semperflorens-cultorum* 'Cinderella'. In : Lord EM, Bernier G (1989) Plant reproduction: from floral induction to pollination. American Society of Plant Physiologists, Rockville, p 199. **Lehmann N,** Sattler R (1992) Am J Bot 79(10):1145–1157. **Leins P,** Erbar C (1989) Flora 182:43–56. **Leins P,** Erbar C (1990) Mitt Inst Allg Bot Hamb 23b:723–729. **Leins P,** Erbar C (1991) Bot Acta 104:336–344. **Liang H-X,** Tucker SC (1989) Am J Bot 76:806–819. **Liang H-X,** Tucker SC (1990) Am J Bot 77:607–623. **Lidén M** (1992) Prog Bot 53:282–317. **Linder HP,** Kurzweil H (1990) Bot J Linn Soc 102:287–302. **Lord EM** (1991) Flowering Newslett 11:4–13. **Lord EM,** Gould KS (1989) In: Lord E, Bernier G (1989) Plant reproduction: from floral induction to pollination. American Society of Plant Physiologists, Rockville, pp 82–94.

**Mabee P** (1991) Science 254:874–875. **Magin N,** Classen R, Gack C (1989) Can J Bot 67:1931–1937. **Marquez-Guzman J,** Engleman M, Martinez-Mena A, Ramos C (1989) Ann Mo Bot Gard 76:124–127. **Martensen HO,** Probst W (1990) Farn- und Samenpflanzen in Europa. Fischer, Stuttgart, 535 pp. **Martinez SE,** Ramos C (1989) Ann Mo Bot Gard 76:128–135. **Mayo SJ** (1989) Bot J Linn Soc 100:139–172. **McDaniel CN** (1989) Floral initiation as a developmental process. In: Lord E, Bernier G (1989) Plant reproduction: from floral induction to pollination. American Society of Plant Physiologists, Rockville, pp 51–57. **McDaniel CN** (1991) Plant Cell 3:431–433. **McKinney ML,** McNamara KJ (1991) The evolution of ontogeny. Plenum, New York, 459 pp. **Meeks-Wagner DR** (1992) Flowering Newslett 13:29–32. **Metzger JD** (1990) Gibberellins and flower initiation in herbaceous angiosperms. In: Pharis RP, Rood S (eds) Plant growth substances 1988. Springer, Berlin Heidelberg New York, pp 476–485. **Meyerowitz EM,** Smyth D, Bowman J (1989) Development 106:209–217. **Meyerowitz EM,** Bowman JL, Brockman LL, Drews GN, Jack T, Sieburth LE, Weigel D (1991) Development (Suppl) 1:157–167. **Mione T,** Bogle AL (1990) Am J Bot 77:77–91. **Moncur MW** (1988) Floral development of tropical and subtropical fruit and nut species. CSIRO, Melbourne, 181 pp.

**Negrutiu I,** Sala F (1991) Plant Sci 80:1–6. **Negrutiu I,** Installé P, Jacobs M (1991) Plant Sci 80:7–18. **Nemoto T,** Ohashi H (1990) Bot Mag (Tokyo) 103:217–231. **Nijhout HF** (1990) Bio Essays 12:441–446.

**Okamoto M** (1989a) Plant Syst Evol 168:7–18. **Okamoto M** (1989b) Am J Bot 76:14–22. **Owens JN,** Hardev V (1990) Crit Rev Plant Sci 9:281–294.

**Penzig O** (1921-1922) Pflanzen-Teratologie, 2nd edn. Borntraeger, Berlin, 3 vols. **Perilleux C,** Bernier G, Kinet J-M (1991) J Exp Bot 42:501–507. **Pharis RP,** Evans LT, King RW, Mander LN (1989) Gibberellins and flowering in higher plants: differing structures yield highly specific effects. In: Lord E, Bernier G (1989) Plant reproduction: from floral induction to pollination. American Society of Plant Physiologists, Rockville, pp 29–41. **Pnueli L,** Abu-Abeid M, Zamir D, Nacken W, Schwarz-Sommer Z, Lifschitz E (1991) Plant J 1:255–266. **Poethig RS** (1990) Science 250:923–930. **Posluszny U,** Gerrath JM (1992) Can J Bot 70:659–660. **Prusinkiewicz P,** Lindenmayer A (1990) The algorithmic beauty of plants. Springer, Berlin Heidelberg New York, 240 pp.

**Raff RA,** Wray GA (1989) J Evol Biol 2:409–434. **Ramirez-Domenech JI,** Tucker SC (1989) Am J Bot 76:1583–1593. **Ramirez-Domenech JI,** Tucker SC (1990) Am J Bot 77:624–635. **Remphrey WR** (1989) Can J Bot 67:1966–1978. **Retallack B,** Willison JHM (1990a) Am J Bot 77:1222–1230. **Retallack B,** Willison JHM (1990b) Acta Bot Neerl 39:247–252. **Richards JH,** Barrett SCH (1992) The development of heterostyly. In: Barrett SCH (ed) Evolution and function of heterostyly. Springer, Berlin Heidelberg New York, pp 85–127. **Rieseberg LH** (1992) Int J Plant Sci 153:V–VI. **Ritterbusch A** (1989) Ann Bot 64:179–183. **Ritterbusch A** (1990) Flora 184:313–323. **Ronse Decraene LP** (1989) Am J Bot 76:1344–1359. **Ronse Decraene LP** (1990) Beitr Biol Pflanzen 65:181–203. **Ronse Decraene LP,** Smets E (1990) Bull Jard Bot Nat Belg 60:229–273. **Ronse Decraene LP,** Smets E (1991a) Plant Syst Evol 178:179–194. **Ronse Decraene LP,** Smets E (1991b) Bot J Linn Soc 105:257–269. **Ronse Decraene LP,** Smets E (1991c) Am J Bot 78:1438–1448. **Ruebsamen-Weustenfeld T** (1991) Bibl Bot 140:1–113. **Rutishauser R,** Huber KA (1991) Plant Syst Evol 178:195–223. **Rutishauser R,** Sattler R (1986) Can J Bot 64:1987-2019.

**Sattler R** (1978) Notes Roy Bot Gard Edinb 36:397–405. **Sattler R** (1988) Am J Bot 75:1606–1617. **Sattler R** (1992) Can J Bot 70:708–714. **Sattler R,** Rutishauser R (1990) Can J Bot 68:1989–2003. **Sawhney VK** (1990) Flowering Newslett 10:3–8. **Sawhney VK** (1992) Can J Bot 70:701–707. **Sawhney VK,** Rastogi R (1990) Acta Hort 280:563–568. **Schick B** (1989) Bot Jahrb Syst 110:289–323. **Schultz EA,** Haughn GW (1991) Plant Cell 3:771–781. **Schwarz-Sommer Z,** Huijser P, Nacken W, Saedler H, Sommer H (1990) Science 250:931–936. **Scribailo RW,** Tomlinson PB (1992) Int J Plant Sci 153:1–13. **Shannon S,** Meeks-Wagner DR (1991) Plant Cell 3:877–892. **Sharman KV,** Sedgley M (1988) Aust J Bot 36:575–587. **Sibony M,** Pinthus MJ (1988) Ann Bot 61:473–479. **Singer SR,** Hsiung LP, Huber SC (1990) Am J Bot 77:1330–1335. **Smyth DR,** Bowman JL, Meyerowitz EM (1990) Plant Cell 2:755–767. **Sommer H,** Nacken W, Beltran P, Huijser P, Pape H, Hansen R, Flor P, Saedler H, Schwarz-Sommer Z (1991) Development (Suppl) 1:169–175. **Steeves MW,** Steeves TA (1990) Can J Bot 68:1680–1688. **Steeves TA,** Steeves MW, Olson AR (1991) Can J Bot 69:844–857. **Steeves TA,** Sussex IM (1989) Patterns in plant development, 2nd edn. Cambridge Univ Press, Cambridge. **Stevens PF** (1991a) Flowering Newslett 12:32–33. **Stevens PF** (1991b) Syst Bot 16:553–583. **Stevenson DW** (1990) Mem N Y Bot Gard 55:8–55. **Stoffberg E** (1991a) Bot J Linn Soc 105:1–19. **Stoffberg E** (1991b) Bot J Linn Soc 105:21–35. **Stuessy TF** (1990) Plant taxonomy: the systematic evaluation of comparative data. Columbia Univ Press, New York, 531 pp. **Stützel T** (1990) Beitr Biol Pflanz 65:275–299. **Sundberg MD,** Orr AR (1990) Am J Bot 77:141–152. **Sussex IM** (1989) Cell 56:225–229.

**Takaso T,** Tomlinson PB (1989a) Am J Bot 76:692–705. **Takaso T,** Tomlinson PB (1989b) Bot Gaz 150:378–390. **Takaso T,** Tomlinson PB (1990) Am J Bot 77:1209–1221. **Takaso T,** Tomlinson PB (1991) Am J Bot 78:417–428. **Taylor P** (1989) The genus *Utricularia* – a taxonomic monograph. Her Majesty's Stationery Office, London, 735 pp. **Tomlinson PB,** Takaso T, Rattenbury JA (1989) Bot J Linn Soc 99:209–221. **Troll W,** Weberling F (1989) Infloreszenzuntersuchungen an monotelen Familien. Fischer, Stuttgart, 497 pp. **Tucker SC** (1989a) Am J Bot 76:714-729. **Tucker SC** (1989b) Monogr Syst Bot Mo Bot Gard 29:59-75. **Tucker SC** (1990) Am J Bot 77:750–761. **Tucker SC** (1991) Am J Bot 78:1130–1149. **Tucker SC** (1992a) Am J Bot 79:318–327. **Tucker SC** (1992b) Flowering Newslett 13:21–28. **Tucker SC** (1992c) Can J Bot 70:692–700. **Tucker SC**, Stirton CH (1991) Bot J Linn Soc 106:209–227. **Turlier M-F,** Alabouvette J (1989) Can J Bot 67:1985–1997. **Turlier M-F,** Alabouvette J, Doulain-Douvier D (1991) Can J Bot 69:866–872. **Turnbull LV** (1991) Trop Agric (Trinidad) 68:139–143.

**Van Heel WA** (1992) Blumea 36:481–499. **Van Loon LC,** Bruinsma J (1992) Acta Bot Neerl 41:1–23. **Van Tunen AJ,** Angenent GC (1991) Flowering Newslett 12:34–37. **Veit B**, Greene B, Lowe B, Mathern J, Sinha N, Vollbrecht E, Walko R, Hake S (1991) Development (Suppl) 1:105–111. **Verbeke JA** (1992a) Annu Rev Plant Physiol Plant Mol Biol 43:583–598. **Verbeke JA** (1992b) Flowering Newslett 13:33–37. **Vlot EC,** Bachmann K (1991) Am J Bot 68:235–241. **Vlot EC**, van Houten WHJ, Mauthe S, Bachmann K (1992) Int J Plant Sci 153:89–97. **Von Baer KE** (1828) Ueber die Entwicklungsgeschichte der Thiere. Borntraeger, Königsberg.

**Weberling F** (1989a) Morphology of flowers and inflorescences. Cambridge Univ Press, Cambridge. **Weberling F** (1989b) Monogr Syst Bot Mo Bot Gard 29:35–58. **Williams DM**, Scotland RW, Blackmore S (1990) Biol J Linn Soc 39:99–108. **Wunderlin U** (1992) Diss Bot 188:1–313.

**Ye D**, Oliveira M, Veuskens J, Wu Y, Installé P, Hinnisdaels S, Truong AT, Brown S, Mouras A, Negrutiu I (1991) Plant Sci 80:93–106.

Dr. Rolf Rutishauser
Botanischer Garten und Institut für
Systematische Botanik der Universität
Zollikerstraße 107
CH-8008 Zürich, Switzerland

# B. Physiology

## I. Plant Water Relations

By Rainer Lösch

As in previous report periods, the total number of papers on plant water relations published during the past 2 years exceeds 2000 titles, of which a fifth is mentioned on the following pages.

*Books*. Baker and Hall (1988): solute transport; Brown (1990): microbial water stress; Calder (1990): evaporation; Feigin et al. (1990): sewage irrigation; Hashimoto et al. (1990): methods; Jackson et al. (1991): oxygen deficiency; Jones et al. (1989), Katterman (1990), Kreeb et al. (1989), Mooney et al. (1991), Rozema and Verkleij (1991), all: plants and environmental stresses; Monteith and Unsworth (1990): environmental physics; Nobel (1991): physicochemical and environmental plant physiology; Stewart and Nielsen (1990): irrigation; Van Hoorn (1988): agrohydrology; Vasilescu (1988): biophysics of water-ion interactions; Weyers and Meidner (1990): methods of stomatal research.

### 1. Water Relations of Cells and Tissues

At the molecular level, osmotic phenomena can be explained in principle by kinetic pressure differences at the porous semipermeable membranes: the concentration difference of solutes brings about a directed drain of water molecules into the holes of the barrier between the compartments tending to lower the concentration gradient. The resulting spurts of water molecules through the pore bring about a time-averaged pressure onto the membrane (Dainty and Ferrier 1989; Kiil 1989), i.e., this process is a statistical phenomenon. Real pressures, however, would give rise to tensions in the surface which are not observed experimentally. Hill (1989a, b), emphasizing this problem for the situation of leaky porous membranes, gives a set of equations which describe the volume and solute flow across such barriers based on general frictional theory. In a membrane system comprising immobile polyelectrolytes, permeable small electrolytes, and water, the osmotic flow is specified by solution-dependent properties resulting from frictional effects (Williams and Comper 1990). This applies to the situation in gels or cells where partition of ions between the gel and the external solution produces different activity coefficients in the two compartments. Calculations based on a normal Gibbs-Donnan equilibrium must fail when describing such a system (Wiggins et al. 1991).

Tensions up to –0.7 MPa can be produced in an artificial osmotic cell constructed with reverse osmotic membranes embedded into rigid metal grids (Zhu et al. 1989). Such a system resembles plant cells with stiff walls like those of sclerophylls which exhibit a considerable resistance to collapse (Oertli et al. 1990). Cells of mesophytic tissues cannot

Progress in Botany, Vol. 54

withstand pressure differences higher than 0.1 MPa. Wheat leaves may reach, with age or under drought, a sufficient degree of stiffness so that tensions (or negative turgor values) can be determined (Jat et al. 1991).

The established methods for the determination of plant water relations became extended also to more difficult objects [application of pressure chamber measurements to chilled plants: Pardossi et al. 1991; determination of water potential ($\psi$) in latex-bearing plants: Milburn et al. 1990; measurements of sugarcane leaf segments with the pressure chamber: Saliendra et al. 1990; use of unsaturated salt solutions to generate conifer needle water-release curves: Livingston and de Jong 1991; determination of the cell wall water content: Hardegree 1989; application of pressure volume (pV) analysis to in-vitro cultures: Auge et al. 1990b; evaluation of pV curves by nonlinear regression software: Andersen et al. 1991; inference of osmotic adaptation from seasonal changes of leaf water contents and osmotic potentials ($\psi_\pi$) using the relations of the general gas equation: Noitsakis and Tsiouvaras 1990; determination of the bulk modulus of elasticity ($\varepsilon$) by vesicular swelling: Rivers and Williams 1990]. Particular attention is paid to the effects of rehydration when pV curves are produced. Artificial rehydration may alter pV-relationships by redistribution of solutes between symplast and apoplast (Evans et al. 1990) and by alteration of the apoplastic water content, which can bring about a plateau in the water content/water potential relations and may mask ontogenetic changes in tissue water relations (Dreyer et al. 1990; Kubiske and Abrams 1990, 1991).

The use of pressure probes to measure single cell water relations has meanwhile been established as a reliable method in several laboratories. Pressure probe measurements have been checked against $\psi$ values of xylem vessels measured with a pressure bomb, with the result that they gave less negative values than the latter. The study came to the conclusion that the water relations of the living parenchyma bordering the xylem vessels play an important role in the maintenance of xylem tension (Balling and Zimmermann 1990). Tensions in maize vessels were measured by Heydt and Steudle (1991). Water flow and hydraulic parameters along a profile through the root cortex were quantified by the same group using a double pressure probe technique (Zhu and Steudle 1991). A similar approach by Rygol and Zimmermann (1990) tested the effects of salinity on *Mesembryanthemum crystallinum* root water relations, and Tyerman et al. 1989 determined root cell ion concentrations by conventional methods and the resulting turgor by pressure probe measurements. Both studies are in agreement, showing a turgor decrease in the cortical cells with increase of salinity. Changes in $\varepsilon$ are correlated with volume changes of the cells, and osmotic adjustment occurs upon sudden changes in the external NaCl concentration. By using differences between initial and final equilibrium turgor pressures associated with pressure-relaxation treatments, the determination of plant cell volumes is possible (Malone and Tomos 1990). Direct measurements of cell turgor by means of an improved pressure probe demonstrated that the intracellular pressure is considerably affected by respiration. The respiration-dependent component of cell turgor in the hypocotyl elongation zone of *Vigna* was 30–40 kPa (Nakahori et al. 1990). According to pressure probe determinations, huge increases in $\varepsilon$ and decreases in membrane hydraulic conductivity ($L_p$) accompany the development of sugarcane stalk storage parenchyma (Moore and Cosgrove 1991) so that it continues growing and serves as a sink for sucrose despite high quantities of water and sugar in the apoplastic spaces (Welbaum and Meinzer 1990).

Pressure-volume analyses indicate the occurrence of osmotic adjustment, which is mostly highly specific with respect to treatment, species, plant organ, and age. In *Lupinus* species under drought, osmotic and water potentials decrease by 1/2 or 1/3 of the initial values and $\varepsilon$ is halved so that turgor can be maintained under drought (Jensen and Henson 1990). Similar changes under such treatment were not enough to prevent turgor loss in cultivars of *Coffea arabica*. Maintenance of protoplasmic volume was more pronounced in this case than maintenance of turgor as soil drying progressed (Meinzer et al. 1990c). Heavy shading can inhibit osmoregulation in rose leaves (Auge et al. 1990a). Root water relations of roses may benefit under drought from mycorrhizal colonization due to an increased symplastic water content of *Glomus*-infected roots as compared with symbiont-free ones (Auge and Stodola 1990). Other means to maintain positive turgor values can be changed elasticities (Blake et al. 1991) or changed osmotic potentials of the tissues (Ranney et al. 1991). Often a complicated interplay of the different effects of morphogenetic and metabolic events takes place (Sharp et al. 1990). Thus, according to Goldstein et al. (1991a), in *Opuntia ficus-indica* in large elastic cells in the water-storage parenchyma, high amounts of mucilage of high water-holding capacity, and polymerization of sugars cooperate in maintaining positive turgor values in the photosynthetic tissue even over a 4-month period of drought (extracellular polysaccharides ameliorating water stress in the composite *Hemizonia luzulifolia*: Morse 1990). The nocturnal increase of osmotically active malate in the chlorenchyma of the *Opuntia* cladodes additionally helps to keep turgor there favorable even in plants under stress (Goldstein et al. 1991b). Age effects upon the degree of osmotic adjustment are reported by Steinberg et al. (1989a) and West et al. (1990), indicating that young leaves are more efficient in osmoregulation than older ones. Among the solutes accumulated during osmotic adjustment, proline plays a prominent role (e.g., Thomas 1990; Jorgensen 1991; Voetberg and Sharp 1991). However, amounts and effects of its accumulation (and that of osmotically active carbohydrates) can be different in different plant organs: Turner (1990); Thomas (1991). The location of these solutes in subcellular compartments and their additional functions (e.g., scavenging of hydroxyl radicals under drought: Smirnoff and Cumbes 1989) will affect also the osmoregulatory efficiency. Although osmotic adjustment is beneficial in delaying and mitigating the consequences of drought stress, it often will not suffice to improve growth or increase productivity under harsh environments (Thomas and Evans 1989; Flower et al. 1990).

Growth is controlled by yield threshold and cell wall extensibility. The latter depends distinctly upon cell water relations, and both are influenced strongly by temperature (Pritchard et al. 1990, 1991; Roden et al. 1990). The primary event during growth inhibition by drought is a change in the growth-induced water potential. Afterwards, low wall extensibility and tissue conductance for water are responsible for the growth limitation. Changes in wall extensibility are due to changes in wall physical properties (Nonami and Boyer 1990a, b). Photoinhibition of stem elongation results similarly from changes in cell wall extensibility; hydraulic properties are unaffected by both blue and red light (Kigel and Cosgrove 1991). Cell expansion is paralleled by solute uptake delivered from the phloem through the symplast (Schmalstig and Cosgrove 1990). In well-watered expanding *Vitis* leaves, 18–40% of the supplied carbohydrates are incorporated into the enlarging cell wall polysaccharides; water stress reduces this process. Cellulose synthesis is particularly strongly affected by water deficits (Sweet et al. 1990). The phytohormones auxin and ethylene increase glucose incorporation and cell wall extensibility (Hoson et al. 1990), and at least the former reduces simultaneously the yield threshold (Okamoto et al. 1990).

## 2. Root Water Uptake and Water Movement Through the Plant

Water flow driven by water potential drops in the immediate periphery of roots needs to be determined with high resolution. This might be achieved by a system of automated time-domain reflectometry developed by Wraith and Baker (1991). NMR imaging (Brown et al. 1990) makes evident a water-depletion region particularly around the taproots of loblolly pine (Macfall et al. 1990) suggesting that water uptake occurs also through suberized root regions. Also in *Pinus elliottii* the importance of the woody roots for water and mineral absorption was demonstrated (Van Rees and Comerford 1990). Local water depletion promotes a root growth in order to exploit further soil water reserves (*inter alia*: Higgs and Jones 1989; Benlaribi et al. 1990; Hodgson et al. 1990; Meyer et al. 1990). The ability for hydrotropism is localized in the root tips, and hydrotropic curvature results from inhibited cell elongation at the surfaces closer to moist environments (Takahashi and Suge 1991). Hydrotropic and gravitropic responses of roots interact with one another (Takahashi and Scott 1991) although the stimulus-perception mechanisms probably are different. Generally, changes in wall extensibility play a major role in controlling gravicurvature (Bagshaw and Cleland 1990). Upon drought stress or abscisic acid (ABA) treatment sunflower roots firstly become reasonably elongated, stop growing after some hours, and start elongation again 3 days after the treatment's commencement (Robertson et al. 1990a). The first phase is accompanied by a rise in $\psi_\pi$ and an inhibition of mitotic activity, the latter lasting until the third phase of growth responses (Robertson et al. 1990b) when a gradual recovery of root growth occurs possibly due to osmotic adjustment. In primary roots of maize, this is brought about chiefly by proline accumulation in the elongation zone (Voetberg and Sharp 1991). Changes in cell wall yielding properties are also important for a continued root growth, despite lowered water potentials (Spollen and Sharp 1991). Under field conditions, the long-delayed maturation of the metaxylem in maize roots may provide an effective barrier partly uncoupling the water relations of the immature root apices from low water potentials in shoots and leaves (Wang et al. 1991).

Effectiveness for water uptake is particularly high during the early period of root growth (Colombo and Asselstine 1989). It decreases exponentially with root age (Malik et al. 1989). Roots are supposed to contribute a reasonable part to the water transport resistance in the soil-plant-atmosphere (SPAC) pathway (Hirasawa and Ishihara 1991). The major part of this resistance is localized in the radial pathway from the root surfaces to the stele, where water and minerals become canalized. This flow has to pass walls of particular cells in the pericycle layer of the root meristematic regions (Huang and Van Steveninck 1989). Within the cortex, the $L_p$ of the cell-to-cell path increases significantly from the first layer towards the inner layers. This is due to increases in the cell-membrane $L_p$ (Zhu and Steudle 1991). Salinization of the growing medium reduces $L_p$ in glycophytes (Joly 1989; Evlagon et al. 1990). In halophytes, this process might be less pronounced due to retention of the overall membrane structure and an increase of the free sterol/phospholipid ratio which also prevents leakage of $K^+$ into the salinized medium (Blits and Gallagher 1990). $L_p$ decreases with lowered temperatures (Ameglio et al. 1990; Nobel et al. 1990; Bassirirad et al. 1991; Udagawa et al. 1991) probably also due to temperature-dependent permeability changes of the cell membranes (Yoshida and Eguchi 1990). Temperature sensitivity of $L_p$ can shift during the early stages of seedling growth

(Lopez and Nobel 1991). Root water and mineral uptake is negatively affected by allelopathic compounds in the soil solution, like ferulic acid (Lyu and Blum 1990), and by phytopathogens (*Phytophthora* root rot: Ristaino and Duniway 1991). There is a diurnal variation in root nutrient but not water uptake, which increases during night (Masuda et al. 1990), and the ratio of nocturnal ion uptake to daily uptake rates varies with the developmental stages of the plants (Terabayashi et al. 1991). During drought, $L_p$ decreases reversibly due to embolism and increase of lacunar spaces in the cortex thereby helping to limit water loss to dry soil (North and Nobel 1991). This applies to established roots as well as to rain roots of cacti. The latter have in principle no higher $L_p$ than the former, but are younger and therefore more efficient in water uptake and transport (Nobel et al. 1990). The drop of $\psi$ along these roots is negligible even over longer distances (Nobel and Lee 1991). Heydt and Steudle (1991) found that $L_p$ is an order of magnitude higher if hydrostatic pressure is applied in the measurements than if osmotic forces bring about the driving gradient.

In many instances, water becomes pushed from the roots to the shoot by root pressure (Singh and Singh 1989), resulting from the release of low molecular weight carbohydrates into the developing xylem (Stasova and Antonova 1989). The intensity of this process often varies with the season reaching peak values of such organic and also of inorganic solutes at bud-break (Glavac et al. 1989). In alder, xylem sap content and composition of amino acids is influenced by the supplied nitrogen sources (Tonin et al. 1990). Seasonal changes with highest levels in spring occur also in the cytokinin composition of the xylem sap (Tromp and Ovaa 1990; Waseem et al. 1991). Their decrease with progress of the season is related to senescence in monocarpic plants (Nooden et al. 1990). Among other phytohormones transported by the transpiration stream ABA is most prominent. It is assumed to be a messenger signaling soil drying (Gowing et al. 1990; Wolf et al. 1990; Zhang and Davies 1991) or hypoxia (Smit et al. 1990).

The conducting tissues in the junction zone of roots and shoots of cereals are constructed either as safety zones where root vessels are separated from those of the shoots by tracheids (rye, wheat, barley) or as unsafe systems where continuous strands of vessels extend from the roots to the shoots (maize, sorghum, oats: Aloni and Griffith 1991). In maize, individual hydraulic pathways exist from nodal roots of a certain level of insertion up to leaves of a corresponding level of insertion oriented into the same direction as the supplying roots (Zhao and Wang 1989). Hydraulic architectures show a high degree of species specifity with respect to Huber values, relative and leaf-area related conductivities (conifers: Sellin 1988, 1990; Ryan 1989; vines: Ewers et al. 1989, 1990; Gartner et al. 1990; Gartner 1991). In jojoba shrubs, the number of vessels decreases but their diameter increases acropetally supporting a reasonable compensation of leaf water losses at low water potentials (Lo Gullo 1989). A correlation may exist between high leaf-specific conductivities and the probability of embolisms (*Schefflera*: Tyree et al. 1991). In many cases, wide and narrow vessels are bundled, forming functional xylem units where the longitudinal water transport is carried mostly by the former and an intensive radial exchange (also to the surrounding symplastic elements) by the latter (Van der Schoot and Van Bel 1989; Canny 1991).

Xylem disfunction can be brought about by vessel occlusion from tyloses and gums or by cavitations (Cochard and Tyree 1990; Bonsen 1991). The latter is monitored using ultrasonic or audible acoustic emissions (Ritman and Milburn 1991 – detection of cavitation in fern sporangia by these methods: Ritman and Milburn 1990). Embolism (review:

Tyree and Sperry 1989) can result in the course of freezing and thawing cycles of trunks and branches (Just and Sauter 1991) or from water stress (Sperry and Tyree 1990). In both cases, recovery of the hydraulic conductivity is possible (Just and Sauter 1991), based particularly on capillary tension forces, not on metabolic processes (Borghetti et al. 1991). A prevention of too many cavitation events can be achieved by sensitive stomatal control of the transpirational flux: obviously feedback loops exist tuning stem hydraulic conductivity and leaf and canopy conductance in accordance with soil water availability (Jones and Sutherland 1991). A plant-internal balance between supply and demand is also achieved by different leaf areas on differently exposed branches (Steinberg et al. 1990; Lilienfein et al. 1991). Velocities and amounts of sap flow are measured by heat-balance methods, the improvement and versatile application of which is reported in many papers (e.g., steady-state heat application: Steinberg et al. 1989b; Dugas 1990; emphasis on multiple-point measurements: Hatton et al. 1990; Dye et al. 1991; combination of heat-balance measurements with computation of potential transpiration from meteorological data: Y. Cohen 1991; with gas exchange measurements: Fichtner and Schulze 1990; with gravimetric estimates of transpiration: Ham and Heilman 1990). By monitoring stem diameter changes (Schoch et al. 1989; Imai et al. 1990) the function of trunks not only as conductors but also as capacitors in the plant water relations is studied. Water stored in the axes will buffer steep increases in transpiration. Dehydration proceeds in this case in three steps related to increasingly lower water potentials: firstly water bound by capillary forces and available in reasonable amounts is released, then the dehydration brings about first cavitation events and a drastic drop of $\psi$ per increment of water loss, finally a further removal of water is possible only by cavitation events in the conducting tissues liberating reasonable amounts of water (Tyree and Yang 1990). Stem succulents, however, are adapted to deliver huge amounts of storage water: Succulent trees from Baja California store up to six times more water per volume than co-occurring nonsucculent trees (Nilsen et al. 1990). An alternative, new approach in order to estimate wood moisture storage by a nearly nondestructive method is the application of time domain reflectometry (Constantz and Murphy 1990).

## 3. Stomatal Behavior and Transpirational Water Loss

The stomatal regulation of transpirational water loss links leaf metabolism to the atmospheric environment. Guard cells as the interfaces between leaf interior and exterior are not only idioblasts in an anatomical sense, but behave also as independent units, partly with metabolic peculiarities differing from those of other leaf tissues. The amount of ribulose-bisphosphate carboxylase/oxygenase (Rubisco) per protein unit in the guard cells is only 1/300 to 1/400 of that in mesophyll cells (Reckmann et al. 1990; Tarczynski et al. 1989). While Calvin-cycle $CO_2$-fixation is thus negligible, phosphoenolpyruvate(PEP)-carboxylation in guard cells occurs. It is stimulated by ammonium (Müller et al. 1991). PEP-carboxylase (PEPcase) activity of guard cell protoplasts is not light-dependent and highly sensitive to malate inhibition (Willmer et al. 1990). NADP-dependent malate dehydrogenase (MDH) is active in darkness and further increases its activity under illumination. It is supplemented additionally by NAD-dependent MDHs (Scheibe et al. 1990). Respiratory $O_2$ turnover does not differ, by contrast, from that of mesophyll cells (Gautier et al. 1991). Adenyl compounds (Curvetto and Delmastro 1990), peroxidases (Pantoja and Willmer 1991), and specific proteins (Park and Thimann 1990) are further involved in the turnover of enzymes, substrates, energy-rich cofactors, and organic solutes. They come from metabolic processes within the guard cell cytoplasm or play a role in transport processes at the plasma membranes.

Plasma membranes of guard cells exhibit higher permeabilities than plasma membranes of mesophyll cells. This applies also for the transport of protonated ABA (Baier et al. 1990). ABA fluxes at the plasma membrane seem to be independent of the cytoplasmatic pH, those at the tonoplast depend upon the pH of the cytoplasm (Baier and Hartung 1991). Ion (particularly $K^+$) fluxes at the membranes occur through specific channels, the gating of which is affected by voltage gradients. Channel closing and opening are different in their voltage dependence (Schroeder 1989a). $Na^+$ depresses the outward rectifying $K^+$ channel current when applied at the cytoplasmic site of the membrane. This effect becomes relieved at extreme positive voltages (Thiel and Blatt 1991). Similarly, $Ca^{2+}$ modulates the activity of time- and voltage-dependent $K^+$ channels through the tonoplast (Colombo et al. 1989). The data suggest that the channels function as single-file pores which are passed by several ions simultaneously (Thiel and Blatt 1991). The redox dye neutral red leads to membrane depolarization and a charge compensating $K^+$ net efflux. These processes bring about an intracellular acidification (Ullrich et al. 1990). Steady-state opening of the stomata is achieved by the balance of both, $K^+$ influx and efflux (Brindley 1990a), closure upon darkness is effected by a large transient increase of efflux while the influx remains nearly unchanged (Brindley 1990b). ABA mobilizes $K^+$ channels or channel conductance but does not affect gating of the $K^+$ channels directly (Blatt 1990). This, however, could be the case with the ABA-dependent activation of $Ca^{2+}$-permeable channels of the plasmalemma (Schroeder and Hagiwara 1990). The ABA effect on $K^+$ efflux out of the guard cells precedes the increase of cytoplasmic $Ca^{2+}$ (Macrobbie 1990). Thus $Ca^{2+}$-dependent and $Ca^{2+}$-independent transduction pathways exist linking ABA perception to stomatal closure (Gilroy et al. 1991). The rise of cytoplasmic $Ca^{2+}$ activates anion channels in the plasma membrane if nucleotides are simultaneously present (Hedrich et al. 1990). The existence of various ways and periods of activation which characterize $Ca^{2+}$ involvement in the regulation of stomatal apertures (review: Mansfield et al. 1990) may explain why an ample supply of $Ca^{2+}$ to plants growing on calcareous soil does not adversely interact with the stomatal regulation of the leaves (Atkinson et al. 1990). Ions involved in the exchange processes during stomatal movements enter the guard cells from the surrounding apoplasmic space but may be supplied or sequestered from and to adjacent subsidiary cells. Turgor changes become intensified by such cell-to-cell ion shifts (Braconnier and D'Auzac 1990). Ionic competition at the exchange sites of the membranes (e.g., between the anions chloride and vanadate) may be important in keeping elemental pool sizes balanced and the ionic transfer rates at appropriate levels (Schwartz et al. 1991).

Long-distance delivery of ions and phytohormones from roots to leaves is a transport not only of substances but also of specific informations (Davies and Zhang 1991). Johnson et al. (1991) focus on the importance of these root/shoot messages modelling water flow in the SPAC. Saab and Sharp (1989) emphasize that nonhydraulic inhibition of leaf growth and reduction of stomatal conductance, both probably evoked by increased ABA, can act as means to conserve water at very early stages of plant deficits. Zhang and Davies (1990a, 1991) bring evidence that the ABA amounts transported in the xylem sap of unwatered plants suffice to trigger stomatal closure. Xylem ABA concentrations reflect the extent of soil drying. It is this ABA the concentration of which corresponds to the degree of increase in leaf resistance. Total leaf ABA levels, by contrast, are much less precisely related to the degree of soil dryness, upon which the stomates respond by closure (Zhang and Davies 1990a; Tardieu et al. 1991). Stomata on wheat ear lemmas and awns are less responsive to ABA than those on leaves (Xu and Ishii 1990). At the microscale, gradients of ABA concentrations exist between guard, subsidiary, and epidermis cells, the mean concentrations being 2.49, 1.25, and 0.86 mM ABA, respectively (Brinckmann et al. 1990). Upon water stress, ABA accumulation within individual guard cells occurs within 15 min (Harris and Outlaw 1991). From comparative studies with artificially produced ABA analogs, the minimum chemical structure which is necessary to bring about ABA-like activity turned out to be a properly substituted cyclohexane unit connected with a six-carbon side chain (Schubert et al. 1991). Atmospheric pollution reduces stomatal responsiveness to ABA (Atkinson et al. 1991).

The hydraulic changes during stomatal opening can be recognized by decrease in leaf thickness due to a loss of epidermal turgor, coinciding in time with the peak of the absorption lag during part of the motor phase (Meidner 1990). Water vapor diffusion from the substomatal cavity through the pore is supplied to approximately 90% from the guard and subsidiary cells, according to model calculations by Yang and Wang (1989b).

Among the environmental factors influencing stomatal apertures light probably directly affects the guard cell metabolism, atmospheric humidity acts via the cell hydraulics, soil and plant water stress are signaled by ABA translocation. Interactions between these factors permit conclusions as to the mechanisms which bring about changed stomatal apertures. Blue light pulses superimposed on the effects on stomata of various vapor pressure differences (VPD) evoke transient increases of conductance. At lower temperatures these increases are smaller at high VPDs, at 30 °C they are higher (Assmann and Grantz 1990a). The velocity of the stomatal opening response to the blue light pulse is higher at high VPDs. Probably, increased epidermal water loss at high VPD decreases the back-pressure from the subsidiary cells and by this way permits a more rapid stomatal opening per unit of guard cell metabolic response to blue light (Assmann and Grantz 1990b). Only under this reduced epidermal turgor can overshoot responses of conductance to blue light pulses be observed (Karlsson and Assmann 1990). Water use efficiency is higher under the combined effects of high VPD and blue light (Karlsson and Assmann 1990; Grantz and Assmann 1991). Patterns of fluorescence transients indicate that blue light modulates photosynthetic activities of guard cell protoplasts (Mawson and Zeiger 1991). According to anoxia experiments by Vavasseur et al. (1990), the blue light effect upon stomatal conductances requires an oxygen supply.

Since epidermal ATP levels become increased under blue light irradiation oxidative phosphorylation may be stimulated by such a treatment (Agbariah and Roth-Bejerano 1990). The hypothetically postulated (Prog. Bot. 46, 42) flavonoid receptor of the blue light irradiation exhibits different absorbances of UV irradiation when stomata are open or closed (Schnabl et al. 1989). Stomatal responses to light are generally quicker in herbaceous plants than in woody plants. Under varying sun and shade regimes, the former therefore save water at the cost of a reduced $CO_2$ uptake, the latter, having a higher buffering capacity with respect to water loss, continue the uptake of $CO_2$ during both sun and shade periods (Knapp and Smith 1990). Different conductances can be observed also on cloudy and on clear days with similar average photon flux densities (Henson et al. 1990). Leaves of different ontogenetic stages normally have different stomatal responses to radiation; in *Populus* the mature leaves are more responsive than the still expanding ones (Ceulemans et al. 1989).

Unlike the stomatal perception of irradiance by means of photoreceptors, no specific sensor for different ambient humidities has been found. Rather, the mechanism for stomatal responses to humidity is based on the rate of water loss from the leaf (Grantz 1990; Mott and Parkhurst 1991). As a consequence of increased evaporative demand outside the epidermis, water potential differences between subsidiary and guard cells change. This can bring about a shortage of water supply to the guard cells that again causes there a higher turgor drop than in the subsidiary cells, and pore narrowing results (Nonami et al. 1991). The appropriate external variable to which response occurs in this case is the vapor saturation deficit between leaf and air, not relative humidity (Aphalo and Jarvis 1991). It is the leaf-air vapor pressure difference at the leaf surface which is sensed by the stomata, not the bulk air vapor pressure deficit. Therefore, humidity responses of stomata can be obscured when canopy conductances are calculated from conventionally determined or remotely sensed meteorological data (Grantz and Meinzer 1990, 1991). On the other hand, due consideration of the stomatal humidity response will considerably reduce the variability otherwise found in calculations of radiation use efficiency for assessing the biomass accumulation (Stockle and Kiniry 1990).

A decided shift in interpretation is presently taking place with regard to water stress effects on stomata. After the classical view of direct hydraulic feedback effects had been replaced during the past years by knowledge about the hydroactive effects of ABA as the messenger affecting guard cell metabolism under water stress, evidence is growing now that soil rather than leaf water shortage causes ABA accumulation. Better statistical correlations between soil (instead of leaf) water potential and gas exchange (Rosa et al. 1991 – compare, however, the statement of Ni and Pallardy (1991) that their experiments do not indicate a more direct influence of soil water potential on conductance as compared with leaf water potential! – and the effect upon stomatal closure of xylem sap ABA concentrations (Zhang and Davies 1990a; Tardieu et al. 1991) bring good evidence for the modern view. Of course, under soil water shortage also leaf water potential decreases and can be correlated with increasing leaf resistances. Both changes, however, seem to be caused independently by the decreasing water potentials in the root zone. Under light water stress an increased leaf resistance brings about decreases in carbon assimilation; under more severe moisture stress the photosynthetic processes themselves are negatively effected as well (Frederick et al. 1989; Ephrath et al. 1990; Kidambi et al. 1990). Nitrogen shortage seriously increases such stress effects (Radin 1990; Heitholt et al. 1991). Threshold values of plant water potential above which no clear increase of stomatal resistance is observed, are different between species (e.g., Jing and Ma 1990) and occur generally in monocotyledons at lower levels of water potential than in dicotyledons. They can change during the season. In several cases they run parallel to the seasonal course of the osmotic potential at turgor loss point (Yoon and Richter 1990).

Plant internal coordination due to morphological and functional feedback relations with aspect to root water uptake, stem water transport, and transpirational water loss becomes evident if these processes are measured synchronously in complex studies. Thus in sugarcane, the coordination of vapor phase and liquid phase conductances results in a balance between water loss and water transport capacity, maintaining leaf water status more or less constant (Meinzer and Grantz 1990). The south- and north-facing branches of hazel growing in a hedgerow are exposed to contrasting microclimatic environments and therefore display quite different transpiration rates per unit leaf area. Total leaf areas of the parts of the crown differ, however, in such a way that the demands for water supply are equal in all main branches regardless of the hedgerow direction they are facing (Lilienfein et al. 1991). Relations between biomass development, leaf conductances, and predawn xylem potentials of thinned *Pinus taeda* stands similarly support the hypothesis that transpiration, leaf area, and water potential interact to form a homeostatic relationship (Cregg et al. 1990). Such a plant-internal coordination probably is very important in order to avoid too high a degree of xylem embolism (Jones and Sutherland 1991). Similar reasons probably lie behind the gradation of bulk water relations and leaf conductances in parallel to the insertion levels of leaves alongside a branch. Osmotic adaptation in terms of solute changes and/or changes in tissue bulk elastic moduli can enable particularly young leaves to keep positive turgor values and to shift the threshold for stomatal closure to lower water potentials (Steinberg et al. 1989a; Hitron and Zur 1990). Local differences in water potential and ABA accumulation also exist between the leaf surfaces of plants with amphistomatic leaves. The greater availability of ABA per stomatal complex on the adaxial compared with the abaxial epidermis of several *Lupinus* species is, for example, consistent with the greater sensitivity to water deficits and ABA in adaxial compared to abaxial stomata (Henson and Turner 1991). Finally, patchiness of stomatal sensitivity over the surface of a leaf is also observed (review: Mansfield et al. 1990) and analyzed by model calculations. It may explain some inconsistencies of gross gas exchange data (Loreto and Sharkey 1990; Cheeseman 1991). This phenomenon also underlines the need for direct observations of stomatal apertures in parallel to their implicit determination by porometric or gas-exchange measurements (visualization of stomatal apertures by low-temperature scanning electron microscopy: Van Gardingen et al. 1989; nondestructively by a remote-control microscope system: Omasa and Maruyama 1990; Kappen and Haeger 1991).

The complex effects of factors influencing stomatal apertures can be best described quantitatively by models. A workshop at Pennsylvania University the contributions of which were published in a special issue of *Agricultural and Forest Meteorology* (Issue Editor: Carlson 1991) dealt with the various facets of the theme. Key topics of the contributions of this publication are: stomatal regulation and the influence of the boundary layer conditions (Collatz et al. 1991), measurements of the environmental influences upon stomata (Turner 1991), stomatal behavior of subalpine conifers (Massman and Kaufmann 1991), or maize (Rochette et al. 1991), of tropical forest plants (Dolman et al. 1991), spatial effects of stomatal control of transpiration (McNaughton and Jarvis 1991), the canopy resistance (Lhomme 1991), the problems of scaling up from individual plants to stands (Baldocchi et al. 1991), the extrapolation of SPAC resistances to regional scales (Hunt et al. 1991; Mascart et al. 1991), the importance of plant stomatal control for mesoscale atmospheric circulations (Avissar and Pielke 1991), transpiration and evapotranspiration (Saugier and Katerji 1991), and the use of complex evapotranspiration models with canopy resistance data in climate models (Dickinson et al. 1991). Other publications from the same journal on the theme deal with stomatal conductance models for grapevines under mediterranean climates (Winkel and Rambal 1990) and with canopy conductances in a temperate grassland (Kim and Verma 1991). Jones and Higgs (1989) modeled leaf conductances in apple orchards, Leuning (1990) stomatal behavior and photosynthesis of *Eucalyptus grandis*, Lösch et al. (1992) stomatal conductance of field-grown barley under the influence of different $K^+$-fertilization regimes.

Resins covering the leaves of *Larrea tridentata* reduce transpirational rates more than assimilation thus acting as a natural antitranspirant (Meinzer et al. 1990b). Other studies were performed with wheat leaves and ears, the cuticular conductance of the latter being considerably higher than that of the former (Araus et al. 1991), with timberline conifers suffering from winter desiccation (Hadley and Smith 1990), with *Eragrostis curvula* provenances (Tischler and Voigt 1990), with *Theobroma* cultivars differing in their wax and lipid fractions (Bhat et al. 1990), and with the efficiency of chemical defoliants as dependent upon cuticle quality (Oosterhuis et al. 1991): All try to determine the importance of cuticular peculiarities for leaf water losses. Santrucek and Slavik (1990a, b) separate by model calculations and by experimental verification the partial contributions of the cuticular and the stomatal component to the total leaf resistance to gas exchange. Glaucous-leaved grass species have a higher leaf reflection of radiation which can reduce midday canopy temperatures, and also higher predawn water potentials (Jefferson et al. 1989). Major compounds of *Citrus* cuticles are l-alkanols, n-alkyl esters, n-alkanolic acids, n-alkanes, and triterpenones (Riederer and Schneider 1990). However, no relationships were found between the particular composition of these soluble lipids and the cuticular water permeability (Geyer and Schönherr 1990), probably since they are not as important for the sorption characteristics of the cuticular membranes as the insoluble components (Chamel et al. 1991). Nonionic surfactants alter amounts and composition of the soluble lipids, but not of the cuticular waxes (Riederer and Schönherr 1990). The latter are probably most important for the cuticular water sorption, which increases very rapidly at high humidities (Chamel et al. 1991). Ion permeability of isolated cuticles is dependent on ionic interactions during the passage (Tyree et al. 1990).

Between ABA concentration levels of the xylem sap and the diurnal courses of transpiration, no consistent relationships were detectable in a study with lysimeter-grown almond trees under desert conditions. Over the seasons, however, an inverse relationship was determined between maximum leaf conductance and the daily average ABA concentration in the xylem sap (Wartinger et al. 1990). The diurnal course of leaf gas exchange is often decoupled from the course of evaporation by the midday depression of transpiration and photosynthesis. Additionally to the amply studied effects of external factors leading to a decreased leaf conductance around noon, leaf internal factors play a role in this phenomenon as well (Han et al. 1988; Wise et al. 1991). These include the photochemical efficiency of photosystem II, the photon yield of $O_2$ evolution, and reversible changes in the leaf's carotenoid amounts and composition (Demmig-Adams et al. 1989). The effects of water stress upon these changes of the photosynthetic gas exchange rates might be mediated primarily by an increased susceptibility to photoinhibition (Epron and Dreyer 1990). An inhibitory effect of high light is probably also responsible for continually reduced gas exchange rates in the afternoon even under constant environmental conditions or when ambient temperature and humidity after the midday extremes have reached levels comparable to the morning (Correia et al. 1990; Bunce 1990a, b).

Transpiration and photosynthetic carbon uptake differ considerably between leaves of different insertion levels alongside a shoot. This is caused not so much by microclimatic differences through the depth of the canopy (Lösch et al. 1992) as by organ-specific differences. In grasses, spikelets of the ears or panicles contribute considerably to the overall plant's gas exchange (Ishihara et al. 1990). Within this context, H.L. Xu et al. (1990) report on different sensitivities of Rubisco activity in ears and flag leaves under the influence of water stress.

Diurnal transpirational water turnover corresponds with measures of the stem flow water transport (Dugas 1990; Steinberg et al. 1990). Approaches now become more and more popular to calculate leaf water vapor loss by Penman-Monteith calculations and by the application of leaf energy budget studies (Leuning and Foster 1990). Exact leaf temperature measurements are essential for this goal; remote measurements by IR thermometry can be usefully applied for this purpose (Inoue 1990; Inoue et al. 1990; Fuchs 1990; Dunin et al. 1991). Crop canopy – air temperature differences reach values up to 8 °C under a high global radiation (Jensen et al. 1990). Plant water content interferes with canopy surface exchange processes (Szlovak et al. 1991), and temperature-affected stomatal regulation of transpirational cooling as a plant-internal homeostatic regulation of the energy exchange rates must be considered as well (Burke and Upchurch 1989).

## 4. Water Relations of Seeds and During Early Germination Stages

Seeds are useful objects to study the properties of tightly bound water in tissues. If water content of embryonic tissue increases to more than 0.1 g $g^{-1}$ dw, both enthalpy and entropy decrease, indicating very tight binding and ordering of water molecules at lower water contents. Unlike in embryonic tissue, a peak at this threshold value of both parameters is not detectable with endosperm tissue (Vertucci, 1990; Bruni and Leopold 1991a). Furthermore, direct current conductivity due to protons of water increases dramatically above this hydration level, indicating that then percollation processes occur within a network of water molecules adsorbed on the surfaces of proteins and membranes (Bruni and Leopold 1991a). From these studies with pea, soybean, and maize seeds, and from sorption curves with barley caryopses (Arabi et al. 1991), it may be generalized that up to approx. 8% of all water molecules are tightly bound as constitutive water, absorption water is present between 8 and 11%, and above this hydration level free water exists in the tissues. Diffusion of water is seriously hindered, however, at such low water contents due to interactions with polymers (Jenner and Jones 1990; Fleischer et al. 1990). Phase transitions including glass formation in desiccation-tolerant seeds (Koster 1991; Bruni and Leopold 1991b) occur and may be responsible for interactions with the temperature regime affecting the seed viability. Freezing and thawing occur only at higher water contents of cotyledonar tissue (Vertucci 1990), and the occurrence of free water, as determined by proton NMR imaging analysis, is paralleled by first signs of respiratory activity (Kano et al. 1990). Special differences exist across the tissues with respect to the metabolically active liquid zones (Aksenov et al. 1991). With increasing hydration, seeds lose their original desiccation tolerance. This is paralleled by an increase of free radicals (Askochenskaya and Chetverikov 1989; Leprince et al. 1990b).

Recalcitrant (= homoiohydrous) seeds lose their viability at much higher water contents than is the case with orthodox seeds. The threshold of viability loss normally lies around 30% in recalcitrant seeds with the embryonic axes being more sensitive (Doley

1990; Pritchard 1991). Recalcitrant seeds are damaged in any case if nonfreezable water is removed, and they may be affected already if freezable water is removed slowly. Orthodox seeds endure this treatment much better (Berjak et al. 1990; Pammenter et al. 1991). The rate of drying and temperature essentially determine the duration of viability of recalcitrant seeds. Quick drying seems to be better in order to maintain relative longevity (Fu et al. 1990; Wang and Fu 1991). The species-specific initial decline in seed moisture content during maturation drying might be crucial for the development of either recalcitrant or orthodox seeds (Hong and Ellis 1990). Whether or not sites of high water affinity are required in order to permit desiccation of the seeds was not unequivocally clear from the determination of water-absorption isotherms of *Cucumis melo* seeds studied by Welbaum and Bradford (1989). Also membrane destruction during water losses of desiccation-intolerant seeds (Pukacka 1989) will contribute to the syndrome of recalcitrant seed behavior. Late developmental stages of recalcitrant seeds (*Zizania palustris*) do not differ in their desiccation sensitivities (Probert and Brierley 1989), but various populations of a species may differ in their desiccation tolerance (Ellis et al. 1991). In orthodox seeds, a log-arithmetic relation exists between longevity and moisture content (fresh-weight related) of seeds. However, a species-specific discontinuity exists breaking this relation at low moisture contents (Ellis et al. 1990).

During the course of development, gain of desiccation tolerance during maturation drying and loss of this tolerance during imbibition are accompanied by structural and metabolic changes. In castor bean and alfalfa seeds a mid-development increase of seed ABA contents is followed by a decrease and concomitant ABA-insensitivity in the dry seeds (Kermode et al. 1989; N.Xu et al. 1990). Generally, ABA seems to play a key role in preventing germination during seed ripening. The high premature germinability of *Sorghum* seeds if affected by intermittent water stress during grain filling is counteracted efficiently by increased ABA contents (Benech et al. 1991). Also in vitro, ABA can inhibit a precocious germination of early stage embryos but not of mature ones (Morris et al. 1991). Additionally to these phytohormonal effects, germinability is governed as well by osmotica which affect the synthesis of specific proteins (Lanfermeijer et al. 1990; Welbaum et al. 1990; N. Xu et al. 1990; Xu and Bewley 1991). The accumulation of proline in imbibing seeds depends upon the water potential regime, not upon the ABA concentration (Palit and Singh 1991). Sugars coming from starch breakdown during maturation drying (Leprince et al. 1990a) may be involved in the production of the osmotica which locally bring about a suitable water potential. Water potential gradients exist between the seeds and the surrounding tissue. In *Cucumis melo* the gradient between both is sufficiently high for a precocious germination to be prevented during the course of development (Welbaum and Bradford 1990, 1991).

Rapid imbibition of dry seeds brings about damage (e.g., Chern and Sung 1991). Particularly the testa plays an important role as a barrier preventing too high a rate of water uptake (McCormac and Keefe 1990). Possibly, the hard seed coats of many Leguminosae have evolved in order to prevent such damage. A hydrophobic layer prevents soaking of the seed and channels water entrance to the chalaza (Serrato-Valenti et al. 1990). Differences in micropyle and strophiole structures of *Cassia* species are correlated with different abilities for water uptake (Bhattacharya and Saha 1990), and strophiole ruptures as a consequence of fluctuating temperature and humidity with time permit a swelling and final germination of the seeds of subterranean clover (Fairbrother 1991). Ion leakage may also occur if dry seeds become soaked. Priming, i.e., controlled wetting and redrying prior to the onset of the germination process, reduces such events (Weges and Karssen 1990). Compartmentation and sensitivity differences of embryo and surrounding tissues (Welbaum and Bradford 1990) again will determine success or failure of the germination process.

## 5. Effects of Water Shortage on Growth and Metabolism

Although plant water potential, or, rather, its components, particularly the turgor potential, essentially govern metabolic processes, water content is often used as a convenient measure of plant water status. Leaf or canopy water contents can be remotely sensed by reflectance parameters in the near- and mid-IR wavelength region. With increasing availability of satellite imagery, expectations and efforts are growing to monitor crop water stress by remote sensing and a sophisticated use of the energy balance equation (e.g., Lepoutre and Vidal 1989; Fuchs 1990; Carlson et al. 1991). However, drought-dependent changes in leaf reflectance are influenced by many factors (Carter 1991). Pilot studies therefore warn that airborne spectroscopic monitoring is far from yet being usable as a routine method to ascertain vegetation water stress unambiguously (e.g., Hunt and Rock 1989; W. B. Cohen 1991; Riggs and Running 1991).

Water stress effects on crop productivity are reported from many investigations which cannot be enumerated here. Comparisons prevail between different cultivars and species, irrigation and fertilization methods, planting schemes, and weeding activities. Most often the effects of reduced water supply are determined by harvest measures, but more detailed studies of $CO_2$- and $H_2O$-gas exchange were performed as well. Nearly 100 papers present information about crop water use efficiencies (WUE) based on such determinations. A relatively modern approach of WUE determinations uses carbon isotope discrimination which is negatively correlated with WUE (Condon et al. 1990; Johnson et al. 1990; Meinzer et al. 1990a; Virgona et al. 1990; Yakir et al. 1990; Ehdaie et al. 1991; Johnson and Bassett 1991).

*Morphogenetic effects* of water shortage at vegetative stages of plant development are changed phyllochron indexes (Ouknider et al. 1991) and leaf sizes and structures (Welander and Hellgren 1990: *Pelargonium hortorum*; Abrams et al. 1990a: *Fraxinus pennsylvanica*; Ristic and Cass 1991a: *Zea mays*). Cell wall growth is affected differently by water stress in developing and mature leaves. Cellulose synthesis is most sensitive to water shortage (Sweet et al. 1990), an increased hemicellulose accumulation occurs in wheat leaves during the plant's flowering time (Rascio et al. 1990). A good water supply is essential also for a normal development of the reproductive organs. Rice pollen production and viability are reasonably affected by lowered water potentials (Ekanayake et al. 1990), maize kernel set suffers from reduced assimilate supply under water stress (Schussler and Westgate 1991), and ABA accumulates in the caryopses, intensifying the accumulation of storage material (Ober and Setter 1990).

Water stress affects *transcription of nucleic acids* so that particular mRNAs and polypeptides appear (Ho and Mishkind 1991). Such dehydrins are similar in barley and wheat and are encoded from regions of the barley chromosomes 6 and 7 (Close and

Chandler 1990). Among the five genes expressed in water-stress-adapted rice, Borkird et al. (1991) identified those which encode the 70 kDa heat shock protein and the ubiquitin. Those and proteins between 71 and 48 kDa were detected also by fluorography after one-dimensional electrophoresis in slash pine hypocotyl segments exposed to –2.5 or –1.8 MPa (Valluri et al. 1989). Possibly, maturation proteins associated with desiccation tolerance of soybean seeds (Blackman et al. 1991) belong to the same group of proteins which have an adaptive value under dry conditions. Transcription of other proteins becomes reduced under water stress. Most prominent is the decrease of mRNA that encodes the small subunit of Rubisco (Ho and Mishkind 1991). Still existing Rubisco in *Lemna minor* fronds becomes converted under osmotic stress to a high-molecular-weight form (Ferreira and Shaw 1989). In *Mesembryanthemum crystallinum*, shifting under water stress from $C_3$ to CAM photosynthesis, concomitantly with a decrease of the Rubisco expression by more than 4/5 of the original amount, an increase of mRNA for PEPCase up to 13-fold occurs. Conversely, upon reimbibition of prestressed leaves the level of PEPcase mRNA falls drastically, that of Rubisco transcript rises (Schmitt 1990). Ribonuclease activity is increased by dehydration (Blekhman and Shelamova 1990).

Key enzymes of $C_4$ *photosynthesis* are generally more stable than those of the Calvin cycle (Yuzbekov et al. 1989). Pool sizes of intermediates of the photosynthetic dark reactions are differentially affected by mild water stress in *Phaseolus vulgaris* leaves; 3-phosphoglycerate is markedly reduced, whilst triosephosphates and hexose monophosphates did not decline as much (Sharkey and Seemann 1989). Rubisco activity was not so much affected by the mild stress applied in this study. Majumdar et al. (1991), by contrast, emphasize the high drought sensitivity of soybean Rubisco (similarly: Santakumari and Berkowitz 1991). Partitioning of newly fixed carbon between starch and sucrose is shifted by water stress in favor of sucrose (Vassey and Sharkey 1989; Bensari et al. 1990). Effects of water stress upon the electron flow through the photosystems are mostly studied by determination of fluorescence signals. With decreasing water contents of the leaves, fluorescence transients become decreased (Bukhov et al. 1989). In parallel with gas-exchange measurements, fluorescence studies permit a screening of intensity of drought-susceptibility (Martiniello and Blum 1989; Scheuermann et al. 1991) and of sites of particular sensitivity within the photosynthetic system. Thus the ratio (difference between maximum and steady-state fluorescence/maximum fluorescence) indicates that the allocation of photosynthetic electrons to $O_2$ is increased during the desiccation of a leaf (Cornic and Briantais 1991). Increased vapor pressure deficits also change the relation between photochemical and nonphotochemical fluorescence quenching (Peterson 1990). While leaf chlorophyll contents remain unaffected by water shortage (Stuhlfauth et al. 1990; Norman et al. 1990), chloroplast ultrastructures are not very much altered in drought-tolerant maize cultivars, but thylacoids were swollen and the outer membrane of the chloroplast envelope disappeared when water stress was imposed on a drought-susceptible line (Ristic and Cass 1991b). Also the overall volume of spinach chloroplasts is better maintained in drought-acclimated than in nonacclimated plants under water stress (Santakumari and Berkowitz 1991). A correlative association can be found also between protoplast volume and photosynthetic capacity in wheat leaves under periodical water stress (Santakumari and Berkowitz 1990). The formation of oxygen radicals, which bring about a destruction of chlorophyll in the presence of iron, is enhanced by desiccation (Price and Hendry 1991). Drought-resistant plants counteract this by increased activities of defensive substances, particularly of ascorbate peroxidase and glutathione reductase (Tanaka et al. 1990; Dhindsa 1991). Photosynthesis of the green stems of *Spartium junceum* turned out to be less sensitive to water stress (Nilsen and Bao 1990), which might confer an ecological advantage of the broom-like habit.

An increase of the specific PEPcase activity due to de novo synthesis upon exposure to water stress was also recorded with the Crassulacean species *Sedum rupestre* (Pilon-Smits et al. 1990). Water stress, however, is not necessarily the only trigger shifting $CO_2$ uptake in $C_3$-CAM intermediates to the prevalence of nocturnal carbon gain. In *Sedum telephium*, CAM is initially enhanced by an increase of the amount of respiratory $CO_2$ which becomes refixed by the PEPcase system (Borland and Griffiths 1990). In well-watered plants, recycled respiratory $CO_2$ completely accounted for all the overnight malic acid formation. Such CAM-cycling was also found in diverse *Talinum* species. The amount of their CAM-cycling was positively correlated with the degree of aridity of their growing sites (Harris and Martin 1991). Under well-watered conditions, shade-adapted specimens of the $C_3$-CAM tree *Clusia minor* have reduced malic acid levels and release $CO_2$ at night, combined with diurnal acidity changes due to increased and decreased citric acid levels. High-PAR-adapted plants showed net $CO_2$-uptake, malate and citrate accumulation during the dark period (Franco et al. 1991). ABA partially can substitute water stress shifting the mode of carbon acquisition from the $C_3$ pathway to CAM in the facultative CAM-species *Mesembryanthemum crystallinum* (Chu et al. 1990). In the obligate CAM-plant *Kalanchoe pinnata* irradiation, nitrogen nutrition, and water availability interact in eliciting flexible responses of gas exchange and $CO_2$ gain (Lüttge et al. 1991). Diurnal amplitudes of chlorenchyma acidity of *Opuntia ficus-indica* became considerably reduced under prolonged drought. Soluble sugars in both compartments, chlorenchyma and water-storage parenchyma, decreased while plants were kept dry for 3–4 months so that water content decreased by nearly 30% in the green, and more than 60% in the chloroplast-free tissue (Nerd and Nobel 1991; Goldstein et al. 1991a, b). In this species, as well as in *Agave deserti* and *Ferrocactus acanthodes* (Tissue et al. 1991), water transfer between water-storage parenchyma and chlorenchyma is of great importance under prolonged drought conditions. Water reserves of the CAM epiphyte *Tillandsia usneoides* are partly external to the leaves where water is kept by capillary forces within the dense indumentum of trichomes (Martin and Schmitt 1989).

Water stress reduces *respiration* (e.g., Hall et al. 1990: root respiration of sunflowers; Yang and Wang 1989a: root respiration of wheat). Under osmotic stress, mitochondrial oxidative and phosphorylative activities decrease, and the cytochrome-mediated electron flow of *Vigna unguiculata* mitochondria is affected negatively. Possibly, by water stress the inner mitochondrial membrane becomes highly permeable and a solubilization of cytochrome C occurs (Schmitt and Dizengremel 1989).

Symbiontic *nitrogen* fixation of *Vigna* becomes reduced by water stress. Nitrogenase activity decreases to 1/5 of its original value when nodule water potential drops from –0.2 MPa to –0.48 MPa; but the drop of nitrogenase activity occurs before the water potential declines and seems to be influenced more by the declining soil water content. A direct link between reduced nitrogenase activity and carbohydrate supply does not exist (Pararajasingham and Knievel 1990a, b). Water stress at a vegetative growth stage of groundnut improves the partitioning of nitrogen into reproductive parts; during pod filling such an allocation by preference ceases (Venkateswarlu et al. 1991). Nitrate reductase activity is also decreased under water shortage. This decrease is enhanced by high temperatures (Il'chukov and Sher 1991); ABA has no direct influence on this process (Larsson et al. 1989). According to studies with horsegram *(Macrotyloma uniflorum)* glutamate synthetase is also reduced under drought, while NAD- and NADP-dependent glutamate dehydrogenase attains higher acitivitiy (or remains unaffected: Argandona and

Pahlich 1991), possibly as a mechanism to assimilate excess ammonia accumulated during water stress (Reddy et al. 1990). Under the influence of both glutamine and glutamate synthetases, the two amino acids are metabolized in the photorespiratory nitrogen cycle, which together with the photosynthetic carbon oxidation cycle in principle remains active at reduced water potentials (Becker and Fock 1989). Arginine decarboxylase is increased under osmotic stress from mannitol solutions; polyamines accumulate under such conditions (Kubis and Krzywanski 1989; Erdei et al. 1990; Foster and Walters 1991).

Many reports on free proline accumulation in plants under stress also appeared during the presence period under report (e.g., Rhizopoulou et al. 1990: seasonal drought and plant proline levels in Mediterranean phrygana plants; Kohl et al. 1991: proline and proline-metabolizing enzymes in drought-stressed soybean nodules). Together with proline, but partially at lower levels, many other pools of free amino acids can also increase under water stress (Cyr et al. 1990).

A higher supply of *mineral nutrients* generally alleviates negative effects of water stress. Cell membrane stability and ability for osmoregulation become improved by higher N-, K-, and P-supply (Premachandra et al. 1990a, b, 1991; Jensen et al. 1992). Potassium mainly acts as an osmoticum, increasing cell size by higher tissue turgor and in this way also changing leaf anatomy. This leads to a better WUE (Andersen et al. 1992a, b), particularly since transpiration becomes lowered due to a more effective stomatal closure per unit of guard cell metabolic response to unfavorable conditions (Lösch et al. 1992). $K^+$ deficiency combined with water shortage reduces photosynthesis overproportionally (Behboudian and Anderson 1990). Under water deficits, $K^+$ is preferentially allocated to leaves at the expense of stem and root $K^+$ (Itoh and Kumura 1990). Phosphorus content in roots, but not in stems and leaves, becomes reduced after water stress is imposed; acid phosphatase increases considerably (Flasinski et al. 1989). An increased phosphorus supply promotes root water exploitation of deeper soil layers and alleviates the sensitivity of photosynthesis and stomatal resistance against decreases of leaf water potentials (Saneoka et al. 1990).

Among *phytohormones* ABA plays a prominent role in plant water stress responses. Its function as a messenger transducing alongside the xylem pathway (Neales et al. 1989; Hartung et al. 1990) the information about root water shortage to the stomata has been quoted already. Also leaf growth is reduced upon such signals (Bunce 1990c) prior to hydraulic restrictions of tissue enlargement (Zhang and Davies 1990b; 1991). ABA is rapidly generated in roots which are at least partially dried. Increases of soil shear strength which accompany soil drying are not required for its production (Neales et al. 1989). The greatest increase of ABA occurs in the root-growing zone (Ribaut and Pilet 1991). Violaxanthin can act as a precursor in ABA biosynthesis (Li and Walton 1990). The amounts of ABA accumulation are variable; some of the more recent studies quote increases up to the tenfold (Robinson and Barritt 1990) and up to the 67-fold (Norman et al. 1990) of the original values. ABA is accumulated in a target tissue even if a positive turgor prevails there (Creelman and Mullet 1991). Chilling temperatures can lead to increased ABA concentrations as a consequence of a low-temperature-induced water stress (Capell and Dörffling 1989). On the other hand, ABA synthesis itself can be reduced by low temperatures (Vernieri et al. 1991). Pretreatment of *Pinus banksiana* plants with ABA considerably increases their drought resistance (Marshall et al. 1991).

Less unequivocal is the effect of water stress upon ethylene production. A rapid bench-drying of detached leaves brings about an increased ethylene evolution, but this might be an artifact, different from the metabolism of intact plants in a drying soil (Morgan et al. 1990; Narayana et al. 1991). Plant IAA contents can be decreased slightly by reduced water potentials. However, IAA concentration changes also depend upon age, plant organs, intermediate metabolism, and stress duration so that up till now generalizations about its involvement in the plant's water stress response are not yet possible (Guinn et al. 1990; Yuan and Jing 1990).

Leaf *lipid composition* is changed under water stress, but large species-specific differences can be found (Navari-Izzo et al. 1990; Norberg and Liljenberg 1991). In susceptible plants, levels of galactolipids and phosphatidylcholine decrease. In drought-resistant cultivars or species they can even be increased under drought (Hubac et al. 1989; Monteiro et al. 1990). An activity increase of lipolytic hydrolases under water stress is probably responsible for the observed breakdown of galactolipids (El-Hafid et al. 1989). Protective substances active under cold or water stress, like substituted amines and soluble carbohydrates, slightly affect the phase behavior of phosphatidylcholine (Rudolph and Goins 1991). Membrane lipids from microsomes of water-imbibed rape roots exhibit a cubic phase which is transformed to a hexagonal$_{II}$ structure ($H_{II}$) under dehydration in stress-sensitive cells. In acclimated cells the cubic phase is transformed to an L2 phase via an intermediate $H_{II}$ phase (Norberg et al. 1990).

## 6. Implications of Waterlogged Conditions

Waterlogging reduces leaf conductance, photosynthesis, and growth by hypoxia affecting both stomatal gas exchange and photosynthetic processes in the mesophyll cells (Oosterhuis et al. 1990; Soldatini et al. 1990; Singh et al. 1991; Vu and Yelenosky 1991). These effects, however, are indirect and perhaps less critical than the immediate damage the root respiration will suffer from anaerobic conditions. Two minutes after onset of hypoxia, root mitochondria swell and change their ultrastructure. These alterations are reversible if aeration occurs after a few hours (Vartapetian and Zakhmilova 1990). Also the capacity of root cells for oxidative phosphorylation is kept as long as hypoxia lasts only few hours (Andreev et al. 1991). Particularly, cortical cells continue to respire in oxygen-free media, if they become supplied with some $O_2$ by aerenchymatous spaces. This supply obviously does not suffice to avoid anoxia in the stelar tissue (Thomson and Greenway 1991). Formation of large aerenchymas therefore is crucial for flooding tolerance (Grineva et al. 1988; Burdick 1989; Koncalova 1990). An extreme and finally not very efficient way to achieve this goal in waterlogging-sensitive plants can be the disintegration of a layer of cortical cells adjacent to the stele (Smith et al. 1990). An ontogenetic adaptation of the roots to imposed anoxia is only possible with relatively young roots, and root lengths are always negatively affected by waterlogging (Thomson et al. 1990; Pardales et al. 1991). The latter implies that root elongation of wetland plants into anaerobic zones of the substrate will preferentially occur during phases of moderate drainage of the soil (Pezeshki et al. 1991). Flood-tolerant and -intolerant species differ in their ability to produce aerenchymatous roots (Blom et al. 1990); but differences exist also among wetland plants: species rooting in organic sediments with a distinct negative redox potential such as nymphaeoid macrophytes under similar conditions develop a larger amount of root aerenchyma than wetland plants of the isoetid type which typically grow in the mineral sediment with an oxidative character (Smits et al. 1990a). Roots of the former group have a reduced radial oxygen permeability. Both types also differ in the number of alcohol dehydrogenase activity (ADH) isozymes which is also higher in the nymphaeoid plants (Smits et al. 1990b). (Large genetically based differences in ADH activity are generally found between species and even cultivars differing in their tolerance to hypoxia: McKee et al. 1989; Roberts et al. 1989. The enzymatic activity increases under anoxia as a typical short-term response of the roots to waterlogging: Burdick and Mendelssohn 1990.) Changes in intercellular spaces and aeration in response to various $O_2$ concentrations also occur in root nodules of legumes (Dakora and Atkins 1991). If they cannot offset the reduced $O_2$ supply, nodule numbers, weights, and physiological activities will become reduced, effects which can be partially alleviated by higher doses of gibberellic acid (Bishnoi and Krishnamoorthy 1990). Roots of desert plants under anoxia seem to be affected negatively more by increased $CO_2$ concentrations than by decreased $O_2$ concentrations (Nobel 1990).

Oxygen supply from shoots to roots and the $O_2$ release to perirhizal zones comes from both atmospheric air and the oxygen produced by photosynthesis. Therefore substantial diurnal and seasonal fluctuations occur (Waters et al. 1989; Yabuki et al. 1990a, b; Gries et al. 1990; Laan and Blom 1990; Caffrey and Kemp 1991). $O_2$ supply to the root zones by diffusion alone is small, but it increases substantially if forced by mass convection (Armstrong and Armstrong 1990). The latter is achieved in several species by thermoosmosis (Schroeder 1989b; Buchel and Grosse 1990) or by hygrometric pressurization (dilution and shifts of the partial pressures of $O_2$ and $N_2$ by water vapor: Hwang and Morris 1991).

Studies on functional differences with respect to water-logging between closely allied species may give insights into the mechanisms which increase the plants' tolerance against hypoxia. *Quercus robur* roots become more adapted to waterlogging than those of *Q. palustris* and *Q. rubra*, so that leaf growth continues and photosynthesis is less affected in the former species (Colin-Belgrand et al. 1991; Dreyer et al. 1991). Competitive performance of the flood-intolerant *Polygonum lapathifolium* is generally higher than that of the wetland species *P. hydropiperoides* and *P. amphibium* which therefore only have a relative superiority in their hypoxic habitat (Carter and Grace 1990). Similarly, a habitat assortment exists from wet to dry sites with *Rumex maritimus*, *R. crispus*, and *R. thyrsiflorus*. *R. maritimus* provides $O_2$ supply to the roots by adaptive shoot elongation above water level, *R. crispus* follows a dormancy-strategy with extremely low consumption of carbohydrates when submerged. Internal aeration and versatility in the production of aerenchyma is quite efficient in these two species as compared with *R. thyrsiflorus*. Metabolic differences (oxidase activities and Fe-immobilization by oxidation) seem to be not so different between the three species than it is the root porosity (Laan and Blom 1990; Laan et al. 1990; 1991a, b).

## 7. Habitat Water Relations and Plant Performance

The regional water balance (actual evapotranspiration and deficit) decisively determines plant distribution and vegetation structure (Stephenson 1990). Based on moisture index data, detailed parallel mapping of climate and vegetation patterns can be realized for areas as large as China (Fang and Yoda 1990). Large-scale drought impact assessment with the aid of remotely sensed data also becomes possible (Sollod 1990; Malo and Nicholson 1990).

Functional studies on plant performance under the local soil and atmospheric water relations and the adaptations to adverse conditions have been conducted during the past years in all vegetation zones of the earth, both concentrating on structure and function of particular species and considerating synecological interrelationships.

Plants of tropical rainforests attain, as a rule, high leaf conductances, but often co-occurring species differ in their particular adaptation to drought – and waterlogging – (Melick 1990a, b), there are gradients in stomatal and boundary layer conductances from the floor up to the canopy (Roberts et al. 1990), and inter- and intraspecific differences occur between plants in the understorey and in forest gaps (De Lillis and Sun 1990; Fisher et al. 1991). Edge effects are clearly measurable if microclimate and plant water relations are measured along transects from clearing sites into the interior of established rainforests (Kapos 1989). Seasonal phenomena, like periodically increased leaf fall, need not to be related to actual leaf water contents (Wright and Cornejo 1990). Stomata of canopy trees of a Columbian elfin forest responded to humidity changes during the course of a day. As a consequence, daily transpiration rates were similar on sunny days (with morning and afternoon peak rates) and on cloudy days when leaf conductances remained high throughout the day. Osmotic potential contributed essentially in order to keep the turgor positive, even when water potentials decreased to –1.8 MPa (Cavelier 1990). Hawaiian *Dubautia* species from a wet forest and a bog site differ in their hydraulic conductances and their ability to attain low osmotic potentials. The reciprocal exclusion from the two contrasting adjacent habitats may be explained by such functional differences (Canfield 1990).

Osmotic adjustment probably enables *Psychotria limonensis* to keep high assimilation rates despite seasonal drought in the understory of a dry tropical forest in Panama (Mulkey et al. 1991). Irregular soil water availability early in the wet season is not the primary cause for mass death of seedlings in such forests; it is rather insufficient light in the understory and seed predation (Blain and Kellman 1991). The important highveld trees *Combretum apiculatum* and *Colophospermum mopane* germinate at water potentials as low as –1 MPa; co-occurring acacias (*A. tortilis* and *A. karroo*) are less drought-adapted (Choinski and Tuohy 1991). Plants of the last-named species growing as isolated savanna trees extract much less water from the soil than the surrounding grass vegetation (Stuart-Hill and Tainton 1989). The deciduous *Berchemia zeyheri* and the evergreen *Olea europaea* ssp. *africana* – other typical elements of the south African savannas – are remarkably similar with respect to seasonal courses of xylem pressure potential, stem circumference changes, and transpiration rates (Van Rooyen 1989). Gas exchange characteristics of savanna grasses and shrubs are clearly influenced by the moisture supply from the different soil horizons reached by the roots of the two contrasting life-forms (Brown and Archer 1990). Wildfires have stronger negative effects upon the woody plants and in this way increase the long-term moisture storage of deeper layers otherwise exploited by the tree roots (Link et al. 1990). Grazing, on the other hand, preferably reduces the herb cover. As a consequence, fire frequency decreases, water contents of the upper soil layers become less depleted, and shrub seedlings can establish (Harrington 1991: investigations with *Dodonaea attenuata* in Australian semi-arid grasslands). Tapping of brackish groundwater by shrubs in semi-desert areas will reduce vitality of such stands; a long-term establishment of the vegetation will be possible only if, additionally to the plant's salt tolerance, a fluctuating water table periodically flushes salts out of the rooting zones (Jarrell and Ross 1990). Alternations between flooded and dry conditions of prairie marshes by changed nutrient, particularly nitrogen, availability affect the biomass production of emergent macrophytes. This again, by altered competition with respect to seedling establishment and plant survival influences the floristic composition of such stands (Neill 1990a, b).

Desert annuals require a high field capacity of the soil for germination thus avoiding precocious germination (Meidan 1990). The Negev geophytes *Bellevalia* spp. are well adapted in phenology and biomass accumulation to the seasonal and regional differences of water availability of their respective growing sites (Boeken 1990; Boeken and Gutterman 1990). The grasses *Digitaria adscendens* and *Eleusine indica*, both occupying different microsites around sand-dunes, differ in their abilities to penetrate deeper into the soil profile by vigorous root growth. *Digitaria* therefore can maintain high water potentials and leaf conductances under water shortage, *Eleusine* remains restricted to sites with sufficient superficial soil water availability (Park 1990). Clonal growth enables *Distichlis spicata* plants growing in the Death Valley to share water among separately rooted but connected ramets (Alpert 1990).

The Moroccan *Argania spinosa* tolerates water potentials down to –3.5 MPa without stomatal closure, and maintains rather high transpiration rates throughout the wet season (Peltier et al. 1990). Under a similar humid mediterranean-type climate in south Portugal, watershed moisture balances were studied in a *Quercus suber* stand and an *Eucalyptus globulus* plantation (Pereira de Almeida and Riekerk 1990). The summer-dry weather depleted soil moisture to such a degree that wintertime recharge of the cork-oak watershed generated only 1% of the annual rainfall into runoff. Temporarily evapotranspiration of the *Eucalyptus* plantation matched that of the cork-oak stand; it was higher at the end of the drought period. Other *Eucalyptus* species in irrigated plantations in the Negev desert displayed considerable species-specific differences of WUE (Herwitz and Gutterman 1990). Evapotranspiration of *Pinus pinaster* stands was measured by Granier et al. (1990) and Berbigier et al. (1991). Macchia and chaparral shrubs respond to seasonal drought by adjusting the amount of leaf area and by reducing gas exchange via stomatal closure mechanisms (model calculations by Tenhunen et al. 1990). From year-round measurements of photosynthetic light utilization efficiency, maximum photosynthetic capacity, and carboxylation efficiency, Beyschlag et al. (1990) conclude that it is primarily reduced $CO_2$ diffusion to the mesophyll which lowers carbon gain during the months of summer drought. Mesophyll photosynthetic capacity, although undergoing seasonal changes of activity as well, is not the limiting factor. Sclerophylly turned out not to be the primary cause adapting the bulk modulus of elasticity so that mediterranean oaks can recover from drought-dependent leaf water deficits (Salleo and Lo Gullo 1990). During the summer drought period, leaves of Greek macchia shrubs were mostly at or near zero turgor with high species-related differences in water potential values (Rhizopoulou and Mitrakos 1990). Three co-occurring *Banksia* species in West Australia displayed comparable seasonal courses of dawn and dusk water potentials and daily patterns of leaf conductances, and their rooting behavior turned out to be not different among each other.

Co-occurrence of these species therefore cannot be attributed to a differential water use (Lamont and Bergl 1991). Fire events in fynbos communities affect soil water availability by producing a reduced vegetation cover. Resprouting and reseeding species benefit from this by different ways during recolonization (Smith and Richardson 1990; Davis and Midgley 1990). Broadleaved evergreens from humid woodlands under the South African mediterranean climate experience, as a rule, no excessive soil water stress conditions. The amplitudes of the diurnal courses of their water potential values are not exceptionally large, and their stomata attain moderate conductances throughout the year (Smith 1990; Richardson and Kruger 1990).

Ash stands of alluvial forests of the Upper Rhone valley seldom experience soil water potentials lower than –0.025 MPa and attain high leaf conductances of 1.6 s $cm^{-1}$ so that transpiration rates become extremely high (Besnard and Carlier 1990). Water relations under drought of established North American deciduous forest trees and shrubs represent, as a rule, the long-term hydric situation of their particular environment (Abrams et al. 1990b). This holds true also for provenances from hydrologically different growing sites of the same species (Abrams and Kubiske 1990: *Acer rubrum*). A successful seedling establishment of such species depends to a high degree on the existing vegetation cover, only in the second place on soil (moisture) and microclimate (temperature) characteristics (Burton and Bazzaz 1991). Different species follow different strategies utilizing various edaphic conditions (*Quercus rubra*: stress-tolerant, *Liriodendron tulipifera*: opportunistic; Kolb et al. 1990). Comparative studies of water relations of evergreen Ericaceae from the southern Appalachian mountains give evidence for peculiar patterns of water potential, gas exchange, and productivity interrelations which could be responsible in part for the local separation to different growing sites (*Rhododendron* and *Kalmia* species: Lipscomb and Nilsen 1990; Muller 1991).

The ericaceous shrub *Arbutus menziesii* as a competitor deteriorates the water supply situation of Douglas fir (Pabst et al. 1990). On the other hand, understory biomass accumulation becomes limited as well by a restricted soil water and nutrient supply in the coniferous forests of Oregon (Riegel et al. 1991). Water stress of *Picea rubens* at high elevation stands is not severe enough to explain the increased decline of this species with higher elevations (Andersen and McLaughlin 1991). *Picea mariana* and *Larix laricina* performed well at both drained and undrained peatlands. At drained sites, however, higher net assimilation rates, WUE, and foliar nitrogen contents could be observed (Macdonald and Lieffers 1990).

Drought affects meadow and grassland perennials less than annual plants at such sites (Bollinger et al. 1991). *Bromus tectorum* as a neophyte in central North America competes very successfully with native species for soil water resources after fire events (Melgoza et al. 1990). *Agropyron cristatum* is inferior to *A. smithii* and *Bouteloua gracilis*, both being dominant species of the shortgrass prairie, in a water-limited environment (Trlica and Biondini 1990). Seedlings of grasses and forbs from slightly acidic grasslands and freshwater marshes in Great Britain differed in their root growth during establishment on dry or wet soils. Wetland species were not able to compensate for reduced soil water availability by intensified root growth (Evans and Etherington 1991). Water-holding capacity of soils covered by *Mesobrometum* communities was up to 20-fold higher than that of adjacent sites were *Xerobrometum* communities developed (comparative measurements by Leuschner (1989) in the Upper Alsace area). Soil moisture has also an important effect upon species composition, growth, biomass allocation, and reproductive efforts of annual plants growing on dune habitats under an atlantic climate (Rozijn et al. 1990). The arctic vascular hemiparasite *Pedicularis dasyantha* has much higher leaf conductances and transpiration rates than its host *Dryas octopetala*, and in this way accumulates much higher nutrient amounts than the host (Gauslaa and Odasz 1990).

Plants of salinity-affected coastal areas must develop very low water potentials in order to cope with the low osmotic potential of their substrate. Mangrove species can use all means of osmoregulation – changed cell wall elasticities, accumulation of solutes, and changes of the symplasmic water fraction – in order to maintain turgor and transpirational water flow (Rada et al. 1989). *Acrostichum aureum* additionally to the accumulation of $Na^+$, $Mg^{2+}$ and sucrose produces the cytoplasmic compatible solute D-1-0-methyl-muco-inositol (Medina et al. 1990). Salinity stress is combined in these habitats, as a rule, with soil flooding. *Avicennia germinans* and *Conocarpus erectus* experience dawn water potentials between –0.4 and –5.78 MPa as a consequence of variable water supply in the rainy and the dry season, respectively (Smith et al. 1989). *Laguncularia* and *Rhizophora* respond to changed waterlogging and salinity conditions by changes in biomass partitioning (Pezeshki et al. 1990). A favorable $K^+/Na^+$ ratio under increased salinity is achieved in the halophyte *Kosteletzkya virginica* by a changed membrane lipid composition (Blits and Gallagher 1990).

## 8. Water Relations in Poikilohydric Plants

Studies on water relations of cryptogams published during the past years continue to deal with various interesting aspects in detail. They are, however, not as numerous as those using higher plants, and are more isolated from each other thematically.

Accumulation in cyanobacteria of sugars (trehalose and sucrose) and amino acids (alanine) as a means of osmoregulation is reported by Hershkovitz et al. (1991) and Thomas and Shanmugasundaram (1991). The pentose-phosphate oxidative pathway plays a major role in *Dunaliella* glycerol synthesis in response to osmotic changes (Chitlaru and Pick 1991). $Na^+$ exchanges under such conditions depend upon pH gradients and the transmembrane electric potential (Weiss and Pick 1990; Oren-Shamir et al. 1990). Changes of ion conductances of the membranes can also result from altered pressure gradients (Kourie and Findlay 1991). Hypotonic turgor regulation of *Lamprothamnium* is retarded by an increased cytoplasmatic $Ca^{2+}$ concentration (Okazaki and Iwasaki 1991). Carbon fixation of some cyanobacteria can continue during desiccation until very low water contents are reached (De Winder et al. 1990). Higher organized algae are much more desiccation-sensitive. Sublitoral red algae under emerse conditions decrease their assimilation rates soon after the beginning of desiccation (Steinke et al. 1989). Seaweeds from the eulitoral are more tolerant to water loss. They can experience even a transient increase of $CO_2$ uptake at water contents just below saturation (Madsen and Maberly 1990). Under desiccation peroxidase activity increases in *Ulva* fronds (Murthy and Sharma 1989). Photosynthesis patterns of various Fucales exhibit $C_4$ attributes to the extent that by an energy-consuming transmembrane transport of inorganic carbon $CO_2$ is accumulated, so that it can be metabolized by Rubisco. The oxigenase activity is suppressed simultaneously (Surif and Raven 1990).

Osmoregulation of lower fungi uses compounds similar to those in unicellular algae (e.g., Kelly and Budd 1990, 1991; Lucas et al. 1990; Jovall et al. 1990; Andre et al. 1991). Hyphal turgor of *Achlya*, measured directly by a pressure probe and indirectly by a plasmolytic technique, is in the order of magnitude of 0.8–1.2 MPa (Money 1990). Fungal decomposition of attached twigs is analyzed with respect to wood and bark water potential/fungus growth relations by Griffith and Boddy (1991a, b, c).

Among lichenological studies, quantifications of water-relation-dependent photosynthesis prevail, both under controlled laboratory conditions and in the field (Negev desert micro- and macrolichens: Palmer and Friedmann 1990a, b; epiphytic lichens from California: Nash et al. 1990; *Teloschistes capensis* of the Namib desert: Lange et al. 1990; Umbilicariaceae in the Spanish Sierra Central: Sancho and Kappen 1989; *Cladonia portentosa* at dune sites under the perhumid climate of the North Sea: Bruns-Strenge and Lange 1991; *Stereocaulon* on the slopes of a tropical volcano: Fritz-Sheridan and Coxson 1988; crustose and fruticose lichens in the antarctic: Kappen et al. 1990; Kappen and Breuer 1991). Under the very extreme conditions of the last named study site continued simultaneous in-situ-recording was for the first time achieved of $CO_2$ gas exchange, thallus water content, actinic chlorophyll fluorescence, and stand microclimate (Schroeter et al. 1991).

Upon desiccation, *Cryptothallus mirabilis* produces multicellular hairs at the dorsal surface of the thallus (Ducket et al. 1990). *Grimmia antarctici* grows either in form of turfs with tall shoots and loosely packed leaves or as small cushions with small shoots covered by tightly packed leaves. The former habit is developed in wet, the latter in dry environments. The morphological difference is an effect of water availability on cell division and differentiation (Wilson 1990). Due to their structure and large biomass, epiphytic bryophytes of a montane tropical forest are very efficient in catching precipitation (Veneklaas et al. 1990). The optimal water content for *Sphagnum fuscum* photosynthesis lies at 600–800% d.w. (Silvola 1990).

Ritman and Milburn (1990), by acoustic detection and visual analysis monitored cavitation events in the anular cells of fern sporangia. Photosynthesis goes to zero at leaf water contents below 25% in the poikilohydric Gesneriaceae *Haberlea rhodopensis* and is restored completely if leaf resaturation occurs under irradiance; not completely, however, in darkness (Kimenov et al. 1989). *Myrothamnus flabellifolius* contains extremely high amounts of sucrose and arginine, substances which might be partially responsible for the high desiccation tolerance of this plant (Suau et al. 1991).

## References

**Abrams MD,** Kubiske ME (1990) Funct Ecol 4:727–734. **Abrams MD,** Kubiske ME, Steiner KC (1990a) Tree Physiol 6:305–316. **Abrams MD,** Schultz JC, Kleiner KW (1990b) For Sci 36:970–981. **Agbariah KT,** Roth-Bejerano N (1990) Physiol Plant 78:100–104. **Aksenov SI,** Gurnina TY, Smirnov AI, Yakimchenko OE (1991) Biofizika 36:318–321. **Aloni R,** Griffith M (1991) Planta (Berl) 184:123–129. **Alpert P** (1990) Am J Bot 77:1648–1651. **Ameglio T,** Morizet J, Cruiziat P, Martignac M (1990) Agronomie (Paris) 10:331–340. **Andersen CP,** McLaughlin SB (1991) Tree Physiol 8:11–22. **Andersen MN,** Jensen CR, Lösch R (1991) J Exp Bot 42:159–166. **Andersen MN,** Jensen CR, Lösch R (1992a) Acta Agric Scand 42:34–44. **Andersen MN,** Jensen CR, Lösch R (1992b) Acta Agric Scand 42:45–56. **Andre L,** Hemming A, Adler L (1991) FEBS Lett 286:13–17. **Andreev VY,** Generozova IP, Vartapetian BB (1991) Plant Physiol Biochem (Paris) 29:171–176. **Aphalo PJ,** Jarvis PG (1991) Plant Cell Environ 14:127–132. **Arabi MI,** Barrault G, Sarrafi A, Albertini L (1991) Can J Bot 69:311–315. **Araus JL,** Febrero A, Vendrell P (1991) Plant Cell Environ 14:545–558. **Argandona V,** Pahlich E (1991) Phytochemistry (Oxf) 30:1093–1094. **Armstrong J,** Armstrong W (1990) New Phytol 114:121–128. **Askochenskaya NA,** Chetverikov AG (1989) Fiziol Biokhim Kul't Rast 21:572–577. **Assmann SM,** Grantz DA (1990a) Plant Physiol (Bethesda) 93:701–707. **Assmann SM,** Grantz DA (1990b) Plant Cell Environ 13:163–170. **Atkinson CJ,** Mansfield TA, Davies WJ (1990) New Phytol 116:19–28. **Atkinson CJ,** Wookey PA, Mansfield TA (1991) New Phytol 117:535–542. **Auge RM,** Stodola AJW (1990) New Phytol 115:285–296. **Auge RM,** Stodola AJW, Pennell BD (1990a) J Am Soc Hortic Sci 115:661–667. **Auge RM,** Stodola AJW, Gupta SD, Hovanesian JC, Conger BV (1990b) J Plant Physiol 136:15–23. **Avissar R,** Pielke RA (1991) Agric For Meteorol 54:353–372.

**Bagshaw SL,** Cleland RE (1990) Plant Cell Environ 13:85–89. **Baier M,** Hartung W (1991) Z Naturforsch Sect C Biosci 46:106–110. **Baier M,** Gimmler H, Hartung W (1990) J Exp Bot 41:351–358. **Baker DA,** Hall JL (1988) Solute transport in plant cells and tissues. Longman, Harlow. **Baldocchi DD,** Luxmore RJ, Hatfield JL (1991) Agric For Meteorol 54:197–226. **Balling A,** Zimmermann U (1990) Planta (Berl) 182:325–338. **Bassirirad H,** Radin JW, Matsuda K (1991) Plant Physiol (Bethesda) 97:426–432. **Becker TW,** Fock HP (1989) Photosynthetica (Prague) 23:655–663. **Behboudian MH,** Anderson DR (1990) Plant Soil 127:137–139. **Benech ARL,** Fenner M, Edwards PJ (1991) New Phytol 118:339–348. **Benlaribi M,** Monneveux P, Grignac P (1990) Agronomie (Paris) 10:305–322. **Bensari M,** Calmes J, Viala G (1990) Acta Oecol 11:843–856. **Berbigier P,** Diawara A, Loustau D (1991) Ann Sci For (Paris) 48:157–178. **Berjak P,** Farrant JM, Mycock DJ, Pammenter NW (1990) Seed Sci Technol 18:297–310. **Besnard G,** Carlier G (1990) Ann Sci For (Paris) 47:353–366. **Beyschlag W,** Lange OL, Tenhunen JD (1990) Flora (Jena) 184:271–289. **Bhat PG,** Daniel EV, Balasimha D (1990) Indian J Exp Biol 28:1171–1173. **Bhattacharya A,** Saha PK (1990) Seed Sci Technol 18:97–102. **Bishnoi NR,** Krishnamoorthy HN (1990) Plant Physiol Biochem (Paris) 28:663–666. **Blackman SA,** Wettlaufer SH, Obendorf RL, Leopold AC (1991) Plant Physiol (Bethesda) 97:868–874. **Blain D,** Kellman M (1991) J Trop Ecol 7:69–83. **Blake TJ,** Bevilacqua E, Zwiazek JJ (1991) Can J For Res 21:1329–1333. **Blatt MR** (1990) Planta (Berl) 180:445–455. **Blekhman GI,** Shelamova NA (1990) Fiziol Rast (Mosc) 37:997–1006. **Blits KC,** Gallagher JL (1990) Plant Cell Environ 13:419–426. **Blom CWPM,** Bogemann GM, Laan P, Van der Sman AJM, Van de Steeg HM, Voesenek LACJ (1990) Aquat Bot 38:29–48. **Boeken B** (1990) Oecologia (Berl) 82:172–179. **Boeken B,** Gutterman Y (1990) Isr J Bot Basic Appl Plant Sci 40:17–31. **Bollinger EK,** Harper SJ, Barrett GW (1991) Am Midl Nat 125:114–125. **Bonsen KJM** (1991) Vierteljahrsschr Naturforsch Ges Zür 136:13–50. **Borghetti M,** Edwards WRN, Grace J, Jarvis PG, Raschi A (1991) Plant Cell Environ 14:357–370. **Borkird C,** Simoens C, Villarroel R, Van Montagu M (1991) Physiol Plant 82:449–457. **Borland AM,** Griffiths H (1990) Funct Ecol 4:33–40. **Braconnier S,** D'Auzac J (1990) Plant Physiol Biochem (Paris) 28:105–112. **Brinckmann E,** Hartung W, Wartinger M (1990) Physiol Plant 80:51–54. **Brindley HM** (1990a) Planta (Berl) 181:432–439. **Brindley HM** (1990b) Planta (Berl) 181:440–447. **Brown AD** (1990) Microbial Water Stress Physiology. Wiley, Chichester. **Brown JD,** Archer S (1990) Oikos 57:366–374. **Brown JM,** Kramer PJ, Cofer GP, Johnson GA (1990) Theor Appl Clim 42:229–236. **Bruni F,** Leopold AC (1991a) Physiol Plant 81:359–366. **Bruni F,** Leopold AC (1991b) Plant Physiol (Bethesda) 96:660–663. **Bruns-Strenge S,** Lange OL (1991) Flora (Jena) 185:73–97. **Buchel HB,** Grosse W (1990) Tree Physiol 6:247–256. **Bukhov NG,** Sabat SC, Mohanty P (1989) Plant Cell Physiol 30:393–398. **Bunce JA** (1990a) Field Crops Res 24:251–260. **Bunce JA** (1990b) Field Crops Res 24:261–272. **Bunce JA** (1990c) Plant Cell Environ 13:295–298. **Burdick DM** (1989) Am J Bot 76:777–780. **Burdick DM,** Mendelssohn IA (1990) J Exp Bot 41:223–228. **Burke J,** Upchurch DR (1989) Environ Exp Bot 29:487–492. **Burton PJ,** Bazzaz FA (1991) Am J Bot 78:131–150.

**Caffrey JM,** Kemp WM (1991) Aquat Bot 40:109–128. **Calder IR** (1990) Evaporation in the Uplands. Wiley, Chichester. **Canfield JE** (1990) Pac Sci 44:449–455. **Canny MJ** (1991) New Phytol 118:367–374. **Capell B,** Dörffling K (1989) J Plant Physiol 135:571–575. **Carlson TN** (1991) Agric For Meteorol 54:103–106. **Carlson TN,** Belles JE, Gillies RR (1991) Remote Sens Environ 35:175–186. **Carter GA** (1991) Am J Bot 78:916–924. **Carter MF,** Grace JB (1990) Am J Bot 77:381–387. **Cavelier J** (1990) Trees (Berl) 4:155–163. **Ceulemans R,** Hinckley TM, Impens I (1989) Physiol Plant 75:174–182. **Chamel A,** Pineri M, Escoubes M (1991) Plant Cell Environ 14:87–96. **Cheeseman JM** (1991) Plant Cell Environ 14:593–600. **Chern GS,** Sung FJM (1991) Seed Sci Technol 19:469–476. **Chitlaru E,** Pick U (1991) Plant Physiol (Bethesda) 96:50–60. **Choinski JS,** Tuohy JM (1991) Ann Bot (Lond) 68:227–234. **Chu C,** Dai Z, Ku MSB, Edwards GE (1990) Plant Physiol (Bethesda) 93:1253–1260. **Close TJ,** Chandler PM (1990) Aust J Plant Physiol 17:333–344. **Cochard H,** Tyree MT (1990) Tree Physiol 6:393–408. **Cohen WB** (1991) Int J Remote Sens 12:1865–1876. **Cohen Y** (1991) Irrig Sci 12:93–98. **Colin-Belgrand M,** Dreyer E, Biron P (1991) Ann Sci For (Paris) 48:193–204. **Collatz GJ,** Grivet C, Ball JT, Berry JA (1991) Agric For Meteorol 54:107–136. **Colombo SJ,** Asselstine MF (1989) Tree Physiol 5:73–82. **Colombo R,** Cerana R, Lado P, Peres A (1989) Plant Physiol Biochem (Paris) 27:557–562. **Con-**

**don AG,** Farquhar GD, Richards RA (1990) Aust J Plant Physiol 17:9–22. **Constantz J,** Murphy F (1990) J Hydrol (Amst) 119:31–42. **Cornic G,** Briantais JM (1991) Planta (Berl) 183:178–184. **Correia MJ,** Chaves MMC, Pereira JS (1990) J Exp Bot 41:417–425. **Creelman RA,** Mullet JE (1991) Plant Physiol (Bethesda) 95:1209–1213. **Cregg BM,** Hennessey TC, Dougherty PM (1990) Can J For Res 20:1508–1513. **Curvetto N,** Delmastro S (1990) Plant Physiol Biochem (Paris) 28:367–378. **Cyr DR,** Buxton GF, Webb DP, Dumbroff EB (1990) Tree Physiol 6:293–304.

**Dainty J,** Ferrier J (1989) Stud Biophys 133:133–140. **Dakora FD,** Atkins CA (1991) Plant Physiol (Bethesda) 96:728–736. **Davies WJ,** Zhang J (1991) Annu Rev Plant Phys Mol Biol 42:55–76. **Davis GW,** Midgley GF (1990) S Afr J Bot 56:199–205. **De Lillis M,** Sun GC (1990) Acta Oecol 11:545–556. **Demmig-Adams B,** Adams WW, Winter K, Meyer A, Schreiber U, Pereira JS, Krüger A, Czygan FC, Lange OL (1989) Planta (Berl) 177:377–387. **De Winder B,** Matthijs HCP, Mur LC (1990) FEMS Microbiol Ecol 74:33–38. **Dhindsa RS** (1991) Plant Physiol (Bethesda) 95:648–651. **Dickinson RE,** Henderson-Sellers A, Rosenzweig C, Sellers PJ (1991) Agric For Meteorol 54:373–388. **Doley D** (1990) Seed Sci Technol 18:33–42. **Dolman AJ,** Gash JHC, Roberts J, Shuttleworth WJ (1991) Agric For Meteorol 54:303–318. **Dreyer E,** Bousquet F, Ducrey M (1990) Ann Sci For (Paris) 47:285–298. **Dreyer E,** Colin-Belgrand M, Biron P (1991) Ann Sci For (Paris) 48:205–214. **Ducket JG,** Renzaglia KS, Pell K (1990) Can J Bot 68:697–706. **Dugas WA** (1990) Theor Appl Clim 42:215–222. **Dunin FX,** Barrs HD, Meyer WS, Trevitt ACF (1991) Agric For Meteorol 55:133–148. **Dye PJ,** Olbrich DW, Poulter AG (1991) J Exp Bot 42:867–870.

**Ehdaie B,** Hall AE, Farquhar GD, Nguyen HT, Waines JG (1991) Crop Sci 31:1282–1288. **Ekanayake IJ,** Steponkus PL, De Datta SK (1990) Crop Sci 30:310–315. **El-Hafid L,** Pham Thi AT, Zuily-Fodil Y, Vieira da Silva J (1989) Plant Physiol Biochem (Paris) 27:495–502. **Ellis RH,** Hong TD, Roberts EH, Tao KL (1990) Ann Bot (Lond) 65:493–504. **Ellis RH,** Hong TD, Roberts EH (1991) J Exp Bot 42:653–658. **Ephrath JE,** Marani A, Bravdo BA (1990) Field Crops Res 23:117–132. **Epron D,** Dreyer E (1990) Ann Sci For (Paris) 47:435–450. **Erdei L,** Trivedi S, Takeda K, Matsumoto H (1990) J Plant Physiol 137:165–168. **Evans CE,** Etherington JR (1991) New Phytol 118:571–580. **Evans RD,** Black RA, Link SO (1990) Plant Cell Environ 13:455–462. **Evlagon D,** Ravina I, Neumann P (1990) Isr J Bot Basic Appl Plant Sci 39:239–248. **Ewers FW,** Fischer JB, Chiu ST (1989) Plant Physiol (Bethesda) 91:1625–1631. **Ewers FW,** Fischer JB, Chiu ST (1990) Oecologia (Berl) 84:544–552.

**Fairbrother TE** (1991) Seed Sci Technol 19:93–106. **Fang JY,** Yoda K (1990) Ecol Res 5:9–24. **Feigin A,** Ravina I, Shalhevet J (1990) Irrigation with Treated Sewage Effluent. Springer, Berlin Heidelberg New York. **Ferreira RB,** Shaw NM (1989) Planta (Berl) 179:456–465. **Fichtner K,** Schulze ED (1990) Oecologia (Berl) 82:355–361. **Fisher BL,** Howe HF, Wright SJ (1991) Oecologia (Berl) 86:292–297. **Flasinski S,** Zamorski R, Kotowska U (1989) Acta Soc Bot Pol 58:47–58. **Fleischer G,** Skirda VD, Werner A (1990) Eur Biophys J 19:25–30. **Flower DJ,** Rani AU, Peacock JM (1990) Aust J Plant Physiol 17:91–105. **Foster S,** Walters DR (1991) Physiol Plant 82:185–190. **Franco AC,** Ball E, Lüttge U (1991) J Exp Bot 42:597–604. **Frederick JR,** Alm DM, Hesketh JD (1989) Photosynthetica (Prague) 23:575–584. **Fritz-Sheridan RP,** Coxson DS (1988) Lichenologist (Oxf) 20:63–82. **Fu JR,** Zhang BZ, Wang XP, Qiao YZ, Huang XL (1990) Seed Sci Technol 18:743–754. **Fuchs M** (1990) Theor Appl Clim 42:253–262.

**Gartner BL** (1991) Oecologia (Berl) 87:180–189. **Gartner BL,** Bullock SH, Mooney HA, Brown VB, Whitbeck JL (1990) Am J Bot 77:742–749. **Gauslaa Y,** Odasz AM (1990) Holarct Ecol 13:112–121. **Gautier H,** Vavasseur A, Gans P, Lasceve G (1991) Plant Physiol (Bethesda) 95:636–641. **Geyer U,** Schönherr J (1990) Planta (Berl) 180:147–153. **Gilroy S,** Fricker MD, Read ND, Trewavas AJ (1991) Plant Cell 3:333–344. **Glavac V,** Koenies H, Jochheim H, Ebben U (1989) Angew Bot 63:471–486. **Goldstein G,** Andrade JL, Nobel PS (1991a) Aust J Plant Physiol 18:95–108. **Goldstein G,** Ortega JK, Nerd A, Nobel PS (1991b) Plant Physiol (Bethesda) 95:274–280. **Gowing DJG,** Davies WJ, Jones HG (1990) J Exp Bot 41:1535–1540. **Granier A,** Bobay V, Gash JHC, Gelpe J, Saugier B, Shuttleworth WJ (1990) Agric For Meteorol 51:309–320. **Grantz DA** (1990) Plant Cell Environ 13:667–680. **Grantz DA,** Assmann SM (1991) Plant Cell

Environ 14:683–690. **Grantz DA,** Meinzer FC (1990) Plant Cell Environ 13:27–38. **Grantz DA,** Meinzer FC (1991) Agric For Meteorol 53:169–184. **Gries C,** Kappen L, Lösch R (1990) New Phytol 114:589–594. **Griffith GS,** Boddy L (1991a) New Phytol 117:251–258. **Griffith GS,** Boddy L (1991b) New Phytol 117:259–270. **Griffith GS,** Boddy L (1991c) New Phytol 117:633–642. **Grineva GM,** Bragina TV, Garkavenkova AF, Polotskii AM (1988) Fiziol Rast (Mosc) 35:1189–1197. **Guinn G,** Dunlap JR, Brummet DL (1990) Plant Physiol (Bethesda) 93:1117–1120.

**Hadley JL,** Smith WK (1990) Can J For Res 20:1306–1311. **Hall AJ,** Connor DJ, Whitfield DM (1990) Plant Soil 121:57–66. **Ham JM,** Heilman JL (1990) Agron J 82:147–152. **Han FS,** Zhao M, Zhao S, Wang M (1988) Acta Agron Sin 14:296–302. **Hardegree SP** (1989) J Exp Bot 40:1099–1104. **Harrington GN** (1991) Ecology 72:1138–1149. **Harris FS,** Martin CE (1991) Oecologia (Berl) 86:575–584. **Harris MJ,** Outlaw WH (1991) Plant Physiol (Bethesda) 95:171–173. **Hartung W,** Heilmeier H, Wartinger A, Kettemann I, Schulze ED (1990) Isr J Bot Basic Appl Plant Sci 39:373–382. **Hashimoto Y,** Kramer PJ, Nonami H, Strain BR (1990) Measurement techniques in plant science. Academic Press, London. **Hatton TJ,** Catchpole EA, Vertessy RA (1990) Tree Physiol 6:201–210. **Hedrich R,** Busch H, Raschke K (1990) EMBO J 9:3889–3892. **Heitholt JJ,** Johnson RC, Ferris DM (1991) Crop Sci 31:135–139. **Henson IE,** Turner NC (1991) New Phytol 117:529–534. **Henson IE,** Jensen CR, Turner NC (1990) Physiol Plant 79:15–22. **Hershkovitz N,** Oren A, Cohen Y (1991) Appl Environ Microbiol 57:645–648. **Herwitz SR,** Gutterman Y (1990) For Ecol Manage 31:81–90. **Heydt H,** Steudle E (1991) Planta (Berl) 184:389–396. **Higgs KH,** Jones HG (1989) J Hortic Sci 64:167–176. **Hill AE** (1989a) Proc R Soc Lond B Biol Sci 237:363–367. **Hill AE** (1989b) Proc R Soc Lond B Biol Sci 237:369–377. **Hirasawa T,** Ishihara K (1991) Jpn J Crop Sci 60:174–183. **Hitron O,** Zur B (1990) Biotronics 19:39–48. **Ho TY,** Mishkind ML (1991) Plant Cell Environ 14:67–76. **Hodgson AS,** Constable GA, Duddy GR, Daniells IG (1990) Irrig Sci 11:143–148. **Hong TD,** Ellis RH (1990) New Phytol 116:589–596. **Hoson T,** Maeda S, Sakaguchi K, Onishi H, Ohta H (1990) Physiol Plant 78:277–284. **Huang CX,** Van Steveninck RFM (1989) J Plant Physiol 135:554–558. **Hubac C,** Guerrier D, Ferran J, Tremolieres A (1989) Plant Physiol Biochem (Paris) 27:737–744. **Hunt ER,** Rock BN (1989) Remote Sens Environ 30:43–54. **Hunt ER,** Running SW, Federer CA (1991) Agric For Meteorol 54:169–196. **Hwang YH,** Morris JT (1991) Plant Physiol (Bethesda) 96:166–171.

**Il'chukov VV,** Sher KN (1991) Fiziol Biokhim Kul't Rast 23:79–82. **Imai S,** Iwao K, Fujiwara T (1990) Environ Control Biol 28:103–108. **Inoue Y** (1990) Jpn J Crop Sci 59:762–768. **Inoue Y,** Kimball BA, Jackson RD, Pinter PJ, Reginato RJ (1990) Agric For Meteorol 51:21–34. **Ishihara K,** Kiyota E, Imaizumi N (1990) Jpn J Crop Sci 59:321–326. **Itoh R,** Kumura A (1990) Jpn J Crop Sci 59:824–829.

**Jackson MB,** Davies DD, Lambers H (1991) Plant Life under Oxygen Deprivation. SPB, The Hague. **Jarrell WM,** Ross AV (1990) J Arid Environ 18:51–58. **Jat KR,** Muralia RN, Kumar A (1991) J Agron Crop Sci 167:73–80. **Jefferson PG,** Johnson DA, Asay KH (1989) Can J Plant Sci 69:513–520. **Jenner CF,** Jones GP (1990) Aust J Plant Physiol 17:107–118. **Jensen CR,** Henson IE (1990) Oecologia (Berl) 82:114–121. **Jensen CR,** Andersen MN, Lösch R (1992) Plant Soil 140:225–239. **Jensen HE,** Svendsen H, Jensen SE, Mogensen VO (1990) Irrig Sci 11:181–188. **Jing J,** Ma S (1990) Acta Agron Sin 16:342–348. **Johnson CR,** Bassett LM (1991) Crop Sci 31:157–162. **Johnson DA,** Asay KH, Tieszen LL, Ehleringer JR, Jefferson PG (1990) Crop Sci 30:338–343. **Johnson IR,** Melkonian JJ, Thornley JHM, Riha SJ (1991) Plant Cell Environ 14:531–544. **Joly RJ** (1989) Plant Physiol (Bethesda) 91:1262–1265. **Jones HG,** Higgs KH (1989) Plant Cell Environ 12:301–308. **Jones HG,** Sutherland R (1991) Plant Cell Environ 14:607–612. **Jones HG,** Flowers TJ, Jones MB (1989) Plants Under Stress. Cambridge Univ Press, Cambridge. **Jorgensen U** (1991) Tidsskr Planteavl 95:181–203. **Jovall PA,** Tunblad-Johansson I, Adler L (1990) Arch Microbiol 154:209–214. **Just J,** Sauter JJ (1991) Trees (Berl) 5:117–121.

**Kano H,** Ishida N, Kobayashi T, Kozumi M (1990) Jpn J Crop Sci 59:503–509. **Kapos V** (1989) J Trop Ecol 5:173–185. **Kappen L,** Breuer M (1991) Antarct Sci 3:273–278. **Kappen L,** Haeger S

(1991) J Exp Bot 42:979–986. **Kappen L,** Schroeter B, Sancho LG (1990) Oecologia (Berl) 82:311–316. **Karlsson PE,** Assmann SM (1990) Plant Physiol (Bethesda) 94:440–447. **Katterman F** (1990) Environmental injury to plants. Academic Press, London. **Kelly DJA,** Budd K (1990) Exp Mycol 14:136–144. **Kelly DJA,** Budd K (1991) Exp Mycol 15:55–64. **Kermode AR,** Dumbroff EB, Bewley JD (1989) J Exp Bot 40:303–314. **Kidambi SP,** Krieg DR, Rosenow DT (1990) Plant Physiol (Bethesda) 92:1211–1214. **Kigel J,** Cosgrove DJ (1991) Plant Physiol (Bethesda) 95:1049–1056. **Kiil F** (1989) Am J Physiol 256:R801–R808. **Kim J,** Verma SB (1991) Agric For Meteorol 55:149–166. **Kimenov GP,** Markovska YA, Tsonev TD (1989) Photosynthetica (Prague) 23:368–371. **Knapp AK,** Smith WK (1990) Physiol Plant 78:160–165. **Kohl DH,** Kennelly EJ, Zhu Y, Schubert K, Shearer G (1991) J Exp Bot 42:831–838. **Kolb TE,** Steiner KC, McCormick LH, Bowersox TW (1990) For Ecol Manage 38:65–78. **Koncalova H** (1990) Aquat Bot 38:127–134. **Koster KL** (1991) Plant Physiol (Bethesda) 96:302–304. **Kourie JI,** Findlay GP (1991) J Exp Bot 42:151–158. **Kreeb KH,** Richter H, Hinckley TM (1989) Structural and functional responses to environmental stresses: water shortage. SPB, The Hague. **Kubis J,** Krzywanski Z (1989) Acta Physiol Plant 11:157–164. **Kubiske ME,** Abrams MD (1990) Plant Cell Environ 13:995–1000. **Kubiske ME,** Abrams MD (1991) Oecologia (Berl) 85:537–542.

**Laan P,** Blom CWPM (1990) J Exp Bot 41:775–784. **Laan P,** Tosserams M, Blom CWPM, Veen BW (1990) Plant Soil 122:39–46. **Laan P,** Tosserams M, Huys P, Bienfait HF (1991a) Plant Cell Environ 14:235–240. **Laan P,** Smolders A, Blom CWPM (1991b) Plant Soil 136:153–162. **Lamont BB,** Bergl SM (1991) Oikos 60:291–298. **Lanfermeijer FC,** Koerselman-Kooij JW, Borstlap AC (1990) Planta (Berl) 181:568–575. **Lange OL,** Meyer A, Zellner H, Ullmann I, Wessels DCJ (1990) Madoqua 17:17–30. **Larsson M,** Larsson CM, Whitford PN, Clarkson DT (1989) J Exp Bot 40:1265–1272. **Lepoutre D,** Vidal A (1989) C R Acad Agric Fr 75:95–102. **Leprince O,** Bronchart R, Deltour R (1990a) Plant Cell Environ 13:539–546. **Leprince O,** Deltour R, Thorpe PC, Atherton NM, Hendry GA (1990b) New Phytol 116:573–580. **Leuning R** (1990) Aust J Plant Physiol 17:159–176. **Leuning R,** Foster IJ (1990) Agric For Meteorol 51:63–86. **Leuschner C** (1989) Phytocoenologia 18:1–54. **Lhomme JP** (1991) Agric For Meteorol 54:227–240. **Li Y,** Walton DC (1990) Plant Physiol (Bethesda) 92:551–559. **Lilienfein U,** Linnenbrink M, Weisheit K, Lösch R, Kappen L (1991) Verh Ges Ökol 20:189–196. **Link SO,** Gee GW, Thiede ME, Beedlow PA (1990) Arid Soil Res Rehabil 4:163–172. **Lipscomb MV,** Nilsen ET (1990) Am J Bot 77:517–526. **Livingston NJ,** de Jong E (1991) Can J For Res 21:53–57. **Lo Gullo MA** (1989) G Bot Ital 123:255–268. **Lopez FB,** Nobel PS (1991) J Exp Bot 42:143–150. **Loreto F,** Sharkey TD (1990) Tree Physiol 6:409–416. **Lösch R,** Jensen CR, Andersen MN (1992) Plant Soil 140:205–224. **Lucas C,** Da Costa M, Van Uden N (1990) Yeast 6:187–192. **Lüttge U,** Ball E, Fetene M, Medina E (1991) J Plant Physiol 137:259–267. **Lyu SW,** Blum U (1990) J Chem Ecol 16:2429–2440.

**Macdonald SE,** Lieffers VJ (1990) Can J For Res 20:995–1000. **Macfall JS,** Johnson GA, Kramer PJ (1990) Proc Natl Acad Sci USA 87:1203–1207. **Macrobbie EAC** (1990) Proc R Soc Lond Ser B Biol Sci 241:214–219. **Madsen TV,** Maberly SC (1990) J Phycol 26:24–30. **Majumdar S,** Ghosh S, Glick BR, Dumbroff EB (1991) Physiol Plant 81:473–480. **Malik RK,** Narda NK, Murty VVN (1989) J Agron Crop Sci 162:135–140. **Malo AR,** Nicholson SE (1990) J Arid Environ 19:1–24. **Malone M,** Tomos AD (1990) Planta (Berl) 182:199–203. **Mansfield TA,** Hetherington AM, Atkinson CJ (1990) Annu Rev Plant Physiol Plant Mol Biol 41:55–75. **Marshall JG,** Scarratt JB, Dumbroff EB (1991) Tree Physiol 8:415–422. **Martin CE,** Schmitt AK (1989) Bot Gaz 150:1–8. **Martiniello P,** Blum A (1989) J Genet Breed 43:7–10. **Mascart P,** Pinty YJP, Taconet O, Ben Mehrez M (1991) Agric For Meteorol 54:319–352. **Massman WJ,** Kaufmann MR (1991) Agric For Meteorol 54:155–168. **Masuda M,** Tanaka T, Matsunari S (1990) J Jpn Soc Hortic Sci 58:951–958. **Mawson BT,** Zeiger E (1991) Plant Physiol (Bethesda) 96:753–760. **McCormac AC,** Keefe PD (1990) J Exp Bot 41:893–900. **McKee KL,** Mendelssohn IA, Burdick DM (1989) Can J Bot 67:3446–3452. **McNaughton KG,** Jarvis PG (1991) Agric For Meteorol 54:279–302. **Medina E,** Cuevas E, Popp M, Lugo AE (1990) Bot Gaz 151:41–49. **Meidan E** (1990) J Arid Environ 19:77–84. **Meidner H** (1990) J Exp Bot 41:1115–1118. **Meinzer FC,** Grantz DA (1990) Plant Cell Environ 13:383–388. **Meinzer FC,** Goldstein G, Grantz DA (1990a) Plant Physiol (Bethesda)

92:130–135. **Meinzer FC,** Wisdom CS, Gonzalez-Coloma A, Rundel PW, Shultz LM (1990b) Funct Ecol 4:579–584. **Meinzer FC,** Grantz DA, Goldstein G, Saliendra NZ (1990c) Plant Physiol (Bethesda) 94:1781–1787. **Melgoza G,** Nowak RS, Tausch RJ (1990) Oecologia (Berl) 83:7–13. **Melick DR** (1990a) Aust J Bot 38:361–370. **Melick DR** (1990b) Aust J Bot 38:371–382. **Meyer WS,** Tan CS, Barrs HD, Smith RCG (1990) Aust J Agric Res 41:253–266. **Milburn JA,** Kallarackal J, Baker DA (1990) Aust J Plant Physiol 17:57–68. **Money NP** (1990) Exp Mycol 14:416–425. **Monteiro PD,** Pham Thi AT, Vieira da Silva J, Justin AM, Demandre C, Mazliak P (1990) Plant Sci (Shannon) 66:185–194. **Monteith JL,** Unsworth MH (1990) Principles of environmental physics. Arnold, London. **Mooney HA,** Winner WE, Pell EJ (1991) Response of plants to multiple stresses. Academic Press, London. **Moore PH,** Cosgrove DJ (1991) Plant Physiol (Bethesda) 96:794–801. **Morgan PW,** He CJ, De Greef J, De Proft M (1990) Plant Physiol (Bethesda) 94:1616–1624. **Morris PC,** Jewer PC, Bowles DJ (1991) Plant Cell Environ 14:444–446. **Morse SR** (1990) Plant Cell Environ 13:39–48. **Mott KA,** Parkhurst DF (1991) Plant Cell Environ 14:509–516. **Mulkey SS,** Wright SJ, Smith AP (1991) Am J Bot 78:579–587. **Muller RN** (1991) Bull Torrey Bot Club 118:123–127. **Müller R,** Baier M, Kaiser WM (1991) J Exp Bot 42:215–220. **Murthy MS,** Sharma CLNS (1989) Bot Mar 32:511–513.

**Nakahori K,** Koizumi K, Muramatsu S, Ohtani H, Masuko M, Nakase S, Katou K, Okamoto H (1990) Plant Cell Physiol 31:859–864. **Narayana I,** Lalonde S, Saini HS (1991) Plant Physiol (Bethesda) 96:406–410. **Nash TH,** Reiner A, Demmig-Adams B, Kilian E, Kaiser WM, Lange OL (1990) New Phytol 116:269–276. **Navari-Izzo F,** Vangioni N, Quartacci MF (1990) Phytochemistry (Oxf) 29:2119–2124. **Neales TF,** Masia A, Zhang J, Davies WJ (1989) J Exp Bot 40:1113–1120. **Neill C** (1990a) Can J Bot 68:1007–1014. **Neill C** (1990b) Can J Bot 68:1015–1020. **Nerd A,** Nobel PS (1991) Physiol Plant 81:495–500. **Ni BR,** Pallardy SG (1991) Tree Physiol 8:1–10. **Nilsen ET,** Bao Y (1990) Am J Bot 77:1007–1015. **Nilsen ET,** Sharifi MR, Rundel PW, Forseth IN, Ehleringer JR (1990) Oecologia (Berl) 82:299–303. **Nobel PS** (1990) J Exp Bot 41:1031–1038. **Nobel PS** (1991) Physicochemical and environmental plant physiology. Academic Press, London. **Nobel PS,** Lee CH (1991) Ann Bot (Lond) 67:549–554. **Nobel PS,** Schulte PJ, North GB (1990) J Exp Bot 41:409–416. **Noitsakis B,** Tsiouvaras C (1990) Acta Oecol 11:419–427. **Nonami H,** Boyer JS (1990a) Plant Physiol (Bethesda) 93:1601–1609. **Nonami H,** Boyer JS (1990b) Plant Physiol (Bethesda) 93:1610–1619. **Nonami H,** Schulze ED, Ziegler H (1991) Planta (Berl) 183:57–64. **Nooden LD,** Singh S, Letham DS (1990) Plant Physiol (Bethesda) 93:33–39. **Norberg P,** Liljenberg C (1991) Plant Physiol (Bethesda) 96:1136–1141. **Norberg P,** Larsson K, Liljenberg C (1990) Biochem Cell Biol 68:102–105. **Norman SM,** Maier VP, Pon DL (1990) J Agric Food Chem 38:1326–1334. **North GB,** Nobel PS (1991) Am J Bot 78:906–915.

**Ober ES,** Setter TL (1990) Ann Bot (Lond) 66:665–672. **Oertli JJ,** Lips SH, Agami M (1990) Acta Oecol 11:281–290. **Okamoto H,** Miwa C, Masuda T, Nakahori K, Katou K (1990) Plant Cell Physiol 31:783–788. **Okazaki Y,** Iwasaki N (1991) Plant Cell Physiol 32:185–194. **Omasa K,** Maruyama S (1990) J Agric Meteorol 45:259–264. **Oosterhuis DM,** Scott HD, Hampton RE, Wullschleger SD (1990) Environ Exp Bot 30:85–92. **Oosterhuis DM,** Hampton RE, Wullschleger SD (1991) J Prod Agric 4:260–265. **Oren-Shamir M,** Pick M, Avron M (1990) Plant Physiol (Bethesda) 93:403–408. **Ouknider M,** Jacquard P, Elkhyari D (1991) Agronomie (Paris) 11:447–456.

**Pabst RJ,** Tappeiner JC, Newton M (1990) For Ecol Manage 37:267–284. **Palit P,** Singh VK (1991) Indian J Exp Biol 29:793–794. **Palmer RJ,** Friedmann EI (1990a) Microb Ecol 19:111–118. **Palmer RJ,** Friedmann EI (1990b) New Phytol 116:597–604. **Pammenter NW,** Vertucci CW, Berjak P (1991) Plant Physiol (Bethesda) 96:1093–1098. **Pantoja O,** Willmer CM (1991) J Exp Bot 42:323–330. **Pararajasingham S,** Knievel DP (1990a) Can J Plant Sci 70:163–172. **Pararajasingham S,** Knievel DP (1990b) Can J Plant Sci 70:1005–1012. **Pardales JR,** Kono Y, Yamauchi A (1991) Environ Exp Bot 31:107–116. **Pardossi A,** Vernieri P, Tognoni F (1991) Plant Cell Environ 14:675–682. **Park YM** (1990) Funct Ecol 4:735–742. **Park J,** Thimann KV (1990) Plant Physiol (Bethesda) 92:696–702. **Peltier JP,** Carlier G, El Aboudi A, Doche B (1990) Acta Oecol 11:643–668. **Pereira de Almeida A,** Riekerk H (1990) For Ecol Manage 38:55–64. **Peter-**

**son RB** (1990) Plant Physiol (Bethesda) 92:608–614. **Pezeshki SR,** Delaune RD, Patrick WH (1990) Can J For Res 20:869–874. **Pezeshki SR,** Matthews SW, Delaune RD (1991) Environ Exp Bot 31:91–98. **Pilon-Smits EA,** THart H, Van Brederode J (1990) J Plant Physiol 136:155–160. **Premachandra GS,** Saneoka H, Fujita K, Ogata S (1990a) Soil Sci Plant Nutr 36:653–660. **Premachandra GS,** Saneoka H, Fujita K, Ogata S (1990b) Soil Sci Plant Nutr 36:661–666. **Premachandra GS,** Saneoka H, Ogata S (1991) J Exp Bot 42:739–746. **Price AH,** Hendry GAF (1991) Plant Cell Environ 14:477–484. **Pritchard HW** (1991) Ann Bot (Lond) 67:43–50. **Pritchard J,** Wyn Jones RG, Tomos AD (1990) J Exp Bot 41:669–676. **Pritchard J,** Wyn Jones RG, Tomos AD (1991) J Exp Bot 42:1043–1050. **Probert RJ,** Brierley ER (1989) Ann Bot (Lond) 64:669–674. **Pukacka S** (1989) Trees (Berl) 3:144–148.

**Rada F,** Goldstein G, Orozco A, Montilla M, Zabala O, Azocar A (1989) Aust J Plant Physiol 16:477–486. **Radin JW** (1990) Plant Physiol (Bethesda) 92:855–857. **Ranney TG,** Bassuk NL, Whitlow TH (1991) J Hortic Sci 66:381–387. **Rascio A,** Cedola MC, Toponi M, Flagella Z, Wittmer G (1990) Physiol Plant 78:462–467. **Reckmann U,** Scheibe R, Raschke K (1990) Plant Physiol (Bethesda) 92:246–253. **Reddy PS,** Sudakar C, Veeranjaneyulu (1990) Indian J Exp Biol 28:273–276. **Rhizopoulou S,** Mitrakos K (1990) Ann Bot (Lond) 65:171–178. **Rhizopoulou S,** Diamantoglou S, Passiakou L (1990) Acta Oecol 11:585–594. **Ribaut JM,** Pilet PE (1991) Physiol Plant 81:156–162. **Richardson DM,** Kruger FJ (1990) S Afr J Bot 56:214–225. **Riederer M,** Schönherr J (1990) Pestic Sci 29:85–94. **Riederer M,** Schneider G (1990) Planta (Berl) 180:154–165. **Riegel GM,** Miller RF, Krueger WC (1991) Northwest Sci 65:10–15. **Riggs GA,** Running SW (1991) Remote Sens Environ 35:51–68. **Ristaino JB,** Duniway JM (1991) J Am Soc Hortic Sci 116:603–608. **Ristic Z,** Cass DD (1991a) Bot Gaz 152:173–185. **Ristic Z,** Cass DD (1991b) Bot Gaz 152:186–194. **Ritman KT,** Milburn JA (1990) J Exp Bot 41:1157–1160. **Ritman KT,** Milburn JA (1991) J Exp Bot 42:123–130. **Rivers RL,** Williams JC (1990) Biophys J 57:627–632. **Roberts JK,** Chang K, Webster C, Callis J, Walbot V (1989) Plant Physiol (Bethesda) 89:1275–1278. **Roberts J,** Cabral OMR, Ferreira de Aguiar L (1990) J Appl Ecol 27:336–353. **Robertson JM,** Hubick KT, Yeung EC, Reid DM (1990a) J Exp Bot 41:325–338. **Robertson JM,** Yeung EC, Reid DM, Hubick KT (1990b) J Exp Bot 41:339–350. **Robinson TL,** Barritt BH (1990) J Am Soc Hortic Sci 115:991–999. **Rochette P,** Desjardins RL, Dwyer LM, Stewart DW, Pattey E, Dube PA (1991) Agric For Meteorol 54:241–262. **Roden J,** van Volkenburgh E, Hinckley TM (1990) Tree Physiol 6:211–220. **Rosa LM,** Dillenberg LR, Forseth IN (1991) Ann Bot (Lond) 67:51–58. **Rozema J,** Verkleij JAC (1991) Ecological responses to environmental stresses. Kluwer, Amsterdam. **Rozijn NAMG,** Ernst WHO, van Andel J, Nelissen HJM (1990) Flora (Jena) 184:303–312. **Rudolph AS,** Goins B (1991) Biochim Biophys Acta 1066:90–94. **Ryan MG** (1989) Can J For Res 19:1397–1401. **Rygol J,** Zimmermann U (1990) Plant Cell Environ 13:15–26.

**Saab IN,** Sharp RE (1989) Planta (Berl) 179:466–474. **Saliendra NZ,** Meinzer FC, Grantz DA (1990) Agron J 82:359–361. **Salleo S,** Lo Gullo MA (1990) Ann Bot (Lond) 65:259–270. **Sancho LG,** Kappen L (1989) Oecologia (Berl) 81:473–480. **Saneoka H,** Fujita K, Ogata S (1990) Soil Sci Plant Nutr 36:267–274. **Santakumari M,** Berkowitz GA (1990) Plant Physiol (Bethesda) 92:733–739. **Santakumari M,** Berkowitz GA (1991) Photosynth Res 28:9–20. **Santrucek J,** Slavik B (1990a) J Exp Bot 41:795–802. **Santrucek J,** Slavik B (1990b) J Exp Bot 41:803–814. **Saugier B,** Katerji N (1991) Agric For Meteorol 54:263–278. **Scheibe R,** Reckman U, Hedrich R, Raschke K (1990) Plant Physiol (Bethesda) 93:1358–1364. **Scheuermann R,** Biehler K, Stuhlfauth T, Fock HP (1991) Photosynth Res 27:189–198. **Schmalstig JG,** Cosgrove DJ (1990) Plant Physiol (Bethesda) 94:1625–1633. **Schmitt JM** (1990) Plant Cell Environ 13:845–850. **Schmitt N,** Dizengremel P (1989) Plant Physiol Biochem (Paris) 27:17–26. **Schnabl H,** Weissenboeck G, Sachs G, Scharf H (1989) J Plant Physiol 135:249–252. **Schoch PG,** L'Hotel JC, Dauple P, Conus G, Fabre MJ (1989) Agronomie (Paris) 9:137–142. **Schroeder JI** (1989a) J Membr Biol 107:229–236. **Schroeder P** (1989b) Trees (Berl) 3:38–44. **Schroeder JI,** Hagiwara S (1990) Proc Natl Acad Sci USA 87:9305–9309. **Schroeter B,** Kappen L, Moldaenke C (1991) Lichenologist 23:253–266. **Schubert J,** Roeser K, Grossmann K, Sauter H, Jung J (1991) J Plant Growth Regul 10:27–32. **Schussler JR,** Westgate ME (1991) Crop Sci 31:1189–1195. **Schwartz A,** Illan N, Assmann SH (1991) Planta (Berl) 183:590–596. **Sellin AA** (1988) Fiziol Rast (Mosc) 35:1099–1107. **Sellin AA**

(1990) Fiziol Rast (Mosc) 37:450–456. **Serrato-Valenti G,** Ferro M, Modenesi P (1990) Ann Bot (Lond) 65:529–532. **Sharkey TD,** Seemann JR (1989) Plant Physiol (Bethesda) 89:1060–1065. **Sharp RE,** Hsiao TC, Kuhn-Silk W (1990) Plant Physiol (Bethesda) 93:1337–1346. **Silvola J** (1990) Holarct Ecol 13:224–228. **Singh G,** Singh TN (1989) Curr Sci (Bangalore) 58:1134–1138. **Singh BP,** Tucker KA, Sutton JD, Bhardwaj HL (1991) HortScience 26:372–373. **Smirnoff N,** Cumbes QJ (1989) Phytochemistry (Oxf) 28:1057–1060. **Smit BA,** Neuman DS, Stachowiak ML (1990) Plant Physiol (Bethesda) 92:1021–1028. **Smith RE** (1990) S Afr J Bot 56:409–412. **Smith RE,** Richardson DM (1990) S Afr J Bot 56:683–694. **Smith JAC,** Popp M, Lüttge U, Cram WJ, Diaz M, Griffiths H, Lee HSJ, Medina E, Schäfer CS et al. (1989) New Phytol 111:293–308. **Smith GS,** Judd MJ, Miller SA, Buwalda JG (1990) New Phytol 115:325–334. **Smits AJM,** Laan P, Thier RH, Van der Velde G (1990a) Aquat Bot 38:3–18. **Smits AJM,** Kleukers RMJC, Kok CJ, Van der Velde G (1990b) Aquat Bot 38:19–28. **Soldatini GF,** Ranieri A, Gerini O (1990) Biochem Physiol Pflanz (BPP) 186:145–152. **Sollod AE** (1990) J Arid Environ 18:97–108. **Sperry JS,** Tyree MT (1990) Plant Cell Environ 13:427–436. **Spollen WG,** Sharp RE (1991) Plant Physiol (Bethesda) 96:438–443. **Stasova VV,** Antonova GF (1989) Lesovedenie O(6):84–89. **Steinberg SL,** McFarland MJ, Miller JC (1989a) Aust J Plant Physiol 16:549–560. **Steinberg SL,** van Bavel CHM, McFarland MJ (1989b) J Am Soc Hortic Sci 114:466–472. **Steinberg SL,** McFarland MJ, Worthington JW (1990) J Exp Bot 41:653–660. **Steinke TD,** Mann FD, Seetal AR (1989) S Afr J Mar Sci 0(8):319–324. **Stephenson NL** (1990) Am Nat 135:649–670. **Stewart B,** Nielsen D (1990) Irrigation of agricultural crops. CRC, Boca Raton. **Stockle CO,** Kiniry JR (1990) Field Crops Res 25:171–182. **Stuart-Hill GC,** Tainton NM (1989) J Grassl Soc South Afr 6:195–204. **Stuhlfauth T,** Steuer B, Fock HP (1990) Photosynthetica (Prague) 24:412–418. **Suau R,** Cuevas A, Valpuesta V, Reid MS (1991) Phytochemistry (Oxf) 30:2555–2556. **Surif MB,** Raven JA (1990) Oecologia (Berl) 82:68–80. **Sweet WJ,** Morrison JC, Labavitch JM, Matthews (1990) Plant Cell Physiol 31:407–414. **Szlovak S,** Szabo T, Koos L (1991) Növenytermeles 40:41–50.

**Takahashi H,** Suge H (1991) Physiol Plant 82:24–31. **Takahashi H,** Scott TK (1991) Plant Physiol (Bethesda) 96:558–564. **Tanaka K,** Masuda R, Sugimoto T, Omasa K, Sakaki T (1990) Agric Biol Chem 54:2629–2634. **Tarczynski MC,** Outlaw WH, Arold N, Neuhoff V, Hampp R (1989) Plant Physiol (Bethesda) 89:1088–1093. **Tardieu F,** Katerji N, Bethenod O, Zhang J, Davies WJ (1991) Plant Cell Environ 14:121–126. **Tenhunen JD,** Sala Serra A, Harley PC, Dougherty RL, Reynolds JF (1990) Oecologia (Berl) 82:381–393. **Terabayashi S,** Takii K, Namiki T (1991) J Jpn Soc Hortic Sci 59:751–756. **Thiel G,** Blatt MR (1991) J Plant Physiol 138:326–334. **Thomas H** (1990) Ann Bot (Lond) 66:521–530. **Thomas H** (1991) New Phytol 118:35–48. **Thomas H,** Evans C (1989) Ann Bot (Lond) 64:581–588. **Thomas SP,** Shanmugasundaram S (1991) Biochem Int 23:93–102. **Thomson CJ,** Greenway H (1991) Plant Physiol (Bethesda) 96:1294–1301. **Thomson CJ,** Armstrong W, Waters I, Geenway H (1990) Plant Cell Environ 13:395–404. **Tischler CR,** Voigt PW (1990) Crop Sci 30:111–117. **Tissue DT,** Yakir D, Nobel PS (1991) Plant Cell Environ 14:407–414. **Tonin GS,** Wheeler CT, Crozier A (1990) Physiol Plant 79:506–511. **Trlica MJ,** Biondini ME (1990) Plant Soil 126:187–202. **Tromp J,** Ovaa JC (1990) J Plant Physiol 136:606–610. **Turner LB** (1990) Ann Bot (Lond) 66:721–728. **Turner NC** (1991) Agric For Meteorol 54:137–154. **Tyerman SD,** Oats P, Gibbs J, Dracup M, Greenway H (1989) Aust J Plant Physiol 16:517–532. **Tyree MT,** Sperry JS (1989) Annu Rev Plant Physiol Mol Biol 40:19–38. **Tyree MT,** Yang S (1990) Planta (Berl) 182:420–426. **Tyree MT,** Scherbatskoy TD, Tabor CA (1990) Plant Physiol (Bethesda) 92:103–109. **Tyree MT,** Snyderman DA, Wilmot TR, Machado JL (1991) Plant Physiol (Bethesda) 96:1105–1113.

**Udagawa Y,** Ito T, Gomi K (1991) J Jpn Soc Hortic Sci 59:711–718. **Ullrich CI,** Koehler K, Baier M, Förster B, Hartung W (1990) Bot Acta 103:214–222.

**Valluri JV,** Castillon J, Newton RJ, Soltes EJ (1989) J Plant Physiol 135:355–360. **Van der Schoot C,** Van Bel AJ (1989) Am J Bot 76:487–503. **Van Gardingen PR,** Jeffree CE, Grace J (1989) Plant Cell Environ 12:887–898. **Van Hoorn JW** (1988) Agrohydrology – recent developments. Elsevier, Amsterdam. **Van Rees KCJ,** Comerford NB (1990) Can J For Res 20:1183–1191. **Van Rooyen N** (1989) S Afr J Sci 85:736–740. **Vartapetian BB,** Zakhmilova NA (1990) Protoplasma 156:39–44. **Vasilescu V** (1988) Water and ions in biological systems. Birkhäuser, Basel. **Vassey TL,** Sharkey TD (1989) Plant Physiol (Bethesda) 89:1066–1070. **Vavasseur A,** Lasceve G, Couchat P (1990) Plant Cell Environ 13:389–394. **Veneklaas EJ,** Zagt RJ, Van Leerdam A, Van Ek R, Broekhoven AJ, Van Genderen M (1990) Vegetatio 89:183–192. **Venkateswarlu B,** Maheswari M, Reddy GS (1991) Indian J Exp Biol 29:272–275. **Vernieri P,** Pardossi A, Tognoni F (1991) Aust J Plant Physiol 18:25–36. **Vertucci CM** (1990) Biophys J 58:1463–1472. **Virgona JM,** Hubick KT, Rawson HM, Farquhar GD, Downes RW (1990) Aust J Plant Physiol 17:207–214. **Voetberg GS,** Sharp RE (1991) Plant Physiol (Bethesda) 96:1125–1130. **Vu JCV,** Yelenosky G (1991) Physiol Plant 81:7–14.

**Wang XF,** Fu JR (1991) Acta Bot Sin 33:118–123. **Wang XL,** Canny MJ, McCully ME (1991) Physiol Plant 82:157–162. **Wartinger A,** Heilmeier H, Hartung W, Schulze ED (1990) New Phytol 116:581–588. **Waseem M,** Phipps J, Carbonneau R, Simmonds J (1991) J Plant Physiol 138:489–493. **Waters I,** Armstrong W, Thompson CJ, Setter TL, Adkins S, Gibbs J, Greenway H (1989) New Phytol 113:439–452. **Weges R,** Karssen CM (1990) Isr J Bot Basic Appl Plant Sci 39:327–336. **Weiss M,** Pick U (1990) J Plant Physiol 136:429–438. **Welander NT,** Hellgren O (1990) Sver Lantbruksuniv Inst Tradgardsvetensk Rapp 0(56):1–17. **Welbaum GE,** Bradford KJ (1989) J Exp Bot 40:1355–1362. **Welbaum GE,** Bradford KJ (1990) Plant Physiol (Bethesda) 92:1038–1045. **Welbaum GE,** Bradford KJ (1991) J Exp Bot 42:393–400. **Welbaum GE,** Meinzer FC (1990) Plant Physiol (Bethesda) 93:1147–1153. **Welbaum GE,** Tissaoui T, Bradford KJ (1990) Plant Physiol (Bethesda) 92:1029–1037. **West CP,** Oosterhuis DM, Wullschleger SD (1990) Environ Exp Bot 30:149–156. **Weyers JDB,** Meidner H (1990) Methods of stomatal research. Longman, Harlow. **Wiggins PM,** Van Ryn RT, Ormrod DGC (1991) Biophys J 60:8–14. **Williams RPW,** Comper WD (1990) Biophys Chem 36:223–234. **Willmer CM,** Petropoulou Y, Manetas Y (1990) J Exp Bot 41:1103–1108. **Wilson ME** (1990) Polar Biol 10:337–342. **Winkel T,** Rambal S (1990) Agric For Meteorol 51:107–122. **Wise RR,** Sparrow DH, Ortiz-Lopez A, Ort DR (1991) Plant Sci (Limerick) 74:45–52. **Wolf O,** Jeschke WD, Hartung W (1990) J Exp Bot 41:593–600. **Wraith JM,** Baker JM (1991) Soil Sci Soc Am J 55:928–932. **Wright SJ,** Cornejo FH (1990) Ecology 71:1165–1175.

**Xu HL,** Ishii R (1990) Jpn J Crop Sci 59:158–161. **Xu HL,** Ishii R, Yamagishi T, Kumura A (1990) Jpn J Crop Sci 59:153–157. **Xu N,** Bewley JD (1991) J Exp Bot 42:821–826. **Xu N,** Coulter KM, Bewley JD (1990) Planta (Berl) 182:382–390.

**Yabuki K,** Kitaya Y, Sugi J (1990a) Environ Control Biol 28:95–98. **Yabuki K,** Kitaya Y, Sugi J (1990b) Environ Control Biol 28:99–102. **Yakir D,** Deniro MJ, Ephrath JE (1990) Plant Cell Environ 13:949–956. **Yang GP,** Wang ST (1989a) Acta Phytophysiol Sin 15:179–184. **Yang SD,** Wang TD (1989b) Acta Bot Sin 31:36–43. **Yoon TM,** Richter H (1990) Physiol Plant 80:520–526. **Yoshida S,** Eguchi H (1990) Biotronics 19:121–128. **Yuan CX,** Jing D (1990) Acta Phytophysiol Sin 16:179–180. **Yuzbekov AK,** Shmat'ko IG, Shedova OE, Latashenko OP (1989) Fiziol Biokhim Kul't Rast 21:523–527.

**Zhang J,** Davies WJ (1990a) Plant Cell Environ 13:277–286. **Zhang J,** Davies WJ (1990b) J Exp Bot 41:1125–1132. **Zhang J,** Davies WJ (1991) J Exp Bot 42:317–322. **Zhao Z,** Wang P (1989) Acta Agron Sin 15:289–296. **Zhu GL,** Steudle E (1991) Plant Physiol (Bethesda) 95:305–315. **Zhu JJ,** Steudle E, Beck E (1989) Plant Physiol (Bethesda) 91:1454–1459.

Prof. Dr. Rainer Lösch
Abteilung Geobotanik
Heinrich Heine-Universität Düsseldorf
Universitätsstraße 1, Geb. 26.13
W-4000 Düsseldorf 1, Germany

## II. The Transport Phloem. Specifics of its Functioning

By Aart J. E. van Bel

### 1. Functions of the Transport Phloem

Translocation of photoassimilates from the source leaves to the heterotrophic parts of the plant demands three different functions executed by the phloem system. The photosynthate is collected in the source leaves (phloem loading), translocated from source to sink (phloem transport), and delivered in the sink tissues (phloem unloading). It is expected that anatomy and physiology of the particular phloem sections reflect the specific tasks to be carried out. Investigations over the past 10 years revealed a diversity of anatomical settings in sources and sinks which seem to correspond with a multiplicity of mechanisms. Several reviews and opinion papers were recently devoted to the phloem sections engaged in loading (Delrot 1987; Van Bel 1987, 1989, 1992; Gamalei 1989, 1990, 1991; Turgeon 1989; Turgeon and Beebe 1991; Van Bel and Gamalei 1991) and unloading (Murray 1987; Turgeon 1989; Patrick 1990; Oparka 1990; Wolswinkel 1990). The functioning of the transport phloem has gained much less attention, although some silent conceptual progress has been made over the past decade. Only aspects such as the photosynthate unloading from the transport phloem (Patrick 1990) and the relation between transport phloem and solute transfer through rays (Van Bel 1990) have been reviewed recently.

In the 1970s, the transport phloem was in the picture due to the attempts to explain the mechanism of phloem translocation. According to the original concept of Münch, the sieve tubes were initially regarded as sugar-impermeable pipes through which a fluid was driven along an osmotic gradient from source to sink (e.g., Canny 1973). Later, it became clear that complementary retrieval of photoassimilates escaping along the path would sustain the osmotic gradient and hence the pressure-driven flow (e.g., Fensom 1981). The continuous retrieval of escaping sugars along the path (Fig. 1) is in concert with a volume flow as the mechanism of phloem transport (Eschrich et al. 1972).

Simultaneously, the awareness grew that the transport phloem has a dual function. Of course, photosynthetic products must be retained in the sieve tubes in order to maintain the pressure gradient and to nourish the terminal sinks, but also maintenance of the mature cells and growth of the sink tissues around the transport pathway (major veins, petioles, stems) ought to be sustained (Fig. 1).

The term sink is mostly attributed to rapidly growing or strongly accumulating organs such as vegetative apices and seeds, fruits, tubers, bulbs, and roots.; but the activity of axial sinks cannot be neglected when considering C- and N-economy in plants. The companion cells with their excep-

Progress in Botany, Vol. 54

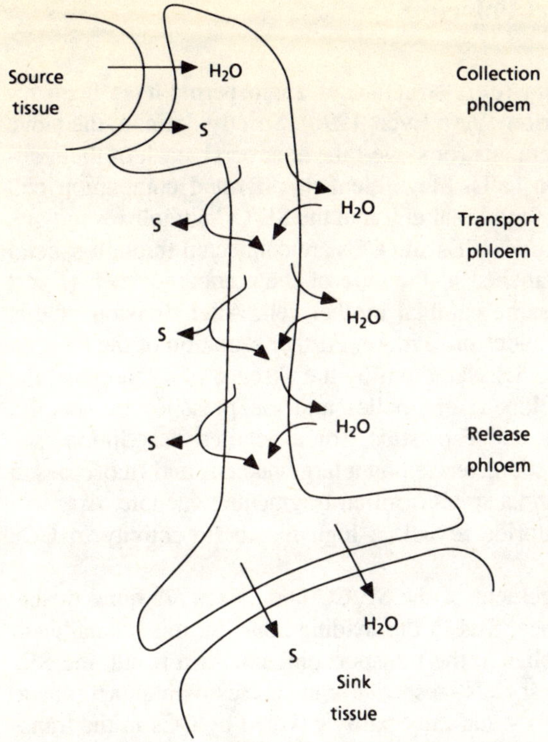


**Fig. 1.** Current view on phloem translocation in higher plants. Photoassimilates are loaded into the collection phloem in source tissues producing an osmotic potential inside the SE/CC complexes sufficient for local attraction of water from the apoplast. The generated turgor gradient gives rise to a source-to-sink pressure gradient creating a mass flow through the sieve tube system. In the transport phloem, solutes escaping from the sieve tubes are partly retrieved by carrier systems located in the SE/CC complexes. The permanent and intense release/retrieval processes are accompanied by the continuous exchange of water. The retrieval serves to maintain the turgor gradient towards the sinks. In the release phloem in the sinks, the turgor potential is lower than in the source, not in the least by intense solute release from the sieve elements sustained by steady accumulation by the sink tissues

tional degree of metabolic activity, the cambial cells with a high rate of division and growth, and the axial and radial parenchyma cells with an occasionally high storage capacity are important sinks for photosynthate.

The dualism in function must be met by a rigorously controlled release/retrieval balance along the transport pathway. The centers of the release/retrieval processes are likely to be the SE/CC complexes. Logically, therefore, this update attempts to structure and combine the present knowledge on the sieve element/companion cell complexes (SE/CC complexes) of the transport phloem and to explore what makes the transport phloem so suitable for photosynthate translocation.

## 2. (Ultra)structure of the Transport Phloem

The basics of sieve element and phloem (ultra)structure of angiosperms have been reviewed recently (Behnke 1989; Eleftheriou 1990; Evert 1990). Strictly defined, the sieve tubes are longitudinal arrays of sieve elements (or sieve-tube members), each of them accompanied by one or a few companion cells. Sieve element (SE) and companion cell (CC) constitute, however, such a strong functional unit that the SE/CC complexes may be considered as the modules of the sieve tubes. SEs and CCs are connected through special plasmodesmatal channels which are branched at the side of the companion cell (Evert 1990). CC and SE originate from the same cambial mother cell. After division of this precursor cell, the immature CC often divides once more. Further septation of the CC can occur by another division. The enucleate SEs stand out by the absence of a tonoplast, the small quantity of (mostly poorly developed) organelles and the presence of specific structures of organelles like P-proteins and P-plastids (for a detailed description, see Eleftheriou 1990; Evert 1990). CCs are characterized by a large nucleus and ribosomes, a dense cytoplasm, many mitochondria, and a small or often fragmented vacuole. Application of thiazolyl blue or tetrazoliumchloride reveals a high metabolic activity in CCs (Van der Schoot and Van Bel 1989).

Structural features and spatial arrangement of the SE/CC complexes are quite different in source, path, and sink. For instance, SEs in the loading zone are much smaller in diameter than the CCs; the opposite applies to the transport phloem. As a result, the SEs in the minor veins are surrounded by their CCs – especially in species with an apoplastic minor vein configuration (Gamalei 1989) – and only partly covered by CCs in the transport phloem (Zahur 1959). The disparate relative size of the CCs probably reflects quantitative differences in the photosynthate collection in the source leaf and the photosynthate retrieval along the path. The collection is supposed to be a more intense process than the retrieval. The parameters of sucrose uptake by isolated vascular tissues from leaves and petioles seem to support this supposition. The affinity for sucrose uptake ($K_m$) is ten times higher in leaf vein networks than in petiole phloem strips, while the uptake capacity ($V_m$) is in the same order (Table 1).

**Table 1.** The parameters of sucrose uptake by isolated vascular bundles and their xylem and phloem constituents. All vascular tissues were isolated from petioles with exception of the *Commelina* tissue. $K_m$ is expressed as mol $m^{-3}$ and $V_m$ as μmol $h^{-1}$ $(gFW)^{-1}$

| | Experimental systems | $K_m$ | $V_m$ | Reference |
|---|---|---|---|---|
| *Heracleum sosnowskyi* | Phloem bundles | 5.9 | 7.6 | Sokolova et al. (1979) |
| | Xylem bundles | 14.3 | 8.8 | Sokolova et al. (1979) |
| *Commelina benghalensis* | Vein networks | 0.5 | 3.2 | Van Bel and Koops (1985) |
| *Apium graveolens* | Vascular bundles | 5 | 5 | Daie (1987a) |
| | Phloem bundles | 10 | 10 | Daie (1987a) |
| | Xylem bundles | 5 | 1.5 | Daie (1987a) |
| *Cyclamen persicum* | Vascular bundles | 5.2 | 4.2 | Grimm et al. (1990) |

## 3. Membrane Potential Mapping of Stem Tissues

Systematic screening of membrane potentials in tomato stem tissues revealed the existence of cell clusters with similar membrane potentials (Van der Schoot and Van Bel 1990). The distinct voltage differences suggest electrical isolation between neighboring clusters (Fig. 2). As plasmodesmata are channels for electric conductance (Overall and Gunning 1982), electrical isolation is indicative of symplastic discontinuity. The existence of voltage difference clusters therefore suggests that the stem tissue is divided in symplastic subunits.

Injection of fluorescent dyes in *Egeria* (Erwee and Goodwin 1985) demonstrated that the symplast is not a protoplastic network with unlimited communication. The symplast appeared to be divided into operational subunits, named symplast domains (Erwee and Goodwin 1985). Intracellular injection of membrane-impermeant fluorescent dyes substantiated the supposition that the stem tissue is also organized in functional symplast domains (Van der Schoot and Van Bel 1990). The domains identified with the injection method largely corresponded with the clusters indicated by the membrane potential mapping (Van der Schoot and Van Bel 1990). Groups of SE/CC complexes with their exceptionally low membrane potentials (–60 to –110 mV) seemingly constitute longitudinally oriented symplast domains in tomato stems (Fig. 2). Similarly, the paratracheal collars around the xylem vessels (Fig. 2) were identified as other longitudinal symplastic strands with a strongly negative membrane potential (–60 to –110 mV). Radial domains with a

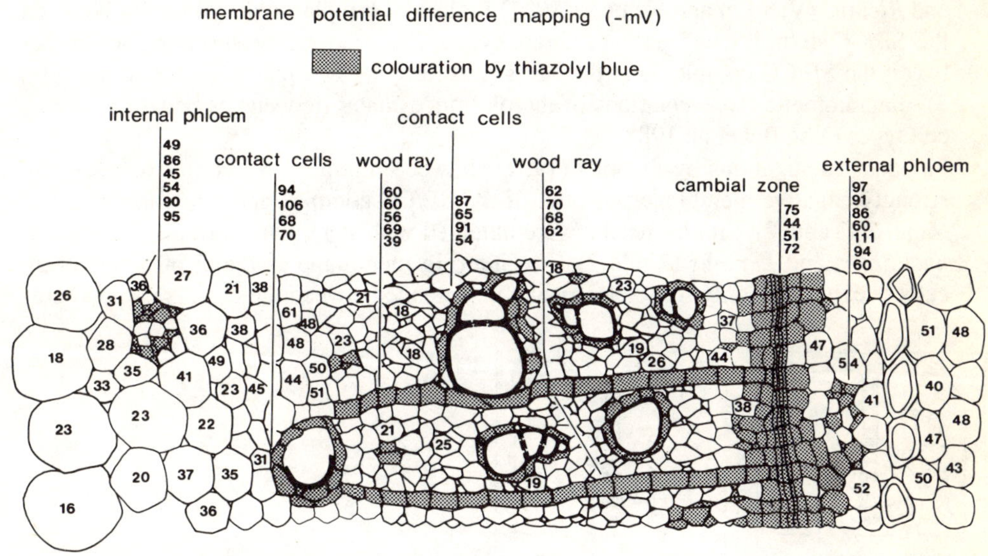


**Fig. 2.** Mapping of membrane potentials and the metabolic activity in tomato internodes. The diagram is a composition of three consecutive transverse sections through the main vascular area of a tomato internode. The map includes the following zones from *left* to *right*: pith with embedded internal phloem, xylem with wood rays, fibers, and xylem vessels surrounded by axial parenchyma, cambial zone, external phloem, and cortex cells (Van der Schoot and Van Bel 1990). Metabolic activity was indicated by staining with thiazolyl blue

relatively negative membrane potential (–50 to –70 mV) in a matrix of living fibers (–20 to –30 mV) link the longitudinal domains (Fig. 2). As the membrane potential ($\Delta\psi$) positively corresponds with the proton-motive force (PMF), domains with negative membrane potential (the SE/CC complexes, in particular) supposedly dominate in retrieving materials from the apoplast.

The relationship between PMF, membrane potential, and the uptake of uncharged substrate molecules ($S_i$ inside, $S_o$ outside concentration) is: $59 \log [S_i]/[S_o] = -PMF$. So, the more negative the PMF, the higher the substrate accumulation capacity. The PMF is the electrochemical free energy of the transmembrane proton gradient which is utilized for transmembrane carriage of the uncharged substrates against their chemical potential. The PMF consists of an electrical (the membrane potential, $\Delta\psi$) and a chemical component (the difference between the outside and inside proton concentration, $\Delta pH$) according to the formula: PMF (in mV) = $\Delta\psi - 59\Delta pH$. The membrane potential, in turn, is the sum of an electrogenic and a diffusional component. The electrogenic potential is created by active extrusion of protons by membrane-bound ATPases; the diffusion potential is the sum of the charge distribution of all ionic species over the plasma membrane.

## 4. Symplastic Discontinuity Between SE/CC Complex and Adjoining Cells

The inferred symplastic discontinuity between SE/CC complexes and adjacent cells (Fig. 2) is supported by a range of other observations:

1. The plasmodesmatal frequencies in the stem phloem of *Phaseolus* (Hayes et al. 1985) and *Ricinus* (Van Bel and Kempers 1991) indicate a symplastic discontinuity between the SE/CC complex and phloem parenchyma (PP). The symplastic constriction between the SE/CC complex and the PP is illustrated by two plasmodesmograms (Fig. 3), diagrammatic representations of the plasmodesmatal frequencies between various cell types (Van Bel et al. 1988).
2. Pioneer investigations by Sibaoka (1962) showed a dramatic difference between the strongly negative membrane potential of the SE/CC complex and the adjacent PP of *Mimosa* (Table 2). Similar results were obtained with stem phloem of *Solanum, Ricinus, Salix*, and *Lupinus* (Table 2). The jump in membrane potential of neighboring cells is explicable only if the electrical conductance between the cells is poor. There-

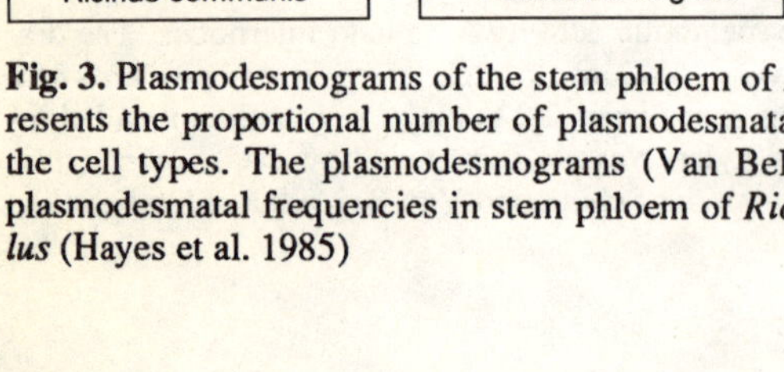


**Fig. 3.** Plasmodesmograms of the stem phloem of *Ricinus* and *Phaseolus*. The striping density represents the proportional number of plasmodesmata (plasmodesmata per unit stem length) between the cell types. The plasmodesmograms (Van Bel and Kempers 1991) were reconstructed using plasmodesmatal frequencies in stem phloem of *Ricinus* (Van Bel and Kempers 1991) and *Phaseolus* (Hayes et al. 1985)

**Table 2.** Membrane potentials of phloem parenchyma and the SE/CC complex in petioles or stems. The membrane potentials of the group above the demarcation line were measured by blind impalement or use of aphids. The second group was measured using glass microelectrodes for impalement into phloem tissue slices under the microscope. Separate data for sieve element and companion cell are given for *Solanum*

| | Phloem parenchyma (mV) | SE/CC complex (mV) | Reference |
|---|---|---|---|
| *Mimosa pudica* | –60 | –160 | Sibaoka (1962) |
| *Salix exigua* | | –160 | Wright and Fisher (1981) |
| *Mimosa pudica* | | –150 | Fromm and Eschrich (1988b) |
| *Cucurbita pepo* | | –170 | Eschrich et al. (1988) |
| *Cucurbita pepo* | | –150 | Opritov and Pyatygin (1989) |
| *Solanum lycopersicum* | –70 | –140/–150 | van der Schoot and van Bel (1990) |
| *Ricinus communis* | –60 | –90 | van Bel and Kempers (1991) |
| *Salix alba* | –50 | –130 | van Bel and Kempers (1991) |
| *Lupinus albus* | –70 | –105 | van Bel and van Rijen (unpubl.) |

fore, the disproportional difference in membrane potential seems to indicate symplastic isolation of the SE/CC complex (Table 2).

3. Fluorescent dyes injected intracellularly into the sieve elements (SEs) were transported longitudinally within the sieve tubes (Van der Schoot and Van Bel 1989; Van Bel and Kempers 1991; Oparka et al. 1992). The dyes always moved quickly to the CCs but not to the PP in stem phloem of *Ricinus* (Van Bel and Kempers 1991) and *Lupinus* (Van Bel and Van Rijen 1992). In these studies, exchange of dyes between SEs, probably via the lateral sieve areas, was often observed. Neighboring SE/CC complexes are most likely clustered, therefore, in a symplast domain, separate from the surrounding PP. In concert with this, dyes injected into PP cells only displaced to other PP cells and never to the SE/CC complexes (Van Bel and Kempers 1991; Van Bel and Van Rijen 1992).
4. The electrical conductance between adjacent PP cells was seven times higher than that between the SE/CC complex and PP (Van Bel and Van Rijen 1992). Since electrical cell-to-cell conductivity and plasmodesmatal frequencies are linearly related (Overall and Gunning 1982), these results underline the relative symplastic isolation of the SE/CC complexes.

How meaningful symplastic isolation is for in vivo functioning remains a matter of debate. Despite the apparent logic of a SE/CC complex with limited symplastic continuity, the results can be misinterpreted quite easily. Moreover, the experimental approach – mostly the use of tissue slices – is prone to artifacts. First, the absence of dye movement does not necessarily imply lack of in vivo symplastic communication, as plasmodesmata tend to shut off in excised tissues (Van der Schoot and Van Bel 1990). Moreover, there is an important pitfall in the interpretation of plasmodesmograms. A paucity of plasmodesmata inevitably constricts the pathway between symplast units, but does not reveal the quantitative consequences for radial transport. This point is crucial for assessing the mode of photosynthate unloading along the pathway (Hayes et al. 1987). For instance, the few plasmodesmatal connections between the SE/CC complex and surrounding cells (Hayes et al. 1985; Van Bel and Kempers 1991) might still account for a symplastic mode of SE unloading in stem tissues.

## 5. Energy Channeling Between Sieve Element and Companion Cell

It is unclear how SE and CC interact metabolically and succeed in maintaining ΔpH and membrane potential of the sieve element. Both values are crucial for setting up the proton gradient for maintenance of the sugar concentration in the sieve tubes. The current idea is that the CCs retrieve photosynthate by a strong PMF (Fig. 4A) and that photosynthate flows back to the SEs (Ziegler 1990). As, however, CCs only partly cover the SEs in the transport phloem (Zahur 1959), such a retrieval mechanism seems to be inefficient. Sugars would easily leak away through the uncovered surface of the SEs.

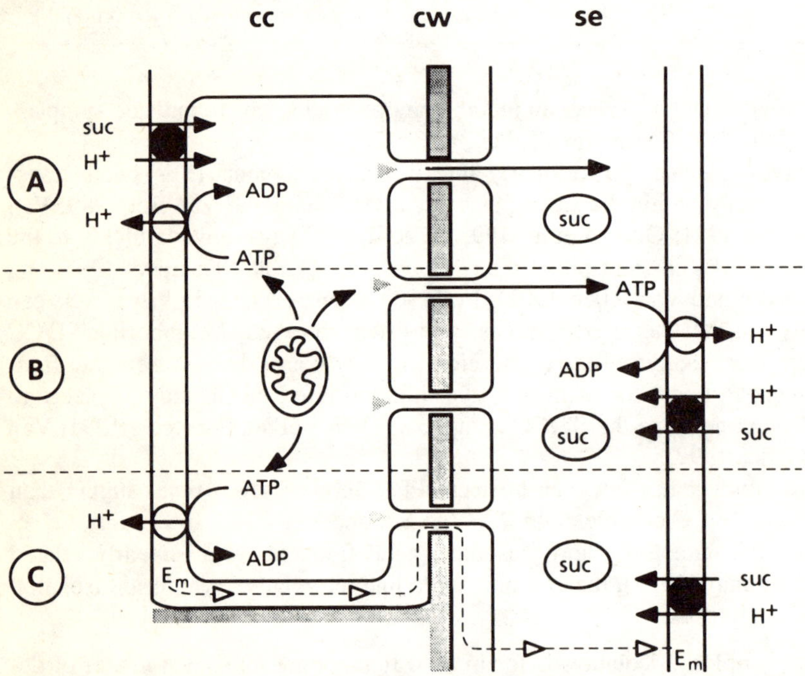


**Fig. 4 A–C.** Potential energy transfer and metabolic interactions between sieve element (*right*) and companion cell (*left*). The model presents three possible ways to maintain the sugar concentration in the sieve elements. **A** Sucrose is retrieved by carriers energized by the excessive PMF of the companion cells and transferred to the sieve tube. (After Ziegler 1990). **B** ATP molecules produced by abundant mitochondria of the companion cell move via the branched plasmodesmata to the sieve element and energize the proton pumps in the energy-deficient sieve elements. The PMF drives the carriers in the sieve-element plasma membrane to retrieve the sugar. **C** The electrogenic potential generated by the proton pumps in the companion-cell plasma membrane propagates via the plasmodesmata to the plasma membrane of the sieve elements. There, the electrogenic potential is part of the PMF which fuels sucrose/proton retrieval mediated by carriers

There are at least two potential alternative or additional mechanisms (Fig. 4B, C). In both models, the CCs – with the abundance of mitochondria – act as the power centers for the SE/CC complex. In the first model, ATP produced by the CCs diffuses to the SEs (Fig. 4B), which would explain the high ATP content of the sieve tube sap (Lehmann 1979). In the second, even more speculative model, the electrogenic potential generated by the CC propagates via the plasma membrane lining of the plasmodesmata to the SE plasma membrane (Fig. 4C). This would match the observation that the membrane potential of the CC is fractionally more negative than that of the SEs (Table 2; Van der Schoot and Van Bel 1989). In the latter models, the sugar carriers of the SE are activated indirectly by the action of the CC (Fig. 4B, C).

In all models, symplastic isolation of the SE/CC complex is functional (Fig. 4). The symplastic coupling between SE and CC guarantees that metabolites are directed preferentially to the SEs (Fig. 4). That the sugar concentration in SE and CC is identical (Fisher 1978) could express the efficiency of metabolite transfer. In addition, symplastic isolation would also serve exclusive channeling of electricity from CC to SE (Fig. 4C). That the sucrose concentration in immature SE/CC complexes is much lower than that of mature ones (Fisher 1978) may show the importance of symplastic isolation for the maintenance of a high sugar content. Young SEs are not symplastically isolated, as shown by intracellular injection of Lucifer Yellow (A. J. E. van Bel and H. V. M. van Rijen, unpubl. results).

The positioning of proton-translocating ATPases may play a key role in the energy channeling between SE and CC. Putative ATPase activity at the SE membrane has been invoked to support apoplastic phloem loading in leaves (Cronshaw 1980). This argument is hardly consistent with the discovery of ATPases associated with the SE/CC complex in the unloading zone of root tips (Eschrich 1983). Moreover, lead deposits also occurred in mesophyll cells (Pesacreta and Lucas 1986), probably resulting from the unspecific action of phosphate-hydrolyzing enzymes (Evert et al. 1988). In freeze-fracture studies, many transmembrane macromolecules appeared to be present in the SE plasma membrane (Sjolund and Shih 1983b). The postulate that a parietal fenestrated ER network in the SEs is a structural prerequisite for ATPase-fueled retrieval (Sjolund and Shih 1983a; Evert 1984) awaits unequivocal determination of ATPase.

## 6. Release/Retrieval Via Pump/Leak Systems in the SE/CC Complex

There is abundant experimental support for pump/leak systems operating along the phloem path (Wolswinkel and Ammerlaan 1983; Minchin et al. 1984; Minchin and Thorpe 1984; Aloni et al. 1986; Hayes et al. 1987; Minchin and Thorpe 1987). The leakage of $^{14}C$-photosynthate from stem segments of broad bean plants, leaves of which had been fed with $^{14}CO_2$, increased in the presence of parachloromercuribenzenesulfonic acid (PCMBS) (Fig. 5). By contrast, PCMBS inhibited the leakage of reducing sugars thought to reside in the PP (Fig. 5). The results were interpreted to evidence substantial passive leakage of photosynthate from the SE/CC complexes along the path. Loss of organic C from the phloem was calculated to be 6% per centimeter stem of bean plant and the resorption 3.2%, resulting in a net loss of 2.8%/cm (Minchin and Thorpe 1987). Thus, to

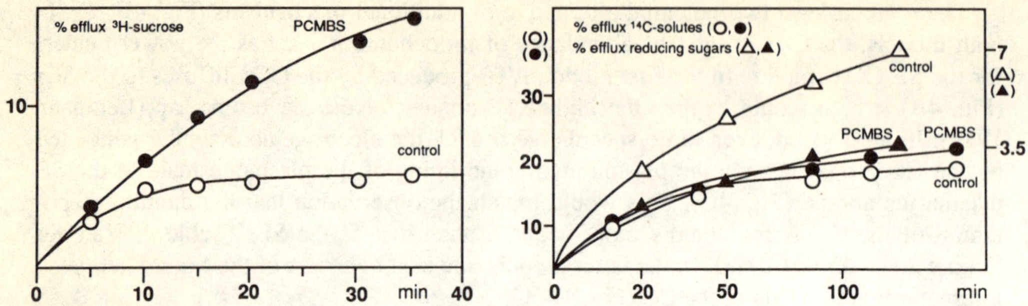


**Fig. 5.** The effect of PCMBS on the sugar release from vascular bundles in stem segments of *Vicia faba* [After Wolswinkel and Ammerlaan 1983 (*left*) and Aloni et al. 1986 (*right* )]. The efflux is related to the label in the stem tissue at the start of the washing period (100%)

maintain a high sugar content, the SE/CC complexes require highly active systems to balance the permanent sugar efflux. These retrieval systems were investigated in uptake studies on isolated bundles containing transport phloem (Daie 1987b; Grimm et al. 1990).

Such a pump/leak system would greatly benefit from symplastic isolation of the SE/CC complexes, i.e., that the SE/CC complexes were symplast domains. Efficient resorption by the SE/CC complex through membrane-bound carrier systems is difficult to reconcile with symplastic exchange of solutes with adjacent cells. Further, maintenance of the extravagant pH of the sieve tube sap may be difficult, if an open communication channel exists between the SE/CC complex and the surrounding cells. Similarly, symplastic discontinuity would assist to create the osmotic disparity between the SE/CC complex and other cells.

## 7. Sugar Uptake by the SE/CC Complex

The kinetics of sucrose uptake by isolated phloem bundles of celery (Daie 1987a) and vascular bundles of *Cyclamen* (Grimm et al. 1990) confirms the presence of carrier systems in SE/CC complexes. The uptake parameters are close: $K_m$ 10 mol $m^{-3}$ and $V_m$ 10 μmol $h^{-1}$ (g FW)$^{-1}$ for celery, and $K_m$ 5.2 mol $m^{-3}$ and $V_m$ 4.2 μmol $m^{-3}$ $h^{-1}$ (g FW)$^{-1}$ for *Cyclamen*. An essential drawback of these experiments is that uptake by SE/CC complexes and other phloem elements interfere with each other. Taking into account the fresh weight, the parameters for uptake by the celery xylem (Daie 1987a) were calculated to be 5 mol $m^{-3}$ ($K_m$) and 1.5 μmol $h^{-1}$ (g FW)$^{-1}$ ($V_m$). A similar computation may rectify the uptake parameters of *Cyclamen* phloem (Grimm et al. 1990). Since, however, the uptake parameters of vascular bundles and vascular phloem alone (Table 1) are close, uptake by the phloem accounts for the bulk uptake by isolated vascular bundles.

That SE/CC complexes monopolize the sugar uptake by isolated vascular tissues (Daie 1987a, b; Grimm et al. 1990) is probably due to their excessive proton-motive force (PMF). Together with the high ΔpH as a result of the exceptionally high pH of the sieve tube sap (see Giaquinta 1983), the membrane potential generates an impressive PMF.

Where sieve tubes are broad enough (in petioles and stem), the membrane potentials can be recorded by use of microelectrodes. Blind impalement into excitable cells in intact stems (Sibaoka 1962; Eschrich et al. 1988; Opritov and Pyatygin 1989) or measurement of the sieve tube content via cut aphid stylets (Wright and Fisher 1981; Fromm and Eschrich 1988b) yielded values between –100 and –150 mV (Table 2). Sieve tube membrane potentials amounted to between –100 and –150 mV in longitudinal phloem strips (Van der Schoot and Van Bel 1989; Van Bel and Kempers 1991). The reaction to high concentrations of exogenous potassium (Wright and Fisher 1981; Van der Schoot and Van Bel 1989) and the effect of inhibitors on the electrogenic proton gradient (Van der Schoot and Van Bel 1989) confirmed that genuine membrane potentials were measured. After membrane potential measurement, the identity of the impaled cells was checked by iontophoretic injection of Lucifer Yellow CH. The exclusive luminescence of the sieve tubes (the sieve plates in particular) proved that the membrane potential of the SE/CC complex had been recorded (Van der Schoot and Van Bel 1989; Van Bel and Kempers 1991).

The values of the membrane potential and the outside-inside pH gradient enable photosynthate accumulation by factors between $10^4$ and $10^6$ by the SE/CC complexes. In the species investigated to date, the PP cells possess less negative membrane potentials than the SE/CC complexes (Table 2). Assuming that the internal pH of PP is close to the value of average parenchyma cells (pH 6.5–7), the PMF of the PP cells is equivalent to an accumulation factor 100 times lower than that of SE/CC complexes.

Direct evidence for proton-driven sugar accumulation by the SE/CC complex is the sucrose-induced depolarization of the sieve tube plasma membrane (Wright and Fisher 1981; Van der Schoot and Van Bel 1989). The electrogenic potential is presumed to be fueled by ATPases at the expense of ATP in the SE/CC complex (Peel 1987).

## 8. Sugar Release from the SE/CC Complex

Photosynthates are probably released into the apoplast from the SE/CC complexes in the stem phloem. Plasmolytic disruption of the plasmodesmata did not inhibit the release of photosynthate and subsequent uptake by adjacent tissues (Minchin and Thorpe 1984; Hayes et al. 1987). The passive character of the release from the SE/CC complex is indicated by the enhanced appearance of radiolabeled photosynthate in the apoplast of stem tissue treated with PCMBS (Wolswinkel and Ammerlaan 1983; Minchin et al. 1984; Aloni et al. 1986).

Rates of phloem unloading into the stem apoplast have been measured by following the efflux of $^{11}C$-labeled substrate (Minchin and Thorpe 1984; Minchin et al. 1984) or by changes in the apoplastic pool of sugars (Hayes and Patrick 1985; Hayes et al. 1987). Given the concentrations of 3 to 100 mol $m^{-3}$ sucrose in the apoplast (Patrick and Turvey 1981; Minchin and Thorpe 1984; Hayes and Patrick 1985) and 500 mol $m^{-3}$ in the sieve tubes, the passive efflux was computed to be between 4 and 5 pmol $cm^{-2}$ $s^{-1}$ (Patrick

1990). On the basis of the passive transmembrane sucrose fluxes in leaf cells and the plasma membrane surface of the phloem cells, Hayes and coworkers (1985) claimed that only the PP cells would be capable of passive release of photosynthate. This would imply symplastic transport from the SE/CC complex to the PP cells (Hayes et al. 1987).

Their considerations were founded on potential symplastic sucrose fluxes (Hayes et al. 1987). With reference to other tissues, they assumed a symplastic standard flux of about $10^4$ pmol $s^{-1}$ $cm^{-2}$ plasmodesmata. However, the putative flux for fully symplastic release from the sieve elements must be ten times higher at the interface between the SE/CC complex and PP than at the other interfaces (Hayes et al. 1987). Therefore, my conclusion is that the connectivity between the SE/CC complex and other cells forms a symplastic bottleneck which is also visualized by the plasmodesmogram of the bean stem phloem (Fig. 3). One could argue that the majority of the photosynthate escapes directly from the SE/CC complex and that only a fraction of the photosynthate is released via the symplast. This conclusion seems to be justified by the finding that the photosynthate release remained constant under osmotic treatments which affected all cells but the SE/CC complexes (Hayes et al. 1987). Under sink-limiting conditions, a partial switch to a symplastic mode of unloading is likely (Hayes et al. 1987).

## 9. Alternative or Parallel Apoplastic and Symplastic Unloading from the SE/CC Complex into Axial Sinks

According to Patrick's (1990) calculations, passive release would account for 60 to 70% of the delivery from the sieve tubes (Hayes et al. 1985). This indicates a potential parallel symplastic route of unloading into the axial sinks. When the prevailing source/sink ratio favors net assimilate storage in stems, unloading indeed seems to switch to a symplastic mode (Hayes et al. 1987).

It is attractive to assume that under source-limiting conditions, the plasmodesmata between the SE/CC complex and PP may be closed to exert ultimate control on the resorption (Hayes et al. 1987; Patrick 1990). Efflux via plasmodesmata would then dissipate the turgor potential of the sieve elements to an undesirable extent. Under sink-limiting conditions (summer-grown plants), the sugar concentration in the sieve tubes will build up and the difference in osmopotential between the SE/CC complex and other phloem cells will further increase (Hayes et al. 1987). Patrick (1990) speculated that the hydrodynamic properties of the plasmodesmata would change with the increasing turgor of the SE/CC complex. Opening of the plasmodesmata would allow the plants to transfer the excess photosynthate to the storage parenchyma. Symplastic transport matches well the glucose flux rates required for starch accumulation in *Populus* ray cells during spring and summer (Sauter and Kloth 1986).

On the other hand, closure of plasmodesmata was observed when the turgor difference was increased by bathing adjacent internode cells of *Chara* in osmotically different media (Zawadzki and Fensom 1986; Côté et al. 1987). Similarly, plasmodesmatal closure was induced by setting up a turgor gradient of 200 kPa between two neighboring cells in *Nicotiana* trichome cells by means of a pressure probe (Oparka and Prior 1992). It is difficult to see these observations as consistent with the opening of plasmodesmata of highly turgescent sieve tubes. Under these conditions, the turgor difference between the SE/CC complex and the PP is expected to be high and plasmodesmata would close in response to the turgor difference between neighboring cells (Zawadski and Fensom 1986; Coté et al. 1987; Oparka and Prior 1992). It is not ruled out, however, that the reaction of the plasmodesmata to turgor may be different between phloem elements and other tissues.

In spite of the confusing evidence so far, reversible closure mechanisms of plasmodesmata may be extremely important for a flexible response to changing conditions (night/day oscillations in hydrostatic potentials, shifts in photosynthate supply from the leaves, shifts of metabolism in the sinks).

## 10. (Un)loading of Storage Carbohydrates in the Transport Phloem

Unloading from the transport phloem is obligatory for the carbohydrates to be stored temporarily (cereals) or seasonally (deciduous trees) in petiole or stem tissues. Cardinal for appreciable storage is that the release/retrieval balance allows major photosynthate loss from the SE/CC complexes. Mechanism and possible regulation of the release were discussed in the preceding section.

An interesting question concerns the entrance of mobilized storage compounds from the storage compartments into the phloem prior to transport to the outgrowing sink. As the release of photosynthate is principally into the apoplast, it is tempting to assume that the carbohydrates released from the storage units follow the reverse route (Van Bel 1990). In principle, the same uptake systems will be employed as are responsible for carrier-mediated retrieval by the SE/CC complex (Van Bel 1990). However, a potential two-way unloading of the SE/CC complex (Hayes et al. 1987) demands reconsideration of the uptake mechanism by the SE/CC complexes in the intact plant.

## 11. Driving Forces of Phloem Transport

According to the original Münch hypothesis, the difference in osmotic potential between the SE/CC complexes in source and sink generates mass flow through the phloem. As discussed above, complementary continuous retrieval of photoassimilates sustains the osmotic gradient and hence a pressure-driven mass flow (Fig. 1). A permanent resorption is in concert with a volume-flow mechanism (Eschrich et al. 1972) which is a dynamic variant of the static pressure-flow hypothesis (Fig. 1). A recent version of the pressure-flow theory proposes that the turgor difference between the sieve tubes in source and sink rather than the difference in osmotic potential along the sieve tubes is responsible for mass flow in the phloem (Wolswinkel 1990). In this concept, the osmotic potential of the phloem sap at the sink end can even exceed that at the source end, provided that the apoplastic osmolarities in source and sink satisfy the osmotic requirements for a source-sink turgor gradient (Lang and Thorpe 1986, 1989).

Computer simulation and analysis of $^{11}C$-experiments are consistent with the mass flow hypothesis (e.g., Goeschl and Magnuson 1986; Magnuson et al. 1986). Some transport physiologists, however, still feel that mass flow is difficult to accept. They doubt if obstructive factors like the small pores in the sieve plates and the abundancy of potentially sealing organelles (P-plastids, P-proteins) are compatible with the measured mass transfer rates. To accommodate the presence of the structural obstacles in the sieve tubes, an alternative to the pressure flow was advanced. This relay hypothesis (Lang 1979) proposes that the phloem is organized in functional units from which assimilates are released at the 'source end' and resorbed by the 'sink end' of the next unit. Like the pump/leak system, the relay (release/resorption) is sensitive to metabolic inhibition. Consequently, unequivocal proof to discriminate between these possibilities was difficult to acquire. An experimental approach with *Ricinus* plants provided proof for a continuous sieve tube system which has distributed capacity to release and resorption (Murphy and Aikman 1989).

## 12. PMF Gradients Along the Phloem Pathway

Of paramount significance for photosynthate distribution in the intact plant is a PMF gradient that drives the photosynthate retrieval along the phloem path to a varying extent. An apoplastic base-to-tip sucrose gradient (from 60 to 25 mol $m^{-3}$) was observed in the stem of bean plants (Minchin and Thorpe 1984). The gradient can be interpreted to ex-

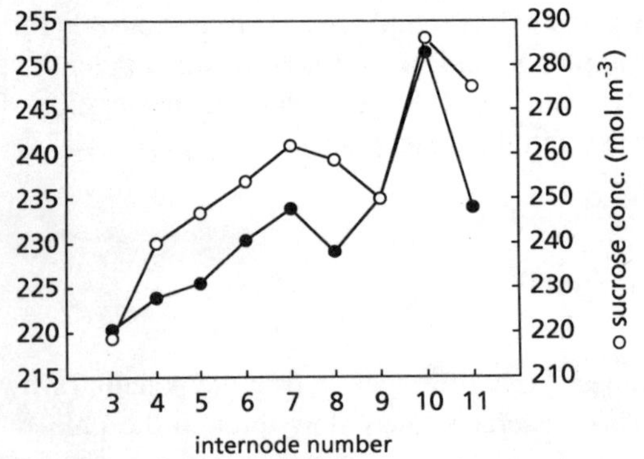


**Fig. 6.** Correlation of calculated PMF and sucrose concentration in the sieve tubes measured in successive internodes of *Ricinus communis* (original data – $H^+$, $K^+$, sucrose concentrations; Vreugdenhil and Koot-Gronsveld 1989). pH and potassium concentration of the phloem bleeding sap in successive internodes were transformed into PMF values (–mV). On the basis of an imaginary apoplast concentration of 5 mol $m^{-3}$ $K^+$, the Nernst potential of potassium ions was computed for each internode. Addition of an estimated electrogenic potential (–60 mV; Van der Schoot and Van Bel 1989) yields a membrane potential of about –130 mV, which is close to the in vivo membrane potential of the SE/CC complex (Van Bel and Kempers 1991). Furthermore, the chemical proton potential, ΔpH (fixed outside pH 5.7, inside pH varying per internode) was calculated and transformed into mV. The sum of electrical and chemical PMF over the plasma membrane of the SE/CC complex is plotted on the *left ordinate* against the sucrose concentration on the *right ordinate* in successive internodes

press a decreasing retrieval capacity with increasing distance from the source. It is expected that the gradient corresponds with a decreasing PMF from source to sink. This supposition conforms with a progressively declining PMF of SE/CC complexes along *Ricinus* stems deduced from the composition of phloem exudate collected at different heights. A decreasing proton extrusion rate inducing an electrophoretic outward potassium flux to reach the Nernst equilibrium would match the basipetal decline of the potassium concentration (Vreugdenhil 1985) and pH (Vreugdenhil and Koot-Gronsveld 1989) in *Ricinus* sieve tube sap. Gradients of PMF over the SE/CC plasma membrane were calculated tentatively by transformation of pH and potassium concentrations into mV (Fig. 6). The corresponding accumulation factors for sucrose are consistent with the observed sucrose concentrations at various distances from the top (Fig. 6). This gives credence to the existence of a longitudinal PMF gradient along the sieve tubes. Projecting the gradient on the *Ricinus* plant illustrates that the highest PMF is reached at the point of attachment of the most mature leaf (Fig. 6).

The decreasing CC/SE volume ratio from source to sink (Offler and Patrick 1984; Hayes et al. 1985; Warmbrodt 1985) supports the concept of a decline in resorption capacity from source to sink. The relative size reduction of the CC amounts from a volume ratio between CC and SE of 2 in the source leaves to 0.25 or less in the sink.

## 13. Potential Consequences of Functioning of the Transport Phloem for the Relative Growth Rate

It has been documented that the sugar composition of the phloem exudate varies considerably with the plant family (Zimmermann and Ziegler 1975; Gamalei 1985). The sugar composition is probably associated with the mode of phloem loading (Gamalei 1989; Turgeon 1991). In symplastic loaders, galactosyl-oligosaccharides are prominent compounds of the phloem sap; in apoplastic loaders, sucrose is the predominant sugar (Gamalei 1985). That phloem loading occurs via symplast or apoplast (Van Bel et al. 1992) may have important implications for the rate of mass flow and the C-distribution over the plant.

Although unequivocal evidence is lacking so far, apoplastic phloem loading driven by the PMF (Giaquinta 1983; Delrot 1987) may set up a higher osmotic gradient in the sieve tubes than symplastic phloem loading would do (Turgeon, 1991). Raffinose solutions become saturated and too viscous (at 250 mol $m^{-3}$) before maximal mass flow is reached (Lang 1978), whereas maximal mass flow of organic C is attained at 750 mol $m^{-3}$ sucrose. One may speculate that, as a result, mass flow rates and pressure gradients are generally (not always!) higher in species loading via the apoplast. There are indications for a higher mass flow rate in apoplastic loaders (Gamalei 1990; Jahnke et al. 1992). These differences in mass flow rate probably correspond with differential escape from the sieve tubes as mass flow through a porous pipe is inversely related to the solute escape from the pipe (Horwitz 1958). A low pressure gradient thus could imply a high rate of release along the sieve tubes. If all these presumptions are valid, loss of assimilates along the stem is likely to be higher in symplastic loaders (Van Bel and Visser 1993) which gives rise to a higher investment into axial tissues. One of the features of plants with a low relative growth rate is a proportionally higher investment of daily fixed carbon in stem tissues (Poorter et al. 1990). In contrast, the apoplastic loaders will direct by preference the photosynthetic products to the terminal sinks (investment in photosynthetic surface), giving rise to higher relative growth rates (Van Bel and Visser 1993).

## 14. Signaling Along the Phloem Pathway

The bioelectric signal in *Mimosa pudica* is accompanied by quick and massive release of photosynthate from the sieve elements (Fromm and Eschrich 1988a, b). This raises the question whether electric signaling controls phloem transport (e.g., Faucher et al. 1982). Another proposal for controlling phloem transport is that uptake and release at the source and sink ends are regulated by turgor signals (Smith and Milburn 1980a, b).

The phloem bundles may also serve as signal conduits themselves and may be used as transmission wires for electric messages. The propagation of action potentials along the phloem would also require symplastic isolation of the SE/CC complex (Fig. 2). If electric signaling is a universal mode of message transfer in plants (Davies 1987), the sieve tubes would be excellent units for this purpose. Some indication for electric messaging along the sieve tubes was obtained in zucchini plants (Eschrich et al. 1988). Further, phytohormones produced in the leaves may join in with the mass flow and then be distributed between the sinks. Considerable amounts of phytohormones with a high mobility were found in phloem exudate of *Ricinus* (Hoad 1980). Wound signals are probably also transported via the phloem, since wound-induced gene expression at sites remote from the wounding is dependent on the direction of the assimilate movement (Davis et al. 1991).

## References

**Aloni B,** Wyse RE, Griffith S (1986) Plant Physiol 81:482–486.

**Behnke HD** (1989) In: Baker DA, Milburn JA (eds) Transport of photoassimilates. Longman, Harlow, pp 79–137.

**Canny MJ** (1973) Phloem translocation. University Press, Cambridge. **Côté R,** Thain JF, Fensom DS (1987) Can J Bot 65:509–511. **Cronshaw J** (1980) Ber Dtsch Bot Ges 93:123–139.

**Daie J** (1987a) Planta 171:474–482. **Daie J** (1987b) Plant Physiol 84:1033–1037. **Davies E** (1987) Plant Cell Environm 10:623–631. **Davis JM,** Gordon MP, Smit BA (1991) Proc Natl Acad Sci USA 88:2393–2396. **Delrot S** (1987) Plant Physiol Biochem 25:667–676.

**Eleftheriou EP** (1990) In: Behnke HD, Sjolund RD (eds) Sieve elements. Springer, Berlin Heidelberg New York, pp 103–137. **Erwee MG,** Goodwin PB (1985) Planta 163:9–19. **Eschrich W** (1983) Planta 157:540–547. **Eschrich W,** Evert RF, Young JH (1972) Planta 107:279–300. **Eschrich W,** Fromm J, Evert RF (1988) Bot Acta 101:327–331. **Evert RF** (1984) In: White RA, Dickison W (eds) Contemporary problems in plant anatomy. Academic Press, Orlando, pp 145–234. **Evert RF** (1990) In: Behnke HD, Sjolund RD (eds) Sieve elements. Springer, Berlin

Heidelberg New York, pp 103–137. **Evert RF,** Mierzwa RJ, Eschrich W (1988) Protoplasma 146:41–51.

**Faucher M,** Bonnemain JL, Doffin M (1982) Physiol Vég 20:395–405. **Fensom DS** (1981) Can J Bot 59:425–432. **Fisher DB** (1978) Planta 139:19–24. **Fromm J,** Eschrich W (1988a) Trees 2:7–17. **Fromm J,** Eschrich W (1988b) Trees 2:18–34.

**Gamalei YV** (1985) Fiziol Rast 32:866–875. **Gamalei YV** (1989) Trees 3:96–110. **Gamalei YV** (1990) Leaf phloem. Nauka, Leningrad (in Russian). **Gamalei YV** (1991) Trees 5:50–63. **Giaquinta RT** (1983) Annu Rev Plant Physiol 34:347–387. **Goeschl JD,** Magnuson CE (1986) Plant Cell Environ 9:95–102. **Grimm E,** Gernot B, Rothe K, Jacob F (1990) Planta 182:480–485.

**Hayes PM,** Patrick JW (1985) Planta 166:371–379. **Hayes PM,** Offler CE, Patrick JW (1985) Ann Bot 56:125–138. **Hayes PM,** Patrick JW, Offler CE (1987) Ann Bot 59:635–642. **Hoad GV** (1980) Planta 150:275–278. **Horwitz L** (1985) Plant Physiol 33:81–93.

**Jahnke S,** Schlesinger U, Knust EJ (1992) Physiol Plant 85:A19.

**Lang A** (1978) Aust J Plant Physiol 5:665–674. **Lang A** (1979) Ann Bot 44:141–145. **Lang A,** Thorpe MR (1986) J Exp Bot 37:495–503. **Lang A,** Thorpe MR (1989) J Exp Bot 40:1069–1078. **Lehmann J** (1979) Z Pflanzenphysiol 94:331–338.

**Magnuson CE,** Goeschl JD, Fares Y (1986) Plant Cell Environ 9:103–109. **Minchin PEH,** Thorpe MR (1984) J Exp Bot 35:538–550. **Minchin PEH,** Ryan KG, Thorpe MR (1984) J Exp Bot 35:1744–1753. **Minchin PEH,** J Exp Bot 13S:11. **Murphy R,** Aikman D (1989) J Exp Bot 40:1079–1088. **Murray DR** (1987) Am J Bot 74:1122–1137.

**Offler CE,** Patrick JW (1984) Aust J Plant Physiol 11:79–99. **Oparka KJ** (1990) Plant Physiol 94:393–396. **Oparka KJ,** Viola R, Wright KM, Prior DAM (1992) In: Farrar JF, Gordon AJ, Pollock CJ (eds) Carbon partitioning within and between organisms. BIOS, Oxford, pp 91–114. **Oparka KJ,** Prior DAM (1992) Plant J 2:741–750. **Opritov VA,** Pyatygin SS (1989) Biochem Physiol Pflanz 184:447–451. **Overall RL,** Gunning BES (1982) Protoplasma 111:151–160.

**Patrick JW** (1990) Physiol Plant 78:298–308. **Patrick JW,** Turvey PM (1981) Ann Bot 47:611–621. **Peel AJ** (1987) Planta 172:209–213. **Pesacreta TC,** Lucas WJ (1986) In: Cronshaw J, Lucas WJ, Giaquinta RT (eds) Phloem transport. Liss, New York, pp 135–144. **Poorter H,** Remkes C, Lambers H (1990) Plant Physiol 94:621–627.

**Sauter JJ,** Kloth S (1986) Planta 168:377–380. **Sibaoka T** (1962) Science 137:226. **Sjolund RD,** Shih CY (1983a) J Ultrastruct Res 82:111–121. **Sjolund RD,** Shih CY (1983b) J Ultrastruct Res 82:189–197. **Smith JAC,** Milburn JA (1980a) Planta 148:28–34. **Smith JAC,** Milburn JA (1980b) Planta 148:42–48. **Sokolova SV,** Krasavina MS, Lushchikov SB (1979) Fiziol Rast 26:721–727.

**Turgeon R** (1989) Annu Rev Plant Physiol Plant Mol Biol 40:119–138. **Turgeon R** (1991) In: Bonnemain JL, Delrot S, Lucas WJ, Dainty J (eds) Recent advances in phloem transport and assimilate compartmentation. Ouest Editions, Nantes, pp 18–22. **Turgeon R,** Beebe DU (1991) Plant Physiol 96:349–354.

**Van Bel AJE** (1987) Plant Physiol Biochem 25:677–686. **Van Bel AJE** (1989) Bot Acta 102:183–185. **Van Bel AJE** (1990) J Exp Bot 41:631–644. **Van Bel AJE** (1992) Acta Bot Neerl 41:121–141. **Van Bel AJE,** Kempers R (1991) Planta 183:69–76. **Van Bel AJE,** Koops AJ (1985) Planta 164:362–369. **Van Bel AJE,** Van Rijen H (1992) J Exp Bot 43S:15. **Van Bel AJE,** Visser AJ (1993) In: Garnier E, Roy J (eds) Carbon-nitrogen interactions: a whole plant perspective. SPB

Academic Publishing, The Hague (in press). **Van Bel AJE,** Van Kesteren WJP, Papenhuijzen C (1988) Planta 176:159–172. **Van Bel AJE,** Gamalei YV, Ammerlaan A, Bik LPM (1992) Planta 186:518–525. **Van der Schoot C,** Van Bel AJE (1989) Protoplasma 149:144–154. **Van der Schoot C,** Van Bel AJE (1990) Planta 182:9–21. **Vreugdenhil D** (1985) Planta 163:238–340. **Vreugdenhil D,** Koot-Gronsveld EAM (1989) Physiol Plant 77:385–388.

**Warmbrodt RD** (1985) Bot Gaz 146:169–180. **Wolswinkel P** (1990) Plant Physiol Biochem 28:399–410. **Wolswinkel P,** Ammerlaan A (1983) J Exp Bot 34:1516–1527. **Wright JP,** Fisher DB (1981) Plant Physiol 67:845–848.

**Zahur MS** (1959) Cornell Univ Ag Exp Mem 358. **Zawadzki T,** Fensom DS (1986) J Exp Bot 37:1353–1363. **Ziegler H** (1990) In: Werner D, Müller P (eds) Fast growing trees and nitrogen fixing trees. Gustav Fischer, Stuttgart, pp 162–170. **Zimmermann MH,** Ziegler H (1975) In: Zimmermann MH, Milburn JA (eds) Encyclopedia of plant physiology. Transport in plants I. Springer, Berlin Heidelberg New York, pp 480–496.

Dr. Aart J.E. van Bel
Transport Physiology
Research Group
Department of Plant Ecology
and Evolutionary Biology
University of Utrecht
Sorbonnelaan 16
3508 TB Utrecht, The Netherlands

# III. Progress in Chlorophyll Fluorescence Research: Major Developments During the Past Years in Retrospect

By Ulrich Schreiber and Wolfgang Bilger

## 1. Introduction

During the past years chlorophyll fluorescence has developed as one of the most frequently used measuring tools in plant science. This somewhat unexpected and remarkable development was triggered by recent progress in instrumentation for measuring fluorescence yield under ambient light conditions and by the increased awareness among plant scientists, as in the general public, of aspects of environmental and stress physiology. Since the discovery of the "Kautsky effect" in 1931, fluorescence had always served as a pioneer tool. However, for more than 50 years it was mainly used by biophysically oriented scientists for basic photosynthesis research. The phenomenology of fluorescence changes in intact cells was considered far too complex to provide more than qualitative information. Because the fluorescence characteristics were known to be strongly affected by preillumination, it appeared necessary to thoroughly dark-adapt a sample before recording dark-light induction curves. Additionally, as the complexity of fluorescence changes increased with increasing illumination time, it was mostly the rapid initial induction kinetics which were analyzed to assess the functioning of the primary reactions. Within less than a decade, a completely different situation has evolved. Along with the availability of new instrumentation and analytical methods for fluorescence analysis under normal daylight conditions, the interest has shifted from the primary reactions to the level of overall electron transport efficiency and photosynthesis regulation, and from induction kinetics to investigations of steady-state reactions. Chlorophyll fluorescence, which used to be a tool preferentially applied in dark laboratories, has made the step into the full sunlight, where in situ photosynthesis takes place with all its intricate and still poorly understood regulatory mechanisms in response to environmental factors. Within a short time this new approach has led to important insights.

It is the aim of this review to summarize the progress in fluorescence research of the last years, focusing on a few selected aspects. The selection of topics and papers is not unbiased, concentrating to some extent on the main research interests of the authors. Besides surveying the major methodological and conceptional developments, new insights into the regulation of photosystem II quantum yield and the protection against photoinhibitory damage are outlined. As chlorophyll fluorescence measurements are applied in many different fields of plant research, a complete survey of the numerous published studies using fluorescence is outside the scope of the present review. Rather it is intended

Progress in Botany, Vol. 54

to present a critical assessment of those studies which have significantly contributed to our present understanding of chlorophyll fluorescence as an indicator of photosynthesis.

The present review complements other review articles on various aspects of chlorophyll fluorescence published during recent years. Fluorescence lifetime measurements were surveyed by Moya et al. (1986), Holzwarth (1987, 1988, 1991) and Geacintov and Breton (1987). Reviews on the role of chlorophyll fluorescence in stress detection were presented by Lichtenthaler et al. (1986), Renger and Schreiber (1986), Schreiber and Bilger (1987), Lichtenthaler and Rinderle (1988), Demmig-Adams (1990) and Karukstis (1991). Van Gorkom (1985, 1986) reviewed work pertaining to the relationship between fluorescence yield and properties of photosystem II. Fluorescence characteristics of algae were surveyed by Fork and Mohanty (1986) as well as Govindjee and Satoh (1986). Current knowledge on the relationship between fluorescence emission and the physiology of photosynthesis is summarized in the reviews by Briantais et al. (1986), Krause and Weis (1984, 1991), Demmig-Adams (1990), Horton and Bowyer (1990), and Walker (1992).

## 2. Progress in Noninvasive Modulation Measuring Techniques

### a) Modulation Fluorometry

Recent progress in chlorophyll fluorescence research is closely linked to the development of modulated chlorophyll fluorometers which allow detection of fluorescence yield in normal daylight (Quick and Horton 1984a, b; Dietz et al. 1985; Ögren and Baker 1985; Schreiber 1986; Schreiber et al. 1986). In principle, it had been long-term practice to select a modulated signal by frequency and phase selective amplification. However, a new generation of such instruments had to be developed to satisfy the requirements of "quenching analysis" by the so-called saturation pulse method (see Sect. 3), which was conceived as the logical consequence of the original "light-doubling" technique of Bradbury and Baker (1981, 1984). Whereas Quick and Horton (1984a, b) and Dietz et al. (1985) were still using mechanical choppers to produce modulated measuring light, Ögren and Baker (1985), as well as Schreiber and co-workers, applied light-emitting diodes, which allow much higher modulation frequencies. Furthermore, a selective pulse-amplitude-modulation system (PAM) was developed (Schreiber 1986; Schreiber et al. 1986) which tolerates nonmodulated background signals more than $10^6$ times stronger than that of the measured fluorescence yield. With such new instruments becoming commercially available, modulation fluorometry gained widespread use in numerous disciplines of plant science. The properties and potential of commercial fluorometers designed for field applications were reviewed in a technical report by Bolhar-Nordenkampf et al. (1989). The performance of a new two-channel fluorometer, which measures fluorescence around 690 and 730 nm, was described by Genty et al. (1990a). Very recently, an exceptionally small but powerful modulation fluorometer based on microprocessor technology was introduced (for examples of application see Figs. 3 and 4), with the help of which Schroeter et al. (1992) studied photosynthetic activity of crustose lichens in Antarctica. Further developments can be foreseen, particularly in the direction of fluorescence image analysis. Following pioneering work of Ellenson and Amundson (1982) and Omasa et al. (1987), Daly et al. (1989) demonstrated application of fluorescence quenching analysis in conjunction with video image analysis. This work confirmed the important previous conclusion by Downton et al. (1988a,b), also made on the basis of quenching analysis, that stomates tend to close in groups ("patchiness phenomenon"), particularly when the stress hormone abscisic acid is applied.

### b) Related Modulation Spectroscopy

The same modulation techniques which were successfully applied in chlorophyll fluorescence studies were also introduced to other closely related fields of photosynthesis research. Weis et al. (1987) and Harbinson and Woodward (1987) used the absorbance change at 820 nm (Havemann and Mathis 1976) to monitor the P700 redox state, providing an ideal indicator for PS I quantum efficiency (see Sect. 5) and thus complementing the PS II indicator chlorophyll fluorescence. Schreiber et al. (1988a) developed an 820-nm emitter-detector unit which can be operated with the same instrument with which modulated fluorescence is measured. An overview on devices and methods is given by Schreiber et al. (1989).

Pulsed light-emitting diodes were also applied for measurement of delayed chlorophyll fluorescence (also called luminescence) which originates from charge recombination at PS II reaction centers (Schreiber and Schliwa 1987; reviews by Malkin 1977, Jursinic 1986, Bilger and Schreiber 1990). Millisecond luminescence strongly depends on membrane energization (i.e., membrane potential and proton gradient), whereas luminescence in the 50 μs time range is favored by a slow-down of the PS II donor-side. Based on these features, parallel luminescence and fluorescence measurements have been useful in the study of energy-dependent and donor-side–dependent fluorescence quenching (Bilger et al. 1988; Schreiber and Neubauer 1989, 1990; Rees et al. 1993).

Another related measuring technique complementing fluorometry, and often used in parallel with fluorescence measurements, is photoacoustic spectroscopy (Bults et al. 1982; Buschmann and Prehn 1983; Canaani and Malkin 1984; Havaux et al. 1986; Malkin 1987). Kolbowski et al. (1990) introduced a computer-controlled pulse modulation system for analysis of photoacoustic signals in the time domain. As the photothermal component of the signal reflects heat dissipation and the photobaric component is indicative of $O_2$-evolution and $O_2$-uptake, parallel measurements of fluorescence and photoacoustics have been useful for the analysis of photochemical and nonphotochemical fluorescence quenching components (Canaani and Malkin 1984; Havaux 1989; Dau and Hansen 1989, 1990; Snel et al. 1990a, b; Reising and Schreiber 1992).

High-frequency pulse-modulated light-emitting diodes were applied by Klughammer et al. (1990) in the development of a so-called LED-array spectrophotometer, with which time resolved absorbance difference spectra in the green spectral region (α-band of cytochrome absorption) can be measured with strongly scattering intact leaves and chloroplasts. Measurements of Cyt b559 absorbance changes may help to clarify the potential role of this component in the mechanism of nonphotochemical quenching (Schreiber and Neubauer 1987; Crofts and Horton 1991) and in protection against photodamage (Thompson and Brudvig 1988).

## 3. The Saturation Pulse Method and Separation of Different Forms of Nonphotochemical Quenching

### a) The Saturation Pulse Method

Under continuous illumination, photosynthetic organisms reach a state of low fluorescence yield which is governed by different quenching mechanisms. On one hand, there is photochemical quenching caused by charge separation at PS II reaction centers. On the other hand, so-called nonphotochemical quenching persists when all PS II centers are closed.

To obtain steady-state information from fluorescence, it is essential to separate photochemical and nonphotochemical quenching components. Krause et al. (1982) succeeded in resolving one component associated with open PS II centers (primary acceptor $Q_A$ oxidized), and another component related to the photo-induced acidification of the intrathylakoid space, by rapid mixing of DCMU to intact chloroplasts and algae. Quick and Horton (1984b) demonstrated with barley protoplasts that equivalent information can be

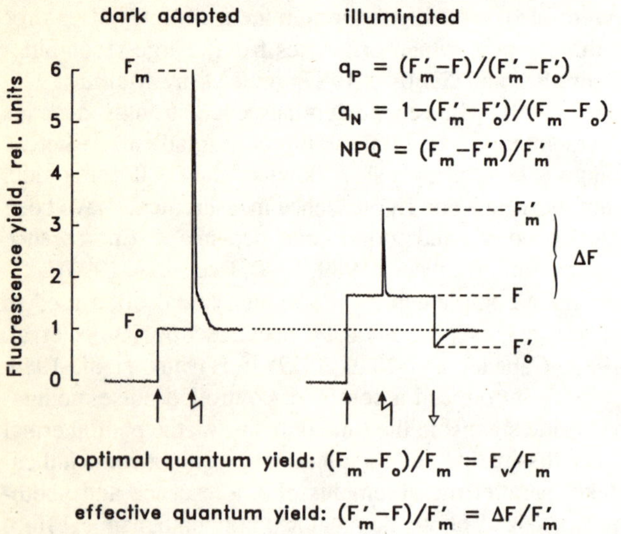


**Fig. 1.** Definition of characteristic fluorescence levels, quenching coefficients, and expressions for PS II quantum yield in dark-adapted and illuminated samples. The fluorescence yield is monitored with a weak modulated measuring beam which by itself does not induce any fluorescence increase. Maximal fluorescence levels, $F_m$ and $F_m'$, are induced by full reduction of $Q_A$ with a saturating light pulse of 0.5–1 s duration. The arrows denote switching-on the measuring light (↑), application of a saturation pulse (ϟ) and turning-off the actinic light (↓). The optimal quantum yield, $F_v/F_m$, is determined using a dark-adapted sample, with all reaction centers open ($q_P = 1$) and at maximal PS II excitation-capture efficiency ($q_N = 0$). The effective quantum yield, which is measured during or after illumination, is determined by the combination of a lowered amount of open centers and a lowered excitation-capture efficiency of open centers (Genty et al. 1989). For determination of the quenching coefficients $q_P$ and $q_N$, knowledge of $F_o'$ is required, i.e., there is the need of sample darkening. This is not the case when the extent of nonphotochemical quenching is expressed as NPQ (Bilger and Björkman 1990). Knowledge of $F_o'$ is also not required for $\Delta F/F_m'$-determination

obtained when a strong pulse of light is applied, which transiently closes all PS II reaction centers. Under the assumption that photochemical quenching can be completely suppressed by a saturating light pulse and that nonphotochemical quenching does not change during such a "saturation pulse", the coefficients for photochemical and nonphotochemical quenching, $q_P$ and $q_N$ (originally called $q_Q$ and $q_E$) were defined (Schreiber et al. 1986). A standard nomenclature for characteristic fluorescence levels and quenching coefficients was proposed at a recent fluorescence workshop (van Kooten and Snel 1990). Following this nomenclature, in Fig. 1 the characteristic fluorescence levels, quenching coefficients and expressions for PS II quantum yield are depicted.

### b) Separation of Different Forms of Nonphotochemical Quenching

Nonphotochemical quenching consists of several components, with the three major ones being so-called energy-dependent quenching, related to the build-up of a transthylakoidal pH-gradient, quenching due to a state transition, a process involving LHC-II phosphorylation (Bennett 1980, 1983; Black et al. 1986; Williams and Allen 1987; Allen 1992), and photoinhibitory quenching (Powles and Björkman 1982; Krause et al. 1985, see also Sect. 4b). It has been proposed to term these three components $q_E$, $q_T$, and $q_I$, respectively (van Kooten and Snel 1990). In isolated chloroplasts $q_E$ can be readily defined as that part of $q_N$ which is sensitive to uncouplers (Krause et al. 1982). Energy-dependent quenching has been shown to be closely correlated with the intrathylakoid proton concentration (Briantais et al. 1979; Laasch 1987; Krause and Weis 1991).

To a certain extent, it is possible to distinguish the different forms of nonphotochemical quenching by their relaxation kinetics upon darkening, a method which can be also used with intact leaves. Schreiber and Bilger (1987) distinguished three relaxation phases in *Arbutus unedo* leaves; that part of $q_N$ which did not relax within 40 min was considered $q_I$, in agreement with previous experiments of Krause et al. (1985) and Samuelsson et al. (1985). Horton and Hague (1988) used NaF, an inhibitor of the thylakoid protein phosphatase (Bennett 1980) to selectively suppress $q_T$-relaxation in isolated barley protoplasts. $q_T$ was found to saturate in low light and to decrease again at high light intensity. Demmig and Winter (1988a, b) studied relaxation of $q_N$ in intact leaves and suggested that a slow component, which appeared only at high light intensities, was correlated with the formation of zeaxanthin. It was inferred that this zeaxanthin-related component reflects photoinhibition (Demmig et al. 1987). However, with the introduction of DTT as an inhibitor of violaxanthin de-epoxidation in leaves (Bilger et al. 1989), it became clear that a pronounced DTT-sensitive component is formed under nonphotoinhibitory conditions, relaxing within less than 5 min (Demmig-Adams et al. 1990). It now appears that DTT-sensitive $q_N$ constitutes a subfraction of $q_E$ which requires the presence of zeaxanthin (see also Sect. 4b). In leaves, DTT-sensitive $q_N$ may comprise more than 90% of the total rapidly relaxing $q_N$ (Demmig-Adams et al. 1990; Bilger and Björkman 1990, 1991).

A detailed analysis of the $q_N$-relaxation kinetics in intact leaves was presented by Quick and Stitt (1989), with a rapid phase ($t_{1/2} \sim 1$ min) identified as $q_E$, a middle phase ($t_{1/2} \sim 5$ min) considered $q_T$, and a slow phase ($t_{1/2}$ of hours) attributed to $q_T$. However,

However, more recent studies have shown that the relaxation of $q_E$ can be rather slow in intact leaves (Bilger and Björkman 1990, 1991; Walters and Horton 1991) and algae cells (Lee et al. 1990). After light activation of the reversible ATP-ase, ATP-hydrolysis can maintain a transthylakoidal proton gradient over extended dark periods (Schreiber 1980; Morita et al. 1981). The extent of such slowly relaxing $q_E$ will be variable, depending amongst other factors on the rate of ATP consumption in the Calvin cycle and of ATP transport from mitochondria to chloroplasts. Besides ATP-hydrolysis, also chlororespiration (Bennoun 1982) may contribute to the maintenance of $q_E$ over extended dark periods. Indications for chlororespiration (electron flow from NAD(P)H to oxygen via the plastoquinone pool) were found in green algae (Peltier et al. 1987), in a diatom (Caron et al. 1987), a xanthophycean alga (Büchel and Wilhelm 1990), and a prasinophycean alga (Wilhelm and Duval 1990). In these studies, fluorescence quenching analysis has been useful to identify plastoquinone reduction (via $q_P$) and membrane energization (via $q_E$) in the dark. One may conclude that relaxation kinetics alone are not sufficient to resolve the different components of $q_N$. In vivo reactions may be operative which can maintain $q_E$ in the dark.

In the original definition of quenching coefficients, only changes of variable fluorescence yield were considered i.e., a constant dark-level fluorescence, $F_o$, was assumed. Bilger and Schreiber (1986) showed that also $F_o$ can be lowered by nonphotochemical quenching. This feature complicates $q_P$ and $q_N$ determination under ambient light conditions and favors analytical expressions, the determination of which do not require knowledge of $F_o$. Genty et al. (1989) introduced the expression $\Delta F/F_m' = (F_m' - F)/F_m'$ (see Fig. 1) as a measure of the effective PS II quantum yield (see Sect. 5). Bilger and Björkman (1990) first used the expression $NPQ = (F_m - F_m')/F_m'$ to characterize the extent of nonphotochemical quenching.

The definitions of $q_P$ and $q_N$ have been questioned by Havaux et al. (1990) by theoretical considerations on the basis of a detailed model of energy migration in the photosynthetic membrane (Strasser 1981; Sironval et al. 1984). However, as pointed out by van Kooten and Snel (1990), our knowledge of the mechanisms of fluorescence quenching is still incomplete and no unequivocal model is at hand yet for rigorous definitions. Meanwhile the saturation pulse method and the empirical terms $q_P$ and $q_N$ have proven quite useful in practical applications in different fields of plant physiological research, the most important of which are:

1. Plant stress investigations (see Sect. 4).
2. Determination of photosynthesis rates in leaves (see Sect. 5).
3. Studies on the mechanism of regulated radiationless energy dissipation (see Sect. 6).
4. Studies on membrane energization and nonassimilatory fluxes (see Sect. 7).

## 4. Plant Stress Investigations

a) Fluorescence as a Nondestructive Tool in Stress Research

Chlorophyll fluorescence is a very sensitive indicator of all parameters which affect the photosynthetic process more or less directly. Hence, the indicator function covers direct effects at the level of the PS II antenna and reaction centers (see Sect. 4b) as well as indirect effects at the "sink-level" of photosynthate metabolism (see e.g., Stitt et al. 1988; Stitt and Grosse 1988; Stitt and Schreiber 1988). Stress may not necessarily lead to a sustained decrease in photosynthetic activity, which would be readily detectable by quenching analysis (see Sect. 3 and 5), as there are regulatory mechanisms at various levels tending to compensate for stress-induced limitations. In this case, however, it is likely that the stress effect and the induced regulatory processes are somehow reflected in the complex course of dark-light fluorescence induction kinetics. In this sense, fluorescence induction patterns and derived indices have long been used as an empirical diagnostic tool in stress physiology (for reviews see Lichtenthaler et al. 1986; Renger and Schreiber 1986; Lichtenthaler and Rinderle 1988; Karukstis 1991). With the advent of quenching analysis, the use of fluorescence in stress investigations has been placed on a more quantitative basis, allowing deeper insights into the molecular mechanisms involved.

In Lichtenthaler (1988) and Snel and van Kooten (1990), some applications of chlorophyll fluorescence and of other nondestructive tools in photosynthesis and plant stress research are described. A general overview was given by Schreiber and Bilger (1987). Because of the complexity of the subject, a detailed survey on recent fluorescence stress studies would be outside the scope of the present review. Some emphasis will be put on photoinhibitory stress, a topic which has found most attention in recent years (see Sect. 4b).

In the other fields of plant stress research the following publications may serve as examples for practical applications of the chlorophyll fluorescence method:

1. Low temperature and freezing stress: Krause and Somersalo (1989), Larcher and Neuner (1989), Terashima et al. (1991), Brüggemann et al. (1992), Rütten and Santarius (1992).
2. Heat stress: Havaux et al. (1991), Bilger et al. (1987).
3. Drought and salinity stress: Cornic and Briantais (1991), Cornic and Ghashghaie (1991), Winter and Gademann (1991), Larcher et al. (1990), Havaux and Lannoye (1985), Stuhlfauth et al. (1988), Stuhlfauth et al. (1990).
4. Environmental pollution: van Kooten et al. (1988), Schmidt et al. (1987, 1990), Veljovic-Jovanovic et al. (1993), Atal et al. (1991).

Primary damage caused by the above-mentioned stress factors is mostly at the level of the enzymic dark reactions (for examples with heat and water stress, see Bilger et al. 1987 and Schreiber and Bilger 1987). This kind of damage is expressed by a changed induction pattern upon continuous illumination and a lowered effective PS II quantum yield, $\Delta F/F_m'$ (see Fig. 1 and Sect. 5), *during* illumination, whereas the PS II quantum yield of dark-adapted samples, $F_v/F_m$, may be unaffected. Generally, due to limitations of the enzymic dark reactions, stressed plants become prone to photoinhibition, the main target of which is PS II, resulting in a lowering of PS II quantum yield, $F_v/F_m$ (see following subsection).

### b) Photoinhibition and Photoinhibitory Quenching

Chlorophyll fluorescence has been particularly useful for the quantification and characterization of the lowering of photosynthetic activity by excess light, which is often referred to as photoinhibition. Actually, this term is somewhat ambiguous, as it is used in the sense of photoinactivation or photodamage (see e.g., Kyle 1987; Ohad et al. 1988; Styring et al. 1990) as well as of a regulatory and even protective process (Barenyi and Krause 1985; Demmig and Björkman 1987; Horton 1989). The amount of work published during recent years on both of these aspects is overwhelming. Most of this work relies on fluorescence measurements. A general overview on various aspects of photoinhibition is given in Kyle et al. (1987). The topic has been extensively reviewed by Powles (1984), Björkman (1987a, b), Kyle (1987), Öquist et al. (1987), Baker and Horton (1987), Krause and Cornic (1987), Krause (1988), Demmig-Adams (1990), Baker (1991), Krause and Weis (1991).

Important insights on the relationship between photoinhibition and chlorophyll fluorescence yield had already been obtained before the advent of pulse modulation fluorometry on the basis of measurements at 77 K (Powles and Björkman 1982; Björkman and Powles 1984; Ögren and Öquist 1984; Barenyi and Krause 1985; Krause et al. 1985; Samuelsson et al. 1985; Greer et al. 1986; Björkman and Demmig 1987; Demmig and Björkman 1987). This extensive work had already demonstrated that the $F_v/F_m$ ratio is an excellent indicator for the PS II quantum yield, which is lowered by photoinhibition, in agreement with a simple fluorescence model of Kitajima and Butler (1975).

Measurements of $F_v/F_m$ have become greatly facilitated by the nondestructive saturation pulse analysis (see Sect. 3). Demmig et al. (1987) and Adams et al. (1990b) demonstrated the equivalence of the 77 K method and the ambient-temperature saturation pulse method. A great advantage of the latter method is that the same sample can be studied under a variety of changing conditions over extended time periods. Only after introduction of the saturation pulse method did it become possible to assess nonphotochemical quenching, an expression of nonradiative energy dissipation, *during* illumination with excessive light. This opened the way for extensive studies of adaptive and protective mechanisms.

Primarily based on measurements with the saturation pulse method, in recent years a large amount of photoinhibition data has been gathered relating to the following major areas:

1. Different phases in the recovery from photoinhibition (see e.g. Greer et al. 1986; Greer and Laing 1988; see also Sect. 3b).
2. Protection against photoinhibition by membrane energization and related nonradiative energy dissipation (see e.g., Krause and Behrend 1986, Krause and Laasch 1987, Oxborough and Horton 1988, Winter and Königer 1989, Bilger and Björkman 1990; see also Sect. 7).

3. Correlation between photoinhibition and zeaxanthin-content (see e.g., Demmig et al. 1987, 1988).
4. Protective and detrimental role of oxygen (see e.g., Barenyi and Krause 1985; Krause et al. 1985; Krause and Cornic 1987; Heber et al. 1989, 1990; Richter et al. 1990; Schöner and Krause 1990; van Wijk and Krause 1991, Wu et al. 1991).
5. Photoinhibition caused by a combination of light and other environmental stress factors (see e.g., Ludlow 1987; Öquist et al. 1987; Björkman et al. 1988; Greer et al. 1986; Smillie et al. 1988; Demmig-Adams et al. 1989; Krause and Somersalo 1989; Somersalo and Krause 1989, 1990).

From the viewpoint of chlorophyll fluorescence as an indicator of photosynthesis regulation, the protection conferred by nonradiative dissipation, as reflected by nonphotochemical quenching, is of particular relevance. That part of nonphotochemical quenching, which is induced during high light treatment and which persists for longer than 5–40 min (depending on authors) after darkening, has been termed $q_I$ (see also Sect. 3b). It has been proposed that at least part of $q_I$ reflects a reversible mechanism to lower the quantum yield of PS II, operating at the antenna pigment level (Demmig and Björkman 1987). On the other hand, $q_I$ or part of it may also be associated with direct damage to PS II, resulting in the formation of energy-dissipating centers (Cleland et al. 1986; Krause et al. 1990). The first mechanism provides "photo-protection", since it leads to safe dissipation of excess light energy, whereas in the latter case the term "photodamage" appears more appropriate, since full recovery requires de novo protein synthesis (Guenther and Melis 1990a, b). As both forms of "photoinhibition" may coexist and vary in their relative extent, depending on plant species and experimental conditions, some confusion has arisen about the notion of "photoprotective $q_I$". Initially, zeaxanthin-dependent $q_N$ had been related to $q_I$ (Demmig et al. 1987, 1988). In later studies, however, it became clear that zeaxanthin-dependent $q_N$ requires the presence of $\Delta pH$ (Demmig-Adams et al. 1989; Bilger et al. 1989; Rees et al. 1989; Gilmore and Yamamoto 1991, 1992; Noctor et al. 1991; Neubauer and Yamamoto 1992). This apparent contradiction could be resolved if it is assumed that after high light stress the dark-relaxation of the $\Delta pH$ and of zeaxanthin related $q_E$ is vastly slowed down (see Sect. 3b). In this sense, the reported "photoprotective $q_I$" in fact could have been slowly reversible $q_E$.

There is general agreement that PS II is a primary target of photoinhibition, which not only lowers PS II quantum yield but also the capacity for PS II-driven electron flow (Powles 1984; Barenyi and Krause 1985; Krause and Cornic 1987). Krause et al. (1990) showed that the decline in PS II capacity was linearly related to the decrease in $F_v/F_m$ and that the rate of $Q_A$-reduction was not affected by photoinhibition. On the basis of these data, it was proposed that photoinhibition is due to transformation of active reaction centers into photochemically inactive centers, which dissipate excitation energy into heat, thus causing nonphotochemical fluorescence quenching. This interpretation, which is analogous to the explanation of $q_N$ (and $q_E$ in particular) by Weis and co-workers (Weis and Berry 1987; Weis and Lechtenberg 1989; Krieger and Weis 1990, Krieger et al. 1992) is in line with earlier suggestions by Cleland and Critchley (1985), Cleland et al. (1986), Critchley (1988), Melis et al. (1988) and Guenther and Melis (1990a, b). Recently, Giersch and Krause (1992) demonstrated that a quasi-linear relationship between $F_v/F_m$ and PS II activity is predicted by a simple model which assumes inactive quenching centers and cooperativity in terms of energy transfer between active and inhibited PS II centers.

## 5. Photosynthetic Rate Determination from Fluorescence Measurements

Following progress in modulation fluorometry (see Sect. 2) and the introduction of the saturation pulse method (see Sect. 3) considerable efforts were expended on the investigation of the relationship between PS II activity, as reflected by fluorescence quenching, and the rate of photosynthetic electron flow. The original studies (Bradbury and Baker 1981, 1984; Quick and Horton 1984a, b; Dietz et al. 1985; Schreiber et al. 1986) had concentrated on the separation of photochemical and nonphotochemical quenching components, to gain selective information both on the redox state of the intersystem electron transport chain and on the energization state of the thylakoid membrane. Under certain conditions, a satisfactory correlation between assimilation rate and the coefficient of photochemical quenching ($q_P$ or $q_Q$) was found (Dietz et al. 1985; Schreiber and Bilger 1987). However, substantial deviations between electron transport rate and $q_P$ were observed at high membrane energization due to a large ΔpH (Krause and Behrend 1986; Krause and Laasch 1987; Weis et al. 1987) or to a weakening of the PS II donor-side (Schreiber and Neubauer 1987; Schreiber and Rienits 1987). Furthermore, it had been established (Björkman and Demmig 1987; Demmig and Björkman 1987; Björkman 1987a, b) that, consistent with fluorescence theory (Butler 1978), in a large variety of leaves the quantum yield of PS II, as determined from the yield of $O_2$-evolution, correlates linearly with the expression $F_v/F_m$ measured at 77K, when different levels of nonphotochemical quenching were created under varying light stress conditions (see Sect. 4b). Hence, the quantum efficiency of PS II is not only determined by the proportion of open reaction centers, but also by the efficiency of excitation capture by open reaction centers.

### a) Expressions for Rate Determination on the Basis of Quenching Analysis

Weis et al. (1987) and Weis and Berry (1987) were the first to present a detailed analysis of simultaneously measured $CO_2$-fixation data and fluorescence quenching parameters and to quantify the observed relationship. This work led to the important conclusion that ΔpH formation and membrane energization result in down-regulation of PS II quantum yield, providing a novel concept of "photosynthetic control" (see reviews by Horton et al. 1989; Foyer et al. 1990; Walker 1992). On the basis of a model which involves "energy-quenched" PS II reaction centers, Weis and co-workers derived an empirical equation, containing only fluorescence parameters, which gave an excellent linear correlation between the electron transport rate calculated from fluorescence and that determined from $CO_2$ exchange:

$$J = I \copyright q_P (m \cdot b \cdot q_N), \quad (1)$$

where J is the rate of electron transport, I is the incident photon flux density, and m and b are empirically derived, plant-specific constants. Sharkey et al. (1988) and Holmes et al. (1990) confirmed the findings of Weis and Berry (1987) under conditions of varying $CO_2$- and $O_2$-partial pressures.

In the original work of Weis and co-workers a linear relationship between $q_N$ (or $q_E$) and the quantum yield of open PS II centers ($\phi_P = J/I{\cdot}q_P$) was postulated, with $\phi_P$ extrapolating to a value above zero with $q_N$ approaching 1. This, however, was not confirmed by Horton and Hague (1988), Peterson et al. (1988) and Weis and Lechtenberg (1989), who found a nonlinear relationship and $\phi_P$ extrapolating to zero for $q_N = 1$. Also Snel et al. (1990b) concluded, from simultaneous measurements of photoacoustically detected $O_2$-evolution and fluorescence quenching, that the quantum efficiency of electron flow in energized open PS II centers is practically zero. Weis and Lechtenberg (1989) suggested that the nonlinearity could be caused by energy transfer between PS II units. Markgraf and Berry (1990) showed that the intensity of the saturation pulses used for $q_N$-determination affects the extent of nonlinearity in the $\phi_P$ vs. $q_N$ relationship (see also Schäfer and Björkman 1989). When extrapolating to infinite intensity, the relationship became linear. These aspects are relevant for a discussion of the molecular mechanisms involved in energy dependent fluorescence quenching and down-regulation of PS II quantum yield (see Sect. 6).

Genty et al. (1989) measured $CO_2$-assimilation rate and fluorescence yield in a variety of plants ($C_3$ and $C_4$) at different light states and $CO_2$-concentrations and found excellent linear correlation (extrapolating to the origin) between the quantum yield of $CO_2$-fixation and the simple fluorescence parameter $\Delta F/F_m'$ (see Fig. 1). This parameter incorporates the extent of open PS II centers and the efficiency of excitation capture of open centers:

$$\phi_{PSII} = q_P \cdot F_v'/F_m' = (F_m'/F) \vee F_m' = \Delta F/F_m'. \quad (2)$$

Notably, for this determination of $\phi_{PSII}$ it is not necessary to know $F_o$ which is of considerable practical importance, since determination of the true $F_o$ often is problematic, particularly in continuous light (Bilger and Schreiber 1986; see also Sect. 3). The findings of Genty et al. (1989) were confirmed and extended by Di Marco et al. (1990), Keiller and Walker (1990), Seaton and Walker (1990), Krall and Edwards (1991) and Krall et al. (1991), who compared the quantum yield of $O_2$-evolution or $CO_2$-fixation and $\Delta F/F_m'$ for leaves of numerous plant species and under a variety of physiological conditions. The relationship held even under circumstances in which photosynthetic electron flow was undergoing marked oscillations (Seaton and Walker 1990) which reflect delicate regulatory mechanisms (Sivak et al. 1985; Laisk et al. 1991).

In Fig. 2 the relationship between $\Delta F/F_m'$ and the quantum yield of $O_2$-evolution is depicted, as measured by Seaton and Walker (1990). These authors extended the analysis to $\phi_{PSII}$ values exceeding those included in Genty et al. (1989) and found the relationship between measured quantum yield and $\Delta F/F_m'$ nonlinear, with curvature increasing towards higher values of $\phi_{PSII}$. The nonlinearity is particularly pronounced in $C_3$-plants in presence of elevated $O_2$-concentration, whereas $C_4$-plants show a close to linear relationship independent of $O_2$-concentration (Krall and Edwards 1990, 1991; Krall et al. 1991). These findings suggest that photorespiratory electron flow is likely to cause at least part

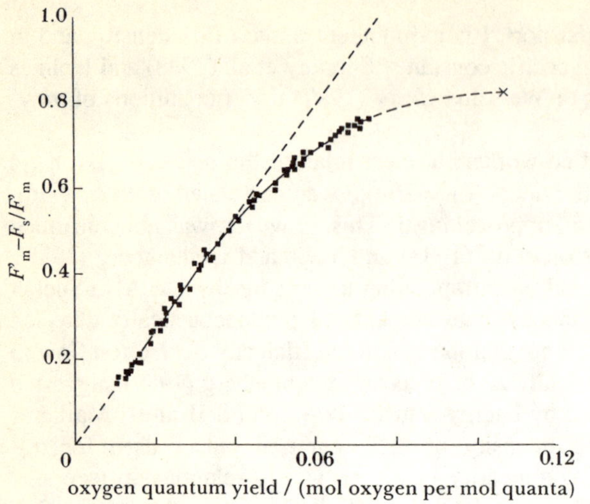


**Fig. 2.** Relationship between the effective quantum yield of PS II, $\Delta F/F_m' = F_m'-F_s/F_m'$, as determined by the saturation pulse method, and the quantum yield of $O_2$-evolution measured with a leaf-disk oxygen electrode (Seaton and Walker 1990). The data points were obtained by variation of actinic photon flux density between 30 and 1040 $\mu E/m^2s$ using leaves of eight different species. $F_s$, steady state fluorescence yield. Point (x) represents estimated maximal values, in practice, of oxygen quantum yield and $F_v/F_m$

of the nonlinearity. Another part could be due to $O_2$- and $H_2O_2$-reduction or cyclic electron flow around PS II (Schreiber and Neubauer 1990; see also Sect. 7).

Irrespective or whether the relationship between $\Delta F/F_m'$ and $\phi_{PSII}$ is linear or curvilinear, and whatever is the cause of non-linearity at high $\phi_{PSII}$ values, it appears that a wide variety of plant species follows the same relationship (Seaton and Walker 1990). Hence, it is now possible to derive a value for overall photosynthetic quantum yield on the basis of fluorescence data alone, and to calculate the relative rate of electron transport at that particular PFD.

Figure 3 shows plots of $\Delta F/F_m'$ vs. PFD for leaves of *Helianthus tuberosus*, grown in full sun light, and of *Fagus sylvatica*, with a varying degree of light exposure. Corresponding $\Delta F/F_m'$-determinations were made under natural field conditions in ambient day light (left) and under controlled laboratory conditions (right). It is apparent that the different light acclimation states are clearly expressed. In Fig. 4 the relative electron transport rates calculated on the basis of the data in Fig. 3 (controlled conditions) are presented. From the depicted curves, electron transport capacities can be evaluated. A comprehensive measure for the degree of light stress is given by the expression of "limitation" (see Schreiber and Bilger 1987), which is equivalent to "excessive PFD" (Demmig and Winter 1988a). Notably, when a certain standard optimal quantum yield of an unstressed plant is assumed (see e.g., Björkman and Demmig 1987; Demmig and Björkman 1987), a single $\Delta F/F_m'$ measurement at a given light intensity is sufficient to determine relative "limitation" or "excessive PFD". Considering that such measurements can be carried out within seconds using portable fluorometers in the field, a great impact of this method on future ecophysiological research can be foreseen.

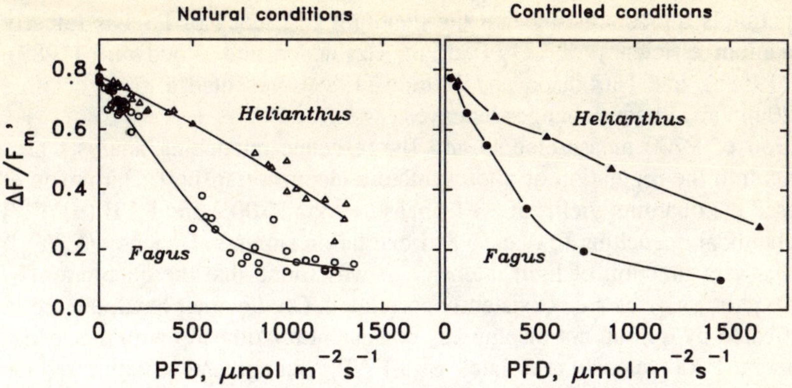


**Fig. 3.** Light-intensity dependence of effective PS II quantum yield, $\Delta F/F_m'$, as determined with a portable chlorophyll fluorometer under natural field conditions (*left*) and controlled laboratory conditions (*right*). With the field measurements, a number of leaves, exposed to different light intensity habitats were assessed, whereas the laboratory experiments were carried out with one sun leaf of *Helianthus* and one shade leaf of *Fagus*, respectively. Measurements were done with a PAM-2000 Portable Fluorometer (Walz). PFD, photon flux density of photosynthetically active radiation

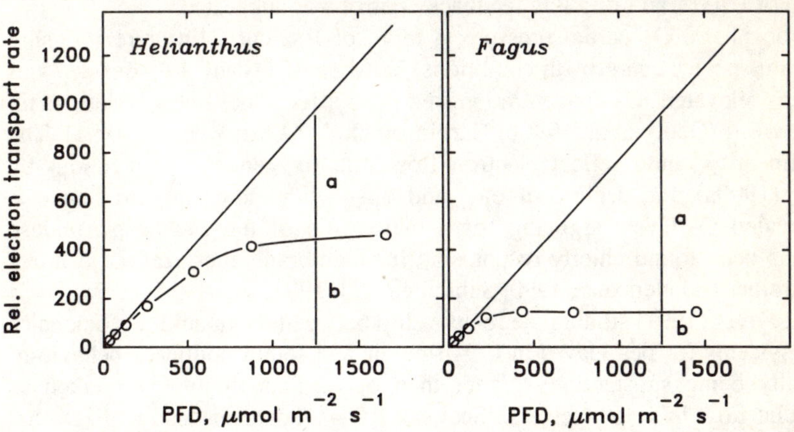


**Fig. 4.** Light saturation curves of relative electron transport rates calculated from fluorescence parameters on the basis of the data in Fig. 3 (controlled conditions). Relative electron transport rates are the product of effective PS II quantum yield, $\Delta F/F_m'$, and incident photon flux density, PFD. The potential rates at maximal quantum efficiency are given by the *initial slope lines*. The expression a/(a+b) is a measure for rate limitation or "excessive light intensity". At 1250 µmol quanta $m^{-2}s^{-1}$ values of 0.54 and 0.84 were obtained for a sun leaf of *Helianthus tuberosus* and a shade leaf of *Fagus sylvatica*, respectively

b) Comparison of Effective Quantum Yields of PS I and PS II

Along with the realization that fluorescence measurements provide a means of quantifying PS II quantum efficiency in vivo, compatible methods for noninvasive determination of PS I quantum efficiency were developed (see Sect. 2). Weis et al. (1987) demonstrated that the absorbance increase at 820 nm caused by $P700^+$ can be monitored in intact leaves and that $P700^+$ parallels $q_E$, concluding that the quantum efficiency of PS I is linearly related to the quantum efficiency of $CO_2$ fixation. Harbinson and Woodward (1987), Schreiber et al. (1988a), and Harbinson and Hedley (1989) presented a more detailed analysis of the 820 nm absorbance changes in leaves.

The combination of P700 measurements and fluorescence quenching analysis provided new insights into the regulation of photosynthetic electron transport. Harbinson et al. (1989) compared the quantum yields of PS I (measured via $P700^+$) and PS II ($\Delta F/F_m'$) as well as photochemical quenching ($q_P$) and PS II excitation capture efficiency ($F_v'/F_m'$) in intact pea leaves as a function of light intensity. It was found that the quantum efficiencies of the two photosystems are consistently correlated. On the other hand, the redox state of PS II (reflected by $q_P$) did not display any consistent relationship with the oxidation of P700, whereas $P700^+$ closely correlated with $F_v'/F_m'$, in agreement with previous findings of Weis et al. (1987). Weis and Lechtenberg (1989), Harbinson et al. (1990 a,b), Genty et al. (1990b), and Peterson (1991) compared simultaneously measured PS I and PS II quantum yields with the quantum efficiency of $CO_2$-fixation in various types of leaves under photorespiratory and nonphotorespiratory conditions. Generally, the cited work revealed a strict control of $CO_2$-fixation by electron donation from PS II when illumination was at low $CO_2$ and 2% $O_2$. In view of an almost parallel decrease in $\phi_{PSII}$ and accumulation of $P700^+$, an intriguing feedback control was suggested.

At normal atmospheric $O_2$ partial pressure, a more or less curvi-linear relationship (depending on plant species and growth conditions) between $\phi CO_2$ and $\phi_{PSI}$ or $\phi_{PSII}$ was observed, whereas elevated $O_2$-concentration did not affect the linear relationship between $\phi_{PSI}$ and $\phi_{PSII}$ (Genty et al. 1990b; Harbinson et al. 1990a; Peterson 1991). The deviation from linearity could reflect electron flow into oxygen. As pointed out by Harbinson et al. (1990a), the decline of $\phi_{PSI}$ and $\phi_{PSII}$ with increasing irradiance is enhanced by elevated $O_2$-levels argueing for a major role of the Mehler-peroxidase reaction which has been found chiefly responsible for membrane energization in intact chloroplasts (Schreiber and Neubauer 1990; Schreiber et al. 1991; see also Sect. 7).

Harbinson and Foyer (1991) studied the relationship between the quantum efficiencies of the two photosystems for pea leaves in $CO_2$-free air and found nonlinear behaviour, with $\phi_{PSI}$ generally being substantially higher than $\phi_{PSII}$. This finding was taken as "evidence for cyclic flow in vivo" (see also Sect. 7). It was, however, also realized that under conditions of PS I acceptor side limitation, there may be a population of reduced P700 which cannot contribute to quantum utilization (Harbinson and Foyer 1991). Recent data by Klughammer (1992) suggest a dissipative cycle at PS I which depends on the presence of $O_2$ and does not necessarily involve the cyt b/f-complex.

## 6. Mechanisms of Regulated Radiationless Energy Dissipation

Since Weis et al. (1987) suggested that nonphotochemical quenching reflects down-regulation of PS II quantum efficiency, numerous laboratories have carried out investigations to elucidate the mechanisms involved. The most controversial question, which has remained unsettled until today, is whether energy dissipation is mediated within the pigment bed or at the reaction centers. Weis and co-workers (Weis and Berry

1987; Weis and Lechtenberg 1989; Weis et al. 1990; Krieger and Weis 1990, Krieger et al. 1992) have presented strong arguments in favor of energy dissipation at PS II centers. Schreiber and Neubauer (1987, 1989, 1990) demonstrated strong control of nonphotochemical quenching by the PS II donor side, and suggested that PS II quantum efficiency for linear flow is lowered by a rapid, dissipative PS II cycle or charge recombination. On the other hand, the work of Demmig and co-workers (Demmig et al. 1987, 1988; Demmig and Winter 1988a, b; Demmig-Adams et al. 1989, 1990; Demmig-Adams 1990; Adams et al. 1990a, b) strongly argued for a decisive role of zeaxanthin at the antenna pigment level. Also the work of Björkman (1987a, b) and Genty and co-workers (Genty et al. 1989, 1990b) led to the conclusion that the exciton capture efficiency of PS II is controlled by nonradiative processes in the pigment matrices of PS II.

Arguments in favor of antenna quenching are generally based on the simple fluorescence model of Kitajima and Butler (1975), which predicts a linear relationship between $F_v/F_m$ and $\phi_{PSII}$, and a certain extent of $F_o$ quenching. However, taking into account energy transfer between PS II units, a quasi-linear relationship can also be obtained on the basis of a model involving inactive, quenching centers (Weis and Berry 1987; Giersch and Krause 1992). Contrary to the original model of Kitajima and Butler (1975), the latter authors make no distinction between the rate constants for nonradiative dissipation at the reaction centers ($k_d$) and in the antenna ($k_D$), which may be justified in view of the work of Schatz et al. (1988) (see also Holzwarth 1991). This work has led to the suggestion, based on fluorescence life time measurements, that PS II reaction centers operate as "shallow traps", inferring excitation equilibration between open reaction centers and antenna pigments. Excitation cycling between reaction centers and antenna was also considered by Butler (1984) in an extension of his original model (Butler and Kitajima 1975; Kitajima and Butler 1975). Rees et al. (1990) applied the extended Butler-model for an analysis of the relationship between $q_N$ and $F_o$ quenching ($q_o$) and came to the conclusion that the observed extent of $q_o$ is more readily explained by antenna quenching than by reaction center quenching.

In an attempt to provide a unifying model incorporating both reaction center and antenna quenching, Schreiber and Neubauer (1990) proposed that, under energized conditions, PS II charge recombination leads to the formation of triplet excited states which are trapped by carotenoids, in particular zeaxanthin, with triplet excited carotenoids causing singlet excitation quenching by an annihilation process. Details of this model have been questioned by very recent work of Krieger et al. (1992), who discovered that energy-dependent quenching correlates with the release of $Ca^{++}$ from the PS II donor side and a redox potential increase of the primary stable acceptor $Q_A$ when the intra-thylakoidal pH drops below pH 5. Whereas these authors in principle confirm the notion of "donor-side dependent" nonphotochemical quenching, they put forward arguments in favor of a direct recombination between $Q_A^-$ amd $P_{680}^+$ which results exclusively in energy dissipation into heat. Notably, Krieger et al. (1992) exclude the possibility of any significant contribution of antenna quenching which, however, is difficult to reconcile with a large body of evidence in favor of a role of zeaxanthin.

Strong support for a quenching role of zeaxanthin comes from experiments with DTT as an inhibitor of violaxanthin de-epoxidation (Bilger et al. 1989; Demmig-Adams et al. 1990; Demmig-Adams and Adams 1990; Winter and Königer 1990; Bilger and Björkman 1990, 1991; Gilmore and Yamamoto 1991, 1992; Noctor et al. 1991; Neubauer and

Yamamoto 1992). In DTT-treated intact leaves zeaxanthin does not accumulate upon illumination and energy-dependent quenching is suppressed. At the same time DTT does not affect the light-response curve of $CO_2$-fixation (Bilger et al. 1989), but at excessive levels of irradiation the presence of DTT increases the susceptibility to irreversible photoinhibitory damage (Winter and Königer 1989; Bilger and Björkman 1990; see also Sect. 4b). Gilmore and Yamamoto (1992) showed that energy-dependent quenching caused by ATP-hydrolysis was not suppressed by DTT when zeaxanthin was formed during a preillumination period. This finding precludes the possible argument that DTT acts by donating to $P680^+$ and, thus, by preventing the proposed recombination quenching (Schreiber and Neubauer 1990; Krieger and Weis 1992).

In all models of regulated radiationless energy dissipation, the transthylakoidal $\Delta pH$ plays a decisive role. Acidification of the thylakoid lumen stimulates the violaxanthin deepoxidase and slows down electron donation from the water splitting enzyme system. Furthermore, it was shown that zeaxanthin requires the presence of $\Delta pH$ for its postulated quenching action (Demmig-Adams et al. 1989). On the other hand, there is evidence from various laboratories that energy-dependent quenching can occur in the absence of zeaxanthin (Bilger et al. 1989; Rees et al. 1989; Adams et al. 1990a; Demmig-Adams et al. 1990; Schreiber and Neubauer 1990). Rees et al. (1989), Gilmore and Yamamoto (1991), and Noctor et al. (1991) showed that chloroplasts from leaves with pre-formed zeaxanthin showed energy-dependent quenching at lower $\Delta pH$-values than chloroplasts prepared from dark-adapted leaves. This finding suggested that zeaxanthin may "amplify" a quenching mechanism which already exists in its absence, although at a lower level. Noctor et al. (1991) also demonstrated that, with or without zeaxanthin present, $q_E$-formation was sensitive to antimycin A, suggesting that in both cases the same molecular mechanism is involved. Early work by Oxborough and Horton (1987a, b; 1988) had revealed that the relationship between $q_E$ and $\Delta pH$ is flexible, i.e., that a reaction which can be inhibited by antimycin A modulates the effect of low internal pH. Further work dealing with this interesting aspect was presented by Laasch and Weis (1988, 1989), Krause et al. (1988), and Rees and Horton (1990).

Experiments with isolated chloroplasts have led to the conclusion that there are two basic mechanisms by which the quantum yield of open PS II centers can be lowered in response to a low internal pH (Rees and Horton 1990). Besides the $q_E$-related mechanism, discussed above, there are also indications for cyclic flow around PS II (Schreiber and Neubauer 1987; Schreiber and Rienits 1987; Falkowski et al. 1988; Oxborough and Horton 1988; Noctor and Horton 1990; Rees and Horton 1990; Crofts and Horton 1991). Crofts and Horton (1991) confirmed previous conclusions by Schreiber and Neubauer (1987, 1989) and Schreiber et al. (1988b) that low-pH-induced cyclic PS II electron transport results from a weakening of the PS II donor side, which can be also induced by ADRY-reagents, like ANT-2p or CCCP. This type of flow, which stimulates $q_P$ much more than $q_N$, is induced between pH 6.5 and 5. At still lower pH values, strong $q_N$ is obtained. In agreement with findings of Krieger and Weis (1990) and Krieger et al. (1992), this type of low-pH quenching, with an apparent pK of 4.8, was found to be light-dependent and requiring oxidizing conditions (Crofts and Horton 1991). According to Krieger et al. (1992), this quenching reflects $Ca^{2+}$ release from PS II centers, which by these authors is considered the basis of pH-dependent down-regulation of PS II in vivo. Horton et al. (1991) argue that the low-pH-induced change at PS II centers is secondary to zeaxanthin-induced quenching. In an extensive review on nonradiative energy dissipation, Demmig-Adams (1990) comes to the conclusion that there are two independent types of nonphotochemical fluorescence quenching, i.e., at the reaction centers, linked to a donor-side limitation, and in the pigment bed, associated with zeaxanthin. Quenching of $F_o$ and DTT-sensitivity are considered indicative of the latter mechanism, while the former mechanism is thought to be selectively suppressed at 77 K.

Very recently, Horton et al. (1991) proposed a new hypothesis to explain the molecular mechanism of energy-dependent quenching. On the basis of an analysis of 77 K fluorescence emission spectra of "dark" and "light" thylakoids (Ruban et al. 1991), arguments were put forward in favor of the idea that a low lumen pH causes an increase in LHC II aggregation, resulting in a stimulation of nonradiative energy dissipation. The aggregation is supposed to be potentiated by the conversion of violaxanthin to zeaxanthin. Although so far the evidence for this model is rather limited, it is appealing for a number of reasons. In particular, it may provide an explanation for the correlation between $q_E$ and light scattering (Sivak et al. 1985; Bilger et al. 1988, 1989).

## 7. Membrane Energization and Nonassimilatory Fluxes

While the molecular mechanisms involved in energy-dependent fluorescence quenching are still under debate (see Sect. 6), there is general agreement that a transthylakoidal proton gradient is indispensable for "membrane energization" and for the resulting downregulation of PS II. Considering that the Calvin cycle consumes at least as much ATP in $CO_2$-assimilation as coupled linear electron flow can produce, non-assimilatory electron flow must be responsible for membrane energization. Surprisingly, only few researchers have attempted to clarify what kind of non-assimilatory flow is involved. Major candidates are $O_2$-dependent flow, cyclic flow and the Q-cycle at the cyt b/f complex.

Fluorescence quenching analysis has been extremely useful in elucidating a decisive role of $O_2$-dependent electron flow in membrane energization (Neubauer and Schreiber 1989a, b; Schreiber and Neubauer 1990; Schreiber et al. 1991). Notably, quenching analysis can reveal the sequence of Mehler reaction and ascorbate peroxidase reaction, the combination of which results in zero net oxygen exchange (Asada and Badger 1984; Badger 1985; Asada and Takahashi 1987). It was found that in intact chloroplasts with high ascorbate peroxidase and low catalase activity $O_2$-dependent flow can maintain a large $q_E$ which is 50% suppressed when $O_2$-concentration is lowered to 60 μM. Schreiber et al. (1991) investigated $q_E$-formation under conditions which allowed only Mehler reaction or only ascorbate peroxidase reaction to occur. Strong energization was found with $H_2O_2$-reduction, but not with $O_2$-reduction as such. The two reactions also displayed contrasting pH-optima: whereas the Mehler reaction peaks at pH 5, the ascorbate peroxidase reaction shows its maximum around pH 8 (Schreiber et al. 1991). Takahashi and Asada (1988) have suggested that the superoxide anion radical may be formed within the aprotic thylakoid membrane and will only leave the membrane after protonation. Protons could be derived from the acidic internal phase. Reising and Schreiber (1992) demonstrated a close correlation between nonphotochemical quenching and a gas uptake component, measured by pulse-modulated photoacoustics, which is likely to reflect $O_2$-uptake.

Bailey and Walker (1992) carried out simultaneous measurements of $q_P$, $q_N$ and $dO_2/dt$ with intact leaves under conditions inducing pronounced oscillatory behavior. As the oscillatory pattern was markedly changed by DTT, these authors suggested a major role of $O_2$-dependent electron flow in ΔpH-related regulation, assuming that DTT not only prevents zeaxanthin formation but also affects the peroxidase reaction (Schreiber and Neubauer 1990). The regulatory aspects of $O_2$-dependent electron flow have been outlined in a recent review by Walker (1992).

Another major candidate for non-assimilatory electron flow responsible for $q_E$ is cyclic flow around PS I, possibly involving a ferredoxine-quinone reductase (Moss and Bendall 1984). The evidence for such flow in $C_3$-plants under in vivo conditions so far is weak. Asada et al. (1990) showed with spinach chloroplasts that cyclic flow, if stimulated by $O_2$-removal, will cause complete reduction of the intersystem chain and total block of PS II activity, even when substrate amounts of methylviologen are available as PS I acceptor. On the other hand, using the same measuring techniques, Mi et al. (1992) found strong cyclic flow in the cyanobacterium Synechocystis PCC 6803. In this context, it appears relevant that the PS II/PS I ratio is low in cyanobacteria (approx. 0.5) and high in spinach (approx. 2) (see review by Melis 1991). Hence, a high PS I/PS II turnover may be prerequisite for effective cyclic flow. LHC II phosphorylation, by causing a shift of excitation distribution in favor of PS I (Bennett 1980, 1983; Horton and Black 1980; Allen et al. 1981; Fernyhough et al. 1984), could promote the relative contribution of cyclic PS I to overall membrane energization.

Virtually nothing is known about a possible contribution of a Q-cycle at the cyt b/f complex to membrane energization. This cycle has the potential of increasing the $H^+/e^-$ ratio from 2 to 3 (Mitchell 1975; Rathenow and Rumberg 1980; Schubert et al. 1990). Klughammer (1992) studied cytochrome redox changes in intact chloroplasts with a kinetic LED array spectrophotometer (see Sect. 2), presenting data which suggest that a Q-cycle is operative irrespective of the cyt f redox state, with the plastoquinone reduction level being the main regulating factor.

## 8. Outlook

Despite the considerable amount of work which has been devoted to the study of chlorophyll fluorescence during recent years, there are still open questions, the most important of which are:

1. What are the molecular mechanisms of energy dependent and photoinhibitory quenching?
2. How does zeaxanthin contribute to quenching and regulation of energy dissipation?
3. What kind of proton translocating electron flow is responsible for membrane energization in vivo?
4. Is the stimulatory role of $O_2$ for membrane energization mainly due to the Mehler-peroxidase reaction sequence or to "redox-posing" of cyclic electron flow?

5. How much are the observed fluorescence characteristics affected by heterogeneities of PS II units and electron transport chains?

Presumably, to solve these questions, other measuring techniques have to complement chlorophyll fluorescence, and further progress in the refinement of fluorescence analysis has to be achieved. However, the time has already come for widespread practical applications of the available fluorescence measuring techniques in many aspects of plant science and, in particular, in ecophysiological field work. At the same time, fluorescence may be expected to remain a "pioneer tool" in basic research, with the potential of revealing still unknown features of the photosynthetic process.

*Acknowledgments*. We wish to thank Murrary Badger for reading the manuscript and Annette Weber for technical assistance. This work was supported by the Deutsche Forschungsgemeinschaft (SFB 176 and 251).

## References

**Adams WW,** Demmig-Adams B, Winter K (1990a) Plant Physiol 92:302–309. **Adams WW,** Demmig-Adams B, Winter K, Schreiber U (1990b) Planta 180:166–174. **Allen JF** (1992) Biochim Biophys Acta 1098:275–335. **Allen JF,** Bennett J, Steinback KE, Arntzen CJ (1981) Nature 291:25–29. **Asada K,** Badger MR (1984) Plant Cell Physiol 25:1169–1179. **Asada K,** Takahashi M (1987) In: Kyle DJ et al. (eds) Photoinhibition. Elsevier, Amsterdam, pp 227–287. **Asada K,** Neubauer C, Heber U, Schreiber U (1990) Plant Cell Physiol 31:557–564. **Atal N,** Saradhi PP, Mohanty P (1991) Plant Cell Physiol 32:943–951.

**Badger MR** (1985) Annu Rev Plant Physiol 36:27–53. **Bailey KJ,** Walker DA (1992) Plant Physiol 99:124–129. **Baker NR** (1991) Physiol Plant 81:563–570. **Baker NR,** Horton P (1987) In: Kyle et al. (eds) Photoinhibition. Elsevier, Amsterdam, pp 145–168. **Barenyi B,** Krause GH (1985) Planta 163:218–226. **Bennett J** (1980) Eur J Biochem 104:85–89. **Bennett J** (1983) Biochem J 212:1–13. **Bennoun P** (1982) Proc Natl Acad Sci USA 79:4352–4356. **Bilger W,** Björkman O (1990) Photosynth Res 25:173–185. **Bilger W,** Björkman O (1991) Planta 184:226–234. **Bilger W,** Schreiber U (1986) Photosynth Res 10:303–308. **Bilger W,** Schreiber U (1990) Photosynth Res 25:161–171. **Bilger W,** Schreiber U, Lange OL (1987) In: Tenhunen J et al. (eds) Plant response to stress. Springer, Berlin Heidelberg New York, pp 391–399. **Bilger W,** Heber U, Schreiber U (1988) Z Naturforsch 43c:877–887. **Bilger W,** Björkman O, Thayer SS (1989) Plant Physiol 92:542–551. **Björkman O** (1987a) In: Biggins J (ed) Progress in photosynthesis research vol 4. Nijhoff, Dordrecht, pp 11–18. **Björkman O** (1987b) In: Kyle et al. (eds) Photoinhibition. Elsevier, Amsterdam, pp 123–144. **Björkman O,** Demmig B (1987) Planta 170:489–504. **Björkman O,** Powles SB (1984) Planta 161:490–504. **Björkman O,** Demmig B, Andrews TJ (1988) Aust J Plant Physiol 15:43–61. **Black MT**, Brearley TH, Horton P (1986) Photosynth Res 8:193–207. **Bolhar-Nordenkampf HR,** Long SP, Baker NR et al. (1989) Funct Ecol 3:497–514. **Bradbury M,** Baker NR (1981) Biochim Biophys Acta 63:542–551. **Bradbury M,** Baker NR (1984) Biochim Biophys Acta 765:275–281. **Briantais JM,** Vernotte C, Picaud M, Krause GH (1979) Biochim Biophys Acta 548:128–138. **Briantais JM,** Vernotte C, Krause GH, Weis E (1986) In: Govindjee et al. (eds) Light emission by plant and bacteria. Academic Press, New York, pp 539–583. **Brüggemann W,** van der Kooig TAW, van Hasselt PR (1992) Planta 186:179–187. **Büchel C,** Wilhelm C (1990) Plant Physiol Biochem 28:307–314. **Bults G,** Horwitz BA, Malkin S, Cahen D (1982) Biochim Biophys Acta 679:452–465. **Buschmann C,** Prehn H (1983) Photobiochem Photobiophys 5:63–69. **Butler WL** (1978) Annu Rev Plant Physiol 29:345–378. **Butler WL** (1984) Photochem Photobiol 40:513–518. **Butler WL,** Kitajima M (1975) Biochim Biophys Acta 376:116–125.

**Canaani O,** Malkin S (1984) Biochim Biophys Acta 766:513–524. **Caron L,** Berkaloff C, Duval JC, Jupin H (1987) Photosynth Res 11:131–139. **Cleland RE,** Critchley C (1985) Photobiochem Photobiophys 10:83–92. **Cleland RE,** Melis A, Neale PJ (1986) Photosynth Res 9:79–88. **Cornic G,** Briantais JM (1991) Planta 183:178–184. **Cornic G,** Ghashghaie J (1991) Planta 185:255–260. **Critchley C** (1988) Aust J Plant Physiol 15:27–41. **Crofts J,** Horton P (1991) Biochim Biophys Acta 1058:187–193.

**Daly PF,** Raschke K, Ball JT, Berry JA (1989) Plant Physiol 90:1233–1238. **Dau H,** Hansen UP (1989) Photosynth Res 20:59–83. **Dau H,** Hansen UP (1990) Photosynth Res 25:269–278. **Demmig B,** Björkman O (1987) Planta 171:171–184. **Demmig B,** Winter K (1988a) Aust J Plant Physiol 15:151–162. **Demmig B,** Winter K (1988b) Aust J Plant Physiol 15:163–177. **Demmig B,** Winter K, Krüger A, Czygan FC (1987) Plant Physiol 84:218–224. **Demmig B,** Winter K, Krüger A, Czygan FC (1988) Plant Physiol 87:17–24. **Demmig-Adams B** (1990) Biochim Biophys Acta 1020:1–24. **Demmig-Adams B,** Adams WW (1990) Photosynth Res 25:187–198. **Demmig-Adams B,** Winter K, Krüger A, Czygan FC (1989) Plant Physiol 90:881-898. **Demmig-Adams B,** Adams WW, Heber U (1990) Plant Physiol 92:293–301. **Dietz KJ,** Schreiber U, Heber U (1985) Planta 166:219–226. **Di Marco G,** Manes F, Tricoli D, Vitale E (1990) J Plant Physiol 136:538–543. **Downton WJS,** Loveys BR, Grant WJR (1988a) New Phytol 108:262–266. **Downton WJS,** Loveys BR, Grant WJR (1988b) New Phytol 110:503–509.

**Ellenson JL,** Amundson RG (1981) Science 215:1104–1106.

**Falkowski PG,** Kolber Z, Fujita Y (1988) Biochim Biophys Acta 933:432–443. **Fernyhough P,** Foyer CH, Horton P (1984) FEBS Lett 176:133–138. **Fork DC,** Mohanty P (1986) In: Govindjee et al. (eds) Light emission by plant and bacteria. Academic Press, New York, pp 451–496. **Foyer C,** Furbank R, Harbinson J, Horton P (1990) Photosynth Res 25:83–100.

**Geacintov NE,** Breton J (1987) In: CRC Crit Rev Plant Sci 55:1–44. Chemical Rubber, Boca Raton. **Genty B,** Briantais JM, Baker NR (1989) Biochim Biophys Acta 990:87–92. **Genty B,** Wonders J, Baker NR (1990a) Photosynth Res 26:133–139. **Genty B,** Harbinson J, Briantais JM, Baker NR (1990b) Photosynth Res 25:249–257. **Giersch C,** Krause GH (1992) Photosynth Res 30:115–121. **Gilmore AM,** Yamamoto HY (1991) Plant Physiol 96:635–643. **Gilmore AM,** Yamamoto HY (1992) Proc Natl Acad Sci USA 89:1899–1903. **Govindjee,** Satoh I (1986) In: Govindjee et al. (eds) Light emission by plant and bacteria. Academic Press, New York, pp 497–537. **Greer DH,** Laing WA (1988) Planta 175:355–363. **Greer DH,** Berry JA, Björkman O (1986) Planta 168:253–260. **Guenther JE,** Melis A (1990a) Photosynth Res 23:105–109. **Guenther JE,** Melis A (1990b) Photosynth Res 23:195–203.

**Harbinson J,** Foyer C (1991) Plant Physiol 97:41–49. **Harbinson J,** Hedley CL (1989) Plant Cell Environ 12:357–369. **Harbinson J,** Woodward FI (1987) Plant Cell Environ 10:131–140. **Harbinson J,** Genty B, Baker NR (1989) Plant Physiol 90:1029–1034. **Harbinson J,** Genty B, Baker NR (1990a) Photosynth Res 25:213–224. **Harbinson J,** Genty B, Foyer CH (1990b) Plant Physiol 94:545–553. **Havaux M** (1989) Plant Physiol 89:286–292. **Havaux M,** Lannoye R (1985) Photosynthetica 19:388–396. **Havaux M,** Canaani O, Malkin S (1986) Plant Physiol 82:827–833. **Havaux M,** Strasser RJ, Greppin H (1990) Photosynth Res 27:41–55. **Havaux M,** Greppin H, Strasser RJ (1991) Planta 186:88–98. **Havemann J,** Mathis P (1976) Biochim Biophys Acta 440: 346–355. **Heber U,** Vijl J, Neimanis S et al. (1989) Z Naturforsch 44c:524–536. **Heber U,** Schreiber U, Siebke K, Dietz KJ (1990) In: Zelitch I (ed) Perspectives in biochemical and genetic regulation of photosynthesis. Plant biology, vol 10. Liss, New York, pp 17–37. **Holmes JJ,** Weger HG, Turpin DH (1990) Plant Physiol 91:331–337. **Holzwarth AR** (1987) In: Barber J (ed) Topics in photosynthesis: the light reactions, vol 8. Elsevier, Amsterdam, pp 95–157. **Holzwarth AR** (1988) In: Lichtenthaler HK (ed) Application of chlorophyll fluorescence. Kluwer, Dordrecht, pp 21–31. **Holzwarth AR** (1991) In: Scheer H (ed) The chlorophylls. CRC, Boca Raton, pp 1125–1151. **Horton P** (1989) In: Briggs WR (ed) Photosynthesis. Liss, New York, pp 393–406.

**Horton P,** Black MT (1980) FEBS Lett 119:141–144. **Horton P,** Bowyer JR (1990) In: Harwood JL, Bowyer JR (eds) Methods in plant biochemistry, vol 4. Academic Press, Orlando, pp 259–296. **Horton P,** Hague A (1988) Biochim Biophys Acta 932:107–115. **Horton P,** Crofts J, Gordon S et al. (1989) Philos Trans R Soc Lond B 323:269–279. **Horton P,** Ruban AV, Rees D et al. (1991) FEBS Lett 292:1–4.

**Jursinic PA** (1986) In: Govindjee et al. (eds) Light emission by plants and bacteria. Academic Press, Orlando, pp 291–328.

**Karukstis KK** (1991) In: Scheer H (ed) Chlorophylls. CRC, Boca Raton, pp 769–795. **Keiller DR,** Walker DA (1990) Proc R Soc Lond B 241:59–64. **Kitajima M,** Butler WL (1975) Biochim Biophys Acta 376:105–115. **Klughammer C** (1992) PhD Thesis, Univ Würzburg, Germany. **Klughammer C,** Kolbowski J, Schreiber U (1990) Photosynth Res 25:317–327. **Kolbowski J,** Reising H, Schreiber U (1990) Photosynth Res 25:309–316. **Krall JP,** Edwards GE (1990) Aust J Plant Physiol 17:579–588. **Krall JP,** Edwards GE (1991) Aust J Plant Physiol 18:267–278. **Krall JP,** Edwards GE, Ku MSB (1991) Aust J Plant Physiol 18:369–383. **Krause GH** (1988) Physiol Plant 74:566–574. **Krause GH,** Behrend U (1986) FEBS Lett 200:298–302. **Krause GH,** Cornic C (1987) In: Kyle et al. (eds) Photoinhibition. Elsevier, Amsterdam, pp 169–196. **Krause GH,** Laasch H (1987) Z Naturforsch 42c:581–584. **Krause GH,** Somersalo S (1989) Philos Trans R Soc Lond B 323:281–293. **Krause GH,** Weis E (1984) Photosynth Res 5:139–157. **Krause GH,** Weis E (1991) Annu Rev Plant Physiol 42:313–349. **Krause GH,** Briantais JM, Vernotte C (1982) Biochim Biophys Acta 679:116–124. **Krause GH,** Köster S, Wong SC (1985) Planta 165:430–438. **Krause GH,** Laasch H, Weis E (1988) Plant Physiol Biochem 26:445–452. **Krause GH,** Somersalo S, Zumbusch E et al. (1990) J Plant Physiol 139:472–479. **Krieger A,** Weis E (1990) In: Baltscheffsky M (ed) Current research in photosynthesis vol 4. Kluwer, Dordrecht, pp 563–569. **Krieger A,** Moya I, Weis E (1992) Biochim Biophys Acta 1102:167–176. **Kyle DJ** (1987) In: Kyle et al. (eds), Photoinhibition. Elsevier, Amsterdam, pp 197–226. **Kyle DJ,** Osmond CB, Arntzen CJ (1987) Photoinhibition. Topics in photosynthesis vol 9. Elsevier, Amsterdam.

**Laasch H** (1987) Planta 171:220–226. **Laasch H,** Weis E (1988) Biochim Biophys Acta 936:99–107. **Laasch H,** Weis E (1989) Photosynth Res 22:137–146. **Laisk A,** Siebke K, Gerst U et al. (1991) Planta 185:554–562. **Larcher W,** Neuner G (1989) Plant Physiol 89:740–742. **Larcher W,** Wagner J, Thammathaworn A (1990) J Plant Physiol 136:92–102. **Lee CB,** Rees D, Horton P (1990) Photosynth Res 24:167–173. **Lichtenthaler HK** (ed) (1988) Application of chlorophyll fluorescence. Kluwer, Dordrecht, 396 pp. **Lichtenthaler HK,** Rinderle U (1988) In: CRC Crit Rev Anal Chem 19, Suppl I:29–85. **Lichtenthaler HK,** Buschmann C, Rinderle U, Schmuck G (1986) Radiat Environ Biophys 25:297–305. **Ludlow MM** (1987) In: Kyle et al. (eds) Photoinhibition. Elsevier, Amsterdam, pp 89–110.

**Malkin S** (1977) In: Barber J (ed) Primary processes of photosynthesis. Elsevier, Amsterdam, pp 349–431. **Malkin S** (1987) Planta 171:65–72. **Markgraf T,** Berry J (1990) In: Baltscheffsky M (ed) Current research in photosynthesis vol 4. Kluwer, Dordrecht, pp 279–282. **Melis A** (1991) Biochim Biopyhs Acta 1058:87–106. **Melis A,** Guenther JE, Morissey PJ, Ghirardi ML (1988) In: Lichtenthaler HK (ed) Application of chlorophyll fluorescence. Kluwer, Dordrecht, pp 33–43. **Mi H,** Endo T, Schreiber U et al. (1992) Plant Cell Physiol 33:1099–1105. **Mitchell P** (1975) FEBS Lett 59:137–139. **Morita S,** Itoh S, Nishimura M (1981) Plant Cell Physiol 22:205–214. **Moss DA,** Bendall DS (1984) Biochim Biophys Acta 767:389–395. **Moya I,** Sebban P, Haehnel W (1986) In: Govindjee et al. (eds) Light emission by plant and bacteria. Academic Press, New York, pp 161–190.

**Neubauer C,** Schreiber U (1989a) Z Naturforsch 44c:262–270. **Neubauer C,** Schreiber U (1989b) Bot Acta 102:314–318. **Neubauer C,** Yamamoto HY (1992) Plant Physiol 99:1354–1361. **Noctor G,** Horton P (1990) Biochim Biophys Acta 1016:228–234. **Noctor G,** Rees D, Young A, Horton P (1991) Biochim Biophys Acta 1057:320–330.

**Ögren E,** Öquist G (1984) Physiol Plant 62:193–200. **Ögren E,** Baker NR (1985) Plant Cell Environ 8:539–547. **Ohad I,** Koike H, Shahat S, Inoue Y (1988) Biochim Biophys Acta 993:288–298. **Omasa K,** Shimazaki KI, Aiga I et al. (1987) Plant Physiol 84:748–752. **Öquist G,** Greer DH, Ögren E (1987) In: Kyle et al. (eds) Photoinhibition. Elsevier, Amsterdam, pp 67–88. **Oxborough K,** Horton P (1987a) Photosynth Res 12:119–128. **Oxborough K,** Horton P (1987b) FEBS Lett 221:211–214. **Oxborough K,** Horton P (1988) Biochim Biophys Acta 934:135–143.

**Peltier G,** Ravenel J, Vermeglio A (1987) Biochim Biophys Acta 893:83–90. **Peterson RB** (1991) Plant Physiol 97:1388–1394. **Peterson RB,** Sivak MN, Walker DA (1988) Plant Physiol 88:158–163. **Powles SB** (1984) Annu Rev Plant Physiol 35:15–44. **Powles SB,** Björkman O (1982) Planta 156:97–107.

**Quick WP,** Horton P (1984a) Proc R Soc Lond B 220:361–370. **Quick WP,** Horton P (1984b) Proc R Soc Lond B 220:371–382. **Quick WP,** Stitt M (1989) Biochim Biophys Acta 977:287–296.

**Rathenow M,** Rumberg B (1980) Ber Bunsenges Phys Chem 84:1059–1062. **Rees D,** Horton P (1990) Biochim Biophys Acta 1016:219–227. **Rees D,** Young A, Noctor G et al. (1989) FEBS Lett 256:85–90. **Rees D,** Noctor GD, Horton P (1990) Photosynth Res 25:199–211. **Rees D,** Horton P, Schreiber U (1993) Photosynth Res (in press). **Reising H,** Schreiber U (1992) Photosynth Res 31:227–238. **Renger G,** Schreiber U (1986) In: Govindjee et al. (eds) Light emission by plant and bacteria. Academic Press, New York, pp 597–619. **Richter M,** Rühle W, Wild A (1990) Photosynth Res 24:237–243. **Ruban AV,** Rees D, Noctor GD et al. (1991) Biochim Biophys Acta 1059:355–360. **Rütten D,** Santarius KA (1992) Planta 187:224–229.

**Samuelsson G,** Lönneborg A, Rosenquist E et al. (1985) Plant Physiol 79:992–995. **Schäfer C,** Björkman O (1989) Planta 178:367–376. **Schatz GH,** Brock H, Holzwarth AR (1988) Biophys J 54:397–405. **Schmidt W,** Schreiber U, Urbach W (1987) Z Naturforsch 43c:269–274. **Schmidt W,** Neubauer C, Kolbowski J et al. (1990) Photosynth Res 25:241–248. **Schöner S,** Krause GH (1990) Planta 180:383–389. **Schreiber U** (1980) FEBS Lett 122:121–124. **Schreiber U** (1986) Photosynth Res 9:261–272. **Schreiber U,** Bilger W (1987) In: Tenhunen DJ et al. (eds) Plant response to stress. Springer, Berlin Heidelberg New York, pp 27–53. **Schreiber U,** Neubauer C (1987) Z Naturforsch 42c:1255–1264. **Schreiber U,** Neubauer C (1989) FEBS Lett 258:339–342. **Schreiber U,** Neubauer C (1990) Photosynth Res 25:279–293. **Schreiber U,** Rienits KG (1987) FEBS Lett 211:99–104. **Schreiber U,** Schliwa U (1987) Photosynth Res 11:173–182. **Schreiber U,** Schliwa U, Bilger W (1986) Photosynth Res 10:51–62. **Schreiber U,** Klughammer C, Neubauer C (1988a) Z Naturforsch 43c:686–698. **Schreiber U,** Neubauer C, Klughammer C (1988b) In: Lichtenthaler HK (ed) Applications of chlorophyll fluorescence. Kluwer, Dordrecht, pp 63–69. **Schreiber U,** Neubauer C, Klughammer C (1989) Philos Trans R Soc Lond B 323:241–251. **Schreiber U,** Reising H, Neubauer C (1991) Z Naturforsch 46c:635–643. **Schroeter B,** Green TGA, Seppelt RD, Kappen L (1992) Oecologia 92:457–462. **Schubert K,** Liese F, Rumberg B (1990) In: Baltscheffsky M (ed) Current research in photosynthesis, vol III. Kluwer, Dordrecht, pp 279–282. **Seaton GGR,** Walker DA (1990) Proc R Soc Lond B 242:29–35. **Sharkey TD,** Berry JA, Sage RF (1988) Planta 176:415–424. **Sironval C,** Strasser RJ, Brouers M (1984) In: Sironval C, Brouers M (eds) Protochlorophyllide reduction and greening. Nijhoff, The Hague, pp 307–316. **Sivak MN,** Dietz KJ, Heber U, Walker DA (1985) Arch Biochem Biophys 237:513–519. **Smillie RM,** Hetherington SE, He J, Nott R (1988) Aust J Plant Physiol 15:207–222. **Snel JFH,** van Kooten O (eds) (1990) The use of chlorophyll fluorescence and other non-invasive spectroscopic techniques in plant stress physiology. Photosynth Res (Spec Iss) 25:146–332. **Snel JFH,** Kooijman M, Vredenberg WJ (1990a) Photosynth Res 25:259–268. **Snel JFH,** van Jeperen W, Vredenberg WJ (1990b) In: Baltscheffsky M (ed) Current research in photosynthesis vol. II Kluwer, Dordrecht, pp 911–914. **Somersalo S,** Krause GH (1989) Planta 177:409–416. **Somersalo S,** Krause GH (1990) Planta 180:181–187. **Stitt M,** Grosse H (1988) J Plant Physiol 133:392–400. **Stitt M,** Schreiber U (1988) J Plant Physiol 133:263–271. **Stitt M,** Grosse H, Woo K-C (1988) J Plant Physiol 133:138–143. **Strasser RJ** (1981) In: Akoyunoglou G (ed) Photosynthesis vol III. Balaban, Philadelphia, pp 727–737. **Stuhlfauth T,** Sültemeyer DF, Weinz S et al. (1988) Plant Physiol

86:246–250. **Stuhlfauth T,** Scheuermann R, Fock HP (1990) Plant Physiol 92:1053–1061. **Styring S,** Virgin I, Ehrenberg A, Andersson B (1990) Biochim Biophys Acta 1015:269–278.

**Takahashi M,** Asada K (1988) Arch Biochem Biophys 267:714–722. **Terashima I,** Sonoike K, Kaweizen T, Katoh S (1991) Plant Cell Physiol 32:1275–1283. **Thompson LK,** Brudvig GW (1988) Biochemistry 27:6653–6658.

**van Gorkom HJ** (1985) Photosynth Res 6:97–112. **van Gorkom HJ** (1986) In: Govindjee et al. (eds) Light emission by plant and bacteria. Academic Press, New York, pp 267–289. **van Kooten O,** Snel JFH (1990) Photosynth Res 25:147–150. **van Kooten O,** van Hove B, van Wijk KJ (1988) In: Lichtenthaler HK (ed) Applications of chlorophyll fluorescence. Kluwer, Dordrecht, pp 203–209. **van Wijk KJ,** Krause GH (1991) Planta 186:135–142. **Veljovic-Jovanovic S,** Bilger W, Heber U (1993) Planta (in press).

**Walker DA** (1992) New Phytol 121:325–345. **Walters RG,** Horton P (1991) Photosynth Res 27:121–133. **Weis E,** Berry JA (1987) Biochim Biophys Acta 894:198–208. **Weis E,** Lechtenberg D (1989) Philos Trans R Soc Lond B 323:253–268. **Weis E,** Ball J, Berry JA (1987) In: Biggins J (ed) Progress in photosynthesis research, vol 2. Nijhoff, Dordrecht, pp 553–556. **Weis E,** Lechtenberg D, Krieger A (1990) In: Baltscheffsky M (ed) Current research in photosynthesis vol 4. Kluwer, Dordrecht, pp 307–312. **Wilhelm C,** Duval JC (1990) Biochim Biophys Acta 1016:197–202. **Williams WP,** Allen JF (1987) Photosynth Res 13:19–45. **Winter K,** Gademann R (1991) Plant Physiol 95:768–776. **Winter K,** Königer M (1989) Planta 180:24–31. **Wu J,** Neimanis S, Heber U (1991) Bot Acta 104:283–291.

Dr. Ulrich Schreiber
Dr. Wolfgang Bilger
Julius-von-Sachs-Institut
für Biowissenschaften mit Botanischem Garten
der Universität Würzburg
Lehrstuhl für Botanik I
Mittlerer Dallenbergweg 64
W-8700 Würzburg, Germany

# IV. Photosynthesis: Carbon Metabolism
# Twenty Years of Following Carbon Cycles in Photosynthetic Cells[1]

By Grahame J. Kelly and Erwin Latzko

"The first man I saw was of a meagre aspect, with sooty hands and face, his hair and beard long, ragged and singed in several places. His clothes, shirt and skin were all of the same colour. He had been eight years upon a project for extracting sunbeams out of cucumbers, which were to be put into vials hermetically sealed, and let out to warm the air in raw inclement summers. He told me, he did not doubt in eight years more, that he should be able to supply the Governor's gardens with sunshine at a reasonable rate..."

Jonathan Swift in *Gulliver's Travels*, ca. 1727

## 1. Introduction

We do not know whether or not the gentleman described above completed his planned 16 years of research on cucumber photosynthesis. If he did, he still would not have matched our efforts for *Progress in Botany*, in which our reviews of photosynthetic carbon metabolism have now reached their 20th year. We began at a time when the Calvin cycle was well established (although sufficient activities of all of its enzymes were not), and the processes of $C_4$ photosynthesis and photorespiration had been only recently discovered. During the ensuing 20 years, many details concerning the operation and regulation of these carbon cycles have been revealed. Seven stand out: (1) the demonstration of adequate activities for all Calvin cycle enzymes; (2) the discovery of the chloroplast envelope's phosphate translocator; (3) the $CO_2$-plus-$Mg^{2+}$ activation of ribulose-1,5-bisphosphate (RuBP) carboxylase and, more recently, this enzyme's regulation by the protein "rubisco activase" and a seemingly endless array of novel sugar phosphates; (4) the demonstration of the light-mediated activation of Calvin cycle enzymes, mediated by the small proteins ferredoxin and thioredoxin; (5) the discoveries that the enzymes phosphoenolpyruvate (PEP) carboxylase, pyruvate Pi dikinase, and sucrose-P synthase are regulated by light-mediated phosphorylation/dephosphorylation; (6) the discoveries of the enzyme PPi-phosphofructokinase (PPi–PFK) and the regulatory compound fructose-2,6-$P_2$ in the cytosol of photosynthetic cells – their roles in regulating sucrose metabolism are

[1] This review is dedicated to M. Gibbs on the occasion of his retirement as Editor-in-Chief of *Plant Physiology*.

Progress in Botany, Vol. 54

still being elucidated; (7) the discovery (reported in this review) that the chloroplast envelope's adenylate translocator accommodates the uptake of ADP-glucose.

Clearly, the 1970s and 1980s were an exciting 20 years, but it is worth noting that no major new metabolic cycles were discovered, as happened in the 1950s and 1960s. It would seem that there are no new cycles to discover. Then what of the future? Will research bc devoted to the acquisition of evermore detailed descriptions of the Calvin cycle, $C_4$ photosynthesis, crassulacean acid metabolism (CAM), sucrose biosynthesis, photorespiration, and the enzymes that catalyze their component reactions? Perhaps it will, but we suspect that the consequences will be anything but boring, if for no other reason than the introduction of the new tools of biotechnology. This is our first review in which we mention transgenic plants, for example: tobacco plants containing yeast invertase (Stitt et al. 1990), tomato plants containing maize sucrose-P synthase (Worrell et al. 1991) and *Synechococcus* containing human carbonic anhydrase (Price and Badger 1991). The construction of such transgenic plants, together with the techniques of gene sequencing and site-directed mutagenesis, will without doubt play a major (perhaps *the* major) role in the next 20 years of research into photosynthetic carbon metabolism.

## 2. Uptake of $CO_2$

It is generally believed that $CO_2$ enters the leaf cells of terrestrial plants by simple diffusion across the cell membrane. However, it is tempting to speculate that one or more transporter proteins may be involved, given that aquatic plants possess these (see below). A similarity between the conductivities to $CO_2$ of algal and higher-plant protoplast membranes has led Gimmler et al. (1990) to suggest that such transporters do indeed exist in higher plants.

A fundamental parameter is the concentration of $CO_2$ in the photosynthesizing chloroplast. This has been calculated by von Caemmerer and Evans (1991) to be about one-half of the ambient partial pressure, i.e., close to 200 ppm for leaves in air.

### a) Cyanobacteria

The cell membranes of cyanobacteria possess mechanisms for the active uptake of both $CO_2$ and $HCO_3^-$. The $CO_2$-uptake system is present irrespective of the level of inorganic carbon ($CO_2$ plus $HCO_3^-$) in the growth medium, but the $HCO_3^-$-uptake system is only present when the level of inorganic carbon is relatively low, e.g., air levels (Espie et al. 1991a, b). In addition, $HCO_3^-$ transport appears in two forms: a $Na^+$-dependent form in cultures bubbled with air, and a $Na^+$-independent form in standing cultures (Espie et al. 1991a; Espie and Kandasamy 1992).

There is some confusion concerning how distinct these mechanisms are. Certainly more than one protein is involved (Ogawa 1990), but whether or not the $CO_2$- and $HCO_3^-$-uptake systems are all part of one multi-protein complex (Badger and Price 1990), and whether or not carbonic anhydrase has a role (Bédu et al. 1990; Price and

Badger 1991), remain to be determined. Several reviews of cyanobacterial inorganic carbon uptake have appeared, including Miller et al. (1990) and Kaplan et al. (1991).

b) Eukaryotes

Studies with unicellular freshwater algae have shown that, like cyanobacteria, they possess plasmalemma-located mechanisms for the active uptake of both $CO_2$ and $HCO_3^-$; the inorganic carbon accumulates internally to a concentration that can suppress photorespiration (Gehl et al. 1990; Palmqvist et al. 1990; Spalding 1990; Rotatore and Colman 1991a, b). Energy to drive the $CO_2$ uptake is derived from photosynthetic, but not respiratory, ATP (Sültemeyer et al. 1991). It seems unlikely that additional uptake mechanisms are located on the chloroplast envelope (Rotatore and Colman 1991c), although a special chloroplast protein does appear when *Chlamydomonas reinhardtii* is grown on a low level of $CO_2$ (Mason et al. 1990).

Marine microalgae and freshwater and marine macrophytes also possess $HCO_3^-$ and $CO_2$-uptake systems able to suppress photorespiration (Beer et al. 1990, 1991; Booth and Beardall 1991; Drechsler and Beer 1991). Extensive investigations with the freshwater macrophyte *Elodea nuttallii* have resulted in the proposal that it utilizes two separate transport systems for $CO_2$ and $HCO_3^-$ (Jahnke et al. 1991). However, other freshwater macrophytes, the marine macrophyte *Ulva rotundata*, and the marine microalga *Emiliania huxleyi* appear to have a limited or even nonexistent ability to transport inorganic carbon (Beer et al. 1991; Levavasseur et al. 1991; Nimer et al. 1992).

c) Carbonic Anhydrase

Investigations of $CO_2/HCO_3^-$ uptake have always been complicated by the enzyme carbonic anhydrase which catalyzes the reversible reaction $H_2O + CO_2 \leftrightarrow H_2CO_3$; the $H_2CO_3$ immediately equilibrates with $H^+$ and $HCO_3^-$ according to the prevailing pH. Unicellular algae contain one or more intracellular carbonic anhydrases, and some species produce an additional, ionically-bound *external* carbonic anhydrase, the amount of which usually increases when cells are transferred from high-$CO_2$ to air-level $CO_2$ media (Husic and Quigley 1990; Sültemeyer et al. 1990; Coleman et al. 1991; Goyal et al. 1992). The induction of this external enzyme seems to be influenced by the intracellular metabolic status (Martínez and Orús 1991; Umino and Shiraiwa 1991). Utilization of $HCO_3^-$ via an extracellular carbonic anhydrase and subsequent entry of the generated $CO_2$ into cells also has been demonstrated for a macroalga, *Ulva rigida* (Björk et al. 1992).

These carbonic anhydrases not only confuse the issue of whether $CO_2$ or $HCO_3^-$ is the species transported across the plasmalemma (indeed, interconversion might be part of the transport process! – see Badger and Price 1990), but also raise the question of what role, if any, they have in the events that occur between the active uptake process at the plasmalemma and the assimilation of $CO_2$ inside the chloroplast. Indications are that the enzyme's importance depends on the species of alga (Gehl et al. 1990; Nara et al. 1990; Palmqvist et al. 1990). This interpretation also applies, in a sense, to terrestrial plants, in that all of the carbonic anhydrase activity present in $C_4$ plant leaves is needed for $C_4$ photosynthesis to proceed at the rates observed (Hatch and Burnell 1990; see Sect. 10). Finally, some studies may need re-evaluation following the report by Hatch (1991) of a problem with the conventional carbonic anhydrase assay system.

## 3. RuBP Carboxylase

The plethora of research on RuBP carboxylase has continued unabated. The carboxylase is the first enzyme of the Calvin cycle to which the tools of so-called molecular biology have been applied. The genes encoding the large and small subunits of the enzyme were the first plant genes to be sequenced; this occurred about 12 years ago (see Raines et al. 1991). Four years ago Rodermel et al. (1988) transformed tobacco plants with a nuclear antisense gene for the small subunit, thereby reducing the level of RuBP carboxylase in the plants. The photosynthesis by similarly transformed tobacco plants now has been characterized by two groups (Quick et al. 1991; Hudson et al. 1992) and the results provide qualified support to the concept that the activity of RuBP carboxylase is one of the prime determinants of photosynthetic capacity. The qualification is that, with these tobacco plants, the amount of carboxylase protein had to be reduced by at least 50% before a drop in photosynthesis, measured at a relatively low irradiance, could be detected (Quick et al. 1991), although with high irradiance the contribution of the carboxylase to control was greater (Stitt et al. 1991; Fig. 1). When the carboxylase content was reduced to only 18% of normal, and irradiance was high, it limited photosynthesis even at high $CO_2$ levels (Hudson et al. 1992). These results infer that natural variation of RuBP carboxylase, such as the range of 15 to 50% of total soluble protein in barley varieties (Metodiev and Demirevska-Kepova 1992), would certainly alter the relative contribution of the enzyme to photosynthetic control. In addition, the results are not unexpected, given that leaves contain sufficient potential RuBP carboxylase activity such that, in situations where it is unlikely to be limiting photosynthesis, it is actually downregulated so that production and consumption of RuBP are balanced (Sage et al. 1990a). Possible situations include high $CO_2$ and/or lower growth temperature (Pi regeneration during starch and/or sucrose synthesis become limiting) and low light (electron transport becomes limiting); the diversion of photorespiratory glycolate away from 3-P-glycerate formation (see Fig. 6 in our 1982 review) modifies these responses (Sage et al. 1990b; Harley and Sharkey 1991).

RuBP carboxylase has been the subject of a range of other studies. One of the more fascinating has been the isolation of a previously undescribed chloroplast protein that catalyzes the addition of methyl groups to the lysyl residue at position 14 of the enzyme's large subunit (Houtz et al. 1991). This residue is normally methylated in some species, e.g., tomato, tobacco, and pea, but not in others, e.g., wheat and spinach; the reason for this difference is not yet known (Houtz et al. 1992). A separate study with a mutant *Chlamydomonas reinhardtii* RuBP carboxylase also provided evidence that some separate protein could influence the carboxylase's structure (Chen et al. 1990).

The enzyme's affinity for RuBP has been halved by either cleaving off the large subunit's C-terminal amino acid (valine; Portis 1990), or by replacing this subunit's lysine 11 with leucine, us-

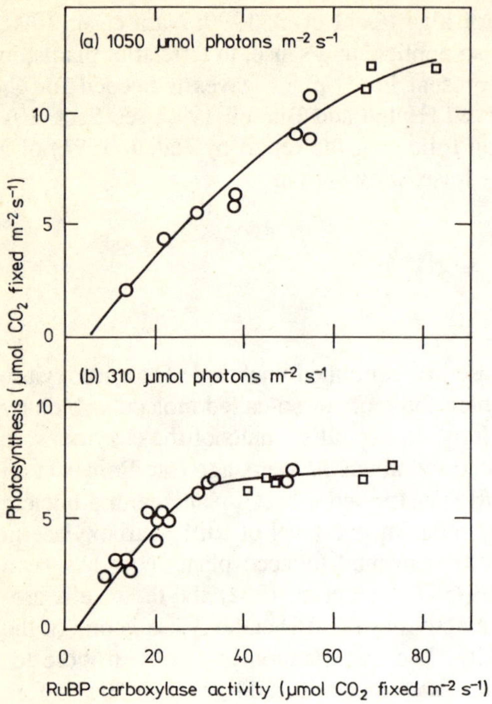


**Fig. 1.** Photosynthesis by tobacco leaves containing varying quantities of RuBP carboxylase. ❐, Wild-type plants; 0, plants transformed to contain reduced levels of the carboxylase. Measurement conditions were: 350 ppm $CO_2$, 22 °C, 65% relative humidity, and irradiances as shown – approximately two-thirds of full sunshine in **(a)**, and one-fifth in **(b)**. (Redrawn from Stitt et al. 1991)

ing site-directed mutagenesis (Kettleborough et al. 1991). An odd observation is that the enzyme blunders 0.7% of the time, and produces pyruvate instead of glycerate-3-P (Andrews and Kane 1991). In storage, the enzyme suffers inactivation in the cold (Heuer and Portis 1990), a phenomenon that also occurs in vivo, such that long-term chilling may lower photosynthetic capacity (Brüggemann et al. 1992).

Evidence for a strong association of a proportion of the chloroplast's RuBP carboxylase with membranes has been reported for red clover, pea, and soybean (Grub and Mächler 1990; Crafts-Brandner et al. 1991; Makino and Osmond 1991). In the latter case, fruit removal acted as a stimulus for the enzyme to move from a soluble to an insoluble form. An obvious candidate membrane for binding the enzyme is the thylakoid membrane, thus the observation by Süss (1990) that the thylakoid-coupling factor forms a specific complex with the carboxylase is of particular interest.

Contemporary environmental issues have persuaded researchers to examine the level of RuBP carboxylase in plants exposed to high levels of $CO_2$, and to increased UV-B radiation. In general, acclimation of plants to a $CO_2$ level more or less double the current atmospheric level tends to involve a decline in the leaf content of RuBP carboxylase, more so if the availability of sinks for increased photosynthate is low (Bowes 1991). Such a response has been documented for tomato (Besford et al. 1990) and rice (Rowland-Bamford et al. 1991). However, this decline, as seen in cultivated species, may be less common in other plants for which consistently higher photosynthetic rates have been measured during prolonged exposure to high $CO_2$ (Arp and Drake 1991; Idso and Kimball 1991; Ziska et al. 1991).

One disturbing report is that the content of RuBP carboxylase in pea leaves dropped dramatically upon exposure of the leaves to supplementary UV-B radiation (Jordon et al. 1992). The dosage was rather high (2.5 times normal at 297 nm), but the response was dramatic: enzyme activity declined by 71% after 3 days, while the mRNAs for the enzyme subunits declined by more than 80% after only 4 h. Similar studies involving lower UV-B irradiances, comparable to those expected next century as a consequence of stratospheric ozone depletion, need to be made available. Another study involving light quality has provided an intriguing insight into fern RuBP carboxylase (Eilenberg et al. 1991): the enzyme from gametophytes is two to three times more active when the gametophytes are grown under blue light than when they are grown under red light, and the small subunits of the enzyme from each of these light qualities differ immunologically. This latter aspect is especially interesting, given that the small subunit is intimately involved in ensuring that the six-carbon intermediate involved in the reaction catalyzed by the carboxylase is not lost (Smrcka et al. 1991).

### a) Regulation of RuBP Carboxylase Activity

No new additions have been made to the exciting discoveries of rubisco activase, carboxyarabinitol-1-P (CA1P), and xylulose-1,5-$P_2$ (XuBP), reported in our last two reviews. However, further details concerning the operation of these regulators of RuBP carboxylase have come to hand. Rubisco activase, which occurs in two isoforms (Shen et al. 1991), is more effective at activating the carboxylase when photosynthetic electron transport through photosystem I is operating, but the reasons for this are not yet clear (Campbell and Ogren 1990a). The effectiveness is decreased by certain sugar-Ps, e.g, RuBP and fructose-1,6-$P_2$ (FBP; Lilley and Portis 1990). It also seems to be decreased when leaves are exposed to prolonged high light intensity (Suzuki et al. 1990).

A further study has confirmed that CA1P inhibition of the carboxylase is confined to certain plants such as bean, but absent from others such as wheat (Moore et al. 1991). The CA1P is degraded by a phosphatase that is, in turn, stimulated by sugar-Ps such as RuBP and FBP (Holbrook et al. 1991), a result in contrast to the effect of these sugar-Ps on the activase (see above). The origin of CA1P is still unknown, but it seems clear that the carboxylase is not involved; an end-product of photosynthesis is the more likely precursor (Moore and Seemann 1990), thus FBP still has a chance (see our preceding review).

With respect to regulation by XuBP, and by the substrate RuBP, a relatively more complex and somewhat unclear picture is emerging. Originally, the phenomenon of "fallover" (initial gradual decrease in carboxylase activity until a final linear rate becomes established) was attributed to XuBP, produced by a side-reaction of the carboxylase and then bound to the catalytic site (Edmondson et al. 1990a). Binding of *RuBP* to a separate *regulatory* site was not believed to be involved (Edmondson et al. 1990b), but Yokota (1991a) has demonstrated the existence of just such a regulatory site, and attributes fallover to a slow conformational change in the carboxylase caused when RuBP occupies this site. More recently, Yokota and Tsujimoto (1992) proposed, from studies of the spinach enzyme, that RuBP needs to be bound to the regulatory site before rubiso activase can be effective. Activase was envisaged to bind to one of the eight large subunits of the carboxylase carrying regulatory RuBP, thereby stimulating the well-known $CO_2$-plus-$Mg^{2+}$ activation. Furthermore, XuBP on the catalytic site can be used as a substrate for $CO_2$ fixation and, consequently, removed – but not before causing some fallover by removing the activator-$CO_2$ (Yokota 1991b) that is involved in the $CO_2$-plus-$Mg^{2+}$ activation. Zhu and Jensen (1991) have also related the binding of XuBP to the absence of this activator $CO_2$. Clearly, more studies are needed in this area. They may be assisted by the novel estimate of Purohit et al. (1991) that the pH of the microenvironment of the catalytic site jumps from <2 to >8 upon activation by $Mg^{2+}$.

Finally, two simple metabolites have been implicated in RuBP carboxylase regulation. These are Pi (Sawada et al. 1990) and glyoxylate (Campbell and Ogren 1990b). In the latter case, it was concluded that glyoxylate inhibited an unidentified component of the RuBP carboxylase activation system. Apparently, the list of regulators is not yet complete.

## 4. Other Calvin Cycle Enzymes

Calvin cycle enzymes other than RuBP carboxylase have attracted a moderate increase in researchers' attention over the past 2 years. The first study of a marine algal FBPase has been made (Yu 1989), and two microalgal ribulose-5-P (Ru5P) kinases, from *Chlamydomonas reinhardtii* (Roesler and Ogren 1990) and *Selenastrum minutum* (Lin and Turpin 1992), have been purified; the latter is somewhat unique in being a heterodimer. The classes of aldolases in algae have been evaluated by Jacobshagen and Schnarrenberger (1990) who thereby obtained molecular evidence that the charophytes form an evolutionary link between chlorophytes and higher plants.

Two unusual effects of metal ions on the higher plant chloroplast bisphosphatases have been detected. Oxidized FBPase (i.e., the "dark-inactivated" form) was found to possess activity in the presence of the substrate FBP and a relatively unphysiological concentration of $Ca^{2+}$ (Chardot and Meunier 1990). On the other hand, oxidized sedoheptulose bisphosphatase (SBPase) in the presence of $Fe^{2+}$ was not only inactive, but could not be activated by dithiothreitol while the $Fe^{2+}$ remained (Queiroz-Claret and Meunier 1992). Other studies, somewhat more physiologically relevant, have shown that the bisphosphatases are sites for the fine control of the Calvin cycle during illumination (being regulated by prevailing levels of glycerate-3-P and sedoheptulose-7-P; Schimkat et al. 1990), and that they are relatively unique in that they were the only two of nine Calvin cycle enzymes to increase in activity in the leaves of soybean plants exposed to low phosphate nutrition – the activities of the other seven enzymes decreased (Fredeen et al. 1990).

A new outlook on chloroplast aldolase is required following the discovery by Iwaki et al. (1991) that the enzyme is strongly inhibited by RuBP when operating in the photosynthetic direction (FBP formation) and with physiological concentrations of its triose-P substrates.

Two groups have simultaneously separated and purified the leaf chloroplast and cytosolic 3-P-glycerate kinases: from spinach (Köpke-Secundo et al. 1990) and from barley

(McMorrow and Bradbeer 1990). In both species at least 90% of the total activity was in the chloroplasts. There is evidence that, within the chloroplast, this kinase interacts with the next enzyme in the Calvin cycle sequence, viz. glyceraldehyde-3-P dehydrogenase (GAPDH) (Macioszek et al. 1990), and that the other kinase, Ru5P kinase, interacts with its nearest neighbor, viz. P-ribose isomerase (Skrukrud et al. 1991). In fact, the latter interaction can occur even when the isomerase is from yeast (Skrukrud and Anderson 1991). Finally, interaction between this kinase and chloroplast GAPDH occurs in a nonreducing environment such that the two enzymes copurify as a complex (Clasper et al. 1991).

a) Light-Mediated Regulation

The chloroplast enzymes FBPase, SBPase, Ru5P kinase, and GAPDH are capable of undergoing light-mediated reductive activation (Fig. 2) because their peptide sequences include "regulatory sections" that have evolved for this purpose; these sections are absent

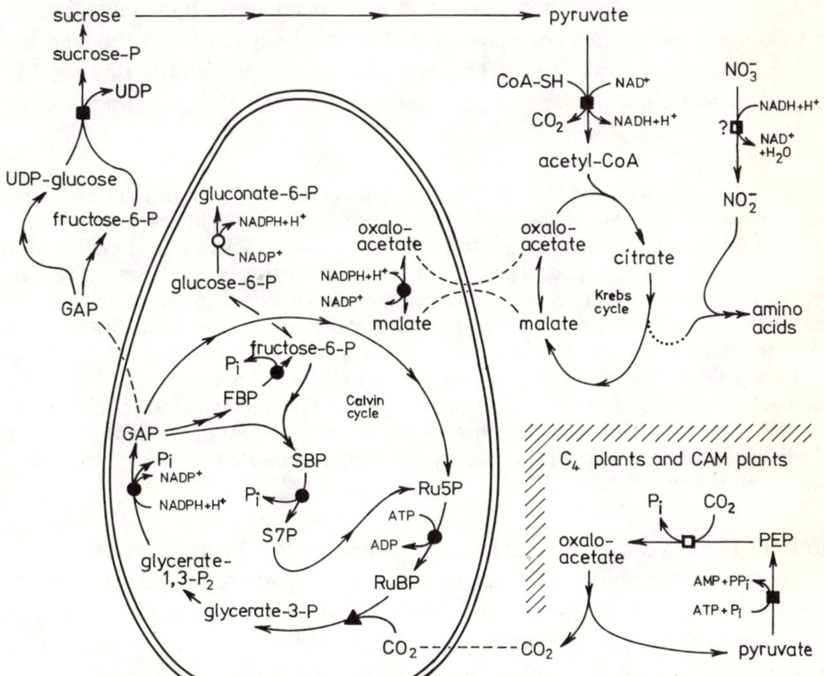


**Fig. 2.** A summary of light-mediated regulation of enzyme activities in photosynthetic cells. Mechanisms: activation (●) or inactivation (○) through reduction of disulfide bonds to sulfhydryls, mediated by thioredoxin; ▲, rubisco activase-promoted carbamylation and/or dissociation of sugar-P inhibitor; ■, phosphorylation resulting in inactivation; □, phosphorylation resulting in activation. *S7P* Sedoheptulose-7-P. (After Budde and Randall 1990; Scheibe 1991)

from nonchloroplast FBPases and GAPDHs (Scheibe 1990, 1991). The in vivo activation of all four enzymes in the cyanobacterium *Nostoc* sp. has recently been demonstrated (Austin et al. 1992). A novel observation is that the activity of three of these enzymes (SBPase was not tested) can be enhanced by high hydrostatic pressure, whereas pressure did not influence nonchloroplast FBPase and GAPDH, or the chloroplast's RuBP carboxylase (Prat-Gay et al. 1991). The authors argue that noncovalent interactions are an important component of light-mediated activation. However, Porter and Hartmann (1990) did not consider that conformational changes were necessary in the case of Ru5P kinase regulation.

Minor attention has been given to the bisphosphatases. Their light-mediated activation in tomato leaves is impaired by chilling, and this seems to be the primary cause of the reduced photosynthetic capacity of these leaves (Sassenrath et al. 1990). An odd observation is that FBPase reduced by dithiothreitol behaves differently to that reduced by the physiological reductant, thioredoxin (Chardot and Meunier 1991).

A collection of studies surrounds the sequence between Ru5P kinase and GAPDH. Light-mediated activation of these two enzymes is more rapid than that of RuBP carboxylase; the latter is not a reductive process, but involves rubisco activase (Servaites et al. 1991; see Sect. 3.a). The most rapid activation seems to be that of GAPDH, which is completely inactive in the dark, but fully activated even in low light (Siebke et al. 1991; Laisk et al. 1992). Ru5P kinase activation may be more dependent on light intensity, as evidenced by its poor activation in vivo when photochemical capacity was reduced by iron deficiency (Arulanantham et al. 1990). This was despite the probability that the kinase was integrated in a multi-enzyme complex, an event that facilitates its activation (Rault et al. 1991).

Five enzymes *not* associated with the Calvin cycle have activities that undergo light-mediated regulation in photosynthetic cells. One of these, viz. the chloroplast $NADP^+$-linked malate dehydrogenase, is activated by reduction of one disulfide bridge per subunit (Scheibe et al. 1991). But the other four are regulated through phosphorylation/dephosphorylation mechanisms (Fig. 2). These are chloroplast pyruvate Pi dikinase (in $C_4$ plants), the mitochondrial pyruvate dehydrogenase complex, and the cytosolic enzymes PEP carboxylase and sucrose-P synthase (Budde and Randall 1990). A fifth contender is the cytosolic nitrate reductase (Kaiser and Spill 1991) that is rapidly inactivated upon darkening, thereby preventing accumulation of toxic nitrite (Riens and Heldt 1992). This mechanism complements another in the light, in which the presence of nitrate indirectly stimulates the quick utilization of nitrite through nitrate enhancement of the light-mediated phosphorylative activation of PEP carboxylase, thereby directing carbon flow towards amino acid synthesis (Van Quy et al. 1991).

Returning to light-mediated reductive activation: electrons for this process are delivered to target enzymes in the chloroplast via ferredoxin, ferredoxin/thioredoxin reductase, and thioredoxin (reviewed by Buchanan 1991). Besides this well-established system, which uses either of two types of thioredoxin, viz. thioredoxin$_f$, (which is present in appreciable amounts; Hermoso et al. 1992) or thioredoxin$_m$, recent evidence points to the existence of a comparable *cytosolic* system of NADPH, NADP/thioredoxin reductase, and a third type of thioredoxin, viz. thioredoxin$_h$, in leaves, algae, and nonphotosynthetic plant cells (Florencio et al. 1988; Huppe et al. 1991; Marcus et al. 1991; Van Langendonckt and Driessche 1992). Its function is unknown.

## 5. Chloroplast Starch

For many years the regulation of starch synthesis in plastids, including chloroplasts, has been considered to take place predominantly at the reaction catalyzed by ADP-glucose (ADPG) pyrophosphorylase, principally through 3-P-glycerate activation and Pi inhibition of this enzyme (reviewed by Kleczkowski et al. 1991). A mutant of *Chlamydomonas reinhardtii* that accumulates very little starch and contains an enzyme with a dramatically lowered sensitivity to 3-P-glycerate supports this view (Ball et al. 1991a). However, by using the wrinkle-seeded and round-seeded genotypes of pea (that differ in leaf content of starch-branching enzyme), and mutants of *Arabidopsis thaliana* constructed to contain very low activities of either ADPG pyrophosphorylase or P-glucomutase in chloroplasts, Smith et al. (1990) and Neuhaus and Stitt (1990) found that the contemporary view of the importance of the pyrophosphorylase was only applicable to plants grown in low light; under high light, the other enzymes became important in controlling the rate of starch synthesis.

Accumulation of starch benefits plant growth mostly when other factors, e.g., nitrogen supply, are not limiting (Schulze et al. 1991). The rate of nocturnal mobilization of starch is somehow determined by the amount finally accumulated at day's end (Hendrix and Grange 1991). The initial steps in mobilization may include phosphorylase, amylases (see our earlier reviews), α-glucosidase (Beers et al. 1990), and an "unidentified regulatory factor" (Casper et al. 1991). Incidentally, a first clue to the role of the *extra*-chloroplastic phosphorylase has been obtained by Yang and Steup (1990) who found that it strongly interacts with a large complex polysaccharide composed of at least four different monosaccharides.

Studies with nonphotosynthetic plastids have had a major repercussion on our view of photosynthetic carbon metabolism. Recent efforts to elucidate the pathway between sucrose (imported into nonphotosynthetic cells) and starch synthesized in the plastids of these cells have added support to the concept that carbon is transported into the plastids in the form of a hexose unit, and not a triose unit (Frehner et al. 1990; Hatzfeld and Stitt 1990; Hill and Smith 1991). The fact that such plastids lack FBPase (Entwistle and ap Rees 1990) gives strength to this concept. Until recently, the import of hexose units in the form of glucose-6-P, and their mediation by a phosphate translocator (Flügge and Heldt 1991), was entertained. Then came the communication from Pozueta-Romero et al. (1991a) that the adenylate translocator of amyloplasts could transport ADPG, thereby supplying to the plastid the immediate substrate of the starch synthase enzyme. Coupled with a demonstration that sucrose synthase might be capable of generating ADPG from sucrose in the cytosol (Pozueta-Romero et al. 1991b), a relatively concise picture of the sucrose-to-starch conversion can be envisaged (Fig. 3). Note, however, that the fructose liberated by the action of sucrose synthase would need to be processed by a hexokinase (Schnarrenberger 1990), and a straightforward pathway for the incorporation of this half of the sucrose molecule into starch would still require utilization of the phosphate translocator for glucose-6-P import (Fig. 3).

The repercussion on photosynthesis came when Pozueta-Romero et al. (1991c) reported that chloroplasts also are capable of importing ADPG. It can be questioned whether or not any significant quantity of ADPG is imported by photosynthesizing chloroplasts already capable of generating hexose-P from $CO_2$ internally. If ADPG, gen-

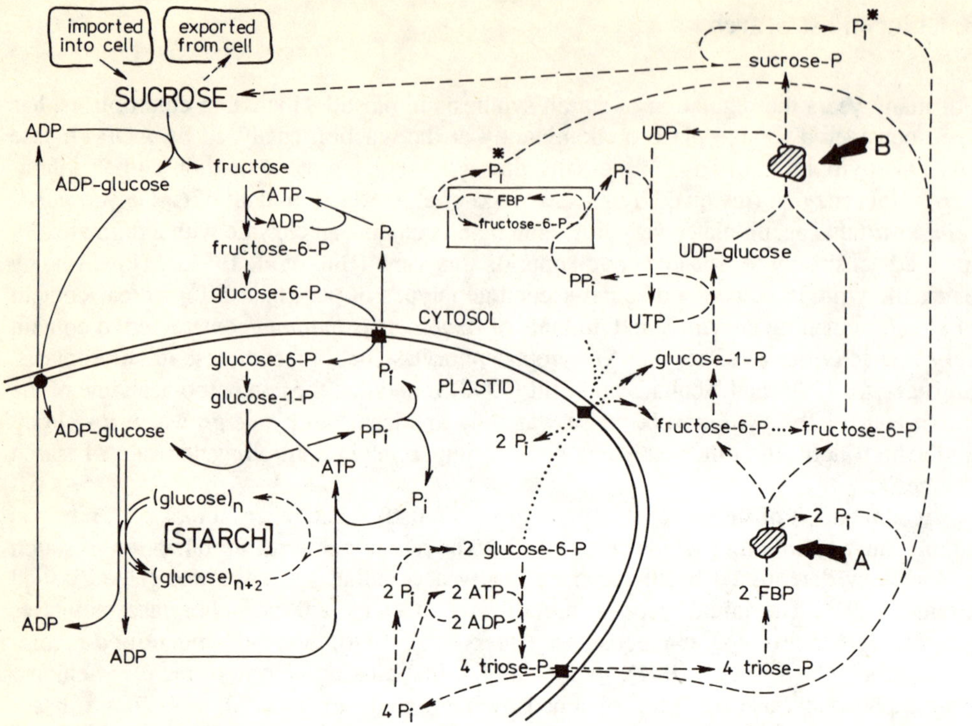


**Fig. 3.** A composite picture of starch and sucrose metabolism in plastid-containing plant cells. —, Plastid starch synthesis from cytosolic sucrose, utilizing ADPG transport via the adenylate translocator (●) (Pozueta-Romero et al. 1991a, c) and glucose-6-P transport via the phosphate translocator (■). The sum of the reactions is: sucrose + $(\text{glucose})_n \rightarrow (\text{glucose})_{n+2}$. This pathway is probably important for starch synthesis by amyloplasts; its activity with chloroplasts might be limited by the lower capacity of the chloroplast's phosphate translocator to transport glucose-6-P (Flügge and Heldt 1991). -----, Cytosolic sucrose synthesis from plastid starch, utilizing triose-P export by the phosphate translocator (■). The sum of the reactions is: $(\text{glucose})_{n+2} \rightarrow (\text{glucose})_n$ + sucrose. Not all of the plastid reactions are shown. This pathway is probably important for the nocturnal mobilization of chloroplast starch. Note that the cytosol lacks pyrophosphatase, but the combination of FBPase and PPi-linked P-fructokinase catalyzes, overall, the same reaction *(boxed small print)*. *A* and *B* point to the cytosolic enzymes FBPase and sucrose-P synthase (respectively), which are the principal regulatory enzymes of sucrose synthesis in photosynthetic cells (Neuhaus et al. 1990a). Mobilization of *amyloplast* starch may involve a modification (and simplification) of this pathway, whereby two glucose-6-P are exported (....) via the phosphate translocator in exchange for the net two Pi liberated (shown *Pi**) when the molecule of sucrose is synthesized

erated from sucrose, was imported by photosynthesizing chloroplasts that were simultaneously exporting triose-P to manufacture the sucrose, a new cycle for photosynthetic carbon metabolism would be set up (Fig. 3), but it might prove difficult to discover whether it achieves anything more significant than the consumption of 4 ATP and 1 UTP per sucrose cycled (Fig. 3). Regulation might prevent the cycle from being quantitatively significant; such regulation could be uncomplicated since the enzymes and membrane proteins involved in the sucrose-to-starch conversion are mostly separate from those that convert starch to sucrose; only chloroplast P-glucomutase, cytosolic P-hexose isomerase, and the chloroplast's phosphate translocator are common to both sectors (Fig. 3).

## 6. Cytosolic Sucrose Biosynthesis: Control by Cytosolic FBPase and Sucrose-P-Synthase

Practically all recent research into sucrose metabolism has centered on two enzymes: cytosolic FBPase and sucrose-P synthase (SPS). The respective locations of these enzymes in the pathway of sucrose synthesis are designated by A and B in Fig. 3.

FBPase activity is controlled via inhibition by the specially manufactured regulator molecule fructose-2,6-$P_2$. Its concentration is greater in tissues not expected to be actively synthesizing sucrose, e.g., the "sink" portion of leaves (Nielsen and Veierskov 1990), and leaves diverting carbohydrate to form the carbon skeletons for amino acid synthesis (Champigny et al. 1992). Blocking the enzyme that normally removes this inhibitor leads, as expected, to a reduction in sucrose synthesis (Brauer and Stitt 1990). However, fructose-2,6-$P_2$ may not be the only regulator of FBPase: inhibition by micromolar concentrations of $Ca^{2+}$ has been reported from two laboratories (Brauer et al. 1990; Prado et al. 1991). Either or both of these inhibitors, acting on the FBPase, may form part of the mechanism whereby sucrose accumulation in leaves inhibits further sucrose synthesis – a phenomenon recently demonstrated with transgenic tobacco expressing yeast invertase in cell walls (Stitt et al. 1990) – but the overall mechanism is still something of a mystery (Krause and Stitt 1992). Perhaps studies of FBPase from plant species evolutionarily distant from spinach and tobacco may provide clues; in this context, the recent discovery of cytosolic FBPase in the green alga *Selenastrum minutum* (Botha and Turpin 1990) may prove valuable.

Another mystery associated with FBPase is the role of the "competitor" enzyme PPi-linked P-fructokinase (which catalyzes the reversible reaction FBP + Pi $\rightleftharpoons$ fructose-6-P + PPi). In the direction of FBP formation it could, in a roundabout way, convert the PPi generated during UDP-glucose synthesis to two Pi, but the significance of such a role seems uncertain (Neuhaus et al. 1990b; Neuhaus and Stitt 1991). On the other hand, it could boost carbon flow from FBP to sucrose (via fructose-6-P) if FBPase activity was low; evidence for such a role in the grass *Lolium temulentum* has been obtained by Collis and Pollock (1991). In studies with nonphotosynthetic tissues, the enzyme was judged to be of value in the direction of FBP formation in potato tubers (and also *Euglena gracilis* under anaerobiosis), but not in banana fruit or carrot root; in the latter, a role comparable to that in leaves, viz. dealing with PPi, was envisaged (Enomoto et al. 1990; Ball et al. 1991b; Hajirezaei and Stitt 1991). Evidence from Wong et al. (1990) that more than one form of the enzyme exists, and that they have different susceptibilities to fructose-2,6-$P_2$ [which activates this enzyme (Kiss et al. 1991), in contrast to its inhibition of FBPase], helps to complicate the picture. Especially enigmatic is the apparent contradiction that fructose-2,6-$P_2$ can retard sucrose synthesis by *inhibiting* FBPase, yet it *stimulates* the PPi-linked P-fructokinase which is capable of achieving the same FBP $\rightarrow$ fructose-6-P conversion that is supposed to be prevented through FBPase inhibition.

The importance of SPS, the other regulatory enzyme of sucrose synthesis, has been highlighted by the report that a transgenic tomato plant, with total SPS activity elevated up to sixfold with maize SPS, synthesized 2.5 times the usual amount of sucrose in its leaves (Worrell et al. 1991). The enzyme has been more highly purified (Salvucci et al. 1990; Bruneau et al. 1991), and its light-mediated activation by dephosphorylation (and corresponding inactivation by phosphorylation in darkness) has been further studied (Budde and Randall 1990). The phosphorylation is probably mediated by a protein kinase, the identity of which is just beginning to appear (Huber and Huber 1991a). Interestingly, it seems that dephosphorylation can go too far, in that highly active (and therefore "dephosphorylated") enzyme is inactivated if further phosphate groups, not normally involved in regulation in vivo, are cleaved from it (Huber et al. 1991). Another unusual observation was that the normal inactive (phosphorylated) enzyme was, in fact, quite active when assayed in the presence of high ionic strength, but only if the enzyme was obtained from the leaves of a $C_4$ plant (Huber and Huber 1991b).

Triose-P for sucrose synthesis is delivered to the cytosol from the chloroplast by the phosphate translocator (Fig. 3). This translocator is the major protein of the chloroplast envelope; it constitutes 12% of the total envelope protein (Flügge et al. 1991). It is possibly universal in species containing chloroplasts: there is now evidence for its presence in the green macroalga *Bryopsis corticulans* (Yu and Pedersén 1991). A review of this and other metabolite translocators has recently appeared (Flügge and Heldt 1991). Operation of the translocators is indirectly regulated by reactions on both sides of the membrane (Heineke et al. 1991).

## 7. Mitochondrial Respiration

The problem of whether or not the mitochondria in photosynthesizing cells respire continues to be intractable. Earlier impressive experiments in which barley protoplast photosynthesis was shown to be inhibited by oligomycin (an inhibitor of mitochondrial electron transport) have been extended by those workers (Krömer and Heldt 1991a, b) to provide additional evidence that ATP generated by mitochondrial respiration and delivered to the cytosol is required for photosynthesis to continue, perhaps by assisting in the disposal of excess photosynthetic energy. It might even be possible for some of the mitochondrial ATP to be delivered to the chloroplast stroma when growth temperatures are high (Tarchevskii and Konyukhova 1989). Supportive evidence from studies with pea protoplasts has been reported by Vani et al. (1990) and Reddy et al. (1991), who described how a preceding period of respiration (dark) stimulated a subsequent short period of photosynthesis (light) and vice versa. However, the complexities of oxygen metabolism in photosynthetic cells have led others to question what is being observed when oxygen uptake in the light is measured: mitochondrial respiration, photosynthetic oxygen reduction (Mehler reaction), or even chlororespiration (Avelange and Rébeillé 1991; Avelange et al. 1991), although the latter is predominantly a dark process (Peltier and Schmidt 1991).

A matter related to mitochondrial respiration is the theoretical possibility that a reversal of the Krebs cycle could substitute for the Calvin cycle in photosynthesis. Such a situation is now more or less established for the primitive green anaerobic photosynthetic sulfur bacterium *Chlorobium limicola* forma *thiosulphatophilum* (Buchanan and Arnon 1990). A first indication that it might also be possible in eukaryotic cells has been provided by Chen and Gibbs (1992) who found that a *Chlamydomonas reihardtii* mutant lacking Ru5P kinase displayed acetate-enhanced $CO_2$ fixation, and contained three of the requisite enzymes.

## 8. Photorespiration

Much of the recent research on photorespiration has been directed at the individual steps between RuBP oxygenation and glycerate entry into the chloroplast. The oxygenase activity of RuBP carboxylase will no doubt receive special attention as biotechnologists dream of using site-directed mutagenesis to reduce or prevent it, thereby increasing plant productivity (Chen and Spreitzer 1989). To date, it appears that the small subunits have little to do with determining the enzyme's $CO_2/O_2$ specificity (Gutteridge 1991), but there may be some other nuclear-encoded protein that does (Chen et al. 1990).

RuBP oxygenation produces glycolate that is transported out of the chloroplast (Howitz and McCarty 1991) and further metabolized in peroxisomes. Acquisition of high quality preparations of peroxisomes is now possible (Turley and Trelease 1990). Although devoid of any visible internal structure, some structural arrangement of enzymes does appear to be present in peroxisomes, so that "metabolite channeling" is possible (Heupel et al. 1991). Nevertheless, variation in enzyme content is possible, as evidenced by the odd glyoxysomal-type peroxisomes in the xanthophycean algae *Bumilleriopsis* and *Vaucheria* (Backeshoff and Stabenau 1990).

The first half of the peroxisome's contribution to photorespiration is to effect the glycolate → glyoxylate → glycine conversion. Glycine is then transported to mitochondria. A barley mutant apparently deficient in the transport system accumulates glycine (Blackwell et al. 1990). Once in the mitochondria, two molecules of glycine are converted to one each of serine, $CO_2$ and $NH_4^+$ by a multi-enzyme complex that constitutes one-third of the total soluble protein of the mitochondrial matrix (Oliver et al. 1990), and possibly exists in a distinct "photorespiratory metabolic domain" of the mitochondrion (Wiskich et al. 1990). Reassimilation of released $NH_4^+$ occurs in the chloroplast (Yu and Woo 1991). If prevented, photosynthesis is inhibited, not so much by accumulated $NH_4^+$, but by accumulated glyoxylate that inhibits RuBP carboxylase (Wendler et al. 1990; see Sect. 3.a).

The second half of the peroxisome's contribution to photorespiration occurs after the serine (generated above) is transported to this organelle. It is deaminated to hydroxypyruvate which is then reduced to glycerate either in the peroxisome, or perhaps in the cytosol (Kleczkowski et al. 1992). The glycerate returns to the chloroplast in exchange for another glycolate (Howitz and McCarty 1991).

The only known, quantitatively important fate of glycolate in plant leaves is its metabolism to glycerate-plus-$CO_2$ via the glycolate pathway. However, in algae, two alternative fates appear to be possible: (1) a short-circuit from glyoxylate to glycerate via tartronic semialdehyde (see our 1978 review), and (2) excretion from the cell. Evidence for a transporter protein that facilitates excretion from *Chlamydomonas reinhardtii* cells has been obtained by Wilson and Tolbert (1991). Algae also differ from plant leaves in that glycolate, when metabolized , is not always oxidized to glyoxylate by a peroxisomal oxidase. Current indications are that brown macroalgae, the unusual *Cyanophora paradoxa*, and chromophyte microalgae other than diatoms utilize the oxidase; but green, prasinophyte, and cryptophyte microalgae, and diatoms, utilize a dehydrogenase located in the mitochondrion (Gross 1990; Suzuki et al. 1991; Betsche et al. 1992). Categories are not clear-cut: the green microalga *Dunaliella primolecta* contains both the oxidase and the dehydrogenase (Nimer et al. 1990). Finally, questions are raised by the apparent total lack of glycolate-oxidizing enzymes in dinoflagellates and prymnesiophytes (Suzuki et al. 1991).

The function of photorespiration in plants has received little recent attention, apart from a commanding study by Wu et al. (1991) in which photorespiration was evaluated to be superior to the photoreduction of oxygen (Mehler reaction and subsequent $H_2O_2$ metabolism) for protection against photoinhibition. The strength of this conclusion will have considerable bearing on the survivability of any future crop plants engineered to contain low (or zero) RuBP oxygenase activity.

## 9. Stress and Photosynthetic Carbon Metabolism

Photosynthetic organisms seldom grow in ideal environments. Temperature, light level, water supply, nutrient availability, and soil pH are five prominent external environmental parameters that plants must deal with. Most commonly, the problem is that one (or more) of these is too low. When the light level is low, plants respond by mimicking shade plants: RuBP carboxylase is down-regulated, and the quantity of light-harvesting machinery is increased. But when temperature or water supply is too low, a major requirement is to deal with an excess of harvested light energy that follows the consequent retardation of photosynthetic carbon metabolism. Failure to deal with this problem is common, and leads to "overreduction"-induced photoinhibition.

Recent studies with the $C_3$ plant *Helianthus annuus* (sunflower) and the CAM plant *Kalanchoe pinnata* indicate that, under normal growth conditions, photosynthetic electron transport and enzyme activities, coupled with temporary engagements of oxygen reduction (Mehler reaction) are sufficient to fully utilize the harvested light energy and thus avoid photoinhibition (Siebke et al. 1990, 1992; Luttge et al. 1991). But a fascinating study of the $C_3$ plant *Gossypium hirsutum* (cotton) by Winter and Königer (1991) modifies this view, and may fundamentally alter our perception of the photosynthetic carbon metabolism of plant leaves. In essence, these authors found that the potential photosynthesis of cotton (a "sun" plant) is underutilized in full sunlight when the temperature is "slightly suboptimal", e.g., 27 °C. When increased to 33–34 °C, extremely high growth rates of up to $0.26\ g \cdot g^{-1} \cdot day^{-1}$ were measured – we calculate that this represents a *field* rate of around 100 $\mu mol\ CO_2$ fixed $\cdot\ mg^{-1}$ chlorophyll $\cdot\ h^{-1}$. Additionally, these authors found that, with irradiance reduced to about one-half of full sunshine, the dramatic difference between 33 and 27 °C plants was not observed. They concluded that high light limits the growth of sun plants because even relatively warm temperatures of 25–27 °C are suboptimal. This proposed "subtle photoinhibition" somehow must have been prevented from increasing to a catastrophic level, otherwise plants would have died. Königer and Winter (1991) obtained evidence that the reversible de-epoxidation of violaxanthin to zeaxanthin was central to the process of prevention, but the exact mechanism by which zeaxanthin assists in the dissipation of excess harvested light energy still has not been experimentally elucidated (Demmig-Adams 1990; Young 1991).

Of course, stress situations can increase to a point where they ultimately overwhelm protective mechanisms, leading to severe photoinhibition. Two recently reported examples are a mutant of *Synechocystis* that contains greatly reduced Ru5P kinase activity (Su and Bogorad 1991), and *Valerianella locusta* protoplasts exposed to a low temperature of 4 °C (van Wijk and Krause 1991). In the latter case, support for the conclusions of Wu et al. (1991), (see Sect. 8) were obtained: energy dissipation by photoreduction of $O_2$ was found to accentuate, rather than relieve, photoinhibition because metabolism of the products of oxygen reduction was too slow, and these reactive products then damaged the photosynthetic apparatus of photosystem II.

Little attention has been given to the induction of photoinhibition by low water supply. Rather, most studies have concentrated on whether the reduction in photosynthesis caused by water stress is due to stomatal closure, or altered metabolism, or a combination of both. Stomatal closure was concluded to cause most of the reduction of photosynthesis with leaves of bean, lupin, sunflower, and *Eucalyptus globulus*, but a metabolic component appeared to be important in grapevine leaves (Cornic and Briantais 1991; Quick et al. 1992), and even with bean leaves a metabolic response to water stress was detected, viz. a reduction in SPS activity (Vassey et al. 1991). Of course, it can be argued that metabolic responses may be a consequence of the reduced availability of $CO_2$ for photosynthesis following stomatal closure. However, four studies with sunflower, and one with soybean (Frederick et al. 1989; Graan and Boyer 1990; Wise et al. 1990, 1991; Gimez et al. 1992) have not supported this interpretation; rather, water stress-induced perturbations of metabolism appeared to be, at least in part, primarily responsible for the reductions in photosynthesis. These perturbations were looked upon as being normal adaptations to accommodate the unpredictable nature of water availability, and may center on a reduction in photophosphorylation and consequent shortfall in the capacity to regenerate RuBP.

The shortfall is less severe with plants acclimatized to water stress (Santakumari and Berkowitz 1991), and the observation that gradual development of water stress had little effect on the photosynthetic *capacity* of soybean leaves (Frederick et al. 1990) again raises questions concerning the relative contributions of metabolism and stomata. Interactions between these may be possible (Wise et al. 1991), and the question of whether or not the partial closure of stomata (due to water stress) results in lowered levels of $CO_2$ in the leaf needs to be critically addressed (Cornic and Briantais 1991).

## 10. $C_4$ Photosynthesis

Two notable events influenced the biosphere about 60 million years ago. One was the catastrophic impact of a large asteroid with the earth, probably upon present-day Mexico, resulting in a temporary but drastic climate change that halted photosynthesis, and caused many species, including dinosaurs, to become extinct (McKinnon 1992). Less well known, and less dramatic, was the adaptation of plants to a gradually declining level of $CO_2$ in the atmosphere, from something like 3000 ppm (120 million years ago) to a value similar to today's, about 350 ppm (60 million years ago). Ehleringer et al. (1991) proposed that the evolutionary appearance of $C_4$ plants probably took place at this time, when the benefits of the $C_4$ pathway's $CO_2$ concentrating mechanism first would have been of benefit to photosynthesis.

Atmospheric $CO_2$ has remained near this low level for the past 60 million years, thus today's $C_4$ plants continue to benefit from possession of the $C_4$ "$CO_2$-pump". This pump is so efficient that, even at very limiting $CO_2$ concentrations, photosynthesis is streamlined in such a way that both photorespiration and photoreduction of $O_2$ are kept low (Krall and Edwards 1990). With such efficiency, one would not expect $C_4$ plants to benefit from a future increase in atmospheric $CO_2$ (as is occurring due to human activities), and indeed this expectation has often been demonstrated (most recently by Ziska et al. 1990, 1991). An exception was *Echinochloea frumentacea*, which did show some response to increased $CO_2$ when grown with a low supply of nitrogen (Wong and Osmond 1991).

When $CO_2$ enters the $C_4$ plant leaf it is initially fixed by PEP carboxylase which, however, uses $HCO_3^-$ as substrate. It is therefore dependent on rapid generation of $HCO_3^-$ from entering $CO_2$ by the enzyme carbonic anhydrase. Measurements have shown that the activity of this enzyme is just sufficient to accommodate the observed carbon flux (Burnell et al. 1990; Hatch and Burnell 1990). One thought was that the unique $Na^+$ requirement of $C_4$ plants might be related to this enzyme, but it was not (Brownell et al. 1991). More likely, $Na^+$ is required to be cotransported with pyruvate during the latter's uptake by mesophyll chloroplasts, but even this explanation is incomplete because it does not seem to apply to one of the three groups of $C_4$ plants (NADP-malic enzyme group) which possibly uses $H^+$/pyruvate cotransport instead (Ohnishi et al. 1990).

A surprise addition to the list of $C_4$ plants is imminent. Reiskind and Bowes (1991) report that the marine green macroalga *Udotea flabellum* displays some of the physiological and biochemical characteristics of a $C_4$ plant, including minimal photorespiration, although it lacks a Kranz anatomy. Also, capture of cytosolic $CO_2$ is not by PEP carboxylase, but by PEP carboxykinase. Nevertheless, it seems that this enzyme participates in the development of a pool of malate that moves to the chloroplast where decarboxylation elevates the $CO_2$ concentration in the vicinity of RuBP carboxylase. Furthermore, it is not inconceivable that a comparable system operates in the brown alga *Ascophyllum nodosum* (Johnston and Raven 1989).

Returning to the standard $C_4$ plant: Doncaster and Leegood (1990) have clarified why cytosolic malate dehydrogenase does not interfere with the substantial flow of malate from mesophyll cells to bundle-sheath cells. Pyruvate produced following malate decarboxylation in the bundle sheath mostly returns to the mesophyll, but a few percent can be diverted for nitrogen assimilation; it is aminated to alanine, which constitutes one-quarter of all amino acids in the phloem of maize (Valle and Heldt 1991).

Enzymes associated with $C_4$ photosynthesis have continued to attract attention. At least six of them undergo reversible polymerization in vitro – the possible significance of this to in vivo regulation has been reviewed by Podestá et al. (1990). The polymerization state of NADP-linked malic enzyme changes noticeably in the physiological pH range of 7 to 8 (Iglesias et al. 1991). PEP carboxylase has received most attention, and is dealt with below (Sect. 10 a). The mesophyll chloroplast's NADP-linked malate dehydrogenase and the bundle sheath's NADP-linked malic enzyme both have been sequenced from clones of their complementary DNAs (Metzler et al. 1989; Börsch and Westhoff 1990). The dehydrogenase, which is reported to exist as isoenzymes in *Echinochloa crusgalli* (Vairinhos and Simon 1990), must undergo at least 50% of its potential reductive activation before any activity appears (Miginiac-Maslow et al. 1990). $C_4$ plants that require NAD-linked malic enzyme apparently require contrasting regulation: in "NAD malic enzyme-type" $C_4$ plants, the enzyme is inhibited by physiological levels of AMP, ADP and especially ATP, but in "PEP carboxykinase-type" $C_4$ plants, ATP activates the enzyme (Furbank et al. 1991).

A characteristic, and possibly limiting (Reddy and Das 1990) enzyme of $C_4$ photosynthesis is pyruvate Pi dikinase which generates the $CO_2$-acceptor PEP (and AMP and PPi) from pyruvate, Pi, and ATP. The form of Pi used has been identified by Nakamoto and Edwards (1990) as $HPO_4^{2-}$. Assay of the enzyme activity is difficult; improvements recommended by Salahas et al. (1990a, b) are inclusion of ample, glucose-6-P-activated coupling PEP carboxylase, and glycerol to stabilize the enzyme. Regulation of the dikinase in vivo is centered on light-mediated activation by dephosphorylation of a threonyl-P residue (reviewed by Budde and Randall 1990; Bennett 1991), a process that is relatively independent of the normal temperatures experienced by $C_4$-plant leaves (Grammatikopoulos and Manetas 1990).

## a) PEP Carboxylase

Almost all investigations of PEP carboxylase over the past 2 years have involved the $C_4$ leaf enzyme. Genetic analysis indicates that its synthesis is directed by a subgroup of genes rather than by one gene, at least in the $C_4$ dicotyledonous *Flaveria trinervia* (Poetsch et al. 1991). Past studies with the enzyme in vitro may need re-evaluation: (1) the *Cynodon dactylon* enzyme was found to be relatively insensitive to glucose-6-P activation but more sensitive to malate inhibition when assayed with glycerol at a concentration that stimulated in vivo concentrations of solutes (Stamatakis et al. 1990), and (2) its specific activity after ion-exchange chromotography that involved sulfate or phosphate anions was more than twice that measured when chloride ions were involved (Angelopoulos and Gavalas 1991). More realistic properties are considered to be seen when the enzyme is assayed immediately after release from protoplasts or single cells. Such tests have confirmed the high sensitivity to feedback inhibition by malate for the maize leaf enzyme (Petropoulou et al. 1990), and for the enzyme from the guard cells of $C_3$-plant stomata (Tarczynski and Outlaw 1990; Willmer et al. 1990).

Maize leaf PEP carboxylase has been the focus of five studies dealing with effects of pH and temperature on hysteretic (i.e., slowly changing) kinetic properties and related reversible enzyme polymerization (Kleczkowski and Edwards 1990, 1991; López-Pozos et al. 1990; Wu et al. 1990). Overall, hysteresis was attributed to slow reversible changes in the degree of polymerization. Cold storage or alkaline pH favored inactive polymers, while neutral pH, warmer temperatures, and the substrate PEP favored the active tetramer. However, the effects of high pH may be complicated by enzyme sensitivity to contaminating heavy metal ions at this pH (Wedding and Black 1990).

Real advances have been made in unraveling the complex mechanism of light-mediated activation of PEP carboxylase, which in turn is a major determinant of the functional properties of the enzyme (Arrio-Dupont et al. 1992). Jiao and Chollet (1991) argue that it is far more important than tetramer/dimer interconversions, although the latter may have some value in CAM plants. As reported in our last review, activation of the $C_4$-plant enzyme involves phosphorylation of a seryl residue, catalyzed by a protein kinase (Fig. 4). The residue involved is near the N-terminus of the polypeptide chain: serine-15 in the maize enzyme, serine-8 in the sorghum enzyme (Jiao et al. 1991a), quite distant from the enzyme's catalytic site (Jiao and Chollet 1990). It seems that the phosphorylation is light-mediated because the protein kinase that catalyzes it is "light-activated" (Echevarría et al. 1990) by a process that, at its simplest, involves light-dependent, short-term degradation and resynthesis of the kinase in the cytosol (Jiao et al. 1991b; Jiao and Chollet 1992; see Fig. 4). The messenger(s) that connects the light signal to kinase resynthesis remains to be found. Other intriguing facets must also be kept in mind: (1) phosphorylative activa-

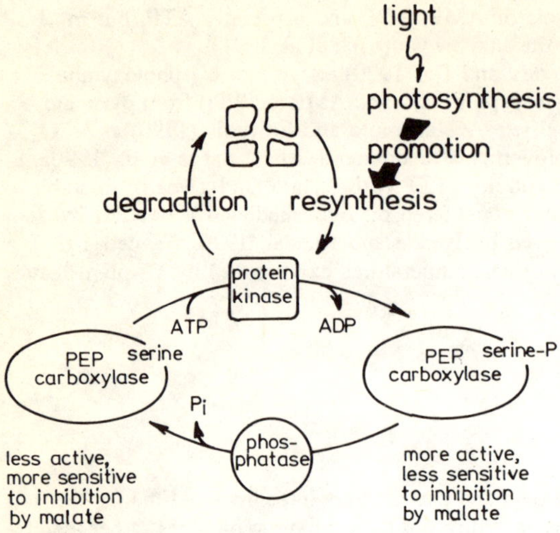


**Fig. 4.** Known aspects of the mechanism of light-mediated activation of PEP carboxylase in $C_4$-plant leaves. (After Echevarría et al. 1990; Jiao et al. 1991b; and Jiao and Chollet 1991, 1992)

tion of PEP carboxylase is *dark*-mediated in CAM plants (see review by Jiao and Chollet 1991), and (2) it does not occur at all in $C_3$-plant guard cells (Willmer et al. 1990).

## 11. $C_3$–$C_4$ Intermediate Photosynthesis

Not a lot has been added to our understanding of those plants that display both $C_3$ and $C_4$ characteristics of photosynthesis. An earlier suggestion that *Manihot esculenta* (cassava) belongs to this group has been contested by Edwards et al. (1990), but the concept that certain photosynthetic tissues in the inflorescence of cereal grasses are $C_3$–$C_4$ has been extended to the rice panicle; it will be difficult to contest this conclusion since substantial activity of the $C_4$-enzyme pyruvate Pi dikinase was found in the panicle (Imaizumi et al. 1990).

The photosynthetic responses of those $C_3$–$C_4$ species more extensively studied during the past decade were monitored by Brown et al. (1991) after leaves were fed a PEP-analog that inhibits PEP carboxylase. Surprisingly, none of *Panicum milioides*, *Moricandia arvensis*, and *Neurachne minor* were affected; these species responded to the inhibitor as if they were $C_3$! However, inhibition of photosynthesis was detected for *Flaveria linearis;* this response was qualitatively similar to those of $C_4$-like and true-$C_4$ species of *Flaveria.* Indeed, the *Flaveria* genus is remarkable in that a continuous gradation from $C_3$ to $C_4$ exists, e.g., *F. pringlei* ($C_3$), *F. sonorensis* ($C_3$–$C_4$, but lacking a $C_4$ cycle), *F. floridana* ($C_3$–$C_4$, with $C_4$ cycle), *F. brownii* ($C_4$-like), *F. bidentis* ($C_4$) (Krall et al. 1991; Ku et al. 1991). In addition, a $C_3$–$C_4$/$C_4$-like hybrid has been reported (Araus et al. 1991).

The gradation of photosynthetic characteristics in *Flaveria* was matched by a gradual decrease in the level of photorespiratory enzymes, leading Ku et al. (1991) to propose that, evolutionarily, photorespiration was first combated by recapture of photorespiratory $CO_2$ (as occurs in $C_3$–$C_4$ plants), and was then repressed when the true $C_4$ pathway's

$CO_2$-pump generated $CO_2$ levels that overwhelmed the oxygenase activity of RuBP carboxylase. Thus, despite the apparent uselessness of PEP carboxylase evidenced from the experiments of Brown et al. (1991; see above), situations where it could efficiently recapture photorespiratory $CO_2$ (e.g., warm leaf temperatures) have been identified (Schuster and Monson 1990), and the high pools of glycine and serine in $C_3$–$C_4$ plant leaves (Rawsthorne and Hylton 1991) attest to a special photorespiratory metabolism in these plants. Indeed, the pivotal discovery of Hylton et al. (1988) that the $CO_2$-releasing enzyme of photorespiration (glycine decarboxylase) is confined to the bundle-sheath cells of $C_3$–$C_4$ plants is now more significant in that it can explain why PEP carboxylase is not necessary for reducing photorespiration in these plants: quite simply, photorespired $CO_2$, being released only in bundle-sheath cells, is directly recaptured by the RuBP carboxylase in these cells (Brown et al. 1991).

## 12. Crassulacean Acid Metabolism (CAM)

Just as more plant species have been discovered with photosynthesis intermediate between $C_3$ and $C_4$, so also more species with photosynthesis between $C_3$ and CAM have been found. Haag-Kerwer et al. (1992) have described a $C_3$-CAM intermediate, *Clusia minor*, and Herrera et al. (1991) report that *Talinum triangular* displays an inducible weak CAM that assists survival during drought. The better-known case of inducible CAM is that shown by *Mesembryanthemum crystallinum*. Environmental factors, such as high salinity, water stress, and photoperiod, are more important than developmental factors for inducing CAM in this species (Cheng and Edwards 1991; Piepenbrock and Schmitt 1991). Interestingly, the induction by water stress could be mimicked by applying abscisic acid to leaves (Chu et al. 1990). After CAM has been induced, *M. crystallinum* retains the capacity for photorespiration (Whitehouse et al. 1991).

In the standard interpretation of CAM, nocturnal $CO_2$ fixation by PEP carboxylase generates a pool of malate that is stored in the vacuole until the next day, when the malate is returned to the cytoplasm and decarboxylated, thereby making the "stored $CO_2$" available for photosynthesis behind closed stomates. Water loss is consequently minimized. The PEP carboxylase activity reflects the degree of CAM: it fluctuates seasonally, and with the degree of water stress (Pilon-Smits et al. 1990, 1991). The fluctuation results form enzyme synthesis (Pilon-Smits et al. 1990). In *Sedum telephium*, which displays inducible CAM, a new molecular species of PEP carboxylase is synthesized (Groenhof et al. 1990). A method for the improved recovery of CAM-plant PEP carboxylase for future studies has become available (Silva et al. 1991).

A variation of CAM that differs from the standard, and may be considered as very $C_3$-like, is "CAM cycling". Species prominently engaged in CAM cycling, such as those in the Portulacaceae, take up atmospheric $CO_2$ during the day, and have stomata open by day and closed by night – these are $C_3$ plant features, and contrast to CAM characteristics. Their CAM attribute derives from their recapture of respiratory $CO_2$ to form malic acid that accumulates nocturnally (Harris and Martin 1991), and can be used to supply $CO_2$ for photosynthesis the next day behind *open* stomates.

However, as may be envisaged, there is no reason why respiratory $CO_2$ may not be recaptured by standard CAM plants, and indeed this does occur, especially if higher nocturnal temperatures enhance respiration, or water stress becomes so severe that stomatal opening even in the dark compromises water economy (Kalt et al. 1990; Fetene and Lüttge 1991). In extreme circumstances, total stomatal closure may occur, but daytime photoinhibition is avoided because sufficient $CO_2$ for photosynthesis is made available from respiratory $CO_2$. Storage of citrate, rather than malate, has been detected, and is considered to be energetically advantageous for this process (Lüttge 1990; Haag-Kerwer et al. 1992).

Two details of CAM biochemistry have been further investigated. The first relates to malate transport in and out of the vacuole. Uptake by the vacuole is active, and is linked to $H^+$ transport which is complex in that three activities are involved: ATP-dependent, PPi-dependent, and pyrophosphatase-activated ATP-dependent (Marquardt-Jarczyk and Lüttge 1990). Efflux of malate is passive, and may be slowed if low temperature early in the morning reduces tonoplast fluidity; this can cause a lag in photosynthesis (Kluge et al. 1991). The second relates to the necessity to cycle one 3C residue through the PEP $\rightarrow$ malate $\rightarrow$ pyruvate $\rightarrow$ reserve polysaccharide $\rightarrow$ PEP circuit for each $CO_2$ captured nocturnally and subsequently fixed diurnally. If the products of net $CO_2$ fixation are not mobilized to sink tissues, excessive storage of the reserve polysaccharide (starch) can distort chloroplasts, ultimately reducing CAM activity (Mayoral et al. 1991). In *Fourcroya humboldtiana* the reserve polysaccharide is not starch, rather it is a low molecular weight fructan (Olivares and Medina 1990).

The requirement of this voluminous traffic of a 3C residue would intuitively suggest that the potential productivity of CAM plants might be less than that of $C_3$ and $C_4$ plants. Indeed, CAM plants have the highest requirement for ATP: about 6 ATP per $CO_2$ fixed, compared to 5 ATP ($C_4$ plants) and 3 ATP ($C_3$ plants). However, a series of calculations and predictions by Nobel (1991), nevertheless, indicates that the potential productivity of CAM plants can equal, or even exceed that of $C_3$ and $C_4$ plants, partly because CAM plants avoid photorespiratory $CO_2$ loss and are better adapted to continue photosynthesis when water stress is substantial. Quite recently, Nobel et al. (1992) have demonstrated this potential by recording productivities averaging 46 t dry $wt.ha^{-1} \cdot year^{-1}$ (which are among the highest ever reported for any plant species) for *Opuntia amyclea* and *Opuntia ficus-indica* growing in Mexico.

## 13. Concluding Comment

In our reviewing of 20 years of literature, we have given much emphasis to enzymes. We make no apologies for maintaining this emphasis during a time when DNA-based biotechnology became "the flavor of the month". After all, the basic aim of this technology is to introduce new enzymes (and other proteins) into cells. Of all the biomolecules, the enzymes are most alive, as Kornberg (1980) pointed out about 12 years ago: "You know a cell's enzymes can sustain it without DNA. The cell may have no future without DNA but it still can exist. By contrast, a cell with DNA but without enzymes cannot exist at all".

There is no doubt that the study of enzymes will continue to hold a prominent place in research on photosynthetic carbon metabolism during the next 20 years. But proteins with other functions, which have received less attention in the past, will also be worth watching. These include: (1) membrane proteins that transport metabolites between cellular compartments – the role of these in photosynthesis has already received detailed attention; (2) proteins designed to directly influence the function of other proteins, e.g., rubisco activase; (3) membrane receptor proteins, about which little is known in relation to photosynthesis; and (4) hormones – the first report of a plant hormone that is a polypeptide was recently published (Pearce et al. 1991).

The currently fascinating picture of how photosynthetic carbon metabolism is run by the molecules of life – proteins – will almost certainly be more intricate, and more wonderful, in 2012.

## References

**Andrews TJ,** Kane HJ (1991) J Biol Chem 266:9447–9452. **Angelopoulos K,** Gavalas NA (1991) J Plant Physiol 137:363–367. **Araus JL,** Brown HR, Byrd GT, Serret MD (1991) Planta 183:497–504. **Arp WJ,** Drake BG (1991) Plant Cell Environ 14:1003–1006. **Arrio-Dupont M,** Bakrim N, Echevarría C, Gadal P, Le Maréchal P, Vidal J (1992) Plant Sci 81:37–46. **Arulanantham AR,** Rao IM, Terry N (1990) Plant Physiol 93:1466–1475. **Austin PA,** Ross IS, Mills JD (1992) Biochim Biophys Acta 1099:226–232. **Avelange M-H,** Rébeillé F (1991) Planta 183:158–163. **Avelange M-H,** Thiéry JM, Sarrey F, Gans P, Rébeillé F (1991) Planta 183:150–157.

**Backeshoff K,** Stabenau H (1990) Bot Acta 103:190–196. **Badger MR,** Price GD (1990) Plant Physiol 94:35–39. **Ball S,** Marianne T, Dirick L, Fresnoy M, Delrue B, Decq A (1991a) Planta 185:17–26. **Ball KL,** Green JH, ap Rees T (1991b) Europ J Biochem 197:265–269. **Bédu S,** Peltier G, Sarrey F, Joset F (1990) Plant Physiol 93:1312–1315. **Beer S,** Israel A, Drechsler Z, Cohen Y (1990) Plant Physiol 94:1542–1546. **Beer S,** Sand-Jensen K, Vindbaek Madsen T, Nielsen SL (1991) Oecologia 87:429–434. **Beers EP,** Duke SH, Henson CA (1990) Plant Physiol 94:738–744. **Bennett J** (1991) Annu Rev Plant Physiol Plant Mol Biol 42:281–311. **Besford RT,** Ludwig LJ, Withers AC (1990) J Exp Bot 41:925–931. **Betsche T,** Schaller D, Melkonian M (1992) Plant Physiol 98:887–893. **Björk M,** Haglund K, Ramazanov Z, Garcia-Reina G, Pedersén M (1992) Planta 187:152–156. **Blackwell RD,** Murray AJS, Lea PJ (1990) Plant Physiol 94:1316–1322. **Booth WA,** Beardall J (1991) Phycologia 30:220–225. **Börsch D,** Westhoff P (1990) FEBS Lett 273:111–115. **Botha FC,** Turpin DH (1990) Plant Physiol 93:1460–1465. **Bowes G** (1991) Plant Cell Environ 14:795–806. **Brauer M,** Stitt M (1990) Physiol Plant 78:568–573. **Brauer M,** Sanders D, Stitt M (1990) Planta 182:236–243. **Brown RH,** Byrd GT, Black CC (1991) Plant Physiol 97:985–989. **Brownell PF,** Bielig LM, Grof CPL (1991) Aust J Plant Physiol 18:589–592. **Brüggemann W,** van der Kooij TAW, van Hasselt PR (1992) Planta 186:179–187. **Bruneau J-M,** Worrell AC, Cambou B, Lando D, Voelker TA (1991) Plant Physiol 96:473–478. **Buchanan BB** (1991) Arch Biochem Biophys 288:1–9. **Buchanan BB,** Arnon DI (1990) Photosyn Res 24:47–53. **Budde RJA,** Randall DD (1990) Plant Physiol 94:1501–1504. **Burnell JN,** Suzuki I, Sugiyama T (1990) Plant Physiol 94:384–387.

**Campbell WJ,** Ogren WL (1990a) Plant Physiol 94:479–484. **Campbell WJ,** Ogren WL (1990b) Photosyn Res 23:257–268. **Casper T,** Lin T-P, Kakefuda G, Benbow L, Preiss J, Somerville C (1991) Plant Physiol 95:1181–1188. **Champigny ML,** Brauer M, Bismuth E, Manh CT, Siegl G, Quy LV, Stitt M (1992) J Plant Physiol 139:361–368. **Chardot T,** Meunier J-C (1990) Plant Sci 70:1–9. **Chardot T,** Meunier J-C (1991) Biochem J 278:787–791. **Chen C,** Gibbs M (1992) Plant

Physiol 98:535–539. **Chen Z,** Spreitzer RJ (1989) J Biol Chem 264:3051–3053. **Chen Z,** Green D, Westhoff C, Spreitzer RJ (1990) Arch Biochem Biophys 283:60–67. **Cheng S-H,** Edwards GE (1991) Plant Cell Environ 14:271–278. **Chu C,** Dai Z, Ku MSB, Edwards GE (1990) Plant Physiol 93:1253–1260. **Clasper S,** Easterby JS, Powls R (1991) Eur J Biochem 202:1239–1246. **Coleman JR,** Rotatore C, Williams TG, Colman B (1991) Plant Physiol 95:331–334. **Collis BE,** Pollock CJ (1991) New Phytol 119:483–489. **Cornic G,** Briantais J-M (1991) Planta 183:178–184. **Crafts-Brandner SJ,** Salvucci ME, Egli DB (1991) Planta 183:300–306.

**Demmig-Adams B** (1990) Biochim Biophys Acta 1020:1–24. **Doncaster HD,** Leegood RC (1990) J Plant Physiol 136:330–335. **Drechsler Z,** Beer S (1991) Plant Physiol 97:1439–1444.

**Echevarría C,** Vidal J, Jiao J, Chollet R (1990) FEBS Lett 275:25–28. **Edmondson DL,** Badger MR, Andrews TJ (1990a) Plant Physiol 93:1390–1397. **Edmondson DL,** Badger MR, Andrews TJ (1990b) Plant Physiol 93:1376–1382. **Edwards GE,** Sheta E, Moore BD, Dai Z, Franceschi VR, Cheng S-H, Lin C-H, Ku MSB (1990) Plant Cell Physiol 31:1199–1206. **Ehleringer JR,** Sage RF, Flanagan LB, Pearcy RW (1991) Trends Ecol Evol 6:95–99. **Eilenberg H,** Beer S, Gepstein S, Geva N, Tadmor O, Zilberstein A (1991) Plant Physiol 95:298–304. **Enomoto T,** Ohyama H, Inui H, Miyatake K, Nakano Y, Kitaoka S (1990) Plant Sci 67:161–167. **Entwistle G,** ap Rees T (1990) Biochem J 271:467–472. **Espie GS,** Kandasamy RA (1992) Plant Physiol 98:560–568. **Espie GS,** Miller AG, Kandasamy RA, Canvin DT (1991a) Can J Bot 69:936–944. **Espie GS,** Miller AG, Canvin DT (1991b) Plant Physiol 97:943–953.

**Fetene M,** Lüttge U (1991) J Exp Bot 42:25–31. **Florencio FJ,** Yee BC, Johnson TC, Buchanan BB (1988) Arch Biochem Biophys 266:496–507. **Flügge U-I,** Heldt HW (1991) Annu Rev Plant Physiol Plant Mol Biol 42:129–144. **Flügge U-I,** Weber A, Fischer K, Lottspeich F, Eckerskorn C, Waegemann K, Soll J (1991) Nature 353:364–367. **Fredeen AL,** Raab TK, Rao IM, Terry N (1990) Planta 181:399–405. **Frederick JR,** Alm DM, Hesketh JD (1989) Photosynthetica 23:575–584. **Frederick JR,** Alm DM, Hesketh JD, Below FE (1990) Photosyn Res 25:49–57. **Frehner M,** Pozueta-Romero J, Akazawa T (1990) Plant Physiol 94:538–544. **Furbank RT,** Agostino A, Hatch MD (1991) Arch Biochem Biophys 289:376–381.

**Gehl KA,** Colman B, Sposato LM (1990) J Exp Bot 41:1385–1391. **Gimenez C,** Mitchell VJ, Lawlor DW (1992) Plant Physiol 98:516–524. **Gimmler H,** Weiss C, Baier M, Hartung W (1990) J Exp Bot 41:785–795. **Goyal A,** Shiraiwa Y, Tolbert NE (1992) Plant Physiol 98:578–583. **Graan T,** Boyer JS (1990) Planta 181:378–384. **Grammatikopoulos G,** Manetas Y (1990) Physiol Plant 80:593–597. **Groenhof AC,** Smirnoff N, Bryant JA (1990) Plant Cell Environ 13:437–445. **Gross W** (1990) J Phycol 26:381–383. **Grub A,** Mächler F (1990) J Exp Bot 41:1293–1301. **Gutteridge S** (1991) J Biol Chem 266:7359–7362.

**Haag-Kerwer A,** Franco AC, Lüttge U (1992) J Exp Bot 43:345–352. **Hajirezaei M,** Stitt M (1991) Plant Sci 77:177–183. **Harley PC,** Sharkey TD (1991) Photosyn Res 27:169–178. **Harris FS,** Martin CE (1991) Plant Physiol 96:1118–1124. **Hatch MD** (1991) Anal Biochem 192:85–89. **Hatch MD,** Burnell JN (1990) Plant Physiol 93:825–828. **Hatzfeld W-D,** Stitt M (1990) Planta 180:198–204. **Heineke D,** Riens B, Grosse H, Hoferichter P, Peter U, Flügge U-I, Heldt HW (1991) Plant Physiol 95:1131–1137. **Hendrix DL,** Grange RI (1991) Plant Physiol 95:228–233. **Hermoso R,** Fonollá J, de Felipe MR, Vivó A, Chueca A, Lázaro JJ, López-Gorgé J (1992) Plant Physiol Biochem 30:39–46. **Herrera A,** Delgado J, Paraguatey I (1991) J Exp Bot 42:493–499. **Heuer B,** Portis AR (1990) Plant Physiol 93:1511–1513. **Heupel R,** Markgraf T, Robinson DG, Heldt HW (1991) Plant Physiol 96:971–979. **Hill LM,** Smith AM (1991) Planta 185:91–96. **Holbrook GP,** Galasinski SC, Salvucci ME (1991) Plant Physiol 97:894–899. **Houtz RL,** Royer M, Salvucci ME (1991) Plant Physiol 97:913–920. **Houtz RL,** Poneleit L, Jones SB, Royer M, Stults JT (1992) Plant Physiol 98:1170–1174. **Howitz KT,** McCarty RE (1991) Plant Physiol 96:1060–1069. **Huber SC,** Huber JL (1990) Arch Biochem Biophys 282:421–426. **Huber SC,** Huber JL (1991a) Plant Cell Physiol 32:319–326. **Huber SC,** Huber JL (1991b) Plant Cell Physiol

32:327–333. **Huber JL,** Hite DRC, Outlaw WH, Huber SC (1991) Plant Physiol 95:291–297. **Hudson GS,** Evans JR, von Caemmerer S, Arvidsson YBC, Andrews TJ (1992) Plant Physiol 98:294–302. **Huppe HC,** Picaud A, Buchanan BB, Miginiac-Maslow M (1991) Planta 186:115–121. **Husic HD,** Quigley EA (1990) Plant Physiol 94:380–383. **Hylton CM,** Rawsthorne S, Smith AM, Jones DA, Woolhouse HW (1988) Planta 175:452–459.

**Idso SB,** Kimball BA (1991) Plant Physiol 96:990–992. **Iglesias AA,** Spampinato CP, Andreo CS (1991) Arch Biochem Biophys 290:272–276. **Imaizumi N,** Usuda H, Nakamoto H, Ishihara K (1990) Plant Cell Physiol 31:835–843. **Iwaki T,** Wadano A, Yokota A, Himeno M (1991) Plant Cell Physiol 32:1083–1091.

**Jacobshagen S,** Schnarrenberger C (1990) J Phycol 26:312–317. **Jahnke LS,** Eighmy TT, Fagerberg WR (1991) Plant Cell Environ 14:147–156. **Jiao J,** Chollet R (1990) Arch Biochem Biophys 283:300–305. **Jiao J,** Chollet R (1991) Plant Physiol 95:981–985. **Jiao J,** Chollet R (1992) Plant Physiol 98:152–156. **Jiao J,** Vidal J, Echevarría C, Chollet R (1991a) Plant Physiol 96:297–301. **Jiao J,** Echevarría C, Vidal J, Chollet R (1991b) Proc Nat Acad Sci USA 88:2712–2715. **Johnston AM,** Raven JA (1989) J Phycol 25:568–576. **Jordan BR,** He J, Chow WS, Anderson JM (1992) Plant Cell Environ 15:91–98.

**Kaiser WM,** Spill D (1991) Plant Physiol 96:368–375. **Kalt W,** Osmond CB, Siedow JN (1990) Plant Physiol 94:826–832. **Kaplan A,** Schwarz R, Lieman-Hurwitz J, Reinhold L (1991) Plant Physiol 97:851–855. **Kettleborough CA,** Phillips AL, Keys AJ, Parry MA (1991) Planta 184:35–39. **Kiss F,** Wu M-X, Wong JH, Balogh Á, Buchanan BB (1991) Arch Biochem Biophys 287:337–340. **Kleczkowski LA,** Edwards GE (1990) Z Naturforsch 45c:42–46. **Kleczkowski LA,** Edwards GE (1991) Plant Physiol Biochem 29:9–17. **Kleczkowski LA,** Villand P, Lönneborg A, Olsen O-A, Lüthi E (1991) Z Naturforsch 46c:605–612. **Kleczkowski LA,** Edwards GE, Randall DD (1992) Phytochemistry 31:51–54. **Kluge M,** Wolf H, Fischer A (1991) Plant Physiol Biochem 29:83–90. **Königer M,** Winter K (1991) Oecologia 87:349–356. **Köpke-Secundo E,** Molnar I, Schnarrenberger C (1990) Plant Physiol 93:40–47. **Kornberg A** (1980) Can J Biochem 58:93–96. **Krall JP,** Edwards GE (1990) Aust J Plant Physiol 17:579–588. **Krall JP,** Edwards GE, Ku MSB (1991) Aust J Plant Physiol 18:369–383. **Krause K-P,** Stitt M (1992) Phytochemistry 31:1143–1146. **Krömer S,** Heldt HW (1991a) Plant Physiol 95:1270–1276. **Krömer S,** Heldt HW (1991b) Biochim Biophys Acta 1057:42–50. **Ku MSB,** Wu J, Dai Z, Scott RA, Chu C, Edwards GE (1991) Plant Physiol 96:518–528.

**Laisk A,** Kiirats O, Oja V, Gerst U, Weis E, Heber U (1992) Planta 186:434–441. **Levavasseur G,** Edwards GE, Osmond CB, Ramus J (1991) J Phycol 27:667–672. **Lilley RMcC,** Portis AR (1990) Plant Physiol 94:245–250. **Lin M,** Turpin DH (1992) Plant Physiol 98:82–88. **López-Pozos R,** Rodríguez-Sotres R, Muñoz-Clares RA (1990) J Plant Physiol 136:451–457. **Lüttge U** (1990) Plant Cell Environ 13:977–982. **Lüttge U,** Ball E, Fetene M (1991) J Plant Physiol 137:268–272.

**Macioszek J,** Anderson JB, Anderson LE (1990) Plant Physiol 94:291–296. **Makino A,** Osmond B (1991) Photosyn Res 29:79–85. **Marcus F,** Chamberlain SH, Chu C, Masiarz FR, Shin S, Yee BC, Buchanan BB (1991) Arch Biochem Biophys 287:195–198. **Marquardt-Jarczyk G,** Lüttge U (1990) Bot Acta 103:203–213. **Martínez F,** Orús MI (1991) Plant Physiol 95:1150–1155. **Mason CB,** Manuel LJ, Moroney JV (1990) Plant Physiol 93:833–836. **Mayoral ML,** Medina E, Garcia V (1991) J Exp Bot 42:1123–1129. **McKinnon WB** (1992) Nature 357:15–16. **McMorrow EM,** Bradbeer JW (1990) Plant Physiol 93:374–383. **Metodiev M,** Demirevska-Kepova K (1992) J Exp Bot 43:155–158. **Metzler MC,** Rothermel BA, Nelson T (1989) Plant Mol Biol 12:713–722. **Miginiac-Maslow M,** Decottignies P, Jacquot J-P, Gadal P (1990) Biochim Biophys Acta 1017:273–279. **Miller AG,** Espie GS, Canvin DT (1990) Can J Bot 68:1291–1302. **Moore Bd,** Seemann JR (1990) Plant Cell Physiol 31:969–974. **Moore Bd,** Kobza J, Seemann JR (1991) Plant Physiol 96:208–213.

**Nakamoto H,** Edwards GE (1990) Plant Physiol Biochem 28:553–559. **Nara M,** Shiraiwa Y, Hirokawa T (1990) Plant Cell Physiol 31:961–967. **Neuhaus HE,** Stitt M (1990) Planta 182:445–454. **Neuhaus HE,** Stitt M (1991) Plant Sci 76:49–55. **Neuhaus HE,** Quick WP, Siegl G, Stitt M (1990a) Planta 181:583–592. **Neuhaus HE,** Krause K–P, Stitt M (1990b) Phytochemistry 29:3411–3415. **Nielsen TH,** Veierskov B (1990) Plant Physiol 93:637–641. **Nimer NA,** Sallal A-KJ, Al-Hasan RH, Merrett MJ (1990) Planta 181:374–377. **Nimer NA,** Dixon GK, Merrett MJ (1992) New Phytol 120:153–158. **Nobel PS** (1991) New Phytol 119:183–205. **Nobel PS,** García-Moya E, Quero E (1992) Plant Cell Environ 15:329–335.

**Ogawa T** (1990) Plant Physiol 94:760–765. **Ohnishi J,** Flügge U-I, Heldt HW, Kanai R (1990) Plant Physiol 94:950–959. **Olivares E,** Medina E (1990) J Exp Bot 41:579–585. **Oliver DJ,** Neuburger M, Bourguignon J, Douce R (1990) Plant Physiol 94:833–839.

**Palmqvist K,** Ramazanov Z, Samuelsson G (1990) Physiol Plant 80:267–276. **Pearce G,** Strydom D, Johnson S, Ryan CA (1991) Science 253:895–898. **Peltier G,** Schmidt GW (1991) Proc Natl Acad Sci USA 88:4791–4795. **Petropoulou Y,** Manetas Y, Gavalas NA (1990) Physiol Plant 80:605–611. **Piepenbrock M,** Schmitt JM (1991) Plant Physiol 97:998–1003. **Pilon-Smits EAH,** 't Hart H, van Brederode J (1990) J Plant Physiol 136:155–160. **Pilon-Smits EAH,** 't Hart H, van Brederode J (1991) J Plant Physiol 138:581–586. **Podestá FE,** Iglesias AA, Andreo CS (1990) Photosyn Res 26:161–170. **Poetsch W,** Hermans J, Westhoff P (1991) FEBS Lett 292:133–136. **Porter MA,** Hartmann FC (1990) Arch Biochem Biophys 281:330–334. **Portis AR** (1990) Arch Biochem Biophys 283:397–400. **Pozueta-Romero J,** Frehner M, Viale A, Akazawa T (1991a) Proc Natl Acad Sci USA 88:5769–5773. **Pozueta-Romero J,** Yamaguchi J, Akazawa T (1991b) FEBS Lett 291:233–237. **Pozueto-Romero J,** Ardila F, Akazawa T (1991c) Plant Physiol 97:1565–1572. **Prado FE,** Lázaro JJ, Gorgé JL (1991) Plant Physiol 96:1026–1033. **Prat-Gay G,** Paladini A, Stein M, Wolosiuk RA (1991) J Biol Chem 266:20913–20921. **Price GD,** Badger MR (1991) Can J Bot 69:963–973. **Purohit US,** Verma NC, Bhagwat AS (1991) FEBS Lett 284:23–26.

**Queiroz-Claret C,** Meunier JC (1992) Phytochemistry 31:751–754. **Quick WP,** Schurr U, Fichtner K, Schulze E-D, Rodermel SR, Bogorad L, Stitt M (1991) Plant J 1:51–58. **Quick WP,** Chaves MM, Wendler R, David M, Rodrigues ML, Passaharinho JA, Pereira JS, Adcock MD, Leegood RC, Stitt M (1992) Plant Cell Environ 15:25–35.

**Raines CA,** Lloyd JC, Dyer TA (1991) Photosyn Res 27:1–14. **Rault M,** Gontero B, Ricard J (1991) Eur J Biochem 197:791–797. **Rawsthorne S,** Hylton CM (1991) Planta 186:122–126. **Reddy AR,** Das VSR (1990) Biochem Intern 20:25–35. **Reddy MM,** Vani T, Raghavendra AS (1991) Plant Physiol 96:1368–1371. **Reiskind JB,** Bowes G (1991) Proc Natl Acad Sci USA 88:2883–2887. **Riens B,** Heldt HW (1992) Plant Physiol 98:573–577. **Rodermel SR,** Abbott MS, Bogorad L (1988) Cell 55:673–681. **Roesler KR,** Ogren WL (1990) Plant Physiol 93:188–193. **Rotatore C,** Colman B (1991a) Plant Cell Environ 14:371–375. **Rotatore C,** Colman B (1991b) Plant Cell Environ 14:377–382. **Rotatore C,** Colman B (1991c) Can J Bot 69:1025–1031. **Rowland-Bamford AJ,** Baker JT, Allen LH, Bowes G (1991) Plant Cell Environ 14:577–583.

**Sage RF,** Sharkey TD, Seemann JR (1990a) Plant Physiol 94:1735–1742. **Sage RF,** Sharkey TD, Pearcy RW (1990b) Aust J Plant Physiol 17:135–148. **Salahas G,** Manetas Y, Gavalas NA (1990a) Photosyn Res 24:183–188. **Salahas G,** Manetas Y, Gavalas NA (1990b) Photosyn Res 26:9–17. **Salvucci ME,** Drake RR, Haley BE (1990) Arch Biochem Biophys 281:212–218. **Santakumari M,** Berkowitz GA (1991) Photosyn Res 28:9–20. **Sassenrath GF,** Ort DR, Portis AR (1990) Arch Biochem Biophys 282:302–308. **Sawada S,** Usuda H, Hasegawa Y, Tsukui T (1990) Plant Cell Physiol 31:697–704. **Scheibe R** (1990) Bot Acta 103:327–334. **Scheibe R** (1991) Plant Physiol 96:1–3. **Scheibe R,** Kampfenkel K, Wessels R, Tripier D (1991) Biochim Biophys Acta 1076:1–8. **Schimkat D,** Heineke D, Heldt HW (1990) Planta 181:97–103. **Schnarrenberger C** (1990) Planta 181:249–255. **Schulze W,** Stitt M, Schulze E-D, Neuhaus HE, Fichtner K (1991) Plant Physiol 95:890–895. **Schuster WS,** Monson RK (1990) Plant Cell Environ 13:903–912. **Servaites JC,**

Shieh W-J, Geiger DR (1991) Plant Physiol 97:1115–1121. **Shen JB,** Orozco EM, Ogren WL (1991) J Biol Chem 266:8963–8968. **Siebke K,** Laisk A, Oja V, Kiirats K, Heber U (1990) Planta 182:513–522. **Siebke K,** Laisk A, Neimanis S, Heber U (1991) Planta 185:337–343. **Siebke K,** Yin Z-H, Raghavendra AS, Heber U (1992) Planta 186:526–531. **Siegl G,** MacKintosh C, Stitt M (1990) FEBS Lett 270:198–202. **Silva AB,** Arrabaca MC, Cabral JMS (1991) Plant Sci 76:29–34. **Skrukrud CL,** Anderson LE (1991) FEBS Lett 280:259–261. **Skrukrud CL,** Gordon IM, Dorwin S, Yuan X-H, Johansson G, Anderson LE (1991) Plant Physiol 97:730–735. **Smith AM,** Neuhaus HE, Stitt M (1990) Planta 181:310–315. **Smrcka AV,** Bohnert HJ, Jensen RG (1991) Arch Biochem Biophys 286:14–19. **Spalding MH** (1990) Photosyn Res 24:245–252. **Stamatakis K,** Skaliora I, Gavalas NA, Manetas Y (1990) Aust J Plant Physiol 17:407–411. **Stitt M,** von Schaewen A, Willmitzer L (1990) Planta 183:40–50. **Stitt M,** Quick WP, Schurr U, Schulze E-D, Rodermel SR, Bogorad L (1991) Planta 183:555–566. **Su X,** Bogorad L (1991) J Biol Chem 266:23698–23705. **Sültemeyer DF,** Fock HP, Canvin DT (1990) Plant Physiol 94:1250–1257. **Sültemeyer DF,** Fock HP, Canvin DT (1991) Can J Bot 69:995–1002. **Süss K-H** (1990) Z Naturforsch 45c:633–637. **Suzuki A,** Lelandais M, Morot-Gaudry J-F (1990) Phytochemistry 29:3067–3072. **Suzuki K,** Iwamoto K, Yokoyama S, Ikawa T (1991) J Phycol 27:492–498.

**Tarchevskii IA,** Konyukhova TM (1989) Photosynthetica 23:472–478. **Tarczynski MC,** Outlaw WH (1990) Arch Biochem Biophys 280:153–158. **Turley RB,** Trelease RN (1990) Physiol Plant 79:570–578.

**Umino Y,** Shiraiwa Y (1991) J Plant Physiol 139:41–44.

**Vairinhos F,** Simon J-P (1990) Plant Sci 71:173–177. **Valle EM,** Heldt HW (1991) Plant Physiol 95:839–845. **Vani T,** Reddy MM, Raghavendra AS (1990) Physiol Plant 80:467–471. **Van Langendonckt A,** Driessche TV (1992) Arch Biochem Biophys 292:156–164. **Van Quy L,** Foyer C, Champigny M-L (1991) Plant Physiol 97:1476–1482. **van Wijk KJ,** Krause GH (1991) Planta 186:135–142. **Vassey TL,** Quick WP, Sharkey TD, Stitt M (1991) Physiol Plant 81:37–44. **von Caemmerer S,** Evans JR (1991) Aust J Plant Physiol 18:287–305.

**Wedding RT,** Black MK (1990) Arch Biochem Biophys 282:284–289. **Wendler C,** Barniske M, Wild A (1990) Photosyn Res 24:55–61. **Whitehouse DG,** Rogers WJ, Tobin AK (1991) J Exp Bot 42:485–492. **Willmer CM,** Petropoulou Y, Manetas Y (1990) J Exp Bot 41:1103–1107. **Wilson BJ,** Tolbert NE (1991) FEBS Lett 279:313–315. **Winter K,** Königer M (1991) Oecologia 87:190–197. **Wise RR,** Frederick JR, Alm DM, Kramer DM, Hesketh JD, Crofts AR, Ort DR (1990) Plant Cell Environ 13:923–931. **Wise RR,** Sparrow DH, Ortiz-Lopez A, Ort DR (1991) Plant Sci 74:45–52. **Wiskich JT,** Bryce JH, Day DA, Dry IB (1990) Plant Physiol 93:611–616. **Wong SC,** Osmond CB (1991) Aust J Plant Physiol 18:137–152. **Wong JH,** Kiss F, Wu M-X, Buchanan BB (1990) Plant Physiol 94:499–506. **Worrell AC,** Bruneau J-M, Summerfelt K, Boersig M, Volker TA (1991) Plant Cell 3:1121–1130. **Wu J,** Neimanis S, Heber U (1991) Bot Acta 104:283–291. **Wu M-X,** Meyer CR, Willeford KO, Wedding RT (1990) Arch Biochem Biophys 281:324–329.

**Yang Y,** Steup M (1990) Plant Physiol 94:960–969. **Yokota A** (1991a) J Biochem 110:246–252. **Yokota A** (1991b) Plant Cell Physiol 32:755–762. **Yokota A,** Tsujimoto N (1992) Eur J Biochem 204:901–909. **Young AJ** (1991) Physiol Plant 83:702–708. **Yu S** (1989) J Phycol 25:608–612. **Yu S,** Pedersén M (1991) Bot Mar 34:323–325. **Yu J,** Woo KC (1991) Aust J Plant Physiol 18:583–588.

**Zhu G,** Jensen RG (1991) Plant Physiol 97:1354–1358. **Ziska LH,** Drake BG, Chamberlain S (1990) Oecologia 83:469–472. **Ziska LH,** Hogan KP, Smith AP, Drake BG (1991) Oecologia 86:383–389.

Dr. Grahame J. Kelly
Centre for Biological Population Management
School of Life Science
Queensland University of Technology
4000 Brisbane, Queensland Australia

Professor Dr. Erwin Latzko
Institut für Allgemeine Chemie und Biochemie
Technische Universität München
W-8050 Freising, Germany.

# V. Metabolism of Inorganic Nitrogen Compounds

By Hermann Bothe

## 1. Assimilatory Nitrate Reduction

Proceedings from a meeting on molecular and genetic aspects of nitrate assimilation are available (Wray and Kinghorn 1989). Most aspects of nitrogen metabolism of plants are covered in the multiauthor books edited by Abrol (1989) and Mengel and Pilbeam (1992).

In most higher plants (e.g. spinach or pea), assimilatory nitrate reduction mainly takes place in the leaves and is largely stimulated by light. Nitrate reduction is controlled by enzyme synthesis and degradation as well as by regulation of enzyme activities. Nitrate and light control *synthesis* of nitrate reductase (= NIA) and of nitrite reductase (= NII). Light acts via phytochrome (Rajasekhar and Oelmüller 1987; Rajasekhar et al. 1988) and an uncharacterized plastidic signal (Oelmüller et al. 1988; Schuster et al. 1988). Alternatively (or additionally), the blue-light photoreceptor may be involved in the modulation of the levels of NIA and NII (Solomonson and Sephar 1977). Recent work with the aurea mutant of tomato deficient in phytochrome (Parks et al. 1987) additionally showed that the expression of NIA and NII genes is light-stimulated and mediated by phytochrome (Becker et al. 1992). In squash cotyledons, the induction of NIA activity requires intact chloroplasts (Oelmüller and Briggs 1990). This is a somewhat surprising result since NIA is a cytoplasmic enzyme and nuclear-encoded.

*Regulation of enzyme activity* had been studied extensively with conflicting results and with no clear picture about the overall pattern until the last review in this series (Bothe 1990). A number of metabolites such as ADP, ATP, $P_i$, NADH, cyanide, hydroxylamine, amino acids, and photosynthetically generated reducing power had been reported to regulate NIA activity (for references, see Riens and Heldt 1992), but a recent study (Sanchez and Heldt 1990) assaying the effects of metabolites on partially purified NIA under physiological conditions suggested that the effects of adenine nucleotides, $P_i$, $Ca^{2+}$, ammonia, and amino acids are insufficient to account for the fluctuations of NIA activity observed in vivo.

Such fluctuations have recently been shown to be larger than believed hitherto. Remmler and Campbell (1986) had observed a decrease in NIA activity by only some 30% within 1 h, but recent determinations showed for spinach leaves that the activity decreased to 15% of the control value with a half-decay time of 2 min during darkening (Riens and Heldt 1992) and to less than 10% with a t 1/2 of 15 min when leaves were exposed to low $CO_2$ in the light (Kaiser and Förster 1989; Kaiser and Brendle-Behnisch 1991). It was known that the levels of NIA activity and protein (the latter measured immunochemically) did not match each other particularly during light/dark transitions (Oaks et al. 1982; Remmler and Campbell 1986). In addition, NIA activity showed considerable fluctuations during such transitions even in transgenic tobacco plants with a high constitutive level of NIA mRNA (Vincentz and Caboche 1991). All this suggests that NIA activity must be subject to metabolic control in intact cells, and it has been pointed out that such a control is mandatory for physiological reasons (Riens and Heldt 1992). Nitrite reduction takes place in chloroplasts and is dependent on reduced ferredoxin generated by the photosynthetic electron transport. When a leaf is suddenly shaded, reduced ferredoxin likely limits the reduction of nitrite transported from the cytosol into chloroplasts. Thus, toxic amounts of nitrite would accumulate unless the cytosolic NIA and the chloroplastic NII activities were not regulated concertedly. Redeterminations of the levels of free nitrite in leaves showed that the concentration is very low under natural conditions (Riens and Heldt 1992; Kaiser et al. 1992).

A new avenue to the problem came from the work of W. M. Kaiser and his group in Würzburg. Kaiser and Brendle-Behnisch (1991) found that NIA is rapidly inactivated in spinach leaves subjected to rapid wilting or exposure to low $CO_2$-concentrations. They also reported that NIA from illuminated leaves could rapidly be inactivated by incubating desalted crude extracts (but not the purified enzyme) with either $Mg^{2+}$, $Ca^{2+}$, ATP, GTP, or UTP (Kaiser and Spill 1991). The stable modulation in crude extracts could be reversed by AMP. These authors hypothesized that NIA is covalently modified by a second enzyme system in the presence of ATP and $Mg^{2+}$, resulting in protein phosphorylation (Kaiser and Spill 1991). Experimental evidence was then forwarded that NIA rapidly responds to changes in the cytosolic adenine nucleotide levels with high activity at low ATP but high AMP amounts and low activity at the opposite (Kaiser et al. 1992). The experimental findings by Kaiser and coworkers were largely confirmed by Huber et al. (1992) with the exception of the stimulation by AMP. Additionally, Huber et al. (1992) indicated that the modification of NIA occurs at the Mo-moiety ($NO_3^-$-reduction activity with reduced methyl viologen) but not at the diaphorase part (NADH-ferricyanide reduction activity). Experiments with $^{32}P$ showed that the enzyme is labelled on seryl residues. Tryptic peptide mapping revealed three major $^{32}P$-labelled polypeptide fragments (Huber et al. 1992). Thus, all the data suggest that *NIA is covalently modified by phosphorylation* which renders it more sensitive to inhibition by $Mg^{2+}$, as also described for approximately 30 other proteins in plants (Budde and Rendall 1990; Huber and Huber 1990), although the evidence for NIA is not stringent as yet. The protein kinase and dephosphorylase for NIA still have to be characterized. In addition, as pointed out by Kaiser et al. (1992), dark inactivation of NIA would require a rapid increase in ATP and decrease in AMP in the cytosol; experimental evidence, however, indicated that the opposite is likely the case under physiological conditions. Thus, further factors in addition to adenylates and $Mg^{2+}$ might regulate NIA activity in the cytosol (Kaiser et al. 1992).

Progress in *molecular biology* is also rapid in the field of nitrate reduction. Complete DNA sequences of at least seven NIAs have been published (Crawford et al. 1988; Cheng et al. 1989; Daniel-Vedele et al. 1989; Gowri and Campbell 1989; Johnstone et al. 1990; Hyde et al. 1991). This work revealed the localization of the binding sites for the internal cofactors FAD, heme-Fe, and molybdopterin within the enzyme (Crawford and Campbell 1990; Campbell and Kingshorn 1990).

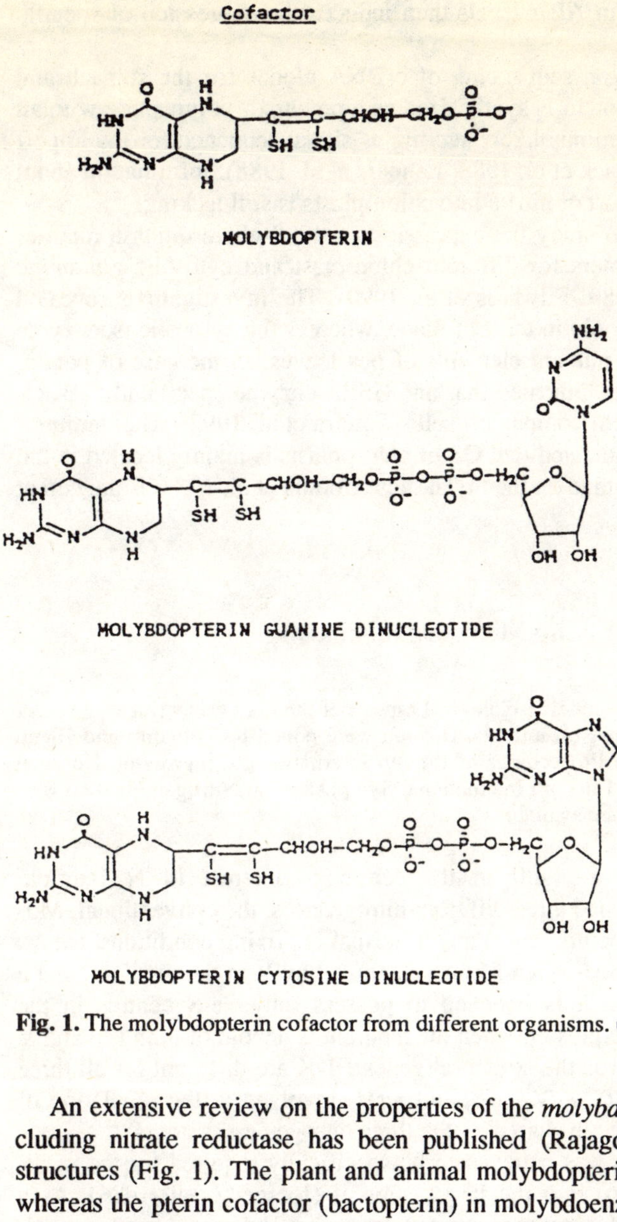


Enzyme

Milk xanthine oxidase
Liver sulfite oxidase
*Chlorella* nitrate reductase
Corn nitrate reductase
*R. sphaeroides* DMSO reductase

*R. sphaeroides* nitrate reductase
*E. coli* nitrate reductase
*E. coli* formate dehydrogenase
*M. formicicum* formate dehydrogenase

*P. carboxydoflava* CO dehydrogenase

**Fig. 1.** The molybdopterin cofactor from different organisms. (from Rajagopalan 1991)

An extensive review on the properties of the *molybdenum cofactor* of all enzymes including nitrate reductase has been published (Rajagopalan 1991) with the chemical structures (Fig. 1). The plant and animal molybdopterins contain one phosphate group, whereas the pterin cofactor (bactopterin) in molybdoenzymes of bacteria is larger by the addition of either a cytosine residue as in *Pseudomonas carboxydoflava* CO-dehydrogenase (Johnson et al. 1990; Tachil et al. 1991) or a guanine moiety as in the dimethyl sulfoxide reductase of *Rhodobacter sphaeroides* (Johnson et al. 1990) and the formylmethanofuran dehydrogenase of *Methanosarcina barkeri* (Karrasch et al. 1990). In all cases, Mo-bridges to the enedithiol group within the cofactor.

Work on barley mutants indicated that the molybdenum cofactor of enzymes, such as aldehyde oxidase, may control ABA-accumulation after water stress (Walker-Simmons et al. 1989). The resultant increase in ABA-levels then induces the expression of specific stress genes.

With regard to *nitrite reductase*, sequencing of c-DNA clones for the spinach and maize enzyme showed that the protein is synthesized as a precursor polypeptide with an amino acid extension at the N-terminus likely serving as signal sequence for the import of the protein into chloroplasts (Back et al. 1988; Lahners et al. 1988). Information about a nitrite carrier catalyzing the import of nitrite into chloroplasts is still lacking.

Transgenic plants were used to study the expression of the ß-glucuronidase reporter genes directed by specific promoters for different chloroplast and cytosolic *glutamine synthetase genes* (Forde et al. 1989; Edwards et al. 1990). The investigations revealed high levels of the chloroplast GS only in the leaf blade, whereas the cytosolic isoenzyme was expressed exclusively in the phloem elements of pea leaves. In the case of potato, immunogold labelling experiments indicated that one GS isoenzyme specifically associated with the plastids of the phloem companion cells (Pereira et al. 1992). The immunogold labelling experiments also indicated that GS of chloroplasts is mainly located in the stroma and only to a minor percentage on the thylakoids (Botella et al. 1988; Brangeon et al. 1989).

## 2. Nitrogen Fixation by Free-Living Microorganisms

A remarkable monograph particularly stressing biological aspects of the field was written by Sprent and Sprent (1990). Multiauthor reviews on nitrogen fixation were edited by Dilworth and Glenn (1991) and Stacey et al. (1992). The Proceedings of the 8th International Congress on Nitrogen Fixation (Gresshoff et al. 1990) and of the 5th International Symposium on Nitrogen Fixation with Non-Legumes (Polsinelli et al. 1991) are available.

The *alternative nitrogenases* are still in the centre of interest in $N_2$-fixation. *Azotobacter vinelandii* can synthesize three different nitrogenases: the conventional, Mo-containing enzyme (encoded by the *nif* genes) under normal $N_2$-fixing conditions, the V-containing protein (genes termed *vnf*) when Mo is substituted by V in the medium, and a third nitrogenase (*anf* genes) which is believed to possess only Fe-S centres in the catalytic active site and which is expressed when the medium is devoid of both Mo and V (Bishop and Joerger 1990). Whereas the structural genes HDK are different for all three nitrogenases, five genes (*nifU*, *nifS*, *nifV*, *nifB*, and *nifM*) involved in the regulation of activity or synthesis of the enzyme are shared by the three nitrogenase systems.

With regard, to *nifB*, this gene was believed to be specifically required for the insertion of the Mo-cofactor (FeMoco) into the larger subunit (Joerger et al. 1986). As a FeVa-cofactor analogous to FeMoco was extracted from the V-nitrogenase using N-methylformamide (Smith et al. 1988), a role of the NifB gene product in the insertion of the FeVa-cofactor with the second nitrogenase is feasible. The role of *nifB* in the formation of the third nitrogenase containing only Fe-S centres, however, remains obscure. The extraction of a specific cofactor from this enzyme has not yet been reported. It remains to be shown whether this protein contains an unusual element or whether NifB has an additional unresolved function in the biosynthesis of the nitrogenases. The gene product of *nifV* is a homocitrate synthase catalyzing the formation of homocitrate which is incorporated into Fe Moco during its formation (Hoover et al. 1987; Shah et al. 1988). The requirement of NifV for the biosynthesis of all three nitrogenases (Kennedy and Dean 1992) suggests that they all contain homocitrate in their prosthetic group. *NifH* is the structural gene for the smaller subunit of Mo-nitrogenase. Its gene product was shown to be additionally required for FeMoco biosynthesis (Robinson et al. 1987). The *vnfH* (encoding the smaller subunit of V-nitrogenase) and *nifH* gene products show 91% amino acid sequence homologies and are thus very similar proteins (Joerger et al. 1990). The *vnfH* gene product is required additionally for the insertion of the FeVa-cofactor into the V-enzyme and, surprisingly, also for the biosynthesis of all products of the structural genes (*anf* genes) of the third nitrogenase (Joerger et al. 1991). Whereas Mo-nitrogenase is encoded by the three structural genes *HDK*, an additional structural gene (*nifG*) is required for the biosynthesis of the two alternative nitrogenases in *A. vinelandii* (Bishop and Joerger 1990). Despite an extensive effort to elucidate the function of this gene, no clear idea about its role has emerged as yet. Expression of Mo-nitrogenase is controlled by the genes *nifLA*. Three different genes *nifA*, *vnf*A, and *anfA* and thus three nitrogenase activator proteins exist in *A. vinelandii* (Joerger et al. 1986, 1989).

Whereas progress in the field of molecular biology of alternative nitrogenases is rapid, information about the *ecological roles of these alternative nitrogenases* is scant. An alternative nitrogenase has not yet been shown to be expressed in natural habitats. Thus, it is not clear whether the possession of an alternative nitrogenase represents a selective advantage for an organism. Several organisms, however, have the genetic capability to synthesize at least one alternative nitrogenase. The photosynthetic bacteria *Rhodobacter capsulatus* (Schneider et al. 1988) and *Rhodospirillum rubrum* (Lehman and Roberts 1991) can synthesize the third nitrogenase, whereas physiological evidence exists that the cyanobacterium *Anabaena variabilis* can express both the V- (Kentemich et al. 1988) and the FeS-protein (Kentemich et al. 1991).

The *electron transport to nitrogenase* has been elucidated only for some microorganisms. In *Klebsiella pneumoniae*, the gene product of *nifF*, flavodoxin, is the immediate electron carrier to nitrogenase (Nieva-Gomez et al. 1980). Flavodoxin itself is reduced by the electron donor pyruvate in the pyruvate phosphoroclastic reaction catalyzed by a pyruvate: flavodoxin oxidoreductase encoded by *nifJ* (Shah et al. 1983). *Klebsiella* flavodoxin possesses a single cysteine residue to which coenzyme A can be attached posttranslationallly (Bagby et al. 1991; Thorneley et al. 1992). The protein modified by coenzyme A is inactive in transferring electrons from pyruvate to the iron protein of nitrogenase. Modification by coenzyme A might represent a novel type of enzyme regulation in prokaryotes (Thorneley et al. 1992). Interestingly, pyruvate: ferredoxin oxidoreductase activity of cyanobacteria had earlier been found to be regulated by coenzyme A (Neuer and Bothe 1985).

The aerobic $N_2$-fixing *Azotobacter* forms constitutively both flavodoxin and ferredoxin. Earlier work indicated that both flavodoxin and ferredoxin are components of an electron transport chain in this bacterium (Benemann et al. 1971); the electron donor (pyruvate, NADH, or NADPH), however, remained unknown. Work with insertion mutants showed that the *nifF*-encoded flavodoxin is required for maximal nitrogenase activity, as *Azotobacter* strains mutated within *nifF* had only 70% of the wild-type nitrogenase activity (Bennett et al. 1988). In contrast, when the ferredoxin gene (= ferredoxin I, according to Yoch and Arnon 1972) of *Azotobacter vinelandii* was specifically interrupted by insertion of a kanamycin cartridge, the cells grew as the wild type under $N_2$-fixing conditions (Morgan et al. 1988), indicating that ferredoxin is not involved in $N_2$-fixation of *A. vinelandii*. It was noted in that publication that the concentration of a small acidic protein with an $M_r = 18\,000$ of unknown function drastically increased in this mutant. An interesting hypothesis has recently been forwarded that ferredoxin from *A. vinelandii* represents a novel class of DNA-binding proteins that regulate gene expression in response to cellular iron(II) levels (Thomson 1991). The protein possesses two cubic iron-sulfur clusters in the prosthetic group. One of the two clusters operates between the $[4Fe-S]^{+2}$ and $[4Fe-4S]^{+1}$ states, has a reduction potential of –645 mV and thus the lowest one so far reported for a biological iron-sulfur centre (Armstrong et al. 1988). When the protein is prepared aerobically, the other cluster is in the $[3Fe-4S]^{+1}$ state where it can be reduced by 1e ($E_0' = -460$ mV; Armstrong et al. 1988). More interestingly, this $[3Fe-4S]$ cluster can be interconverted to a $[4Fe-4S]$ centre by the uptake of iron which requires the presence of nucleic acids and pH-values above 8 (Thomson 1991). The *Azotobacter* ferredoxin possesses α surface α-helix capable of DNA recognition. Such a ferredoxin could be a widespread DNA-binding protein occurring also in non-$N_2$-fixing microorganisms. It could act as activator/repressor and regulate protein expression depending upon the iron and redox status of the cells. The conversion of $[3Fe-4S] \rightarrow [4Fe-4S]$ may be the chemical sensor of iron(II) levels. Ferredoxin could control the expression of bacterioferretin, which is possibly identical to the 18 000 $M_r$-protein overexpressed in the ferredoxin mutant (Morgan et al. 1988).

In bacteria at least three other DNA-binding proteins have been described whose activity is controlled by the redox state of the cells and which bind iron. The first one is FNR (extensively discussed in the previous review of this series, cf. Bothe 52:122–137, 1990) which regulates the expression of genes of anaerobic metabolism, particularly of denitrification. Iron was suggested to function as cofactor of FNR for gene activation and $O_2$-sensing in *E. coli* (Unden and Trageser 1991; Engel et al. 1991)

| FNR · $Fe^{2+}$ | $O_2$ | FNR · $Fe^{3+}$ |
|---|---|---|
| gene activator | $N_2$ | |
| (anaerobic) | | (aerobic) |

More recent evidence, however, casts some doubts that the functional state of FNR during aerobic/anaerobic switch is regulated by the iron content of the cells and by reversible binding of $Fe^{2+}$ under physiological conditions (Niehaus et al. 1991). The second example from *E. coli* is the *fur* gene, whose product negatively controls the transcription of genes coding for proteins involved in iron uptake. The product of the *fur* gene acts as an aporepressor which is converted to the active repressor by binding $Fe^{2+}$ (Schaffer et al. 1985; Bagg and Neilands 1987; Saito and Williams 1991). The third example is the NifA protein from *Bradyrhizobium japonicum* which positively controls the expression of the *nif* and *fix* genes in this bacterium. It had been shown that metal ions such as iron (II) are absolutely required for NifA to activate *nif* transcription in vivo (Fischer et al. 1988; Morett et al. 1991).

*In cyanobacteria*, ferredoxin is synthesized constitutively, whereas the formation of flavodoxin is induced by Fe-deficiency in some species (Bothe 1977). Flavodoxin from *Anabaena* was recently cloned after amplification by PCR and expressed in *E. coli* (Fillat et al. 1991). In vitro, flavodoxin and ferredoxin replace each other with more or less the same activity in all reactions assayed (Trebst and Bothe 1966; Fillat et al. 1988). A heterocyst-specific ferredoxin from *Anabaena variabilis* had been described which was claimed to be particularly active in transferring electrons to nitrogenase (Schrautemeier and Böhme 1985; Böhme and Schrautemeier 1987). This ferredoxin was shown to be expressed only under $N_2$-fixing conditions (Böhme and Haselkorn 1988). Recent immunoquantitations using flavodoxin antibodies, however, showed that this protein is constitutively synthesized in heterocysts (but not in vegetative cells) of *Anabaena* ATCC 29151 even at high concentrations of iron in the medium (Sandmann et al. 1990). The pyruvate clastic reaction had been described to occur in cyanobacteria (Neuer and Bothe 1985), Thus, it appears that the electron transport to nitrogenase involves pyruvate and flavodoxin in cyanobacteria as in *Klebsiella.* It is somewhat tempting to speculate that the heterocyst-specific ferredoxin could be a DNA-binding and regulatory protein as ferredoxin from *Azotobacter*.

Evidence is now accumulating that nitrogenase activity is regulated by *covalent modification of the smaller subunit* (the Fe-protein, nitrogenase reductase) in cyanobacteria (Reich and Böger 1989; Ernst et al. 1990) similar to that reported for *Rhodospirillum rubrum* (Preston and Ludden 1982). In the latter organism, nitrogenase is regulated by reversible ADP-ribosylation at a specific arginine residue (arg 101) of one of the Fe-protein polypeptides (Pope et al. 1985). Modification to an inactive form can clearly be shown to occur at the smaller subunit of nitrogenase in cultures of the cyanobacterium *Anabaena variabilis* by supplementing the medium with $NH_4Cl$ or by growing them under C-starvation (Ernst et al. 1990).

It was generally believed that organisms smaller than 2 µm (the picoplankton) are most important in carbon and *nitrogen cycling of the open oceans* (Fogg 1986). A recent reevaluation (Carpenter and Romans 1991) indicates that large cyanobacteria such the filamentous *Trichodesmium* and *Richelia intracellularis* (the latter occurring endosymbiotically in the diatoms *Rhizoselenia* and *Hemiaulis*) play a major role in both C- and N-fixation in the tropical North Atlantic Ocean and presumably also in other tropical seas.

## 3. Nitrogen Fixation in Symbiosis

D. Werner's remarkable monograph on *Symbiosis of Plants and Microbes* is now available in English (Werner 1992). Proceedings from two meetings have recently been published (Hennecke and Verma 1991; Werner and Müller 1990).

In the *legume-Rhizobium symbiosis*, the recognition of the partners is mediated by *signal substances* which have been identified recently (reviews: Long 1989; Kondorosi 1989, 1991; Djordjevic and Weinman 1991). From the plant side, they are all components of the phenylpropanoid pathway (flavanones, flavones, isoflavones/anthocyanidins, and chalcones) which specifically induce the expression of *Rhizobium* genes required for the formation of nodules (the *nod* genes). Each plant apparently releases its specific "cocktail" mixture of flavonoids from roots or seeds, and antiinducers such as umbelliferone (a coumarin) and formononetin (an isoflavone) have been described to modulate *nod* gene expression (Djordjevic and Weinman 1991). In addition to their function in *nod* gene activation, flavonoids cause a significant increase in the production of indole acetic acid by *Rhizobium* (Prinsen et al. 1991), induce resistance to potentially bactericidal concentrations of the phytoalexin glyceollin (Parniske et al. 1991), and promote spore germination of vesicular-arbuscular fungi of the genus *Glomus* (Tsai and Phillips 1991).

The *nod* genes in *Rhizobium* are localized on large indigenous plasmids, often clustered in the vicinity of the *nif* genes, whereas they are on the chromosome in the case of *Bradyrhizobium* and *Azorhizobium* (Kondorosi 1989, 1991). The *nod* genes of all rhizobia necessary for host recognition and nodule induction can be subdivided into three categories:

1. the regulatory *nodD* gene, which can occur in multiple forms in the different rhizobia;
2. the common *nodABC* genes, which occur everywhere and which are genetically interchangeable among the different rhizobia;
3. the host-specific genes such as *nodFEG, HQ*, and others which determine the specificity of the symbiosis between a legume and its rhizobial partner.

The *nodD* gene product is a 33-kDa protein which has two domains (one at each end) apparently involved in flavonoid recognition (Djordjevic and Weinman 1991). The presence of the correct flavonoid cocktail enables NodD to become a potent DNA-binding transcriptional activator. NodD binds to highly conserved promoter elements preceding *nodA, nodF, nodM,* and the *nodO* genes in *Rhizobium leguminosarum* bv. *viciae* thereby enabling the synthesis of the products of the common and of the host-specific genes (Downie et al. 1991; Downie 1991). A major breakthrough in the field has now come from the identification of the product of the common *nodABC* and the specific *H* and *Q* genes as an acylated and sulfated *lipo-oligosaccharide* in *Rhizobium meliloti* (Lerouge et al. 1990; see Fig. 2). It was shown that this component causes root hair deformation (Lerouge et al. 1990) and cortical cell division (Truchet et al. 1991) in alfalfa (*Medicago sativa*). The *nodABC* genes determine the glucosamine backbone of the molecule. Synthesis of this backbone in *Rhizobium leguminosarum* bv. *viciae* might also require NodM, a glucosamine synthase. This strain has two different glucosamine synthases of which only one is induced by flavonoids (Marie et al. 1992), possibly indicating that the biosynthesis of this signal molecule and of the peptidoglycan components of the bacterial cell wall follow different pathways. In the case of *R. meliloti* specifically, a sulfo-group is attached to the molecule which is determined by the genes *nodH* and *nodQ* (Roche et al. 1991). The long-chain unsaturated fatty acid is apparently different from strain to strain. In the case of *R. meliloti*, the molecule carries a C16:2 N-acyl group (Lerouge et al. 1990), whereas it is a C18:4 chain genetically determined by *nodFE* in *R. leguminosarum* bv. *viciae* (Spaink et al. 1991). An additional O-acetyl group at the non-reducing end sugar (*nodL* gene product) can occur in the signal molecule of both *R. leguminosarum* bv. *viciae and R. meliloti* (Spaink et al. 1991; Truchet et al. 1991) which makes the molecule more active than the non-acetylated form (Truchet et al. 1991). Additionally, the number of N-acetylglucosamine molecules appears to be variable (n = 2 or 3; see Fig. 2) (Spaink et al. 1991, 1992). The studies on the structure-function relationship of this lipo-oligosaccharide signal molecule are just at the beginning. The molecule apparently acts on plants in minute concentrations comparable to those of phytohormones. Remarkably, nitrate completely suppresses the induction of nodule formation by the lipo-oligosaccharide in *R. meliloti* (Truchet et al. 1991). The receptor sides of the signal molecule in the plant

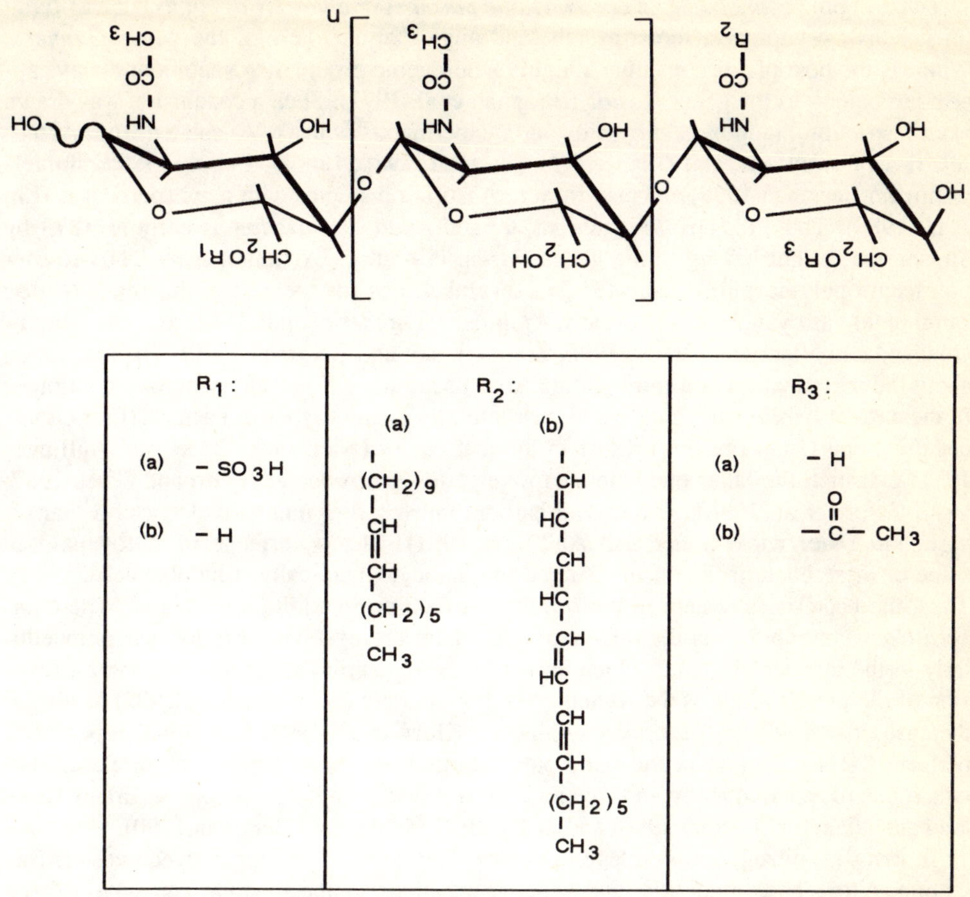

**Fig. 2.** The *Rhizobium* signal substances (*nod* gene product). Substituents (*a*) for *Rhizobium meliloti* (from Lerouge et al. 1990; Truchet et al. 1991); (*b*) for *Rhizobium leguminosarum* bv. *viciae.* (Spaink et al. 1991; Downie et al. 1991)

roots are unknown. Recent evidence supports the old hypothesis (Bohlool and Smidt 1974; Dazzo and Hubbell 1975) that lectins are also major host specificity determinants of the *Rhizobium*-legume symbiosis (Diaz et al. 1989; Kijne et al. 1990; Lugtenberg et al. 1991). It is not yet clear whether the bacterial lipo-oligosaccharide interacts directly with the plant lectins, causes the synthesis of early plant nodulins such as the proline-rich glycoprotein N-15 of soybeans (Franssen et al. 1987; Bisseling et al. 1991), and/or has other target sides. It is noteworthy in this context that the formation of empty nodules of alfalfa clones can occur in total absence of *Rhizobium* (Truchet et al. 1989; Joshi et al. 1990; Kapp et al. 1990), indicating that the legume possesses the entire genetic set-up for nodule organogenesis and that *Rhizobium* provides only a trigger to switch on this program. Recent evidence (Campbell and Meeks 1992) suggests that the bryophyte *Anthoceros punctatus* directly controls nitrogenase expression of its symbiotic partner *Nostoc* similar as in the legume-*Rhizobium* symbiosis.

Diazotrophic *cyanobacteria are symbiotic partners of a variety of plants* such as fungi (in lichens), bryophytes, ferns, cycads, and angiosperms solely of the genus *Gunnera.* Whereas the host plants constitute a highly heterogenic group, the cyanobiont always appears to belong to the genus *Nostoc* (Bergman et al. 1993). Such a conclusion was drawn from restriction mapping and hybridization experiments with DNA isolated from *Anthoceros*, *Azolla*, cycads, and *Gunnera* (Meeks et al. 1988; Lindblad et al. 1989; Zimmermann and Bergman 1990) and also from reconstitution studies with megasporocarps (Lin et al. 1988). Thus, the *Azolla* symbiont, termed *Anabaena azollae* as early as 1873 by Strasburger, might belong to the genus *Nostoc* in reality. Examination by DNA-restriction length polymorphism indicated that several strains (or species) of the genus *Nostoc* can function as cyanobionts in cycads, *Azolla*, and *Gunnera* (Lindblad et al. 1989; Braun-Howland and Nierzwicki-Bauer 1990; Zimmermann and Bergman 1990). The discussion about the participation of a third partner (the "bacteriobiont") in the symbiosis continues. In the case of *Azolla carolina*, five ultrastructurally distinct types of bacteria (four Gram-negative, one Gram-positive) reside in the leaf cavity (Nierzwicki-Bauer and Aulfinger 1991) of which the latter one belongs to the genus *Arthrobacter* (Petro and Gates 1987; Grilli-Caiola et al. 1988). *A. carolina* harbors much larger numbers of bacteria than *A. mexicana* (Nierzwicki-Bauer and Aulfinger 1991). The occurrence of nitrogenase in some of these bacteria was demonstrated immunocytochemically (Lindblad et al. 1991). Thus, the bacteria may help in performing $N_2$-fixation in addition to *Nostoc*. The *Gunnera-Nostoc* symbiosis is the only example where the cyanobiont is located intracellularly in the interior of glands which do not harbor bacteria (Bergman et al. 1992). Invasion of the glands requires the formation of hormogonia (Bergman et al. 1992) as also in the case of the *Nostoc-Geosiphon* symbiosis (Kluge et al. 1991) which was reported to perform $C_2H_2$-reduction in the same publication. There is, however, no correlation between heterocyst frequency and maximal nitrogenase activity in *Gunnera* (Braun-Howland and Nierzwicki-Bauer 1990) and in cycads (Lindblad and Bergman 1990).

In *Frankia*, nitrogenase resides in vesicles. The interrelationship between vesicle formation, nitrogenase induction, and $O_2$-concentrations in the medium can now also be studied in continuous cultures of *Frankia* (Harris and Silvester 1992). The vesicle envelope has been shown to largely consist of bacteriohopanetetrol in nodule tissues of *Alnus* and *Ceanothus* and in solvent extracts of *Frankia* cells (Berry et al. 1991). It is possible that this unusual amphiphilic-lipid compound functions by altering cell permeability.

## 4. Denitrification

The proceedings of a symposium on denitrification in soil and sediment are available (Revsbech and Sörensen 1990).

In the reduction of $NO_3^-$ to $N_2$ by denitrifying bacteria $NO_2^-$ and $N_2O$ are established intermediates. The recent discussion in denitrification focuses on *a functional role of NO*. Nitric oxide could be a free and obligatory intermediate according to the following equation:

$$\text{I. } NO_3^- \rightarrow NO_2^- \xrightarrow{\text{NIR}} NO \xrightarrow{\text{NOR}} N_2O \rightarrow N_2$$

$$\left(\text{NIR} = \text{nitrite reductase, NOR} = \frac{\text{nitric oxide}}{\text{reductase}}\right)$$

Alternatively, $NO_2^-$ could be directly reduced to $N_2O$ by a single: nitrite reductase

$$\text{II. } NO_3^- \rightarrow NO_2^- \xrightarrow{\text{NIR}} N_2O \rightarrow N_2,$$

and NO could be produced and reduced by a side reaction of enzyme(s) catalyzing something else in reality.

Two different dissimilatory nitrite reductases containing either cytochrome $cd_1$ or solely Cu in the prosthetic group exist in microorganisms. A distinct enzyme catalyzing the reduction of NO to $N_2O$ has now been characterized. In both *Pseudomonas stutzeri* (Heiss et al. 1989) and *Paracoccus denitrificans* (Carr and Ferguson 1990; Dermastia et al. 1991) the enzyme was shown to contain both heme b and c in the prosthetic group. Mutants defective in cytochrome $cd_1$ nitrite reductase of *P. stutzeri* could reduce NO unimpaired (Zumft et al. 1988; Jüngst et al. 1991a, b). Such a finding did not rule out the possibility that $NO_2^-$-reductase and NO-reductase are working in parallel. After a long search, a mutant of *P. stutzeri* has recently been isolated which had lost the capability to reduce NO (Braun and Zumft 1991). This mutant reduced $NO_2^-$ to NO which proceeded without any production of $N_2O$ or $N_2$. Production of NO by this mutant strongly affected cell viability under anaerobiosis. A double mutant defective in both $NO_2^-$- and NO-reductases neutralized the conditionally lethal mutation in the NO-reductase genes. Thus, NO is apparently toxic to the cells and $NO_2^-$- and NO-reductases must act concertedly in denitrification following equation I. In order to keep NO at a very low steady state concentration in the cells, NO-reductase has a very low $K_m$-value and is a very active enzyme, probably the most active one in denitrification (Goretski et al. 1990). The release of NO by intact cells can, however, be demonstrated by trapping extracellularly with hemoglobin (Goretski and Hollocher 1988; Carr et al. 1989). Likewise, very high external concentrations of $NO_2^-$ and low pH-values between 5.5–6.0 favor NO-formation in intact cells of *P. denitrificans* (Ji and Hollocher 1988) and *Pseudomonas aeruginosa* (Voßwinkel et al. 1991). NO is extensively produced in nature in some environments (Remde et al. 1989), where it can be a product either of microbial denitrification or nitrification (Goretski and Hollocher 1988; Carr et al. 1989; Zafiriou et al. 1989).

The statement that the reduction of $NO_2^-$ to $N_2O$ involves two different enzymes with the obligatory intermediate NO according to Equation I is now proven for *P. stutzeri* and is likely true for all denitrifying bacteria which possess the cytochrome $cd_1$ containing nitrite reductase. Exchange studies with the $H_2{}^{18}O$ isotope indicate that $NO_2^-$-reduction proceeds directly to $N_2O$ without the intermediate formation of free NO in microorganisms which express the Cu-nitrite reductase but not in those with the cytochrome $cd_1$ enzyme (Ye et al. 1991). The Cu-nitrite reductase catalyzes the formation of an enzyme-bound $Cu^+ - NO^+$ species from $NO_2^-$ which is then reduced either by a second $NO_2^-$-molecule or by free NO to $N_2O$ by the same enzyme (Jackson et al. 1991). Thus, for bacteria containing a Cu-nitrite reductase equation II is true.

Denitrifying bacteria with the cytochrome $cd_1$ nitrite reductase largely differ in their *activity for NO-utilization.* The expression of NO-reductase requires anaerobic conditions (Braun and Zumft 1991) and presumably other, unresolved factors. Nitric oxide is reduced at the periplasmic surface of the cytoplasmic membrane (Bell et al. 1992). The electron transport pathway to nitric oxide does not involve cytochrome $c_2$ and is different from that to nitrous oxide in this respect in *Rhodobacter capsulatus* (Richardson et al. 1991). NO-reduction to $N_2O$ is not coupled with the generation of an $H^+$-gradient in *Rhodobacter capsulatus* (Bell et al. 1992), *Pseudomonas aeruginosa*, and *Azospirillum brasilense* (Voßwinkel et al. 1991). Thus, these bacteria cannot grow with NO as sole respiratory electron acceptor. In contrast, *Paracoccus denitrificans* forms a gradient of 3.5 $H^+$ per NO-molecule reduced to 1/2 $N_2O$ (Garber et al. 1982; Shapleigh and Payne 1985; Voßwinkel et al. 1991) but does not grow anaerobically with NO due to the toxicity of this gas.

*Nitrous oxide reductase* is a homodimeric multi-Cu protein. The structural gene as well as three other genes specific for Cu acquisition and processing have recently been identified (Zumft et al. 1990). The product of one of the genes involved belongs to the family of nucleotide-binding proteins (Zumft et al. 1990). An unusual $N_2O$-reductase was described to occur in *Wolinella succinogenes* (Teraguchi and Hollocher 1989; Zhang et al. 1991). The protein contains only three Cu atoms (instead of four) but additionally a basic cytochrome c subunit of $M_r$ 88 000. An $N_2O$-reductase catalyzing the reduction of $N_2O$ to $N_2$ apparently occurs also in *E. coli* (Kaldorf et al. 1993). This protein is strongly affected by nitrite and requires very high amounts of the substrate $N_2O$. It is suggested that $N_2O$-reduction is an artificial side reaction of an enzyme catalyzing something else in *E. coli*. Enteric bacteria thus contain two different enzymes catalyzing the reduction of $N_2O$ to $N_2$, namely, nitrogenase in *Klebsiella pneumoniae* (Burns and Hardy 1975) and the protein of unidentified function in *E. coli* (Kaldorf et al. 1993).

Determination of three nucleotide sequences of the structural genes for cytochrome $cd_1$ nitrite reductase (Jüngst et al. 1991a) and $N_2O$-reductase (Viebrock and Zumft 1988) indicated the presence of a *signal peptide for both enzymes.* Immunogold-labelling established the periplasmic location of both $N_2O$-reductase and nitrite reductase of *Pseudomonas stutzeri* (Körner and Mayer 1991) and supported the same conclusion obtained for *P. aeruginosa* (Coyne et al. 1990). The copper chromophore assembly for nitrous oxide reductase might take place in the periplasm after export of the protein (Zumft et al. 1990). In contrast, nitrate reductase and NO-reductase are integral membrane proteins.

*DNA probing* has now been employed to track the distribution of microorganisms in natural environments. A gene probe of part of the cytochrome $cd_1$ containing nitrite reductase of *P. stutzeri* allowed the detection of denitrifying bacteria by DNA hybridizations in samples such as a bioreactor consortium, aquifer microcosms, and denitrifying toluene-degrading enrichments (Smith and Tiedje 1992). In a similar approach, a screening of denitrifying bacteria in different soil types was performed using probes for dissimilatory nitrite reductase from *Alcaligenes eutrophus* and nitrous oxide reductase from *A. eutrophus* and *P. stutzeri* (Linne von Berg and Bothe 1992). The DNA-DNA hybridization studies with these probes indicated that denitrifying bacteria predominantly occur at the vicinity of plant roots, whereas non-denitrifiers have the tendency to settle in the bulk, plant-free soil. These studies are obviously very much at the beginning, but could be the start of a new research area in modern ecology.

## 5. Nitrification

Despite the importance of these organisms, progress in this field appears to be slow probably due to the fact that these bacteria can be grown in larger quantities only with difficulties. The *ecology of the organisms* is the centre of interest.

In waterlogged soils, nitrification stops due to the lack of $O_2$ (Belser 1979), but it may occur there in the vicinity of such plants which have aerenchyma in their roots (Blacquière 1986). In such soils the availability of $NH_4^+$ may be more limiting for nitrification than the $O_2$-supply (Engelaar et al. 1991). Nitrification also occurs in acid forest soils at a substantial rate due to the action of fungi and chemolithoautotrophic bacteria. All the ammonium-oxidizing bacteria isolated from such soils were acid-sensitive (Hankinson and Schmidt 1984; Prosser 1989). It is therefore believed that the bacteria in the soils perform nitrification in microsites such as clay particles with relatively high pH. *Nitrosomonas europaea*, when colonizing clay minerals, was found to be protected against nitrapyrin as the most commonly used commercial inhibitor of ammonium oxidation (Powell and Prosser 1991). Alternatively, the bacteria may form aggregates where the cells residing inside may be protected against deleterious nitrous acid (De Boer et al. 1991). Nitrifying bacteria have the potential to remove ammonium nitrogen from wastewater. The search for the optimal material to immobilize the bacteria without loss of their $NH_4^+$-oxidizing capability continues (Sumino et al. 1992).

*Acknowledgment*. A grant from Fonds der Chemischen Industrie greatly aided in writing this article.

## References

**Abrol YP** (ed) (1989) Nitrogen in higher plants. Research Studies Press, Taunton, GB Wiley, New York, pp. 492 p. **Armstrong FA,** George SJ, Thomson AJ, Yates MG (1988) FEBS Lett 234:107–110.

**Back E,** Burckhardt W, Meyer M, Privalle L, Rothstein S (1988) Mol Gen Genet 212:20–26. **Bagby S,** Barker PD, Hill HAO, Sanghera GS, Dunbar B, Ashby GA, Eady RR, Thorneley RN (1991) Biochem J 277:313–319. **Bagg A,** Neilands JB (1987) Microbiol Rev 51:509–518. **Becker TW,** Foyer C, Caboche M (1992) Planta 188:39–47. **Bell LC,** Richardson DJ, Ferguson SJ (1992) J Gen Microbiol 138:437–443. **Belser LW** (1979) Annu Rev Microbiol 33:309–333. **Benemann JR,** Yoch DC, Valentine RC, Arnon DI (1971) Biochim Biophys Acta 226:205–212. **Bennett LT,** Jacobson MR, Dean DR (1988) J Biol Chem 263:1364–1369. **Bergman B,** Rai AN, Johansson C, Söderbäck E (1993) Symbiosis (in press). **Berry AM,** Moreau RA, Jones ADC (1991) Plant Physiol 95:111–115. **Bishop PE,** Joerger RD (1990) Annu Rev Plant Physiol Plant Mol Biol 41:109–125. **Bisseling T,** Franssen HJ, Govers F, Horvath B, Moerman M, Scheres B, van de Wiel C, Yang WC (1991) In: Hennecke H, Verma DPS (eds) Advances in molecular genetics of plant-microbe interactions, vol 1. Kluwer, Dordrecht, pp 300–303. **Blacquière T** (1986) Plant Soil 91:377–380. **Bohlool BB,** Smidt EL (1974) Science 185:269–271. **Böhme H,** Haselkorn R (1988) Mol Gen Genet 214:278–285. **Böhme H,** Schrautemeier B (1987) Biochim Biophys Acta 891:1–7. **Botella JR,** Verhelen JP, Valpuesta V (1988) Plant Physiol 88:943–946. **Bothe H** (1977) In: Trebst A, Avron M (eds) Encyclopedia of plant physiol, vol 5. Springer, Berlin Heidelberg New York, pp 217–221. **Bothe H** (1990) Prog Bot 52:122–137. **Brangeon J,** Hirel B, Forchioni A (1989) Protoplasma 151:88–97. **Braun C,** Zumft WG (1991) J Biol Chem 266:22785–22788. **Braun-Howland EB,** Nierzwicki-Bauer SA (1990) In: Rai AN (ed) Handbook of symbiotic cyanobacteria. CRC Press, Boca Raton, pp 65–117. **Budde JA,** Randall DD (1990) Plant Physiol 94:1501–1504. **Burns RC,** Hardy RWF (1975) Nitrogen fixation. Springer, Berlin Heidelberg New York, 189 p.

**Campbell EL,** Meeks JC (1992) J Gen Microbiol 138:473–480. **Campbell WH,** Kingshorn JR (1990) Trends Biochem Sci 15:315–319. **Carpenter EJ,** Romans K (1991) Science 254:1356–1358. **Carr GJ,** Ferguson SJ (1990) Biochem J 269:423–429. **Carr GJ,** Page MD, Ferguson SJ (1989) Eur J Biochem 179:683–692. **Cheng C,** Dewdney J, Nam H, Den Boer BGW, Goodman HM, Choi HK, Kleinhofs A, An G (1989) Plant Mol Biol 13:731–733. **Coyne MS,** Arunakumari A, Pankratz HS, Tiedje JM (1990) J Bacteriol 172:2558–2562. **Crawford NM,** Campbell WH (1990) Plant Cell 2:829–835. **Crawford NM,** Smith M, Bellissimo D, Davis RM (1988) Proc Natl Acad Sci USA 85:5006-5010.

**Daniel-Vedele F,** Dorbe MF, Caboche M, Rouzé P (1989) Gene 85:371–380. **Dazzo FB,** Hubbell DH (1975) Appl Microbiol 30:1017–1033. **De Boer W,** Klein Gunnewiek PJA, Veenhuis M, Bock E, Laanbroek HJ (1991) Appl Environ Microbiol 57:3600–3604. **Dermastia M,** Turk T, Hollocher TC (1991) J Biol Chem 266:10899–10905. **Diaz CL,** Melchers LS, Hooykaas PJJ, Lugtenberg BJJ, Kijne JW (1989) Nature 338:579–581. **Dilworth MJ,** Glenn AR (eds) (1991) Biology and biochemistry of nitrogen fixation. Elsevier, Amsterdam, 438 p. **Djordjevic MA,** Weinman JJ (1991) Aust J Plant Physiol 18:543–557. **Downie JA** (1991) Curr Biol 1:382–384. **Downie JA,** Marie C, Scheu AK, Firmin JL, Wilson KE, Davies AE, Cubo TM, Mavridou A, Johnston AWB, Economou A (1991) In: Henecke H, Verma DPS (eds) Advances in molecular genetics of plant-microbe interactions, vol 1. Kluwer, Dordrecht, pp 134–141.

**Edwards JW,** Walker EL, Coruzzi GM (1990) Proc Natl Acad Sci USA 87:3459–3463. **Engel P,** Trageser M, Unden G (1991) Arch Microbiol 156:463–470. **Engelaar WMHG,** Bodelier PLE, Laanbroek HJ, Blom CWPM (1991) FEMS Microbiol Ecol 86:33–42. **Ernst A,** Reich S, Böger P (1990) J Bacteriol 172:748–755.

**Fillat MF,** Sandmann G, Gómez-Moreno C (1988) Arch Microbiol 150:160–164. **Fillat MF,** Borrias WE, Weisbeek PJ (1991) Biochem J 280:187–191. **Fischer HM,** Bruderer T, Hennecke H (1988) Nucleic Acids Res 16:2207–2224. **Fogg GE** (1986) Proc R Soc Lond Ser B 228:1–24. **Forde BG,** Day HM, Turton JF, Wen-jun S, Cullimore JV, Oliver JE (1989) Plant Cell 1:391–401. **Franssen HJ,** Nap JP, Gloudemans T, Stiekema W, van Dam H, Govers F, Louwerse J, van Kammen A, Bisseling T (1987) Proc Natl Acad Sci USA 84:4495–4499.

**Garber AE,** Castignetti D, Hollocher TC (1982) Biochem Biophys Res Comm 107:1504–1507. **Goretski J,** Hollocher TC (1988) J Biol Chem 263:2316–2323. **Goretski J,** Zafiriou OC, Hollocher TC (1990) J Biol Chem 165:11535–11538. **Gowri G,** Campbell WH (1989) Plant Physiol 90:792–798. **Gresshoff PM,** Roth LE, Stacey G, Newton WE (eds) (1990) Nitrogen fixation: achievements and objectives. Proc 8th Int Congr on Nitrogen fixation, Knoxville, TN. Chapman & Hall, New York, 869 p. **Grilli-Caiola M,** Forni C, Castagnola M (1988) Symbiosis 5:185–198.

**Hankinson TR,** Schmidt EL (1984) Can J Microbiol 30:1125–1132. **Harris SL,** Silvester WB (1992) New Phytol 121:43–48. **Heiss B,** Frunzke K, Zumft WG (1989) J Bacteriol 171:3288–3297. **Hennecke H,** Verma DPS (eds) (1991) Advances in molecular genetics of plant-microbe interactions, vol 1. Kluwer, Dordrecht, 482 p. **Hoover TR,** Robertson AD, Cerny RL, Hayes RN, Imperial J, Shah VK, Ludden PW (1987) Nature 329:855-857. **Huber SC,** Huber JLA (1990) Current Top Plant Biochem Physiol 9:329–343. **Huber JL,** Huber SC, Campbell WH, Redinbaugh MG (1992) Arch Biochem Biophys 296 (in press). **Hyde GE,** Crawford NG, Campbell WH (1991) J Biol Chem 266:23542–23547.

**Jackson MA,** Tiedje JM, Averill BA (1991) FEBS Lett 291:41–44. **Ji XB,** Hollocher TC (1988) Biochem Biophys Res Comm 157:106–108. **Joerger RD,** Premakumar R, Bishop PE (1986) J Bacteriol 168:673–682. **Joerger RD,** Jacobson MR, Bishop PE (1989) J Bacteriol 171:3258–3267. **Joerger RD,** Loveless TM, Pau RN, Mitchenall LA, Simon BH, Bishop PE (1990) J Bacteriol 172:3400–3408. **Joerger RD,** Wolfinger ED, Bishop PE (1991) J Bacteriol 173:4440–4446. **Johnson JL,** Rajagopalan KV, Meyer O (1990) Arch Biochem Biophys 283:542–545. **Johnstone JI,** McCabe PC, Greaves P, Curr SJ, Cole GE, Brow MAD, Unkles SE, Clutterbuck AJ, Kinghorn JR, Innis MA (1990) Gene 90:181–192. **Joshi PA,** Caetano-Anollés G, Graham ET, Gresshoff PM (1990) In: Gresshoff PM, Roth LE, Stacey G, Newton WE (eds) Nitrogen fixation: achievements and objectives. Proc 8th Int Congr on Nitrogen fixation, Knoxville, TN. Chapman Hall, New York, pp 744–744. **Jüngst A,** Braun C, Zumft WG (1991a) Mol Gen Genet 225:241–248. **Jüngst A,** Wakabajashi S, Matsubara M, Zumft WG (1991b) FEBS Lett 279:205–209.

**Kaiser WM,** Brendle-Behnisch E (1991) Plant Physiol 96:363–367. **Kaiser WM,** Förster J (1989) Plant Physiol 91:970–974. **Kaiser WM,** Spill D (1991) Plant Physiol 96:368–375. **Kaiser WM,** Spill D, Brendle-Behnisch E (1992) Planta 186:236–240. **Kaldorf M,** Linne von Berg K-H, Bothe H (1993) (submitted). **Kapp D,** Niehaus K, Quandt J, Müller P, Pühler (1990) Plant Cell 2:139–151. **Karrasch M,** Börner G, Thauer RK (1990) FEBS Lett 274:48–52. **Kennedy C,** Dean D (1992) Mol Gen Genet (in press). **Kentemich T,** Danneberg G, Hundeshagen B, Bothe H (1988) FEMS Microbial Lett 51:19–24. **Kentemich T,** Haverkamp G, Bothe H (1991) Z Naturfosch 46c:217–222. **Kijne JM,** Diaz CL, dePater BS, Smit G, Bakhuizen R, Lugtenberg BJJ (1990) In: Gresshoff BM, Roth LE, Stacey G, Newton WE (eds) Nitrogen fixation: achievements and objectives. Chapman & Hall, New York, pp 187–192. **Kluge M,** Mollenhauer D, Mollenhauer R (1991) Planta 185:311–315. **Kondorosi A** (1989) In: Kosuge T, Nester EW (eds) Plant-microbe interactions molecular and genetic perspectives, vol 3. McGraw Hill, New York, pp 383–420. **Kondorosi A** (1991) In: Hennecke H, Verma DPS (eds) Advances in molecular genetics of plant-microbe interactions, vol 1. Kluwer, Dordrecht, pp 111–118. **Körner H,** Mayer F (1991) Meeting of the German Vereinigung für Allgemeine und Angewandte Microbiologie (VAAM), March 1991, Freiburg, Abstr Book, Poster No 65.

**Lahners K,** Kramer V, Back E, Privalle L, Rothstein S (1988) Plant Physiol 88:741–746. **Lehman LJ,** Roberts GP (1991) J Bacteriol 173:5705–5711. **Lerouge P,** Roche P, Faucher C, Maillet F, Truchet G, Promé JC, Dénarié J (1990) Nature 344:781–784. **Lin C,** Watanabe I, Liku CC, Zheng DY, Tang LF (1988) In: Bothe H, deBruijn FJ, Newton WE (eds) Nitrogen fixation: hundred years after. Gustav Fischer, Stuttgart, pp 223–227. **Lindblad P,** Bergman B (1990) In: Rai AN (ed) Handbook of symbiotic cyanobacteria. CRC Press, Boca Raton, pp 137–159. **Lindblad P,** Haselkorn R, Bergman B, Nierzwicki-Bauer SA (1989) Arch Microbiol 152:20–24. **Lindblad P,** Bergman B, Nierzwicki-Bauer SA (1991) Appl Environ Microbiol 57:3637–3640. **Linne von Berg KH,** Bothe H (1992) FEMS Microbiol Ecol 86:331–340. **Long SR** (1989) Cell 56:203–214. **Lugtenberg BJJ,** Diaz C, Smit G, dePater S, Kijne JW (1991) In: Hennecke H, Verma DPS (eds) Advances in molecular genetics of plant-microbe interactions, vol 1. Kluwer, Dordrecht, pp 174–181.

**Marie C,** Barny MA, Downie JA (1992) Mol Microbiol (in press). **Meeks JC,** Joseph CM, Haselkorn R (1988) Arch Microbiol 150:61–71. **Mengel K,** Pilbeam DJ (eds) (1992) Nitrogen metabolism of plants. Proceedings of the Phytochemical Society of Europe, vol 33. Clarendon Press, Oxford, 289 p. **Morett E,** Fischer HM, Hennecke H (1991) J Bacteriol 173:3478–3487. **Morgan TV,** Lundell DJ, Burgess BK (1988) J Biol Chem 263:1370–1375.

**Neuer G,** Bothe H (1985) Arch Microbiol 143:185–191. **Niehaus F,** Handtke K, Unden G (1991) FEMS Microbiol Lett 84:319–324. **Nierzwicki-Bauer SA,** Aulfinger H (1991) Appl Environ Microbiol 57:3629–3636. **Nieva-Gomez D,** Roberts GP, Klevickis S, Brill WJ (1980) Proc Natl Acad Sci USA 77:2555–2558.

**Oaks A,** Poulle M, Goodfellow VJ, Cass LA, Deising H (1982) Plant Physiol 88:442–447. **Oelmüller R,** Schuster C, Mohr H (1988) Planta 174:75–83. **Oelmüller R,** Briggs WR (1990) Plant Physiol 92:434–439.

**Parks BM,** Jones AM, Adamse P, Koornneef M, Kendrick RE, Quail PH (1987) Plant Mol Biol 9:97–107. **Parniske M,** Ahlborn B, Werner D (1991) J Bacteriol 173:3432–3439. **Pereira S,** Carvalho H, Sunkel C, Salema S (1992) Protoplasma, 167:66–73. **Petro MJ,** Gates JE (1987) Symbiosis 3:41–48. **Polsinelli M,** Materassi M, Vincenzini M (1991) Nitrogen fixation. Proc 5th Int Symp on Nitrogen fixation with non-legumes, Florence, Sept 1990. Kluwer, Dordrecht, 678 p. **Pope MR,** Murrell SA, Ludden PW (1985) Proc Natl Acad Sci USA 82:3173–3177. **Powell SJ,** Prosser JI (1991) J Gen Microbiol 137:1923–1929. **Preston GG,** Ludden PW (1982) Biochem J 205:489–494. **Prinsen E,** Chauvaux N, Schmidt J, John M, Wieneke U, De Greef J, Schell J, Van Onckelen H (1991) FEBS Lett 282:53–55. **Prosser JI** (1989) Adv Microbiol Physiol 30:125–181. **Prosser JI,** Lazarus LM (1990) Plant Mol Biol 15:187–190.

**Rajagopalan KV** (1991) In: Meister A (ed) Advances in enzymology and related areas of molecular biology. Wiley, London, pp 215–290. **Rajasekhar VK,** Oelmüller R (1987) Plant Physiol 71:517–521. **Rajasekhar VK,** Gowri G, Campbell WH (1988) Plant Physiol 88:242–254. **Reich S,** Böger P (1989) FEMS Microbiol Lett 58:81–86. **Remde A,** Slemr F, Conrad R (1989) FEMS Microb Ecol 62:221–230. **Remmler JL,** Campbell WH (1986) Plant Physiol 80:442–447. **Revsbech NP,** Sörensen J (eds) (1990) Denitrification in soil and sediment. FEMS Symp no 56. Plenum Press, New York, 349 p. **Richardson DJ,** Bell LC, McEwan AG, Jackson JB, Ferguson SJ (1991) Eur J Biochem 199:677–683. **Riens B,** Heldt HW (1992) Plant Physiol 98:537–577. **Robinson AC,** Dean DR, Burgess RK (1987) J Biol Chem 262:14327–14332. **Roche P,** Lerouge P, Promé JC, Faucher C, Vasse J, Maillet F, Camut S, de Billy F, Dénarié J, Truchet G (1991) In: Hennecke H, Verma DPS (eds) Advances in molecular genetics of plant-microbe interactions, vol 1. Kluwer, Dordrecht, pp 119–126.

**Saito T,** Williams RJP (1991) Eur J Biochem 197:29–38. **Sanchez J,** Heldt HW (1990) Plant Physiol 92:684–689. **Sandmann G,** Peleato ML, Fillat MF, Lázaro MC, Gómez-Moreno C (1990) Photosynth Res 26:119–125. **Schaffer S,** Handtke K, Braun V (1985) Mol Gen Genet 201:204–212. **Schneider K,** Müller A, Schramm U, Klipp W (1988) 195:653–661, 303–314. **Schrautemeier B,** Böhme H (1985) FEBS Lett 184:304–308. **Schuster C,** Oelmüller R, Bergfeld R, Mohr H (1988) Planta 174:289–297. **Shah VK,** Stacey G, Brill WJ (1983) J Biol Chem 258:12064–12068. **Shah VK,** Hoover TR, Imperial J, Paustian TD, Roberts GP, Ludden PW (1988) In: Bothe H, de Bruijn FJ, Newton WE (eds) Nitrogen fixation: hundred years after. Gustav Fischer, Stuttgart, pp 115–120. **Shapleigh JP,** Payne WJ (1985) J Bacteriol 163:837–840. **Smith BE,** Eady RR, Lowe DJ, Gormal C (1988) Biochem J 250:299–302. **Smith GB,** Tiedje JM (1992) Appl Environ Microbiol 58:376–384. **Söderbäck E,** Lindblad P, Bergman B (1990) Planta 182:355–362. **Solomonson LP,** Sephar AM (1977) Nature 265:373–375. **Spaink HP,** Sheeley DP, van Brussel AAN, Glushka J, York WS, Tak T, Geiger O, Kennedy EP, Reinhold VN, Lugtenberg BJJ (1991) Nature 354:125–130. **Spaink HP,** Aarts A, Stacey G, Bloemberg GV, Lugtenberg BJJ, Kennedy EP (1992) Molecular Plant Microbe Interactions 5:72–80. **Sprent JI,** Sprent P (1990) Nitrogen fixing organisms–pure and applied aspects. Chapman & Hall, London, 256 p. **Stacey G,** Burris RH, Evan HJ (eds) (1992) Biological nitrogen fixation. Chapman & Hall, New York, 943 pp. **Strasburger E** (1873) Über *Azolla*. Hermann Dobis Verlag, Jena. **Sumino T,** Nakamura N, Kawaguchi Y, Tada M (1992) Appl Microbiol Biotechnol 36:556–560.

**Tachil J,** Frunzke K, Meyer O (1991) Meeting of the German Vereinigung für Allgemeine und Angewandte Mikrobiologie (VAAM), Freiburg, März; Abstract Book, Poster 277. **Teraguchi S,** Hollocher TC (1989) J Biol Chem 264:1972–1979. **Thomson AJ** (1991) FEBS Lett 285:230–236. **Thorneley RNE,** Abell C, Ashby GA, Drummond MH, Eady RR, Huff S, Macdonald C (1992) Biochemistry 31:1216–1224. **Trebst A,** Bothe H (1966) Ber Dtsch Bot Ges 79:44–47. **Truchet G,** Barger DG, Camut S, de Billy F, Vasse J, Huguet T (1989) Mol Gen Genet 219:65–68. **Truchet G,**

Roche P, Lerouge P, Vasse J, Camut S, de Billy F, Promé JC, Dénarié J (1991) Nature 351:670–673. **Tsai SM,** Phillips DA (1991) Appl Environ Microbiol 57:1485–1488.

**Unden G,** Trageser M (1991) Antonie Leeuwenhoek J Microbiol 59:65–76.

**Viebrock A,** Zumft WG (1988) J Bacteriol 170:4658–4668. **Vincentz M,** Caboche M (1991) Embo J 10:1027–1035. **Voßwinkel R,** Neidt I, Bothe H (1991) Arch Microbiol 156:62–69.

**Walker-Simmons M,** Kudona DA, Warner RL (1989) Plant Physiol 90:728–733. **Werner D** (1992) Symbiosis of plants and microbes. Chapman & Hall, London. **Werner D,** Müller P (eds) (1990) Fast growing trees and nitrogen fixing trees. Int Conf, Marburg, 1989. Fischer, Stuttgart, 396 p. **Wray J,** Kinghorn J (eds) (1989) Molecular and genetic aspects of nitrate assimilation. Oxford Sci Publ, Oxford.

**Ye RW,** Suarez-Toro J, Tiedje JM, Averill BC (1991) J Biol Chem 266:12848–12851. **Yoch DC,** Arnon DI (1972) J Biol Chem 247:4514–4520.

**Zafiriou OC,** Hanley QS, Snyder G (1989) J Biol Chem 264:5694–5699. **Zhang CS,** Hollocher TC, Kolidzij AF, Orme-Johnson WH (1991) J Biol Chem 266:2199–2202. **Zimmermann WJ,** Bergman B (1990) Microb Ecol 19:291–302. **Zumft WG,** Döhler K, Körner H, Löchelt S, Viebrock A, Frunzke K (1988) Arch Microbiol 149:492–498. **Zumft WG,** Viebrock-Sambale A, Braun C (1990) Eur J Biochem 192:591–599.

Professor Dr. Hermann Bothe
Botanisches Institut
Universität zu Köln
Gyrhofstraße 15
W-5000 Köln 41, Germany

# VI. Secondary Plant Substances
# Further Topics of the Phenylpropanoid Metabolism

By Horst-Robert Schütte

## 1. Introduction

Some of the many plant-specific phenylpropanoid branch pathways and the corresponding functional diversity of their products were discussed in the last volume (Schütte 1992). Last year it was mentioned that special topics of the phenylpropanoid metabolism had been discussed by Schütte (1978, 1979, 1985) and that recent developments in this area involve major advances in elucidating the structural organization, mode of expression and functional relationships of genes encoding enzymes of phenylpropanoid metabolism in dicotyledonous plants. In this Volume, special interest is given to the quinones, lignans and to capsaicin.

## 2. Quinones

Most naturally occurring quinones are p-quinones and their structure is based on the benzoquinone, naphthoquinone, and anthraquinone ring systems (Thomson 1987). The naphthoquinones of the vitamin K type have a long isoprenoid side chain. Other naphthoquinones are devoid of a side chain, e.g., juglone (XIII). A typical representative of the anthraquinone derivatives is alizarin (XXII). Naphthoquinones with a long isoprenoid side chain are formed in microorganisms (menaquinones, vitamin $K_2$) and in higher plants (phylloquinone, vitamin $K_1$). Naphthoquinone derivatives with short or absent side chains, as well as anthraqinones of the alizarin type, are produced in a few families of higher plants, e.g., Balsaminaceae, Juglandaceae, Rubiaceae (Wijnsma and Verpoorte 1986).

One of the most interesting features of quinone biosynthesis in higher plants is that it is derived from a variety of different precursors and by different pathways (Leistner 1981; van den Berg and Labadie 1989; Dewick 1988, 1989, 1990, 1991; van den Berg 1991). We can differentiate between acetate/malonate-derived quinones and the isochorismate/o-succinylbenzoic acid-derived quinones as the most important pathways to be mentioned. For example, in the field of anthraquinones, emodin and islandicin in lower plants (Gatenbeck 1960) and emodin and chrysophanol in higher plants (Leistner 1971) have been shown to be derived from acetic acid, alizarin (XXII) and purpurin carboxylic acid (XXIII) from *Rubia tinctorum*, however, are derived from shikimic acid (Conn 1986) by way of o-succinylbenzoic acid.

Progress in Botany, Vol. 54

In the field of isochorismate-derived quinones (Leistner 1985), work on the biosynthesis of vitamin $K_2$ (IX) (menaquinone) is most advanced when compared to the investigation on the biosynthesis of other quinones (Bentley and Meganathan 1981, 1983). Shikimic acid is incorporated nonsymmetrically into menaquinone (Baldwin et al. 1974). Earlier, it was believed that chorismic acid (II) is the branch point compound that links the shikimate (I) pathway to the vitamin K biosynthetic pathway. Recent results, however, show that in the case of vitamin $K_2$, isochorismic acid (III) and not chorismic acid undergoes aromatization (Weische and Leistner 1985; Weische et al. 1987a).

Isochorismic acid synthase, the enzyme catalyzing the formation of isochorismic acid from chorismic acid (Young et al. 1969; Young and Gibson 1969; Johanni et al. 1989; Poulson et al. 1991) is encoded in *Echerichia coli* by the entC gene which is part of the enterobactin gene cluster (Elkins and Earhart 1988; Ozenberger et al. 1989; Tummuru et al. 1989). This cluster maps at 13.8 min on the *E. coli* linkage map. Enterobactin is an ionophore which is essential for iron uptake during aerobic growth; but isochorismic acid synthase and the entC gene not only play a role in enterobactin biosynthesis but also are functionally related to men genes encoding enzymes which catalyze the formation of menaquinones (Kaiser and Leistner 1990). These genes are located at 49 and 88 min on the *E. coli* linkage map.

The key reaction in vitamin $K_2$ biosynthesis is the formation of o-succinylbenzoic acid (VI) as the first aromatic intermediate (Fig. 1) (Dansette and Azerad 1970). It has been isolated from a men B mutant strain of *E. coli* which is blocked in menaquinone biosynthesis (Young 1975) and is formed by a new and hitherto unprecedented aromatization process from isochorismic acid (III) and 2-ketoglutaric acid (IV) in the presence of thiamine pyrophosphate (Campbell et al. 1971; Baldwin et al. 1974; Meganathan and Bentley 1983; Weische and Leistner 1985; Weische et al. 1987a). Protein extracts from menaquinone-producing bacteria such as *E. coli*, as well as *Aerobacter aerogenes* mutants are known to catalyze this formation (Meganathan 1981). 2-Ketoglutaric acid is decarboxylated by the o-succinylbenzoic acid synthase system in the presence of thiamine pyrophosphate. The anion formed probably attacks isochorismic acid in a Michael-type reaction. The aromatization leading to o-succinylbenzoic acid (VI) is likely to be a multistep reaction sequence and proceeds via 2-succinyl-6-hydroxy-2,4-cyclohexadiene-carboxylate (V) as an intermediate (Emmons et al. 1985; Popp et al. 1989; Breuer et al. 1991). The o-succinylbenzoic acid synthase, which has been characterized (Weische et al. 1987b), dissociates, and decarboxylating activity can be separated from fractions containing this synthase (Marley et al. 1986). It is a subunit of the o-succinylbenzoic acid synthase. This decarboxylating enzyme, 2-succinyl-6-hydroxy-2,4-cyclohexadiene-carboxylate synthase, has a molecular weight of 66500, whereas the intact synthase has a molecular weight of 190 000. The gene (men D) believed to encode this enzyme was sequenced (Hawkins et al. 1989; Popp 1989). The men D gene is linked to the men C gene on the *E. coli* chromosome. The latter gene encodes the enzyme that converts 2-succinyl-6-hydroxy-2,4-cyclohexadiene-carboxylate (V) to o-succinylbenzoic acid (VI). o-Succinylbenzoic acid is the first aromatic intermediate not only in menaquinone biosynthesis in bacteria but also in phylloquinone and anthraquinone biosynthesis in higher plants, see below (Dansette and Azerad 1970).

The next step in menaquinone biosynthesis is the conversion of o-succinylbenzoic acid (VI) into 1,4-dihydroxy-2-naphthoic acid (VIII) (McGovern and Bentley 1978;

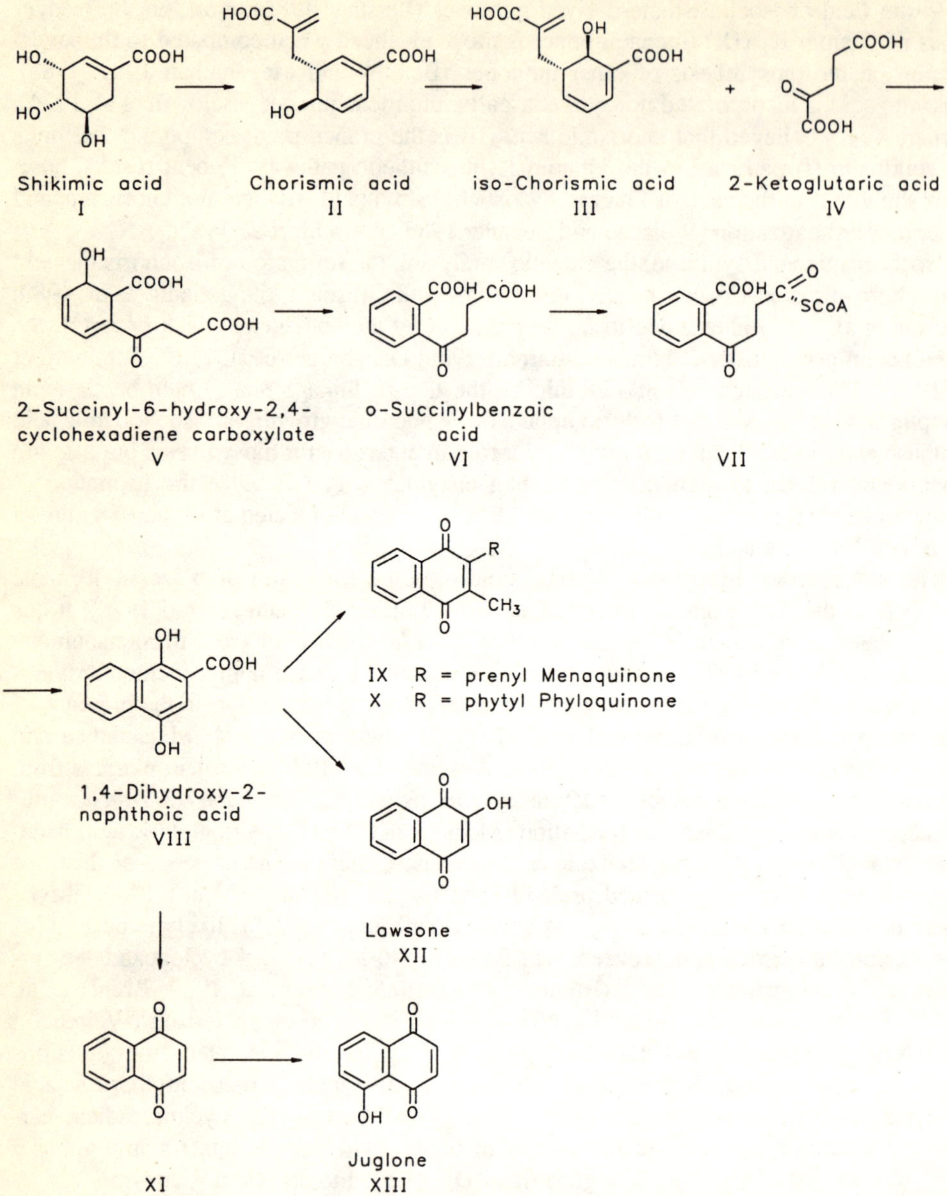


**Fig. 1.** Biosynthesis of o-succinylbenzoic acid and derived quinones

Meganathan and Bentley 1979; Meganathan et al. 1980; Shaw et al. 1982). Cell-free protein extracts from different menaquinone producing bacterial strains including *E. coli* and *Mycobacterium phlei* catalyze the coenzyme A, ATP and $MG^{2+}$-dependent conversion. The intermediate is a mono coenzyme A ester (VII) (Heide et al. 1982a) which has been characterized (Kolkmann and Leistner 1987a) and converted enzymically to 1,4-dihydroxy-2-naphthoic acid (VIII) (Kolkmann and Leistner 1987b). The conversion of o-succinylbenzoic acid (VI) to 1,4-dihydroxy-2-naphthoic acid (VIII) is catalyzed by two enzymes, the o-succinylbenzoate: coenzym A ligase and 1,4-dihydoxy-2-naphtoate synthase which is the product of the men B gene in *E. coli* (Meganathan and Bentley 1979; Meganathan et al. 1980; Shaw et al. 1983). o-Succinlybenzoic acid: coenzyme A ligase was purified from *M. phlei* and characterized with respect to isoelectric point, molecular weight, pH optimum, temperature optimum and kinetic data (Sieweke and Leistner 1991). The enzyme hydrolyses ATP to AMP and appears to possess broad specificity for substrates and cofactors. Of the two possible carboxyl groups, the aliphatic one is activated (Heide and Leistner 1981b; Kolkmann et al. 1982; Kolkmann and Leistner 1985). The conversion of o-succinylbenzoic acid-CoA ester (VII) to 1,4-dihydroxy-2-naphthoic acid (VIII) is a two-step process (Igbavboa and Leistner 1990). In the first step the $2H_{Re}$ atom of (VII) is equilibrated with solvent protons in a fast and reversible reaction. Subsequently, the $2H_{Si}$ and $3H_{Si}$ atoms are removed whereas the $3H_{Re}$ atom becomes the proton at C–3 of 1,4-dihydroxy-2-naphthoic acid (VIII). The second step in this ring closure reaction is the rate-limiting step.

The final reactions in the biosynthesis of menaquinone involve prenylation and methylation (Samuel and Azerad 1972). A 1,4-dihydroxy-2-naphthoate: polyprenyltransferase was detected in the membrane fraction from *Micrococcus luteus* (Saito and Ogura 1981). The results show that menaquinone biosynthesis involves prenylation of 1,4-dihydroxy-2-naphthoate (VIII) prior to decarboxylation and methylation.

Phylloquinone (X) (vitamin $K_1$) is a constituent of higher plants. Its structure is similar to vitamin $K_2$ in that it also contains the menadione nucleus to which, however, a phytyl residue is attached. Phylloquinone occurs in chloroplasts of all green plants. It was shown to be contained in photosystem I reaction center core polypeptides (Takahashi et al. 1985), and has been suggested to be the primary stable electron acceptor (Brettel et al. 1986; Schoeder and Lockau 1986; Andreasson and Vängard 1988; Itoh and Iwaki 1989), which mediates electron transfer between the electron acceptor intermediate chlorophyll a and the iron sulfur center. Although less information is available, phylloquinone biosynthesis in plants apparently follows the same general pattern as for menaquinone biosynthesis except that phytylation replaces prenylation. Strong support for this conclusion is the asymmetric incorporation of o-succinylbenzoic acid (VI) into phylloquinone (X) in *Zea mays* shoots (Thomas and Threlfall 1974; Hutson and Threlfall 1980); this asymmetry is exactly that observed with bacterial menaquinone.

The detailed investigation of phylloquinone (X) biosynthesis in plants is made difficult by the small amounts of this material that are present. The average concentration in plants is the lowest of all the terpenoid quinones and chromanols: thus for every 100 molecules of chlorophyll a, there are about two of phylloquinone (Lichtenthaler 1968). Some work has been achieved with the more easily manipulated phytoflagellate, *Euglena gracilis*, which contains phylloquinone in amounts comparable to those of higher plants and also contains 20–30 times more 5'-monohydroxyphylloquinone than phylloquinone

(Law et al. 1973; Ziegler et al. 1989). This novel hydroxylated material may carry out the function normally performed by phylloquinone in other organisms. It has also been found in *Anacystis nidulans* (Allen et al. 1967) and *Chlorella pyrenoidosa* (Whistance and Threlfall 1970).

The pathway for synthesis in *Euglena* is likely to be similar to that proposed for higher plants. This organism uses the same early precursors as do bacteria. The pathway has been reported to be located in the chloroplast. Whole cell extracts, purified organelles, and organelle subfractions from organotrophically grown *Euglena gracilis* were assayed for the synthesis of the phylloquinone precursors, o-succinylbenzoic acid (VI) and 1,4-dihydroxy-2-naphthoic acid (VIII) (Seeger and Bentley 1991). The enzymes responsible for the synthesis of o-succinylbenzoate from isochorismate and 2-ketoglutarate were localized in the photosynthetic apparatus. Tentative evidence was obtained for the synthesis of 1,4-dihydroxy-2-naphthoate from o-succinylbenzoate; the enzymes appeared to be associated with the chloroplasts and synthesis occurred at the chloroplast envelope membrane.

The final steps in phylloquinone (X) biosynthesis have been localized in subplastidic fractions. Thus cell-free systems catalyzing the phytylation of 1,4-dihydroxy-2-naphthoic acid (VIII) by phytylpyrophosphate have been obtained from plant chloroplast membrane envelopes, whereas the final methylation reaction occurs in isolated thylakoids provided protein of the stroma is added (Threlfall 1972; Schultz et al. 1981; Kaiping et al. 1984). Although chromoplasts of *Capsicum* do not contain vitamin $K_1$, enzymes involved in its biosynthesis seem to be present. Addition of 1,4-dihydroxy-2-naphthoic acid (VIII) and (S)-adenosyl-L-methionine to chromoplasts resulted in formation of phylloquinone (X) (Gaudilliere et al. 1984).

Studies on the biosynthesis of plant naphthoquinones derived from the shikimic acid-o-succinylbenzoic acid pathway have been carried out with plants of three species, namely *Juglans regia* (Juglandaceae) containing preferably juglone (XIII), *Impatiens balsamina* (Balsaminaceae) containing preferably lawsone (XII), and *Catalpa ovata* (Bignoniaceae) containing preferably catalponol (XVI) and catalponone (XVII).

Early work with *J. regia* had shown that shikimic acid (I) is a direct precursor of juglone (XIII) with its carboxylic group via the symmetrical intermediate 1,4-naphthoquinone, that metabolites of the tricarboxylic acid cycle, particularly 2-ketoglutaric acid, were specifically incorporated into naphthoquinones and that o-succinylbenzoic acid is also a key intermediate in the pathway (Fig. 1) (Dansette and Azerad 1970; Floß et al. 1971). The incorporation of shikimic acid (I) into juglone (XIII) involves loss of the pro R6H and retention of the pro 6SH of shikimic acid. Similar results were obtained for the biosynthesis of lawsone (XII) in *I. balsamina* (Campbell 1969; Grotzinger and Campbell 1972a, 1974). In analogy to bacterial systems, it is suggested that o-succinylbenzoic acid (VI) was converted to 1,4-dihydroxy-2-naphthoic acid (VIII). Application of this acid to *J. regia* and *I. balsamina* plants resulted in a specific and rather high incorporation into juglone (XIII) and lawsone (XII) (Fig. 1); but whereas incorporation of this acid into juglone proceeded via 1,4-naphthoquinone (XI) as symmetrical intermediate, incorporation into lawsone took place directly without symmetrical intermediates being involved in the pathway (Leistner and Zenk 1968; Grotzinger and Campbell 1972b; Müller and Leistner 1976).

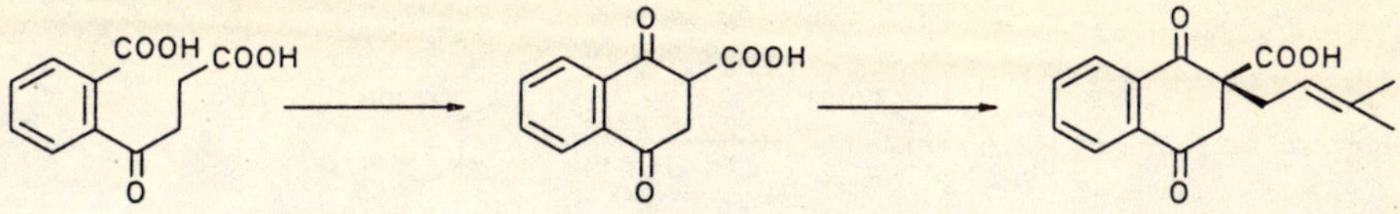

VI o-Succinylbenzoic acid  XIV 2-Carboxy-4-oxo-tetralone (COT)  XV (2S)2-Prenyl-COT

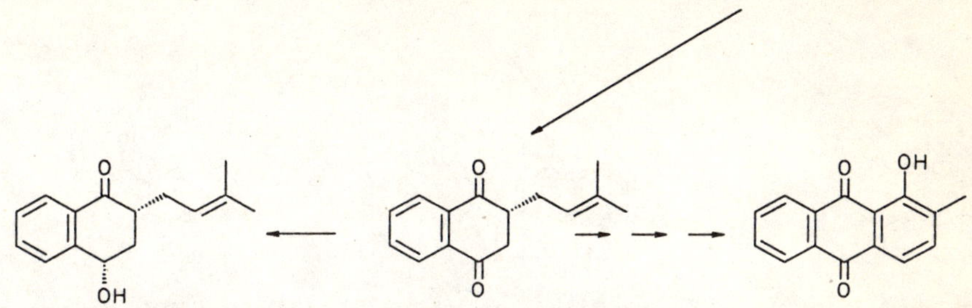

XVI Catalponol  XVII (2R)Catalponone  XVIII

**Fig. 2.** Biosynthesis of naphthalene derivatives in *Catalpa*

However, for the incorporation of o-succinylbenzoic acid into catalponol (XVI) and catalponone (XVII) in *Catalpa ovata*, a pathway without the intermediacy of an aromatic compound is assumed. It is suggested that 2-carboxy-4-oxotetralone (XIV) or 2-carboxy-4-hydroxytetralone rather than 1,4-dihydroxy-2-naphthoic-acid (VIII) were involved in this case (Fig. 2) (Inouye et al. 1975, 1978, 1981; Ueda et al. 1975; Inoue et al. 1977, 1980, 1981). The main biosynthetic pathway to catalponol (XVI) and related compounds in callus tissue of *C. ovata* stereospecifically pass through o-succinylbenzoic acid (VI), 2-carboxy-4-oxo-1-tetralone (COT) (XIV), (2S)2-prenyl-COT (XV) and (2R) catalponone (XVII), but there also exists a subsidiary route from COT (XIV) to catalponol (XVI) via 2-carboxy-4-hydroxy-1-tetralone (CHT) and 2-prenyl-CHT. The assumption that nonaromatic naphthalene derivatives are involved in the pathway between o-succinylbenzoic acid (VI) and quinones also in *J. regia* is in agreement with the observation that oxotetralone and β-hydrojuglone are likely present in these plants (Müller and Leistner 1978).

Naphthoquinones and anthraquinones have been repeatedly shown to occur in the same plant. For example, *Catalpa ovata*, additionally to the naphthalene derivatives mentioned (Fig. 2), contains the anthraquinone derivative (XVIII). In many cases an isoprene unit is attached to these naphthalene derivatives. In earlier work it has been shown that anthraquinones of the alizarin (XXII) type are derived from naphthalene derivatives via o-succinylbenzoic acid (VI) and prenylation. Thus in *Rubia tinctorum* alizarin (XXII) and purpurincarboxylic acid (XXIII) are derived from shikimic acid, glutamic acid, and mevalonic acid via o-succinylbenzoic acid (VI) (Leistner and Zenk 1967, 1971; Leistner 1973a), which is then transformed to a nonsymmetrical intermediate such as (VIII).

Dimethylallylpyrophosphate is then attached to (XIX) following ring closure and further modification to alizarin (XXII) (Fig. 3). Morindone (XXIV), the most abundant anthraquinone in *Morinda citrifolia*, another anthraquinone containing Rubiaceae, is formed by the same pathway as alizarin in *R. tinctorum* (Leistner 1973b, 1975).

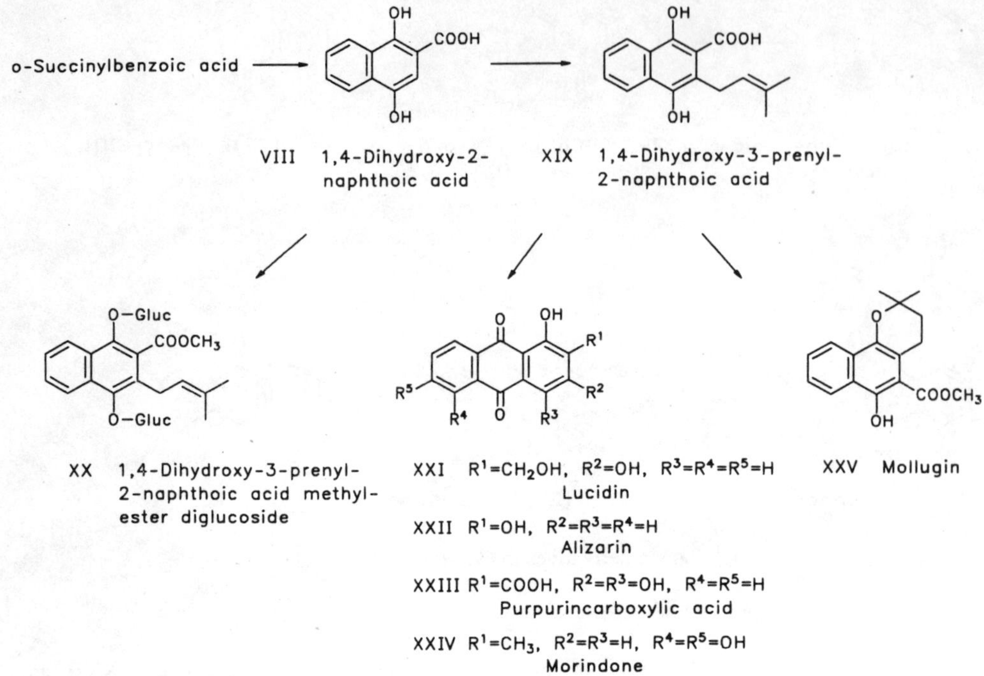


**Fig. 3.** Biosynthesis of anthraquinones and related compounds

Isolation of the enzymes from higher plants catalyzing the synthesis of o-succinylbenzoic acid seems to pose great problems because anthraquinones occur predominantly in lignified tissue of the root bark (Zenk et al. 1975), whereas phylloquinone-producing young green leaves contain only insignificant amounts of the enzyme protein. Cells cultivated in suspensions represent, however, a more suitable enzyme source (Stöckigt and Schübel 1989; Zenk 1991). Hence cells of a phylloquinone-producing photoautotrophic suspension culture raised from *Morinda lucida* and cells of anthraquinone producing heterotrophic suspension cultures raised from *Galium* species (Bauch and Leistner 1978a) were used, which were optimized from different Rubiaceae species for the production of anthraquinones (Schulte et al. 1984).

Also the anthraquinones of *Galium mollugo* cell cultures such as lucidin-(XXI)-3-primveroside, which could be isolated along with 1,4-dihydroxy-3-prenyl-2-naphthoic acid methylester diglucoside (XX) and mollugin (XXV), are derived from shikimic acid (I), 2-ketoglutaric acid (IV), and mevalonic acid with o-succinylbenzoic acid (VI) being a key intermediate (Bauch and Leistner 1978b; Inoue et al. 1979, 1984a). The latter could be isolated from such cell cultures (Simantiras et al. 1991).

Enzyme preparations from such cell cultures were shown to catalyze the synthesis of o-succinylbenzoic acid from isochorismic acid and 2-oxoglutaric acid in the presence of thiamine diphosphate and $Mn^{2+}$ (Simantiras and Leistner 1989, 1991). One of the intermediates is likely to be identical to 2-succinyl-6-hydroxy-2,4-cyclohexadiene-1-carboxylate, a metabolite known to occur in the bacterial pathway (Fig. 1). An active enzyme preparation could be obtained from this source which has led to the detection of o-succinylbenzoic acid: coenzyme A ligase (Heide et al. 1982b; Heide and Leistner 1983). The CoA-ester (VII) was converted to dihydroxynaphthoic acid by an enzyme extract from *Mycobacterium phlei*, indicating that the bacterial CoA-ester and the ester from *G. mollugo* are identical.

1,4-Dihydroxy-3-prenyl-2-naphthoic acid (XIX) has been postulated as an intermediate and the methylester (as glucoside XX) is a natural constituent of *G. mollugo* (Heide and Leistner 1981a, 1982). The prenylation during the biosynthesis of anthraquinone (Fig. 3) occurring at the position corresponding to C–3 of o-succinylbenzoic acid is in contrast to naphthoquinone derivatives in *Catalpa* (Fig. 2) and to menaquinone (IX) which are known to be formed by prenylation at the position corresponding to C–2 of o-succinylbenzoic acid.

The genus *Streptocarpus* belongs to the family Gesneriaceae. *S. dunnii* contains several pigments in its roots and leaves, some rare prenylated naphthoquinones such as dunnione (XXIX) and δ-dunnione (XXVIII), together with two anthraquinones (XXXII and XXXIII) (Stöckigt et al. 1973; Inoue et al. 1983). In *S. dunnii* suspension cultures the naphthoquinones were formed through a unique prenylation mode via o-succinylbenzoic acid (VI), dihydroxynaphthoic acid (VIII), lawsone (XII) and the O-prenylated lawsone (XXVI) (Fig. 4). By the Claisen-type rearrangement of the latter compound, the key intermediate 2-hydroxy-3-(1,1-dimethylallyl)-1,4-naphthoquinone (XXVII) was formed. The key intermediate in the formation of anthraquinones (Fig. 5), 2-prenyl-1,4-naphthohydroquinone (XXX), which gives (XXXII) and (XXXIII) via (XXXI), was derived from o-succinylbenzoic acid (VI), either via dihydroxynaphthoic acid (VIII), or via 2-carboxy-2,3-dihydro-1,4-naphthoquinone, 2-prenyl-2-carboxy-2,3-dihydro-1,4-naphthoquinone (XV) and catalponone (XVII) (Inoue et al. 1982, 1984b). During the biosynthesis of these

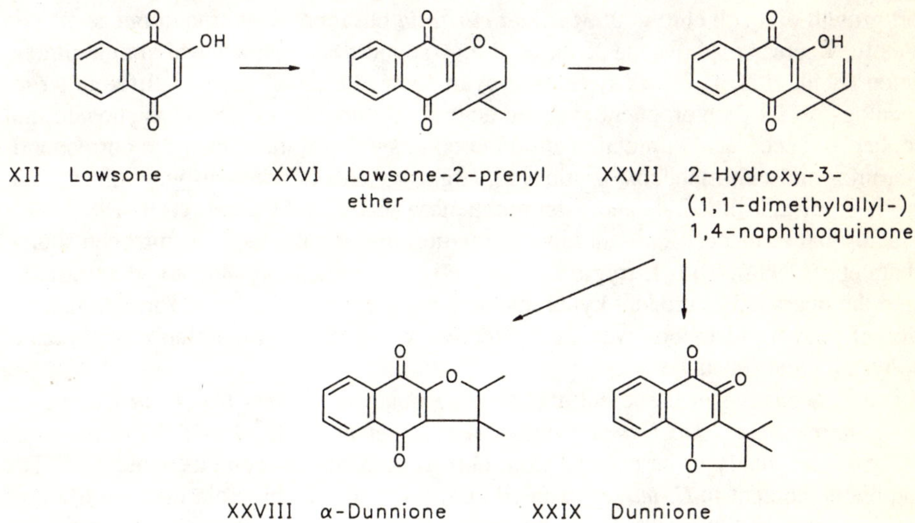


**Fig. 4.** Formation of dunnione and related compounds in *Streptocorpus dunnii*

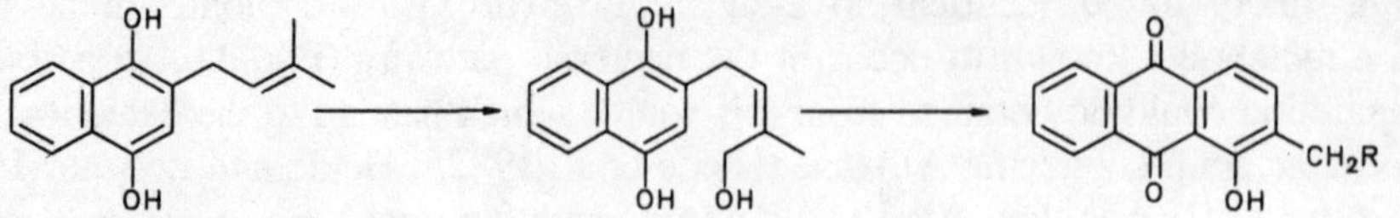

XXX 2-Prenyl-1,4-naphthohydroquinone — XXXI — XXXII R=H 1-Hydroxy-2-methylantrquinone — XXXIII R=OH 1-Hydroxy-2-hydroxymethylanthraquinone

**Fig. 5.** Biosynthesis of anthraquinones in *Streptocorpus dunnii*

anthraquinones, prenylation occurs in a manner different than that observed with *Rubia, Galium* and *Morinda.* The C–2 position (rather than C–3) of the hypothetical naphthalene derivative was prenylated (comparing with *Catalpa* Fig. 2) during the reactions leading to anthraquinones.

The green photosynthetic cell suspension culture of *Morinda lucida* contains also phylloquinone. In these cultures the biosynthesis of anthraquinones and its glycosides can be triggered. Upon transfer of cells into darkness and a medium containing sucrose, the chlorophyll decreases and the lipoquinones, including phylloquinone, disappear from the cells (Igbavboa et al. 1985). Simultaneously, anthraquinone pigments are formed. Thus, in the suspension culture, photoautotrophy correlates with lipoquinone synthesis while heterotrophy correlates with anthraquinone synthesis. This reflects the situation in the intact plants where lipoquinones are associated with chloroplasts, whereas anthraquinones occur in the roots.

Various observations indicate that anthraquinone biosynthesis can be repressed, induced or increased in cell suspension cultures. Thus 2,4-dichlorophenoxyacetic acid may support growth of a cell culture of *Morinda citrifolia* but repress anthraquinone synthesis completely, whereas naphthalene acetic acid may support both growth and anthraquinone formation (Zenk et al. 1975). A systematic evaluation of the influence of different structural analogs of 2,4-dichlorophenoxyacetic acid on anthraquinone synthesis showed that 4-iodophenoxyacetic acid stimulated anthraquinone synthesis more than the corresponding 4-bromo- or 4-chlorophenoxyacetic acid. Highest yields of pigments were obtained in the presence of 2,3-dimethyl- and 2-bromophenoxyacetic acids (Zenk et al. 1984). Another factor that influences anthraquinone formation in cell cultures is the nitrogen source (El-Shagi et al. 1984). Thus L-tryptophan as well as its precursors indole and anthranilic acid and the degradation product kynurenic acid suppress anthraquinone formation. Suppression of pigment formation was most effective, however, with 2-aminobenzothiazole, a nonphysiological substance.

Quinones occurring in tissue cultures of the Rubiaceous plants *Cinchona ledgeriana* and *C. pubescens* = *C. succirubra* (Mulder-Krieger et al. 1982, 1984; Wijnsma et al. 1984, 1986a) are likely to be derived from chorismic acid /o-succinylbenzoic acid. The anthraquinone content in *C. ledgeriana* cell cultures increased ninefold upon addition of sterilized mycelia of Phytophthora cinnamomi, a fungus pathogenic to *Cinchona* species (Wijnsma et al. 1985). Infected bark material from an intact *Cinchona ledgeriana* tree

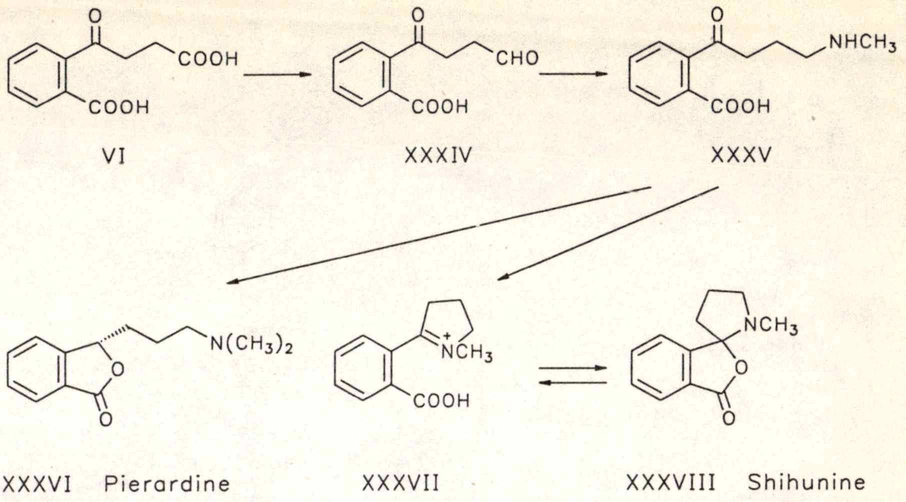

Fig. 6. Biosynthesis of shihunine and pierardine

contained anthraquinones at levels of 0.08%, whereas no anthraquinones could be detected in healthy bark (Wijnsma et al. 1986b). Thus the anthraquinones in *Cinchona* are phytoalexins. From suspension cultures of *C. succirubra* five distinct anthraquinone-specific O-glucosyl-transferases have been isolated (Khouri and Ibrahim 1987).

Another example for the precursorship of o-succinylbenzoic acid is shihunine (XXXVIII). This alkaloid is a phthalidopyrrolidine, no other representatives of this class being known. It was first isolated from the orchid *Dendrobium lohohense* (Inubushi et al. 1964, 1968), and later from *D. pierardii* (Elander et al. 1971), which also yielded the related alkaloid, pierardine (XXXVI) (Elander et al. 1969).

It was considered that these alkaloids would have a common biosynthetic origin and initially they were regarded as polyketides (Leete and Bodem 1976), but after feeding *D. pierardii* with (1–$^{14}$C) acetate, the bulk of the activity was located in the pyrrolidine ring at C–5. On the other hand, o-succinylbenzoic acid (VI) is incorporated with a high rate and specifically into shihunine (XXXVIII) and pierardine (XXXVI) by *D. pierardii* (Fig. 6). It is suggested that (VI) is reduced to the aldehyde (XXXIV), which on transamination and N-methylation yields the ketoamine (XXXV). Cyclization results in the formation of the betaine (XXXVII), which is the predominant state of shihunine (XXXVIII) in polar solvents. This scheme would be consistent with the results of the acetate feeding experiments since (1–$^{14}$C) acetate would yield 2-ketoglutaric acid labeled on its carboxyl groups by operation of the Krebs cycle. The o-succinylbenzoic acid would then be labeled at C–1 and would ultimately afford shihunine labeled at C–5.

## 3. Lignans

Lignans consist of two phenylpropane units by a C-C-bond between the carbon atoms 2' and 2" of the side chains (e.g. Fig. 7). The presence of further o-heterocyclic rings is

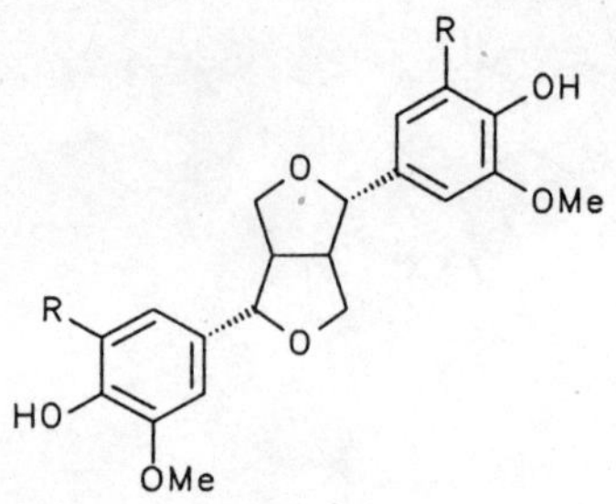

XXXIX R=H
(+)-Pinoresinol

XL R=OCH$_3$
Syringaresinol

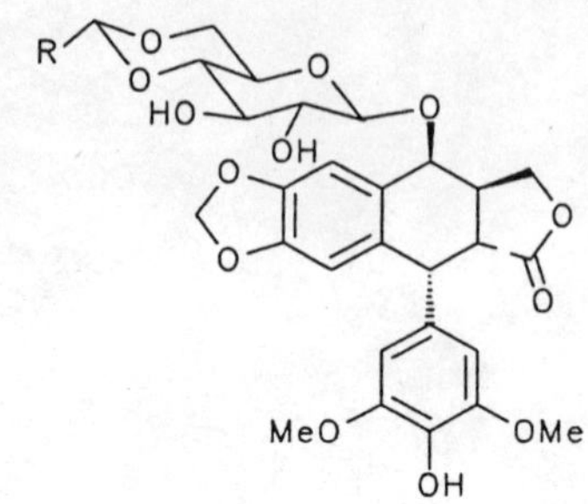

XLI R=Me
Etoposide

XLII R= 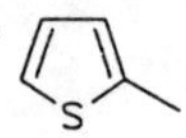

Teniposide

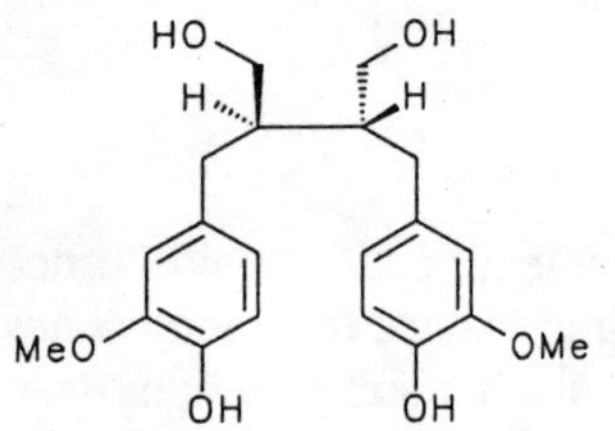

XLIII (-)Secoisolariciresinol

**Fig. 7.** Different lignans and derivatives

common (Rao 1978; Whiting 1985, 1987, 1990; Pelter 1986; Dewick 1990). Lignans are widely distributed plant products. Particularly common are the dimers derived from two units at the coniferyl alcohol oxidation level. Lignans are of enormous structural diversity, and their formation has long been proposed to proceed in a manner analogous to that of lignin, i.e., via an $H_2O_2$-requiring peroxidase-catalyzed oxidative coupling of monolignols (Lewis and Yamamoto 1990). However, such an oxidative free-radical polymerization would be expected to form a randomly linked polymer with no measurable optical activity. Lignans are, however, normally enantiomerically pure, although the optical rotation of each may differ depending upon the plant species, e.g., (+)-pinoresinol (XXXIX) from *Forsythia suspensa* (Chiba et al. 1978, 1979) or (–)-pinoresinol from *Xanthoxylum ailanthoides* (Ishii et al. 1983). Therefore the reactions for the formation of lignans must be somehow enzyme-mediated.

Species of Podophyllum produce a range of aryltetralin lactone lignans (Fig. 8) possessing antitumor activity which are exploited commercially as raw material for the semisynthesis of the anticancer drugs etoposide (XLI) and teniposide (XLII) (Dewick and Jackson 1981; Jackson and Dewick 1984c, 1985; MacRae and Towers 1984; Broomhead and Dewick 1990a). Podophyllotoxin (LV) derivatives are obtained also from *Juniperus salina* leaves (San Feliciano et al. 1990). The antitumor agents from *Bursera* are

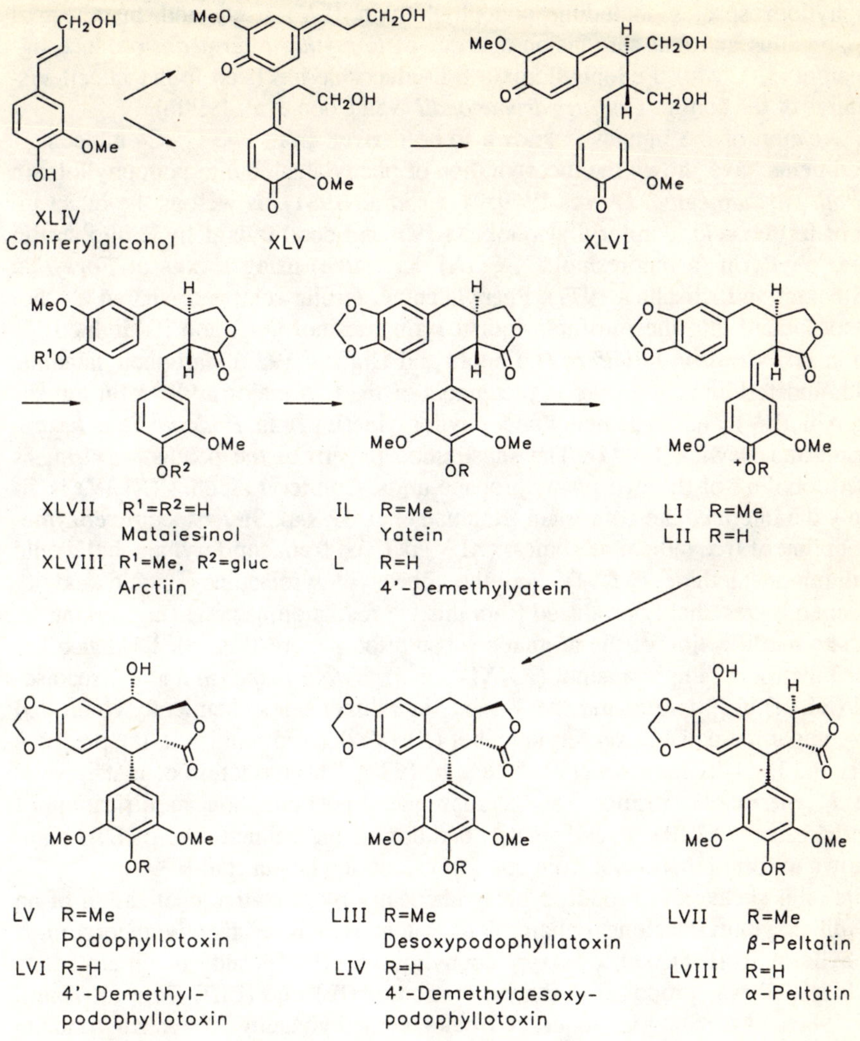


**Fig. 8.** Proposed biosynthetic pathway to *Podophyllum* lignans

ß-peltatin (LVII) methylether and desmethoxy-ß-peltatin methylether (Bianchi et al. 1969).

Lignan production has been investigated also in cultured plant cells. Thus, *Podophyllum peltatum* and *P. hexandrum* tissue cultures produce podophyllotoxin (LV) and derivatives (Kadkade 1981, 1982; Heyenga et al. 1990). Two lignans, the furofuran derivative pinoresinol and 2,3-dihydro-2-(4-hydroxy-3-methoxyphenyl)-3-hydroxymethyl-5-(ω-hydroxypropyl)-7-methoxybenzofuran, have been isolated from callus tissues of *Larix leptolepis* (Nabeta et al. 1991). *Linum flavum* cell cultures and plants are well known for their ability to produce cytotoxic aryltetralin lignans (Berlin et al. 1986, 1988; Broomhead and Dewick 1990b; van Uden et al. 1990a; Wichers et al. 1990, 1991). These lignans show many structural similarities to the podophyllotoxin-like compounds from Podophyllum species including podophyllotoxin (LV) itself and matairesinol (XLVII). Also callus and cell suspension cultures of *Forsythia intermedia* produce lignans (Rahman et al. 1990b). Podophyllotoxin-β-D-glucoside has been found in cell suspension cultures of the conifer *Callitris drummondii* (van Uden et al. 1990b).

The $C_{18}$ skeleton of the lignans is known to be derived from two $C_6$ $C_3$ molecules. Earlier experiments have shown the incorporation of phenylalanine into podophyllotoxin (LV) with *Podophyllum emodi* (Ayres 1969; Ayres et al. 1981), as well as the intact incorporation of ferulic acid, coniferyl alcohol (XLIV), and coniferylaldehyde into arctiin (XLVIII) and phyllyrin (a pinoresinol (XXXIX) derivative) using shoots of *Forsythia suspensa* (Stöckigt and Klischies 1977). Phenylalanine, ferulic acid, and sinapyl alcohol could be incorporated into the furofuran lignan syringaresinol (XL) and its diglucoside liriodendrin in *Liriodendron tulipifera* (Fujimoto and Higuchi 1977), and phenylalanine, cinamic acid, and ferulic acid are good precursors of the two major aryltetralin lignans podophyllotoxin (LV) and 4'-demethylpodophyllotoxin (LVI) in *Podophyllum hexandrum* (Jackson and Dewick 1984a). The substitution pattern of the pendent arylring is built up after coupling of the two phenylpropane units. Coniferyl alcohol (XLIV) is almost certainly the intermediate concerned (Rahman et al. 1990a). Stereospecific enzyme-catalyzed coupling of free radical mesomers (XLV) derived from coniferyl alcohol would lead to a diquinonemethide (XLVI), and the dibenzylbutyrolactones yatein and 4'-demethylyatein are presumably produced from this via reduction, lactone ring formation, and appropriate modification of the aromatic substitution pattern (Fig. 8). Evidence has been obtained that both (+) pinoresinol (XXXIX) in *Forsythia suspensa* and (–) secoisolariciresinol (XLIII) in *F. intermedia* are formed via a direct stereochemically controlled coupling of compartmentalized coniferyl alcohol (XLIV) derived moities in the presence of NAD(P)H and $H_2O_2$ (Umezawa et al. 1990a, b, 1991). The formation of matairesinol (XLVII) in *F. intermedia* results from a subsequent post-coupling modification of (–)secoisolariciresinol (XLIII). A cell-free preparation of the callus tissus of *Larix leptolepis* is shown to form pinoresinol from coniferyl alcohol (Nabeta et al. 1991).

The aryltetralin skeleton is produced biosynthetically by oxidative cyclization of an appropriate dibenzylbutyrolactone lignan. Thus yatein (IL) is efficiently incorporated into podophyllotoxin (LV) via desoxypodophyllotoxin (LIII) and an intermediate quinonemethide (LI) was proposed in the conversion of (IL) into (LIII) (Fig. 8) (Kamil and Dewick 1986a). An analogous sequence from 4'-demethylyatein (L) was suggested to account for biosynthesis of the 4'-demethyl series of lignans, such as 4'-demethyldesoxypodophpyllotoxin (LIV) and 4'-demethylpodophyllotoxin (LVI). The peltatins, ß-peltatin (LVII) and α-peltatin (LVIII) were shown to be derived from (LIII) and (LIV) respectively by aromatic hydroxylation at C–5 (Kamil and Dewick 1986b).

Matairesinol (XLVII) is likely to be a common intermediate on the pathway to (IL) and (L), and the probable branch-point to the two series of *Podophyllum* lignans, which has been confirmed by feeding matairesinol to *Podophyllum hexandrum, P. peltatum,* and *Diphylleia cymosa* (Broomhead et al. 1991).

In *P. hexandrum* plants, desoxypodophyllotoxin (LIII) is converted into podophyllotoxin (LV) which in turn is oxidized to podophyllotoxone, although this latter step appears to be reversible (Jackson and Dewick 1984b). A similar sequence is proposed for the corresponding 4'-demethyl derivatives. Although 4'-demethyldesoxyposophyllotoxin (LIV) is readily converted into demethylpodophyllotoxin (LVI), neither compound is incorporated into lignans of the 4'-methyl series such as podophyllotoxin (LV). The *Podophyllum* lignans may be subdivided biogenetically into two groups, those with 3,4,5-trimethoxy substitution in the pendent aryl ring, and those with a 4-hydroxy-3,5 dimethoxy substituted pendent ring although these probably arise from a common precursor. Biotransformation studies using cell suspension cultures of *Forsythia intermedia* showed that these are capable of rapidly oxidizing podophyllotoxin (LV) into podophyllotoxone (Broomhead and Dewick 1991). A *Penicillium* strain was found to convert desoxypodophyllotoxin (LIII) to epipodophyllotoxin which is the basic skeleton of etoposide (XLI) (Kondo et al. 1988, 1989a, b).

## 4. Capsaicinoids

Capsaicinoids such as capsaicin (LXI) are responsible for the pungent property of red peppers. It has been reported that phenylalanine, cinnamate derivatives, and vanillylamine (LIX) are incorporated in vivo into the vanillylamine moiety of capsaicinoids and that valine is incorporated via isobutyric acid and 8-methylnonenoic acid (LX) into the acyl moiety of capsaicin (LXI) and dihydrocapsaicin, the acyl constituents of which are evennumbered, branched-chain fatty acids, while leucine was incorporated via isopentanoic acid and 9-methyldecanoic acid into nordihydrocapsaicin and homodihydrocapsaicin, which contain odd-no., branched-chain fatty acids (Fig. 9) (Bennett and Kirby 1968; Leete and Louden 1968; Kopp and Jurenitisch 1981, 1982; Suzuki et al. 1981a; Fujiwake et al. 1982a, Kopp et al. 1983). Furthermore dihydrocapsaicin, capsaicin, and nordihydrocapsaicin are formed and accumulated in the fruits of the sweet pepper, *Capsicum annuum*, when the peduncle is put into an aqueous solution of vanillylamine and isocapric acid and the whole fruit aged for several days under continuous light (Iwai et al. 1978). The content of capsaicinoid reached its maximum about 25 days after flowering (Suzuki et al. 1981b). The fluctuation of the capsaicinoid content and composition is related to those of leucine and valine. Dihydrocapsaicin was also formed by cell-free extracts of the fruits in a reaction mixture containing vanillylamine and isocapric acid, and corresponding extracts of placenta of *C. annuum* catalyzed the formation of capsaicinoid from vanillylamine and different acyl-CoA compounds (Iwai et al. 1977; Fujiwake et al. 1980b). The capsaicinoid-synthesizing enzyme activity in sweet pepper fruits was found in both mitochondrial and microsomal fractions of placenta and pericarp. The capsaicinoid synthetase was located in the vacuole (Fujiwake et al. 1980a, 1982b).

In chilli pepper, *Capsicum frutescens*, the capsaicnoids are only synthesized and accumulated in the maturing fruits at one particular stage of development, and have not been detected in other parts of the plant (Hall et al. 1987). Once capsaicin is synthesized, it is not turned over. It has been proposed that as capsaicin synthesis in vivo coincides with fruit softening; such a softening may provide a huge supply of substrates for capsaicin synthesis (Gross et al. 1986). The activity of capsaicin synthase, the ultimate enzyme in the sequence, can be detected in the developing fruits long before capsaicin is synthesized and therefore the role of compartmentation may be an important issue (Yeoman et al. 1989). More progress is coming from cell suspension culture of *C.*

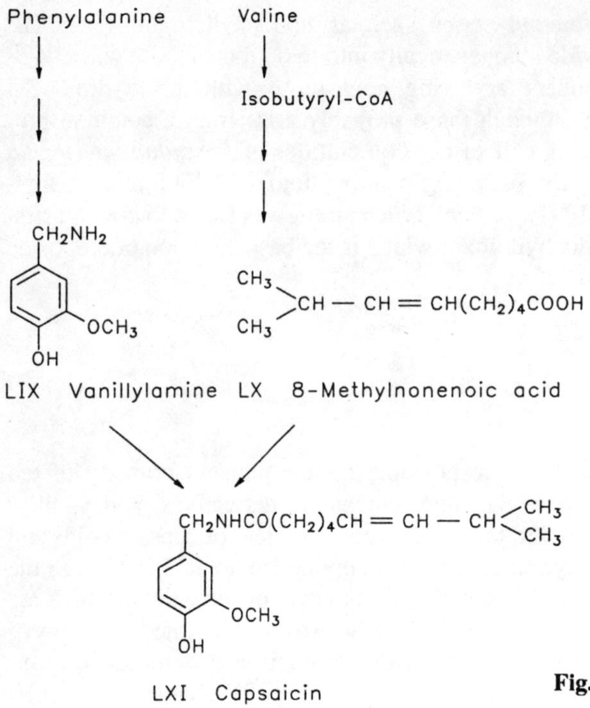


**Fig. 9.** Biosynthesis of capsoicin

*frutescens* for which the existence and operation of the complete pathway could be shown.

A series of immobilized cell systems were successful in raising the yield of capsaicin from culture cells and in providing excellent systems for exploring the regulation of the biosynthetic pathway leading to capsaicin (Lindsey and Yeoman 1984a). Immobilized cells of *C. frutescens* with high viability were obtained using polyurethane-foam particles (Lindsey et al. 1983). Such immobilized cells produced higher yields of capsaicin than did freely suspended cells, and generally incorporated higher levels of phenylalanine and cinnamic acid than free suspensions (Lindsey 1986). Using iso-capric acid, which is very close to the end of the biosynthetic pathway and gives rise to 8-methylnonenoic acid, which condenses with vanillylamine to produce capsaicin, quite interesting results were obtained (Lindsey and Yeoman 1984b). Enhanced capsaicin accumulation in immobilized cells is associated with reduced growth, increased organization of the cultures and decreased primary metabolic processes (Lindsey 1985). These conditions can conveniently be induced by the passive immobilisation of the cells in an inert support, with or without the use of suboptimal growth conditions. Capsaicin, through a feedback-inhibition mechanism, negatively influences its own synthesis (Hall and Yeoman 1991). Furthermore, capsaicin synthesis in these cells is not controlled via the activity of the enzymes phenylalanine ammonia-lyase and cinnamate 4-hydroxylase which may determine the rate of entry of metabolites into the phenylpropanoid pathway. In addition, p-fluorophenylalanineresistant cell lines (phenylalanine over-producer) of the chilli pepper have also been found to accumulate enhanced levels of capsaicin (Salgado-Garciglia and Ochoa-Alejo 1990).

## References

**Allen CF,** Franke H, Hirayama O (1967) Biochem Biophys Res Commun 26:562–568. **Andreasson AW,** Vängard T (1988) Annu Rev Plant Physiol 39:379–411. **Ayres DC** (1969) Tetrahedron Lett 883–886. **Ayres DC,** Farrow A, Carpenter BG (1981) J Chem Soc Perkin Trans I 2134–2136.

**Baldwin RM,** Snyder CD, Rapoport H (1974) Biochemistry 13:1523–1530. **Bauch HJ,** Leistner E (1978a) Planta Med 33:105–123. **Bauch HJ,** Leistner E (1978b) Planta Med 33:124–127. **Bennett DJ,** Kirby GW (1968) J Chem Soc C 442–446. **Bentley R,** Meganathan R (1981) Microbiol Rev 46:241–280. **Bentley R,** Meganathan R (1983) J Nat Prod 46:44–59. **Berlin J,** Wray V, Mollenschott C, Sasse F (1986) J Nat Prod 49:435–439. **Berlin J,** Bedorf N, Mollenschott C, Wray V, Sasse F, Höfle G (1988) Planta Med 54:204–206. **Bianchi E,** Sheth K, Cole JR (1969) Tetrahedron Lett 2759–2762. **Brettel P,** Setif P, Mathis P (1986) FEBS Lett 203:220–224. **Breuer M,** Igbavboa U, Kaiser A, Leistner E, Simantiras M, Weische A (1991) In: Biswanger H, Ullrich J (eds) Biochemistry and physiology of thiamin diphosphate enzymes. VHC, Weinheim. **Broomhead AJ,** Dewick PM (1990a) Phytochemistry 29:3831–3837. **Broomhead AJ,** Dewick PM (1990b) Phytochemistry 29:3839–3844. **Broomhead AJ,** Dewick PM (1991) Phytochemistry 30:1511–1517. **Broomhead AJ,** Rahman MMA, Dewick PM, Jackson DE, Lucas YA (1991) Phytochemistry 30:1489–1492.

**Campbell IM** (1969) Tetrahedron Lett 4777–4780. **Campbell IM,** Robins DJ, Kelsey M, Bentley R (1971) Biochemistry 10:3069–3078. **Chiba M,** Hisada S, Nishibe S (1978) Shoyakugaku Zasshi 32:194–197. **Chiba M,** Tsukamoto H, Hisada S, Nishibe S (1979) Shoyakugaku Zasshi 33:150–154. **Conn EE** (ed) (1986) The shikimic acid pathway. Rec Adv Phytochem, vol 20. Plenum, New York.

**Dansette P,** Azerad R (1970) Biochem Biophys Res Commun 40:1090–1095. **Dewick PM** (1988, 1989, 1990, 1991) Nat Prod Rep 5:73–97, 6:263–290, 7:165–189, 8:149–170. **Dewick PM** (1990) In: Atta-ur-Rahman A (ed) Studies in natural products chemistry, vol 5 Structure Elucitation, part B. Elsevier, Amsterdam, pp 459–503. **Dewick PM,** Jackson DE (1981) Phytochemistry 20:2277–2280.

**Elander M,** Leander K, Lüning B (1969) Acta Chem Scand 23:2177–2178. **Elander M,** Gawell L, Leander K (1971) Acta Chem Scand 25:21–24. **Elkins MF,** Earhart CF (1988) FEMS Microbiol Lett 56:35–40. **El-Shagi H,** Schulte U, Zenk MH (1984) Naturwissenschaften 71:267. **Emmons GT,** Campbell IM, Bentley R (1985) Biochem Biophys Res Commun 131:956–960.

**Floß HG,** Scharf KH, Zenk MH, Onderka DK, Carroll M (1971) J Chem Soc D 576–577. **Fujimoto H,** Higuchi T (1977) Wood Res 62:1–10. **Fujiwake H,** Suzuki T, Iwai K (1980a) Plant Cell Physiol 21:1023–1030. **Fujiwake H,** Suzuki T, Oka S, Iwai K (1980b) Agric Biol Chem 44:2907–2912. **Fujiwake H,** Suzuki T, Iwai K (1982a) Agric Biol Chem 46:2591–2592. **Fujiwake H,** Suzuki T, Iwai K (1982b) Agric Biol Chem 46:2685–2689.

**Gatenbeck S** (1960) Acta Chem Scand 14:102–106. **Gaudilliere JP,** d'Harlingue A, Camara B, Moneger R (1984) Plant Cell Rep 3:240–242. **Gross KC,** Watada AE, Kang MS, Kim SD, Kim KS, Lee SW (1986) Physiol Plant 66:31–36. **Grotzinger E,** Campbell IM (1972a) Tetrahedron Lett 4685–4686. **Grotzinger E,** Campbell IM (1972b) Phytochemistry 11:675–679. **Grotzinger E,** Campbell IM (1974) Phytochemistry 13:923–926.

**Hall RD,**Yeoman MM (1991) Planta 185:72–80. **Hall RD,** Holden MA, Yeoman MM (1987) Plant Cell Tissue Organ Cult 8:163–176. **Hawkins CF,** Borges A, Perham RN (1989) FEBS Lett 255:77–82. **Heide L,** Leistner E (1981a) Chem Commun 334–336. **Heide L,** Leistner E (1981b) FEBS Lett 128:201–204. **Heide L,** Leistner E (1982) Z Naturforsch 37c:354–362. **Heide L,** Leistner E (1983) Phytochemistry 22:659–662. **Heide L,** Arendt S, Leistner E (1982a) J Biol Chem 257:7396–7400. **Heide L,** Kolkmann R, Arendt S, Leistner E (1982b) Plant Cell Rep 1:180–182. **Heyenga AG,** Lucas JA, Dewick PM (1990) Plant Cell Rep 9:382–385. **Hutson KG,** Threlfall DR (1980) Phytochemistry 19:535–537.

**Igbavboa U,** Leistner E (1990) Eur J Biochem 192:441–449. **Igbavboa U,** Sieweke HJ, Leistner E, Röwer I, Hüsemann W, Barz W (1985) Planta 166:537–544. **Inoue K,** Ueda S, Shiobara Y, Inouye H (1977) Phytochemistry 16:1689–1694. **Inoue K,** Shiobara Y, Nayeshiro H, Inouye H, Wilson G, Zenk MH (1979) Chem Commun 957–959. **Inoue K,** Ueda S, Shiobara Y, Inouye H (1980) Tetrahedron Lett 21:621–622. **Inoue K,** Ueda S, Shiobara Y, Kimura J, Inouye H (1981) J Chem Soc Perkin Trans I 1246–1258. **Inoue K,** Ueda S, Nayeshiro H, Inouye H (1982) Chem Commun 993–994. **Inoue K,** Ueda S, Nayeshiro H, Inouye H (1983) Phytochemistry 22:737–741. **Inoue K,** Shiobara Y, Nayeshiro H, Inouye H, Wilson G, Zenk MH (1984a) Phytochemistry 23:307–311. **Inoue K,** Ueda S, Nayeshiro H, Moritome N, Inouye H (1984b) Phytochemistry 23:313–318. **Inouye H,** Ueda S, Inoue K, Hayashi T, Hibi T (1975) Chem Pharm Bull 23:2523–2533. **Inouye H,** Ueda S, Inoue K (1978) Tetrahedron Lett 4551–4554. **Inouye H,** Ueda S, Inoue K, Shiobara Y (1981) Phytochemistry 20:1707–1710. **Inubushi Y,** Tsuda Y, Konita T, Matsumoto S (1964) Chem Pharm Bull 12:749–750. **Inubushi Y,** Tsuda Y, Konita T, Matsumoto S (1968) Chem Pharm Bull 16:1014–1018. Ishii H, Ishikawa T, Mihara M, Akaike M (1983) Yakugaku Zasshi 103:279–292. Itoh S, Iwaki M (1989) FEBS Lett 243:47–52. **Iwai K,** Suzuki T, Lee KR, Kobashi M, Oka S (1977) Agric Biol Chem 41:1877–1882. **Iwai K,** Lee KR, Kobashi M, Suzuki T, Oka S (1978) Agric Biol Chem 42:201–202.

**Jackson DE,** Dewick PM (1984a) Phytochemistry 23:1029–1035. **Jackson DE,** Dewick PM (1984b) Phytochemistry 23:1037–1042. **Jackson DE,** Dewick PM (1984c) Phytochemistry 23:1147–1152. **Jackson DE,** Dewick PM (1985) Phytochemistry 24:2407–2409. **Johanni M,** Hofmann P, Leistner E (1989) Arch Biochem Biophys 271:495–501.

**Kadkade PG** (1981) Naturwissenschaften 68:481–482. **Kadkade PG** (1982) Plant Sci Lett 25:107–115. **Kaiping A,** Soll J, Schultz G (1984) Phytochemistry 23:89–91. **Kaiser A,** Leistner E (1990) Arch Biochem Biophys 276:331–335. **Kamil WM,** Dewick PM (1986a) Phytochemistry 25:2093–2102. **Kamil WM,** Dewick PM (1986b) Phytochemistry 25:2089–2092. **Khouri HE,** Ibrahim RK (1987) Phytochemistry 26:2531–2535. **Kolkmann R,** Leistner E (1985) Tetrahedron Lett 26:1703–1704. **Kolkmann R,** Leistner E (1987a) Z Naturforsch 42c:542–552. **Kolkmann R,** Leistner E (1987b) Z Naturforsch 42c:1207–1214. **Kolkmann R,** Knauel G, Arendt S, Leistner E (1982) FEBS Lett 137:53–56. **Kondo K,** Ogura M, Midorikawa Y, Kozawa M, Tsujibo H, Inamori Y (1988) Agric Biol Chem 52:2673–2674. **Kondo K,** Ogura M, Midorikawa Y, Kozawa M, Tsujibo H, Baba K, Inamori Y (1989a) Agric Biol Chem 53:777–782. **Kondo K,** Ogura M, Midorikawa Y, Kozawa M, Tsujibo H, Inamori Y (1989b) Agric Biol Chem 53:1171–1172. **Kopp B,** Jurenitsch J (1981) Planta Med 43:272–279. **Kopp B,** Jurenitsch J (1982) Sci Pharm 50:150–157. **Kopp B,** Jurenitsch J, Woeginger R, Haunold E (1983) Sci Pharm 51:274–277.

**Law A,** Thomas G, Threlfall DR (1973) Phytochemistry 12:1999–2004. **Leete E,** Bodem GB (1976) J Am Chem Soc 98:6321–6325. **Leete E,** Louden LCL (1968) J Am Chem Soc 90:6837–6841. **Leistner E** (1971) Phytochemistry 10:3015–3020. **Leistner E** (1973a) Phytochemistry 12:337–345. **Leistner E** (1973b) Phytochemistry 12:1669–1674. **Leistner E** (1975) Planta Med Suppl 214–224. **Leistner E** (1981) Biosynthesis of Plant Quinones. In: Conn EE (ed) The biochemistry of plants, vol 7. Academic Press, New York, pp 403–423. **Leistner E** (1985) Rec Adv Phytochem 20:243–261. **Leistner E,** Zenk MH (1967) Z Naturforsch 22b:865–868. **Leistner E,** Zenk MH (1968) Z Naturforsch 23b: 259–268. **Leistner E,** Zenk MH (1971) Tetrahedron Lett 1677–1681. **Lewis NG,** Yamamoto E (1990) Annu Rev Plant Physiol Mol Biol 41:455–496. **Lichtenthaler HK** (1968) Planta 81:140–152. **Lindsey K** (1985) Planta 165:126–133. **Lindsey K** (1986) Phytochemistry 25:2793–2801. **Lindsey K,** Yeoman MM (1984a) Planta 162:495–501. **Lindsey K,** Yeoman MM (1984b) J Exp Bot 35:1684–1696. **Lindsey K,** Yeoman MM, Black GM, Mavituna F (1983) FEBS Lett 155:143–149.

**MacRae WD,** Towers GHN (1984) Phytochemistry 23:1207–1220. **Marley MG,** Meganathan R, Bentley R (1986) Biochemistry 25:1304–1307. **McGovern EP,** Bentley R (1978) Arch Biochem Biophys 188:56–63. **Meganathan R** (1981) J Biol Chem 256:9386–9388. **Meganathan R,** Bentley R (1979) J Bacteriol 140:92–98. **Meganathan R,** Bentley R (1983) J Bacteriol 153:739–746. **Meganathan R,** Folger T, Bentley R (1980) Biochemistry 19:785–789. **Mulder-Krieger T,** Verpoorte R, de Water A, van Gessel M, van Oeveren BCJA, Baerheim Svendsen A (1982) Planta Med 46:19–24. **Mulder-Krieger T,** Verpoorte R, van der Kreek M, Baerheim Svendsen A (1984) Planta Med 50:17–20. **Müller WU,** Leistner E (1976) Phytochemistry 15:407–410. **Müller WU,** Leistner E (1978) Phytochemistry 17:1735–1738.

**Nabeta K,** Nakahara K, Yonekubo J, Okuyama H, Sasaya T (1991) Phytochemistry 30:3591–3593.

**Ozenberger BA,** Brickman TJ, McIntosh MA (1989) J Bacteriol 171:775–783.

**Pelter A** (1986) Rec Adv Phytochem 20:201–241. **Popp JL** (1989) J Bacteriol 171:4349–4354. **Popp JL,** Berliner C, Bentley R (1989) Anal Biochem 178:306–310. **Poulson C,** van der Heijden R, Verpoorte R (1991) Phytochemistry 30:2873–2876.

**Rahman MMA,** Dewick PM, Jackson DE, Lucas JA (1990a) Phytochemistry 29:1841–1846. **Rahman MMA,** Dewick PM, Jackson DE, Lucas JA (1990b) Phytochemistry 29:1861–1866. **Rao CBS** (1978) Chemistry of Lignans. Andhra University Press, Visakhapatnam.

**Saito Y,** Ogura K (1981) J Biochem (Tokyo) 89:1445–1452. **Salgado-Garciglia R,** Ochoa-Alejo N (1990) Plant Cell Rep 8:617–620. **Samuel O,** Azerad R (1972) Biochemie 54:305–317. **San Feliciano A,** Miguel del Corral JM, Gordaliza M, Castro A (1990) Phytochemistry 29:1335–1338. **Schoeder HU,** Lockau W (1986) FEBS Lett 199:23–27. **Schulte U,** El-Shagi H, Zenk MH (1984) Plant Cell Rep 3:51–54. **Schultz G,** Ellerbrock BH, Soll J (1981) Eur J Biochem 117:329–332. **Schütte HR** (1978) Progr Bot 40:126–149. **Schütte HR** (1979) Progr Bot 41:93–107. **Schütte HR** (1985) Progr Bot 47:118–141. **Schütte HR** (1992) Progr Bot 53:78–98. **Seeger JW,** Bentley R (1991) Phytochemistry 30:3585–3589. **Shaw DJ,** Guest JR, Meganathan R, Bentley R (1982) J Bacteriol 152:1132–1137. **Shaw DJ,** Robinson EC, Meganathan R, Bentley R, Guest JR (1983) FEMS Microbiol Lett 17:63–67. **Sieweke HJ,** Leistner E (1991) Z Naturforsch 46c:585–590. **Simantiras M,** Leistner E (1989) Phytochemistry 28:1381–1382. **Simantiras M,** Leistner E (1991) Z Naturforsch 46c:364–370. **Simantiras M,** Schmidt K, Leistner E (1991) Phytochemistry 30:823–824. **Stöckigt J,** Klischies M (1977) Holzforschung 31:41–44. **Stöckigt J,** Schübel H (1989) Dtsch Apoth Ztg 129:1187–1192. **Stöckigt J,** Srocka U, Zenk MH (1973) Phytochemistry 12:2389–2391. **Suzuki T,** Kawada T, Iwai K (1981a) Plant Cell Physiol 22:23–32. **Suzuki T,** Kawada T, Iwai K (1981b) Agric Biol Chem 45:535–537.

**Takahashi Y,** Hirota K, Katoh S (1985) Photosynth Res 6:183–192. **Thomas G,** Threlfall DR (1974) Phytochemistry 13:807–813. **Thomson RH** (1987) Naturally occurring quinones. Chapman and Hall, London. **Threlfall DR** (1972) Biochim Biophys Acta 280:472–480. **Tummuru NKR,** Brickman TJ, McIntosh MA (1989) J Biol Chem 264:20547–20551.

**Ueda S,** Inoue K, Hayashi T, Inouye H (1975) Tetrahedron Lett 2399–2401. **Umezawa T,** Davin LB, Yamamoto E, Kingston DGI, Lewis NG (1990a) Chem Commun 1405–1408. **Umezawa T,** Davin LB, Lewis NG (1990b) Biochem Biophys Res Commun 171:1008–1014. **Umezawa T,** Davin LB, Lewis NG (1991) J Biol Chem 266:10210–10217.

**van den Berg AJJ** (1991) Pharm Weekbl Sci Ed 13:74–77. **van den Berg AJJ,** Labadie RP (1989) In: Harborne JB (ed) Methods in Plant Biochemistry. Academic Press, London, pp 451–491. **van Uden W,** Pras N, Vossebeld EM, Mol JNM, Malingre TM (1990a) Plant Cell Tissue Organ Cult 20:81–87. **van Uden W,** Pras N, Malingre TM (1990b) Plant Cell Rep 9:257–260.

**Weische A,** Leistner E (1985) Tetrahedron Lett 26:1487–1490. **Weische A,** Johanni M, Leistner E (1987a) Arch Biochem Biophys 256:212–222. **Weische A,** Garvert W, Leistner E (1987b) Arch Biochem Biophys 256:223–231. **Whistance GR,** Threlfall DR (1970) Biochem J 117:593–600. **Whiting DA** (1985, 1987, 1990) Nat Prod Rep 2:191–211, 4:499–525, 7:349–364. **Wichers HJ,** Harkes MP, Arroo RRJ (1990) Plant Cell Tissue Organ Cult 23:93–100. **Wichers HJ,** Versluis-DeHaan GG, Marsman JW, Harkes MP (1991) Phytochemistry 30:3601–3604. **Wijnsma R,** Verpoorte R (1986) Prog Chem Org Nat Prod 49:79–149. **Wijnsma R,** Verpoorte R, Mulder-Krieger T, Baerheim-Svendsen A (1984) Phytochemistry 23:2307–2311. **Wijnsma R,** Go JTKA, van Weerden IN, Harkes PAA, Verpoorte R, Baerheim-Svendsen A (1985) Plant Cell Rep 4:241–244. **Wijnsma R,** Go JTKA, Harkes PAA, Verpoorte R, Baerheim-Svendsen A (1986a) Phytochemistry 25:1123–1126. **Wijnsma R,** van Weerden IN, Verpoorte R, Harkes PAA, Lugt CB, Scheffer JJC, Baerheim-Svendsen A (1986b) Planta Med 52:211–212.

**Yeoman MM,** Holden MA, Hall RD, Holden PR, Holland SS (1989) Regulation of capsaicin synthesis in cultured cells of Capsicum frutescens. In: Kurz WGW (ed) Primary and secondary metabolism of plant cell cultures. Springer, Berlin Heidelberg New York, pp 162–174. **Young IG** (1975) Biochemistry 14:399–406. **Young IG,** Gibson F (1969) Biochim Biophys Acta 177:401–411. **Young IG,** Batterham TJ, Gibson F (1969) Biochim Biophys Acta 177:389–400.

**Zenk MH** (1991) Phytochemistry 30:3861–3863. **Zenk MH,** El-Shagi H, Schulte U (1975) Planta Med Suppl 79–101. **Zenk MH,** Schulte U, El-Shagi H (1984) Naturwissenschaften 71:266. **Ziegler K,** Maldener I, Lockau W (1989) Z Naturforsch 44c:468–472.

Prof. Dr. Horst-Robert Schütte
Institut für Pflanzenbiochemie
Weinberg 3
O-4050 Halle/Saale, Germany

# VII. Mineral Nutrition: Tropical Savannas

By Ernesto Medina

## 1. Introduction: The Concept of Savanna

Tropical savannas are ecosystems characterized by a continuous herbaceous cover of heliophilous bunch grasses and sedges, almost exclusively with the C4 photosynthetic pathway, that show a clear seasonality related to water stress (Sarmiento 1984; Cole 1986; Frost et al. 1986). Woody species (shrubs, trees, and palms) occur in varying density depending on soil depth and water availability (Cole 1986; Medina and Silva 1990). Legumes are commonly found in savanna vegetation as herbaceous, shrubby and tree species, and particularly in South American savannas, they can reach high specific diversity, although their biomass seldom surpasses 1% of the total aboveground biomass of the grass layer (Medina and Bilbao 1991). In contrast to the relative homogeneity in form and function of the grass layer, the woody components vary in their phenology and structure between Africa, Australia, and South America. While in Africa most woody components are drought deciduous (with the exception of a few species in western African savannas), most shrub and tree species in Australia and South America are evergreen, changing their leaves during the dry season. Large differences are also observable in the leaf structure. Most woody species in Australia and South America are scleromorphic due to the abundance of mechanical tissue and silica, while typical African tree species are mesophyllous in structure (Cole 1986).

Savanna systems are burned regularly, presently mostly as a result of human influences. However, the occurrence of fire probably modified the distribution and species composition of savannas before man, particularly in tropical South America (Walter 1973; Medina 1982a; Sarmiento 1984; Medina and Silva 1990).

The concept of savanna has been widened by different authors and includes a large array of different vegetation physiognomies (Eiten 1972; Walter 1973; Huntley 1982; Sarmiento 1984; Cole 1986) from savanna woodlands and parklands, to pure grasslands. One extreme view has been to include all those ecosystems which lie between the tropical rainforests and the deserts and semideserts of Africa, Australia, and South America (Huntley and Walker 1982, p. 1). Another extreme has been to assimilate dissimilar tropical grasslands within the savanna concept (Jones et al. 1992). Such extreme views contribute little to understanding savanna ecology.

This review is restricted to dystrophic and slightly mesotrophic tropical savannas. Savannas in the African highlands characterized by the occurrence of brown, eutrophic soils of alkaline pH have not been considered here. The nutrient relations of these savannas

Progress in Botany, Vol. 54

have been reviewed by several authors, and the reader is referred to the work of McNaughton and co-workers (McNaughton 1985, 1988; Ruess 1987).

## 2. Recent Reviews on Savannas in General and Nutrients in Particular

The increasing recognition of the ecological and economic importance of savanna ecosystem has stimulated intensive research leading to the publication of an important number of books during the last decade. Emphasis has been laid on human aspects, ecology, and management (UNESCO 1979; Harris 1980; Huntley and Walker 1982; Bourliére 1983; Sarmiento 1984; Tothill and Mott 1985; Cole 1986; Proctor 1989). Several collective books on savannas resulted from a research project launched by the International Union of Biological Sciences in the middle 1980s (Research on Savanna Stress and Disturbance). This project attempted a functional analysis of savanna ecosystems, emphasizing the environmental factors determining both structure and function. These collective books contain a large amount of information on nutritional relationships of savanna ecosystems (Frost et al. 1986; Walker 1987; Walker and Menaut 1988; Sarmiento 1990; Werner 1991).

## 3. Savanna Soils and the Concept of Nutrient Availability

Soil fertility is a concept applied to the potential productivity of a certain soil, and has been developed within the context of agricultural crops. Fertility is defined on the basis of certain physicochemical properties of the upper soil layers. These properties include texture, pH, cation exchange capacity, extractable bases, availability of P, and content of potentially mineralizable nitrogen. However, soil fertility cannot be defined independently of the plants growing on it. Different crops or wild plant species develop root systems which vary widely in their ability to extract nutrients from the soil. These differences are associated with root morphology, and the production of root exudates capable of increasing the nutrient availability in the rhizosphere environment (Marschner 1986). Nevertheless, soil properties alone have been used to evaluate variations in soil fertility within different savanna regions of the world. Frequently only the physicochemical characteristics of the upper soil layers is taken into account to evaluate the fertility status of the soils. This is understandable when savanna soils are going to be used for agricultural purposes. However, the upper soil layers are those mostly modified by the existing vegetation.

Predominant soil types in seasonal, nonflooded tropical savannas of South America, Africa, and Australia are oxisols, ultisols, and alfisols (Sánchez and Isbell 1978; Montgomery and Askew 1983; Acquaye and Kang 1987; Møberg and Esu 1991). These soils have in general low fertility, but the oxisols and ultisols, particularly in South America, have relatively higher P fixation capacity, lower K availability, and higher Al percentage saturation (Lópes and Cox 1977a; Sánchez and Isbell 1978; Ayodele 1986; Furley and Ratter 1988; Cochrane 1989). These low fertility levels are the main constraint for agricultural utilization of savanna soils on a large scale. Savanna soils require substantial amendements with Ca to reduce P fixation and Al toxicity (Alvim and Araujo 1952; Montgomery 1988); P (for establishment of legumes), and S for production of cereals (McClung and de Freitas 1959; Acquaye and Kang 1987).

Defining a synthetic parameter to evaluate soil fertility has proved to be difficult. The actual significance of all chemical factors associated with soil fertility is modified by soil texture and water availability. Alvim (1979) was able to show clear fertility gradients existing across soil types in the Amazonian region using an index based on the integration of all chemical fertility factors, calculated as the area of a regular polygon defined by the axes organic matter (0–3.5%), P (0–20 ppm), K (0–0.5 mEq/100 g), Ca+Mg (0–15 mEq/100 g), percent Al saturation (20–0%) and pH (0–7) (Alvim and Cabala 1974). This index has been criticized because of (1) the arbitrary order and magnitude of the axes; (2) that plant growth responds linearly to each factor, and; (3) that all factors affect growth in a similar way (Medina in Walker and Menaut 1988). The sum of bases expressed as equivalents of $K^+$, $Ca^{2+}$, $Mg^{2+}$, $Na^+$ per unit dry soil weight (Sarmiento 1990) or per unit clay weight (Huntley 1982) appears to be a simpler index correlated with soil fertility. With this index savanna soils have been classified as dystrophic [$\Sigma$bases < 5 cmol(+) $kg^{-1}$ soil], eutrophic [$\Sigma$bases > 20 cmol(+) $kg^{-1}$ soil] and mesotrophic between the two extremes (+ indicates positive charges, this unit substitutes the usual mEq per 100 g soil weight). Sarmiento (1990) even used the term hyperdystrophic for those soils with $\Sigma$bases less than 1 cmol(+) $kg^{-1}$ soil. The inclusion of P availability or N content in fertility indices might be misleading when applied to those soil layers under the influence of the vegetation. The value of $\Sigma$bases was shown to be strongly correlated with available P in soil layers below the grass root level, while the correlation was weak within the root level (Medina and Silva 1990).

Bioassays have been successfully applied to estimate fertility of savanna soils in South Africa (Ölsvig-Whittaker and Morris 1982), however there is no general consensus on the use of indicator species, or the standardization of the climatic conditions under which the assays are conducted (Medina in Walker and Menaut 1988).

## 4. Correlations Between Soil Fertility and Savanna Composition and Structure

Plant available moisture and nutrients have been hypothesized to be the main environmental variables determining savanna structure and function in the tropics (Frost et al. 1986). While moisture appears to influence mainly the frequency, height, and density of the shrubby/tree components, nutrient availability seems to regulate the composition of the vegetation units within each savanna category (Cole 1986; Medina and Silva 1990).

In Africa, the distribution of tropical and subtropical savannas and the distribution of large mammal populations have been clearly associated with soil nutrient status, which regulates the nutritional value of the vegetation growing on them (Bell 1982; Huntley 1982; Frost et al. 1986; McNaughton 1988). Nutrient availability influences the slope of the relationship between rainfall and aboveground productivity of the herbaceous component (Scholes 1990). Nutrient-poor savannas in southern Africa [$\Sigma$bases < 20 cmol(+)

$kg^{-1}$] are dominated by Combretaceae and Caesalpinioidae in the tree stratum, have a conspicuous litter layer indicating slow decomposition rates, and the herbivore biomass, both of insects and mammals, is lower compared to savannas on nutrient-rich substrates [$\Sigma$bases > 20 cmol(+) $kg^{-1}$] (Scholes 1990). The nutrient-rich savannas are dominated by Mimosoidae in the tree stratum and have in general grasses with higher nutritional value.

In Brazil, savanna vegetation constitutes a continuum from savanna woodland, shrubby savannas (Cerrado, Campos Cerrados, Campos sujos) and open grasslands (Campo Limpo) (Eiten 1972). The woody components of the brazilian Cerrado, and also of the Llanos in northern South America, are typically sclerophyllous, resistant to annual fires, evergreen, changing their leaves during the dry season, and seem to have access to the water table, therefore remain photosythetically active throughout the year (Medina 1982a, 1987; Sarmiento 1983; Goldstein et al. 1986; Medina and Silva 1990). This sequence has been hypothesized to be the result of gradients in soil fertility, particularly of Ca and P availability and high Al percent saturation (Goodland and Pollard 1973). Open grasslands are found on the least fertile soils, while the sequence leading to closed sclerophyllous forests, Cerradão, follows a gradient of increasing soil fertility. This sequence was shown in large-scale sampling of top soils (30 cm depth) in the Brazilian Cerrado by Lópes and Cox (1977b). On a smaller scale, García-Miragaya and Cáceres (1990) showed the same pattern of higher nutrient content in the upper 20 cm of forest soils compared to open savannas in Calabozo, Venezuela. The contention that Al toxicity is the main edaphic factor regulating tree density in the Brazilian savannas is only of a correlative nature. Extractable P and Ca are negatively correlated with extractable Al. Percent Al saturation, but not extractable Al, is correlated with vegetation types (Lópes and Cox 1977b). More dramatic changes along these sequences are observed in the Al/Ca ratio. While the Al% saturation decreases from 66 to 44 from Campo Limpo to Cerradão (33% reduction), the Al/Ca ratios decrease from 3.7 in Campo Limpo to 0.88 in Cerradão (76% reduction). Similar relationships for the A1 horizon were demonstrated by Cochrane (1989) when comparing Cerrado and forest soils in Central Brazil. The percent Al saturation was about three times higher in Cerrado as compared to forest soils, while the Al/Ca ratio was 56 times larger in Cerrado soils. These analyses of the upper soil layers are difficult to interpret because the chemical properties might be strongly affected by the activity of vegetation (litter accumulation above and below ground). The calculation of relative accumulation of biogenic elements in the upper layers of Cerrado soils indicate factors of 3.9 for Ca, 2.3 for K and 1.9 for Mg, while Al and Mn are accumulated only by factors of 1.6 and 1.3 respectively, indicating the nutrient selection exerted by the vegetation (calculations with data from Cochrane 1989).

## 5. Grassland Productivity and Nutrient Availability

The nutrient content of plants growing in savanna soils is comparatively low. Several patterns may be discerned among savanna components: (a) trees have higher nutrient content than the xeromorphic C4 grasses dominating the herbaceous layer; (b) grasses and woody plants from semiarid savannas in general have higher nutrient content than species from dystrophic humid tropical savannas; (c) nutrient content of leaves depends on leaf duration, deciduous trees produce leaves with higher nutrient content than evergreen trees; (d) leguminous trees have usually higher N and P contens than nonlegumes (Norman 1963, 1966; Sen and Mabey 1966; Ahmad and Jones 1969; Montes and Medina 1977; Medina et al. 1978; Villecourt et al. 1979; Sarmiento 1984; Frost 1985; Medeiros and Haridasan 1985; Tolsma et al. 1987; García-Miragaya and Cáceres 1990). Nitrogen, sulfur, and phosphorus concentrations decrease, while Ca, Mg, and heavy metals increase with leaf age (Montes and Medina 1977; Medina et al. 1978; Medeiros and Haridasan 1985; Tolsma et al. 1987).

Utilization of savanna environments for agricultural production has always been associated with the problem of nutrient availability. Agrobusiness has simply got around the problem by using high fertilizer inputs, given economic rentability, for establishment of highly productive crops or introducing African grasses in combination with herbaceous legumes (Goedert 1983, 1990; Winter et al. 1989a; Thomas et al. 1990). Fertilization of native grasslands in savanna areas has been performed to improve their nutritional value as animal fodder. These experiments give insight into the problem of nutrient limitation of savanna soils for native savanna plants. Fertilization of native grasslands in dystrophic savannas resulted in higher aboveground biomass production adding N or P alone, but more commonly significant effects have been observed when both nutrients are added together (Hall et al. 1941; Norman 1963, 1966; Medina et al. 1978; San José and García-Miragaya 1981; García-Miragaya et al. 1983). All these experiments have been conducted in native grasslands with grass species of particularly low nutritional requirements. In all cases, in spite of the increased production of aboveground organic matter, the P concentration in biomass remained higher than the controls at the end of the experiment, while the N concentration turned back to levels little above control (Medina 1987). This indicates that % retranslocation of P measured at the end of the growing season is relatively smaller than that of N in grasses.

When P is not limiting, the proportional response to N supply is similar for many tropical grass species (Wilson and Haydock 1971). Responses to P supply vary considerably among grass species from environments differing in soil fertility. Under nonlimiting N conditions grass species from nutrient-rich soils always have larger proportional responses to P than those species from dystrophic soils. This is clear when comparing *Themeda australis* and *Heteropogon contortus* from southern Queensland, Australia, with African grasses such as *Panicum maximum* and *Sorghum almum* (Wilson and Haydock 1971). A similar pattern is obtained by comparing *Andropogon gayanus* introduced in central Venezuela and *Paspalum plicatulum*, a native grass (Bilbao and Medina 1990).

The evaluation of the fertility conditions of a given savanna soil, particularly with respect to the nutrients P and N, can be made using the P/N ratios of plant matter as proposed by Penning de Vries et al. (1980). Being those two elements directly involved in all enzymatic and energetic pathways within a living cell, the hypothesis has been put forward that the total P/organic N ratios may be used to determine if plant productivity is limited by either one of the two elements (Penning de Vries et al. 1980). A similar hypothesis has been applied to tropical forests (Medina and Cuevas 1989). P/N ratios vary within certain limits which can be established experimentally. For semiarid grasslands in the Sahel region of Africa, the ratios vary in average from 0.038 g P/ g N in P-deficient plants to 0.15 g P/ g N in P-sufficient plants (about 17 to 68 mmol P/mol N). The range between those two extreme values defines the response to P fertilization. The extreme

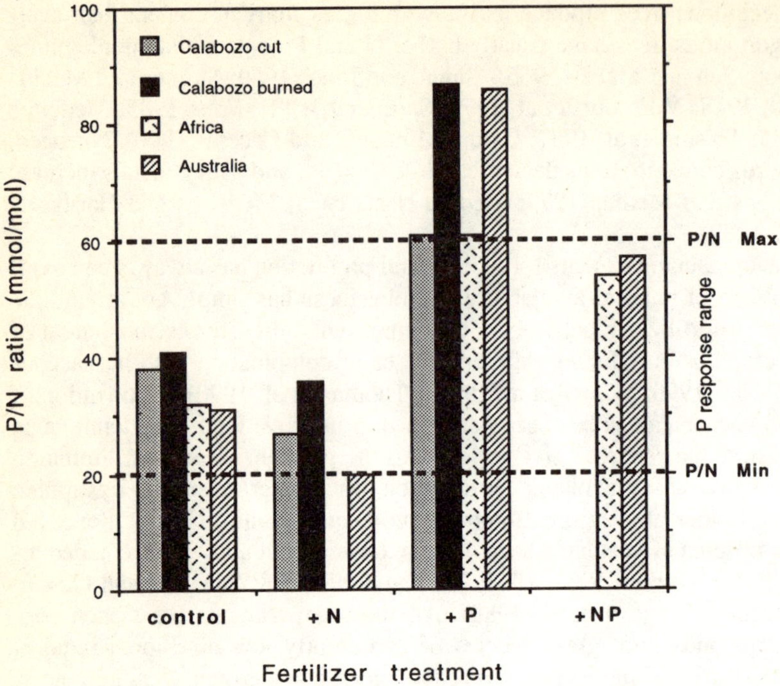


**Fig. 1.** Changes in the P/N ratios (mmol P/mol N) in the aboveground biomass of different grasslands in Africa (Hall et al. 1941), Australia (Norman 1963, 1966), and South America. (*cut* and *burned* refer to treatments applied to *Trachypogon*-savannas in central Venezuela at the end of the dry season. (Medina et al. 1978)

values measured experimentally in nine tropical grass species were 20 mmol/mol for P limited plants to 60 mmol/mol for plants receiving plenty of P supply (Wilson and Haydock 1971). If this index proves to be of general validity, it may be useful to assess the nutritional status of savanna vegetation. Their applicability can be seen in the variation induced by fertilization in native grasslands in Venezuela, South Africa, and Australia (Fig. 1). A brief look at published data on N and P contents of the herbaceous layers in different savanna areas shows that the grasslands in the Brazilian Cerrado, in central Venezuela, at the Nylsvley research site in South Africa, and at the Katherine research station in Australia are strongly deficient in P, with P/N ratios well below 60 mmol/mol (Medina et al. 1978; Coutinho 1979; Frost 1985), while grasslands in the savanna station of Lamto in Ivory Coast (Villecourt et al. 1979) and in western Venezuela in Barinas state (Sarmiento 1984), with P/N ratios above 60, probably would not respond to P additions.

## 6. Biological Interactions Regulating Nutrient Availability in Savanna Environments

Savanna soils are generally of low nutrient availability because of the predominance of oxisols, ultisols and alfisols of low nutrient status. Under these circumstances biological interactions are fundamental to guarantee availability of nutrients. These biological interactions include grass-tree interactions, herbivore-plant interactions, particularly leaf-cutting ants in South America, and mound-building termites throughout the seasonal dry tropics. Several of those processes contribute to create a large degree of patchiness in nutrient availability in savanna soils.

### a) Grass-Tree Interactions

The structure of savannas characterized by a continuous grassland cover interspersed with more or less isolated trees and shrubs constitutes a mosaic of contrasting resources utilization. Trees and shrubs generally exploit deeper soil layers compared to grasses and other herbaceous species (Walter 1973). Scattered trees and shrubs constitute areas of nutrient enrichment, at least in the upper soil layers. This may account partially for the differences in floristic composition, density, and productivity of the vegetation underneath woody perennials in several savanna areas of the world. The presence of trees within a grassland also results in changes in light conditions, temperature, and evapotranspiration, factors powerful enough to explain the variations observed in the herbaceous vegetation (Belsky et al. 1989; Muoghalu and Isichei 1991). Availability of K, Ca, and Mg in the upper soil layers is always higher under the tree canopy in semiarid savannas in Kenya (Belsky et al. 1989), in humid, probably derived savannas in Central America (Kellman 1979), and in humid savannas in the Orinoco plains in Central Venezuela (Susach 1984). The levels of organic carbon and available P were not significantly different in the Central American savanna. The richer nutrient environment underneath the tree components may allow the establishment of other woody species with higher nutrient requirements leading to the development of "tree groups" (Blydenstein 1962; Kellman 1979). In South American savannas, deciduous trees, presumably more nutrient-demanding, are restricted to forest groves, while evergreen trees, lower in nutrient content, are found scattered within the grass layer of the savannas (Montes and Medina 1977). However, this distribution may result simply from differences in fire tolerance.

The mechanism of soil enrichment underneath tree canopies is probably associated with the accumulation of litter, produced by trees concentrating nutrients from large soil volumes (Bernhard-Reversat 1982; Belsky et al. 1989). In addition, bird droppings, because of the use of trees as perching objects, might play a role in nutrient enrichment (Belsky et al. 1989). Kellman (1979) suggested that the main source of nutrients is the precipitation inputs, because enrichment has been observed in trees without deep root systems (see also García-Miragaya and Cáceres 1990). However, in Kellman's study *Pinus caribaea* is less efficient than *Byrsonima crassifolia* (Malpighiaceae) in increasing the nutrient content underneath their canopy, and this may be better explained by the higher nutrient content of the *Byrsonima* litter than by a higher nutrient capture efficiency of its canopy compared to *Pinus*.

### b) Mound-Building Termites and Leaf-Cutting Ants

Mound-building termites are a common characteristic of all savanna regions of the world, although they reach their most significant development in northern Australia and western Africa. During mound building, termites explore and excavate soil between 0.5 to 1 m depth (Pomeroy 1983). In this process termites modify the physicochemical characteristics of the soils, in texture by selecting finer particles more appropriate for mound building, and in nutrient content because of the mixing with food residues (Arshad 1981; Pomeroy 1983). The enrichment in Ca, K, and Mg is particularly noticeable when the subsoil is nutrient-poor (Pomeroy 1983). In the Ivory Coast and Venezuela, it was shown that mounds of soil-eating termites are enriched in easily extractable P, while that is not the case in grass-eating termites (López-Hernández et al. 1989; López-Hernández and Febres 1984). Several termite species are efficient foragers of organic debris (nearly 1000 kg $ha^{-1}$ $a^{-1}$), therefore they can deprive the surroundings of organic carbon input and the nutrients contained in the plant litter (Goodland 1965). This author found an inverse correlation between nutrient status of the soil and the occurrence of termitaria in the Rupununi savannas. The effect of sequestering nutrients in termite mound is probably minor, at least in the semiarid savannas of Kenya (Pomeroy 1983). In Australia, *Amitermes* and *Nasutitermes* mineralize approximately 250 kg $ha^{-1}$ $a^{-1}$ of organic matter, and this may be an important contribution to nutrient recycling in nutrient-poor soils (Holt and Coventry 1990).

On the American continent, ants of the genera *Atta* and *Acromyrmex* occur in a range of tropical ecosystems, from the rainforest to strongly seasonal savannas (Salick et al. 1983; Jaffe and Delascio 1991). In savanna environments, these ants form mounds of variable size, occurring both in open grasslands, and in forest groves interspersed with savanna vegetation (Coutinho 1982; Farji 1992). In these areas they reach dimensions comparable to termites in Africa and Australia. These ant species forage actively on grass and tree leaves, which are transported to complicated underground chamber systems where the material is used for the cultivation of fungi. Therefore they act as nutrient concentrators comparable to termites. The litter from the plant material is deposited in deep chambers where it remains immobilized for prolonged periods contributing to the patchiness in nutrient distribution in South American savannas (Haines 1975; Salick et al. 1983). Coutinho (1984) postulates that leaf-cutting ants play a significant role in nutrient cycling in some Brazilian savannas. Burying large amounts of nutrient-enriched litter may provide a source of nutrients for deep-rooted woody plants, which can thereby evade competition with grass roots in the upper soil layers. Brazilian authors calculated that one well-developed ant nest can consume about 1 ton of fresh plant weight per year (Coutinho 1984). Therefore, this foraging activity would also sequester significant amounts of nutrients from above ground, avoiding their volatilization through fire at the end of the rainy season. There are no adequate studies to document quantitatively the significance of leaf-cutting ant colonies in the nutrient budget of savannas in South America.

### c) Mycorrhiza

Mycorrhizas are the most common symbioses in terrestrial ecosystems and in the tropics vesicular-arbuscular mycorrhizas (VA-mycorrhizas) are almost of universal occurrence (Harley and Smith 1983). These symbioses play a fundamental role in phosphorus acquisition in nutrient-poor soils, as clearly demonstrated in cultivated plants (Sanni 1976). The effects on legume forage plants are quite strong because the improvement in P nutrition brought about by mycorrhiza has a substantial impact on nodulation efficiency with *Rhizobium* (Crush 1974; Lynd et al. 1985). Management of mycorrhizal symbioses in nutrient-poor savanna soils in Colombia is improving the growth and productivity of forage species (*Brachyaria dycteroneura* and *Pueraria phaseoloides*), and important crop species such as cassava (*Manihot esculenta*) and *Sorghum* sp. (Dodd et al. 1990; Arias et al. 1991).

Quantitative assessments of the amount of P incorporated through mycorrhiza in natural vegetation are not available. South American legume species from savannas which are being selected and improved were shown to be infected with VA-mycorrhiza in experimental plots in the field in Colombia (Saif 1986). In natural savannas including *Trachypogon* sp., mycorrhizal infection was lower than that of cultivated forage plants, both grasses and legumes. Phosphorus fertilization decreased infection while its deficiency stimulated it. Calcium fertilization stimulated mycorrhizal infection in some legume and grasses. K fertilization stimulated infection in legumes but not in grasses. All the legumes were nodulated and had significant mycorrhizal infection. Grasses and legumes are both dependent on mycorrhiza for P supply. Native populations of mycorrhizal fungi are low in savanna soils, therefore inoculation with appropiate mycorrhizal endophytes might improve the establishment of introduced grasses and legumes.

The widespread occurrence of mycorrhizal symbioses in tropical savannas has been well documented. In semiarid savannas in Kenya, five grass species were shown to be highly infected with mycorrhizal fungi in different soil types, including seasonal flooded soils (Newman et al. 1986). Shrubby and herbaceous legume species of the Papilionoidae and Caesalpinioidae subfamilies were shown to develop tripartite symbioses (legume-*Rhizobium*-mycorrhiza) in Venezuelan savannas (Medina and Bilbao 1991). Trees from savanna woodlands in Tanzania were mostly endomycorrhizal (41 species out of 47 investigated), but the legume trees characteristic of Miombo woodlands, *Julbernardia globiflora* and *Brachystegia* spp., were all ectomycorrhizal (Högberg 1982).

### d) $N_2$-Fixation by Legume-*Rhizobia* Associations

A characteristic feature of South American savannas is the high diversity of herbaceous and shrubby legume species. In fact, many of the forage legume species cultivated today in the tropical world originate from South America (Winter et al. 1989b). Most legumes in tropical savannas live in association with native *Rhizobia* strains. In a large survey of legume species of the savannas in Venezuela, Barrios and González (1971) found 109 nodulated species out of a total of 127 species investigated. Most of the nodulated species were within the subfamily Papilionoidae (73 species). Many of these species develop thick underground organs (lignotubers and xylopodia, Rizzini and Heringer 1961) accumulating carbohydrates and nutrients (Medina and Bilbao 1991). Nutrient and energy-rich compounds of these organs support active resprouting after fire or prolonged drought.

Frequency of legume species in Venezuelan savannas has been associated with low levels of exchangeable Al and high levels of exchangeable Ca (these two parameters are inversely correlated) (Medina and Bilbao 1991). In natural savannas legumes have leaf nitrogen concentrations above the critical level for normal growth, however, the levels of P and of K are clearly deficient (Andrews and Robins 1969). This nutrient status leads to reduced plant growth because of the inhibition of effective nodulation. The grasses appear to be limited both by N and P (Norman 1966; Medina et al. 1978; García-Miragaya et al. 1983; Medina and Bilbao 1991).

There are few data documenting nitrogen fixation by native legumes under natural conditions in South American savannas. Indirect evidence has been reported for a few species based on determination of natural abundance of $^{15}N$ and the concentration of ureids and $\alpha$-amino compounds in the xylem sap of young shoots (Medina and Bilbao 1991). In Nylsvley, South Africa, significant rates of acetylene redution were measured in soil cores, equivalent to $N_2$ fixation rates of 6.6 to 8.5 kg $ha^{-1}$ $a^{-1}$. This activity was attributed exclusively to nodulated legumes (Zietsman et al. 1988). In savanna woodlands of Tanzania, Högberg (1986) compared 27 species of legume trees separated in groups of potentially $N_2$-fixing and non-$N_2$-fixing species. Nodulation and nitrogenase activity (acetylene reduction) was demonstrated in 12 of 14 potentially $N_2$-fixing legumes. In these species natural abundance of $^{15}N$ was 0.0007% lower, and the leaf N content was considerably higher than that of the non-$N_2$-fixing species. The use of natural abundance of $^{15}N$ to determine the contribution of $N_2$ fixation through symbiosis to the N budget of the whole plant is expanding (Shearer and Kohl 1986). The method proved to be useful to detect $N_2$ fixation by grasses and legumes of the herbaceous layers in savannas of the Ivory Coast and Venezuela (Abbadie et al. 1992; Medina and Bilbao 1991). In subtropical savanna-like vegetation in Namibia, Schulze et al. (1991) showed that $\delta^{15}N$ values of potentially $N_2$-fixing species was 3% lower on average compared to non-$N_2$-fixing species. It was calculated that the symbiosis contributed an average of 30% to the total N budget of the species studied. There is a large degree of variation of $\delta^{15}N$ values published for tropical plants, and this may be associated with the differences in $^{15}N$ abundance of the substrate where the plants are growing (Heaton 1987).

### e) $N_2$-Fixation by Free-Living Microorganisms

Cyanobacteria can make important contributions to soil N in most humid savannas. In Nigerian savannas with cyanobacterial crust cover of around 30%, rates recorded during the rainy season average 2.3 mg N $m^{-2}$ $day^{-1}$, while in the *Trachypogon*-savannas of central Venezuela, with 10–12% cyanobacterial coverage, rates during the rainy season varied between 0.4 and 0.7 mg N $m^{-2}$ $day^{-1}$ (Santaella 1985). These rates are not enough to compensate potential N losses trough fire (Medina 1982b).

Free-living bacteria in the rhizosphere of grass roots have been reported to fix significant amounts of atmospheric nitrogen (Dobereiner and Day 1975; De-Polly et al. 1977). The reports are somewhat inconsistent, and disagreement persists regarding the significance of this activity for the nitrogen budget of natural savannas.

Reported rates for these root-*Azospirillum* associations vary in different savanna areas, with natural and introduced pastures from 8 to 800 g N $ha^{-1}$ $day^{-1}$ (Balandreau et al. 1976 and literature therein). With field-grown Brazilian grasses, rates ranging from 14.7 to 51.4 g N $ha^{-1}$ day $^{-1}$ have been reported (van Berkum and Day 1980). In Zimbabwe, nitrogen fixation activity associated with grass roots was probably restricted to low-lying, seasonally water-logged areas; in these areas rates of $N_2$ fixation by *Paspalum urvillei* amounting to 76 kg N $ha^{-1}$ $a^{-1}$ have been reported (Maasdorp 1987). It seems that rhizospheric nitrogen fixation is a slow but continuous process contributing small amounts of organic nitrogen to savanna ecosystems. These amounts are significant in areas where plant production is severely limited by nitrogen supply (Giller and Day 1985).

In grasses of savannas within the Lamto station in the Ivory Coast, Abbadie et al. (1992) calculated that about 17% of the nitrogen requirements of the grass layer is derived from fixation of atmospheric $N_2$ operated by free-living micoorganisms associated with the rhizosphere. In these savannas legumes contribute little to total biomass of the herbaceous layer. In order to explain the low $\delta^{15}N$ of the grasses, the authors proposed that most of the nitrogen used for growth every year originates from the decay of root material and not from the surrounding soil. In this process the $\delta^{15}N$ value of the plant would decrease slowly with plant age. The same argument could be applied to the legumes reported in the paper, since their $\delta^{15}N$ values are within the same range of the grasses.

## 7. Nutrient Cycling and the Effect of Fire: Are Savannas Sources or Sinks for Biogenic Nutrients?

Recurrent fires are probably responsible for increasing nutrient depletion of savanna soils, particularly those nutrients volatilized during burning, namely nitrogen and sulfur (Enwezor 1976; Robertson and Rosswall 1986). To my knowledge, no complete nutrient balance study is available for savanna ecosystems, even at the grasslayer level. Most emphasis has been laid upon the cycling of nitrogen, because of its intrinsic metabolic importance, and also because it is easier to measure compared to S and P. It has been established that there are two main nutrient cycling pathways in parkland savannas, a faster and variable cycle through the herbaceous layer and a slower and more stable cycle through the woody plants (Bate 1981; Bernhard-Reversat 1982; Frost 1985).

Nutrient inputs through rains to savanna regions of the world have not been evaluated comprehensively, a job which has to be done if the process of environmental change and its potential impact in savanna environments is to be understood. The input of K, Ca, and

**Table 1.** Nutrient inputs through rainfall in different savanna regions of the world. Values in kg $ha^{-1}$ $a^{-1}$

| Region[a] | P | N inorg | S | K | Ca | Mg |
|---|---|---|---|---|---|---|
| Australia (1) | 0.2–0.4 | 1.4–2.0 | 1.0–6.0 | 0.3–11.0 | 0.5–12 | 0.2–3.0 |
| Lamto (2) Africa | 3.6 | 4.5 | n.d. | 3.4 | 26.0 | 4.8 |
| Calabozo (3) South America | 0.3–0.5 | 1.3–3.4 | 1.3–3.4 | 2.7–5.1 | 2.1–4.6 | 5.9–5.2 |
| Emas (4) South America | 0.94 | n.d. | 12.0 | 2.6 | 5.6 | 0.9 |

[a] (1) Holt and Coventry (1990); (2) Villecourt and Roose (1978); (3) Montes and San José (1989); (4) Coutinho (1979).

Mg varies considerably among regions and years, a fact probably associated with dust transportation, or the stochastic differences in the distribution and timing of fires in savanna regions (Villecourt and Roose 1978; Coutinho 1979; Montes and San José 1989; Holt and Coventry 1990; Table 1). In some areas, the input of nitrogen in organic form has been reported to be larger than the input in inorganic form (Villecourt and Roose 1978).

Nutrients in rainfall are fixed efficiently in the upper layers of oligotrophic savanna soils (Kellman and Sanmugadas 1985; Kellman et al. 1985). The nutrients most efficiently fixed are Ca and P, followed by Mg and K. The mechanisms of nutrient retention is not dependent on increased root uptake, but on the physicochemical properties of the soil. The efficiency of nutrient capture is associated with the cation exchange capacity of the soil layers through which the rain water percolates. Experiments increasing the nutrient load of incoming water in the same savanna region showed higher leaching to deeper soil layers. Leaching was more pronounced in soils partially covered by trees than in open savannas (Kellman and Sanmugadas 1985).

### a) Nitrogen Losses Through Denitrification and Ammonification

Enzymatic denitrification in soils of seasonal savannas has been assumed to be insignificant because of the lack of favorable conditions for denitrifiers (Woodmansee et al. 1981). In addition, substrate for denitrification ($NO_3^-$) is produced only during the rainy season, when the vegetation is active and presumably rapidly absorbs all the $NO_3^-$ produced.

Losses of $NH_3$ in natural savannas have seldom been measured. In the *Trachypogon*-savannas they amount to 11 kg $ha^{-1}$ during the reproductive period (San José et al. 1991) and are therefore quite significant for the nitrogen budget of these ecosystems.

Savanna vegetation appears to be an important source of nitrogen oxides, particularly during the rainy season. Values ranging from 3 to 13 kg N $ha^{-1}$ $a^{-1}$ have been measured in savannas of the Venezuelan Llanos (Johansson et al. 1988). These high estimates may not be typical for the whole range of savannas in northern Venezuela; values around 2.6 kg $ha^{-1}$ $a^{-1}$ for NO-N and 0.65 kg $ha^{-1}$ $a^{-1}$ for $N_2O$-N are probably more accurate (Sanhueza et al. 1990). The effect of burning increasing the NOx release from savanna soils appears to be very clear (Johansson and Sanhueza 1988). These amounts constitute a significant loss of N from savanna soils and show that tropical savannas may play an important role in tropospheric photochemistry.

b) Nitrogen Losses Through Fire

Nitrogen losses through fire depend on the above-ground biomass present at the time of burning, the N content of the biomass, and the efficiency of burning. The calculated losses through fire range from 4.5–5.6 kg $ha^{-1}$ $a^{-1}$ in grasslands near Katherine, Australia (Norman and Wetselaar 1960) and 8–10 kg N $ha^{-1}$ $a^{-1}$ in Africa (Villecourt et al. 1980; Isichei and Sanford 1980) (much higher values have been estimated for the total above-ground biomass losses in South Africa, see Frost 1985); to around 8 kg N $ha^{-1}$ $a^{-1}$ in central Venezuelan savannas (Medina 1982b; Chacón et al. 1991).

c) Nitrogen Balances

Several of the nitrogen balances published for tropical savannas indicate a net gain of N in spite of the recurrent annual fires (Villecourt et al. 1980; Chacón et al. 1991). However, these estimations are based on the assumption of large N inputs through biological fixation (~10 kg $ha^{-1}$ $a^{-1}$), particularly those corresponding to bacteria-rhizosphere associations. In some savannas this activity has not been detected but the contribution of the nodulated legumes is of the same order (Zietsman et al. 1988). On the other hand, savannas appear to be important sources of NOx and $NH_3$, representing net nitrogen losses of the same order of the biological fixation (Johansson et al. 1988; San José et al. 1991). These points have to be further documented with intensive and more precise evaluation of nutrient inputs and outputs regulated by microbiological activities in the field.

## 8. Conclusions

There is a great bulk of information on nutrient relations in savanna environments. The physiological characteristics of mineral nutrition of native savanna plants are reasonably well understood. Many studies on the biological interactions facilitating nutrient acquisition in dystrophic savanna soils will help considerably in the management of agricultural production in the tropics using microbiological techniques such as improved *Rhizobium* and mycorrhizal fungi strains.

Essential information regarding the nutrient balances of most savanna regions of the world is still lacking (see the study of Robertson and Rosswall 1986). Judging from studies on biomass production in tropical grasslands concluding that these systems constitute important carbon sinks (Long et al. 1992), it could be predicted that this tendency is also valid for nutrients such as nitrogen and sulfur (because of their association with organic matter), but no such statement can be made regarding the status of P, K, and Ca. Savannas are extremely stressed systems, much more than tropical rainforests, partially because they are the center of dense human populations throughout the tropical world (Harris 1980; Werner 1991), but also because they are perceived as natural ecosystems of relatively lower ecological value. Studies of nutrient dynamics on a regional scale are required to clarify the role of savanna ecosystems in atmosphere dynamics, and their capability to tolerate changes in chemical composition of rainfall.

## References

**Abbadie L,** Mariotti A, Menaut JC (1992) Ecology 73:608–613. **Acquaye DK,** Kang BT (1987) Soil Sci 144:43–52. **Ahmad N,** Jones RL (1969) Soil Sci Soc Am Proc 33:762–765. **Alvim P de T** (1979) Potencial de la producción agrícola en la región amazónica. In: Tergas LE, Sánchez PA (eds) Potencial de pastos en suelos ácidos de los trópicos. Centro Internacional de Agricultura Tropical, Colombia, pp 17–28. **Alvim P de T,** Araujo W (1952) Turrialba 2:153–160. **Alvim P de T,** Cabala FP (1974) Cacau Atual (Ilheus/Bahia) 11:2–6. **Andrews CS,** Robins MF (1969) Aust J Agric Res 20:665–674. **Arias I,** Koomen I, Dodd JC, White RP, Hayman DS (1991) Plant Soil 132:253–260. **Arshad MA** (1981) Soil Sci 132:161–174. **Ayodele OJ** (1986) Trop Agric (Trinidad) 63:297–300.

**Balandreau J,** N'Dri Allou R, Villemin G, Weinhard P, Villecourt P (1976) Rev Ecol Biol Sol 13:529–544. **Barrios S,** González V (1971) Plant Soil 34:707–719. **Bate GC** (1981) Nitrogen cycling in savanna ecosystems. In: Clark EE, Rosswall T (eds) Terrestrial nitrogen cycles. Ecol Bull (Stockholm) 33:463–475. **Bell RHV** (1982) The effect of soil nutrient availability on community structure in savanna ecosystems. In: Huntley BJ, Walker BH (eds) Ecology of tropical savannas. Ecological Studies 42. Springer, Berlin Heidelberg New York, pp 193–216. **Belsky AJ,** Amundson RG, Duxbury JM, Riha SJ, Ali AR, Mwonga SM (1989) J Appl Ecol 26:1005–1024. **Bernhard-Reversat F** (1982) Oikos 38:321–332. **Bilbao B,** Medina E (1990) J Biogeogr 17:421–425. **Blydenstein J** (1962) BolSoc Ven Cienc Nat 23(102):139–206. **Bourliére F** (ed) (1983) Tropical savannas. Ecosystems of the world, vol 13. Elsevier, Amsterdam.

**Chacón P,** López-Hernández ID, Lamotte M (1991) Rev Ecol Biol Sol 28:67–75. **Cochrane TT** (1989) Soil Sci Soc Am J 53:139–141. **Cole MM** (1986) The savannas, biogeography and geobotany. Academic Press, London. **Coutinho LM** (1979) Rev Bras Biol 2:97–101. **Coutinho LM** (1982) Rev Bras Biol 42:147–153. **Coutinho LM** (1984) Bol Zool, Univ São Paulo 8:1–9. **Crush JR** (1974) New Phytol 73:743–752.

**De-Polli H,** Matsui E, Dobereiner J, Salati E (1977) Soil Biol Biochem 9:119–123. **Dobereiner J,** Day JM (1975) Nitrogen fixation in the rhizosphere of tropical grasses. In: Stewart WDP (ed) Nitrogen fixation by free living microorganisms. IBP Series 6. Cambridge Univ Press, Cambridge, pp 39–56. **Dodd JC,** Arias I, Koomen I, Hayman DS (1990) Plant Soil 122:229–240.

**Eiten G** (1972) Bot Rev 38:201–241. **Enwezor WO** (1976) Geoderma 15:401–411.

**Farji AG** (1992) Dinámica de islas boscosas en una sabana-parque: la actividad de las Attini (Hymenoptera, Formicidae) y su relación con la invasión de leñosas. Thesis Mag Sci, Univ Andes,

Fac Cienc, Mérida Venezuela. **Frost P** (1985) Organic matter and nutrient dynamics in a broadleafed African savanna. In: Tothill JC, Mott JJ (eds) Ecology and management of the world's savannas. Australian Academy of Sciences, Canberra, pp 200–206. **Frost P,** Medina E, Menaut JC, Solbrig O, Swift M, Walker BJ (1986) Responses of savannas to stress and disturbance. Biol Int Spec Issue 10. IUBS, Paris. **Furley PA,** Ratter JA (1988) J Biogeogr 15:97–108.

**García-Miragaya J,** Cáceres A (1990) Acta Oecol 11:775–781. **García-Miragaya J,** San José JJ, Hernández J (1983) Trop Ecol 24:33–42. **Giller KE,** Day JM (1985) Nitrogen fixation in the rhizosphere: significance in natural and agricultural systems. In: Fitter AH (ed) Ecological interactions in soil. Blackwell, London, pp 127–147. **Goedert WJ** (1983) J Soil Sci 34:405–428. **Goedert WJ** (1990) Estratégias de manejo das savanas. In: Sarmiento G (ed) Las sabanas americanas. Fondo Editorial Acta Cient Ven, Caracas, pp 191–218. **Goldstein G,** Sarmiento G, Meinzer F (1986) Acta Oecol, Oecol Plant 7:107–119. **Goodland R** (1965) Can J Zool 43:641–650. **Goodland R,** Pollard R (1973) J Ecol 61:219–224.

**Haines BL** (1975) Impact of leaf-cutting ants on vegetation development at Barro Colorado island. In: Golley FB, Medina E (eds) Tropical ecological systems. Ecological Studies 11. Springer, Berlin Heidelberg New York, pp 99–111. **Hall TD,** Meredith T, Murray SM (1941) S Afr J Sci 37:111–129. **Harley JL,** Smith SE (1983) Mycorrhizal symbiosis. Academic Press, London. **Harris H** (ed) (1980) Human ecology in savanna environments. Academic Press, London. **Heaton THE** (1987) Oecologia 74:236–246. **Holt JA,** Coventry RJ (1990) J Biogeogr 17:427–432. **Högberg P** (1982) New Phytol 92:407–415. **Högberg P** (1986) J Appl Ecol 23:675–688. **Huntley BJ** (1982) Southern African savannas. In: Huntley BJ, Walker BH (eds) Ecology of tropical savannas. Ecological Studies 42. Springer, Berlin Heidelberg New York, pp 101–109. **Huntley BJ,** Morris JW (1982) Structure of the Nylsvlei savanna. In: Huntley BJ, Walker BH (eds) Ecology of tropical savannas. Ecological Studies 42. Springer, Berlin Heidelberg New York, pp 432–455. **Huntley BJ,** Walker BH (eds) (1982) Ecology of tropical savannas. Ecological Studies 42. Springer, Berlin Heidelberg New York.

**Isichei AO** (1980) Nitrogen fixation by blue-green algal crusts in a Nigerian savanna. In: Rosswall T (ed) Nitrogen cycling in West-African ecosystems. SCOPE-UNEP Int Nitrogen Unit, R Swed Acad Sci, Stockholm, pp 191–198. **Isichei AO,** Sanford WW (1980) Nitrogen losses by burning from Nigerian grasslands ecosystems. In: Rosswall T (ed) Nitrogen cycling in West-African ecosystems. SCOPE-UNEP Int Nitrogen Unit, R Swed Acad Sci, Stockholm, pp 325–331.

**Jaffe K,** Delascio F (1991) Ecotropicos 4:60–67. **Johansson C,** Rhode H, Sanhueza E (1988) J Geophys Res 93:7180–7192. **Johansson C,** Sanhueza E (1988) J Geophys Res 93:14193–14198. **Johnson RW,** Tothill JC (1985) Definition and broad geographic outline of savanna lands. In: Tothill JC, Mott JJ (eds) Ecology and management of the world's savannas. Aust Acad Sci, Canberra, pp 1–13. **Jones MB,** Long SP, Roberts MJ (1992) Synthesis and conclusions. In: Long SP, Jones MB, Roberts MJ (eds) Primary productivity of grass ecosystems of the tropics and subtropics. Chapman & Hall, London, pp 212–255.

**Kellman M** (1979) J Ecol 67:565–577. **Kellman M,** Sanmugadas K (1985) J Ecol 73:935–951. **Kellman M,** Miyanishi K, Hiebert P (1985) J Ecol 73:953–962.

**Long SP,** Jones MB, Robert MS (eds) (1992) Primary productivity of grass ecosystems of the tropics and the subtropics. Chapman & Hall, London. **Lópes AS,** Cox FR (1977a) Soil Sci Soc Am J 41:742–747. **Lópes AS,** Cox FR (1977b) Agron J 69:828–831. **López-Hernández D,** Fardeau JC, Niño M, Nannipieri P, Chacón P (1989) J Soil Sci 40:635–640. **López-Hernández D,** Febres A (1984) Rev Ecol Biol Sol 21:477–489. **Lynd JQ,** Tyrl RJ, Purcino AAC (1985) J Plant Nutr 8:1047–1059.

**Maasdorp BV** (1987) Soil Biol Biochem 19:7–12. **Marschner H** (1986) Mineral nutrition of higher plants. Academic Press, London. **McClung AC,** de Freitas LM (1959) Ecology 40:315–317. **McNaughton SJ** (1985) Ecol Monogr 55:259–294. **McNaughton SJ** (1988) Nature 334:343–345. **Medeiros RA,** Haridasan M (1985) Plant Soil 88:433–436. **Medina E** (1982a) Physiological ecology of neotropical savanna plants. In: Huntley BJ, Walker BH (eds) Ecology of tropical savannas. Ecological Studies 42. Springer, Berlin Heidelberg New York, pp 308–335. **Medina E** (1982b) Plant Soil 67:305–314. **Medina E** (1987) Nutrients: requirements, conservation and cycles in the herbaceous layer. In: Walker B (ed) Determinants of savannas. IUBS Monogr Series No. 3. IRL Oxford, pp 39–65. **Medina E,** Bilbao B (1991) Significance of nutrient relations and symbiosis for the competitive interaction between grasses and legumes in tropical savannas. In: Esser G, Overdieck D (eds) Modern Ecology, basic and applied aspects. Chapt 5. Elsevier, Amsterdam, pp 295–319. **Medina E,** Cuevas E (1989) Patterns of nutrient accumulation and release in Amazonian forests of the upper Río Negro basin. In: Proctor J (ed) Mineral nutrients in tropical forest and savanna ecosystems. Blackwell, Oxford, pp 217–240. **Medina E,** Silva J (1990) J Biogeogr 17:403–413. **Medina E,** Mendoza A, Montes R (1978) Trop Agric 55:243–253. **Møberg JP,** Esu IE (1991) Geoderma 48:113–129. **Montes R,** Medina E (1977) Geo-Eco-Trop 4:295–307. **Montes R,** San José JJ (1989) Biogeochemistry 7:241–256. **Montgomery RF** (1988) J Biogeogr 15:11–18. **Montgomery RF,** Askew GP (1983) Soils of tropical savannas. In: Bourliére F (ed) Tropical savannas. Ecosystems of the world, vol 13. Elsevier, Amsterdam, pp 63–78. **Mott JJ,** Williams J, Andrew MH, Gillison AN (1985) Australian savanna ecosystems. In: Tothill JC, Mott JJ (eds) Ecology and management of the world's savannas. Aust Acad Sci, Canberra, pp 56–82. **Muoghalu JI,** Isichei AO (1991) Vegetatio 95:167–175.

**Newman EI,** Child RD, Patrick CM (1986) J Ecol 74:1179–1183. **Norman M,** Wetselaar R (1960) NTJ Aust Inst Agric Sci 26:272–273. **Norman MJT** (1963) Aust J Exp Agric Anim Husb 3:119–124. **Norman MJT** (1966) Katherine research station 1956-64: a review of published work. Div Land Res, Tech Pap 28, CSIRO, Australia.

**Ölsvig-Whittaker L,** Morris JW (1982) S AfrJ Bot 1:91–96.

**Penning de Vries FWT,** Krul JM, van Keulen H (1980) Productivity of Sahelian rangelands in relation to the availability of nitrogen and phosphorus. In: Rosswall T (ed) Nitrogen cycling in West-African ecosystems. SCOPE-UNEP Int Nitrogen Unit, R Swed Acad Sci, pp 95–114. **Pomeroy DE** (1983) J Soil Sci 34:555–570. **Proctor J** (ed) (1989) Mineral nutrients in tropical forest and savanna ecosystems. British Ecological Society No. 9. Blackwell, Oxford.

**Rizzini CT,** Heringer EP (1961) Phyton (Argentina) 17:105–124. **Robertson GP,** Rosswall T (1986) Ecol Monogr 56:43–72. **Ruess RW** (1987) Herbivory: the role of large herbivores in nutrient cycling of tropical savannas. In: Walker BH (ed) Determinants of tropical savannas. IUBS Monogr Series 3. IRL, Oxford, pp 67–91.

**Saif SR** (1986) Angew Bot 60:125–139. **Salick J,** Herrera R, Jordan CF (1983) Biotropica 15:1–7. **Sánchez PA,** Isbell RF (1978) Comparación entre los suelos de los trópicos de América latina y Australia. In: Tergas LE, Sánchez PA (eds) Producción de pastos en suelos ácidos de los trópicos. Cent IntAgricTrop, Medellin, Colombia, pp 17–28. **Sanford WW,** Usman S, Obot EO, Isichei AO, Wari M (1982) Trop Agric (Trinidad) 59:315–318. **Sanhueza E,** Hao WM, Scharffe D, Donoso L, Crutzen PJ (1990) J Geophys Res 95:22481–22488. **Sanhueza E,** Arias MC, Donoso L, Graterol N, Hermoso M, Martí I, Romero J, Rondón A, Santana M (1992) Tellus 44B:54–62. **San José JJ,** García-Miragaya J (1981) Bol Soc Ven Cienc Nat 36(139):347–374. **San José JJ,** Montes R, Nikonova-Crespo N (1991) Ann Bot 68:321–328. **Sanni SO** (1976) New Phytol 77:667–671. **Santaella S** (1985) Fijación biológica de nitrógeno en sabanas de Trachypogon spp. Trabajo Especial de Grado. Esc Biol, Univ Central Venez, Caracas. **Sarmiento G** (1983) The savannas of tropical America. In: Bourliére F (ed) Tropical savannas. Ecosystems of the world, vol 13. Elsevier, Amsterdam, pp 245–288. **Sarmiento G** (1984) The ecology of neotropical savannas. Harvard Univ

Press, Cambridge. **Sarmiento G** (1990) Ecología comparada de ecosistemas de sabanas de América del Sur. In: Sarmiento G (ed) Las sabanas americanas. Aspectos de su biogeografía, ecologia y utilización. Fondo Editorial Acta Cient Ven, Caracas, pp 15–56. **Scholes RJ** (1990) J Biogeogr 17:415–419. **Schulze ED,** Gebauer G, Ziegler H, Lange OL (1991) Oecologia 88:451–455. **Sen KM,** Mabey GL (1966) The chemical composition of some indigenoius grasses of coastal savanna of Ghana at different stages of growth. In: Proc 9th Int Grasslands Congr, vol 1. São Paulo, Brasil, pp 763–771. **Shearer G,** Kohl DH (1986) Aust J Plant Physiol 13:699–756. **Susach F** (1984) Caracterización ecológica de las sabanas de un sector de los Llanos Centrales bajos de Venezuela. PhD Thesis, Univ Cent Venez, Fac Cienc, Caracas.

**Thomas D,** Vera RR, Lascano C, Fisher MJ (1990) Use and improvement of pastures in neotropical savannas. In: Sarmiento G (ed) Las sabanas americanas. Fondo Editorial Acta Cient Ven, Caracas, pp 141–161. **Tolsma DJ,** Ernst WHO, Vermeij RA, Vooijs R (1987) J Ecol 75:755–770. **Tothill JC,** Mott JJ (eds) (1985) Ecology and management of the world's savannas. Aust Acad Sci, Canberra.

**UNESCO** (1979) Tropical grazingland ecosystems. Nat Res Res XVI UNESCO, Paris.

**van Berkum P,** Day JM (1980) Soil Biol Biochem 12:137–140. **Villecourt P,** Roose E (1978) Rev Ecol Biol Sol 15:1–20. **Villecourt P,** Schmidt W, César J (1979) Rev Ecol Biol Sol 16:9–15. **Villecourt P,** Schmidt W, César J (1980) Rev Ecol Biol Sol 17:7–12.

**Walker BH** (ed) (1987) Determinants of tropical Savannas. IRL, Oxford. **Walker BH,** Menaut JC (1988) Research procedure and experimental design for ecology and management. RSSD Aust, Publ 1, CSIRO Printing Center, Melbourne. **Walter H** (1973) Die Vegetation der Erde. B I. Die tropischen und suptropischen Zonen. VEB Gustav Fischer, Jena. **Werner PA** (ed) (1991) Savanna ecology and management. Blackwell, London. **Wilson JR,** Haydock KP (1971) Aust J Agric Res 22:573–587. **Winter WH,** Mott JJ, McLean RW, Ratcliff D (1989a) Aust J Exp Agric 29:613–622. **Winter WH,** Mott JJ, McLean RW (1989b) Aust J Exp Agric 29:623–630. **Woodmansee RG,** Vallis I, Mott JJ (1981) Grassland nitrogen. In: Clark EE, Rosswall T (eds) Terrestrial nitrogen cycles. Ecol Bull (Stockholm) 33:443–462.

**Zietsman PC,** Grobbelaar N, van Rooyen N (1988) S Afr J Bot 54:21–27.

Professor Dr. Ernesto Medina
Centro de Ecología,
Instituto Venezolano de
Investigaciones Científicas
Aptdo. 21827,
Caracas 1020-A, Venezuela

## VIII. Developmental Physiology: Signal Transduction

By Hubert H. Felle

### 1. Introduction

Ever since researchers have become aware of the general importance of apparently trivial signals for the plant development, signal transduction in the plant kingdom has become a rapidly expanding field. If we venture to compare signal transduction in plants to that in animals, we quickly realize that individual plant cells must sense a much greater variety of signals than animals. This is because plants are essentially sessile organisms and, apart from their protected embryological development, encounter a wide variety of signals to which they respond. Furthermore, many different signals such as light, temperature, gravity, $pH_2O$, $pCO_2$, $pO_2$, growth substances, wounding, pathogens, amino acids, sugars, and possibly laboratory chemicals, some of which may arrive at the same time, must be integrated and converted into a coherent response. Developmental plasticity is the result. Of course, the precondition for such behavior is lack of specialization, which clearly defines the difference to most animal cells, which often are highly specialized, in that they respond to one single signal or stimulus only. Individual plant cells, on the other hand, have many more compartments than animal cells and have retained a multiplicity of functions responding to a variety of signals. In spite of this apparent variety, very few pathways now begin to emerge in plant cells, of which the socalled PI-response (see below) has seemingly drawn the most attention.

### 2. Perception of the Signal

A stimulus must be perceived, modulated, and transduced to initiate a (cellular) response. Apart from light and gravitropic stimuli (not reviewed here), which cross the plasma membrane before they find their perceiving structures somewhere inside the cells, to many (external) signals the plasma membrane is a first barrier. Since any preformed intracellular cascade of reactions would not be triggered by a primary stimulus, a plasma membrane bound receptor/transducer system mediates the functional gap.

Progress in Botany, Vol. 54

a) Growth Substances

*Auxin*. Remarkable progress has been made in identifying an auxin receptor. Since the reports on the three classical auxin binding sites (Dohrmann et al. 1978), many more binding proteins have been identified. One of these, most intensively investigated, led to the identification of a 22 kDa protein (Löbler and Klämbt 1985; Shimomura et al. 1986; Venis 1987; Hesse et al. 1989; Palme et al. 1990). Several cDNA clones corresponding to this auxin-binding protein have recently been isolated, the primary amino acid sequences of which defines a protein (ABP1), apparently a luminal component of the endoplasmic reticulum (Hesse et al. 1989; Inohara et al. 1989; Tillmann et al. 1989). Recently, Palme et al. (1992 and refs. therein) identified another auxin-reactive 23 kDa (pm 23) protein which specifically associates with plasma membranes, but not with membranes from intracellular compartments.

Although several auxin-binding proteins (ABPs) have been identified in the ER and the plasma membrane (Cross 1985 and refs. therein; Löbler and Klämbt 1985; Palme et al. 1990, 1992), functional evidence is accumulating that the auxin receptor, responsible for the rapid stimulation of the plasma membrane $H^+$ ATPase, is located at the plasma membrane. Applying classical electrophysiology, Barbier-Brygoo et al. (1990, 1991) were able to demonstrate that antibodies raised against maize ABPs (Napier et al. 1988) block an electrical auxin response in tobacco protoplasts. Only recently, Rück et al. (1993), using the patch-clamp technique, could show that in maize protoplasts the same antibody indeed blocked the response of the plasma membrane proton pump to auxins (i.e., the generation of an outwardly directed ATP-dependent electrical current). Finally, tests with membrane-impermeant auxin conjugates, which elicit a strong auxin response, indicate that this auxin receptor functions at the cell surface (Venis et al. 1990). Since the same antibody was effective in both monocots (maize) and dicots (tobacco), it appears that this receptor is ubiquitous. An interesting variation of the auxin response, viz. a deactivation of the proton pump through submicromolar auxin concentrations has been demonstrated in root hairs of *Sinapis* (Tretyn et al. 1991). Since the reaction of the cells to auxin was delayed by about 30 s, the authors assume the involvement of an auxin receptor, without defining its location.

*Fusicoccin*. Although fusicoccin (FC) is a fungal toxin, its mode of action traditionally has been compared with that of auxin. Especially because of its "stimulatory" effect on the plasma membrane $H^+$ ATPase, FC was referred to as "super auxin". After two decades of intensive research, it is clear now that the signal chain(s) of the auxin and the FC effects are different in every way. Apart from the finding that FC also affects $K^+$ transport (Clint and Blatt 1989), there is a body of evidence for a specific FC-binding site/receptor at the plasma membrane (Aducci et al. 1982, 1988, 1989, 1992; Blum et al. 1988; Feyerabend and Weiler 1988a, b; Stout 1988; De Boer et al. 1989; De Michelis et al. 1989; Meyer et al. 1989) which indicates a "deregulation" of the $H^+$ ATPase rather than a stimulation. Recently, Marra et al. (1992) demonstrated that FC stimulated the ATP-driven intravesicular acidification of liposomes which were reconstituted with crude FC receptors and the $H^+$ ATPase from maize plasma membrane. FC stimulation was also maintained, when receptors from maize were substituted by those from spinach.

b) Pathogens

An important part of plant resistance is the induced formation of phytoalexins, antifungal hydrolases, and the hypersensitive response. Many observations indicate that specific molecules of the *pathogens*, the so-called elicitors, are the primary signals for the induction of the defense reaction. Such signals can be extracellular components released by pathogens (polysaccharides, glycoconjugates, enzymes, toxins), structural elements of pathogens (components of cell wall and membranes), or products of avirulence genes. Two examples: (1) Lee and West (1981) have examined the elicitor activity of the endopolygalacturonase purified from *Rhizopus stolonifer* more closely and found that elicitor activity of this enzyme is based on its catalytic activity. Subsequently, it was shown that the heat-labile pectic enzymes formed heat-stable elicitors when incubated with plant cell wall preparations (Bruce and West 1982; Davis et al. 1984). These results are compatible with the following model: the pathogen-produced extracellular pectic enzymes are the primary signals for recognition. The pectin in the plant cell is the receptor for these primary signals. (2) Isolated cell walls of microorganisms have nonspecific elicitor activity (Hadwiger and Beckman 1980; Kurosaki et al. 1986). How can insoluble pathogen cell walls be recognized? It is assumed that the receptor of this signal is an extracellular plant enzyme directed against structures of microorganisms. Signal-receptor interaction results in the release of soluble pathogen wall fragments that act as secondary extracellular signals, similar to the endogenous elicitors. For further information the interested reader is referred to the review by Boller (1989).

## 3. Transformation of the Signal

After perception by a specific receptor, the primary signal (first messenger) is transformed into a more general one. As the bacterial chemotaxis demonstrates, this can go along with a dramatic reduction from a wide variety of stimuli to a few (transducers), or it may be a one-to-one transformation, necessary to overcome the plasma membrane barrier. However, some of the signal molecules may also be transported into the cells to meet their respective receptor. In bacterial chemotaxis the PTS-sugars would be an example of this, and in plants the auxin. IAAH (indole-acetic acid), as a weak acid, easily crosses the plasma membrane, but is also symported as $nH^+/IAA^-$ (Felle et al. 1991). Cross (1991) discusses the probability of a primary auxin receptor, located at the endoplasmic reticulum (ER), and explains the plasma membrane reactions through endo- and exocytosis. In the cases where a receptor is involved, ligand binding causes conformation changes on the respective receptor which will be transferred onto a neighboring protein. As we know from animal systems, such proteins may be GTP-binding proteins (G-proteins) and/or phospholipases, both of which are involved in the production of so-called second messengers which then serve signal amplification to act as activators of protein kinases (see below).

Whereas in animal cells G-proteins are well established and known as an integral part of many signal chains, not much is known yet about G-proteins in plants, although they have been shown to exist in plants (Hasunuma et al. 1987; Millner 1987). Recently, Zbell et al. (1989, 1990) demonstrate high-affinity binding of GTP to microsomal membranes of *Funaria* and soybean, and tentatively discuss this effect with respect to auxin. Also Palme et al. (1992) discuss a potential role of G-proteins for auxin action in maize. Rodriguez-Rosales et al. (1992) report the identification of low molecular mass G-proteins in membranes of *Dunaliella salina*.

## 4. Amplification of the Signal: Secondary Messengers

Since the original stimulus (first messenger) may only be a weak signal, it has to be amplified. During this process, which may proceed for as long as the respective receptor remains engaged or until some sort of regulatory inhibition takes place, so-called second messengers are formed (molecules) or released (ions). Apart from $Ca^{2+}$ and $H^+$, a growing number of metabolites which meet the minimum requirement for a second messenger, i.e., that they are well regulated to a low free concentration, are discussed here.

### a) Phosphoinositides

Current interest in the metabolism of phosphoinositides in plants has arisen from studies of animal cells, where it has been demonstrated that these lipids are a source of second messengers (Hokin and Hokin 1953; Berridge and Irvine 1984). Although we should assume that the regulation of this metabolism may be different in plants and animals, the animal pathway provides a framework for investigations in plants: the receptor activates a phosphatidylinositol-4,4-bisphosphate ($PIP_2$)-specific phospholipase C or phosphodiesterase. In animal cells, activation of the phospholipase C is enhanced by GTP (see above), and it has been proposed that a G-protein may increase the binding for soluble lipases to the membrane, thereby enhancing $PIP_2$ hydrolysis. The $PIP_2$ present on the cytosolic side of the plasma membrane is hydrolyzed to inositol-1,4,5-trisphosphate ($IP_3$) and diacylglycerol (DAG). Each of these metabolites is a potential second messenger, initiating a cascade of events.

What about plants? Whereas there is enough evidence that phosphoinositide metabolism exists in plants (Boss 1989 and refs. therein), few reports give evidence that this metabolism is part of a signal chain. If $IP_3$ is a second messenger in plants, it should mobilize intracellular calcium stores and cause increase in cytosolic free $Ca^{2+}$. The first indication that inositol phosphates might affect $Ca^{2+}$ fluxes in plants was a report by Droeback and Ferguson (1985). They showed a small $Ca^{2+}$ release from zuccini microsomes after treatment with $IP_3$. Although Schumaker and Sze (1987), Reddy and Poovaiah (1987), Ranjeva et al. (1988), Brosnan and Sanders (1990), and Alexandre et al. (1990) reported $IP_3$-induced $Ca^{2+}$ release from the vacuole or vacuolar vesicles, the ER is still under debate as $IP_3$-sensitive $Ca^{2+}$ store, especially since highly vacuolated permeabilized protoplasts did not respond to $IP_3$, whereas protoplasts densely filled with cytoplasm did (see review Boss 1989). Few reports claim isolation of $IP_3$ from plant cells (Dillenschneider et al. 1986; Morse et al. 1987). Whereas most of these reports simply document that inositol phosphates in principal can cause $Ca^{2+}$ release from intracellular stores, few were able to connect this reaction with physiological events: Morré et al. (1984a), and Zbell and Walter (1987) were able to show that auxins affect the phosphorylation of membrane lipids; Morré et al. (1984a) demonstrated that the synthetic auxin 2,4-dichlorophenoxy-acetic acid (2,4-D) stimulated PI-turnover in soybean membranes. Gilroy et al. (1990b) show that $Ca^{2+}$ released by caged $IP_3$ causes stomatal closure, whereas Blatt et al. (1990), also using caged $IP_3$, demonstrate reversible inactivation of $K^+$ channels of *Vicia faba* guard cells. Kurosaki et al. (1987) report breakdown of phosphatidylinositol during the elicitation of phytoalexin production in cultured carrot cells. Kamada and Muto (1991) show that the phosphatidylinositol turnover in the plasma membrane of tobacco suspension culture cells is $Ca^{2+}$-regulated.

### b) $Ca^{2+}$

Calcium ions exist in free concentrations of 0.1 to 1 μM in the cytoplasm of plant cells (Hepler and Wayne 1985; Felle 1988, 1989a and refs. therein; Gilroy and Trewavas 1990 and refs. therein). Increased levels of $Ca^{2+}$ are assumed to amplify the original stimulus-receptor signal that induced it, and to modulate a variety of $Ca^{2+}$-dependent cellular processes, such as protein phosphorylation, either directly, or by forming a complex with other regulatory proteins, such as calmodulin (Marmé 1989 and refs. therein; Hsieh et al. 1991; Rasi-Caldogno et al. 1992).

As to regulation of cytosolic $Ca^{2+}$ in plant cells, the information is still rather patchy. Nevertheless, recent work indicates that cytosolic free $Ca^{2+}$ is well regulated, although the recovery times in the second range appear clearly slower as known from specialized animal cells (Felle 1991). The maintenance of the steep electrochemical $Ca^{2+}$ gradient across the plasma membrane requires the $Ca^{2+}$ ATPase to be connected to the electrochemical $H^+$ gradient; hence a $nH^+/Ca^{2+}$-ATPase is discussed (Rasi-Caldogno et al. 1987, 1989a, b; Miller and Sanders 1989; Felle 1991; Felle et al. 1992). Increase in cytosolic free $Ca^{2+}$ may arise from extracellular $Ca^{2+}$ through $Ca^{2+}$ channels or from intracellular stores. $Ca^{2+}$ channels in plants appear not as specific as those reported from animal cells and usually display quite substantial $K^+$ conductances (Tester 1990 and refs. therein). The activation of $Ca^{2+}$ from internal stores has been proposed for the ER, mitochondria and vacuoles (see above). Two $Ca^{2+}$-conducting channels in the tonoplast membrane of red beet were described by Brosnan and Sanders (1990), and by Johannes et al. (1991).

Although a wide range of physiological processes in plants have been found to require $Ca^{2+}$ (Hanson 1984; Hepler and Wayne 1985; Gilroy and Trewavas 1990 and refs. therein), its uniquivocal role as a typical second messenger is still poorly documented. Miller and Sanders (1987) have shown that light depletes cytosolic $Ca^{2+}$ in *Nitellopsis* and may affect sucrose synthesis. Investigation of the charophyte action potential has a long history. The importance of $Ca^{2+}$ for the action potential has been demonstrated repeatedly, but so far its role remains speculative (Beilby 1984). Willamson and Ashley (1982) relate the rapid $Ca^{2+}$ transients in *Chara* to cytoplasmic streaming. Shiina and Tazawa (1987) show that increased cytoplasmic $Ca^{2+}$ opens Cl-channels. An interesting physiological view comes from work on *Lamprothamnium*, where $Ca^{2+}$ increase interferes with turgor regulation (Okazaki and Tazawa 1990). In tip-growing cells it could be demonstrated that perturbations of the steady-state $Ca^{2+}$ gradient (Brownlee and Wood 1986) by $Ca^{2+}$ channel antagonists lead to inhibition of growth (Nobiling and Reiss 1987). Felle (1988), and Gehring et al. (1990) demonstrate $Ca^{2+}$ increase in maize coleoptiles as a response to auxin; Tretyn et al. (1991) report a rapid and massive decrease in cytosolic free $Ca^{2+}$ in *Sinapis* root hairs in the presence of submicromolar auxin concentrations. Since this $Ca^{2+}$ change occurs after a lag phase of about 30 s, followed by a decrease in $H^+$ pump activity, its involvement in a signal chain as a second messenger appears obvious. Convincing evidence also comes from work on guard cells. Gilroy et al. (1990a), McAinsh et al. (1990), and Schroeder and Hagiwara (1990) report transient $Ca^{2+}$ increase in response to externally added abscisic acid (ABA). On the other hand, Wang et al. (1991) show a decrease in response to ABA. Using caged $IP_3$, Gilroy et al. (1990b) demonstrate that upon illumination with UV-light, which breaks the "cage" and sets $IP_3$ free, the released $Ca^{2+}$ causes stomatal opening. Bush et al. (1989) and Bush (1992) present evidence that in barley aleurone protoplasts gibberellic acid ($GA_3$) induces a two- to threefold rise in free cytosolic and ER [$Ca^{2+}$]. These increases seemingly depend on extracellular [$Ca^{2+}$] (Bush and Jones 1987), are not transient, but represent a steady-state rise which is maintained as long as GA is present. For the same cells, Gilroy and Jones (1992) discuss the regulation of $Ca^{2+}$ buffering in the cytosol and the ER by ABA and $GA_3$, and show that 5 μM $GA_3$ induces an increase in cytosolic $Ca^{2+}$ by a factor of about 4, whereas ABA reduces this again (cf. the reports of Wang et al. 1991). For the green alga *Eremosphera viridis*, Förster et al. (1989) report oscillating opening of calcium-dependent potassium channels, which appears also connected with cytosolic pH changes (Thaler et al. 1989). In *Fucus*, $Ca^{2+}$ has been traditionally regarded as a cytosolic signal during egg development (Jaffe 1983). Recently, Speksnider et al. (1989) demonstrated that injected BAPTA buffers block fucoid egg development by facilitating calcium diffusion. Brownlee and Taylor (1992) show that calcium influx increased significantly (but slowly) following the addition of sperm to unfertilized eggs.

### c) Fatty Acid-Derived Second Messengers

A review on membrane-derived fatty acids as precursors to second messengers is given by Anderson (1989). Diacylglycerol (DAG) is the other product formed by phospholipase C hydrolysis of phospholipids. Several reports indicate that DAG or DAG analogs, in concert with phosphatidylserine or phorbol esters, will promote ATP-dependent protein phosphorylation (Morré et al. 1984b; Schäfer et al. 1985; Elliot and Skinner 1986; Olah and Kiss 1986; Favre and Turian 1987). Lee and Assmann (1991), applying the patch-clamp technique, report that diacylglycerols induce ATP-dependent proton pumping in guard-cell protoplasts. The green alga *Dunaliella salina*, when subjected to osmotic shock, responds with an immediate 50% increase in cell volume followed directly by a substantial loss of its plasma membrane localized $PIP_2$ (Einspahr et al. 1988, 1989; Einspahr and Thompson 1990) and increase in DAG, whereas the DAG content of the microsomal fraction was unchanged (Ha and Thompson 1991). DAG-kinase has been solubilized and purified from membrane fractions of *Catharanthus roseus* (Wissing et al. 1989; Biffen and Hanke 1990; Wissing and Wagner 1992), and tobacco suspension culture cells (Kamada and Muto 1991). In testing the $Ca^{2+}$-regulation of the PI-turnover, they found that optimal [$Ca^{2+}$] exeeded 10 μM, and they rejected a cytoplasmic regulation. Wood et al. (1990) give some evidence that IAA, together with phorbolesters, can replace cytokinin as a promoter of cell proliferation. Many other lipids or lipid-derived products generated by phospholipases acting on phospholipids in membranes are also implicated as mediators and second messengers in signal transduction. Apart from the phospholipase C (Melin et al. 1987; Pfaffmann et al. 1987; Tate et al. 1989), which is known to split $PIP_2$ into $IP_3$ and DAG, in plants the activation of phospholipase $A_2$ ($PLA_2$) is discussed. Recently, Scherer et al. (1990) and André and Scherer (1991) reported stimulation of phospholipase $A_2$ by auxins. From animal systems we know that arachidonic acid and a lysophospholipid are the two products of $PLA_2$; however, in the few reports on plant cells, the production of lysophospholipids only was found so far. When this lipid is an alkylether-containing phosphatidylcholine (PC), the lysoPC forms platelet-activation factor (PAF), a lipid which is a potent cellular mediator (Dennis et al. 1990). Indeed, as Scherer and coworkers could demonstrate, PAF stimulates protein kinase and $H^+$ ATPases in cucumber (Scherer et al. 1988), soybean cells (Scherer and Nickel 1988), and microsomes of zuccini (Martiny-Baron and Scherer 1988). Brightman et al. (1991) show activation of plasma membrane NADH oxidase activity by lysophospholipids.

A candidate for long-distance signaling was recently reported to be an 18-amino acid polypeptide (systemin; Pearce et al. 1991). The recent finding that methyl-jasmonate and jasmonic acid are also powerful inducers of proteinase inhibitor synthesis (Farmer and Ryan 1990, 1992; Farmer et al. 1992) raised the possibility that jasmonates could be intracellular signal intermediates in the induction of proteinase inhibitor gene expression by wounding or through the action of systemin. Several intermediates in the pathway from linolenic acid to jasmonic acid were all found to be powerful inducers of proteinase inhibitor synthesis. It seems possible, therefore, that octadecanoids, as well as jasmonate, may be involved in complex signaling networks that regulate defensive or developmental genes in plants.

### d) Sphingolipids

Sphingolipid breakdown products, sphingosine and lysosphingolipids, may serve as surface receptors, as modifiers of the properties of cell receptors for growth factors, and as activators and inhibitors of protein kinases, ion transporters, and other proteins, and as such have the potential to regulate cell behavior at many levels of signal transduction (Hannun and Bell 1989; Merrill 1991). In spite of this wide range of potential activity in cellular regulation, in plant cells sphingolipids have not gained much attention yet. Recently, however, Bille et al. (1992) reported the effect of sphingosine on the pyrophosphate-driven, but not on the ATP-dependent $H^+$ transport into tonoplast vesicles, and by means of patch-clamp experiments, on the corresponding clamp current into isolated vacuoles of *Chenopodium rubrum*.

### e) pH

As cytosolic $Ca^{2+}$, protons fulfill the basic requirements of being well regulated at a low cytosolic concentration. Still, the role of $H^+$ as second messenger, or better as cellular messenger, is less widely accepted, but it is clearly a highly important part of any signal chain: besides being both the substrate and product of metabolic pathways, protons have the apparent potential to communicate information about the cellular energy balance to enzymes and structures that may share no common effector. Since small changes in cytosolic pH may cause large effects on the activity of enzymes, the potential impacts on intracellular processes are manifold and, in principle, need not be mediated by a specific receptor (Felle 1989b and refs. therein).

### f) Electrical Fields and Waves

Are changes in membrane potential secondary signals? Probably not in the classical sense, because the transferred information is very general. Membrane potential changes occur wherever the steady-state situation across a given membrane is disturbed. Such a change may be only local and cause ion flow, e.g., through opening of a voltage-gated ion channel, or the electrical signal may propagate, e.g., as action potential. We also know that wounding elicits electrical waves which will travel through the whole plant, i.e., from the root to the leaves or vice versa (Roblin and Bonnemain 1985; Frachisse and Desbiez 1989; Julien et al. 1991). This electrical wave, the nature of which is not understood yet, will cause metabolic and possible developmental changes in target organs. Apparently, electrical signals may also play a role in gravitropism. As Behrens et al. (1983, 1985) could show, dislocating roots of *Lepidium* causes differential changes in membrane potential, but also changes the extracellular electrical field. Whereas it appears questionable whether the observed membrane potential changes travel along the root to trigger a physiological process in the growing zone, the changes in electrical field could well be sufficient to cause instance channel opening or dislocation of charged molecules (growth substances) in that zone and thus induce the bending process.

## 5. The Transduction of the Signal: Protein Kinases

Numerous calcium-dependent protein kinases can be identified (Blowers and Trewavas 1989 and refs. therein; Guo and Roux 1990; Harper et al. 1991). However, for the primary stimulus and second messenger production to be meaningful, they must be directly related to a cellular response.

Protein kinase C (PKC) is a phospholipid- and $Ca^{2+}$-dependent protein kinase, which can be activated by DAG (Nishisuka 1984). This activation is a consequence of a DAG-dependent increase in PKC's affinity for $Ca^{2+}$ and thereby a reduction in the amount of $Ca^{2+}$ needed for activity. In animal systems, phorbol esters have been used as extracellular activators of PKC, which has been shown to be required for transduction of a large number of chemical messengers, including growth factors, hormones, and neurotransmitters. The response to most of these messengers has been linked to phospholipase C-dependent turnover of phosphoinositides (see above).

The first evidence for a $Ca^{2+}$- and DAG-stimulated protein kinase in plants came from Morré et al. (1984a, b), and Schäfer et al. (1985) partially purified such an enzyme from zuccini. Also, a putative PKC has been isolated from wheat by Olah and Kiss (1986). A similar protein kinase was isolated from *Amaranthus* by Elliott and Skinner (1986); however, in none of the cases did phorbol esters cause the typical stimulation; and, although Wood et al. (1990) report effects of phorbol esters on plant cell proliferation in connection with auxin, a possible role of PKC as well as for tyrosine kinase in the control of plant cell development remains open.

An interesting alternative comes from experiments with platelet-activating factor (PAF; Martiny-Baron and Scherer 1989; Scherer 1989 and refs. therein). PAF, a molecule similar to oleylacetyldiglyceride, and a known artificial stimulator of protein kinase C in animal cells (Nishizuka 1984) stimulates ATP-dependent proton transport on both plasmalemma and tonoplast in zuccini microsomes. Whereas in animal cells only picomolar concentrations of PAF are sufficient to elicit a physiological reaction, in plant vesicles micromolar concentrations are required. It was suggested that PAF could replace a natural activator for a plant $Ca^{2+}$/phospholipid-stimulated protein kinase which then would be involved in the regulation of both $H^+$ ATPases.

Blowers and Trewavas (1987, 1989) describe the attractive model of protein kinase autophosphorylation. Since protein kinases are proteins, they are potential substrates for their own enzyme activity. Whereas this is common in animal systems, for plant cells only few cases have been reported so far: Wong et al. (1986) found that purified preparations of phytochrome contain protein kinase activity leading to phosphorylation of phytochrome. Moreover, there are the redox-state-dependent thylacoid protein kinase of Bennet (1983), and Coughlan and Hind (1986), the plasma membrane calcium- and Cam-dependent protein kinase of Blowers et al. (1985), and the soluble calcium-dependent protein kinase of Harmon et al. (1987). Three effects of autophosphorylation have been noted: autophosphorylation can (1) modify the catalytic activity to make the PK independent of associated $Ca^{2+}$, (2) increase the sensitivity of the PK to $Ca^{2+}$, (3) accelerate catalytic activity (positive feedback). As such, these systems may act as metabolic switches.

## 6. The Target(s) and the Cellular Response

The response to auxins comprises as physiological response the elongation of the coleoptile (cells), whereas one of the targets of the activated signal chain(s) is the stimulation of the plasma membrane $H^+$ ATPase (proton extrusion, hyperpolarization). Of course, auxin may involve other targets: recently, Marten et al. (1991) demonstrated the auxin-induced activation of a $Cl^-$-channel in the plasma membrane of guard cells, and since cytosolic $Ca^{2+}$ increases in response to auxin (Felle 1988), $Ca^{2+}$ channels may also be a direct target. Hager et al. (1991) report rapid exocytosis in *Zea mays*, and suggest increase in pump density as primary action of IAA-dependent acidification. In general, membrane (transport) proteins, receptor proteins, cytoskeletal proteins, enzymes, and gene activation are potential targets of activation/deactivation through phosphorylation by kinases.

In plant cells, the best-investigated target is the plasma membrane $H^+$ ATPase. This for very obvious reasons: (a) this transport enzyme is functionally available at all experimental levels and therefore relatively easy to handle; (b) since this ion pump is the central energy mediator, many changes in metabolism and growth (development) are expressed, at least in part, through an activity change of this ATPase.

On the other hand, gene activation has also been subject to intensive investigation. Many of the effects of growth substances on plant tissues are very complex, so in many cases it has been difficult to identify the involvement of individual genes in these processes. However, in a few cases, it has been possible to identify gene products, to identify their role in the hormonal response and to study their control: auxins (Theologis 1986; Key 1989; Guilfoyle et al. 1992); ABA (Quatrano 1987; Mundy et al. 1990; Shriver and Mundy 1990); ethylene (Lincoln et al. 1987; Holdsworth and Laties 1989; Hamilton et al. 1992; White et al. 1992); gibberellic acid (Jones and Jacobsen 1991; Rogers et al. 1992).

## 7. Outlook

The auxin signal apparently has drawn the most attendance, and researchers have tried to focus on different aspects of auxin action. Therefore, Fig. 1 serves as a still rather hypo-

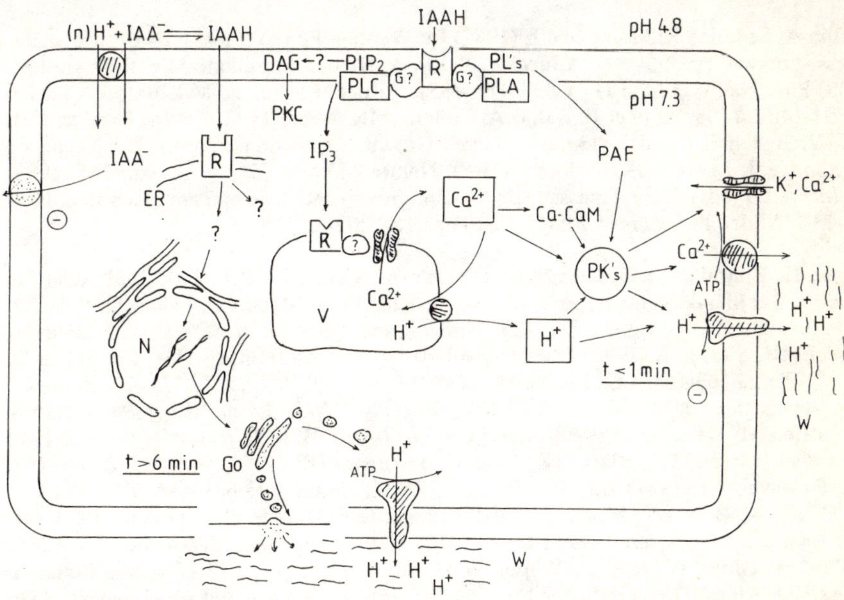


**Fig. 1.** Integrated model of auxin action. Included are hard scientific findings as cited in the text, together with ideas frequently discussed in the literature. *Cam* Calmodulin; *DAG* diacylglycerol; *ER* endoplasmc reticulum; *G* GTP-binding protein; *Go* Golgi apparatus; *$IP_3$* inositol-1,4,5-trisphosphate; *N* nucleus; *PAF* platelet-activating factor; *$PIP_2$* phosphatidylinositol-4,4-bisphosphate; *PK's* proteinkinases; *PL's* phospholipids (e.g., phosphatidylcholine); *PLA, PLC* phospholipases A and C, respectively; *R* receptor; *V* vacuole; *W* wall

thetical model of primary auxin action. This scheme gives a vague impression of the minimum complexity of signal transduction in plant cells. Many of the findings cited in the text have been included, however, for simplicity, by far not all. It also demonstrates the interesting case of a forking signal chain at the end of which stands the stimulation of one target, the plasma membrane $H^+$ ATPase: as Barbier-Brygoo et al. (1990), and Rück et al. (1993) demonstrate for protoplasts in tobacco and maize, respectively, that the auxin-induced stimulation of this ATPase occurs within less than 1 min after external auxin application, but it takes 6 to 10 min in grass coleoptiles (Cleland et al. 1977; Felle et al. 1986, 1991; Hager et al. 1991), as does the onset of elongation growth. Since auxin receptors at the plasma membrane but also at the ER are presently discussed, it appears possible that their activation leads to separate responses which are integrated to a final response; but, be it gravity, light, elicitors, growth substances, or other stimuli, from the suspected signal chain, we know only fragments. And, regardless how different the stimuli and their activated cascades may be, in many cases growth is the cellular response. This means, regardless of how many different stimuli the plant cell has to deal with, and regardless of how many of the different pathways the respective signals may travel along at first, at the end they all may funnel into one or two end chains, as long as growth is the final response.

## References

**Aducci P,** Ballio A, Federico R, Montesano L (1982) In: Wareing PF (ed) Plant growth substances. Academic Press, London, pp 395–404. **Aducci P,** Ballio A, Blein J-P, Fullone MR, Rossignol M, Scalla R (1988) Proc Natl Acad Sci USA 85:7849–7851. **Aducci P,** Fullone MR, Ballio A (1989) Plant Physiol 91:1402–1406. **Aducci P,** Ballio A, Fullone MR, Marra M (1992) In: Karssen CM, van Loon LC, Vreugdenhil D (eds) Progress in Plant Growth Regulation. Kluwer, Dordrecht, pp 686–690. **Alexandre J,** Lasalles JP, Kado RT (1990) Nature 343:567–570. **Anderson JM** (1989) In: Boss WF, Morré DJ (eds) Second messengers in plant growth and development. Alan Liss, New York, pp 181–212. **André B,** Scherer GFE (1991) Planta 185:209–214.

**Barbier-Brygoo H,** Ephritikhine G, Klämbt D, Ghislain M, Guern J (1990) Proc Natl Acad Sci USA 86:891–896. **Barbier-Brygoo H,** Ephritikhine G, Klämbt D, Maurel C, Palme K, Schell J, Guern J (1991) Plant J 1:83–93. **Bennet J** (1983) Philos Trans R Soc Lond 302:113–125. **Behrens HM,** Weisenseel MH, Sievers A (1983) Plant Physiol 70:1079–1083. **Behrens HM,** Gradmann D, Sievers A (1985) Planta 163:463–472. **Beilby MJ** (1984) Plant Cell Environ 7:415–421. **Bennet J** (1983) Philos Trans R Soc Lond 302:113–125. **Berridge MJ,** Irvine RF (1984) Nature 312:315–321. **Biffen M,** Hanke DE (1990) Plant Sci 69:147–155. **Bille J,** Weiser T, Bentrup F-W (1992) Physiol Plant 84:250–254. **Blatt MR,** Thiel G, Trentham DR (1990) Nature 346:766–769. **Blowers DP,** Trewavas AJ (1987) Biochem Biophys Res Commun 143:691–696. **Blowers DP,** Trewavas AJ (1989) In: Boss WF, Morré DJ (eds) Second messengers in plant growth and development. Alan Liss, New York, pp 1–28. **Blowers DP,** Hetherington AM, Trewavas AJ (1985) Planta 166:208–215. **Blum W,** Key G, Weiler W (1988) Physiol Plant 72:279–287. **Boller T** (1989) In: Boss WF, Morré DJ (eds) Second messengers in plant growth and development. Alan Liss, New York, pp 227–255. Boss WF (1989) In: Boss WF, Morré DJ (eds) Second messengers in plant growth and development. Alan Liss, New York, pp 29–56. **Brightman AO,** Zhu XZ, Morré DJ (1991) Plant Physiol 96:1314–1320. **Brosnan JM,** Sanders D (1990) FEBS Lett 260:70–72. **Brownlee C,** Taylor AR (1992) In: Karssen CM, Van Loon LC, Vreugdenhil D (eds) Progress in plant growth regulation. Kluwer, Dordrecht, pp 651–662. **Brownlee C,** Wood JW (1986) Nature 320:624–628. **Bruce RJ,** West CA (1982) Plant Physiol 69:1181–1188. **Bush DS** (1992) In:

Karssen CM, Van Loon LC, Vreugdenhil D (eds) Progress in plant growth regulation. Kluwer, Dordrecht, pp 96–104. **Bush DS,** Jones RL (1987) Eur J Cell Biol 46:466–469. **Bush DS,** Biswas AK, Jones RL (1989) Planta 178:411–420.

**Cleland RE,** Prins HBA, Harper JR, Higinbotham N (1977) Plant Physiol 59:395–397. **Clint GM,** Blatt M (1989) Planta 178:495–508. **Coughlan SJ,** Hind G (1986) J Biol Chem 261:11378–11385. **Cross JW** (1985) Plant Cell Environ 8:351–359. **Cross JW** (1991) New Biol 3:813–819.

**Davis KR,** Lyon GD, Darvill AG, Albersheim P (1984) Plant Physiol 74:52–60. **De Boer AH,** Watson BA, Cleland RE (1989) Plant Physiol 89:250–259. **De Michelis IM,** Pugliarello MC, Rasi-Caldogno F (1989) Plant Physiol 90:133–139. **Dennis EA,** Rhee SG, Billah MM, Hannun YA (1990) FASEB J 5:2068–2077. **Dillenschneider A,** Hetherington A, Graziana G, Alibert P, Berta J, Ranjeva R (1986) Fed Eur Biochem Soc 208:413–417. **Dohrmann U,** Hertel R, Kowalik H (1978) Planta 140:97–106. **Droeback BK,** Ferguson IB (1985) Biochem Biophys Res Commun 130:1241–1246.

**Einspahr KJ,** Peeler TC, Thompson GA Jr (1988) J Biol Chem 263:5775–5779. **Einspahr KJ,** Peeler TC, Thompson GA Jr (1989) Plant Physiol 90:1115–1120. **Einspahr KJ,** Thompson GA (1990) Plant Physiol 93:361–366. **Elliott DC,** Skinner JD (1986) Phytochemistry 25:39–44.

**Farmer EE,** Ryan CA (1990) Proc Natl Acad Sci USA 87:7713–7716. **Farmer EE,** Ryan CA (1992) Plant Cell 4:129–134. **Farmer EE,** Johnson RR, Ryan CA (1992) Plant Physiol 98:995–1002. **Favre B,** Turian G (1987) Plant Sci 49:15–21. **Felle H** (1988) Planta 174:495–499. **Felle H** (1989a) Plant Physiol 91:1239–1242. **Felle H** (1989b) In: Boss WF, Morré DJ (eds) Second messengers in plant growth and development. Alan Liss, New York, pp 145–166. **Felle H** (1991) Plant Sci 74:27–33. **Felle H,** Brummer B, Bertl A, Parish R (1986) Proc Natl Acad Sci USA 83:8992–8995. **Felle H,** Peters WS, Palme K (1991) Biochim Biophys Acta 1064:199–204. **Felle H,** Wagner G, Tretyn A (1992) Planta 188:306–313. **Feyerabend M,** Weiler EW (1988a) Planta 174:115–122. **Feyerabend M,** Weiler EW (1988b) Planta 178:282–290. **Förster B,** Thaler M, Urbach W, Simonis W (1989) FEBS Lett 256:159–162. **Frachisse J-M,** Desbiez M-O (1989) Biochem Physiol Pflanz 185:357–368.

**Gehring CA,** Williams D, Cody SH, Parish RW (1990) Nature 345:528–530. **Gilroy S,** Jones RL (1992) In: Karssen CM, Van Loon LC, Vreugdenhil D (eds) Progress in plant growth regulation. Kluwer, Dordrecht, pp 643–650. **Gilroy S,** Trewavas A (1990) In: Larsson C, Moeller IM (eds) The plant plasma membrane. Springer, Berlin Heidelberg New York, pp 203–232. **Gilroy S,** Read ND, Trewavas AJ (1990a) Nature 343:186–188. **Gilroy S,** Read ND, Trewavas AJ (1990b) Nature 346:769–771. **Guilfoyle TJ,** Hagen G, Li Y, Gee MA, Martin G, Ulmasov TN (1992) In: Karssen CM, Van Loon LC, Vreugdenhil D (eds) Progress in plant growth regulation. Kluwer, Dordrecht, pp 128–135. **Guo Y-L,** Roux SJ (1990) Plant Physiol 94:143–150.

**Ha KS,** Thompson GA Jr (1991) Plant Physiol 97:921–927. **Hadwiger LA,** Beckman LM (1980) Plant Physiol 66:205–211. **Hager A,** Debus G, Edel H-G, Stransky H, Serrano R (1991) Planta 185:527–537. **Hamilton A, Bouzayen M,** Grierson D (1992) In: Karssen CM, Van Loon LC, Vreugdenhil D (eds) Progress in plant growth regulation. Kluwer, Dordrecht, pp 156–159. **Hannun YA,** Bell RM (1989) Science 243:500–507. **Hanson JB** (1984) In: Tinker PB, Läuchli A (eds) Advances in plant nutrition. Praeger, Eastbourne, pp 149–208. **Harmon AC,** Putnam-Evans C, Courier MJ (1987) Plant Physiol 83:830–837. **Harper JF,** Sussman MR, Schaller GE, Putnam-Evans C, Charbonneau H, Harmon AC (1991) Science 252:951–954. **Hasunuma K,** Funadera K, Kubota M, Watanabe M (1987) Photochem Photobiol 46:531–535. **Hepler PK,** Wayne RO (1985) Annu Rev Plant Physiol 36:397–439. **Hesse T,** Feldwisch J, Balshüsemann D, Bauw G, Puype M, Vandekerckhove J, Löbler M, Klämbt D, Schell J, Palme K (1989) EMBO J 8:2453–2461. **Hokin M,** Hokin LE (1953) J Biol Chem 203:967–977. **Holdsworth MJ,** Laties GG (1989) Planta 179:17–23. **Hsieh W-L,** Pierce WS, Sze H (1991) Plant Physiol 97:1535–1544.

**Inohara N,** Shimomura S, Fukui T, Futai M (1989) Proc Natl Acad Sci USA 86:3564–3568.

**Jacobsen JV,** Gubler F (1992) In: Karssen CM, Van Loon LC, Vreugdenhil D (eds) Progress in plant growth regulation. Kluwer, Dordrecht, pp 116–127. **Jaffe LF** (1983) Dev Biol 99:265–276. **Johannes E,** Brosnan JM, Sanders D (1991) Bioessays 13:331–336. **Jones RL,** Jacobsen JV (1991) Int Rev Cytol 126:49–88. **Julien JL,** Desbies M-O, De Jaegher G, Frachisse L-M (1991) J Exp Bot 42:131–137.

**Kamada Y,** Muto S (1991) Biochim Biophys Acta 1093:72–79. **Key JL** (1989) Bioassays 11:52–58. **Kurosaki F,** Amin M, Nishi A (1986) Physiol Mol Plant Pathol 28:359–370. **Kurosaki F,** Tsurusawa Y, Nishi A (1987) Plant Physiol 85:601–604.

**Lee S-C,** West CA (1981) Plant Physiol 67:633–639. **Lee Y,** Assmann SM (1991) Proc Natl Acad Sci USA 88:2127–2131. **Lincoln JE,** Cordes S, Read E, Fisher RL (1987) Proc Natl Acad Sci USA 84:2793–2797. **Löbler M,** Klämbt D (1985) J Biol Chem 260:9854–9859

**Marmé D** (1989) In: Boss WF, Morré DJ (eds) Second messengers in plant growth and development. Alan Liss, New York, pp 57–80. **Marra M,** Ballio A, Fullone MR, Aducci P (1992) 98:1029–1034. **Marten I,** Lohse G, Hedrich R (1991) Nature 353:759–762. **Martiny-Baron G,** Scherer GFE (1988) Plant Cell Rep 7:579–582. **Martiny-Baron G,** Scherer GFE (1989) J Biol Chem 264:18052–18059. **McAinsh MR,** Brownlee C, Hetherington AM (1990) Nature 343:186–188. **Melin P-M,** Sommarin M, Sandelius AS, Jergil B (1987) FEBS Lett 2232:87–91. **Merrill AH** (1991) J Bioenerg Biomembr 23:83–104. **Meyer C,** Feyerabend M, Weiler EW (1989) Plant Physiol 89:692–699. **Miller AJ,** Sanders D (1987) Nature 326:397–400. **Miller AJ,** Sanders D (1989) Plant Physiol Biochem 27:551–556. **Millner PA** (1987) FEBS Lett 226:155–160. **Morré DJ,** Gripshover B, Monroe A, Morré JT (1984a) J Biol Chem 259:15364–15368. **Morré DJ,** Morré JT, Varnold RL (1984b) Plant Physiol 75:265–268. **Morse MJ,** Crain RC, Satter RL (1987) Plant Physiol 83:640–644. **Mundy J,** Yamaguchi-Shinogazi K, Chua N-H (1990) Proc Natl Acad Sci USA 87:1406–1410.

**Napier R,** Venis MA, Bolton MA, Richardson LI, Butcher GW (1988) Planta 176:519–526. **Nishizuka Y** (1984) Science 225:1365–1370. **Nobiling R,** Reiss H-D (1987) Protoplasma 139:20–24.

**Okazaki Y,** Tazawa M (1990) J Membr Biol 114:189–194. **Olah Z,** Kiss Z (1986) Fed Eur Biochem Soc Lett 195:33–37.

**Palme K,** Feldwisch J, Hesse T, Bauw G, Puype M, Vanderkerkehove J, Schell J (1990) In: Roberts JA, Kirk C, Venis MA (eds) Hormone perception and signal transduction in animals and plants. Symp Soc Exp Biol XLIV:299–313. **Palme K,** Feldwisch J, Peters W, Schell J, Zettl R, Campos N, Felle H (1992) In: Karssen CM, Van Loon LC, Vreugdenhil D (eds) Progress in plant growth regulation. Kluwer, pp 73–81. **Pearce G,** Strydon D, Johnson S, Ryan CA (1991) Science 253:895–898. **Pfaffmann H,** Hartmann E, Brightman AO, Morre DJ (1987) Plant Physiol 85:1151–1155.

**Quatrano RS** (1987) In: Davies PJ (ed) Plant hormones and their role in plant growth and development. Martinus Nijhoff, Dordrecht, pp 494–514.

**Ranjeva R,** Carreso A, Boudet AM (1988) FEBS Lett 230:137–141. **Rasi-Caldogno F,** Pugliarello MC, De Michelis MI (1987) Plant Physiol 83:994–1000. **Rasi-Caldogno F,** Pugliarello MC, De Michelis MI (1989a) In: Dainty J, De Michelis MI, Marré E, Rasi-Caldogno F (eds) Plant membrane transport, the current position. Elsevier, Amsterdam, pp 225–230. **Rasi-Caldogno F,** Pugliarello MC, De Michelis MI (1989b) Plant Physiol 90:1429–1434. **Rasi-Caldogno F,** Carnelli A, De Michelis MI (1992) Plant Physiol 98:1202–1206. **Rayle DL,** Nowbar S, Cleland RE (1991) Plant Physiol 97:449–452. **Reddy ASN,** Poovaiah BW (1987) J Biochem 101:569–573. **Roblin G,**

Bonnemain JL (1985) Plant Cell Physiol 26:1273–1283. **Rodriguez-Rosales M-P,** Herrin DL, Thompson GA Jr (1992) Plant Physiol 98:446–451. **Rogers JC,** Lanahan MB, Rogers SW, Mundy J (1992) In: Karssen CM, Van Loon LC, Vreugdenhil D (eds) Progress in plant growth regulation. Kluwer, Dordrecht, pp 136–146. **Rück A,** Palme K, Venis AM, Napier R, Felle HH (1993) Plant J (in press).

**Schäfer A,** Bygrave F, Matzenauer S, Marmé D (1985) FEBS Lett 186:25–28. **Scherer GFE** (1989) In: Boss WF, Morré DJ (eds) Second messengers in plant growth and development. Alan Liss, New York, pp 167–179. **Scherer GFE,** Nickel R (1988) Plant Cell Rep 7:575–578. **Scherer GFE,** Martiny-Baron G, Stoffel B (1988) Planta 175:241–253. **Scherer GFE,** André B, Martigny-Baron G (1990) In: Randall DD, Blevins D (eds) Current topics in plant biochemistry and physiology. University of Columbia-Missouri, pp 190–218. **Schroeder JI,** Hagiwara S (1990) Proc Natl Acad Sci USA 87:9305–9309. **Schumaker KS,** Sze H (1987) J Biol Chem 262:3944–3946. **Shiina T,** Tazawa M (1987) J Membr Biol 99:137–146. **Shimomura S,** Sotobayashi T, Futai M, Fukui T (1986) J Biochem 99:1513–1524. **Shriver K,** Mundy J (1990) Plant Cell 2:503–512. **Speksnider JA,** Miller AL, Weissenseel MH, Chen T-H, Jaffe LF (1989) Proc Natl Acad Sci USA 86:6607–6611. **Stout RG** (1988) Plant Physiol 88:999–1001.

**Tate BF,** Schaller GE, Sussman MR, Crain RC (1989) Plant Physiol 91:1275–1279. **Tester M** (1990) New Phytol 114:305–340. **Thaler M,** Steigner W, Förster B, Köhler K, Simonis W, Urbach W (1989) J Exp Bot 40:1195–1203. **Theologis A** (1986) Annu Rev Plant Physiol 37:407–438. **Tillmann U,** Viola G, Kayser B, Siemeister G, Hesse T, Palme K, Löbler M, Klämbt D (1989) EMBO J 8:2463–2467. **Tretyn A,** Wagner G, Felle HH (1991) J Physiol 139:187–193.

**Venis AM** (1987) In: Klämbt D (ed) Plant hormone receptors. Springer, Berlin Heidelberg New York, pp 27–39. **Venis AM,** Thomas EW, Barbier-Brygoo H, Ephritikhine G, Guern J (1990) Planta 182:232–235.

**Wang M,** Van Duijn B, Schram AW (1991) FEBS Lett 278:69–74. **White JA,** Olsen DC, Kende H (1992) In: Karssen CM, Van Loon LC, Vreugdenhil D (eds) Progress in plant growth regulation. Kluwer, Dordrecht, pp 147–155. **Williamson RE,** Ashley CC (1982) Nature 297:647–650. **Wissing JB,** Wagner KG (1992) Plant Physiol 98:1148–1153. **Wissing JB,** Heim S, Wagner KG (1989) Plant Physiol 90:1546–1551. **Wong Y-S,** Cheng H-C, Waksh DA, Lagarias JC (1986) J Biol Chem 261:12089–12097. **Wood HN,** Alves LM, Sterner R, Basile DV (1990) In Vitro Cell Dev Biol 26:1125–1127.

**Zbell B,** Walter C (1987) In: Klämbt D (ed) Plant hormone receptors. Springer, Berlin Heidelberg New York, pp 141–153. **Zbell B,** Schwendemann I, Bopp M (1989) J Plant Physiol 134:639–641. **Zbell B,** Paulik M, Morré DJ (1990) Protoplasma 154:74–79.

Professor Dr. Hubert Felle
Botanisches Institut
Justus-Liebig-Universität
Senckenbergstraße 17–21
W-6300 Gießen, Germany

# C. Genetics

## I. Replication

By Rudolf Eichenlaub

### 1. Introduction

Previous reviews of this series on prokaryotic DNA replication dealt only with the mechanism of vegetative replication which is a prerequisite for the genetic continuity of a cell and its extrachromosomal elements when division occurs. However, many bacterial plasmids carry genetic information allowing for another type of DNA replication which is employed during conjugal transfer of DNA molecules between bacterial cells. This plasmid property, conjugal plasmid transfer, which is governed by a mechanism distinctly different from general DNA replication, will be discussed in the present review, since it is of general importance for the understanding of plasmid biology.

Further information on conjugation and related subjects can be found in reviews by Manning and Achtmann (1979), Willetts and Skurray (1980), Willetts and Wilkins (1984), Clark (1985), and in more recent papers by Ippen-Ihler and Minkley (1986), Low (1987), Sanderson and MacLachlan (1987), and Willets and Skurray (1987).

### 2. Bacterial Conjugation

The discovery of bacterial conjugation is linked to the identification of the F factor confering sexuality to *Escherichia coli* (Lederberg and Tatum 1946). During conjugation, which requires a close cell to cell contact, DNA is transferred from the donor to the recipient cell. Studies of this highly specific process began with the F factor, although this important genetic trait was also found to exist in a number of other plasmids which confer antibiotic resistance and colicin production to a bacterial cell (reviewed in Willetts 1977a). Although the basic mechanisms which govern conjugation processes among plasmids seem to be quite similar, one has to anticipate distinct genetic differences. Despite the fact that conjugation is one of the main reasons for the genetic evolution of microorganisms, the process itself is still not understood in all detail. Most investigations have been performed on the F factor and closely related plasmids like R1 and R100, therefore most of facts presented in this chapter are based on studies of these systems.

The genes engaged in the conjugation of F, probably more than 25, are clustered in the transfer region (*tra*) which has a size of a approximately 33 kb and is organized in the form of huge operon transcribed into a 32 kb mRNA (Helmuth and Achtman 1975). The

Progress in Botany, Vol. 54

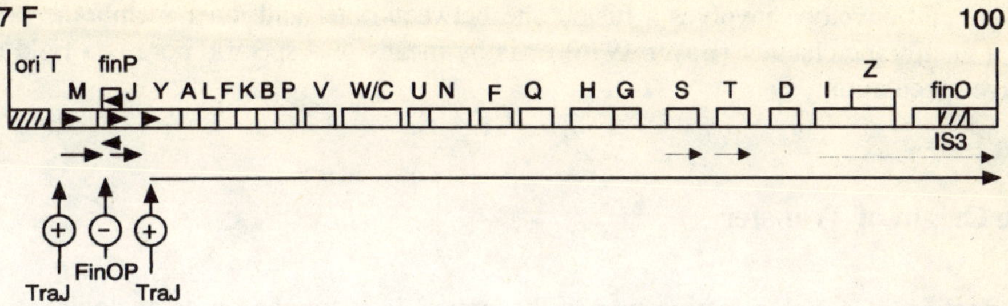


**Fig. 1.** Genetic map of the transfer region of plasmid F. *67F* and *100F* refer to F coordinates in kb, length of gene loci is not proportional to physical length. *Arrowheads in boxes* indicate promoters and their direction of transcription also the transcripts are shown by *arrows*. *Dashed arrows* indicate hypothetical transcripts for *tra*S, *tra*T, and *tra*I. (+) and (–) refer to positive or negative control exerted by TraJ or FinOP, respectively

structure of this region is shown in Fig. 1. Most of these genes have been identified by complementation studies with transfer-deficient point or deletion mutants and polar insertions in the *tra* region. However, some genes are only characterized by their gene products while the position of the gene is still unclear due to not available point mutations. Recently, genetic studies have been extended by a more detailed molecular analysis and the nucleotide sequence of a number of genes has been determined.

The process of conjugation can be divided into two successive stages (1) pilus-mediated intercellular contact between donor and recipient cell which leads to a stable mating pair formation and (2) the process of DNA transfer. An *E. coli* cell harboring the F-factor usually has only one to two F pili, thin threads of 1–2 μm in length, which protrude from the cell surface. They consist of a single protein, termed pilin, encoded by the gene *tra*A (Minkley et al. 1976; Date et al. 1977). In the pilus, the pilin molecules are arranged in helical order forming a hollow cylinder with a diameter of about 8 nm and a central hole of 2 nm (Folkhard et al. 1979). Pilin is synthesized as a preproprotein with a molecular weight of about 13 000–14 000. This preprotein seems to be processed by the *tra*Q gene product, possibly in two succesive steps involving also the removal of a 51 amino acid signal peptide to give rise to the mature pilin molecule, a polypetide of 7000 $M_r$ (Laine et al. 1985). Mature pilin is further modified by acetylation at the amino terminus and may be also phosphorylated as well as glycosylated (Frost et al. 1984). Mutations in *tra* genes B, C, E, F, G, H, K, L, U, V, and W prevent the production of F pili without effecting pilin synthesis (Moore et al. 1981; Laine et al. 1985); however, information on their function in pilus morphopoesis is lacking in most points.

The pilus mediates a specific donor-recipient surface interaction by a contact between the tip of the pilus and a receptor structure on the recipient cell. The initial unstable aggregation is stabilized by a direct cell to cell contact (Achtman et al. 1978). For stable aggregate formation, the *tra*N and *tra*G gene products of the donor and the major outer membrane protein *omp*A of the recipient are required, as indicated by the deficiency of mutants to form mating pairs (Manning et al. 1981; Manoil and Rosenbusch 1982); however, their function and their kind of interaction remain unclear. The disassembly of the pilus, possibly at its base in the donor cell, leads to docking of the recipient at the surface of the donor cell. Whether the cell-to-cell interaction required for transport of DNA across the cell envelope involves a fusion site between outer and inner membrane providing a membrane channel (Bayer 1976) or is by means of a specific pore is a field of extensive speculation.

## 3. The Origin of Transfer

The formation of a stable mating pair is the prerequisite for the onset of conjugative DNA transfer. It was shown already in early studies that the 5'-end of a unique linear DNA strand of F is tranferred while the complementary strand remains in the donor (Ohki and Tomizawa 1968). Such kind of DNA transfer requires nicking of the F DNA which was shown to occur at a specific site, the origin of transfer (*ori*T) (Thompson et al. 1984) (Fig. 2).

*Ori*T is located at the 5'-end of the *tra*-region adjacent to the promoter of *tra*M. The *ori*T region is highly conserved among plasmids related to F and features direct as well as inverted repeats and an AT-rich sequence indicating a locus for protein-DNA interactions. In fact, several mutations which reduce but do not abolish *ori*T function have been isolated, and interestingly map within the stem region of inverted repeats (Everett and Willetts 1982; Thompson et al. 1984). More recently, binding of the *tra*M gene product to *ori*T has been demonstrated in F-related plasmids R1 and R100 (Abo et al. 1991; Schwab et al. 1991).

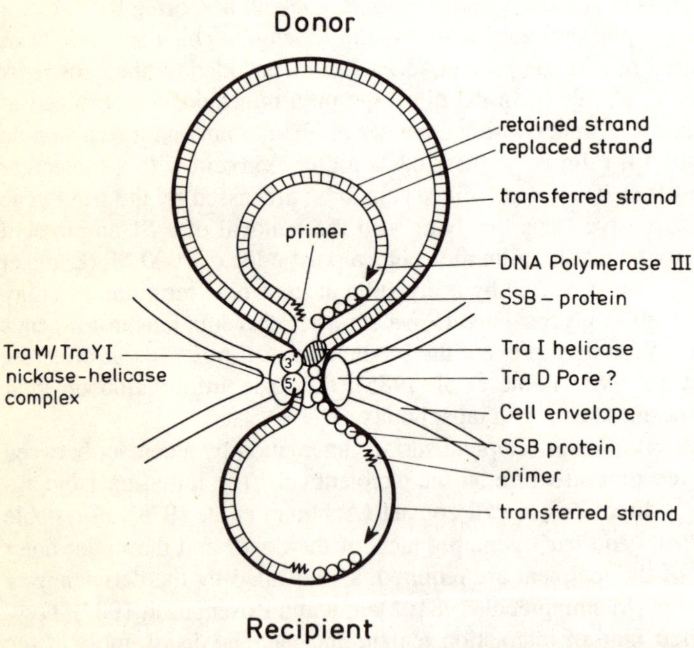


**Fig. 2.** Model for conjugative DNA transfer of F. (After Willetts and Skurray (1987). Details for DNA transfer and function of components involved are described in the text

It is no doubt that site- and strand-specific nicking of the DNA requires a specific enzyme. Three major nicking sites have been detected in the *ori*T region, but they may not represent the actual nicking site, since under experimental conditions nick translation may change the location of the position of the nick. Relating to the before-mentioned binding of the *tra*M gene product to the *ori* region, this binding site is distinctly separated from the nicking site (Abo et al. 1991; Schwab et al. 1991).

There is even less information on the gene product which represents the nickase. A final answer to these questions can only be given by an in vitro nicking system with purified components which awaits to be established.

An in vivo nicking assay with a lambda-F *ori*T hybrid phage in cells carrying various point and deletion transfer-deficient mutants suggested that the products of genes, *tra*Z and *tra*Y, constitute the endonuclease required for nicking (Everett and Willetts 1980). It is quite unfortunate that no point mutations of *tra*Z and *tra*Y exist which would allow analysis of their phenotype.

The gene product of *tra*Y has the characteristic properties of a membrane protein (Fowler et al. 1983). The origin of protein TraZ is not clear and the gene locus *tra*Z is undefined. There is some evidence from expression studies in *E. coli* minicells that the *tra*Z protein may be a product of reinitiation of translation within the *tra*I mRNA (cited in Willetts and Skurray 1987). Therefore, one may speculate that instead of *tra*Z, the *tra*I gene product may interact with TraY in the nicking of *ori*T or consecutive reaction steps. Since an unwinding of the DNA strands and the product of *tra*I has been shown to be DNA helicase I (Abdel-Monem et al. 1983), this seems indeed a very attractive idea. 70–80 molecules of DNA helicase I are bound processively to a single-stranded DNA gap of about 200 nucleotides and move along the DNA in 5' to 3' direction by unwinding about 1000 bp per second. This is in good agreement with the observed transfer rate of about 750 bases per second. The energy for unwinding is provided by hydrolysis of ATP (Abdel-Monem and Hoffmann-Berling 1976). Calculations show that several molecules of ATP per bp unwound per helicase monomer are hydrolyzed. It is highly unlikely that unwinding alone consumes so much energy, therefore, some part may be used for the translocation of the helicase along the DNA molecule and another part may drive the DNA transfer process itself. To accomplish directed transfer of the displaced DNA strand into the recipient, TraI has to be immobilized at the cell fusion site. This membrane anchoring is possibly mediated by the *tra*Y or *tra*D gene products (Panicker and Minkley 1985). TraD is a highly hydrophobic inner membrane protein which can also bind to single-stranded and double-stranded DNA cellulose (cited in Ippen-Ihler and Minkley 1986). Due to these two properties, it is feasible that TraD may form a pore for DNA export in the donor cell.

## 4. Transfer Replication

The desciption of events during early stages of DNA transfer which are followed by transfer replication are summarized in Fig. 2. This model, as proposed by Willetts and Skurray (1987), suggests a rolling circle type replication where the 3'- and 5'-end of the nick are kept in close contact at the fusion site. As a consequence, synthesis of the replacement strand in the donor is according to the leading strand mechanism requiring a single priming event, while synthesis of the complementary strand in the recipient follows the type of lagging strand synthesis with repeated priming. In contrast to vegetative replication of the F factor, which is bidirectional (Eichenlaub et al. 1977), transfer replication is unidirectional. While DNA helicase I unwinds in the 5' → 3' direction bound to the transferred strand, it is possible that *rep* protein (helicase II) is also engaged in this process which would then proceed into the 3' → 5' direction on the retained nontransferred strand (Yarranton and Gefter 1979). Experimental data which clearly demonstrate that *rep* helicase is required for DNA transfer are lacking, as is the requirement for gyrase (Bouck and Adelberg 1970). Gyrase could facilitate unwinding of the DNA strands by providing negative superhelicity in the unreplicated portion of the DNA molecule and furthermore prevent topological constraints by positive superhelicity. It is also possible that the nicked site at *ori*T when stabilized by proteins has the inherent property of a "swivel" removing positive superhelical turns in the DNA.

Strand separation is accompanied by binding of single-strand-binding protein (SSB) to stabilize ssDNA regions and also to protect the DNA during transfer. Besides the possibility that *E. coli* may provide SSB protein the F factor possesses an *ssb* gene of its own (Kolodkin et al. 1983; Golub and Low 1985), which maps ouside the *tra* region and is transferred early during conjugation.

It is quite clear that DNA polymerase III the product of the *E. coli* gene *dna*E is responsible for replacement synthesis in the donor and for complementary strand synthesis in the recipient (Wilkins and Hollom 1974; Kingsman and Willetts 1978). However, as is the case in many replication systems, there is no final answer to the question of which enzyme is responsible for RNA primer synthesis.

It is not consistent with the model for transfer replication as presented in Fig. 2 that the 3'-end at the nick is used for the start of replacement synthesis, an option which rolling circle type replication generally provides. Instead, an RNA primer molecule has to be synthesized de novo either by RNA polymerase, by *dna*G primase, or the primosome complex. Inhibition of conjugative DNA synthesis in the donor by rifampicin suggests the requirement of RNA polymerase for priming of replacement strand synthesis. A potential promoter allowing transcription into the right direction is located in the *ori*T region. The priming of complementary strand synthesis in the recipient may either be by RNA polymerase or *dna*G, since neither rifampicin nor a defect in *dna*B alone, the latter rendering the primosome inactive, leads to a complete inhibition of DNA synthesis (Willetts and Skurray 1987).

In some systems other than F, a plasmid encoded primase has been identified which is engaged in primer synthesis in the recipient (Lanka and Barth 1981).

The final step in transfer replication is the recircularization of the transferred DNA molecule. The possibility that this occurs by recombination between homologous sequences at the ends if more than one unit length is synthesized can be ruled out, since circularization is a *rec*A-independent process (Clark and Margulies 1965) and only precisely one unit length molecule is transferred. It is possible that the membrane-associated proteins which are responsible for the initial nick in the *ori*T region can also ligate the 3'- and 5'-ends.

Such a process closely resembling the one postulated above is found in bacteriophage θX174 replication and in the rolling circle replication of certain plasmids (see Eichenlaub 1991). Here the Rep protein, which has a nicking and a nicking-closing activity, catalyzes circularization of the displaced (+) strand (Koepsel et al. 1985, 1986). Rep protein is covalently bound to the 5'-end of the nick in the (+) strand (Thomas et al. 1988) and when displacement synthesis reaches the origin, the protein will recognize a termination sequence and introduce a second nick into the (+) strand and join the ends to form a circular DNA molecule (Gros et al. 1987; Iordanescu and Projan 1988).

## 5. Regulation of DNA Transfer

In a mixture of donor and recipient cells conjugational transfer of F can be as high as 100% while other plasmids closely related to F like R1 and R100 transfer with a 1000 times lower frequency. This indicated that the *tra* genes are generally not expressed constitutively but are usually repressed. This may represent a selective advantage by rendering the cells resistant to pilus specific bacteriophages but will still allow spreading of the plasmid within a bacterial population at a biologically sufficient rate.

Three transcription products originate in the *tra* operon. Single transcription is observed for genes *tra*M and *tra*J, the third transcript covers the genes *tra*Y-Z (see Fig. 1). Promoters in front of these genes have been confirmed by RNA polymerase binding studies which also suggested the possibility for further promoters for *tra*S, *tra*T and within the *tra*F-H region (Manning et al. 1984). Also *tra*I can be transcribed from a weak internal promoter, as shown by a *tra*I-*gal*K fusion (Mullineaux and Willetts 1985).

The observation that an amber mutant in *traJ* has a pleiotropic effect by preventing expression of the *tra*Y-Z genes suggested a role of *traJ* as a transcription activator for *tra* operon expression (Willetts 1977b). *TraJ* is expressed from a strong promoter resulting in relatively high concentrations of this protein in the cell. Furthermore, TraJ is associated with the outher membrane (Kennedy et al. 1977; Mullineaux and Willetts 1985), although the protein is lacking a signal sequence (Cuozzo et al. 1984). However, these unusual properties for a regulatory protein are only observed when *traJ* is cloned on a high copy number vector. In a singly-copy situation like in a $F^+$ cell, the concentration of TraJ protein is only about 2000 molecules and it is present in the cytoplasm (Silverman 1986).

As mode of action of TraJ in activation of *tra* gene expression one may expect an interaction as for cAMP-CRP complex or for a sigma-factor with core RNA polymerase. It has also been suggested that TraJ may function in antitermination of *tra* operon transcription (Ipper-Ihler and Minkley 1986). This alternative seems quite persuasive since the *tra*Y-Z transcript is extremely long, possesses several interal promoters within the operon and a termination signal between *tra*T and *tra*D which all might contribute to premature arrest of transcription.

The 5'end of the *traJ* mRNA has a 105 base noncoding leader sequence (Mullineaux and Willetts 1985). Located in the amino-terminal region of the *traJ* coding region is the promoter for *fin*P from which a 80 base transcript is synthesized into the anti-*traJ* direction (Fee and Dempsey 1986). Thus the transcript is complementary to the nontranscribed leader sequence of the *traJ* mRNA resembling an antisense-RNA which might prevent transcription or translation by interacting with *traJ* mRNA (Eguchi et al. 1991).

The study of F-related R factors has identified another element involved in regulation, the product if *fin*O. The gene *fin*O located at the distal end of the *tra* operon which encodes a 22 kDa protein has been shown to be essential for repression of *tra*J (Finnegan and Willetts 1972, 1973). In the F factor, which is naturally derepressed in conjugational transfer, *fin*O is inactivated by the integration of an *IS*3 element (Chea and Skurray 1986; Yoshioka et al. 1987). Complementation by *fin*O supplied in trans, for example from plasmid R100, is possible.

There is some evidence that FinO may act by stabilizing FinP RNA (Frost et al. 1989) but it is also conceivable that FinO is directly engaged in duplex formation between the FinP RNA and the 5' untranslated portion of *tra*J RNA. Such a mechanism, a protein required for the interaction of an antisense RNA with its target RNA, is known in the control of initiation of replication in plasmid ColE1 where the Rom protein catalyzes the interaction of RNAI and RNAII (Tomizawa 1987).

In repressed systems the regulation of DNA transfer follows a two stage process. Early during plasmid transfer *tra*J is expressed in the recipient converting this cell into an efficient donor for consecutive transfer events. Thus rapid spreading of a transmissible plasmid in a bacterial population is possible since it takes several hours until the typically low transfer rate is established in the cell (Willetts 1977c). The latter repression occurs by the combined action in FinOP products which prevent production of TraJ and thus shut off expression of all other *tra* genes.

## 6. Summary

The transfer of DNA between bacteria by conjugation requires direct cell contact and DNA replication. To accomplish this, transmissible plasmids carry *tra* genes which encode cell-surface components as well as proteins which promote DNA transfer and a specific site, *ori*T, the origin of transfer. The analysis of the large number of genes clustered in the *tra* region of F which all effect in some way the transfer efficiency presents an extensive problem. Only for some genes, like *tra*A and *tra*I, the function of the gene product is clear. Most of the other genes are defined only by their phenotype, suggesting involvement in pilus formation, stabilization of mating pairs, and conjugal DNA metabolism. A more detailed molecular analysis will certainly fill in many existing gaps but it is doubtful whether the mechanism of DNA transfer through the membrane will be completely understood in the near future. Also the role of functions determined by the host has been largely neglected. Since bacterial cells can be manipulated to take up DNA molecules, the same basic questions can be asked here: is there a DNA membrane channel and what process is driving the uptake of long linear DNA molecules?

Finally, it has to be pointed out that conjugative DNA transfer is not restricted to bacteria. In the pathogenic interaction between *Agrobacterium tumefaciens* and the host plant the T-DNA, a segment of the Ti plasmid, is transferred to plant cells (reviewed by Bevan and Chilton 1982) and recently it has been observed that plasmid DNA transfer can occur between bacteria and yeast (Heinemann and Sprague 1989).

## References

**Abdel-Monem M,** Hoffmann-Berling H (1976) Eur J Biochem 65:431–440. **Abdel-Monem M,** Taucher-Scholz G, Klinkert MQ (1983) Proc Natl Acad Sci USA 80:4659–4663. **Abo T,** Inamoto S, Ohtsubo E (1991) J Bacteriol 173:6347–6354. **Achtman M,** Morelli G, Schwuchow S (1978) J Bacteriol 135:1053–1061.

**Bayer ME** (1976) Role of adhesion zones in bacterial cell-surface function and biogenesis. In: Tzagaloff A (ed) Membrane biogenesis. Plenum, New York, pp 393–427. **Bevan MW,** Chilton MD (1982) Annu Rev Genet 16:357–384. **Bouck N,** Adelberg EA (1970) J Bacteriol 102:688–710.

**Chea K-C,** Skurray R (1986) J Gen Microbiol 132:3269–3275. **Clark AJ** (1985) Conjugation and its after effects in *Escherichia coli.* In: Halvorson HO, Monroy A (eds) The origins and evolution of sex. Alan Liss, New York, pp 47–68. **Clark AJ,** Margulies AD (1965) Proc Natl Acad Sci USA 53:451–459. **Cuozzo M,** Silverman PM, Minkley EG (1984) J Biol Chem 259:6659–6666.

**Date T,** Inuzuka M, Tomoeda M (1977) Biochem 16:5579–5585.

**Eguchi Y,** Itoh T, Tomizawa J-I (1991) Annu Rev Biochem 60:631–652. **Eichenlaub R** (1991) Prog Bot 52:190–200. **Eichenlaub R,** Figurski D, Helinski DR (1977) Proc Natl Acad Sci USA 74:1139–1141. **Everett R,** Willetts NS (1980) J Mol Biol 136:129–150. **Everett R,** Willetts NS (1982) EMBO J 1:747–753.

**Fee BE,** Dempsey WB (1986) J Bacteriol 167:336–345. **Finnegan D,** Willetts NS (1972) Mol Gen Genet 119:57–66. **Finnegan D,** Willetts NS (1973) Mol Gen Genet 127:307–316. **Folkhard W,** Leonard KR, Malsey S, Marvin DA, Dubochet J, Engel A, Achtmann M, Helmuth R (1979) J Mol Biol 130:145–160. **Fowler T,** Taylor L, Thompson R (1983) Gene 26:79–89. **Frost LS,** Paranchych W, Willetts NS (1984) J Bacteriol 160:395–401. **Frost LS,** Lee S, Yanchar N, Paranchych W (1989) Mol Gen Genet 218:152–160.

**Golub EJ,** Low KB (1985) J Bacteriol 162:235–241. **Gros MF,** te Riele H, Ehrlich SD (1987) EMBO J 6:3863–3869.

**Heinemann JA,** Sprague GF (1989) Nature 340:205–209. **Helmuth R,** Achtman M (1975) Nature 257:652–656.

**Iordanescu S,** Projan SJ (1988) J Bacteriol 170:256–260. **Ippen-Ihler KA,** Minkley EG (1986) Annu Rev Genet 20:593–624.

**Kennedy N,** Beutin L, Achtman M, Skurray R, Rahmnsdorf U, Herrlich P (1977) Nature 270:580–585. **Kingsman A,** Willetts NS (1978) J Mol Biol 122:287–300. **Koepsel RR,** Murray RW, Rosenblum WD, Khan SA (1985) Proc Natl Acad Sci USA 82:6845–6849. **Koepsel RR,** Murray RW, Khan SA (1986) Proc Natl Acad Sci USA 83:5484–5488. **Kolodkin AL,** Capage MA, Golub EJ, Low KB (1983) Proc Natl Acad Sci USA 80:4422–4426.

**Laine S,** Moore D, Kathir P, Ippen-Ihler KA (1985) Genes and gene products involved in the synthesis of F-pilus. In: Helinski DR et al. (eds) Plasmids in bacteria. Plenum, New York, pp535–553. **Lanka E,** Barth PT (1981) J Bacteriol 148:769–781. **Lederberg J,** Tatum E (1946) Nature 158:558. **Low KB** (1987) Hfr strains of *Escherichia coli* K12. In: Neidhardt FC (ed) *Escherichia coli* amd *Salmonella typhimurium* : cellular and molecular biology. Am Soc Microbiol, Washingon DC, pp 1134–1137.

**Manning PA,** Achtman M (1979) Cell to cell interactions in conjugative *Escherichia coli* : The involvement of the cell envelope. In: Inouye M (ed) Bacterial outer membrane biogenesis and function. Wiley, New York, pp 409–447. **Manning PA,** Morelli G, Achtman M (1981) Proc Natl Acad Sci USA (1981) 78:7487–7491. **Manning PA,** Morelli G, Fisseau C (1984) Gene 27:121–123. **Manoil C,** Rosenbusch JP (1982) Mol Gen Genet 187:148–157. **Minkley EG,** Polen S, Brinton CC, Ippen-Ihler KA (1976) J Mol Biol 108:111–121. **Moore D,** Sowa BA, Ippen-Ihler KA (1981) Mol Gen Genet 184:260–264. **Mullineaux P,** Willetts NS (1985) Promoters in the transfer region of plasmid F. In: Helinksli DR et al. (eds) Plasmids in bacteria. Plenum, New York, pp 605–614.

**Ohki M,** Tomizawa J-I (1968) Cold Spring Harbor Symp Quant Biol 33:651–657.

**Panicker MM,** Minkley EG (1985) J Bacteriol 162:584–590.

**Sanderson KE,** MacLachlan PR (1987) F-mediated conjugation, F$^+$ strains, and Hfr strains of *Salmonella typhimurium,* and *Salmonella abony*. In: Neidhardt FC (ed) *Escherichia coli* and *Salonella typhimurium* : cellular and molecular biology. Am Soc Microbiol, Washington DC, pp 1138–1144. **Schwab M,** Gruber H, Högenauer G (1991) Mol Biol 5:439–446. **Silverman PM** (1986) Bioessays 2:254–259.

**Thomas CD,** Balson D, Shaw WV (1988) Biochem Soc Trans 16:758–759. **Thompson R,** Taylor L, Kelly K, Everett R, Willetts NS (1984) EMBO J 3:1175–1180. Tomizawa J-I (1987) Regulation of ColE1 DNA replication by antisense RNA. In: Inouye M, Dudock BS (eds) Molecular biology of RNA – new perspectives. Harcourt, San Diego, pp 249–260.

**Wilkins BM,** Hollom SE (1974) Mol Gen Genet 134:143–156. **Willetts NS** (1977a) The genetics of conjugation. In: Mitsuhashi S (ed) R factor: drug resistance plasmid. University of Tokyo Press, Tokyo, pp 89–107. **Willetts NS** (1977b) J Mol Biol 112:141–148. **Willetts NS** (1977c) Mol Gen Genet 129:123–130. **Willetts NS,** Skurray R (1980) Structure and function of the F factor and mechanism of conjugation. In: Neidhardt FC (ed) *Escherichia coli* amd *Salmonella typhirmurium* : cellular and molecular biology. Am Soc Microbiol, Washington DC, pp 1110–1133. **Willetts NS,** Wilkins B (1984) Microbiol Rev 48:24–41.

**Yarranton GT,** Gefter ML (1979) Proc Natl Acad Sci USA 76:1658–1662. *Yoshioka Y,* Ohtsubo H, Ohtsubo E (1987) J Bacteriol 169:619–623.

Professor Dr. Rudolf Eichenlaub
Universität Bielefeld
Fakultät für Biologie
Gentechnologie/Mikrobiologie
Postfach 8640
W-4800 Bielefeld 1, Germany

# II. Recombination: Sexual Reproduction – a Tool for Outcrossing and Recombination of Genetic Material

By Ursula Kües and Ulf Stahl

## 1. Introduction: The Benefits of Sexual Reproduction

Sexual reproduction is characterized by karyogamy and meiosis and is associated with an alternation of haploid and diploid nuclear phases. Amongst evolutionary biologists and geneticists there is a continuing controversy concerning the reasons why sexual reproduction has been evolved (e.g., see E. Darwin 1974; C. Darwin 1839, 1876; Weismann 1891; Muller 1932; Mayr 1963; Ghiselin 1974; Maynard Smith 1978; Bell 1985; Stearns 1987; Michod and Levin 1988; Bernstein and Bernstein 1991; Haig and Grafen 1991). In terms of proliferation or improving the mean fitness of a population, sexual reproduction is not very efficient. From the bioeconomic point of view, sexual reproduction is seen to be energetically expensive, that is, the need to find a mate, special cell processes such as fertilization and meiosis, and additional costs in the development of special sex organs (Maynard Smith 1978; Lewis 1983; Uyenoyama 1984; Krieber and Rose 1986; Michod and Levin 1988). Sexual reproduction does, however, seem to be advantageous but there is no consensus as to what these advantages may be. In fact, the arguments just given indicate a clear advantage in asexual over sexual reproduction – why then do we find such elaborate and expensive sexual reproduction mechanisms so ubiquitoulsy in nature?

Crow (in Michod and Levin 1988, pp 56–73) defines three classical arguments for the evolutionary advantages of sexual reproduction in conjunction with DNA repair and recombination: (1) adjustment to a changing environment, (2) combination of beneficial mutations, and (3) loss of deleterius mutations.

The genetic material (DNA, is unstable under physiological conditions and is permanently accessible to damage by reactive chemicals found in the intracellular environment. If DNA damage remains unrepaired a loss of genetic information will occur with often deleterious or even lethal consequences (Frese and Stahl 1992 and references therein).

Rates for endogenous DNA damage in mammalian cells are estimated as 12 000–14 000 depurinations, 600–700 depyrimidations and more than 55 000 single-strand breaks per day per cell (Lindahl 1977; Tice and Setlow 1985; further information in Bernstein and Bernstein 1991).

Sexual reproduction has often been connected with aging of individual organisms and may be seen as a kind of rejuvenation process (Holliday, in Michod and Levin 1988, pp 45–55).

Progress in Botany, Vol. 54

For example, the ascomycete *Podospora anserina* has a life span of about 24 days. Thereafter, the vegetative mycelia become invariably senescent and finally die. Survival of the fungus is dependent on sexual reproduction (Rizet 1953; Marcou 1961). *Paramecium aurelia* has a limited ability for asexual division before it has to be rejuvenated by meiosis and subsequent conjugation (Siegel 1967).

In sexual offspring, DNA damage can be removed by outcrossing and recombination (DNA repair hypothesis of sex, see Bernstein and Bernstein 1991).

In this context, mixis, the generation of new combination of genes by mixing genomes or parts of genomes from different individuals, may be seen as the more coincidental effect of sexual reproduction. However, diploidy could, in principle, be sufficient for eliminaton of deletereous mutations and DNA damage by double strand DNA repair (Maynard Smith, in Michod and Levin 1988; pp 106–125).

When sexual exchange is unnecessary for error repair, recombination in the sense of mixing of genomes or parts of genomes (mixis) would seem to be the rationale for sexual reproduction (Crow, Michod and Levin 1988, pp 56–73). Holliday (1984) sees a function for sexual reproduction and meiosis in the correction of errors in DNA methylation: non-methylated sequences can be recognized as stimulators for homologous recombination when brought together with a methylated homolog.

Attention has to be paid to the special roles of inbreeding and outbreeding in sexual reproduction. Inbreeding conserves adaptation by reconstituting parental genomes in the face of damage and mutation, outbreeding is a more creative approach with possible advantages in strongly varying environments (Shield in Michod and Levin 1988; pp 253–269). In addition, some authors suggest that sexual reproduction may be a means of promoting transfer of parasitic genetic elements (Hickey 1982, 1984).

Whatever the value of sexual reproduction, in each case there will always be several different arguments as to the combined aspects of costs and benefits, short- and long-term adaptions, individual and species advantages all play a role in its evolution and maintenance. Recombination must surely be considered to be a major advantage of sexual reproduction.

Special genes are necessary for sexual reproduction. In bacteria, where a true diploid stage is missing, there are sets of "sexual" genes, in general plasmid encoded, which are responsible for mediating cell-cell contacts (conjugation) and intracellular DNA transfer (Willetts and Wilkins 1984; Ippen-Ihler and Minkley 1986; Willetts and Skurray 1987; Thomas 1989). These bacterial "sexual" systems are relatively well studied, but an understanding of the genes that control sexual reproduction in eukaryotes is only just starting. Recent studies have revealed the structural organization and functions of the fungal genes regulating sexual reproduction, the mating type genes (see recent review of Kües and Casselton 1992a). These genes will now be described and discussed with respect to their role in promoting recombination.

## 2. Genetic Control of Sexual Reproduction – the Mating Type Genes

For sexual reproduction it is generally necessary to recognize and discriminate between putative sexual partners. Sexual differentiation into morphologically distinct sexual organs or sex cells (i.e., male and female) is one approach to recognition of sexual partners. However, not all organisms have morphologically distinguishable sexes (Meinhardt and Esser 1990; Stahl and Esser 1992). Where sexual differentiation is absent, incompatibility barriers between different individuals decide the success of sexual reproduction. Incompatibility can be determined extracellularly as in the ascomycetes *Saccharomyces cerevisiae* (see Fields 1990; Marsh et al. 1991) and *Neurospora crassa* (Bistis 1981, 1983) and the hemibasidiomycetes *Ustilago maydis* (Bölker et al. 1992) and *Tremella* species (Ishibashi et al. 1984) or intracellularly, as in holobasidiomycetes such as *Coprinus cinereus* (Kües and Casselton 1992b).

Incompatibility barriers between fungal strains occur at the vegetative as well as the sexual stage of development. The first occurs between strains of the same species from different areas. It restricts the coexistence of nuclei and hence transfer of genetic material between individuals with nonidentical mating types (heterogenic incompatibility, Esser 1971; also called heterokaryon incompatibility, Turcq et al. 1990, 1991). In nature, this mechanism acts to keep different gene pools separated and may be responsible for speciation. The second prevents sexual reproduction between putative maters if their mating types are identical; this phenomenon is called homogenic incompatibility (Esser 1971).

The genes that control sexual incompatibility within species are the mating type genes. Molecular analysis has revealed two classes of mating type: one class codes for transcription factor proteins, that bind specific DNA sequences in order to regulate gene transcription; the other class encodes pheromones and their receptors. Interestingly, in some species more than two mating types exist and these are determined by several genes with multiple alleles (for review see Kües and Casselton 1992a). As a consequence, there may be more than 20 000 different "sexes" in a single fungal species (Raper 1966; Raper and Raper 1968). With several mating type genes and multiple alleles of these genes an outbreeding potential of > 95% can be achieved (Stamberg and Koltin 1973).

Multiple mating types have been described in myxomycetales (*Didymium iridis*; Collins 1980; *Physarum polycephalum*; Youngman et al. 1981; Kawano et al. 1987), in acrasiales (*Dictyostelium discoideum*; Robson and Williams 1990) and in basidiomycetes (e.g., *Agrocybe aegerita*, Meinhardt and Leslie 1982; Marmeisse 1988, 1991; *Armilliaria mellea*; Ullrich and Anderson 1978; Guillaumin et al. 1991; *C. cinereus*; Casselton 1978, *Fomes cajanderi*; Neuhauser and Gilbertson 1971; *Polyporus betulinus*, Burnett 1965; *Polyporus palustris*, Flexer 1965; *Schizophyllum commune*, J. R. Raper 1966; C. Raper 1983; *U. maydis*; Puhalla 1970; Wong and Wells 1985; *Tremella mesenterica*, Wong and Wells 1985).

### a) The Ascomycetous Yeast *Saccharomyces cerevisiae*

The best-studied mating type genes are those of *S. cerevisiae* (Fig. 1) and there are several reviews which focus on these genes and related subject areas (e.g., Nasmyth 1982;

### *Saccharomyces cerevisiae*

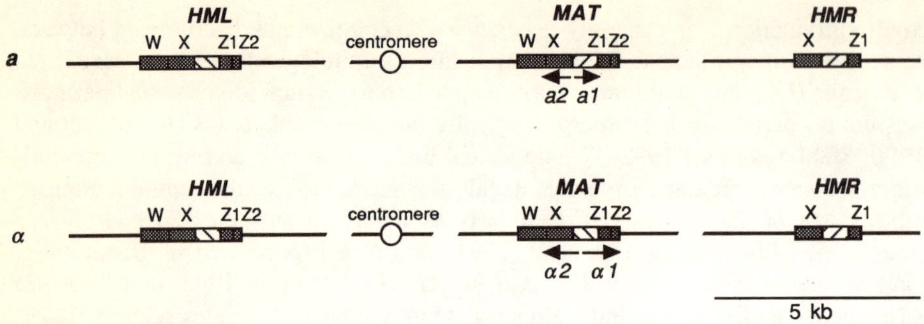


**Fig. 1.** Arrangement of the mating type loci and the *HML* and *HMR* elements in chromosome III of *Saccharomyces cerevisiae* **a** and α cells. *Gray boxes* are homologous sequences within mating type DNA whereas *hatched regions* indicate unique DNA sequences. *Arrows* denote direction of transcription

Cross et al. 1987; Herskowitz 1988, 1989; Klar 1989; Fields 1990; Sprague 1990; Dolan and Fields 1991; Glass and Lorimer 1991; Marsh et al. 1991; Lipke and Kurjan 1992). A haploid *S. cerevisiae* cell contains three copies of mating type information on chromosome III (Fig. 1). Only the genes at the *MAT* locus are transcribed and therefore active. The two distinguishable types of haploid cells, **a** and α, differ at this locus by having either the *MAT**a*** or the *MAT*α mating type specific genes. The two inactive loci, *HML* and *HMR*, are located at either side of the *MAT* locus 200 and 150 kb apart. The *HML* and *HMR* loci may have alternate alleles either with copies of the **a** or α genes. Most *S. cerevisiae* strains contain the silent α allele at *HML* and the silent **a** allele at *HMR* (Herskowitz 1988, 1989). *HML* and *HMR* function as storage loci for mating type conversions by transposition to the *MAT* locus (see Sect. 3).

The transcriptional silencing of mating type information requires a number of other genes. The products of the four SIR (*S*ilent *I*information *R*egulator) genes repress transcription of the **a** or α information at the *HML* or *HMR* loci (Ivy et al. 1986; Rine and Herskowitz 1987; Pillus and Rine 1989). Histone 4 and the DNA binding protein RAP1 are involved (Kayne et al. 1988; Kurtz and Shore 1991) as well as NAT1 and ARD1, components of a protein acetylase (Whiteway et al. 1987; Mullen et al. 1989). The transcriptional repression of *HML* and *HMR* determines that these loci act as donors of information in mating type switching (Klar et al. 1981b; Nasmyth 1982, see Sect.3).

The sequences that confer **a** and α mating type are completely dissimilar: The **a** sequence is composed of 642 bp of unique sequence, the α sequence of 747 bp. These unique sequences are sandwiched between homologous regions X (704 bp) and $Z_1$ (239 bp) in all of the *MAT*, *HML*, and *HMR* loci. Sequence W (723 bp) and $Z_2$ (88 bp) lie outside the X and $Z_1$ regions in the *MAT* and *HML* loci but not in *HMR* (Strathern et al. 1980; Astell et al. 1981; Nasmyth et al. 1981a, see Fig. 2). To denote the discovery of dissimilar sequences at the same locus on a chromosome, the term idiomorph has been introduced (Metzenberg and Glass 1990).

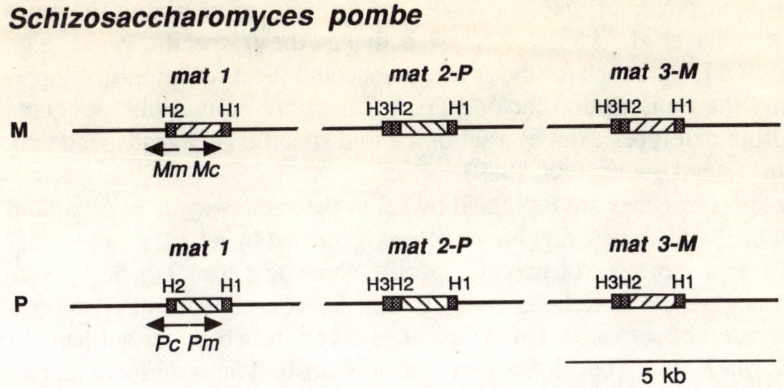


**Fig. 2.** Structural organization of the mating type loci and the silent copies of mating type information in *Schizosaccharomyces pombe* M and P cells. *Gray boxes* are homologous sequences within mating type DNA whereas *hatched regions* indicate unique DNA sequences. *Arrows* denote direction of transcription

The term idiomorph has been developed from the Greek word *idio* (singular, unique) and *morphos* (form) analogous to allelomorph, the original version of allele. Alleles are related forms of a gene which have arisen by gene mutation and therefore are highly similar to each other. In contrast, the highly dissimilar idiomorph may not have arisen from the same ancestor (Metzenberg 1990; Metzenberg and Glass 1990).

*MATa* and *MAT*α both contain two divergently transcribed genes designated ***a1*** and ***a2*** and α*1* and α*2*, respectively. ***a2*** has been shown to be dispensable, the three other genes are essential for developmental gene regulation either in haploid or in diploid cells (Herskowitz 1988, 1989; Dolan and Fields 1991). The **a1**, α1, and α2 proteins (126 aa, 175 aa, and 210 aa, respectively, Astell et al. 1981; Miller 1984; Ner and Smith 1989) have all been shown to bind DNA and to be transcription factors regulating the expression of different sets of genes (Johnson and Herskowitz 1985; Goutte and Johnson 1988; Jarvis et al. 1988, Passmore et al. 1989; Dranginis 1990; Tan and Richmond 1990; Dolan and Fields 1991; Sengupta and Cochran 1991; Keheler et al. 1992; Smith and Johnson 1992). **a1** and α2 belong to the homeodomain family of transcription factors with the typical trihelical DNA binding structure of the homodomain (Laughon and Scott 1984; Shepherd et al. 1984; Phillips et al. 1991; Wolberger et al. 1991). The DNA-binding domain of α1 is as yet uncharacterized, but amino acid homology exists to a domain in the mating type proteins of *Neurospora crassa* and *Podospora anserina*, suggesting that this may contain the DNA binding motif (Glass et al. 1988; Debuchy and Coppin 1992).

Haploid cells of *S. cerevisiae* contain all the information necessary for establishing either the **a** or the α character (e.g., the two types of pheromone genes and the two types of pheromone receptor) (Herskowitz 1988, 1989), but depending on the mating type genes integrated at the *MAT* locus only one set of cell-specific genes is expressed (Nasmyth et al. 1981b; Klar et al. 1981a; Siliciano and Tatchell 1984). In haploid **a** cells the **a1** transcription factor is not necessary and the **a**-specific functions, e.g., the genes for the **a** factor pheromone (*MFA1*, *MFA2*) and the α factor receptor (STE2), are constitutively expressed with the help of the general transcription factor MCM1 (Elble and Tye 1991). In α haploid cells, α2 forms dimers and acts in conjunction with MCM1 and two other general transcription factors, SSN6 and TUP1, as a repressor of **a** specific genes (Sauer et al. 1988; Keheler et al. 1988, 1992). α1 activates the α specific genes, e.g., *MFAα1*, *MFAα2* and *STE3*, the genes for the α pheromone and the **a** receptor, in cooperation with MCM1 and the transcription factor STE12 (Passmore et al. 1989; Sengupta and Cochran 1991). Both cell types express a set of haploid specific genes independently of mating type proteins (Herskowitz 1988, 1989).

In diploid cells, **a**-specific genes are repressed by α2 in the same way as in α haploid cells (Herskowitz 1988, 1989). A new regulatory complex formed by **a1**-α2 acts together with SSN6 and TUP1 as a repressor of haploid specific genes and the a1 gene (Goutte and Johnson 1988; Dranginis 1990; Keheler et al. 1992). As a consequence of α1 repression, α-specific genes are not activated. It is relevant to mention two important haploid specific genes. One is *rme1*, which codes for repressor of meiosis. Transcriptional repression of *rme1* in diploid cells initiates meiosis and sporulation. The other, the *HO* gene (*ho*mothallism) encodes a site-specific endonuclease that initates mating-type interconversion (Nasmyth and Shore 1987; Herskowitz 1988, 1989, Sect. 3).

### b) The Ascomycetous Yeast *Schizosaccharomyces pompe*

Haploid cells of *Schizosaccharomyces pombe* exhibit one of two mutually exclusive mating types, P (plus) and M (minus) (Leupold 1950, 1958; Egel et al. 1990). The mating type information in *S. pombe* is arranged similarly to that of *S. cerevisiae*. Each cell contains three copies of mating type information (*mat1*, *mat2*, *mat3*) residing on linkage group II (Egel et al. 1980; Klar 1989, see Fig. 2). Only the copy present at the *mat1* locus is active which can have either the P or the M information. The two copies located at the *mat2* and *mat3* loci are unexpressed, contain the mutually opposite mating type, and serve as storage loci for mating type switching (Klar 1989; see Sect. 3). The possible alternative mating type information at *mat1*, *mat1-P*, and *mat1-M*, are not similar in sequence (Kelly et al. 1988), seem to be of separate ancestry, and therefore idiomorphs (Metzenberg 1990). The unique mating type sequences are accompanied by short homologous regions called H1 (59 bp), H2 (135 bp) and H3 (57 bp). *mat1* is flanked by H2 and H1, whereas *mat2* and *mat3* have a copy of H3 at one end and a copy of H1 and H2 at the other (Kelly et al. 1988; see Fig. 2).

The unique P information extends over 1104 bp and that of M over 1128 bp (Kelly et al. 1988). Each *mat1* idiomorph codes for two transcripts, *Pc* and *Pm* in *mat1-P*, *Mc* and *Mm* in *mat1-M*, whose products are thought to be regulatory proteins. The protein products Pc (118 aa) and Mc (181 aa), (c for *c*onjugation) are necessary for mating and conjugation, Pm (159 aa) and Mm (42 aa), (m for *m*eiosis) have combined regulatory functions in meiosis (Egel et al. 1990; Kües and Casselton 1992a). As yet, none of these proteins has been shown to bind DNA and to be involved directly in transcriptional activation and repression. However, the Pm product has a domain with homology to the α2 homeodomain (Kelly et al. 1988), and Mc contains an HMG motif (HMG = *H*igh *M*obility *G*roup proteins), a DNA binding motif found in one class of transcription factors (Ner 1992). Amongst the proteins containing an HMG motif is the product of the

SRY gene (sex-determining region Y) which determines maleness in humans and mice (Gubbay et al. 1990; Sinclair et al. 1990; Harley et al. 1992).

### c) Filamentous Ascomycetes: *Cochliobolus heterostrophus, Neurospora crassa* and *Podospora anserina*

Filamentous ascomycetes possess alternative mating types at a homologous chromosomal site in different strains, called *mtA* and *mta* in *Neurospora crassa*, *mat* + (*mt*+) and mat – (*mt*–) in *Podospora anserina*, *MAT-1* and *MAT-2* in *Cochliobolus heterostrophus*. Cloning and molecular analysis of the alternate mating type information has revealed the existence of idiomorphs in all three species (Glass et al. 1988, 1990; Yoder et al. 1989; Staben and Yanofsky 1990; Picard et al. 1991). The unique sequences of the *A* mating type locus of *N. crassa* is 5.3 kb in length and that of the *a* mating type is 3.2 kb (Glass et al. 1990; Staben and Yanofsky 1990, see Fig. 3). In *P. anserina*, the length of the unrelated *mat*– and *mat*+ sequences are 4.7 and 3.8 kb, respectively (Debuchy and Coppin 1992; Fig. 3). One transcript responsible for regulating sexual functions, *mt a-1* and *mt A-1*, has been identified from each *N. crassa* idiomorph (Glass et al. 1990; Staben and Yanofsky 1990). Analogous genes for fertilization in *P. anserina*, the putative genes *FPR1* (*F*ertilization *P*lus *R*egulator) and *FMR2* (*F*ertilization *M*inus *R*egulator) have been found by sequencing (Debuchy and Copping 1992). *mt a-1* and *FPR1* encode related polypeptides (382 aa and 365 aa, respectively) with a putative DNA-binding domain, the HMG motif. mt A-1 (288 aa) and FMR1 (305 or 349 aa) proteins share a region of amino acid homology with the α1 protein of *S. cerevisiae*, which could be responsible for DNA binding, and it is proposed to call this common motif the α1-domain (Debuchy and Coppin 1992). A role as DNA binding proteins and transcription factors has yet to be demonstrated at the molecular level for the *N. crassa* and *P. anserina* mating type proteins.

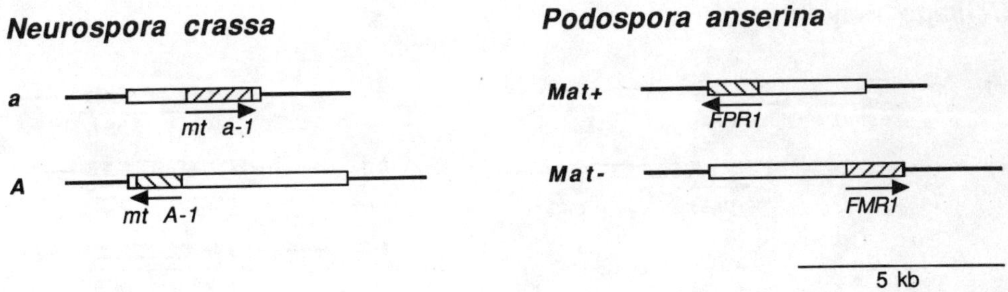


**Fig. 3.** Structural organization of the mating types *a* and *A* of *Neurospora crassa* and *Mat*+ and *Mat*– of *Podospora anserina*. Unique sequences at the mating type loci are shown as *open boxes*, coding regions within these are *hatched*. *Arrows* denote direction of transcription

d) The Hemibasidiomycete *Ustilago maydis*

Two different loci determine the mating types in *Ustilago maydis* (see also recent reviews of Banuett 1991, 1992): The *a*-locus exists in two forms (Rowell and DeVay 1954) and the *b*-locus is multiallelic (≥25 alleles: Puhalla 1968, 1970).

By cloning, hybridization, and sequencing it was found that the *a* loci are typical idiomorphs, having unique sequences extending over a length of 4.5 (*a1*) and 8 kb (*a2*) (Froeliger and Leong 1991; Bölker et al. 1992; Fig. 4). Within each locus, *a1* and *a2*, two different genes have been described: one gene, *mfa*, appears to encode a mating type specific pheromone (*mfa1*, mating factor encoded by the *a1* idiomorph: 40 aa; *mfa2*, mating factor encoded by the *a2* idiomorph: 38 aa). The other, *pra*, encodes the pheromone receptor (*pra1*, pheromone receptor encoded by *a1*: 357 aa; *pra2*, pheromone receptor encoded by *a2*: 346 aa), (Bölker et al. 1992).

The multiallelic *b*-locus presents a regulatory system for self/nonself cell recognition. In each *b*-factor two divergently transcribed genes, *bW* and *bE*, have been identified. The products for both genes (bE: 473 aa; bW:626 aa) contain homoedomain motifs, suggesting that the polypeptides act as transcriptional regulators. It has been shown that for a compatible sexual reaction a bW-protein has to react with an heterologous bE-protein (Gillissen et al. 1992). Each *b* gene is composed of a short variable 5' end and a common 3' end and each *bW-bE* gene pair is separated by a unique sequence of 260 bp (Kronstad and Leong 1990; Schulz et al. 1990; Gillissen et al. 1992). The coding sequence of the *bE* genes differs in the first 110 amino acids (63% identity) and has 90% homology at the 3' end (Kronstad and Leong 1990, Schulz et al. 1990). The coding sequence of *bW* genes differs in the first 130 amino acids (46% identity) and shares 96% sequences at their 3' ends (Gillissen et al. 1992). Due to the extended homology between the different *b* factors, their different *bW* and *bE* genes may be referred to as alleles rather than as idiomorphs. Nevertheless, *bW* and *bE* alleles of different *b* factors and their products are the analogs to the genes and products of the two different mating type idiomorphs of *S. cerevisiae*, *S. pombe* and in the filamentous ascomycetes (e.g., *bW* could correspond to *Mat*α*2* in *S. cerevisiae*, *bE* to *Mata1*), (discussed further in Kües and Casselton 1992a).

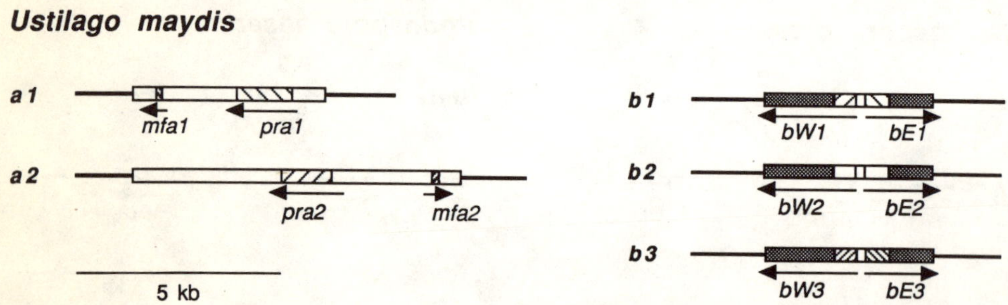


**Fig. 4.** Structural organization of the *a* and *b* mating type loci of *Ustilago maydis*. Unique sequences at the mating type loci are shown as *open boxes*, coding regions within these are *hatched*. *Gray boxes* are homologous sequences within mating type DNA. *Arrows* denote direction of transcription

### e) Holobasidiomycetes: *Coprinus cinereus* and *Schizophillum commune*

In the hymenobasidiomycetes *Coprinus cinereus* and *Schizophyllum commune*, two multiallelic mating type factors, *A* and *B*, each with two loci, α and β, have been defined by classical genetics (Raper 1966; Casselton 1978). The structure of the two *B* loci is not yet known, but molecular cloning has resolved the structure of the entire *A*-factor of *C. cinereus* (May et al. 1991; Kües et al. 1992a,b; Kües and Casselton 1992b) and of the *A*α locus of *S. commune* (Novotny et al. 1991; Ullrich et al. 1991).

The well-studied *A*42 factor of *C. cinereus* has been shown to consist of two gene subcomplexes of two and five genes, respectively, which are separated by 7 kb of noncoding sequences (Fig. 5). The subcomplexes correspond to the classical α and β loci and the 7 kb distance between them to the 0.07 map units shown to separate them. Two genes, one in each complex (orignally called *α1* and *β5*, now renamed *α-fg*, *β-fg*, Kües and Casselton 1992a), are found in all *A* factors. At least four of the remaining five *A*42 genes determine *A* factor specifictiy. They are functionally redundant in triggering sexual development (Kües e al. 1992a). Another cloned *A* factor, *A*43, similarly has five genes at the β locus but has an additional one at the α locus (May et al. 1991; Kües and Casselton 1992a, b; Fig. 5). Recent studies with other *A* factors shows that in general, five to six genes seem to determine the specificity. These are likely arranged in divergently transcribed gene pairs. It can be concluded that an ancient *A* factor may have had four pairs of these functionally redundant specificity genes and that one or more genes of different pairs have been lost in the collection of *A* factors studied (Kües and Casselton 1992a).

The different forms of the *A* factor specificity genes have unique sequences, as judged from hybridization studies and their protein products contain homeodomain motifs (Mutasa et al. 1990; May et al. 1991; Kües et al. 1992a; Tymon et al. 1992). These homeodomain motifs can be divided into two classes, HD1 and HD2 (Kües and Casselton 1992c). Proteins with alternate HD-motifs (putative transcription factors of distinct classes) seem to react with each other to regulate sexual development (Kües and Casselton 1992a).

Each *A* factor specificity gene pair of *C. cinereus* appears to contain one gene with a 2.4 kb transcript that encodes the HD1 protein and one gene with a 2.1 kb transcript that encodes the HD2 protein. Each pair is by itself comparable to the divergently transcribed *bW* and *bE* genes of *U. maydis* (Kües and Casselton 1992a). The *Ustilago* proteins can also be classified as HD1 and HD2 proteins on the basis of their homeodomains (Kües and Casselton 1992c). Similarly, a pair of divergently transcribed genes with unique DNA sequences have been found in the *A*α factor of *S. commune* (Fig. 6) and their predicted protein products also have HD1 and HD2 homeodomain motifs (Giasson et al. 1989; Kües and Casselton 1992c; Stankis et al. 1992). As in *U. maydis* and obviously in *C. cinereus*, an HD1-protein (e.g., from the *A*α3 factor) has to react with an HD2 protein that is encoded by a compatible gene pair (e.g., in the *A*α4 factor) (Specht et al. 1992).

In the *A*α factor of *S. commune*, a third gene has been detected that is common to all *A*α factors of *S. commune* (Stankis et al. 1992). This may be the analog of α-fg in the *A*α factor of *C. cinereus.*

As mentioned above for the *b* loci products of *U. maydis* (Sect. 2e), the HD1 and HD2 proteins of *C. cinereus* and *S. commune* may have some analogy to the **a1** and α2 pro-

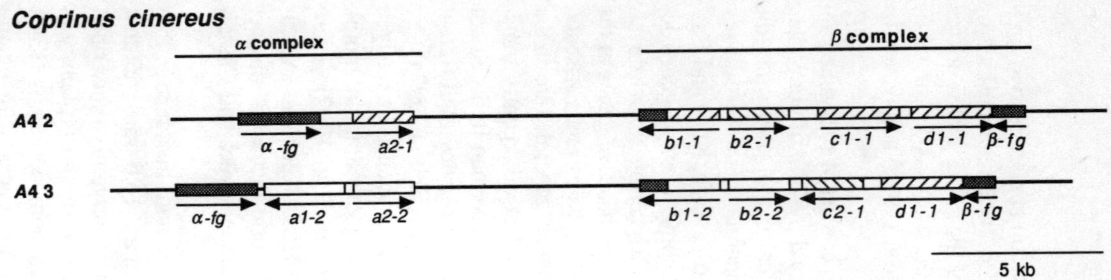


**Fig. 5.** Structural organization of the *A42* and *A43* mating type factors of *Coprinus cinereus*. Unique sequences at the mating type loci are shown as *open boxes*, coding regions within these are *hatched*. *Gray boxes* are homologous sequences within mating type DNA. *Arrows* denote direction of transcription

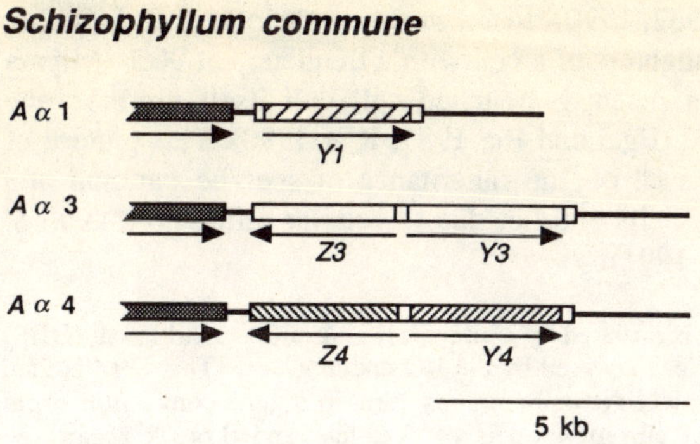


**Fig. 6.** Structural organization of different $A\alpha$ factors of *Schizophyllum commune*. Unique sequences at the mating type loci are shown as *open boxes*, coding regions within these are *hatched*. *Gray boxes* are homologous sequences within mating type DNA. *Arrows* denote direction of transcription

ducts of the mating type idiomorphs of *S. cerevisiae* (Herskowitz 1992; Kües and Casselton 1992a). Hydridization studies revealed no extensive homology between different specificity gene alleles of *C. cinereus* and *S. commune* (Giasson et al. 1989; Mutasa et al. 1990; May et al. 1991; Kües et al. 1992a). Sequence data for different forms of the $A\alpha$ specificity genes of *S. commune* revealed an actual sequence identity of 50–80% (Ullrich et al. 1991; Stankis et al. 1992) and the different gene forms seem to have developed from a common ancestor and are therefore true alleles. This is likely be true for the different forms of the *C. cinereus* specificity genes (Kües and Casselton 1992a), but sequence data for different alleles is still unavailable.

## 3. The Significance of Mating Type Switching

Mating type switching, a recombination process by which cells change between the alternate mating types, is a well-established phenomenon in *S. cerevisiae* and *S. pombe* (Klar 1989).

Mating type switching has also been observed in some filamentous ascomycetous fungi. However, nothing is known about the molecular mechanism (Perkins 1987). In basidiomycetes, comparable recombination events in changing the mating behavior and incompatibility are unknown.

In homothallic strains of both yeasts, cells of either mating type can switch to the other type efficiently and regularly.

Homothallism implies a capability for self-fertilization (Blakeslee 1904). Heterothallic haploid *S. cerevisiae* clones contain stable mating types, either **a** or α. Such isolates contain an inactive allele (*ho*) of the *HO* gene (Russell et al. 1986), the gene encoding the site-specific endonuclease that initiates mating-type interconversion (Kostriken and Heffron 1984).

The pattern of switching is quite different in the two yeasts: in *S. cerevisiae*, there are three types of switching control: the cell-type control, the mother-daughter control and the cell-cycle control. Only **a** or α *haploid mother cells* can change mating types with 73% frequency and only during the *late G1 phase* of the cell cycle because only under these conditions is the *HO* gene expressed (Strathern and Herskowitz 1979; Jensen et al.

1983; Nasmyth 1983; Nasmyth et al. 1987, 1990). In *S. pombe*, on the other hand, mating types switch in one of the four granddaughters of a cell with a frequency of 90% (Miyata and Miyata 1981) and the sister of a recently switched cell will itself produce one switched cell with a probability of 90% (Egel and Eie 1987; Klar 1990). This pattern of switching in *S. pombe* seems to be based on the inheritance of specific parental and grandparental DNA strands in progeny cells and not due to cellular differences as in *S. cerevisiae* (Klar 1987, 1990; Klar et al. 1991).

Mating type conversion in *S. cerevisiae* is initiated by a site-specific double-strand break (DBS) at the Y/Z junction in the *MAT* locus (see Fig. 1) caused by the HO endonuclease. This DBS is then repaired by copying information from an intact donor locus, resulting in a gene conversion event (McGill et al. 1989; for further details of current models of double-stranded-break repair see Szostak et al. 1983; Klar 1989; White and Haber 1990). Similarly, the recombination event that mediates mating type switching in *S. pombe* is initiated by a site-specific DBS at the *smt* cut site (*smt* = *s*witching of *m*ating *t*ype) located between the *mat1* allelic sequences and the H1 region (Nielsen and Egel 1989, see Fig. 2). A copy of mating type information from a silent locus is then transported to the *mat1* locus (Klar 1989). The molecular mechanism of mating type switching seems to be different from that in *S. cerevisiae*, e.g., in that in *S. pombe* a machinery must exist to seal the DBS without switching (Klar and Miglio 1986; Klar et al. 1991) and that mating type switching continues in diploid cells (Beach 1983; Egel 1984; Egel et al. 1984).

Mating type switching in *S. cerevisiae* and *S. pombe* makes sense in that the genetic information specific for **a** and α and P and M cells is present in all haploid cells independently of which mating type idiomorph is expressed. Switching simply results in another set of mating type-controlled genes being expressed. This situation is quite different from the one in the *a1* and *a2* haploid cells of the hemibasidiomycete *Ustilago maydis* and probably also in the basidiomycetous yeast, *Rhodotorula toruloides* (Akada et al. 1989a, b). Neither fungus switches mating type because haploid cells do not contain all the information for both mating type specific functions. In *Ustilago*, the mating type-specific genes encoding the mfa1 and mfa2 pheromones and their receptors have become mating type genes themselves (Sect. 2d) – this is probably true for *Rhodotorula* also (Akada et al. 1989a, b).

In species such as *U. maydis* and *R. toruloides* with only two different mating types (the *a* locus with the *a1* and *a2* alternates in *U. maydis*, the single locus of *R. toruloides* with the A and *a* alternates; Banno 1967) an inbreeding-outbreeding potential of ~50% exists.

It is interesting to note that because there are only two alleles at the *a* locus in *U. maydis*, it is essential to have multiple alleles of the unlinked *b* locus in order to approach 50% outbreeding potential.

In *S. cerevisiae* and *S. pombe* where the mating type locus has only two forms, the inbreeding potential rises with mating type switching (Herskowitz 1988; Klar 1989). If inbreeding increases, the chance of gene repair and loss of deleterius mutations by simultaneous conservation of the original gene pool improves.

## 4. Unlike Other Genes, Intragenic Recombination May Be Prevented in Mating Type Genes

In contrast to other types of recombination (e.g., illegitimate recombination), general or homologous recombination occurs with a relatively high frequency between homologous DNA segments (Franklin 1967, 1971). Such recombination may occur in the vegetative phase during mitosis (Stahl and Esser 1992), but is rare in comparison to the frequency during meiosis.

Mitotic crossing over in vegetative cells occurs with a frequency of $10^{-4}$–$10^{-5}$ per cell division (Günther 1984).

Meiotic recombination events are not uniformly distributed along the chromosome. So-called "hot spots" serves as localized stimulators of general recombination; "cold spots" are sites which depress the level of recombination (Szostak et al. 1983; Klar and Bonaduce 1991).

Examples of such cold spots are centromere regions as shown in *S. pombe* (Nakaseko et al. 1986). The 15 kb region between *mat2-P* and *mat3-M* is equally inaccessible to meiotic recombination, for which the chromatin structure in this region is thought to be responsible (Egel 1984; Klar and Bonaduce 1991). Suppression of homologous recombination in this region is advantageous because DNA exchange between the two different silent copies of the mating type genes would inevitably lead to the loss of one and the duplication of the other copy. It is interesting to note that in *S. cerevisiae* occasionally chromosomal deletions occur as a result of recombination between the silent mating type genes (Hawthorne 1983).

In *S. commune* variable recombination rates have been detected between the α and β loci of both the *A* mating type factor (3.2–22.8%; Raper et al. 1960) and the *B* mating type factor (0.1–8.0% or no recombination at all; Koltin et al. 1967). This variability is under the control of specific genes (Simchen 1967; Simchen and Stamberg 1969a,b) and affects the level of self-fertility amongst mycelia from sister basidiospores (Stamberg 1968).

As mentioned in Section 1, one of the main functions of sexual reproduction may be recombination during meiosis. The prerequisite for this is that DNA sequences are largely homologous. Idiomorphs of mating type genes, as occurs in the ascomycetous fungi and in the *a* locus of *U. maydis*, represent regions of nonhomology within which "intragenic" recombination is not possible. This explains the observed stability of the mating types in *N. crassa* (Glass and Lorimer 1991) and the *a* locus of *U. maydis* (Froeliger and Leong 1991). In species such as *C. cinereus* and *S. commune*, with a low level of homology between alleles of mating type genes, homologous recombination is likely to be rare.

Where short stretches of homology exist, as in the coding region for the homeodomains, it is found that these sequences are interrupted by introns having no DNA homology (Stankis et al. 1992; Kües and Casselton, unpubl.), and thus reducing the change of recombination. The existence of introns may strongly influence intergenic recombination.

In the *U. maydis b* locus, in contrast, there is considerable DNA homology between *b* gene alleles and recombination within genes is possible. However, one would not expect this to generate new mating type gene specificities: The specificity of the *bE* gene alleles (at least) has been localized to the 5' end of the gene, the region where there is no homology (Banuett 1991).

Are there benefits in limiting recombination within mating type loci? The conservation of mating type genes guarantees the potential for sexual development. With sexual reproduction there is a high outbreeding protential as high as >95% for the holobasidiomycetes (Sect. 2). If recombination would be possible within the complex mating type loci described, the consequences would be clearly deleterious and may lead to nonfunctional genes. In the case of the basidiomycetes, there is the interesting situation that in each haploid organism a pair(s) of incompatible mating type genes encodes the two transcription factor proteins necessary for sexual development. Recombination between alleles of the two genes that code for these proteins would lead to an arrangement of two compatible genes in one chromosome and, in consequence, the haploid monokaryons would be self-fertile. With self-fertilization the benefits of outbreeding are lost.

## 5. Conclusions

The contrasting arguments presented in the introduction show that there is no consensus as to why sexual reproduction has evolved. There are systems (e.g. mating type switching in *S. cerevisiae* and *S. pombe*) which facilitate inbreeding whereas others clearly promote a preference for outbreeding (e.g., the holobasidiomycetes with multi-allele mating type organization). On the one hand, the conservation of a given genetic pool is favored (inbreeding), and on the other, its permanent reorganization (outbreeding). There is thus no generalized answer as to which of these phenomena is most favorable. There seems to be a general trend towards outbreeding. However, comparisons of breeding systems of more or less related species indicate how finely balanced the respective advantages of outbreeding and inbreeding may sometimes be (Fincham 1983). The life cycle of an organism and the nature of its mating type gene obviously play an important part in the decision between in- and outbreeding. There may be the additional influence of environment, such as the accessibility of genetically different mating partners.

*Acknowledgments.* We are very grateful to Dr. Lorna Casselton for discussion and critical reading of the manuscript and to Dr. David Haig for discussion and supplying suitable literature about the evolution of sexual reproduction.

## References

**Akada R,** Kai J, Yamashita I, Miyakawa T, Fului S (1989a) Arch Microbiol 152:484–487. **Akada R,** Minomi K, Kai J, Yamashita I, Miyakawa T, Fukui S (1989b) Mol Cell Biol 9:3491–3498. **Astell CR,** Ahlstrom-Jonasson L, Smith M, Tatchell K, Nasmyth KA, Hall B (1981) Cell 27:15–23.

**Banno I** (1967) J Gen Appl Microbiol 13:167–196. **Banuett F** (1991) In: Bennett JW, Lasure LL, (eds) More manipulations in fungi. Academic Press, San Diego, CA, pp 217–234. **Banuett F** (1992) Trends Genet 8:174–180. **Beach DH** (1983) Nature 305:682–688. **Bell G** (1985) Experientia 41:1235–1245. **Bernstein C,** Bernstein H (1991) Aging, sex and DNA repair. Academic Press, San Diego, CA. **Bistis GN** (1981) Mycologia 73:959–975. **Bistis GN** (1983) Exp Mycol 7:292–295. **Blakeslee AF** (1904) Proc Acad Sci 40:203–319. **Bölker M,** Urban M, Kahmann R (1992) Cell 68:441–450. **Burnett JH** (1965) In: Esser K, Raper JR (eds) Incompatibility in fungi. Springer, Berlin Heidelberg New York, pp 98–113.

**Casselton LA** (1978) In: Smith JE, Berry DR (eds) The filamentous fungi, vol 3. Arnold, London, UK, pp 275–297. **Collins RR** (1980) Mycologia 72:1109. **Cross F,** Hartwell LH, Jackson C, Konopka JB (1987) Annu Rev Cell Biol 4:429–457.

**Darwin C** (1839) Journal of researches into the geology and natural history of the various countries visited by the H.M.S. Beagle, under the command of captain Fitzroy, R.N. from 1832 to 1836. Henry Colburn, London. **Darwin C** (1876) The effects of cross and self fertilization in the vegetable kingdom. John Murray, London. **Darwin E** (1794) Zoomania; or, The laws of organic life, vol 1. J. Johnson, London. **Debuchy R,** Coppin E (1992) Mol Gen Genet 233:113–121. **Dolan JW,** Fields S (1991) Biochim Biophys Acta 1088:155–169. **Dranginis AM** (1990) Nature 347:682–685.

**Egel R** (1984) Curr Genet 8:205–210. **Egel R,** Eie B (1987) Curr Genet 12:429–433. **Egel R,** Kohli J, Thuriaux P, Wolf K (1980) Annu Rev Genet 14:77–108. **Egel R,** Beach DH, Klar AJS (1984) Proc Natl Acad Sci USA 81:3481–3485. **Egel R,** Nielson O, Weilguny D (1990) Trends Genet 6:369–373. **Elble R,** Tye B-K (1991) Proc Natl Acad Sci USA 88:10966–10970. **Esser K** (1971) Mol Gen Genet 110:86–100.

**Fields S** (1990) TIBS 15:270–273. **Fincham JRS** (1983) Genetics. John Wright and Sons Ltd. Bristol, pp 518–522. **Flexer A** (1965) Am J Bot 52:634. **Franklin NC** (1967) Genetics 55:699–707. **Franklin NC** (1971) In: Hershey AD (ed) The bacteriophage lambda. Cold Spring Harbor Lab. Cold Spring Harbor New York, pp 175–194. **Frese D,** Stahl U (1992) Mechanisms of aging and development 65:277–288. **Froeliger EH,** Leong SA (1991) Gene 100:113–122.

**Ghiselin MT** (1974) The economy of nature and the evolution of sex. Univ California Press, Berkeley, Ca. **Giasson L,** Specht CA, Milgrim C, Novotny CP, Ullrich RC (1989) Mol Gen Genet 218:72–77. **Gillissen B,** Bergemann J, Sandmann C, Schroer B, Bölker M, Kahmann R (1992) Cell 68:647–658. **Glass NL,** Lorimer IAJ (1991) In: Bennett JW, Lasure LL (eds) More manipulations in fungi. Academic Press, San Diego, CA, pp 194–216. **Glass NL,** Vollmer SJ, Staben C, Groteleuschen J, Metzenberg RL, Yanofsky C (1988) Science 241:570–573. **Glass NL,** Groteleuschen J, Metzenberg RL (1990) Proc Natl Acad Sci USA 87:4912–4916. **Goutte C,** Johnson AD (1988) Cell 52:875–882. **Gubbay J,** Collignon J, Koopmam P, Chapel B, Economou A, Munsterberg A, Vivian N, Goodfellow P, Lovell-Badge R (1990) Nature 346:245–250. **Guillaumin JJ,** Anderson JB, Korhonen K (1991) In: Shaw III CG, Kile GA (eds) *Armillaria* root disease. Agric Handbook No 691, Forest Service, United States Department of Agriculture Washington, DC, pp 10–20. **Günther E** (1984) Lehrbuch der Genetik. Fischer, Stuttgart, 349 pp.

**Haig D,** Grafen A (1991) J Theor Biol 153:531–558. **Harley VR,** Jackson DI, Hextal PJ, Hawkins JR, Berkowitz GD, Sockanathan S, Lovell-Badge R, Goodfellow PN (1992) Science 255:463–456. **Hawthorne DC** (1983) Genetics 48:1727–1729. **Herskowitz I** (1988) Microbiol Rev 52:536–553. **Herskowitz I** (1989) Nature 342:749–757. **Herskowitz I** (1992) Nature 357:190–191. **Hickey DA** (1982) Genetics 101:519–531. **Hickey DA** (1984) Nature 311:417–418. **Holliday R** (1984) In: Evans CE, Dickenson HG (eds) Controlling events in meiosis. Soc Exp Biol Symp 38, Cambridge Univ Press, Cambridge, pp 381–394.

**Ippen-Ihler KA,** Minkley EG Jr (1986) Annu Rev Genet 20:593–624. **Ishibashi Y,** Sakagami Y, Isogai A, Suzuki A (1984) Biochemistry 23:1399–1404. **Ivy JM,** Klar AJS, Hicks JB (1986) Mol Cell Biol 6:688 702.

**Jarvis EE,** Clark KL, Sprague GF JR (1988) Mol Cell Biol 8:309–320. **Jensen R,** Sprague F, Herskowitz I (1983) Proc Natl Acad Sci USA 80:3035–3039. **Johnson AD,** Herskowitz I (1985) Cell 42:237 247.

**Kawano S,** Kuroiwa T, Anderson RW (1987) J Gen Microbiol 133:2539–2546. **Kayne PS,** Kim U-J, Han M, Mullen JR, Yoshizaki F, Grunstein M (1988) Cell 55:27 39. **Keheler CA,** Goutte C, Johnson AD (1988) Cell 53:927–936. **Keheler CA,** Redd MJ, Schultz J, Carlson M, Johnson AD (1992) Cell 68:709–719. **Kelly M,** Burke J, Smith M, Klar A, Beach D (1988) EMBO J 7:1537–1547. **Klar AJS** (1987) Nature 326:466–470. **Klar AJS** (1989) In: Berg DE, Howe MM, Mobile DNA (eds) Am Soc Microbiol, Washington, DC, pp 671–692. **Klar AJS** (1990) EMBO J 9:1407–1415. **Klar AJS,** Bonaduce M3 (1991) Genetics 129:1033–1042. **Klar AJS,** Miglio LM (1986) Cell 46:725–731. **Klar AJS,** Strathern JN, Broach Jr, Hicks JB (1981a) Nature 289:239–244. **Klar AJS,** Strathern JN, Hicks JB (1981b) Cell 25:517–524. **Klar AJS,** Bonaduce MJ, Cafferkey R (1991) Genetics 127:489–496. **Koltin Y,** Raper JR, Simchen G (1967) Proc Natl Acad Sci USA 57:55–62. **Kostriken R,** Heffron F (1984) Cold Spring Harbor Symp Quant Biol 49:89–96. **Krieber M,** Rose MR (1986) J Theror Biol 122:421–440. **Kronstad JW,** Leong SA (1990) Genes Dev 4:1348–1395. **Kües U,** Casselton LA (1992a) Mycol Res 96:993–1006. **Kües U,** Casselton LA (1992b) In: Setlow JK (ed) Genetic engineering, principles and methods, vol 14, Plenum, New York, pp 251–268. **Kües U,** Casselton LA (1992c) Trends Genet 8:154–155. **Kües U,** Richardson WVJ, Mutasa ES, Tymon AM, Gaubatz S, Gregoriades A, Casselton LA (1992a) Genes Dev 6:568–577. **Kües U,** Richardson WVJ, Mutasa ES, Tymon AM, Gaubatz S, Casselton LA (1992b) In: Stahl U, Tudzynski P (eds) Proc. of the EMBO Workshop Molecular Biolgoy of filamentous fungi. Verlag Chemie, Weinheim, pp 241–351. **Kurtz S,** Shore D (1991) Genes Dev 5:616–628.

**Laughon A,** Scott MP (1984) Nature 310:25–31. **Leupold U** (1950) C R Lab Carlsberg Ser Physiol 24:381–480. **Leupold U** (1958) Cold Spring Harbor Symp Quant Biol 23:161–170. **Lewis JW Jr** (1983) J Theor Biol 93:927–951. **Lindahl T** (1977) In: Nichols WW, Murphy DG (eds) DNA repair processes. Symposia Specialists, Miami, pp 225–240. **Lipke PN,** Kurjan J (1992) Microbiol Rev 56:180–194.

**Marcou D** (1961) Ann Sci Nat Bot 12:653–764. **Marmeisse R** (1988) Sex Plant Reprod 1:195–201. **Marmeisse R** (1991) Mycol Res 95:465–468. **Marsh L,** Neiman AM, Herskowitz I (1991) Annu Rev Cell Biol 7:699–728. **May G,** Le Chavanton L, Pukkila P (1991) Genetics 128:529–538. **Maynard Smith J** (1978) The evolution of sex. Cambridge Univ Press, Cambridge. **Mayr E** (1963) Animal species and evolution. Harward Univ Press, Cambridge. **McGill C,** Shafer B, Strathern JN (1989) Cell 57:459–467. **Meinhardt F,** Esser K (1990) Crc Rev Plant Sci 9:329–341. **Meinhardt F,** Leslie JF (1982) Curr Genet 5:65–68. **Metzenberg RL** (1990) Genetics 125:457–462. **Metzenberg RL,** Glass NL (1990) BioEssays 12:53–68. **Michod RE,** Levin BR (eds) (1988) The evolution of sex. An examination of current ideas. Sinauer, Sunderland, Ma. **Miller AM** (1984) EMBO J 3:1061–1065. **Miyata H,** Miyata M (1981) J Gen Appl Microbiol 27:365–371. **Mullen JR,** Kayne PS, Moerschell RP, Tsunsasawa S, Gribskow M, Colavito-Shepanski M, Grunstein M, Sherman F, Sternglanz R (1989) EMBO J 8:2067–2075. **Muller HJ** (1932) Am Natur 66:118–138. **Mutasa ES,** Tymon AM, Göttgens B, Mellon FM, Little PFR, Casselton LA (1990) Curr Genet 18:223–229.

**Nakaseko Y,** Adachi Y, Funahashi S, Niwa O, Yanagida M (1986) EMBO J 5:1011–1021. **Nasmyth KA** (1982) Annu Rev Genet 16:439–500. **Nasmyth KA** (1983) Nature 302:670–676. **Nasmyth K,** Shore D (1987) Science 237:1162–1170. **Nasmyth KA,** Tatchell K, Hall BD, Astell C, Smith M (1981a) Cold Spring Harbor Symp Quant Biol 45:961–981. **Nasmyth KA,** Tatchell K, Hall BD, Astell C, Smith M (1981b) Nature 289:244–250. **Nasmyth KA,** Stillman D, Kipling D (1987) Cell 48:579–587. **Nasmyth K,** Adolf G, Lydall D, Seddon A (1990) Cell 62:631–647. **Ner SS** (1992) Curr Biol 2:208–210. **Ner SS,** Smith M (1989) Mol Cell Biol 9:4623–4620. **Neuhauser KS,** Gilbertson RL (1971) Mycologica 63:722–735. **Nielsen O,** Egel R (1989) EMBO J 8:269–276.

**Novotny CP,** Stankis MM, Speckt CA, Yang H, Giasson L, Ullrich RC (1991) In: Bennett JW, Lasure LL (eds) More manipulations in fungi. Academic Press, San Diego, CA, pp 235–257.

**Passmore S,** Elble R, Tye B-K (1989) Genes Dev 3:921–935. **Perkins DD** (1987) Genetics 115:215–216. **Phillips CL,** Vershon AK, Johnson AD, Dahlquist FW (1991) Genes Dev 5:764–772. **Picard M,** Debuchy R, Coppin E (1991) Genetics 128:539–547. **Pillus L,** Rine J (1989) Cell 59:637–647. **Puhalla JE** (1968) Genetics 60:461–474. **Puhalla JE** (1970) Genet Res 16:229–232.

**Raper C** (1983) In: Bennett JW, Ciegler A (eds) Seconadary metabolism and differentiation in fungi. Marcel Dekker, New York, pp 195–238. **Raper JR** (1966) Genetics of sexuality in higher fungi. Ronald, New York. **Raper JR,** Raper CA (1968) J Elisha Mitchell Sci Soc 84:267–273. **Raper JR,** Baxter MG, Ellingboe AH (1960) Proc Natl Acad Sci USA 46:833–842. **Rine J,** Herskowitz I (1987) Genetics 116:9–22. **Rizet G** (1953) C R Acad Sci 237:1106–1109. **Robson GE,** Williams KL (1990) Curr Genet 1:299. **Rowell JB,** DeVay JE (1954) Phytopathology 44:356–362. **Russell DW,** Jensen R, Zoller M, Burke J, Errede B, Smith M, Herskowitz I (1986) Mol Cell Biol 6:4281–4294.

**Sakagami Y,** Yoshida M, Isogai A, Suzuki A (1981) Science 212:1525–1527. **Sauer RT,** Smith DL, Johnson AD (1988) Genes Dev 2:807–816. **Schulz B,** Banuett F, Dahl M, Schlesinger R, Schäfer W, Martin T, Herskowitz I, Kahmann R (1990) Cell 60:295–306. **Sengupta P,** Cochran BH (1991) Genes Dev 5:1924–1934. **Shepherd JCW,** McGinnis W, Carrasco AE, De Robertis EM, Gehring WJ (1984) Nature 310:70–71. **Siegel RW** (1967) In: Woolhouse HW (ed) Aspects of the biology of aging. Soc Exp Biol Symp 21. Cambridge Univ Press, Cambridge, pp 127–148. **Siliciano PG,** Tatchell K (1984) Cell 37:969–978. **Simchen J** (1967) Genet Res 9:195–210. **Simchen G,** Stamberg J (1969a) Heredity 24:369–381. **Simchen G,** Stamberg J (1969b) Nature 222:329–332. **Sinclair AH,** Berta P, Palmer MS, Hawkins JR, Griffiths BL, Smith MJ, Foster JW, Frischauf A-M, Lovell-Badge R, Goodfellow PN (1990) Nature 346:240–244. **Smith DL,** Johnson AD (1992) Cell 68:133–142. **Specht CA,** Stankis MM, Giasson L, Novotny CP, Ullrich RC (1992) Proc Natl Acad Sci USA 89:7174–7178. **Sprague GF Jr** (1990) Adv Genet 27:33–62. **Staben C,** Yanofsky C (1990) Proc Natl Acad Sci USA 87:4917–4921. **Stahl U,** Esser K (1992) In: Rehm HJ, Reed G (eds) Biotechnology, vol 2, 2nd edn. VCH, Weinheim, pp 73–92. **Stamberg J** (1968) Mol Gen Genet 102:221–228. **Stamberg J,** Koltin Y (1973) Heredity 30:15–26. **Stankis MM,** Specht CA, Yang H, Giasson L, Ullrich RC, Novotny CP (1992) Proc Natl Acad Sci USA 89:7169–7173. **Stearns SC** (ed) (1987) The evolution of sex and its consequences. Birkhäuser, Basel. **Strathern JN,** Herskowitz I (1979) Cell 17:371–381. **Strathern JN,** Spatola E, McGill C, Hicks JB (1980) Proc Natl Acad Sci USA 77:2839–2843. **Szostak JW,** Orr-Weaver TL, Rothstein RJ, Stahl F (1983) Cell 33:25–35.

**Tan S,** Richmond TJ (1990) Cell 62:367–377. **Thomas CM** (ed) (1989) Promiscous plasmids of Gram-negative bacteria. Academic Press, London. **Tice RR,** Setlow RB (1985) In: Finch CE, Schneider EL (eds) Handbook of the biology of aging. Van Nostrand Reinhold, New York, pp 173–224. **Turcq B,** Denayrolles M, Begueret J (1990) Curr Genet 17:297–303. **Turcq B,** Deleu C, Denayrolles M, Begueret J (1991) Mol Gen Genet 228:265–269. **Tymon A,** Kües U, Richardson WVJ, Casselton LA (1992) EMBO J 11:1805–1813.

**Ullrich RC,** Anderson JB (1978) Exp Mycol 2:119–129. **Ullrich RC,** Specht CA, Stankis MM, Yang H, Giasson L, Novotny CP (1991) In: Setlow JK (ed) Genetic engineering, principles and methods, vol 13. Plenum, New York, pp 279–306. **Uyenoyama MK** (1984) Evolution 38:87–102.

**Weismann A** (1891) Essays upon heredity and kindred biological problems, 2nd edn, vols I and II. Clarendon, Oxford. **White CI,** Haber JE (1990) EMBO J 9:663–673. **Whiteway M,** Freeman R, van Arsdell S, Szostak JW, Thorner J (1987) Mol Cell Biol 7:3713–3722. **Willetts N,** Skurray R (1987) In: Neudhardt FC (ed) *Escherichia coli* and *Salmonella typhimurium*. Am Soc Microbiol,

Washington DC, vol 2, pp 1110. **Willetts N,** Wilkins B (1984) Microbiol Rev 48:24–41. **Wolberger C,** Vershon AK, Liu B, Johnson AD, Pabo CO (1991) Cell 67:517–528. **Wong GJ,** Wells K (1985) Trans Br Mycol Soc 84:95–109.

**Yoder OC,** Turgeon BG, Schäfer W, Ciuffetti L, Bohlmann H, van Etten HD (1989) Found Biotechnol Ind Ferment Res 6:189–196. **Youngman PJ,** Anderson RW, Holt CE (1981) Genetics 97:513–530.

Professor Dipl.-Ing. Dr. Ulf Stahl
Institut für Biotechnologie
Fachgebiet Mikrobiologie und Genetik
Technische Universität Berlin
Gustav-Meyer-Allee 25
W-1000 Berlin 65, Germany

Dr. Ursula Kües
Department of Plant Sciences
University of Oxford
South Parks Road
Oxford OX1 3RB, UK

# III. Isolation of Plant Genes by T-DNA and Transposon Mutagenesis – Gene Tagging

By Anders Lönneborg and Christer Jansson

## 1. Introduction

Modern biology depends to a large extent on the identification, isolation and characterization of genes. Describing the expression, function and regulation of these genes is crucial in understanding many of the fundamental questions in development, morphology, pathogen resistance, stress responses, and other phenomena. The classical approaches to cloning require information about the gene product, i.e., the transcript or polypeptide. This information can be used to identify cDNA or genomic clones from a DNA library employing a variety of strategies including different hybridization procedures, synthetic oligonucleotides, expression vectors, specific antibodies or heterologous DNA probes, etc. However, the majority of genes is not easily cloned with these techniques. For most of the agronomically important plant genes very limited information is available about their polypeptide products. Often the only character at hand is a phenotype, a feature difficult to utilize in most of the cloning techniques listed above. Here, gene tagging offers an attractive alternative. The rationale for gene tagging as a technique to identify and isolate genes is that random or directed insertion of DNA sequences into a gene both mutates and tags the gene. The insertional element in plant systems is either T-DNA or transposons (or a combination thereoff). T-DNA and transposon-mediated gene tagging have been used successfully in the isolation of several plant genes in *Arabidopsis*, snapdragon (*Antirrhinum majus*) and maize (*Zea mays*) genes, respectively. In this review we present the basic techniques and results using the gene tagging approach for gene isolation.

## 2. T-DNA Mutagenesis

### a) T-DNA Integration

The basis for *Agrobacterium*-mediated transformation of plants is covered in several recent reviews (e.g. Zambryski 1988) and in the following we describe only the most salient features of this process and then discuss the applications of T-DNA insertion as a gene-tagging technique.

Progress in Botany, Vol. 54

*Agrobacterium tumefaciens* contains a large, ~200 kilobase pair (kb), tumor-inducing (Ti) plasmid. A ~22 kb segment of the Ti-plasmid, called transfer DNA (T-DNA), is excised during infection and transferred in single-stranded form to the plant cell where it can integrate into the nuclear DNA. The only T-DNA sequences required for its mobilization and integration are 25 base-pair (bp) at each end. These border sequences are direct, albeit imperfect, repeats and are termed left border (LB) and right border (RB). The stuffer region of the T-DNA contains genes for auxin, cytokinin and opine synthesis. These genes are controlled by plant-like promoters and are expressed in the plant cell. Integration of T-DNA into the plant genome therefore results in overproduction of auxin and cytokinin and, as a consequence, causes uncontrolled proliferation of the plant cells, i.e. tumor formation. The opines, e.g. nopalin and octopine, are amino acid or keto acid derivatives and are transported out of the plant cell where they serve as carbon and nitrogen sources for the *A. tumefaciens* cells, thus further promoting the infection. Since the stuffer region with the oncogenic auxin and cytokinin genes is not required for the transformation process, it can be deleted from the T-DNA, generating so-called disarmed Ti-plasmids. Another segment of the Ti-plasmid, the ~35-kb *vir* region, is located outside the T-DNA and contains loci essential for T-DNA mobilization and transfer. The *vir* D gene encodes endonucleases that excise the T-DNA at the border sequences. The *vir* E gene encodes single-stranded DNA-binding proteins required to stabilize the single-stranded T-DNA during transfer. Other *vir* products include translocated proteins and a *vir* transcriptional activator.

A wide variety of Ti-plasmids have now been constructed for plant transformation. The two general types of vector system used are the cointegrative vector system and the binary vector system. In the cointegrative system (Fig. 1) a disarmed Ti-plasmid contains a sequence from the *E. coli* plasmid pBR322 and a marker gene functional in plants (to allow for detection of transformed plant cells) cloned beween LB and RB. The desired gene for integration into the plant genome is cloned in pBR322 or a pBR322-based vector. This construct is then transferred to *A. tumefaciens* with the aid of a helper plasmid providing mobilization and transfer functions. In the *A. tumefaciens* cell the desired gene is cointegrated into the Ti-plasmid through homologous recombination between the pBR322 sequences. The binary system makes use of a cloning T-DNA vector consisting of a disarmed Ti-plasmid with a deleted *vir* region and containing a broad host range origin of replication functional in *E. coli* and *A. tumefaciens*. It also contains cloning sites and a marker gene inserted between LB and RB. The desired gene is cloned between LB and RB in the cloning T-DNA vector and the construct is shuttled to *A. tumefaciens* using a helper plasmid as described above. The recipient *A. tumefaciens* cells harbor a *vir*-containing Ti-plasmid, perferably lacking T-DNA. Thus, in the binary vector system the *vir* functions are provided in *trans* to the T-DNA. The major advantage with the cointegrative vector system is that it does not require conveniantly located restriction sites in the Ti-plasmid. One advantage of the binary system is that is does not depend on a recombination event in *A. tumefaciens*. Thus, each vector system has its benefits and drawbacks and they are both widely employed in plant transformation protocols.

As indicated above, transformation of plants with T-DNA vectors requires that the vectors carry a marker gene that is expressed in the plant cell. Such markers can be (1) genes conferring antibiotic resistance (antibiotic resistance genes), (2) color indicators, such as the *gus* gene encoding β-glucuronidase, or (3) light elicitors, such as the bacterial

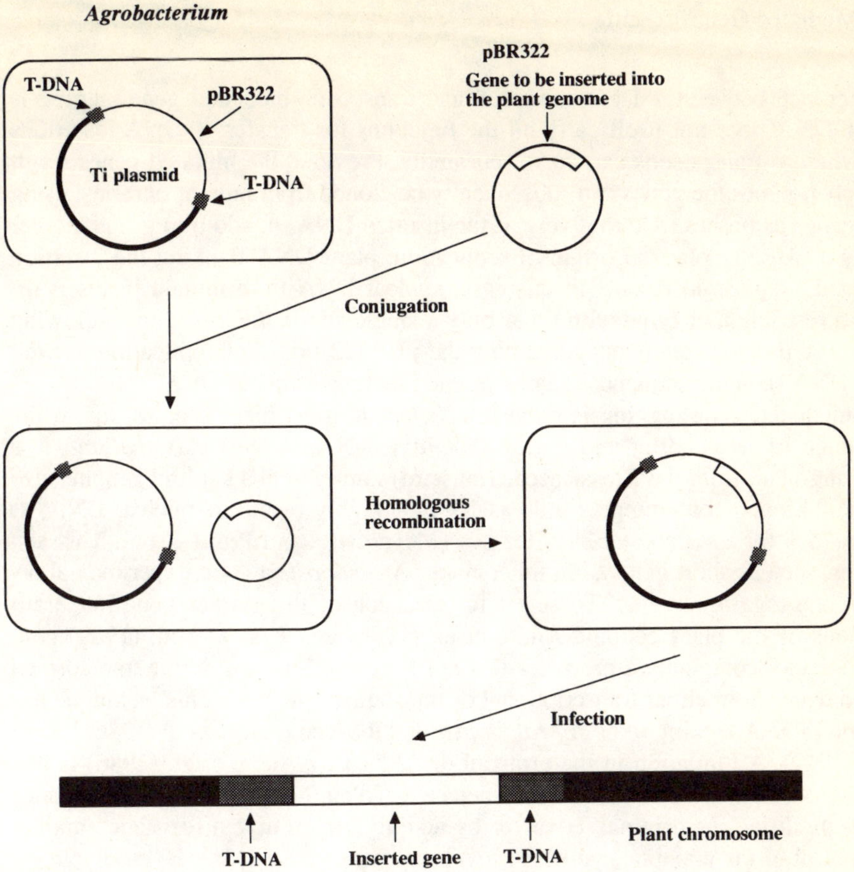


**Fig. 1.** Transformation of plants with *Agrobacterium tumefaciens* using the cointegrative vector system. A portion of the *E. coli* plasmid pBR322 is cloned together with a marker gene (not shown) between the left and right border sequences of the T-DNA in a disarmed Ti-plasmid. The recombinant Ti plasmid is used to transform *A. tumefaciens* cells. The gene of interest is cloned in the pBR322 vector and the construct is transferred to *A. tumefaciens* by conjugation. Homologous recombination between the pBR322 sequences in the Ti plasmid and the cloning vector results in the cointegrative vector, which contains the cloned gene plus all elements required for infection of the plant cells

*lux* or firefly *luc* genes. By far the most commonly used selectable plant markers in T-DNA-mediated transformation is the *npt* II gene, conferring kanamycin resistance ($km^r$).

During plant transformation, the T-DNA inserts seemingly at random in the nuclear chromosomes and it has not been possible to direct the insertion to a particular locus. This is in contrast to insertional mutagenesis in prokaryotic systems such as cyanobacteria (Jansson and Lönneborg 1991) where a modified construct can be used to replace a wild-type gene through homologous recombination. Recent developments in mouse genetics (Mansour et al. 1988) may hold promise for a more general exploitation of homologous recombination in eukariotic systems, including plants.

### b) T-DNA-Mediated Gene Tagging

A main difference between T-DNA-mediated and transposon-mediated gene tagging is that, since T-DNA does not itself carry all the functions for transfer, T-DNA insertions are stable, whereas transposons can move repeatedly. Provided the mutated genes result in scorable phenotypes the genes can subsequently be cloned from mutant libraries, using the marker genes as probes. Alternatively, if the inserted DNA, in addition to the marker gene, carries a pBR322 plasmid origin of replication, plant DNA flanking the insertion can be isolated by plasmid rescue. In this case, nuclear DNA from mutant plants is digested with a restriction enzyme with no or only a single site in the insertion. Following self-ligation circularized fragments containing the pBR322 origin of replication as well as the plant DNA flanking sequences can be rescued by transformation in *E. coli*.

T-DNA-mediated gene tagging is rather inefficient in most plant species due to the high percentage of non-coding and highly repetitive DNA. *Arabidopsis thaliana* is a small flowering plant from the Brassicaceae (mustard) family with a haploid genome size of only 70 000 kb and containing as little as 10–14% highly repetitive nuclear DNA, as compared to 75% for tobacco and 84% for pea (Meyerowitz and Pruitt 1985). This and other features, such as short generation time, makes *Arabidopsis* a good experimental organism for gene tagging studies. To select for insertion of the marker gene into transcribed regions of the plant genome André et al. (1986) and Koncz et al. (1989) constructed a Ti-vector containing a promoter-less *npt* II gene. Thus, $km^r$ in the transformed plants would result from either transcriptional or translational fusions. This technique has been used for T-DNA tagging in both *Arabidopsis* and tobacco (André et al. 1986; Koncz et al. 1989, 1990). A limitation in the protocol developed by André et al. (1986) is that only genes transcribed at the callus level are detected. To circumvent this problem Koncz et al. (1989) modified the original Ti-vector by adding a hygromycin resistance marker under the control of the nopaline synthase (*nos*) promoter (see Sect. 2a). Hygromycin-resistant transgenic plants were then screened for $km^r$ in vegetative tissues. The Ti-vector was also provided with a pBR322 origin of replication to allow for plasmid rescue of the tagged genetic material. Using this vector system Koncz et al. (1990) identified a recessive *pale* mutation in *Arabidopsis* and isolated the tagged gene responsible for the mutation. DNA fragments of rescued clones were then used as probes to isolate wild-type cDNA and genomic clones.

Tissue culture procedures often induce high frequencies of somaclonal variation. This causes a potential problem in gene tagging experiments in that most of the transformation systems developed for *Arabidopsis* involve tissue cultures. To avoid this problem Feldmann and Marks (1987) developed a nontissue culture transformation system for *Arabidopsis*, which involves infecting germinating seeds with *Agrobacterium.* Using this whole seed transformation technique and a Ti-vector containing the *npt* II gene, Feldmann et al. (1989) and Feldmann (1991) constructed large collections of transgenic *Arabidopsis* lines. Several of these transformants displayed a visible mutant phenotype, such as arrested or altered development, floral abnormalities, size reduction or altered pigmentation. Analysis of the progeny showed that the majority of the mutants segregated in a Mendelian manner and suggested that in these cases the phenotype was caused by T-DNA insertion mutagenesis. For one of the floral mutants tagged DNA sequences were isolated by plasmid rescue and the corresponding wild-type cDNA cloned by screening a library with one of the rescued plasmids (Yanofsky et al. 1990). For another mutant, with an altered trichome phenotype, flanking sequences to the T-DNA insert were isolated by screening a mutant library with a region of the insert as a probe (Marks and Feldmann 1989). The obtained clone was subsequently used as a probe to isolate the wild-type gene from a genomic library (Herman and Marks 1989). A third mutant, displaying a dwarf (*dwf*) phenotype, has been examined in some detail (Feldmann et al. 1989).

## 3. Transposon Mutagenesis

### a) Transposons

Transposons were first discovered and described by Barbara McClintock in the late 1940s and early 1950s (McClintock 1948, 1951, 1953). This work provided the foundation for much of today's genetics and molecular genetics. The majority of information has been obtained from studies on transposon behavior in maize (McClintock 1947, 1948, 1951a, b, 1953, 1958, 1965; Peterson 1953) and snapdragon *(Antirrhinum majus)* (Harrison and Fincham 1964; Fincham and Harrison 1967; Harrison and Carpenter 1973) and is now being used to elucidate the mechanisms of transposon activity at the molecular level.

A transposon is defined as a DNA sequence that has the capability of moving from one place in the genome to another trough a process of excision and reintegration. There is genetic evidence for the existence of transposable element systems in a variety of plant species. The best-characterized systems are the *Ac*/*Ds* elements and *Spm*/*dSpm* elements in maize and the *Tam* elements in snapdragon, all of which have been isolated and characterized at the molecular level (Barker 1984; Behrens et al. 1984; Bennetzen et al. 1984; Bonas et al. 1984a; Müller-Neumann et al. 1984; Pohlmann et al. 1984a; Schwarz-Sommer et al. 1984; Pereira et al. 1985).

Of the most well-characterized transposons, all occur as families of related elements. In each family two types of transposons can be identified. Autonomous elements, such as *Ac* and *Spm* carry the genes necessary for excision and reintegration and are thus capable of regulating their own transposition (Fedoroff et al. 1983; Pereira et al. 1986; Frey et al. 1990). These elements can be rendered inactive through methylation of cytosine residues (Schwartz and Dennis 1991; Kunze et al. 1988). Non-autonomous elements, such as *Ds* and *dSpm*, also called cryptic or defective elements, are not by themselves able to transpose but when an autonomous element is present in the genome its gene products can activate the non-autonomous element in *trans*, resulting in transposition of the non-autonomous element (McClintock 1948, 1954). These non-autonomous elements are devoid of smaller or larger regions within the terminal inverted repeats and possibly lack the transposase gene necessary for transcription. Multiple copies of related autonomous or non-autonomous sequences within a family are often scattered throughout the genome so that the family represents middle repetitive DNA (Fedoroff et al. 1983; Barker 1984; Sutton et al. 1984; Peacock et al. 1985; Upadhyaya et al. 1985).

If the site of transposon integration is in or in the vicinity of a particular gene, the expression of that gene can be altered in a variety of ways depending on the proximity of the element to the structural or regulatory regions of the gene. Most transposable elements have been characterized for their activity on genes with an easily scorable phenotype like flower or seed pigmentation or seed structure (McClintock 1948, 1954, 1953, 1958; Harrison and Fincham 1964; Fincham and Harrison 1967; Harrison and Carpenter 1973). For example, if a transposon is inserted in a gene involved in anthocyanin pigmentation of the outermost layer of the maize endosperm, the aleurone, the normal purple color is not produced and the aleurone becomes unpigmented. Mutations generated by transposons are commonly more or less unstable. Therefore, if the element has been excised from the gene, the purple pigmentation can be restored. Somatic mutations are recognized by such variegated phenotypes. With an active element inserted in a gene involved in anthocyanin synthesis purple spots on a yellow background will indicate that the transposon has been excised from the gene during somatic tissue growth and transposed to another place in the genome. A germinal excision is much more stable and the revertant progeny breeds true for the new phenotype.

Upon insertion, a small sequence of the target site is duplicated. The length of this sequence is characteristic for the type of transposon that has inserted. For *Ac/Ds* it is an 8-bp duplication (Pohlmann et al. 1984a; Sutton et al. 1984; Weck et al. 1984; Peacock et al. 1985), for *Spm/dSpm* it is a 3-bp duplication (Schwarz-Sommer et al. 1984, 1985; Gierl et al. 1985; Pereira et al. 1985) and for *Tam* 1 and *Tam* 2 it is also a 3-bp duplication (Bonas et al. 1984a; Uphadhyaya et al. 1985). When a transposon is excised, the unstable phenotype is not always reverted to the wild-type phenotype, but instead the mutated phenotype has become stable or a new phenotype is created. This is because the duplication is retained, usually in a mutated form (Dooner and Nelson 1979; Tuschall and Hannah 1982). If the footprint from an excised transposon is in an intron or in a leader sequence of a gene, it may not be detected. However, if the insertion is in an exon, the footprint may cause a frameshift mutation.

### b) Transposon-Mediated Gene Tagging

Employing a transposon in gene tagging experiments requires that the element be cloned and characterized at the molecular level. Representatives of the *Ac/Ds*, *En/Spm/dSpm* and *Mu*1 families of maize and *Tam* elements from snapdragon have been cloned (Barker 1984; Behrens et al. 1984; Bennetzen et al. 1984; Bonas et al. 1984a; Pohlmann et al. 1984a, b; Schwarz-Sommer et al. 1984; Pereira et al. 1985; Upadhyaya et al. 1985) and all of them have been used in successful gene tagging experiments (Fedoroff et al. 1984; Dooner et al. 1985; Martin et al. 1985; O'Reilly et al. 1985; Cone et al. 1986; Paz-Ares et al. 1986; Wienand et al. 1986; Schmidt et al. 1987; Theres et al. 1987; Martienssen et al. 1989; McCarty et al. 1989; McLaughlin and Walbot 1987; Hake et al. 1989). The principle of transposon tagging is to cross a plant known to harbor an active transposable element system to another, genetically distinct, plant and screen the progeny for the presence of new mutations.

When an attempt is being made to tag a specific gene, a targeted approach can be used (Fig. 2; McClintock 1953; Motto et al. 1986; Schmidt et al. 1987; described in Chandlee 1990, 1991), in which one parent is a homozygous recessive for the particular gene and the other homozygous dominant with an active transposable element located in the vicinity of the gene. When a recessive phenotype is found in the progeny, there is good indi-

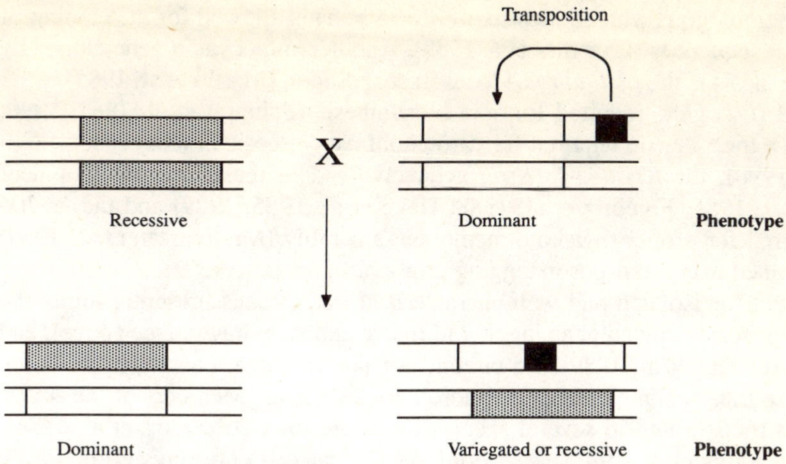


**Fig. 2.** Targeted transposon tagging. A parent that is homozygous recessive for the targeted gene is crossed with a parent that is homozygous dominant for the gene and that harbors a transposon element elsewhere in the genome. At a low frequency, transposition will place the transposable element in the targeted gene. Such a transposition will be uncovered in the $F_1$ progeny as a variegated or recessive phenotype. (After Chandlee 1990)

cation that the gene has been tagged by the transposon. If an attempt is being made to find and recover new mutations for a variety of different plant characteristics, the non-targeted approach should be used (described in Chandlee 1990, 1991). In this approach a standard line is crossed with a line harboring an active transposable element. The $F_1$ progeny is selfed and recessive mutations are detected in the $F_2$ generation. In rare instances a wild-type and a variegated (or stable recessive) phenotype will segregate 3:1 in the $F_2$ progeny of an $F_1$ individual, indicating the insertion of a transposable element. Once the homozygous mutant is recovered, a genomic library can be made. Since the mutation is created by a transposon that has been characterized, a labelled probe of the transposon can be used to screen for the mutated gene. A transposon with a flanking sequence from the mutated gene can then be isolated and the flanking sequence subsequently used to screen a wild-type genomic library for the corresponding non-mutated gene.

In contrast to the large number of characterized maize (and snapdragon) mutants caused by transposable elements (for review, see Döring and Starlinger 1986; Weil and Wessler 1990), a relatively small number of genes have so far been isolated with transposon tagging protocols. With few exceptions these genes are involved in pigmentation. The *C1*, *C2*, *R*, *B*, *P*, *A1*, *Bronze-1 (Bz1)* and *Bz2* loci in maize are part of a network controlling anthocyanin biosynthesis (Chandlee 1990, 1991). This network also includes the *Vp1* locus, which apart from its role in anthocyanin production also regulates embryo dormancy (Chandlee 1991; Neill et al. 1987). The tagged *C1* (Cone et al. 1986; Paz-Arez et al. 1986), *C2* (Wienand et al. 1986), *R* (Dellaporta et al. 1988), *P* (Chen et al. 1987; Lechelt et al. 1989), *A1* (O'Reilly et al. 1985), *Bz1* (Fedoroff et al. 1984), *Bz2* (McLaughlin and Walbot 1987; Theres et al. 1987) and *Vp1* (McCarty et al. 1989a, b) genes and their wild-type progenitors have all been cloned by transposon tagging. The *R* and *B* genes are homologous with an identity of 85–95%. This allowed for the cloning of the *B* gene using *R* sequences (Chandler et a. 1989). Another anthocyanin gene cloned by transposon mutagenesis is the *pallida (pal)* locus in snapdragon (Martin et al. 1985).

The *Opaque-2* (*O2*) gene required for zein biosynthesis (Schmidt et al. 1987, 1990; Motto et al. 1988), the *Y1* gene required for carotenoid biosynthesis in maize endosperm (Buckner et al. 1990), the *Knotted-1* (*Kn1*) gene involved in maize leaf development (Freeling and Hake 1985; Freeling et al. 1988; Hake et al. 1985, 1989) and the *hcf106* maize gene required for proper thylakoid membrane assembly (Martienssen et al. 1989) have also been cloned using transposon tagging protocols.

The requirement for isolated and well-characterized transposons presently limits the approach of transposon-mediated gene tagging to maize and snapdragon (see above), and possibly petunia (Gerats et al. 1989). To circumvent this restriction attempts are being made to use maize transposon elements in heterologous systems. Members of the *Ac/Ds* family have been transformed to several species including tobacco (Baker et al. 1986), *Arabidopsis* (Van Sluys et al. 1987; Dean et al. 1992), carrot (Van Sluys et al. 1987), potato (Knapp et al. 1988), tomato (Yoder et al. 1988).

Another problem encountered with transposon tagging is that the transposons characterized so far all have a tendency to insert relatively close to the excision site, making it difficult to tag a gene at some distance from the transposon-associated region in the parent. There is also the risk of losing a mutation by a second transposition of the same element. Also, here, the approach with heterologous systems should prove useful. The strategy would be to introduce non-autonomous elements, like *Ds* or *dSpm*, at various locations in the genome of a plant. These elements are all stable as long as no autonomous element is introduced in the plant. When this plant is crossed with a plant containing an active element, like *Ac* or *Spm*, this element will activate the defective elements and induce new mutations. The progeny is then selfed and new transposon-tagged mutations are detected. Once the mutation of interest has been identified the autonomous element is outcrossed, thereby stabilizing the mutation. With backcrossing, the defective element cosegregating with the phenotype can be identified and the tagged gene isolated. This approach is now being tested in *Arabidopsis* (Bancroft et al. 1992) and tobacco (Rommens 1992). *Arabidopsis* was transformed with a stabilized *Ac (sAc)* lacking the 3' terminus. Due to the 3' deletion, this element is unable to transpose but retains the capability to produce the transposase and can induce transposition of the *Ds* element in *trans*. In the tobacco system only the transposase gene of the *Ac* element was used. The gene was fused to the flower-specific chalcone synthase A promoter (*chsA*), the anther-specific chalcone isomerase B promoter (*chiB*) or the anther-specific chalcone isomerase A2 promoter ($chiA_2$). The modified transposase genes were introduced into tobacco plants containing a *Ds* element, which was subsequently activated in *trans* by the transposase. Somatic excision frequencies increased with the *chsA* construct, while it decreased with *chiB* and was unaffected with $chiA_2$.

An additional advantage with heterologous systems is that the species receiving a transposable element do not contain any element of the same family in its genome. This makes the identification of a transposable element cosegregating with the mutant phenotype easier. Since the well-characterized transposon families of maize and snapdragon are found as middle repetitive DNA in the normal host plant, the identification of the correct transposon in these plants is much more cumbersome. However, the observation of undermethylation of sequences in transposable elements and their flanking regions (Chandler and Walbot 1986; Dellaporta and Chomet (1985) has been exploited in gene tagging experiments to distinguish between active elements and related, inactive sequences (Schmidt et al. 1987; Cone et al. 1986). This is an especially useful strategy in maize that generally displays a high degree of methylation.

## 4. Prospects

Gene tagging offers a very powerful means to delineate the genetic regulation of phenotypically scorable plant processes. Non-targeted gene tagging, to which also T-DNA tagging belongs, is the more general approach and allows the construction of a large number of novel mutations. In addition, since non-targeted transposon tagging does not rely on homozygous recessive lines, essential genes can be also tagged. The advantage of gene tagging is best demonstrated in the cloning of genes expressed at very low levels, and therefore difficult to clone using other approaches. However, the potential of the technique should make gene tagging an increasingly important step in the cloning, characterization and exploitation of virtually any functional gene, including those of agronomic interest.

## References

**André D,** Colau D, Schell J, van Montagu M, Hernalsteens J-P (1986) Mol Gen Genet 204:512–518.

**Baker B,** Schell J, Lörz H, Fedoroff N (1986) Proc Natl Acad Sci USA 83:4844–4848. **Bancroft I,** Bhatt A, Sjödin C, Scofield S, Jones J, Dean C (1992) Mol Gen Genet (in press). **Barker R** (1984) Nucleic Acid Res 12:5955–5967. **Behrens U,** Fedoroff N, Laird A, Muller-Neumann M, Starlinger P, Yoder J (1984) Mol Gen Genet 194:346–349. **Bennetzen J,** Swanson J, Taylor C, Freeling M (1984) Proc Natl Acad Sci USA 81:4125–4129. **Bonas U,** Sommer H, Saedler H (1984a) Mol Gen Genet 1984:138–143. **Bonas U,** Sommer H, Saedler H (1984b) EMBO J 3:1015–1019. **Buckner B,** Kelson T, Robertson DS (1990) Plant Cell 2:867–876.

**Chandlee JM** (1990) Physiol Plant 79:105–115. Chandlee JM (1991) Dev Genet 12:261–271. **Chandler VL,** Walbot V (1986) Proc Natl Acad Sci USA 83:1767–1771. **Chandler VL,** Radicella JP, Robbins TP, Chen J, Turks D (1989) Plant Cell 1:1175–1183. Chen J, Greenblatt IM, Dellaporta S (1987) Genetics 117:109–116. **Cone KC,** Burr FB, Burr B (1986) Proc Natl Acad Sci USA 83:9631–9635.

**Dean S,** Sjödin C, Page T, Jones J, Lister C (1992) Plant J 2:69–82. **Dellaporta SL,** Chomet PS (1985) The activation of maize controlling elements. In: Hohn B, Dennis ES (eds) Genetic flux in plants, Springer, Berlin Heidelberg New York, pp 169–216. **Dellaporta SL,** Greenblatt I, Kermicle JL, Hicks JB, Wessler SR (1987) Stadler Symp 18:263–282. **Dooner HK,** Nelson O (1979) Proc Natl Acad Sci USA 76:2369–2371. **Dooner HK,** Weck E, Adams S, Ralston E, Favreau M, English J (1985) Mol Gen Genet 200:240–246. **Döring H-P,** Starlinger P (1986) Annu Rev Genet 20:175–200.

**Fedoroff F,** Wessler S, Shure M (1983) Cell 35:235–242. **Fedoroff N,** Furtek DB, Nelson OE (1984) Proc Natl Acad Sci USA 81:3825–3829. **Feldmann KA** (1991) Plant J 1:71–82. **Feldman KA,** Marks MD (1987) Mol Gen Genet 208:1–9. **Feldmann KA,** Marks MD, Christianson ML, Quatrano RS (1989) Science 243:1351–1354. **Fincham JR,** Harrison BR (1967) Heredity 22:211–224. **Freeling M,** Bongard-Pierce DK, Harberd N, Lane B, Hake S (1988) Genes involved in the patterns of maize leaf cell division. In: Verma P, Goldberg R (eds) Plant gene research – temporal and spatial regulation of plant genes. Springer, Berlin Heidelberg New York, pp 41–62. **Frey M,** Reinecke J, Grant S, Saedler H, Gierl A (1990) EMBO J 9:4037–4044.

**Gerats AGM,** Beld M, Huits H, Prescott A (1989) Dev Genet 10:561–568. **Gierl A,** Schwarz-Sommer Z, Saedler H (1985) EMBO J 4:579–583.

**Hake S,** Bird R, Neuffer MG, Freeling M (1985) Development of the maize ligule and mutants that affect it. In: Freeling M (ed) Plant genetics. Alan R Liss, New York, pp 61–71. **Hake S,** Vollbrecht E, Freeling M (1989) EMBO J 8:15–22. **Harrison BJ,** Carpenter R (1973) Heredity 31:309–323. **Harrison BJ,** Fincham JRS (1964) Heredity 19:237–258. **Herman PL,** Marks MD (1989) Plant Cell 1:1051–1055.

**Janson C,** Lönneborg A (1991) Construction and analysis of cyanobacterial mutants in photosynthesis research. In: Behnke H-D, Esser K, Kubitzki K, Runge M, Ziegler H (eds) Progress in Botany, vol 52. Springer, Berlin Heidelberg New York, pp 226–233.

**Knapp S,** Coupland G, Uhrig U, Starlinger P, Salamini F (1988) Mol Gen Genet 213:285–290. **Koncz C,** Martini N, Mayerhofer R, Koncz-Kalman Z, Körber H, Redei GP, Schell J (1989) Proc Natl Acad Sci USA 86:8467–8471. **Koncz C,** Mayerhofer R, Koncz-Kalman Z, Nawrath C, Reiss B, Redei GP, Schell J (1990) EMBO J 9:1357–1346. Kunze R, Starlinger P, Schwartz D (1988) Mol Gen Genet 214:325–327.

**Lechelt C,** Peterson T, Laird A, Chen J, Dellaporta S, Dennis E, Peacock W, Starlinger P (1989) Mol Gen Genet 219:225–234.

**Mansour SL,** Thomas KR, Capecchi MR (1988) Nature 336:348–352. **Marks MD,** Feldmann KA (1989) Plant Cell 1:1043–1050. **Martienssen RA,** Barkan A, Freeling M, Taylor WC (1989) EMBO J 8:1633–1639. **Martin C,** Carpenter R, Sommer H, Saedler H, Coen ES (1985) EMBO J 4:1625–1630. **McCarty DR,** Carson CB, Stinard PS, Robertson DS (1989a) Plant Cell 1:523–532. **McCarty DR,** Carson CB, Lazar M, Simonds SC (1989b) Dev Genet 10:473–481. **McClintock B** (1947) Carnegie Inst Wash Year Book 47:155–169. **McClintock B** (1948) Carnegie Inst Wash Year Book 48:142–154. **McClintock B** (1951a) Carnegie Inst Wash Year Book 50:174–181. **McClintock B** (1951b) Cold Spring Harbor Symp Quant Biol 16:13–47. **McClintock B** (1953) Genetics 38:579–599. **McClintock B** (1954) Carnegie Inst Wash Year Book 54:254–260. **McClintock B** (1958) Carnegie Inst Wash Year Book 58:452–456. **McClintock B** (1965) Brookhaven Symp Biol 18:162–184. **McLaughlin M,** Walbot V (1987) Genetics 117:771–776. **Meyerowitz EM,** Pruitt RE (1985) Science 229:1214–1218. **Motto M,** Marotta R, DiFonzo N, Soave C, Salamini F (1986) Genetics 112:121–133. **Motto M,** Maddaloni M, Ponziani G, Brembilla M, Martta R, DiFonzo N, Soave C, Thompson R, Salamini F (1988) Mol Gen Genet 212:488–494. **Müller-Neumann M,** Yoder J, Starlinger P (1984) Mol Gen Genet 198:19–24.

**Neill JJ,** Horgan R, Rees AF (1987) Planta 171:358–354.

**O'Reilly C,** Shepherd N, Pereira A, Schwarz-Sommer Z, Bertram I, Robertson D, Peterson P, Saedler H (1985) EMBO J 4:877–882.

**Paz-Arez J,** Wienand U, Peterson P, Saedler H (1986) EMBO J 5:829–833. **Peacock WJ,** Dennis ES, Gerlach WL, Sachs MM, Schwarz D (1985) Cold Spring Harbor Symp Quant Biol

19:347–354. **Pereira A,** Schwarz-Sommer Z, Gierl A, Bertran I, Peterson P, Saedler H (1985) EMBO J 4:17–23. **Pereira A,** Cuypers H, Gierl A, Schwarz-Sommer Z, Saedler H (1986) EMBO J 5:835–841. **Peterson PA** (1953) Genetics 38:682–683. **Pohlmann R,** Fedoroff N, Messing J (1984a) Cell 37:635–643. **Pohlmann R,** Fedoroff N, Messing J (1984b) Cell 39:417.

**Rommens CMT,** van Haaren MJJ, Buchel AS, Mol JNM, van Tunen AJ, Nijkamp HJJ, Hille J (1992) Mol Gen Genet 231:433–441.

**Schmidt RJ,** Burr FA, Burr B (1987) Science 238:960–963. **Schmidt RJ,** Burr FA, Aukerman MJ, Burr B (1990) Proc Natl Acad Sci USA 87:46–50. **Schwartz D,** Dennis ES (1986) Mol Gen Genet 205:476–482. **Schwarz-Sommer Z,** Gierl A, Klosgen R, Wienand U, Peterson P, Saedler H (1984) EMBO J 3:1021–1028. **Schwarz-Sommer Z,** Gierl A, Berntgen R, Saedler H (1985) EMBO J 4:2439–2443. **Sutton WD,** Gerlach WL, Schwartz D, Peacock WJ (1984) Science 223:1265–1268.

**Theres N,** Scheele T, Starlinger P (1987) Mol Gen Genet 109:193–197. **Tuschall DM,** Hannah LC (1982) Genetics 100:105–111.

**Upadhyaya K,** Sommer H, Krebbers U, Saedler H (1985) Mol Gen Genet 199:201–207.

**Van Sluys MA,** Tempe J, Fedoroff N (1987) EMBO J 6:3881–3889.

**Weck E,** Courage U, Döring H, Fedoroff N, Starlinger P (1984) EMBO J 3:1713–1716. **Weil CF,** Wessler SR (1990) Annu Rev Plant Physiol Plant Mol Biol 41:527–552. **Wienand U,** Weydeman V, Niesbach-Klosgen U, Peterson P, Saedler H (1986) Mol Gen Genet 203:202–207.

**Yanofsky MF,** Ma H, Bowman J, Drews GN, Feldmann KA, Meyerowitz EM (1990) Nature 346:35–39. **Yoder JI,** Palys J, Alpert K, Lassner M (1988) Mol Gen Genet 213:291–296.

**Zambryski P** (1988) Annu Rev Genet 22:1–30.

Dr. Christer Jansson
The Arrhenius Laboratories
Department of Biochemistry
University of Stockholm
S-10691 Stockholm, Sweden

Dr. Anders Lönneborg
Norwegian Forest Research Institute
Plant Molecular Biology Laboratory
P.O. Box 5051
N-1432 Ås-NLH, Norway

# IV. Function of the Genetic Material: Transposable Elements in Lower Eukaryotes

By Michael Ciriacy

## 1. Introduction

In the past 10 years increasing evidence has been provided that eukaryotic genomes are inhabited by a variety of mobile genetic elements. The occurrence of such elements in forms of plasmids or transposons has changed our view of the eukaryotic genome in terms of its stability and its evolutionary origins quite considerably. Even though our knowledge of the spectrum of mobile elements and their occurrence in the various taxa is still limited, there is now sufficient information available on their molecular nature, their replication strategies, and their cellular function to outline some common and general features. This review will be focused on lower eukaryotes such as yeast, fungi, slime molds, and algae. Research on this group of organisms has contributed, although to a quite different degree, very much to our current knowledge on the nature of eukaryotic mobile elements. Furthermore, I will restrict this review to mobile sequences which interact physically in some way with the nuclear or extranuclear genomes. Thus, the reader is referred to the comprehensive treatise by (Berg and Howe 1989) which covers the entire field of "mobile DNA". The presentation of material in this review is not only biased by my own research interests, but also by extreme variability in knowledge in the various systems investigated. Starting with an overview of transposons in lower eukaryotes and their possible relationship to each other and to other entities of cellular life, I will discuss the major strategies for expression of genetic information carried by transposons and for replication of the genome.

## 2. LTR-Transposons

One of the first hints for the existence of mobile DNA in lower eukaryotes was provided by (Cameron et al. 1979). They showed that baker's yeast cells, *Saccharomyces cerevisiae*, contain a moderately repetitive sequence in the nuclear genome. This sequence, called Ty (*t*ransposon in *y*east), apparently was inserted at various locations within the genome and, because there were considerable variations between different strains, it was anticipated that the Ty sequence is a transposable element. Almost at the same time it was discovered that a number of yeast regulatory mutations at the *HIS4*, *CYC7*, and *ADH2* loci were in fact caused by insertions of a Ty sequence within the promotor region

Progress in Botany, Vol. 54

of these genes (Farabaugh and Fink 1980; Errede et al. 1980; Williamson et al. 1981). It also has been observed that a Ty element can disappear from its location by homologous recombination between the terminal, 340-bp direct repeats, leaving behind a single LTR sequence (Ciriacy and Williamson 1981). Such sequences (delta, sigma, tau) have fortuitously been detected in the vicinity of tRNA genes (Gafner and Philippsen 1980); approximately 100 copies are dispersed over the entire genome, while usually 30 to 40 copies of the complete Ty element have been found per haploid genome in laboratory strains, and significantly less in natural isolates.

The first sequence published by Clare and Farabaugh (1985) revealed its striking similarity in structure to the retroviral proviruses of mammals and to the mobile element *copia* of *Drosophila melanogaster* (Mount and Rubin 1985). The sequence consists of two directly repeated terminal elements (long terminal repeats, LTR) flanking a 5.2-kb central coding region (Fig. 1). Due to the similarities deduced in the protein sequences it has been predicted that Ty as well as *copia* encode a reverse transcriptase, an integrase, and a protease activity. As revealed later the protein encoded by the 5'proximal part (TYA) has functional equivalence to retroviral gag proteins (group-specific antigen) which form the virion capsid.

At the same time (Boeke et al. 1985) demonstrated in an ingenious way that transposition of Ty elements occurs via an RNA intermediate. Upon insertion of a small yeast intron into a genetically marked Ty1 sequence and subsequent analysis of the transposed sequence, it could be demonstrated that in all cases the intron had been precisely removed; this could only be explained by assuming that the processed RNA was used as a template for DNA synthesis. In this context the term "retrotransposon" has been introduced in order to classify transposable elements which replicate via a retrovirus-like strategie, i.e., reverse transcripition and integration into the host chromosomal DNA.

The functional homology between retrotransposons (Ty, *copia*) could even be extended when it was shown by several groups that Ty proteins and Ty-RNA/DNA are assembled in cytoplasmic virus-like particles of approximately 60 nm diameter (Garfinkel et al. 1985; Müller et al. 1987) Normally, yeast cells contain only a few such particles (~10 per haploid cell). Upon overexpression of Ty-RNA and proteins by appropriate fu-

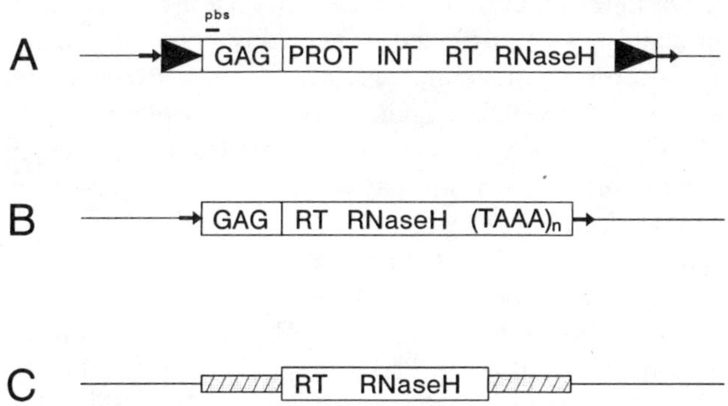


**Fig. 1.** Structural features of three major classes of retrotransposons. **A** LTR elements: **B** poly(A) retrotransposons; **C** retroelements of organellar class II introns. *Arrows* indicate target-site duplications of chromosomal DNA. *Abbrevations* denote domains with structural and functional homology to retroviral proteins. *GAG* Nucleic acid-binding capsid protein; *PROT* protease; *INT* integrase; *RT* reverse transcriptase; *RNaseH* ribonuclease H; *pbs* primer binding site

sion of a Ty sequence to a strong promotor, several thousand Ty particles per cell have been observed. This experimental approach has been extremely useful in a detailed analysis of Ty-protein functions and the integration process (see below).

Since the discovery of the retrovirus-like life-style of Ty1 and Ty2 elements in *S. cerevisiae*, several structurally related elements have been isolated in various lower eukaryotes. In *S. cerevisiae* itself, two other transposons became known. One of them, Ty3, resembles retroviruses even more than Ty1/Ty2 in the order of functional domains of the *TYB* and *pol* genes, respectively (protease – reverse transcriptase – integrase; cf Fig. 1). The peculiarity of this element is its integration specificity for tRNA genes (Chalker and Sandmeyer 1992). Approximately 10% of the estimated 360 yeast tRNA genes is associated with Ty3 or its single LTR sequence, the sigma element. Most, if not all, of them are inserted 16–18 bp upstream of the mature 5'end of the tRNA coding sequence. A fourth element, Ty4, is like Ty3 a low-copy number element (Sandmeyer et al. 1990).

Two retrotransposons, Tf1 and Tf2, belonging to the "gypsy" family (according to a retrotransposon from *D. melanogaster*), have recently been isolated from the fission yeast, *Schizosaccharomyces pombe*. Although Tf1 shows the overall structure of the gypsy/Ty3 elements, it is only 4.7 kb in length. Surprisingly, the nucleotide sequence of Tf1 revealed a unique feature. It contains only one open reading frame (ORF) of 1340 amino acids (Levin et al. 1990). Typically, all known retroelements or retroviruses contain two or more ORFs which allow differential expression of the gene products. The first ORF, the *gag* homolog, produces abundant virion structural protein, whereas the second ORF, the *pol* homolog, encodes the enzymatic components of the virus-like particle (see also below). In the *Sch. pombe* Tf1 element apparently only one primary translation product is formed. It has recently been shown that the element used was able to transpose, although at much lower frequencies, as known for the Ty1 element in a similar transposition assay (Levin and Boeke 1992).

An LTR retrotransposon has been very recently identified in the phytophathogenic fungus *Cladiosporium fulvum*, the causative agent of tomato leaf mold (McHale et al. 1992). This is the first report on a retrotransposable element in hyphal fungi. The element, CfT-1, shows clear resemblance to the *gypsy* group of retrotransposons in the order of the domains of the *pol* gene. Interestingly, the element carries a third ORF 3' to the *pol* gene, although a function could not be attributed to it by any homology of the protein sequence predicted. CfT-1 occurs in moderate repetitive copies. Remarkably, intracellular virus-like particles were isolated which contain reverse transcriptase activity and nucleic acid homologous to CfT-1 DNA. It has been speculated that fungal retrotransposons are involved in promoting variations in pathogen populations, however, clear evidence has not been obtained as yet.

A 5.7-kb transposable element called TOC1 has been isolated from the nuclear genome of the unicellular green alga *Chlamydomonas reinhardtii*. It was discovered as an insertion mutant in the *OEE1* gene (Day and Rochaix 1991). The classification as an LTR retrotransposon is complicated by the fact that the 3'LTR is split by a unique 123-bp sequence and the 5'LTR is split by the central 4.6-kb coding region. Since this LTR structure is apparently not unique to the particular element isolated, it is very likely not a result of some abberant reverse transcription of a TOC1 element with a normal LTR arrangement. The split LTR structure is seemingly not unique to the *Chlamydomonas* element because a family of retrotransposons from the nematode *Panagrellus redivivus* resembles TOC1 in its terminal repeat organization (Day and Rochaix 1991). It should also be mentioned in this context that the reverse transcriptase required for cDNA synthesis has not yet been detected in this example or in many other cases where the retrotransposon of mobile elements is only inferred from some homology to reverse transcriptase sequences or from the structual arrangement of LTRs and putative domains or ORFs with some homologics to the *pol* genes. Unfortunately, the TOC1 element sequenced did not contain large ORFs; the element isolated could be degenerated as it is the case with many repetitive transposons from other organisms. It is also possible that replication of this element is dependent on a yet undetected reverse transcriptase activity.

A similar situation with regard to the lack of functional ORF structure has been observed with a putative retrotransposon from the slime mold *Physarum polycephalum* (Rothnie et al. 1991). The genome of this organism contains a highly repetitive fraction (approximately 20% of the genome) previously referred to as HpaII repeats which appears to be arranged in scrambled clusters up to 50 kb in length. These sequences are hypermethylated and sequence analysis revealed the retrotransposon nature (Tp1) of this fraction of the *Physarum* genome. As with the TOC1 element no continuous ORFs could be detected. However, limited homologies to the nucleic acid binding motif, protease, integrase, and reverse transcriptase domains were found on the predicted sense strand of Tp1. Because the sequence is interspersed with stop codons and putative frameshift mutations, it was not possible to predict the functional organization of the Tp1 reading frames (Rothnie et al. 1991). The most likely explanation for the large, scrambled clusters of Tp1 sequences assumes a specific, directed mechanism of integration comparable to the directed integration of yeast Ty3 in RNA polymerase III promotors (Sandmeyer et al. 1990). It is also likely that homologous recombination between Tp1 sequences within such a cluster contributes to specific sequence arrangements. The preferential integration of a retrotransposon into copies of its own could lie in the limitation of active elements and, hence, of transposition into other sites of the genome. Methylation may also contribute to inactivation of Tp1 gene expression and may limit the number of transposon-competent copies. Once inactivated, the lack of selective pressure would result in the accumulation of mutations in these Tp1 copies. This would ensure that the inactivation is irreversible.

In the slime mold *Dictyostelium discoideum* an LTR retrotransposon has been characterized in detail (Hofmann et al. 1991; Marschalek et al. 1992). This so-called DRE element occurs at quite variable copy numbers in different strains (up to 200 per genome) and in size variants (5.7–6.4 kb). Remarkably, these elements are inserted exclusively some 50 bp upstream of tRNA genes and always in a constant orientation. The LTRs deviate from the standard structure in that DRE-LTRs are nonsymmetric, although a tripartite modular structure is apparent. The entire structure appears to be similar to other retrotransposons like Ty in that two overlapping ORFs might be responsible for differential expression of proteins. Whether the DRE gene products and RNA are assembled in virus-like particles has not yet been demonstrated. In *D. discoideum* cells transcription of the DRE genes into a full length RNA (which is possibly processed) is rather poor despite the high copy number.

## 3. Non-LTR Retrotransposons

A number of retroelements do not show the typical, retrovirus-like structural arrangement of genes flanked by LTRs (Fig. 1B). These elements contain either a 3'poly(A) or oligo(A) stretch [poly(A) retrotransposons according to Boeke and Corces 1989]; other elements have been identified in ORFs with homology to retroviral reverse transcriptases in organellar genomes. The non-LTR elements were initially discovered in *Drosophila* and mammals; recently, a maize element of this type has been described (for reviews, see Boeke and Corces 1989; Grandbastien 1992). In lower eukaryotes this type of element is apparently much less abundant than the LTR type. Examples have been described in the trypanosomes *Trypanosoma brucei* and *Crithidia fasciculata* (Gabriel et al. 1990; Boeke and Corces 1989). In a *Neurospora crassa* natural isolate a transposon of this type has been identified (Kinsey and Helber 1989), although this element is obviously not present in laboratory strains. The general functional organization comprises a 5'proximal ORF which is thought to have functional or structural similarities to *gag*, and a *pol* ORF which contains reverse transcriptase/RNaseH domains only.

A more abundant class of retroelements in yeasts, fungi, and algae has been discovered within organellar genomes. Osiewacz and Esser (1984) revealed first the existence of a circular, 2.5-kb mitochondrial plasmid (plDNA or α-SEN DNA) of *Podospora anserina*. This fungus is to some extent a model system to study the molecular aspects of aging processes. Interestingly, the appearance of this plasmid is correlated with the onset of senescence. In juvenile hyphae, plDNA is an integral part of the cytochrome oxidase subunit I gene and represents a 5'proximal group II intron of that gene. Michel and Lang (1985) revealed that some yeast and fungal group I introns contain reading frames with homologies to reverse transcriptases. A similar situation has been found for an intron of a plasmid gene from a green algae (Kück 1989). It should be noted that the verification of reverse transcriptase activity and its functional significance in the mobility of introns is difficult to achieve. First, the expression of such proteins is quite low, secondly, expression of mitochondrial or plastid genes in heterologous systems poses specific problems. Nevertheless, recent data suggest the the *Podospora* plDNA-encoded protein has indeed reverse transcriptase activity in vitro (U. Kück, pers. comm.). Amplification of the plDNA and hence the onset of the aging syndrome is controlled by a number of nuclear genes (for review of this issue, cf. Osiewacz 1990).

The most elaborate example of an extranuclear retroelement is provided by the Mauriceville and Varkud plasmids of *Neurospora* mitochondria. These plasmids can integrate at various locations in the mitochondrial DNA and cause other gross alterations in the mtDNA (Akins et al. 1986). The plasmids contain a long ORF with homology to reverse transcriptases (Akins et al. 1988) and produce a major transcript which initiates and terminates at the same site. Mutants have been isolated which carry insertion of tRNA sequences precisely at the transcription start site on the plasmid. Such insertions can most plausibly be explained by a process involving reverse transcription. Kuiper and Lambowitz (1988) have isolated a particulate complex of reverse transcriptase activity and full-length plasmid transcripts. Whether reverse transcription is the normal mode of replication of these plasmids is yet unclear.

## 4. Origin and Evolution of Retroelements

The obvious similarities in the structural and functional organization of several transposable elements first discovered in yeast and *Drosophila* and of retroviruses of vertebrates suggested some close evolutionary relationship between these mobile genetic elements. The discovery of reverse transcriptase-like sequences in extranuclear genomes, non-LTR transposons, and retroelements even in bacteria (for review, see Inouye and Inouye 1991) required a broader view of the phylogenetic relationships between retroelements. The most reliable analysis is based on the reverse transcriptase (RT) region since it is the only region common to all elements, thus it can be used for a comprehensive phylogenetic analysis. Upon comparison of more than 80 RT sequences Xiong and Eickbusch (1990) divided retroelements into two major branches (cf. Fig. 2); one branch contains the non-LTR elements, the group II mitochondrial and plastid introns, and bacterial reverse transcriptase (the so-called ms-DNA associated RT). The second major branch of the RT tree encompasses retroviruses, LTR retrotransposons, and several viruses which contain an RT function (for instance, hepadnaviruses, and cauliflower mosaic virus). This distinction is supported by the strong correlation between the RT sequence and the terminal (LTR) structure of the elements. Indeed, there is no example of an LTR retrotransposon whose RT falls on the non-LTR branch and vice versa (Xiong and Eickbusch 1990). Other sequence similarities referring to putative RNaseH, protease, and integrase func-

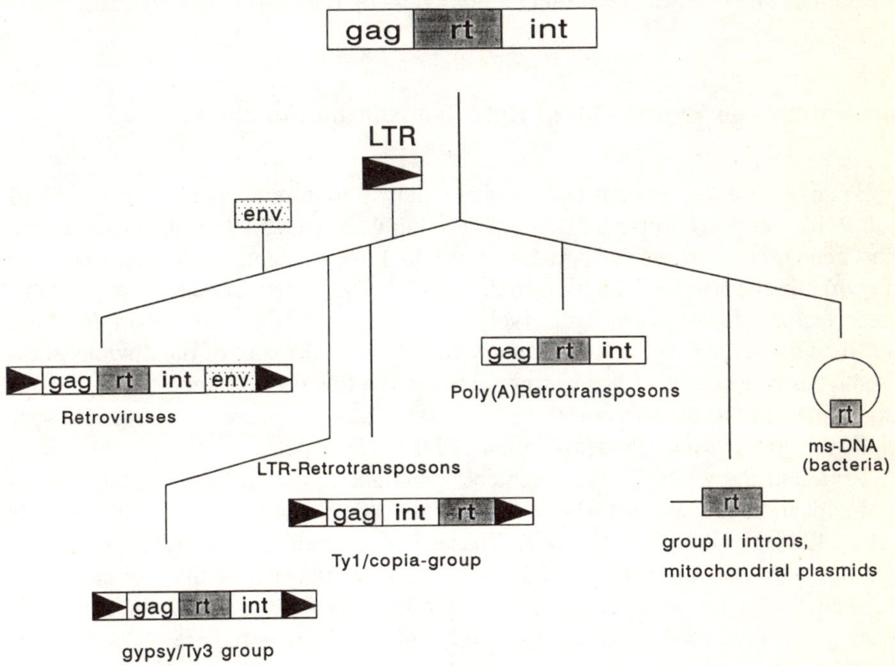


**Fig. 2.** Hypothetical relationships of retroelements according to Xiong and Eickbusch (1990). Designation of functional domains as in Fig. 1. *env* retroviral envelope gene. *Top* Putative structure of an ancestral retroelement

tions have also been used in such analyses (Doolittle et al. 1989). However, these similarities are not consistently detected within the non-LTR branch of retroelements.

It is generally assumed that a progenitor element did not contain LTRs; thus, the non-LTR retrotransposons are presumably the oldest group of retroelements. The coding capacity of the ancestral element would consist of *gag* and *pol* genes. The similarity in *gag* sequences is based on a series of cysteine motifs with putative features of a zinc finger, suggesting that the gag function may be essential to sequester the retroelement's RNA. Given this putative structure of an ancestral element the various categories of retroelements can be explained by gain or loss of various functions. In this view, retroviruses would have evolved from LTR retrotransposons by acquisition of an envelope (*env*) gene allowing the virion particle to bud across the cell membrane. Thus, the yeast and fungal LTR elements would be older than retroviruses and should not be regarded as defective retroviruses. There are also several indications that LTR retrotransposons have been transferred horizontally across major taxonomic groups of organisms (Xiong and Eickbusch 1990). The presence of two different types of elements in *S. cerevisiae* (Ty1/Ty3) and in *D. melanogaster* (copia/gypsy) could be taken as a hint for such a horizontal transfer since both organisms are ecologically closely related (Doolittle et al. 1989).

The evolutionary significance of RT-like sequences of organellar group II introns is not yet clearly understood. While the precise removal of introns can most plausibly be explained by reverse transcription of spliced RNA and subsequent recombination events, the involvement of reverse transcription in intron insertion is less understood. An extensive discussion of this issue can be found in the review by Lambowitz (1989).

## 5. Transcription and Expression of Retrotransposon Functions

Transcription of retrotransposons is critical since transposition requires the availability of full-length RNA which is reverse transcribed into double-stranded DNA, the substrate for integration into chromosomal DNA. At least with the LTR elements reverse transcription occurs in virus-like particles which consists exclusively of element-encoded proteins. These basic features have almost exclusively been established in a few examples from yeast and some higher eukaryotes (*Drosophila* and mouse). Because of the obvious similarities in the basic strategies of gene expression and the functional organization of retrotransposons between these taxonomically quite distant organisms, these findings are very likely applicable to other retroelements, at least of the LTR type.

Transcription of the yeast Ty1/Ty2 elements is unique in that it is dependent on transcriptional activation sequences 150–500 bp downstream of the transcriptional start site (Fulton et al. 1988; Farabaugh et al. 1989). These sites are similar to some eukaryotic enhancers in that they can stimulate transcription from a position within the transcribed region of a gene. These Ty enhancers are also responsible for the transcriptional activation of adjacent genes (see above). Not surprisingly, this activation requires host proteins only, a number of which have been identified by mutants defective in Ty transcription (Boeke and Corces 1989; Ciriacy et al. 1991; Happel et al. 1991). These factors are by no means specific for Ty transcription but are required for expression of very diverse and unrelated cellular functions. Similarily, a DNA-binding protein (STE12), normally implicated in the transcription of genes functionally required for mating, binds to the Ty enhancer and stimulates transcription of Ty-adjacent genes (Company et al. 1988); by contrast, transcription of the Ty element is seemingly unaffected by the STE12 protein. At least two other proteins bind to Ty-enhancer regions, but these have not yet been identified. Another interesting feature of Ty transcription is its differential regulation in haploid versus diploid cells. In the latter type of cells Ty transcription is suppressed about tenfold. This control is presumably exerted through a region of the Ty enhancer with homology to the *MAT*α gene the expression of which is also repressed in diploid cells (Errede et al. 1987). Another yet unexplained regulatory effect which attenuates expression of Ty elements is probably effective at the posttranscriptional level. Despite the abundance of Ty-mRNA [in haploid cells more than 2% of the poly(A)RNA] Ty proteins are synthesized at a surprisingly low level in normal yeast cells. The basis of this down-regulation is not obvious since several different promotor fusions enhance Ty transcription levels only by a factor of 2, while Ty protein levels are increased by a factor of 50 to 100. While the molecular mechanism of Ty transcription remains to be elucidated, the above-mentioned mating-type control has clear implications for the abundance of Ty insertions in natural strains. Since such strains are mostly a/α diploids, this condition keeps Ty insertions at low levels. This coincides with the lower copy number in natural isolates in comparison to haploid laboratory strains.

Translation of the Ty-mRNA into proteins is complicated by the fact that the two ORFs (*TYA* and *TYB* homologous to retroviral *gag* and *pol* genes, respectively) overlap by about 40 bp. In order to express the *TYB* ORF a + 1 frameshift is required relative to *TYA*. This is achieved by a translational frameshift within a 7-bp sequence which is necessary and sufficient to mediate this process. The underlying mechanism has recently been dissected by mutational analysis (Clare et al. 1988; Belcourt and Farabaugh 1990). Ribosomal frameshifting is a result of translational pausing upon low abundance of a specific tRNA-ARG. This pause during translation enhances the probability of a + 1 frameshift by an isoaccepting tRNA-LEU and subsequent expression of TYB proteins. As a result, TYA is made without frameshifting, while TYB is made as a large ~190-kDa TYA-TYB fusion protein. Since frameshifting obviously does not occur very often, TYA is made in approximately 20-fold excess over TYA-TYB. In this way, the relative amounts of TYA and TYA-TYB proteins are intrinsically regulated.

These precursors assemble together with Ty-RNA to virus-like particles. This assemblage is followed by proteolytic processing which is a prerequisite for reverse transcription and transposition (Youngren et al. 1988). The process is at least initially autocatalytic because the protease activity is part of the TYA-TYB fusion protein to be processed. Mutants with defects in the protease domain accumulate unprocessed particles (Müller et al. 1987). The various mature TYB proteins have recently been mapped; the major components are the 60-kDa RT/RNaseH protein and the 90-kDa integrase along with larger processing intermediates (Brühl 1991; Garfinkel et al. 1991). The protease is probably a low-abundant, 23-kDa protein. The gag-homolog TYA is also proteolytically processed by the TYB protease. By contrast to retroviral gag proteins, which are usually cleaved into two or three mature proteins, maturation of the TYA is achieved by removal of a C-terminal 6-kDa peptide the function of which is unknown. However, processing of TYA is required for the maturation of TYB proteins and, hence, for DNA synthesis (Brühl 1991). Figure 2 depicts the essential steps in the Ty life cycle.

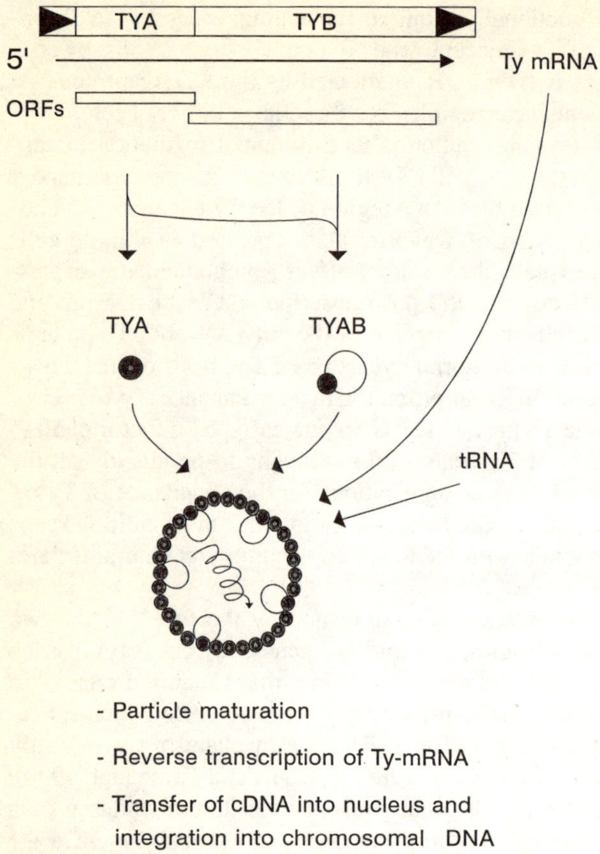


**Fig. 3.** Expression of genes *TYA* (*gag*) and *TYB* (*pol*) of the yeast retrotransposon Ty and formation of retrovirus-like particles. Expression of *TYB* requires a translational frameshift within the overlap between ORF A and B leading to the formation of a fusion protein which is autocatalytically processed during particle assembly

Recent examination of the structure of Ty particles by electron micrographs and neutron scattering revealed some interesting features (Burns et al. 1992). These concern the size and symmetry of Ty particles which turned out to be quite flexible. On average, the presumably icosahedral Ty particle consists of approximately 330 TYA subunits. In addition, these particles seem to form a rather porous structure that does not protect the encapsulated RNA as it is thought for authentic viral capsids. This feature of the Ty particle would suggest some basic differences in structure and function in relation to retroviral particles. The remarkable ability of the TYA protein to form particles even as large fusion proteins has been used in the production and purification of powerful antigens with possible implications for vaccine technology. Up to 40 kDa of heterologous protein can be added to TYA without affecting efficient particle formation (Adams et al. 1987).

## 6. Reverse Transcription and Integration

Due to the similarity between LTR retrotransposons and retroviruses in the arrangement of functional domains required for reverse transcription of a full-length, plus-strand RNA into double-stranded DNA, it has generally been assumed that replication of LTR elements is quite similar to the retroviral replication mode. Again, there are only a few experimental data available for the yeast Ty element. The process can be divided into two steps, i.e., minus-strand and plus-strand synthesis. Crucial in both steps are specific priming events. The (–)strand primer is always a tRNA complementary to a sequence just 3' to the 5'LTR. In the case of Ty elements, the primer is the initiator tRNA-MET. Consequently, the (–)strand would initially contain a covalently linked RNA. Such an intermediate has indeed been observed in nucleic acid preparations from Ty particles (Müller et al. 1991). Plus-strand synthesis is not very well understood. In contrast to the retroviral model, several priming sites have been verified. In addition, it appears that the majority of the Ty-DNA molecules synthesized in vivo contain incomplete (+)strands (Eichinger and Boeke 1990; Müller et al. 1991). These aberrant reverse transcripts are unlikely to be used as substrates for integration. It is possible that the low abundance of full-length, double-stranded Ty-DNA provides another regulatory mechanism which keeps transposition frequencies low. On the other hand, recent work on retrovirus replication indicates that such unpredicted replication intermediates occur in retroviruses as well and may have some specific functions in retroviral replication (J. Boeke, pers. comm.).

Recent work on retroviral integration and on Ty1 transposition in vitro (Brown et al. 1987; Eichinger and Boeke 1988) provided evidence that the linear, double-stranded DNA is an intermediate in transposition. It is not known how the transfer of the cDNA-integrase complex from the cytoplasmic particle into the nucleus is achieved. The surprising finding that in yeast an RNA-debranching enzyme (Chapman and Boeke 1991), usually required in the processing of introns, is also necessary for Ty transposition has not yet been explained. A possible target could be the RNA primers.

Relatively little is known about replication strategies of non-LTR elements. The *Neurospora* Mauriceville and Varkud plasmids probably do not utilize a tRNA primer for (–) strand synthesis (Kuiper and Lambowitz 1988). Hypothetically, priming is achieved by proteins or by some sort of self-priming. For further details concerning this issue, cf. the review of Boeke and Corces (1989).

## 7. Conclusions

Recent work on lower eukaryotes from diverse phylogenetic taxa has strongly supported the previous assumption that mobile retroelements are ubiquitous in nature. Although the evolutionary relationships are still hypothetical for the two major classes of retroelements (LTR and non-LTR elements), the structural and functional homologies within the LTR class are striking. As in retroviruses, replication by reverse transcription is bound to cytoplasmic particles. Some details of the replication mode have been worked out, and it is tempting to speculate that the elucidation of several steps which are still obscure will contribute to novel insights into the retroviral and retrotransposon replication machinery. Another interesting aspect of retrotransposon replication concerns the broad spectrum of integration specificity. This issue is quite relevant to the propagation of retroelements within the genome and, consequently, for the impact of insertions on genome stability.

The structural similarities among non-LTR elements are less obvious, and little is known about the biochemistry of reverse transcription. Nevertheless, the ubiquitous occurrence of retroelements in organellar class II introns or even as mitochondrial plasmids may reflect a peculiar role for the stability of organellar genomes, especially with regard to intron removal or intron insertion. Therefore, viewing retroelements as parasitic or selfish genes is less obvious for this type of element. With regard to the LTR class of retrotransposons it remains to be seen in which way expression of the chromosomal copies is achieved using exclusively host transcription factors. In the case of yeast Ty elements it appears that transcription of retroelements can be quite complex and can serve as a model for downstream activation of eukaryotic genes.

*Ackowledgments.* I wish to thank Kerstin Freidel and Elke Bärenfänger for art work, Drs. K. Esser and H. D. Osiewaczs for helpful suggestions, and the DFG for support of my research work.

## References

**Adams SE,** Dawson KM, Kingsman SM, Kingsman AJ (1987) Nature 329:68–70. **Akins RA,** Kelley RL, Lambowitz AM (1986) Cell 47:505–516. **Akins RA,** Grant DM, Stohl LL, Bottorff DA, Nargang FE, Lambowitz AM (1988) J Mol Biol 204:1–25.

**Belcourt MF,** Farabaugh PJ (1990) Cell 62:339–352. **Berg DE,** Howe MM (1989) Mobile DNA. American Society of Microbiology, Washington, DC, pp 1–972. **Boeke JD,** Corces VG (1989) Annu Rev Microbiol 43:403–434. **Boeke JD,** Garfinkel DJ, Styles CA, Fink GR (1985) Cell 40:491–500. **Brown PO,** Bowermann B, Varmus HE, Bishop JM (1987) Cell 49:347–356. **Brühl KG** (1991) PhD Thesis, Heinrich-Heine-Universität, Düsseldorf. **Burns NR,** Saibil HR, White NS, Pardon JF, Timmins PA, Richardson SMH, Richards BM, Adams SE, Kingsman SM, Kingsman AJ (1992) EMBO J 11:1155–1164.

**Cameron JR,** Loh LY, Davis RW (1979) Cell 16:739–751. **Chalker DL,** Sandmeyer SB (1992) Genes Dev 6:117–128. **Chapman KB,** Boeke JD (1991) Cell 65:483–492. **Ciriacy M,** Williamson VM (1981) Mol Gen Genet 182:159–163. **Ciriacy M,** Freidel K, Löhning C (1991) Curr Genet 20:441–448. **Clare J,** Farabaugh PJ (1985) Proc Natl Acad Sci USA 82:2829–2833. **Clare JJ,** Belcourt M, Farabaugh PJ (1988) Proc Natl Acad Sci USA 85:6816–6820. **Company M,** Adler C, Errede B (1988) Mol Cell Biol 8:2545–2554.

**Day A,** Rochaix JD (1991) Nucleic Acids Res 19:1259–1266. **Doolittle RF,** Feng DF, Johnson MS, McClure MA (1989) Q Rev Biol 64:1–30.

**Eichinger DJ,** Boeke JD (1988) Cell 54:955–966. **Eichinger DJ,** Boeke JD (1990) Genes Dev 4:324–330. **Errede B,** Cardillo TS, Wever B, Sherman F (1980) Cold Spring Harbor Symp Quant Biol 45:593–602. **Errede B,** Company M, Hutchinson CA (1987) Mol Cell Biol 7:258–265.

**Farabaugh PJ,** Fink GR (1980) Nature 286:352–356. **Farabaugh P,** Liao X-B, Belcourt M, Zhao H, Kapakos J, Clare J (1989) Mol Cell Biol 9:4824–4834. **Fulton AM,** Rathjen PD, Kingsman SM, Kingsman AJ (1988) Nucleic Acid Res 16:5439–5458.

**Gabriel A,** Yen TJ, Schwartz DC, Smith CL, Boeke JD, Sollner-Webb B, Cleveland DW (1990) Mol Cell Biol 10:615–624. **Gafner J,** Philippsen P (1980) Nature 286:414–418. **Garfinkel DJ,**

Boeke JD, Fink GR (1985) Cell 42:507–517. **Garfinkel DJ,** Hedge A-M, Youngren SD, Copeland TD (1991) J Virol 65:4573–4581. **Grandbastien M-A** (1992) TIG 8:103–108.

**Happel AM,** Swanson MS, Winston F (1991) Genetics 128:69–77. **Hofmann J,** Schumann G, Borschet G, Gosseringer R, Bach M, Bertling WM, Marschalek R, Dingermann T (1991) J Mol Biol 222:537–552.

**Inouye M,** Inouye S (1991) TIBS 16:18–21

**Kinsey JA,** Helber J (1989) Proc Natl Acad Sci USA 86:1929–1933. **Kuiper MT,** Lambowitz AM (1988) Cell 55:693–704. **Kück U** (1989) Mol Gen Genet 218:257–265.

**Lambowitz AM** (1989) Cell 56:323–326. **Levin HL,** Boeke JD (1992) EMBO J 11:1145–1153. **Levin HL,** Weaver DC, Boeke JD (1990) Mol Cell Biol 10:6791–6798.

**Marschalek R,** Hofmann J, Schumann G, Gosseringer R, Dingermann T (1992) Mol Cell Biol 12:229–239. **McHale MT,** Roberts IN, Noble SM, Beaumont C, Whitehead MP, Seth D, Oliver RP (1992) Mol Gen Genet 233:337–347. **Michel F,** Lang BF (1985) Nature 316:641–643. **Mount SM,** Rubin GM (1985) Mol Cell Biol 5:1630–1638. **Müller F,** Brühl KH, Freidel K, Kowallik KV, Ciriacy M (1987) Mol Gen Genet 207:421–429. **Müller F,** Laufer W, Pott U, Ciriacy M (1991) Mol Gen Genet 226:145–153.

**Osiewacz HD** (1990) Mutat Res 237:1–8. **Osiewacz HD,** Esser K (1984) Curr Genet 8:299–305.

**Rothnie HM,** McCurrach KJ, Glover LA, Hardman N (1991) Nucleic Acids Res 19:279–286.

**Sandmeyer SB,** Hansen LJ, Chalker DL (1990) Annu Rev Genet 24:491–518.

**Williamson VM,** Young ET, Ciriacy M (1981) Cell 23:605–614.

**Xiong Y,** Eickbusch TH (1990) EMBO J 9:3353–3362.

**Youngren SD,** Boeke JD, Sanders NJ, Garfinkel DJ (1988) Mol Cell Biol 8:1421–1431.

Professor Dr. Michael Ciriacy
Institut für Mikrobiologie
Heinrich-Heine-Universität
Universitätsstraße 1
W-4000 Düsseldorf 1, Germany

# V. Extranuclear Inheritance: Mitochondrial Genetics

By Gudula Riemen, Thomas Lisowsky, Frideriki Maggouta, Georg Michaelis, and Elke Pratje

## 1. Introduction

Continuing the previous articles of Ecke et al. (1989) and Dohmen et al. (1991) in this series, the present chapter covers mainly the literature of the past 2 years. The following topics were selected: (a) mitochondrial genomes, especially those of the green alga *Chlamydomonas reinhardtii* and the liverwort *Marchantia polymorpha*, (b) mitochondrial mutations in higher plants, (c) replication and transcription of mitochondrial DNA, (d) processing, splicing, and editing of RNA, (e) regulation of mitochondrial translation, and (f) protein import into mitochondria.

For additional topics not treated in this chapter, the reader is referred to the following reviews on small repeats and recombination of plant mtDNA (André et al. 1992), organelle transformation (Butow and Fox 1990), evolution of mitochondrial DNA (Gray 1989; Palmer 1990; Avise 1991), replication and transcription of vertebrate mitochondrial DNA (Clayton 1991), interactions between mitochondria and the nucleus (Forsburg and Guarente 1989), tissue-specific genes of respiratory proteins (Lomax and Grossman 1989), aminoacyl tRNA synthetases (Tzagoloff et al. 1990), the respiratory chain NADH-dehydrogenase (Weiss et al. 1991), cytochrome c oxidase (Capaldi 1990), alternative oxidase of plant mitochondria (Moore and Siedow 1991), ATP synthase (Glaser and Norling 1991), and movements of mitochondria in living cells (Bereiter-Hahn 1990).

## 2. Mitochondrial Genomes

### a) Complete DNA Sequence of Mitochondrial DNA from *Chlamydomonas reinhardtii* and *Marchantia polymorpha*

Recently, the complete DNA sequence of two mitochondrial (mt) genomes has been determined: that of the unicellular green alga *Chlamydomonas reinhardtii* and the liverwort *Marchantia polymorpha*. We will concentrate mainly on these two genomes and only a few aspects from other mtDNA will be mentioned.

The mt genome of *Chlamydomonas reinhardtii* (Fig. 1) is small, linear, and displays a compact organization of 13 genes (Michaelis et al. 1990). The 15.8 kb mtDNA encodes

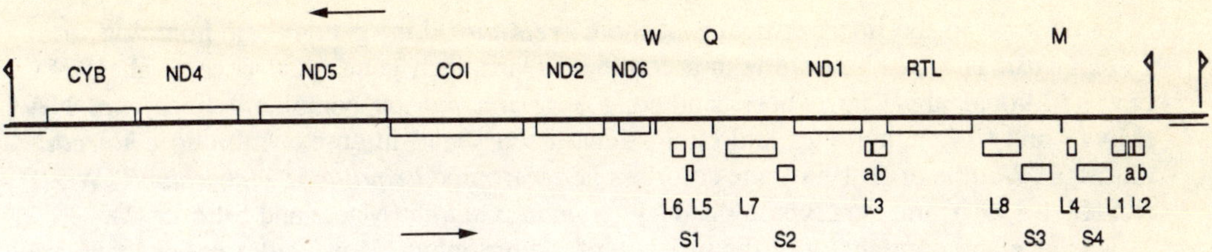


**Fig. 1.** Gene organization of the mitochondrial DNA from *Chlamydomonas reinhardtii*. Genes for apocytochrome b *CYB*, subunits of NADH dehydrogenase *ND*, subunit 1 of cytochrome oxidase *COI*, and a reverse transcriptase-like protein *RTL* are indicated. *W* (tryptophan), *Q* (glutamine), and *M* (methionine) denote tRNA genes. Regions of the large (*L*) and small (*S*) scrambled rRNA genes are indicated. The long terminal inverted repeats are characterized by a *double line*. Three copies of a small repeat are represented by *flags*. Directions of transcription are shown by *arrows*. (Boer and Gray 1989; Ma et al. 1989; Michaelis et al. 1990; Vahrenholz, Riemen, Pratje, Dujon, Michaelis, submitted)

genes for subunit I of cytochrome oxidase, apocytochrome b, five subunits of NADH-dehydrogenase, and a reading frame showing some homology to a reverse transcriptase. Genes for any subunit of mitochondrial ATPase and for subunit II and III of cytochrome oxidase, normally present in mitochondria, are lacking on the mtDNA of *C. reinhardtii*. A similar situation is found in the legumes mung bean (*Vigna radiata*) and cowpea (*Vigna unguiculata*) for the coxII genes. In these two legumes, the coxII genes are transferred to the nucleus probably via RNA transfer (Nugent and Palmer 1991). The coxII gene of mung bean contains an intron at the junction of the transit peptide sequence and the mature protein coding sequence, suggesting a function of the intron in establishing a functional coxII gene. Most legumes have coxII genes in both the mitochondria and in the nucleus, but the gene is expressed in one compartment only.

A minimum of 22 tRNAs is needed to translate all codons of the *Chlamydomonas* mitochondrial genome assuming a "2 out of 3" codon-anticodon pairing. In the 15.8 kb mtDNA of *C. reinhardtii* only genes for the tryptophane, glutamine, and methionine tRNAs could be identified (Boer and Gray 1989; Ma et al. 1989). All other tRNAs have to be imported. A tRNA import might occur in liverwort (see below) and was demonstrated in mitochondria of higher plants (Marechal-Drouard et al. 1988; Small et al. 1992).

The genes for the two ribosomal rRNAs of *C. reinhardtii* have unusual structures. They are divided in separate coding regions and scrambled with one another and tRNA- and protein genes (Boer and Gray 1988). A gene for a mitochondrial 5S rRNA is absent, in contrast to the situation in higher plants.

The linear mtDNA molecules contain long terminal inverted repeats, 3' extensions and the outmost nucleotides are repeated a third time inside the molecule (Vahrenholz, Riemen, Pratje, Dujon, Michaelis, submitted). This organization is probably necessary to replicate the linear molecule.

The mitochondrial DNA of the interfertile alga *Chlamydomonas smithii* is colinear with that of *C. reinhardtii* except for a 1 kb insertion in the apocytochrome b gene. The insert corresponds to a typical group I intron and contains an 237 codon open reading frame with homology to the SceI endonuclease encoded by the omega locus of yeast mitochondrial DNA (Colleaux et al. 1990). This situation is an argument for a horizontal gene transfer between fungi and algae.

In contrast to the linear structure of the *C. reinhardtii* mt genome the homolog from *C. moewusii* is a 22 kb circle with a complete rearranged gene order (Lee et al. 1991). Few additional algae have been studied. *Pandorina morum* contains a linear mtDNA (Moore and Coleman 1989), and circles have been found in an exsymbiotic *Chlorella*-like alga (Waddle et al. 1990), the colorless heterotrophic *Prototheca wickerhamii* (Wolff and Kück 1990), and the cryptomonad *Pyrenomonas salina* (Maerz and Sitte 1991).

The size and complexity of the mtDNA of the bryophyte *Marchantia polymorpha* are much greater. In the complete sequence of the 186.6 kb mtDNA (Oda et al. 1992), 94 possible genes were identified (Fig. 2). These include genes for three species of ribosomal RNA (26S, 18S, 5S rRNA), 29 genes for 27 tRNAs, genes for 16 ribosomal proteins, three subunits of mitochondrial ATPase, three subunits of cytochrome oxidase, apocy-

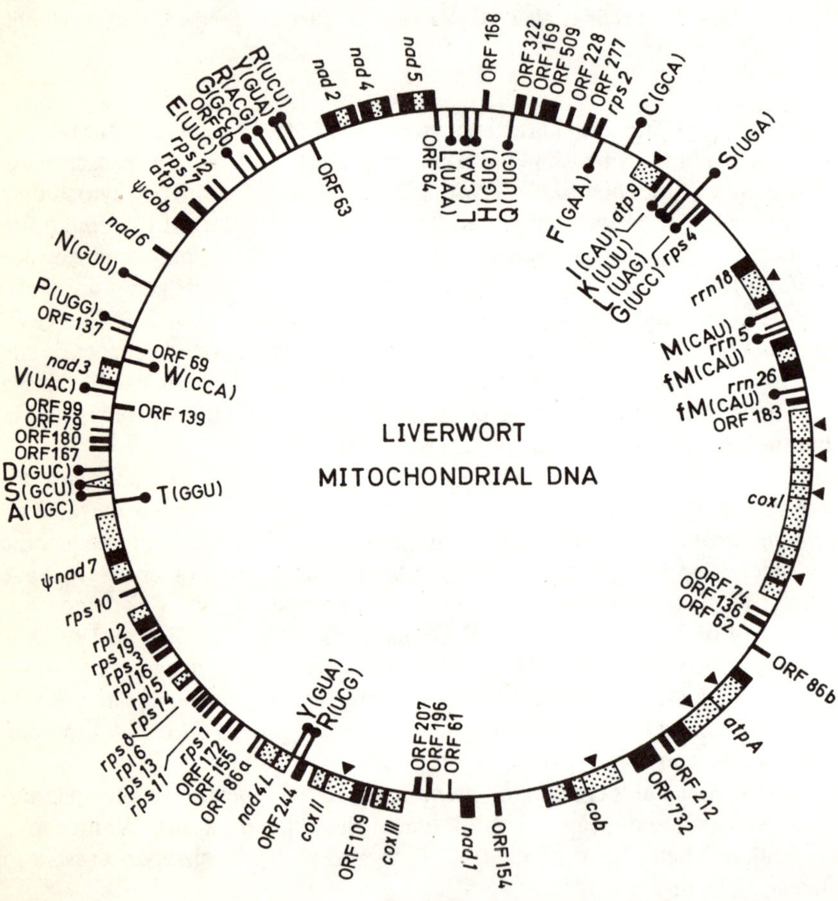


**Fig. 2.** Gene organization of the mitochondrial DNA from the liverwort *Marchantia polymorpha*. Genes for the small subunit *rps* and for the large subunit *rpl* ribosomal proteins, for subunits of ATPase *atp*, NADH dehydrogenase *nad*, cytochrome oxidase *cox*, and apocytochrome b *cob* are indicated. *Filled boxes* indicate exons, *shaded boxes* introns. *Arrowheads* denote introns that contain an ORF homologous to RNA maturase. Genes encoding tRNAs were identified by the *single-letter amino acid code* with their anticodons in parentheses. Genes shown *outside the map* are transcribed anticlockwise, and those *inside* are transcribed clockwise. (After Oda et al. 1992)

tochrome b, and seven subunits of NADH dehydrogenase. Ten open reading frames (ORFs) showing similarity to RNA maturases of yeast occur in introns. Interestingly, 31 ORFs of unknown functions encode putative proteins of at least 60 amino acid residues. The sequence reveals that plant mt genomes encode more functions than their fungal and animal homolog.

The genes for a $tRNA^{ile}$ and a $tRNA^{thr}$ that are required for reading all codons of the liverwort mt genome have not been detected. A nuclear location is one possibility to explain this situation. The genes for ribosomal proteins are organized in a similar manner in mitochondrial and chloroplast genomes of *Marchantia polymorpha*. This indicates that in this organism similar ribosomal protein genes were transferred from the two organelles into the nucleus.

Seventeen genes are interrupted by 32 introns of group I or group II. The striking similarity between several intron sequences is interpreted as intron migration from gene to gene within liverwort mitochondria.

CoxIII belongs to the class of genes containing an intron. This intron has not been found in the coxIII gene of another moss, *Physcomitrella patens* (Marienfeld et al. 1991). The sequence of *Marchantia* mtDNA does not reveal incorporation of foreign genes, the generation of subgenomic molecules by recombination, or the existence of an RNA editing system as described for higher plants. The mtDNA of the horsetail *Equisetum arvense* is at least 200 kb in size, while that of the fern *Onoclea sensibilis* is approximately 300 kb (Palmer et al. 1992). A third of the *Onoclea* mt genome is composed of two large (> 8 kb) repeated sequences which are present in about five copies each and contain the 26S rRNA gene.

### b) Paternal Inheritance of Mitochondrial DNA

Mitochondrial DNA was thought to be strictly maternally inherited in higher plants. Exceptions were first described for some gymnosperms. Paternal inheritance has been inferred in *Sequoia sempervirens* (Neale et al. 1989) and in *Calocedrus decurrens* (Neale et al. 1991) by analyzing restriction fragment length polymorphism. In the genus *Pinus*, maternal inheritance of mtDNA is predominant but paternal leakage occurs in crosses of *Pinus banksiana* and *Pinus contorta* (Wagner et al. 1991). Paternal mtDNA was exclusively detected in 6 out of 125 seedlings obtained from 23 controlled matings.

Paternal inheritance of mitochondrial genomes does not seem to be restricted to gymnosperms. Deviations from the maternal inheritance have been reported in the angiosperm *Brassica napus* (Erickson and Kemble 1990). *Brassica napus* plants carrying the chloroplast marker of triazin resistance were crossed to plants carrying the polima cytoplasm. In the absence of restorer genes, this cytoplasm confers the trait of cytoplasmic male sterility. Approximately 10% of the F1 progeny displayed the male sterility and flower morphology of the male parent. Mixed maternal and paternal mtDNA molecules were found in these F1 plants.

## 3. Mitochondrial Mutants of Higher Plants

Mitochondrial genetics was initiated with the yeast *Saccharomyces cerevisiae* and different classes of mutations are known in this system. In higher plants, an abnormal striping of leaves and the failure to produce functional pollen have been correlated with mitochondrial mutations. The description of nonchromosomal stripe mutants (NCS) is limited to maize. These NCS mutants show variable leaf striping, poor growth, and often kernel abortions. Mitochondrial DNA rearrangements and deletions were demonstrated in the four mutants NCS2, NCS3, NCS5, and NCS6. Because the mutations are lethal, they can only be found in a heteroplasmic state together with normal mitochondrial genomes. A similar situation has recently been reported for some human diseases (for review see Shoffner and Wallace 1990). NCS mutations of *Zea mays* seem to be generated by homologous recombination between very short repeats. In NCS2 mutants, a set of transcripts homologous to the NCS2-altered DNA region is aberrant. In addition, reduced amounts of a 24 kDa polypeptide of unknown function are synthesized by isolated mitochondria from these NCS2 mutants (Feiler and Newton 1987). The deletions found in NCS3, NCS5, and NCS6 mutants could be assigned to known genes for mitochondrial proteins. In the NCS3 mutant a deletion eliminates part of a region encoding two proteins with similarity to the ribosomal proteins S3 and L16 of *E. coli* (Hunt and Newton 1991). NCS5 and NCS6 striped plants carry normal and partially deleted versions of the mitochondrial cytochrome oxidase subunit 2 gene (coxII), resulting in greatly reduced levels of coxII transcripts and the coxII protein (Lauer et al. 1990; Newton et al. 1990).

Cytoplasmic male sterility (CMS) is a maternally inherited alteration in the development of the male gametophyt. Such CMS mutations have been reported in numerous species, including many crop plants. The interest in CMS plants is due to the fact that CMS provides a convenient method for hybrid seed production. The CMS phenotype is thought to be caused by intra- and intermolecular rearrangements of mitochondrial DNA and nuclear-mitochondrial interactions. Chimeric genes have been shown to be directly involved in the CMS trait of maize (T-cytoplasm), *Petunia*, and sorghum (for a review see Hanson 1991).

Recent reports indicate that chimeric genes are also involved in CMS of other plants. Aberrant recombinations between the mitochondrial atp9, atp6, coxII genes and sequences of unknown origin have created unique chimeric sequences in the CMS-C cytoplasm of maize (Dewey et al. 1991).

In CMS of rice the 5' portion of the mitochondrial atp6 gene is involved in the chimeric reading frame urf-rmc (Kadowaki et al. 1990). Investigations on the Polima CMS of *Brassica napus* have shown that rearrangements in the mitochondrial genome upstream of the atp6 gene have generated a chimeric 224 codon open reading frame designated orf224 (Singh and Brown 1991). Orf224 is cotranscribed with atp6 in the Polima male sterile plants. When nuclear restorer genes are present, the transcript pattern is altered and monocistronic atp6 mRNA predominates.

One CMS system of sunflower results from interspecific crosses of *Helianthus petiolaris* with *H. annuus*. Sterile and fertile lines only differ by a 12 kb DNA inversion and a 5 kb insertion/deletion (Siculella and Palmer 1988). This rearrangement has created a new open reading frame which is cotranscribed with atpA in sterile sunflower. As shown by in vitro translation experiments with isolated organelles, a 16 kDa protein is expressed in male sterile lines carrying the *H. petiolaris* cytoplasm (Horn et al. 1991; Laver et al. 1991). A direct relation of the 16 kDa protein to CMS remains to be established because the 16 kDa polypeptide is found in similar amounts in fertility-restored lines.

In *Nicotiana sylvestris*, CMS is associated with large deletions (up to 50 kb) of mitochondrial DNA (Chetrit et al. 1992). These deletions were probably generated by two homologous recombination events. The *Nicotiana sylvestris* CMS plants have been obtained by protoplast culture.

In *Petunia*, *Zea mays* with T cytoplasm, *Brassica napus* and rice, nuclear restorer genes affect the expression of mitochondrial genes at the RNA level. For example, the abundance of *Petunia* transcripts specific for the CMS associated S-Pcf locus is strongly reduced by the presence of a nuclear restorer gene (Pruitt and Hanson 1991). Transcription of the chimeric gene urf-rmc of rice (see above) is also altered by a dominant nuclear fertility restorer gene (Kadowaki et al. 1990). This is demonstrated by an urf-rmc-specific probe which hybridizes to a 2.7 kb RNA in the male sterile plant and to a 2.8 kb RNA in the restored plant.

Quite a different restoration mechanism has been characterized in common bean (Mackenzie and Chase 1990; Johns et al. 1992). In this case, restoration of pollen fertility is associated with the loss of at least 25 kb DNA from the mitochondrial genome. A similar loss is observed in spontaneous fertile revertants. The deleted DNA region encodes at least two open reading frames with portions derived from the chloroplast genome.

In *Petunia*, normal and CMS lines differ in the activity of their cyanide-sensitive and cyanide-insensitive pathways of electron transport (Connett and Hanson 1990). It has still to be shown whether the same mitochondrial DNA locus which specifies CMS also affects alternative oxidase activity. A normal alternative oxidase activity was found in restored plants. The reported differences in electron transport between fertile and CMS lines of *Petunia* do not seem to be a general rule, because no such differences were detected in *Nicotiana* (Hakansson et al. 1990).

## 4. Components for Mitochondrial Transcription and Replication: an Evolutionary Mosaic in Yeast

The recent major achievements in this field of research have been made by the identification and characterization of key components for mitochondrial transcription and replication in the yeast *Saccharomyces cerevisiae*. Therefore this organism will serve as a model system for the following discussion. The results for yeast are summarized in Table 1. It is important to notice that the genes for all these components are localized in the nucleus of the cell. This seems to be a general rule for all the mitochondrial systems analyzed so far. The available sequence data from mitochondria including the latest complete sequence from *Marchantia polymorpha* (Oda et al. 1992) demonstrate that none of the known factors are encoded by the genome of the organelle. The most surprising picture which emerges from the comparison of the respective factors is the completely different evolutionary relationship among them (see Table 1). The transcription and replication machineries are build up by factors related to viral-type proteins and factors similar to prokaryotic or eukaryotic components. A closer look at the transcription apparatus

**Table 1.** Components for mitochondrial transcription and replication: an evolutionary mosaic in yeast

| Component | Gene symbol | Evolutionary relation | Reference |
|---|---|---|---|
| Core enzyme of RNA polymerase | RPO41 | Homologous to *viral* SP6/T7 RNA polymerase | Masters et al. (1987) |
| Specificity factor for transcription | MTF1 | Similarities to *prokaryotic* sigma factors and *viral* proteins | Jang and Jaehning (1991); Lisowsky and Michaelis (1988) |
| Transcription stimulation factor | ABFII | Homologous to *eukaryotic* high mobility group proteins (HMG1) | Diffley and Stillman (1991) |
| Core enzyme of DNA polymerase | MIP1 | Homologous to *eukaryotic* DNA polymerase I and similarities to *viral* reverse transcriptases | Foury (1989) |
| Single strand DNA binding protein | RIM1 | Homologous to *prokaryotic* single strand DNA binding protein (SSB) | Foury (pers. comm.) |

shows that the core enzyme for mitochondrial RNA polymerase is homologous to the viral SP6/T7 enzymes (Masters et al. 1987); but in contrast to the viral enzymes, the mitochondrial RNA polymerase is dependent on a second protein factor for accurate initiation at the promoter (Schinkel et al. 1987). This specificity factor (MTF1) shows structural similarities to prokaryotic sigma factors and viral proteins (Lisowsky and Michaelis 1988, 1989; Jang and Jaehning 1991). A third component with influence on mitochondrial transcription was identified as a homolog of the nuclear high mobility group 1 protein (Diffley and Stillman 1991) which is a typical eukaryotic factor normally localized in the nucleus of the cell (Mayes 1982). This is especially surprising as one always expected to find prokaryotic-type histone-like proteins (HU) inside mitochondria (Drlica and Rouviere-Yaniv 1987). From the latest data it is clear now that yeast and human mitochondria do not contain histone-like proteins, but that instead HMG1-related eukaryotic factors are present in mitochondria (Diffley and Stillman 1991; Parisi and Clayton 1991).

On the other hand, one finds that the core enzyme for mitochondrial DNA synthesis is homologous to eukaryotic DNA polymerases I, and in addition shows structural similarities to viral reverse transcriptases (Foury 1989). Another component essential for the replication of mitochondrial DNA was recently identified as a single-strand DNA-binding protein (SSB) homologous to the prokaryotic SSB protein of *Escherichia coli* (Foury, pers. comm.). In that respect it is interesting that the latest data about replication of mitochondrial DNA in yeast hint at an unusual rolling circle type mechanism for this genome (Malesska et al. 1991). In conclusion, all these data show that it is really justified to speak of an "evolutionary mosaic" in respect of the components of the replication and transcription systems inside yeast mitochondria. From the evolutionary point of view, two major processes could be responsible for the present situation. First the transfer of prokaryotic-related genes from the genome of the former endosymbiont to the nucleus of the host, and second the adaptation of nuclear host factors for the special requirements of the organelle genome. Later, these eukaryotic factors replaced the former prokaryotic proteins of the endosymbiont. Today, the important question is if this is a general picture for all the mitochondrial systems. In this respect, the major tasks for the future will be to bridge the conceptual gap between yeast and higher plants.

A first attempt for such a comparison is based on the detailed analysis of plant mitochondrial promoters. Recent data from *Zea mays* mitochondria exhibit some structural similarities with yeast. As in yeast, one finds several independent start sites for transcription initiation on the mitochondrial genome of *Zea mays*, with the possibility of multiple initiation sites for some transcription units (Costanzo and Fox 1990; Mulligan et al. 1991). The in vitro analysis of different independent promoters identifies for the first time a special consensus sequence motif associated with these initiation sites (Rapp and Stern 1992). In spite of no direct sequence homologies with yeast, these elements are functionally comparable to the well-defined consensus sequences of yeast mitochondrial promoters (Baldacci and Bernardi 1982). Further functional analysis of additional plant promoters will have to define the extent of similarity.

The major problem is, however, that up to now no nuclear gene for mitochondrial transcription or replication has been identified in higher plants, and it will be interesting to find out what kind of proteins are essential for the key processes of gene expression in the mitochondria of these organisms.

## 5. RNA Processing and Splicing

Transcription of mitochondrial genes is followed by numerous RNA-processing events such as cleavage of multigenic precursors, trimming of the 5' and 3' termini and the removal of introns.

The 5' leaders of tRNA precursors are removed by the enzyme RNase P (for review see Altman 1990; Pace and Smith 1990). The catalytic subunit of this ribonucleoprotein is an RNA molecule encoded by the mitochondrial genome in different yeast species (Wise and Martin 1991) and possibly by a nuclear gene in human (Doersen et al. 1985).

The coding sequences of several mitochondrial genes are interrupted by introns which are excised from the pre-mRNA by RNA splicing. A number of introns have characteristics in common with mobile elements and exhibit self-splicing activities in vitro (for review see Cech 1990). Efficient splicing in vivo, however, depends on protein factors. Some of these proteins, called RNA maturases, are encoded by the introns themselves, others are products of nuclear genes. Components of the translation system such as aminoacyl-tRNA synthetases belong to the latter group (reviewed by Lambowitz and Perlman 1990).

Recent studies in higher plants have revealed the unusual organization of some chloroplast and mitochondrial genes whose coding sequences are widely separated on the genome. The exons are independently transcribed, suggesting that mature mRNAs are created by a process termed *trans* splicing. The *nad1* (NADH dehydrogenase subunit 1) gene of *Oenothera* (Wissinger et al. 1991) as well as the *nad5* genes of *Arabidopsis* and *Oenothera* (Knoop et al. 1991) are split into five exons which are located at three distant genomic sites. For maturation of the mRNAs, two *trans* splicing events are required respectively. The *nad1* gene of wheat is also distributed in five exons, which are, however, scattered over four genomic regions (Chapdelaine and Bonen 1991). The assembly of exons by *trans* splicing presumably involves intermolecular base pairing of flanking intron sequences to generate an apparently complete group II type intron structure. For this reason, the separation of coding sequences in plant mitochondria is thought to be the result of DNA rearrangements which occurred within intron sequences during evolution (Chapdelaine and Bonen 1991).

## 6. RNA Editing

RNA editing is a post-transcriptional alteration of mRNA coding sequences. This process has been shown to occur in the mitochondria of trypanosomes (for review see Stuart 1991a, b), the slime mold *Physarum polycephalum* (Mahendran et al. 1991), and higher plants (reviewed by Walbot 1991).

In trypanosomes, mRNA editing involves the addition and deletion of uridines at defined sites in primary transcripts of at least seven different mitochondrial genes. RNA editing can be more or less extensive. Up to hundreds of uridines are inserted in transcripts of *T. brucei* (Feagin et al. 1988; Bhat et al. 1990; Koslowsky et al. 1990). The editing proceeds in a 3' to 5' direction along the primary transcript. The information for the sequence alterations is provided by small RNAs, termed guide RNAs, which are encoded by intergenic regions of maxicircle DNA as well as by minicircle DNA. The recent detection of chimeric guide RNA-mRNA molecules (Blum et al. 1991) suggests a *trans*-esterification mechanism similar to that of RNA splicing/reverse splicing for RNA editing in kinetoplast mitochondria (Cech 1991). According to this model, the editing reaction results in the transfer of uridine residues from the 3' oligo(U) tail of the guide RNA to the editing site of the transcript.

The editing system of *Physarum polycephalum* is characterized by insertions of cytosines, uridines, and adenosines. Interestingly, insertion sites have been identified in mRNAs as well as in a structural RNA (for review see Cattaneo 1991).

RNA editing in mitochondria of higher plants affects all protein-coding transcripts examined to date (for review see Walbot 1991). Furthermore, editing of 5' and 3' untranslated sequences and of the 26S rRNA has been reported (Schuster et al. 1991). Nearly all editing events involve C to U conversions and, to a lesser extent, U to C changes. The alterations result in the creation of start and stop codons in open reading frames or in the exchange of the amino acid specified by the codon. RNA editing thus restores translation according to the universal genetic code and improves the degree of similarity of the polypeptide with the respective sequences of other species. Recent results indicate that editing can precede intron splicing (Sutton et al. 1991; Yang and Mulligan 1991) and that mRNAs associated with polysomes are rather advanced in editing. The specificity determinants, as well as the precise mechanism of RNA editing in plant mitochondria, are still unknown.

The RNA editing process seems not to be limited to mitochondria. Similar alterations of primary transcripts have also been reported for paramyxoviruses (see Cattaneo 1990 for review) as well as for some nuclear and chloroplast transcripts (reviewed by Cattaneo 1992; Hodges and Scott 1992).

## 7. Regulation of Mitochondrial Translation by Nuclear Genes

Genetic studies in the yeast *Saccharomyces cerevisiae* have shown that the expression of a number of mitochondrial genes is controlled by nuclear encoded proteins. These *trans*-acting factors exert their effects post-transcriptionally, affecting either RNA maturation, RNA stability, or translation of individual mRNAs (reviewed by Costanzo and Fox 1990).

Gene-specific translational activator proteins have been identified for mRNAs of cytochrome c oxidase subunits I, II, and III, apo-cytochrome b and ATPase subunit 9 (see Table 2). Mutational defects of an activator protein result in the accumulation of its mitochondrial target mRNA, the disappearance of the corresponding protein and a respiratory deficient *pet* phenotype. The described defects can be suppressed by mitochondrial mutations which replace the 5'- untranslated leader sequence of a specific mRNA by that of another mitochondrial gene. This suppression suggests direct interactions of the *PET* gene products with the 5' leader sequence of their target mRNA (see Costanzo and Fox 1990 for a review). Sequence similarities between the translational activator *PET494* as well as *CBP6* and eukaryotic initiation factors are strong support for the idea that these proteins function as mRNA-specific mitochondrial initiation factors (Grivell 1989). From extensive studies of the nuclear genes *CBS1* and *CBS2* which are involved in the translation of apocytochrome b, a model for translational activation has been presented that could apply to other mitochondrial mRNAs as well (Michaelis et al. 1991). The *CBS1* and *CBS2* gene products have been identified as integral or associated components of the inner mitochondrial membrane. The *CBS2* gene product may, in addition, be associated with the mitochondrial small ribosomal subunit. Translation of apocytochrome b should thus occur in direct vicinity of the mitochondrial membrane possibly facilitating membrane insertion of this hydrophobic protein. The function of the *CBS1* and *CBS2* gene

**Table 2.** Nuclear genes involved in translation of specific yeast mitochondrial mRNAs

| Mitochondrial mRNA | Nuclear gene | Reference |
|---|---|---|
| Cytochrome oxidase subunit I (*coxI*) | *MSS51* | Decoster et al. (1990) |
| Cytochrome oxidase subunit II (*coxII*) | *PET111* | Poutre and Fox (1987) |
| Cytochrome oxidase subunit III (*coxIII*) | *PET54* | Costanzo et al. (1989) |
| | *PET494* | Costanzo and Fox (1986) |
| | *PET122* | Ohmen et al. (1988) |
| Apocytochrome b | *CBS1, CBS2* | Rödel (1986), Rödel et al. (1986) |
| | *CBP6* | Dieckmann and Tzagoloff (1985) |
| ATPase subunit 9 | *ATP13* | Ackerman et al. (1991) |

products is to target the cytochrome b mRNA to the membrane by interacting with the 5' leader sequence and to connect the mRNA with the mitochondrial ribosome.

A similar model has been suggested for translation activation of cytochrome oxidase subunit III (*coxIII*) mRNA. Based on genetic data, the *PET122*, *PET494*, and *PET54* gene products are assumed to interact with the 5' leader of the *coxIII* mRNA (Costanzo and Fox 1988). Interestingly, certain carboxy-terminal alterations of the *PET122* gene product are specifically suppressed by mutations in two nuclear genes, *PET123* and *MRP1*, that encode proteins of the small ribosomal subunit. This observation suggests a functional interaction of the *PET122* gene product with the mitochondrial ribosome (Haffter et al. 1990, 1991; McMullin et al. 1990).

## 8. Protein Import into Mitochondria

Most mitochondrial proteins are nuclear encoded and translated on cytosolic ribosomes as larger precursors with amino-terminal presequences of 10–70 amino acid residues. The mechanism of protein import is best analyzed in the fungi *Saccharomyces cerevisiae* and *Neurospora crassa*. Therefore this section will focus on these two organisms (for reviews see also Glick and Schatz 1991; Horwich et al. 1991; Pfanner et al. 1991).

The precursor proteins bind to receptors on the cytosolic side of the outer mitochondrial membrane. The receptor proteins MAS70 and p32 of yeast (Hines et al. 1990; Murakami et al. 1990) and MOM19 and MOM72 of *Neurospora crassa* (Söllner et al. 1990; H. Schneider et al. 1991) have been identified. Disruption of the yeast MAS70 gene reduces the import of only some precursor proteins. The receptor function of MOM72 seems to be limited to the ADP/ATP translocator. After binding to the receptor, most proteins are transferred to translocation sites, termed GIP (general insertion proteins) in *Neurospora crassa*. Apocytochrome c crosses the outer membrane independently of GIP (Stuart and Neupert 1990). The yeast ISP42 (Baker et al. 1990) and the *N. crassa* MOM38, MOM7, MOM8, and MOM30 (Söllner et al. 1992) proteins are components of the translocation sites. Passage across the two mitochondrial membranes occurs at contact sites between the outer and inner membranes. Nevertheless, translocation sites are not exclusively located in morphological contact sites but distributed over the entire outer membrane. Precursor proteins are exposed to the intermembrane space while being transported from the outer to the inner membrane, suggesting that contact sites are dynamic structures that can be formed and disassembled (Geli and Glick 1990; Rassow and Pfanner 1991).

Protein transport requires correct folding of the precursors. The cytosolic chaperone hsp70 catalyzes the correct folding in the cytoplasm, whereas the matrix-located chaperones mhsp70 and hsp60 have similar functions in the mitochondria (Kang et al. 1990; Scherer et al. 1990). In the matrix imported precursors are first bound to mitochondrial mhsp70 and then to hsp60 (Manning-Krieg et al. 1991). Hsp60 is composed of 14 identical 60 kDa subunits and mediates refolding of the imported proteins. ATP is required for the dissociation of the two chaperone-precursor complexes.

Passage of proteins through the inner membrane requires a membrane potential. The matrix targeting signals of most mitochondrial precursors are cleaved off by the matrix protease. The two subunits MAS1 and MAS2 of the yeast enzyme form a stable 100 kDa heterodimer. The MAS2 subunit of the assembled holoenzyme contains the binding site for the precursor proteins (Yang et al. 1991). In *Neurospora* the two subunits MPP and PEP of the matrix protease form a less stable complex and the PEP subunit was reported to be associated with the cytochrome $bc_1$ complex (Schulte et al. 1989).

At least three different pathways can be used for protein sorting to the intermembrane space. Cytochrome c crosses the outer membrane independently of translocation sites and lacks a cleavable signal sequence (Stuart and Neupert 1990).

Cytochrome c heme lyase, an enzyme that catalyzes the covalent attachment of heme to apocytochrome c, uses another pathway. The enzyme, lacking also a cleavable signal sequence, takes the route through a receptor-GIP complex in the outer membrane (Lill et al. 1992). In contrast to most proteins of the intermembrane space, cytochrome c heme lyase does not transverse the inner membrane.

The third group of intermembrane proteins enter the mitochondrial matrix via a receptor-translocation complex and are then reexported across the inner membrane. This pathway was termed "conservative sorting", because the process resembles bacterial protein secretion, thus supporting the theory of endosymbiotic mitochondrial origin (Hartl and Neupert 1990). Precursor proteins of the intermembrane space (ims) like cytochrome c peroxidase, cytochrome $b_2$ and cytochrome $c_1$ contain a bipartite presequence with an amino-terminal matrix-targeting signal followed by a hydrophobic ims-sorting signal. When the matrix-targeting signals are removed by the matrix protease intermediate precursor proteins are formed. At present, it is not clear whether all intermembrane space proteins of the last group follow the conservative import pathway and are completely translocated into the matrix. Arguments exist that the ims-sorting signal acts as a "stop-transfer" sequence (van Loon and Schatz 1987). In this model only the matrix-targeting sequence is entering the matrix, the ims-sorting sequence crosses the inner membrane and the rest of the proteins remains in the intermembrane space. After removal of the targeting-signal the intermediates are transiently bound to the outer face of the inner membrane as demonstrated for cytochrome $b_2$ (Pratje and Guiard 1986). The processing of the intermediate cytochrome $b_2$ ($cytb_2$) to the mature protein is catalyzed by the inner membrane protease 1 (Behrens et al. 1991; A. Schneider et al. 1991). The inner membrane protease 1 (IMP1) is bound to the outer face of the inner membrane and contains a 21.4 kDa subunit. The sequence of this 21.4 kDa protein shows similarity to the bacterial leader peptidase. This similarity is a strong argument for a conservative sorting mechanism (Behrens et al. 1991). IMP1 is not only involved in processing of cytochrome $b_2$, but also in processing of cytochrome oxidase subunit II. Cytochrome oxidase subunit II (coxII) is encoded by mitochondrial DNA, synthesized with a transient presequence in the mitochondrial matrix and exported to the outer face of the inner membrane. The processing of coxII and $cytb_2$ precursors is prevented in an IMP1 mutant (pet-ts2858). Maturation of cytochrome c peroxidase and cytochrome $c_1$ is unaffected in the IMP1 mutant, indicating that additional proteases exist in the inner membrane to process these precursor proteins.

The "conservative sorting" pathway can also be used by proteins of the inner membrane. Some proteins, like the ADP/ATP translocator, enter the inner membrane directly from receptor-translocation sites without passing the matrix. The ADP/ATP translocator

**Table 3.** Components of the mitochondrial protein import system in yeast and *Neurospora crassa*

| Components *S. cerevisiae* | *N. crassa* | Function |
|---|---|---|
| p32 | MOM19 | Receptor of the outer membrane |
| MAS70 | MOM72 | Receptor of the outer membrane |
| ISP42 | MOM38 | Translocation site (GIP) |
| | MOM7, MOM8, MOM30 | Translocation site (GIP) |
| MAS1 | PEP | Subunit of matrix protease |
| MAS2 | MPP | Subunit of matrix protease |
| hsp70 | hsp70 | antifolding of precursors in the cytosol |
| mhsp70 | mhsp70 | Import and folding of precursors in the matrix |
| hsp60 | hsp60 | Folding of the imported proteins in the matrix |
| IMP1 | | inner membrane protease 1 |

For references see text.

lacks a matrix-targeting signal and its import is independent of hsp60, which is in agreement with such an import mechanism (Mahlke et al. 1990).

The components of the mitochondrial protein import system identified so far are summarized in Table 3.

## References


**Ackerman SH,** Gatti DL, Gellefors P, Douglas MG, Tzagoloff A (1991) FEBS Lett 278:234–238. **Altman S** (1990) J Biol Chem 265:20053–20056. **André C,** Levy A, Walbot V (1992) Trends Genet 8:128–132. **Avise JC** (1991) Annu Rev Genet 25:45–69.

**Baker KP,** Schaniel A, Vestweber D, Schatz G (1990) Nature 348:605–609. **Baldacci G,** Bernardi G (1982) EMBO J 1:987–994. **Behrens M,** Michaelis G, Pratje E (1991) Mol Gen Genet 228:167–176. **Bereiter-Hahn J** (1990) Int Rev Cytol 122:1–63. **Bhat GJ,** Koslowsky DJ, Feagin JE, Smiley BL, Stuart K (1990) Cell 61:886–894. **Blum B,** Sturm NR, Simpson AM, Simpson L (1991) Cell 65:543–550. **Boer PH,** Gray MW (1988) Cell 55:399–411. **Boer PH,** Gray MW (1989) Nucleic Acids Res 17:3993. **Butow RA,** Fox TD (1990) Trends Biochem Sci 15:465–468.

**Capaldi RA** (1990) Annu Rev Biochem 59:569–596. **Cattaneo R** (1990) Experientia 46:1142–1148. **Cattaneo R** (1991) Annu Rev Genet 25:71–88. **Cattaneo R** (1992) Trends Biochem Sci 17:4–5. **Cech TR** (1990) Annu Rev Biochem 59:543–568. **Cech TR** (1991) Cell 64:667–669. **Chapdelaine Y,** Bonen L (1991) Cell 65:465–472. **Chetrit P,** Rios R, Paepe RD, Vitart V, Gutierres S, Vedel F (1992) Curr Genet 21:131–137. **Clayton DA** (1991) Annu Rev Cell Biol 7:453–478. **Colleaux L,** Michel-Molwertz MR, Matagne RF, Dujon B (1990) Mol Gen Genet 223:288–296. **Connett MB,** Hanson MR (1990) Plant Physiol 93:1634–1640. **Costanzo MC,** Fox TD (1986) Mol Cell Biol 6:3694–3703. **Costanzo MC,** Fox TD (1988) Proc Natl Acad Sci USA 85:2677–2681. **Costanzo MC,** Fox TD (1990) Annu Rev Genet 24:91–113. **Costanzo MC,** Seaver EC, Fox TD (1989) Genetics 122:297–305.

**Decoster E,** Simon M, Hatat D, Faye G (1990) Mol Gen Genet 224:111–118. **Dewey RE,** Timothy DH, Levings III CS (1991) Curr Genet 20:475–482. **Dieckmann CL,** Tzagoloff A (1985) J Biol Chem 260:1513–1520. **Diffley JFX,** Stillman B (1991) Proc Natl Acad Sci USA 88:7846–7868. **Doersen CJ,** Guerrier-Takada C, Altman S, Attardi G (1985) J Biol Chem 260:5942–5949. **Dohmen G,** Gessner-Ullrich K, Tudzynski P (1991) Prog Bot 52:244–263. **Drlica K,** Rouviere-Yaniv J (1987) Microbiol Rev 51:301–319.

**Ecke W,** Pannenbecker G, Wasmund O, Michaelis G (1989) Prog Bot 50:198–206. **Erickson L,** Kemble R (1990) Mol Gen Genet 222:135–139.

**Feagin JE,** Abraham JM, Stuart K (1988) Cell 53:413–422. **Feiler HS,** Newton KJ (1987) EMBO J 6:1535–1539. **Forsburg SL,** Guarente L (1989) Annu Rev Cell Biol 5:153–180. **Foury F** (1989) J Biol Chem 264:20552–20560.

**Geli V,** Glick B (1990) J Bioenerg Biomembr 22:725–751. **Glaser E,** Norling B (1991) Curr Top Bioenerg 16:223–263. **Glick B,** Schatz G (1991) Annu Rev Genet 25:21–44. **Gray MW** (1989) Annu Rev Cell Biol 5:25–50. **Grivell LA** (1989) Eur J Biochem 182:477–493.

**Haffter P,** McMullin TW, Fox TD (1990) Genetics 125:495–503. **Haffter P,** McMullin TW, Fox TD (1991) Genetics 127:319–326. **Hakansson G,** Glimelius K, Bonnett HT (1990) Plant Physiol 93:367–373. **Hanson MR** (1991) Annu Rev Genet 25:461–486. **Hartl FU,** Neupert W (1990) Science 247:930–938. **Hines V,** Brandt A, Griffiths G, Horstmann H, Brütsch H, Schatz G (1990) EMBO J 9:3191–3200. **Hodges P,** Scott J (1992) Trends Biochem Sci 17:77–81. **Horn R,** Köhler RH, Zetsche K (1991) Plant Mol Biol 17:29–36. **Horwich AL,** Cheng M, West A, Pollock RA (1991) Curr Top Microbiol Immunol 170:1–42. **Hunt MD,** Newton KJ (1991) EMBO J 10:1045–1052.

**Jang SH,** Jaehning JA (1991) J Biol Chem 266:22671–22677. **Johns C,** Lu M, Lyznik A, Mackenzie S (1992) Plant Cell 4:435–449.

**Kadowaki K,** Suzuki T, Kazama S (1990) Mol Gen Genet 224:10–16. **Kang PJ,** Ostermann J, Shilling J, Neupert W, Craig EA, Pfanner N (1990) Nature 348:137–143. **Knoop V,** Schuster W, Wissinger B, Brennicke A (1991) EMBO J 10:3483–3493. **Koslowsky DJ,** Bhat GJ, Perollaz AL, Feagin JE, Stuart K (1990) Cell 62:901–911.

**Lambowitz AM,** Perlman PS (1990) Trends Biochem Sci 15:440–444. **Lauer M,** Knudsen C, Newton KJ, Gabay-Laughnan S, Laughnan JR (1990) New Biol 2:179–186. **Laver HK,** Reynolds SJ, Moneger F, Leaver CJ (1991) Plant J 1(2):185–193. **Lill R,** Stuart RA, Drygas ME, Nargang FE, Neupert W (1992) EMBO J 11:449–456. **Lisowsky T,** Michaelis G (1988) Mol Gen Genet 214:218–223. **Lisowsky T,** Michaelis G (1989) Mol Gen Genet 219:125–128. **Lee RW,** Dumas C, Lemieux C, Turmel M (1991) Mol Gen Genet 231:53–58. **Lomax MI,** Grossman LI (1989) Trends Biochem Sci 14:501–503.

**Ma D-P,** Yang Y-W, Hasnain SE (1989) Nucleic Acids Res 17:1256. **Mackenzie SA,** Chase CD (1990) Plant Cell 2:905–912. **Maerz M,** Sitte P (1991) Plant Mol Biol 16:593–600. **Mahendran R,** Spottswood MR, Miller DL (1991) Nature 349:434–438. **Mahlke K,** Pfanner N, Martin J, Horwich AL, Hartl FU, Neupert W (1990) Eur J Biochem 192:551–555. **Malesska R,** Skelly PJ, Clark-Walker GD (1991) EMBO J 10:3923–3930. **Manning-Krieg UC,** Scherer PE, Schatz G (1991) EMBO J 10:3273–3280. **Maréchal-Drouard L,** Weil J-H, Guillemaut P (1988) Nucleic Acids Res 16:4777–4788. **Marienfeld JR,** Reski R, Abel WO (1991) Curr Genet 20:319–329. **Masters B,** Stohl LL, Clayton DA (1987) Cell 51:89–99. **Mayes ELV** (1982) In: John EW (ed) The HMG chromosomal proteins. Academic Press, London, pp 9–40. **McMullin TW,** Haffter P, Fox TD (1990) Mol Cell Biol 10:4590–4595. **Michaelis G,** Vahrenholz C, Pratje E (1990) Mol Gen Genet 223:211–216. **Michaelis U,** Körte A, Rödel G (1991) Mol Gen Genet 230:177–185. **Moore LJ,**

Coleman AW (1989) Plant Mol Biol 13:459–465. **Moore AL,** Siedow JN (1991) Biochem Biophys Acta 1059:121–140. **Mulligan RM,** Leon P, Walbot V (1991) Mol Cell Biol 11:533–543. **Murakami H,** Blobel G, Pain D (1990) Nature 347:488–491.

**Neale DB,** Marshall KA, Sederoff RR (1989) Proc Natl Acad Sci USA 86:9347–9349. **Neale DB,** Marshall KA, Harry DE (1991) Can J For Res 21:717–720. **Newton KJ,** Knudsen C, Gabay-Laughnan S, Laughnan JR (1990) Plant Cell 2:107–113. **Nugent JM,** Palmer JD (1991) Cell 66:473–481.

**Oda K,** Yamato K, Ohta E, Nakamura Y, Takemura M, Nozato N, Akashi K, Kanegae T, Ogura Y, Kohchi T, Ohyama K (1992) J Mol Biol 223:1–7. **Ohmen JD,** Kloeckener-Gruissem B, McEwen JE (1988) Nucleic Acids Res 16:10783–10802.

**Pace NR,** Smith D (1990) J Biol Chem 265:3587–3590. **Palmer JD** (1990) Trends Genet 6:115–121. **Palmer JD,** Soltis D, Soltis P (1992) Curr Genet 21:125–129. **Parisi MA,** Clayton DA (1991) Science 252:965–969. **Pfanner N,** Söllner T, Neupert W (1991) Trends Biochem Sci 16:63–67. **Poutre CG,** Fox TD (1987) Genetics 115:637–647. **Pratje E,** Guiard B (1986) EMBO J 5:1313–1317. **Pruitt KD,** Hanson MR (1991) Mol Gen Genet 227:348–355.

**Rapp WD,** Stern DB (1992) EMBO J 11:1065–1073. **Rassow J,** Pfanner N (1991) FEBS Lett 293:85–88. **Rödel G** (1986) Curr Genet 11:41–45. **Rödel G,** Michaelis U, Forsbach V, Kreike J, Kaudewitz F (1986) Curr Genet 11:47–53.

**Scherer PE,** Krieg UC, Hwang ST, Vestweber D, Schatz G (1990) EMBO J 9:4315–4322. **Schinkel AH,** Groot-Koerkamp MJA, Tow EPW, Tabak HF (1987) J Biol Chem 262:12785–12791. **Schneider A,** Behrens M, Scherer P, Pratje E, Michaelis G, Schatz G (1991) EMBO J 10:247–254. **Schneider H,** Söllner T, Dietmeier K, Eckerskorn C, Lottspeich F, Trülzsch B, Neupert W, Pfanner N (1991) Science 254:1659–1662. **Schulte U,** Arretz M, Schneider H, Tropschug M, Wachter E, Neupert W, Weiss H (1989) Nature 339:147–149. **Schulze M,** Rödel G (1988) Mol Gen Genet 211:492–498. **Schuster W,** Ternes R, Knoop V, Hiesel R, Wissinger B, Brennicke A (1991) Curr Genet 20:379–404. **Shoffner JM,** Wallace DC (1990) Adv Human Genet 19:267–330. **Siculella L,** Palmer JD (1988) Nucleic Acids Res 16:3787–3799. **Singh M,** Brown GG (1991) Plant Cell 3:1349–1362. **Small I,** Maréchal-Drouard L, Masson J, Pelletier G, Cosset A, Weil JH, Dietrich A (1992) EMBO J 11:1291–1296. **Söllner T,** Pfaller R, Griffiths G, Pfanner N, Neupert W (1990) Cell 62:107–115. **Söllner T,** Rassow J, Wiedmann M, Schlossmann J, Keil P, Neupert W, Pfanner N (1992) Nature 355:84–87. **Stuart K** (1991a) Annu Rev Microbiol 45:327–344. **Stuart K** (1991b) Trends Biochem Sci 16:68–72. **Stuart RA,** Neupert W (1990) Biochemie 72:115–121. **Sutton CA,** Conklin PL, Pruitt KD, Hanson MR (1991) Mol Cell Biol 11:4274–4277.

**Tzagoloff A,** Gatti D, Gampel A (1990) Prog Nucleic Acids Res Mol Biol 39:129–158. **van Loon APGM,** Schatz G (1987) EMBO J 6:2441–2448.

**Waddle JA,** Schuster AM, Lee KW, Meints RH (1990) Plant Mol Biol 14:187–195. **Wagner DB,** Dong J, Carlson MR, Yanchuk AD (1991) Theor Appl Genet 82:510–514. **Walbot V** (1991) Trends Genet 7:37–39. **Weiss H,** Friedrich T, Hofhaus G, Preis D (1991) Eur J Biochem 197:563–576. **Wise CA,** Martin NC (1991) J Biol Chem 266:19154–19157. **Wissinger B,** Schuster W, Brennicke A (1991) Cell 65:473–482. **Wolff G,** Kück U (1990) Curr Genet 17:347–351.

**Yang M,** Geli V, Oppliger W, Suda K, James P, Schatz G (1991) J Biol Chem 266:6416–6423.
**Yang AJ,** Mulligan RM (1991) Mol Cell Biol 11:4278–4281.

Elke Pratje
Institut für Allgemeine Botanik
der Universität Hamburg
Ohnhorststraße 18
W-2000 Hamburg 52, Germany

Gudula Riemen
Thomas Lisowsky
Frideriki Maggouta
Georg Michaelis
Botanisches Institut
der Universität Düsseldorf
Universitätsstraße 1
W-4000 Düsseldorf 1, Germany

# VI. Extranuclear Inheritance: Linear Protein-Primed Replicating Genomes in Plants and Microorganisms

By Friedhelm Meinhardt and Matthias Rohe

## 1. Introduction

Plasmids were primarily defined as extrachromosomal genetic elements replicating autonomously (Lederberg 1952). They were first described in the 1950s, subsequently, however, plasmids were detected in numerous prokaryotes. They all are covalently closed circular molecules, which carried genes being cryptic or at best advantageous to their hosts under special conditions (Günther 1984). These common structural properties, as well as their occurrence in prokaryotic microorganisms, were considered as characteristic features and strongly influenced the definition of the term "plasmid" (Bresch and Hausmann 1972; Kaudewitz 1983; Lewin 1988). When plasmids were detected in yeast (Sinclair et al. 1967), this was thought to be an exception from a general rule. However, in the meantime, a large number of plasmids could be detected and characterized in plants, yeasts, and in filamentous fungi. A compilation of these plasmids detected to the year 1986 is given in a booklet by Esser et al. (1986) *Plasmids of Eukaryotes*.

The above-mentioned yeast plasmids reflect their prokaryotic counterparts as far as the structure is concerned, i.e., they are covalently closed circular DNA molecules. Subsequently, plasmids were also detected in higher eukaryotes, such as *Zea mays* (Pring et al. 1977) and the ascomycete *Podospora anserina* (Stahl et al. 1978). The plasmids found in the fungus fulfilled the generally accepted criteria as being covalently closed circular molecules. The corn plasmid, however, was the first linear extrachromosomal genetic element described.

Similar genetic elements have been detected in a great number of plants and fungi since then. Evidently the occurrence of circular plasmids represents exceptions among eukaryotes, restricted to the above-mentioned examples and a further few in plants and fungi (Compilation in Esser et al. 1986). In prokaryotes, however, the reverse situation can be observed. Circular plasmids are quite common in bacteria, but linear extrachromosomal elements have so far only been described for *Streptomyces* sp. (Hayakawa et al. 1979; Hirochika et al. 1984) and for *Borrelia* sp. (Plastkerk et al. 1985; Bergström et al. 1989). This review focuses on linear plasmids, their common properties, localization in their hosts, applied aspects, and their phylogeny. Since the role of these genetic elements in phytopathogenic interactions is at least debatable, the reader who is interested in this field is referred to a mini-review (Meinhardt et al. 1990) concerning this subject. Basic principles and relationships of these peculiar genetic elements will be outlined in this

Progress in Botany, Vol. 54

chapter; thus, specific aspects and functions are not given in detail but the reader is referred to original papers and reviews covering the special fields.

## 2. Occurrence and Structure

Linear extrachromosomal genetic elements, also known as linear plasmids, exhibit some common structural properties. Besides their linearity they have terminal inverted repeats (TIRs) ranging from 50 bp in pFOXC2 from *Fusarium conglutinans* (Kistler and Leong 1986) to 1360 bp in kalilo of *Neurospora intermedia* (Chan et al. 1991) and additionally, they have proteins covalently attached to their 5'-ends (terminal Protein = TP).

Representatives of *Borrelia* (Spirochaetaceae), blood parasites of man and mammals, harbor linear plasmids. However, they differ structurally from all other elements known as linear plasmids (Plastkerk et al. 1985; Bergström et al. 1989). The *Borrelia* plasmids do not possess TIRs nor 5'TPs, but have covalently closed hairpin ends (Barbour and Garron 1987). A *Borrelia* bacterium harbors two storage plasmids, each carrying a different nontranscribed gene coding for a variable major surface protein (VMP). On a third plasmid, there is a copy of one of these genes downstream to an expression sequence. Recombinational events between the resident linear plasmids cause antigenic variation of the variable major surface proteins allowing the bacterium to escape the immunological response of its host. Thus, these linear plasmids are the physical basis of the relapsing fever caused by *Borrelia* sp. Similar mechanisms of antigenic variation are known from Trypanosomes, which harbour 50–100 kb minichromosomes. The ends of these minichromosomes do not possess TIRs and 5'TPs but telomeres (for a review see Borst 1986; Lewin 1988). Additionally, a linear plasmid with telomeric end-structures were recently reported to exist in *Physarum polycephalum* (Takano et al. 1991). In this chapter we will not deal with these genetic traits, but focus on the linear dsDNA plasmid with TIRs and TPs.

Because of the high copy number as compared to other DNA molecules, linear elements are considered to replicate autonomously. Further evidence for autonomous replication is derived from molecular analysis as discussed in detail later. All the linear plasmids known so far are never localized in the nucleus, but either in the cytoplasm (the yeast plasmids), or in organelles (the fungal and plant plasmids). The latter are predominantly localized within the mitochondria, there is only one exception in the green alga *Chlamydomonas moewusii* (Turmel et al. 1986), where a linear plasmid was found in the chloroplasts. A compilation of all linear plasmids know thus far is presented in Table 1.

Some of the plasmids listed in Table 1 were subjected to molecular analyses, but for only few of them were the nucleotide sequences determined. From sequencing data, however, it became evident that these elements code for a DNA- and/or RNA-Polymerase (see also Fig. 1).

Only for some of the plasmids has functionality of ORFs been demonstrated: Transcription analysis was performed for pGKL1 (Stark et al. 1984; Sor and Fukuhara 1985) and pGKL2 (Tommasino et al. 1988) of *Kluyveromyces lactis* indicating 4 and 8 transcripts respectively pCLK1 of *Claviceps pupurea* (Düvell et al. 1988; Gessner-Ulrich and Tudzynski 1992). Direct detection of corresponding proteins was done for the plasmids of *Zea mays* (Manson et al. 1986; Zabala and Walbot 1988) and pCLK1 (Gessner-Ulrich and Tudzynski 1992) by immunological methods.

**Table 1.** Compilation of linear plasmids (After Meinhardt et al. 1990)

| Organsim | Plasmid name | Size (kb) | TIR[a] (kb) | Localization | Reference |
|---|---|---|---|---|---|
| Prokaryotes | | | | | |
| *Streptomyces clavuligerus* | pSCL | 12.0 | | | Keen et al. 1988 |
| *Streptomyces rimosus* | pSRM | 43.0 | | | Chardon-Loriaux et al. 1986 |
| *Streptomyces rochei* | pSLA1 | 17.0 | 0.61 | | Hayakawa et al. 1979 |
| | pSLA2 | 17.0 | 0.61 | | Hirochika and Sakaguchi 1982 |
| | | | | | Hirochika et al. 1984 |
| Yeasts | | | | | |
| *Kluyveromyces lactis* | pGKL1 (k1) | 8.9 | 0.20 | Cytoplasm | Gunge et al. 1981 |
| | pGKL2 (k2) | 13.4 | 0.18 | Cytoplasm | Sor et al. 1983 |
| | | | | | Stam et al. 1986 |
| *Phaffia rhodozyma* | | 8.0 | | | Wilber and Profitt 1987 |
| | | 2.0 | | | |
| *Pichia acaciae* | pPac1-1 | 13.6 | | | Worsham and Bolen 1990 |
| | pPac1-2 | 7.3 | | | Worsham and Bolen 1990 |
| *Pichia inositovora* | pPIN11 | 18.0 | | Cytoplasm | Ligon et al. 1989 |
| | pPIN12 | 13.0 | | Cytoplasm | Ligon et al. 1989 |
| | pPIN13 | 10.0 | | Cytoplasm | Ligon et al. 1989 |
| *Saccharomyces kluyveri* | pSKL | 14.2 | 0.48 | Cytoplasm | Kitada and Hishinuma 1984 |
| | | | | | Hishinuma and Hirai 1991 |
| *Saccharomycopsis crataegensis* | pScr1 | 15.0 | | Cytoplasm | Sheperd et al. 1987 |
| | pScr2 | 7.0 | | Cytoplasm | Sheperd et al. 1987 |
| | pScr3 | 5.0 | | Cytoplasm | Sheperd et al. 1987 |
| Filamentous fungi | | | | | |
| *Agaricus bitorquis* | pEM | 7.4 | 1.0 | Mitochondria | Robison et al. 1991 |
| | pMPJ | 3.7 | | Mitochondria | Mohan et al. 1984 |

| | | | | | |
|---|---|---|---|---|---|
| *Ascobulus immersus* | pA1 | 6.4 | 1.2 | | Francou 1981 |
| | pAI1 | 7.9 | | | Meinhardt et al. 1986 |
| | pAI2 | 5.6 | 0.7 | Mitochondria | Kempken et al. 1989 |
| | pAI3 | 2.8 | | Mitochondria | Kempken et al. 1989 |
| *Asochyta rabiei* | | 13.0 | | Mitochondria | Barz and Klein-Bölting pers. comm. |
| *Balansia obtecta* | | 6.0 | | Mitochondria | Mogen et al. 1991 |
| *Ceratocystis fimbriata* | pCF637 | 8.2 | | | Giasson and Lalonde 1987 |
| | pFQ501 | 6.0 | 0.75 | | Normand et al. 1987 |
| *Claviceps purpurea* | pCLK1 | 6.7 | 0.33 | Mitochondria | Tudzynski et al. 1983 |
| | pCLK2 | 5.5 | | Mitochondria | Tudzynski and Düvell 1985 |
| | pCLK3 | 1.1 | | Mitochondria | Tudzynski and Düvell 1985 |
| | pCLT9 | 6.7 | | Mitochondria | Düvell et al. 1988 |
| | pCLB4 | 6.7 | | Mitochondria | Düvell et al. 1988 |
| | pCLT5 | 7.1 | | Mitochondria | Düvell et al. 1988 |
| *Epichloe typhrina* | Et2.0L | 2.0 | | Mitochondria | Mogen et al. 1991 |
| | Et2.1L | 2.1 | | Mitochondria | Mogen et al. 1991 |
| | Et7.5L | 7.5 | | Mitochondria | Mogen et al. 1991 |
| *Erysiphe graminis* f.sp. *hordei* | | 9.0 | | Mitochondria | Giese et al. 1990 |
| *Fusarium oxysporum* | pFOXC1 | 1.9 | | | Kistler and Leong 1986 |
| f. sp. *conglutinans* | pFOXC2 | 1.9 | | Mitochondria | Kistler and Leong 1986 |
| *Fusarium solani* | pFSC1 | 9.2 | 1.21 | Mitochondria | Samac and Leong 1986/1988 |
| f. sp. *curcurbitae* | pFSC2 | 8.3 | 1.02 | Mitochondria | Samac and Leong 1986/1988 |
| *Fusarium sporotrichoides* | | 2.1 | | Mitochondria | Cullen et al. 1985 |
| *Fusarium merismoides* | | 2.1 | | | Rubidge 1986 |
| | | 1.8 | | | Rubidge 1986 |
| *Gaeumannomyces graminis* | E1 | 8.4 | | Mitochondria | Honeyman and Currier 1983/1986 |
| *var. tritici* | E2 | 7.2 | | Mitochondria | Honeyman and Currier 1983/1986 |
| *Hebeloma circinans* | pHC1 | 10.3 | | | Schründer et al. 1991, 1992 |
| | pHC2 | 9.3 | | | Schründer et al. 1991, 1992 |
| *Lentinus edodes* | pLLE1 | 11.0 | | Mitochondria | Katayose et al. 1990 |
| *Morchella conica* | pMC3-1 | 8.0 | | Mitochondria | Meinhardt and Esser 1984/1987 |
| | pMC3-2 | 6.0 | 0.71 | Mitochondria | Meinhardt and Esser 1984/1987 |
| | | | | | Rohe et al. 1991 |

**Table 1** (cont.)

| Organsim | Plasmid name | Size (kb) | TIR[a] (kb) | Localization | Reference |
|---|---|---|---|---|---|
| *Morchella elata* | pME141-1 | 6.7 | | | Meinhardt and Esser 1987 |
| | pME141-2 | 6.0 | | | Meinhardt and Esser 1987 |
| *Morchella hortensis* | pMH1 | 7.8 | | | Meinhardt and Esser 1987 |
| *Morchella esculenta* | pMS2 | 7.5 | | | Meinhardt and Esser 1987 |
| | pMS1 | 4.0 | | | Meinhardt and Esser 1987 |
| *Neurospora intermedia* | kalilo | 9.0 | 1.36 | Mitochondria | Bertrand et al. 1985/1986<br>Chan et al. 1991 |
| *Neurospora crassa* | maranhar | 7.1 | 0.35 | Mitochondria | Bertrand and Griffith 1988<br>Court et al. 1991 |
| *Podospora anserina* | pAL2-1 | 8.4 | | Mitochondria | Osiewacz et al. 1989,<br>Hermanns and Osiewacz 1990 |
| *Pleurotus ostreatus* | pLPO1 | 10.0 | | Mitochondria | Yui et al. 1988 |
| | pLPO2 | 9.4 | | Mitochondria | Yui et al. 1988 |
| *Rhizoctonia solani* | pRS64 | 2.6 | | | Hashiba et al. 1984 |
| *Tilletia caries* | pTCC | 7.2-7.6 | | Mitochondria | Kim et al. 1990 |
| *Tilletia laevis* | pTCL | 7.2/7.4 | | Mitochondria | Kim et al. 1990 |
| *Tilletia controversa* | pTCT | 7.4/7.6 | | Mitochondria | Kim et al. 1990 |
| Green algae | | | | | |
| *Chlamydomonas moewusii* | | 5.9 | | Chloroplasts | Turmel et al. 1986 |
| Higher Plants | | | | | |
| *Brassica campestris/napus* | | 11.3 | | Mitochondria | Palmer et al. 1983<br>Erikson et al. 1986 |
| *Beta maritima* | | 10.4 | | Mitochondria | Samitou-Laprade et al. 1989 |
| *Sorghum bicolor* | N1 | 5.7 | | Mitochondria | Pring et al. 1982 |
| | N2 | 5.3 | | Mitochondria | Chase and Pring 1986 |
| *Zea diploperennis* | D1 | 7.5 | | Mitochondria | Timothy et al. 1982 |
| | D2 | 5.4 | | Mitochondria | Timothy et al. 1982 |

| | | | | | |
|---|---|---|---|---|---|
| *Zea mays* N line | | 2.3 | 0.05 | Mitochondria | Pring et al. 1977<br>Leon et al. 1989 |
| *Zea mays* C line | | 2.3 | 0.05 | Mitochondria | Kemble et al. 1980<br>Leon et al. 1989 |
| *Zea mays* S line | S1 | 6.4 | 0.2 | Mitochondria | Pring et al. 1977 |
| | S2 | 5.5 | 0.2 | Mitochondria | Levings and Sederoff 1983 |
| | | 2.3 | 0.05 | Mitochondria | Kemble and Thompson 1982;<br>Paillard et al. 1985;<br>Leon et al. 1989 |
| *Zea mays* R line | R1 | 7.5 | 0.2 | Mitochondria | Weissinger et al. 1982 |
| | R2 | 5.5 | 0.2 | Mitochondria | Weissinger et al. 1982 |
| *Zea mays* T line | | 2.1 | 0.05 | Mitochondria | Bedinger et al. 1986 |

[a] TIR = terminal inverted repeats.

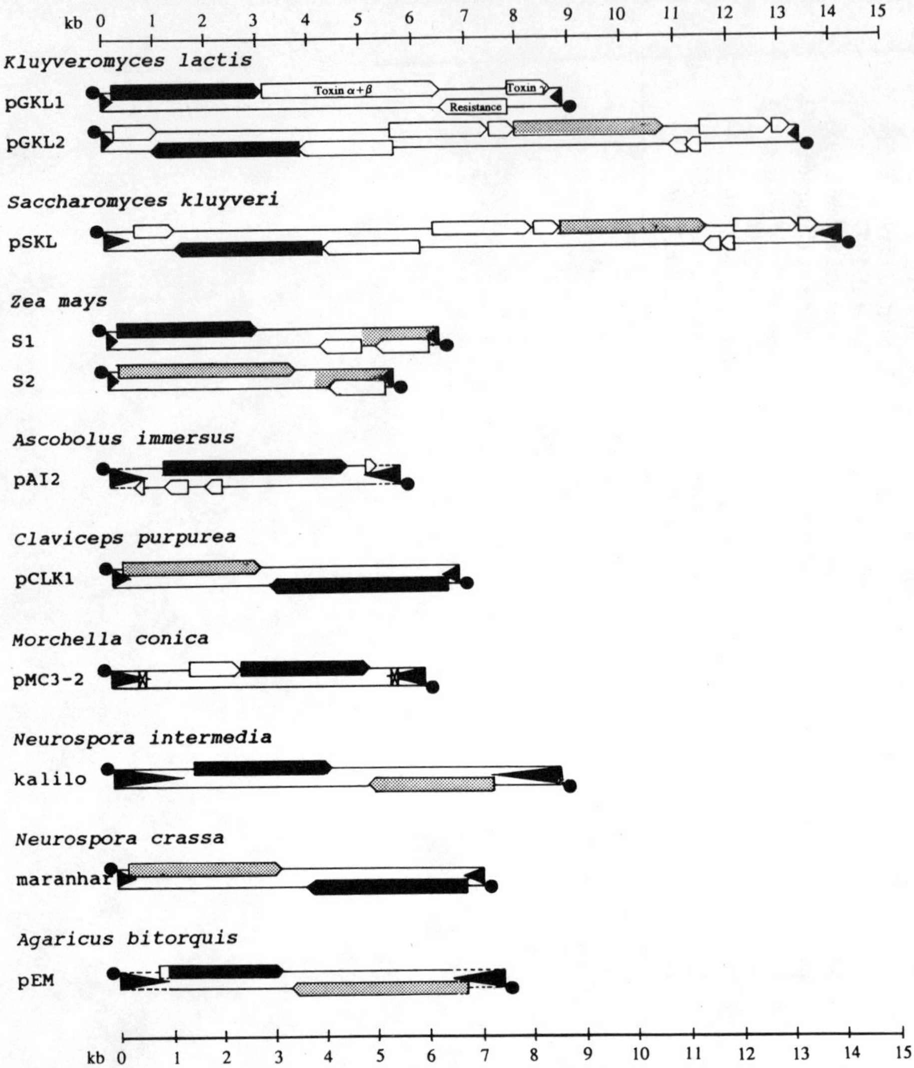


**Fig. 1.** Scheme of linear plasmids showing the localization of open reading frames (ORFs) and length of terminal inverted repeats (TIRs) (adopted from Meinhardt et al. 1990, updated). Symbols: *white arrow* ORF, the direction of the arrow gives the orientation; *dotted arrow* ORF presumably coding for a RNA polymerase; *black arrow* ORF presumably coding for a DNA polymerase; *dotted areas* homologous parts of plasmids S1 and S2; *black triangles* terminal inverted repeats; *filled circles* terminal proteins. Terminal regions of pAI2 and pEM which were not completely sequenced are drawn with *broken lines*. References. *Kluyveromyces lactis* pGKL1: Hishinuma et al. 1984; Stark et al. 1984; Sor and Fukuhara 1985; Stam et al. 1986; Jung et al. 1987a; *Kluyveromyves lactis* pGKL2: Stam et al. 1986; Tommasino et al. 1988; Wilson and Meacock 1988; *Saccharomcyes kluyveri* pSKL: Hishinuma and Hirai 1991; *Zea mays* S1: Kemble and Thompson 1982, Levings et al. 1984; Paillard et al. 1985; Kuzmin and Levchenko 1987; *Zea mays* S2: Kemble and Thompson 1982; Levings and Sederoff 1983; Levings et al. 1984; Kuzmin et al. 1988; Oeser 1988. *Ascobolus immersus* pAI2: Meinhardt et al. 1986; Kempken et al. 1989; *Claviceps purpurea* pCLK1: Düvell et al. 1988; Oeser and Tudzynski 1989; *Morchella conica* pMC3-2: Rohe et al. 1991; *Neurospora intermedia* kalilo: Chan et al. 1991; *Neurospora crassa* maranhar: Court et al. 1991; unpublished data. (Complete sequence of maranhar is not published but released to EMBL database.) *Agaricus bitorquis* pEM: Mohan et al. 1984; Robison et al. 1991. These references comprise all publications on which information in the figure is based

Remarkably, a linear plasmid encodes both a DNA- and RNA-polymerase when found as a single element in an organism, whereas there is more than one element when either a DNA- *or* an RNA-polymerase is encoded. Thus, linear elements presumably represent autonomous systems in terms of replication and transcription. With the exception of the yeast plasmids, there is no other ORF to which a function can be attributed, a fact which is significant with respect to the conception of a selective advantage provided by the plasmids and thereby ensuring maintainance of the elements. The yeast plasmids evidently constitute a special system as outlined below.

## 3. Replication Mechanism

The common structural properties of linear plasmids listed above suggest that all these elements belong to a group of related genetic traits, which replicate via the same mechanism. Because the overall structure of the linear plasmids with its terminal inverted repeats and the terminal proteins is reminiscent of some DNA viruses, i.e., the adenoviruses, a viral mode of replication has repeatedly been suggested (Paillard et al. 1985; Meinhardt et al. 1986; Düvell et al. 1988).

For pAI2 of *Ascobolus immersus*, an adenoviral mode of replication could be domonstrated using an in organello synthesis system. It was shown that the radioactivity is incorporated into the replication-product from the 5'-termini when adding radioactively labeled nucleotides, as would be expected when the replication starts from the termini of

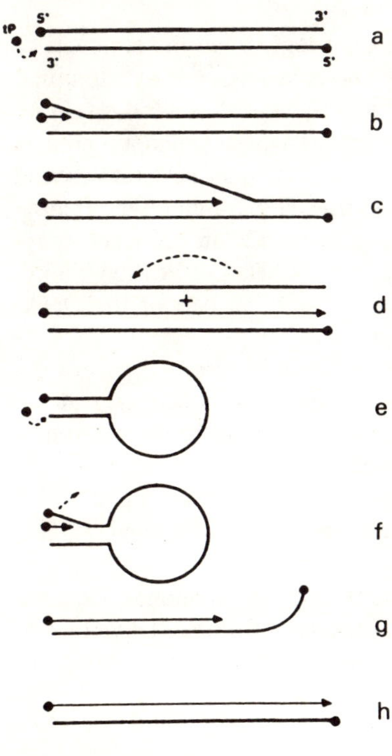


**Fig. 2.** Replication mechanism of the adenoviruses (after Challberg and Kelly 1982). Replication is initiated by the terminal protein (tP) and the DNA polymerase (**a**). DNA replication occurs in 5'-3'direction while the parental strand of the same oriantation is displaced (**b, c, d**). The displaced parental strand creates a partial double strand by homologous base paring in the region of TIRs (**e**) where the second-strand replication starts (**f, g**). Replication continues until the second strand is also completely replicated (**h**)

the plasmid in a mode as it is known from the adenoviruses (Kempken et al. 1989). The supposed model of replication is demonstrated in Fig. 2.

Keen et al. (1988) suggested bacteriophages to be the progenitors of linear plasmids from *Streptomyces*. Since the linear plasmids of the eukaryotes are predominantly located in organelles, generally believed to be descendants of prokaryotes, namely in the mitochondria, it can be assumed that prokaryotic phages are their progenitors (Rohe et al. 1992). Phages with linear DNA and 5'-proteins at least do exist in *E. coli*, e.g., PRD1 (Bamford et al. 1983; Savilahti and Bamford 1986), *Streptococcus pneumoniae*; Cp-1 (Garcia et al. 1983) and *Bacillus*, phi29 (Harding et al. 1978; Hermoso and Salas 1980). The possible relationships of linear plasmids and prokaryotic phages are discussed in a later section of this chapter.The proposed adenoviral-like replication mechanism requires a terminal protein acting as a primer in the initiation of replication, as was also demonstrated for the adenoviruses (Desiderio and Kelly 1981) and linear prokaryotic viruses phi 29 (Hermoso and Salas 1980) and PRD1 (Bamford et al. 1981, 1983).

In linear phages the replication mechanism is well studied. The presence and – in several cases – even the structure and size of the terminal protein in known. Especially for phi29, which, together with the adenoviruses, is the best-investigated linear genome, genetic organization and proteins required for replication are known (for a review, see Salas 1988). In contrast to this, the presence of a terminal protein has been demonstrated for most linear plasmids by indirect methods only (i.e., retardation of terminal fragments in gel-electrophoresis or protection against exonuclease attack). Direct detection of the terminal proteins by ionidation and determination of its molecular weight was done only in *Kluyveromyces lactis*, pGKL1: 28 kDa, pGKL2: 36 kDa (possibly degraded) (Stam et al. 1986), *Fusarium solani*, pFSC1: 80 kDa (Samac and Leong 1989b) and *Neurospora intermedia*, kalilo: 120 kDa (Vierula et al. 1990).

However, until today, only speculations about the coding region and the structure of terminal proteins of linear plasmids exist, because no amino acid sequence of a terminal protein is available up to now. Since the known amino acid sequences of viral origin do not show significant similarities, a detection of coding regions via sequence alignments was not possible. However, there seems to be a characteristic secondary structure around the DNA-binding amino acid (usually Thr, Tyr or Ser) in virus-TPs. Based on the finding that a similar region might be present in pCLK1 around serine 327 in the DNA-polymerase ORF of pCLK1 and the amino terminal region of the 33 kDa ORF of the 2.3 kb-plasmid of *Zea mays*, it has been postulated that the aminoterminal part of the DNA-polymerase ORF codes for this primer (Oeser and Tudzynski 1989).

Thus, conserved motifs were expected to be present in the aminoterminal part of the DNA-polymerases of the closely related linear plasmids. A suggestion concerning such a region has been put forward by Chan et al. (1991). The conservation of the supposed motif, however, is quite poor (see Fig. 3). The significance of these observations awaits further elucidation, since in pMC3-2 of *Morchella conica* two regions can be found which fit into this aligment (see Fig. 3), making evident the problem arising from the comparison of poorly conserved amino acid sequences.

In some of the TIRs of linear elements, sequence similarities to functional domains, responsible for replication in adenoviruses (Hay 1985a, b) can be found (Düvell et al. 1988; Rohe et al. 1991). The significance of these observations, however, remains unclear, because other elements do not fit this model.

```
phi29 TP   242aa NVYKLISYLEQYRRGDFDLSL--KGF-COOH
phi29 TP   150aa SKVRSYSR-LRTLEESMEM----RTDP-QYYEK
                    | ||| ||*| |       *    *
Ad2 pTP     80aa FDSRAYSR-LRYTELSQPG----HQ-TVNWSVM
                      |||  ||          *  |   *
PRD1 TP    156aa TFNGNYSRRT-YTSFDEAF--N-KFMTYDIIID
                      ||   |      |  * *
pGKL1      201aa FVSETYSTVIFYAKSIY-F-PQPRC-VNNWGNN
                      |||  || |*       * *  * |  |
pCLK        65aa RELRFYST-SAYTKTHW----NNKDRI-NKWMQ
                    |  |   |  *           |    ||
pMC3-2     211aa IFTRELSGVKSYICSNTK-----KVLYYNKPLS
                   |   |   |     *        |   | ||
kalilo     168aa KPTSNPSIKTKYEAHKS----NIK-RNKNINLS
                      | | *|              |  ||  |
S1         181aa PKMGRRS-KRRYQSYIP----VDKTEMKNKTLF
                  |    *   |   |    | |  |   || *
pAI2       380aa LKTIEGTKYANYTFPIK-KDIVVKD--INKKIN
                       |   |    *|  |   ||  *|*
pHC2        82aa FNLNMETDSL-YNLSLKNKRYFVKVSLLNRDNE
                   *   |   | ||*        |   **|
pMC3-2     286aa KLIISQTKRLWYNVINSF-----KRFIINLPGY

CONS             ------S----Y-----------K----N----
                       T                R
```

aa considered to be conserved: (I,L,V);(D,E);(K,R,H);(S,T);(N,Q) and (Y,F)

**Fig. 3.** A moderately conserved amino-acid sequence that occurs in the terminal proteins of linear viruses and bacteriophages and in the cryptic amino-terminal domains of the DNA polymerases of eukaryotic linear plasmids. (After Chan et al. 1991). Terminal proteins are identified at the *left* by *TP*, and preterminal protein by pTP, all of the other proteins are from DNA-sequence data of putative DNA polymerases. A consensus sequence is shown at the bottom of the figure. Source of sequences: *phi29* Escarmis and Salas 1982; *Ad2* Aleström et al. 1982; Petit et al. 1988; *PRD1* Savilahti and Bamford 1987; *pGKL1* Stark et al. 1984; *pCLK1* Oeser and Tudzynski 1989; *kalilo* Chan et al. 1991; *S1* Paillard et al. 1985; *pAI2* Kempken et al. 1989; pMC3-*2* Rohe et al. 1991; *pHC2* Schründer and Meinhardt, unpubl. Note that there are two alignments for pMC3-2 of *Morchella conica*, for details and discussion see text

Remarkably, there is no sequence similarity even between TIRs of relatively closely related elements such as the yeast plasmids pGKL1 and pGKL2 (Hishinuma et al. 1984), and that is why efforts aiming to identify those regions by comparion to the adenoviruses must fail.

## 4. Prokaryotic Linear Elements

In prokaryotes, two types of linear elements replicating with a protein primer can be found, i.e., linear plasmids and linear phages. Linear plasmids have only been detected in representatives of the Gram-positive genus *Streptomyces* (see Table 1). The plasmids pSLA1 and pSLA2 of *Streptomyces rochei* were analyzed in detail. They have almost the same size of 17 kb and are thought to be closely related, because they differ only in a small deletion or insertion (Hirochika and Sakaguchi 1982). Curing experiments suggest that these linear plasmids are involved in the production of the lankacidin group antibiotics (Hayakawa et al. 1979). However, sequencing data only exist for the terminal parts of pSLA2 (Hirochika et al. 1984), which was shown to have TIRs of 614 bp and further interrupted homologous sequences beyond this area up to 800 bp.

Compared to linear plasmids, linear phages seem to be more common in prokaryotes: from Gram-negative bacteria, linear phages (i.e., PRD1, PR4, PR5, L17, and PR722) were isolated. These phages are closely related as is evident from similar genome sizes, quite uniform restriction maps, and protein pattern (Coetzee and Bekker 1979; Bamford et al. 1981) as well as the nucleotide sequences of the 110–111 bp long TIRs which differ only in very few positions (Savilahti and Bamford 1986). They infect a variety of different bacteria, including *Escherichia coli* and *Salmonella typhimurium* (Olsen et al. 1974). Detailed analysis was done for PRD1 and PR4. The phage particles consist of an outer protein coat and an inside membrane vesicle, haboring a linear dsDNA of about 14.7 kb (Bamford and Mindich 1982; Davies et al. 1982; McGraw et al. 1983).

Another group of phages infecting the Gram-positive bacterium *Streptococcus pneumoniae* also possesses terminal proteins and TIRs (Ronda et al. 1981; Lopez et al. 1984). Cp-1, its best investigated representative, has a hexagonal head and a short tail (Ronda et al. 1981). Its double-stranded linear DNA genome is about 18 kb in size and has a 28-kD protein covalently bound to its 5'termini (Garcia et al. 1983). The TIRs of Cp-1, and the related phages Cp-5, and Cp-7 were sequenced and shown to be 236 bp (Cp-1) 343 and 347 bp in size respectively (Escarmis et al. 1984, 1985). Biochemical characterization as well as sequencing data of the termini show striking similarities (highest divergence (Cp-1 and Cp-7): 8.4%) between these proteins but clear differences to the *Bacillus subtilis* phage phi29 (Escarmis et al. 1985).

In the Gram-positive bacterium *Bacillus subtilis*, a number of similar phages possessing linear genomes with 5'bound proteins have been described. For some of them, especially phi29, detailed analysis was performed, and knowledge of the replication mechanism including the corresponding proteins exists (for a review see Salas 1988).

Concerning replication of the phi29 genome, mutant studies revealed that several genes are instrumental in viral DNA replication. Products of gene 2 and 3, coding for the viral DNA-polymerase and the terminal protein respectively are sufficient for DNA-replication in vitro (Blanco et al. 1984; Blanco and Salas 1984, 1985; Prieto et al. 1984; Watabe et al. 1984). However, in vivo, gene products of at least three other phage genes are necessary for replication (Talavera et al. 1971; Carrascosa et al. 1976; Hagen et al. 1976).

By comparing the terminal proteins, restriction maps, and nucleotide sequences of the termini from phages phi29, phi15, Nf, M2Y, GA-1, PZA, PZE, and B103, these phages could be subdivided into three classes (A: phi29, phi15 and PZA; B: PZE, Nf and M2Y, and C: GA-1, B103). However, all these phages possess similar, very short TIRs of only 6–8 bp (Yoshikawa and Ito 1981; Gutierrez et al. 1986).

As outlined above, PRD1, Cp-1, and phi29 represent different groups of closley related phages. Sequence similarities between TIR regions of PRD1 and Cp-1 indicate a relationship between these groups, too (Savilahti and Bamford 1986). However, as mentioned above, phi29 group-phages differ from the other elements in having only short

TIRs of 6–8 bp, though there are some motifs at the phi 29 termini outside the short TIR regions similar to PRD1 and Cp-1 regions.

For the adenoviruses (Lechner and Kelly 1977) and the linear plasmids (Hirochika et al. 1984; Sakaguchi 1990), a replication mechanism was postulated, in which a panhandle structure is formed by base pairing of the TIRs, stabilized by different proteins. Such a mechanism is considered to lead to coevolution of opposite long terminal inverted repeats. However, in the phage φ29, extended base pairing is not possible (Savilahti and Bamford 1986). Additionally, for most linear elements there are no ORFs for the postulated proteins and a much simpler replication mechanism can be postulated (see also previous section).

The overall similarity of the genomes with TIRs and 5'bound proteins might suggest a common replication machinery. Based on this assumption, the replication mechanism of phages can be considered to be valid for all the linear elements. Apart from the general rule, however, striking differences with respect to replication do exist: there are TIRs of different size in linear elements and the occurrence of conserved functional domains cannot be generalized.

## 5. Cytoplasmic Elements

### a) Fundamentals

The linear plasmids from the yeast *Kluyveromyces lactis* differ from all the other linear plasmids of eukaryotes in various aspects. They are located in the cytoplasm and not in the organelles, they encode other proteins in addition to DNA- and RNA-polymerases, and they confer an advantageous phenotype to their host, i.e., the ability to kill other sensitive yeast strains. Cells carrying the killer plasmids exhibit immunity to the toxin. The genome organization of the smaller plasmid pGKL1 is largely understood, and specific functions can be attributed to all of its four open reading frames (Stark and Boyd 1986; Jung et al. 1987a, Kuzmin and Levchenko 1987; Tokunaga et al. 1987; Gunge and Kitada 1988) (see also Fig. 1). For the larger plasmid pGKL2 a function can be attributed to only some of the ORFs, most of them are deduced from nucleotide sequence data only: ORF4 (helicase?), ORF2 (DNA-polymerase) (Tommasino et al. 1988), ORF6 (RNA-polymerase) (Wilson and Meacock 1988), ORF10 (DNA-binding protein) and ORF1 was discussed as being instrumental in immunity expression (Tommasino et al. 1988; Stark et al. 1990; Tommasino 1991). ORF1, however, was shown to be dispensable for the expression of the immunity phenotype and for replication and maintenance of the plasmids (Schaffrath et al. 1992).

The killer system of *Kluyveromyces lactis* based on pGKL1/2 differs from the long-known killer system first detected in *Saccharomyces cerevisiae*, which is dependent on virus-like particles containing dsRNA. For further information concerning the *Kluyveromyces* and *Saccharomyces* killer systems, the reader is referred to reviews written by Stark et al. (1990) and Tipper and Bostian (1984) respectively.

Recently, it was reported that in the yeast *Pichia inositovora* (Hayman and Bolen 1991) as well as in *Pichia acaciae* (Worsham and Bolen 1990), different linear plasmids are associated with a killer toxin activity, indicating that the linear plasmid killer system of *K. lactis* is not a unique phenomenon.

b) Applied Aspects

Linear plasmids are ubiquitious. They represent a reservoir of autonomously replicating vectors for genetic manipulation of various organisms. Most of these plasmids are located in the mitochondria, which obviously is an impediment for transformation. The use of microprojectiles, which allows introduction of DNA into organelles (Boynton et al. 1988; Johnston et al. 1988), may, however, facilitate the development of reliable transformation protocols for linear plasmids in the future.

The cytoplasmically localized killer plasmids of *Kluyveromyces lactis* are ideally suited for vector development, because they can be transferred to a number of different yeasts including *Saccharomyces cerevisiae* (Gunge and Sakaguchi et al. 1981; Gunge et al. 1982; Sugisaki et al. 1985), they are stably maintained without selective pressure and have a copy number of about 100 per cell. Vectors based on the pGKL-plasmids are therefore potentially advantagous to conventional yeast vectors.

First attempts at transformation of *Kluyveromyces lactis* (DeLouvencourt et al. 1983) and *Saccharomyces cerevisiae* (Fujimura et al. 1987; Kämper et al. 1989a) with in vitro recombined pGKL1 derived molecules resulted in the formation of circular hybrid vectors only.

Attempts to integrate the LEU2 coding region from *Saccharomyces cerevisiae* into the smaller plasmid pGKL1 by in vivo recombination resulted in the construction of linear vectors. However, the hybrid molecules obtained had no proteins attached to their 5'ends but telomere-specific sequences were added to the ends of pGKL1. In contrast to the cytoplasmically localized pGKL plasmids, the newly obtained linear hybrid vector (pLS1) probably replicates within the nucleus and provides evidence that the nuclear LEU2-gene cannot be expressed in the cytoplasm (Kämper et al. 1989a). Thus, several lines of evidence suggest that linear plasmids constitute an independent transcription system. To make use of this, and, attempting to escape nuclear control mechanisms, the LEU2 gene was flanked with promotor and terminator sequences from pGKL1. Subsequently this was introduced into a pGKL1/2 bearing *S. cerevisiae* strain eventually leading to integration via homologous recombination. Like the killer plasmids pGKL1, the hybrid molecules are located in the cytoplasm, have terminal inverted repeats, and possess covalently linked proteins at their 5'-ends (Kämper et al. 1989b, 1991). The copy number of the hybrid plasmids is comparable to that of the native plasmids, as is also the stability (Kämper et al. 1989b).

The methodology developed by Kämper et al. (1989b, 1991) offered for the first time the opportunity to manipulate these genetic elements, making possible functional analysis of putative ORFs of the larger pGKL2 also. For this purpose, an in vitro modified version of the cytoplasmically expressible LEU2 gene flanked by appropriate pGLK2 segments was used to replace the central part of the ORF1 coding region of pGKL2. Transforma-

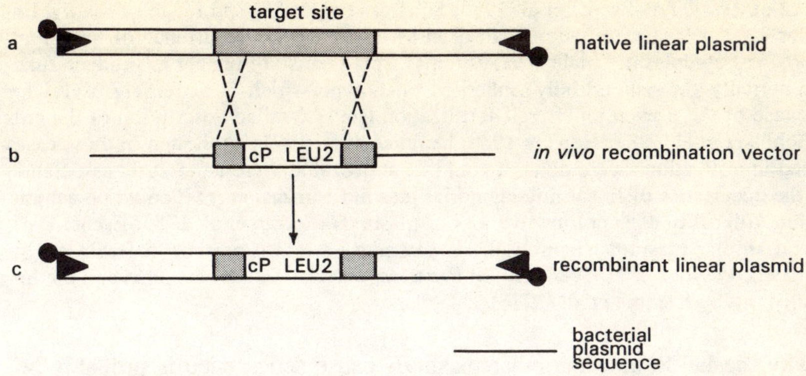


**Fig. 4.** Diagram showing use of vectors for transformation and in vivo recombination with a linear killer-plasmid. **a** In vivo-transformation vector is constructed by flanking the LEU-cartridge (LEU2-gene of yeast under control of a cytoplasmic promoter) with regions homologous to the target site on the killer plasmid. **b** The recombinant plasmid is introduced into the killer-plasmid-harboring strain and the region flanked by the killer-plasmid sequences integrate into the killer plasmid by homologous recombination into the target region. **c** Under selective conditions, segregation of the native and recombinant plasmids leads to total loss of the native plasmid, leading to plasmids harboring recombinant linear plasmids only

tion of a Leu⁻ *Kluyveromyces lactis* killer strain resulted in the expected disruption of ORF1 in the resident pGKL2. No differences concerning the killer and immunity phenotype between a strain harboring this new hybrid plasmid and the native pGKL2-carrying recipient could be detected (Schaffrath et al. 1992). Since there is no detectable influence on stability and maintenance of the plasmids when ORF1 is disrupted, this locus is the ideal target site to integrate foreign DNA into pGKL2. The integration of foreign DNA into a dispensable locus of pGKL2 is of great importance for gene technology, because it is no longer necessary to use a system with two plasmids. In previous experiments, integration of foreign DNA was done into pGKL1 exclusively (Kämper et al. 1989b, 1991; Tanguy-Rougeau et al. 1990). However, replication and maintenance of pGKL1 is strictly dependent on pGKL2. Application can therefore now be performed in a much simpler system based solely on the larger killer plasmid PGKL2. The developed system offers in addition the possibility of adressing basic problems concerning the biology of the linear plasmids as well. A scheme showing the principle of the killer-plasmid-based expression system with vectors using the LEU-marker is shown in Fig. 4.

## 6. Mitochondrial Plasmids

The majority of the linear plasmids from which detailed information is available are located within the mitochondria. The location of these plasmids causes serious problems with respect to biotechnology: every attempt at transformation faces the problem of the barrier of the mitochondrial membrane. All the linear plasmids are cryptic, except the kalilo and maranhar plasmids of *Neurospora intermedia* and *Neurospora crassa*, respectively, which cause senescence in their hosts.

In the case of S2 of *Zea mays* (Escote et al. 1985; Schardl et al. 1985) and for the *Brassica* linear plasmid (Palmer et al. 1983), correlations had been assumed for the occurrence of mitochondrial linear plasmids and cytoplasmic male sterility (cms). cms plants are unable to produce functional pollen, a maternally (mitochondrially) inherited phenotype which is extremely useful for plant breeding because of the prevention of self-fertilization (for a detailed description of the cms phenomenon see Schuster et al. 1987; Newton 1988; Dohmen et al. 1991). Although in those cases where more detailed investigations were done, it could be shown that there is no strict association between cms and the occurrence of linear mitochondrial plasmids: in *Zea mays* it could be demonstrated that reversion to fertility does not involve loss of plasmids (Escote et al. 1985; Escote-Carlson et al. 1988) and similar plasmids (namely the R plasmids) are also present in fertile strains (Weissinger et al. 1982; Levings et al. 1984) and in *Brassica*, correlation between cms and the linear plasmid was ruled out by Kemble et al. (1986).

The reason why the kalilo and maranhar plasmids cause senescence is probably because they mutate mitochondrial DNA by integration into the mitochondrial genome (Bertrand et al. 1986; Chan et al. 1991). However, the integration seems to be a very rare event. Thus, senescence is presumably the result of the suppressive accumulation of defective mtDNAs generated by only one or few integration events, rather than being caused by several separate integration events into many individual mitochondrial genomes (Chan et al. 1991).

It is considered that plasmid bearing mitochondria displace other mitochondria in a given strain in a few cell cycles after an integration occurred which causes functional impairment. This separation is brought about by a control mechanism in the cell which leads to increased proliferation of defective mitochondria. The existence of such a control mechanism can be demonstrated with antibiotics affecting the respiratory chain (Antimycin for example) which result in a much higher amount of mitochondria than found in wild type cells. The existence of such mechanism can be demonstrated also by the fact that mutations affecting mitochondrial function have the same result of leading to senescence and accumulation during the degenerative process (Chan et al. 1991; Court et al. 1991; Bertrand pers. comm.). Thus, kalilo and maranhar act not only as selfish but as true parasitic DNA, spreading by contact transmission like a molecular disease (Griffith et al. 1990). Remarkably, other linear plasmids show a similar mobility. Integration of plasmid-DNA into the mitochondrial genome was also found in the case of pCLK1/*Claviceps purpurea* (Tudzynski and Esser 1986), pEM/*Agaricus bitorquis* (Robison et al. 1991) and pAL2-1/*Podospora anserina* (Hermanns and Osiewacz 1990). In *Brassica napus* and *Neurospora* (Griffith et al. 1990) transferability of plasmids was demonstrated. Possibly the capability of integration is a general feature of all linear plasmids, and this raises the question of whether the senescence-causing phenotype is restricted to kalilo and maranhar, which, however, show no significant structural differences to the other mitochondrial plasmids.

The crypticism, as well as the senescence phenomenon, raises the question as to the reason of the maintenance of the plasmids without any selective pressure. The coding capacity seems to be sufficient only for a DNA-and an RNA polymerase. For the very little ORFs which can be found in addition to the DNA- and/or RNA polymerases, a function is unknown, and no similarity to any other protein could be detected. Thus, it seems that the plasmids do not contribute a selective advantage to their hosts.

One may suppose that this is a secondary feature and the plasmids could have lost their selection marker during evolution. Another possibility is that they descend from viral ancestors presenting a minimized viral system. Evidence for this hypothesis will be presented in the next section.

## 7. Phylogenetic Relationships

As already mentioned, a relationship of the linear plasmids to the adenoviruses has been repeatedly suggested because of basic structural features as well as similarity of the sequences of the DNA polymerase and/or RNA polymerases coding regions (Meinhardt and Esser 1984 and 1987; Düvell et al. 1988; Mohr and Esser 1990).

However, estimation of relationships among the linear plasmids themselves and relative relationships to the different groups of viruses replicating via protein primers were not possible for a long time, because only molecular data from very few plasmids were available. Recently, sequence data of several plasmids have been published almost simultaneously, namely pSKL (Hishinuma and Hirai 1991), pEM (Robison et al. 1991), kalilo (Chan et al. 1991), pMC3-2 (Rohe et al. 1991), pHC2 (Schründer et al. 1991), maranhar (Court et al. 1991), pCLT5 (Tudzynski and Deing, in prep.). The number of plasmids from which sequence data were available was thereby doubled within 3 months. When comparing the amino acid sequences of the putative DNA polymerases from these linear plasmids and from various adenoviruses and prokaryotic protein-primed replicating phages, not only striking similarities could be observed, it was also noticed that some of the amino acids in the conserved regions of the DNA polymerases were not conserved in all elements but in specific groups (Rohe et al. 1991).

Three conserved motifs called PolI-III were shown to be present in different DNA polymerases (Bernad et al. 1987). Later, three conserved 3'-5'exonuclease active-regions designated Exo1-3 were additionally described for different polymerases (Bernad et al. 1989). With the data available, structural relationships between DNA polymerases of different types were determined including those from the other terminal protein-primed replicating genomes. This data suggested, that the DNA polymerases from the terminal protein-primed replicating genomes constitute a separate group of closely related elements among the other polymerases (Bernad et al. 1989). However, the amount of data from this type of polymerases was not sufficient to determine closer relationships within this group.

A comparative alignment of the conserved regions of DNA-polymerases of the linear plasmids from which data are known as well as some viruses with protein primed replication is given in Fig. 5.

Because the DNA polymerase is the only structural element in the linear plasmids which is common to all the linear elements replicating via a protein primed mechanism (except S2 of *Zea mays*) and which additionally shows enough conservation for comparative analyses, it was chosen by Rohe et al. (1992) for a calculation of a distance matrix for a phylogenetic tree according to the method of Feng and Doolittle (1987). Because the N-terminal part of the DNA polymerases shows great variation in length, and the real start of transcription is in most cases unknown, only the region of the ORFs spanning the conserved motifs from the Exo1 to Pol3 was taken for the analysis. The derived phylogenetic tree is presented in Fig. 6.

```
           Exo1              Exo2              Exo3
pMC3-2     ISTFDIETYTTPSGI   IYVHNLTKFDAYFIL   KETIKYLEQDIKIL
pCLK1      LGTIDLETTTGAQPK   FYAHNLGKFDAVFLL   DETLEYLNLDLISL
pCLT5      IGTFDIETTSNEQPK   FYAHNLGKFDAVFLL   VETLKYLVLDLRSL
maranhar   ILAFDIETFQVPTGN   VYCHNFSKFDINFII   KETIIYLEKDIKAL
kalilo     IITLDLETRSVPIHP   VYTHNFSYFDGIFII   KELIKYCEIDTIAL
pAI2       ISTLDLETRMDTNNR   FYVHNLAHFDSVFIL   KELKNYCEIDCIAL
pEM        IGSLDLETYGDNCFE   IYAHNLGRFDSVFII   KECINYLDKDIIGL
S1         FFVADLETLLLKRRD   VYFHNLSQFDGIMIL   EDSLTYLKQDILIT
pGKL1      TVCFDIESYFDPEKE   VIG-NIVDFEGRDCV   EWSKDYCANDVLVL
pGKL2      EVFFDIESFSDETKH   FNVINKGEFPHSITH   ENAIEYCKVDVLAM
pSKL       EIFFDIESFTNEKNF   FNVINKGSFPNSLIK   EKAIEYCSIDVIAM
Ad2        FUTYDVETITWMGAF   V-GHNINGFDEIFLA   KETLDYCALDVKVT
Ad5        FVTYDVETYTWMGAF   V-GHNINGFDEIVLA   KETLDYCALDVQVT
Ad7        FLTYDVETYTWMGSF   V-GHNINGFDEIVLA   QETLNYCALDVLVT
Ad12       FITYDVETYTWMGAF   V-GHNINGFDEIVLA   GETLNYCALDVLVT
phi29      MYSCDFETTTKVEDC   LIFHNL-KFDGAFII   DSLKGY--KDIITT
PRD1       IAAFDFETDPFKHDR   IYAHNGGKFDFLFLM   E-LKQFMPYDPVRK

Cons mt    I-T-DLET-------   -Y-HNL--FD--FIL   KE---YL--D---L
Cons cyt   ---FDIES-------   ----N---F------   E----YC--DV---
```

```
           Pol1                                                 Pol2                  Pol3
pMC3-2     IWEAYYGGRCEVFSYPCIK----TKS---YDCNSIYPICMLK-DMPVEN   PTAKLLLNGL-YGRFGMNP   KVFMTDTDCIWMNGSLS
pCLK1      IHEAYYGGRVEVFN-PIIMADS-TKSYYYYDVNSLYPFASIN-DIPGLK   NIAKLILNSL-IGRFGMNI   TLYYTDTDSIVTDLKLP
pCLT5      IHQSYYGGRVEVFN-PIISKDS-SKPCLYYDVNSLYPFASLN-DLPGLN   NIAKLILNSL-IGRFGMSI   TLYYTDTDSIVTDIRLP
maranhar   VRAAYFGGRNEIFI-PIINNISF-------DFNSLYPTAMMM-PMPVGI   QMAKLLLNTL-YGRTGMND   NSAYTDTDSIFVEKPLD
kalilo     IKLSYFGGICELYK-PFGVNI---KS---YDVNSLYPFAMKYFKMPSGI   YTSKLLMNSL-YGRFGLNS   NIYYIDTDGLKVDIDLD
pAI2       LKQAYTGGYCDVFK-PEGKNI---HS---YDINSLYPSAMAKFDMPTGT   FIAKLLMNSL-YGRFGMDP   NLYAVDTDGLKVDTEID
pEM        IRSSYFGGNSDIFVSGEERLVKNG---YHYDMNSQYPYAMLQ-SMPTGN   YIAKLSLNSL-YGKFGQKE   LAIASNTDSLILRKPLE
pHC2       IRKKYYGGIVDVFNLNTQUIFET----YKYDVNSHYPCAMLK-PMPGGI   LLNKLLMNSL-YGRLGIKD   EYLGGDTDSIILTHPLE
S1         IREGYYGGHTDVYK-PYGENL------YYYDVNSLYPSSMLD-DMPIGK   FIYKIIMNSL-YGRFGISP   CYYYTDTDSVVVERELP
pGKL1      MREALIGGRCISVNGIYKDVL-------CLDVKSLYPASMAFYDQPYGS   NVIKIIMNSL-WGKFAQKW   ECIYSDTDSIFVVHKEH
pGKL2      IRNAIYGGRVISKNGIYEEN------IVYADVVSLYPSAMKLLKHSYGK   MVAKIALNGGGYGKFVQKP   DIIYSDTDSIFVKQKSV
pSKL       IRNSIYGGR-AAKNGTYNEE------IVYADVVSLYPSAMKLLEHSYGK   MVAKIALMGGGYGKFVQKP   EIIYSDTDSIFVRKNTI
Ad2        VRASIRGGRCYP---TYLGIL-REPLYVY-DICGMYASALTH-PMPYGE   SIAKLLSNAL-YGSFATKL   KSVYGDTDSLFVTERGH
Ad5        VRASIRGGRCYP---TYLGIL-REPLYVY-DICGMYASALTH-PMPWGP   SIAKLLSNAL-YGSFATKL   KSVYGDTDSLFVTERGH
Ad7        VRASIRGGRCYP---TYIGIL-EEPLYVY-DICGMYASALTH-PMPWGT   SIAKLLSNAL-YGSFATKL   KSVYGDTDSLFVTQRGH
Ad12       VRASIRGGRCYP---TYIGVL-KEPIYVY-DICGMYASALTH-PMPWGP   SIAKLLSNAL-YGSFATKL   KSVYGDTDSLFVTEKGR
phi 29     VRYAYRGGFTWLNDRFKEKEI--GEGMVF-DVNSLYPAQMYSRLLPWGP   QLAKLMLNSL-YGKFASNP   RIIYCDTDSIHLTGTEI
PRD1       MRPFYFGGRCQAF----EKGIIEDDIKVY-DVNSMYPMAMRNFRHPFSD   IFYKLILNSS-YGKFAQNP   RPLYCDTDSIICRDLKN

Cons mt    -R----GGR-EVF--P------------YDVNSLYP-AML--DMPYG-   -IAKLL-NSL-YGRFGM--   --YYTDTDSI-----L-
Cons cyt   -R----GGR----NG-Y-------------DV-SLYP--M------YG-   -V-KI--N----GKF-QK-   --IYSDTDSIFV-----
```

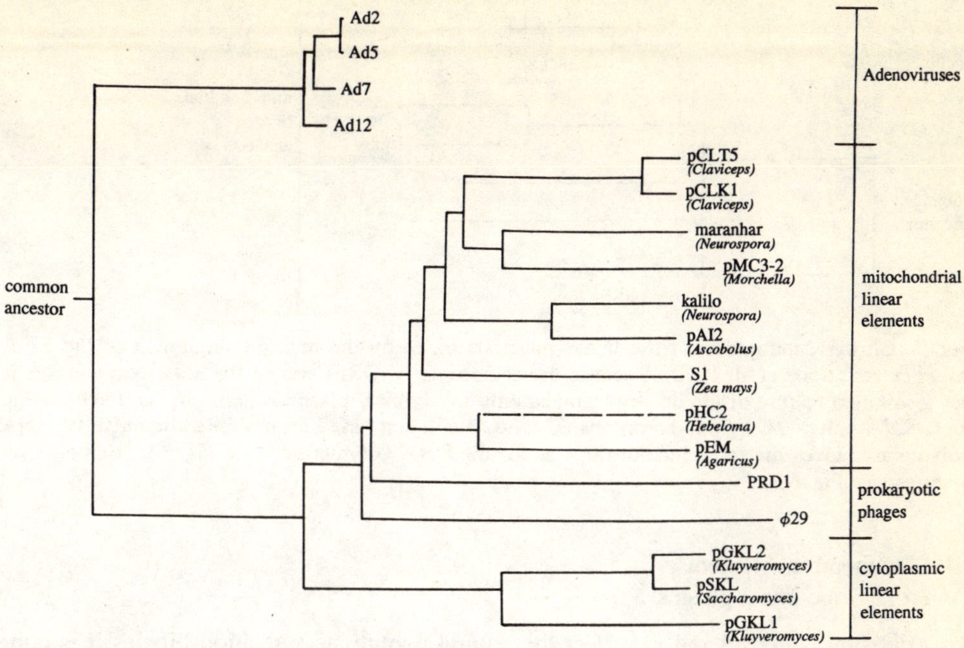


**Fig. 6.** Cluster dendogram for the linear plasmids and viruses replicating with protein primer derived from comparison of putative DNA-polymerases (Rohe et al. 1992). Source of sequences: see Fig. 5

In order to test the reliability of this phylogenetic tree, and additionally for determining the position of S2, which, as mentioned above, does not possess a DNA polymerase, the calculation was also done with the available RNA polymerases, leading to the dendogram presented in Fig. 7.

As can be seen, the relationships based on the RNA polymerase comparisons are consistant with those derived from DNA polymerase analysis.

The observations made in the analyses led to the identification of four major groups within the protein primed replicating genomes, i.e., the

I. Adenoviruses,
II. Cytoplasmic linear elements,

---

**Fig. 5.** Comparison of conserved sites of DNA polymerases of linear genetic elements with protein primed replication after Rohe et al. 1992: Source of sequences: pMC3-2 Rohe et al. 1991; *pCLK1* Oeser and Tudzynski 1989; *pCLT5* Tudzynski and Deing, in prep; *maranhar* Court et al. 1991; *kalilo* Chan et al. 1991; *pAI2* Kempken et al. 1989; *pEM* Robison et al. 1991; *pHC2* Meinhardt et al. unpubl. data; *S1* Paillard et al. 1985; *pGKL1* Stark et al. 1984; Jung et al. 1987a; *pGKL2* Tommansino et al. 1988; *pSKL* Hishinuma and Hirai 1991; *Ad2* Gingeras et al. 1982; *Ad5* Dekker and van Ormondt 1984; *Ad7* Engeler et al. 1983; *Ad12* Shu et al. 1986; *phi29* Yoshikawa and Ito 1982; *PRD1* Jung et al. 1987b. *Below* a consensus sequence for the mitochondrial linear plasmids (*Cons mt*) and the cytoplasmic linear plasmids (*Cons cyt*) is shown. Amino acids, which are selectively conserved indicate differences for the mitochondrial or cytoplasmic group of linear plasmids

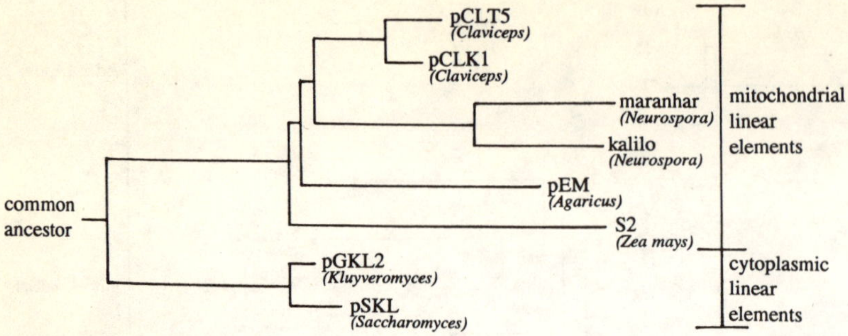


**Fig. 7.** Cluster dendogram for the linear plasmids based on the multiple alignment of the RNA polymerases (Rohe et al. 1992). Because the occurrence of ORFs coding for RNA polymerases is not a common feature of the linear plasmids, only those linear plasmids harboring an ORF coding for a RNA polymerase could be compared. The calculation based on multiple alignment of RNA polymerases gives the same relationships as for the DNA polymerases (see Fig. 6). Source of sequences: see Fig. 6

III. Mitochondrial elements,
IV. Prokaryotic linear phages.

This division is considered to reflect the natural evolutionary relationships as it is consistant with other group specific characteristics such as the host organism, codon usage, and compartmental specifity in the host organisms as far as the linear plasmids are concerned (see also Table 2).

The cytoplasmically localized yeast plasmids exhibiting the universal genetic code represent a relatively isolated group, which is much more closely related to the elements

**Table 2.** Characteristics of linear elements (After Rohe et al. 1992)

| Designation | Localization | TGA-Codon | Host | Host-taxon |
|---|---|---|---|---|
| pGKL1 | Cytoplasm | Stop[a] | *Kluyveromyces lactis* | Endo- |
| pGKL2 | Cytoplasm | Not used | *Kluyveromyces lactis* | mycetes |
| pSKL | Cytoplasm | Stop | *Saccharomyces kluyveri* | |
| pAI2 | Mitochondria | Trp | *Ascobolus immersus* | Asco- |
| pCLK1 | Mitochondria | Trp | *Claviceps purpurea* | mycetes |
| pCLT5 | Mitochondria | Trp | *Claviceps purpurea* | |
| pMC3-2 | Mitochondria | Trp | *Morchella conica* | |
| kalilo | Mitochondria | Trp | *Neurospora intermedia* | |
| maranhar | Mitochondria | Trp | *Neurospora crassa* | |
| pEM | Mitochondria | Trp | *Agaricus bitorquis* | Basidio- |
| pHC2 | Mitochondria | Trp | *Hebeloma circinans* | mycetes |
| S1 | Mitochondria | Not used | *Zea mays* | Higher |
| S2 | Mitochondria | Not used | *Zea mays* | plants |

[a] The mitochondrial genetic code of filamentous fungi differs from the universal code in using TGA as a Trp codon instead of a stop codon (Fox 1987). While some of the linear plasmids (i.e., those of the ascomycetes) also use TGA as a codon for Trp, other circumvent the problem by omitting the TGA codon. The actual use of the universal genetic code can only be assumed for those plasmids carrying ORFs ending with a TGA codon.

from filamentous fungi and prokaryotes than to adenoviruses. The phylogenetic distance of the linear plasmids from filamentous fungi and plants to prokaryotic phages is obviously much less than to the cytoplasmic yeast linear plasmids, indicating a prokaryotic origin of these plasmids.

Because the phylogenetic trees group the linear plasmids according to the phylogenetic relationships of their host organisms, one can rule out that the plasmids originate from different independent infections, but they have to be considered as very old elements in evolutionary terms. The descendence from a viral (prokaryotic) ancestor can easily explain the lack of selective pressure. For propagation within their hosts, infectivity is no longer necessary for those plasmids which are located in organelles, which are inherited cytoplasmically. In yeast, propagation may occur via fusion of haploid cells in the course of sexual reproduction. Thus, linear plasmids may represent a specialized "selfish DNA" having lost all viral functions, thus existing as a minimal intracellular parasitic system replicating efficiently and thereby ensuring survival in their host organisms.

## 8. Prospects

Today, the main scientific questions concerning the general structure, the evolutionary descendence, and the relation between the plasmids are in principle answerable. In our opinion, there are two questions left to be solved. The first is the origin of the terminal protein and the second concerns integration of linear plasmids, i.e., is this a general phenomenon or restricted to some of them? An application of mitochondrial plasmids for vector development is at least uncertain because of the disadvantages caused by the organelle location.

In contrast to this, vector development seems to be advantageous and more practicable for the yeast systems. First important steps have already been made as described before. Possibly, because of the evident advantages of the system, yeast vectors derived from cytoplasmic plasmids will become a powerful alternative tool to the commonly used circular vectors, especially when additional research provides more knowledge of the function of these plasmids. The developed system now offers the possibility of addressing basic problems concerning the biology of the linear plasmids as well as the expression of heterologous genes.

## References

**Aleström P,** Akusjärvi G, Petterson M, Petterson U (1982) J Biol Chem 257:13492–13489.

**Barbour AG,** Garron CF (1987) Science 237:409–411. **Bamford DH,** Mindich L (1982) J Virol 44:1031–1038. **Bamford DH,** Rouhiainen L, Takkinen K, Söderlund H (1981) J Gen Virol 57:365–373. **Bamford DH,** McGraw T, Mackenzie G, Mindich L (1983) J Virol 311–316. **Bedinger P,** de Hostos EL, Leon P, Walbot V (1986) Mol Gen Genet 205:206–212. **Bergström S,** Bundoc VG, Barbour AG (1989) Mol Microbiol 3:479–486. **Bernad A,** Zaballos A, Salas M, Blanco L (1987) EMBO J 6:4219–4225. **Bernad A,** Blanco L, Lazaro JM, Martin G, Salas M

(1989) Cell 59:219–228. **Bertrand H,** Griffith AJF (1988) Abstr 14th EMBO Annu Symp, pp 7–8. **Bertrand H,** Chan BSS, Griffith AJF (1985) Cell 41:877–884. **Bertrand H,** Griffith AJF, Court DA, Cheng CK (1986) Cell 47:829–837. **Blanco L,** Salas M (1984) Proc Natl Acad Sci USA 81:5325–5329. **Blanco L,** Salas M (1985) Proc Natl Acad Sci USA 82:6404–6408. **Blanco L,** Garcia JA, Salas M (1984) Gene 29:33–40. **Borst** (1986) Annu Rev Biochem 55:701–732. **Boynton JE,** Gillham NW, Harris EH, Hosler JP, Johnson AM, Jones AR, Randolph-Anderson BL, Robertson D, Klein TM, Shark K, Sanford JC (1988) Science 240:1534–1538. **Bresch,** Hausmann (1972) Klassische und molekulare Genetik. 3. Aufl. Springer, Berlin Heidelberg New York.

**Carrascosa JL,** Camacho A, Moreno F, Jimenez F, Mellado RP, Vinuela E, Salas M (1976) Eur J Biochem 66:229–241. **Challberg MD,** Kelly TJ (1982) Annu Rev Biochem 51:901–934. **Chan BSS,** Court DA, Vierula JP, Bertrand H (1991) Curr Genet 20:225–237. **Chardon-Loriaux I,** Charpentier M, Percheron F (1986) FEMS Microbiol Lett 35:151–155. **Chase CD,** Pring DR (1986) Plant Mol Biol 6:53–64. **Coetzee WF,** Bekker PJ (1979) J Gen Virol 45:195–200. **Court DA,** Griffith AJF, Kraus SR, Russel PJ, Bertrand H (1991) Curr Genet 19:129–137. **Cullen D,** Budde A, Kistler H, Samac D, Leong S (1985) J Cell Biochem 9C:169.

**Davies TN,** Muller ED, Cronan TJ Jr (1982) Virology 120:287–306. **Dekker BMM,** Van Ormondt H (1984) Gene 27:115–120. **DeLouvencourt L,** Fukuhara H, Heslot H, Wesolowski M (1983) J Bacteriol 154:737–742. **Desiderio SV,** Kelly TJ Jr (1981) J Mol Biol 145:319–337. **Dohmen G,** Gessner-Ulrich K, Tudzynski P (1991) Prog Bot 52:244–263. **Düvell A,** Hessberg-Stutzke H, Oeser B, Rogmann-Backwinkel P, Tudzynski P (1988) Mol Gen Genet 214:128–134.

**Engeler JA,** Hoppe MS, Van Bree MP (1983) Gene 21:145–159. **Erikson L,** Beversdorf WD, Pauls KP (1986) Curr Genet 9:410–415. **Escarmis C,** Salas M (1982) Nucleic Acids Res 10:5785–5798. **Escarmis C,** Gomez A, Garcia E, Ronda C, Lopez R, Salas M (1984) Virology 133:166–171. **Escarmis G,** Garcia P, Mendez E, Lopez R, Salas M, Garcia E (1985) Gene 36:341–248. **Escote LJ,** Gabay-Laughnan SJ, Gabay-Laughnan JR (1985) Plasmid 14:264–267. **Escote-Carlson LJ,** Gabay-Laughnan SJ, Gabay-Laughnan JR (1988) Theor Appl Genet 75:659–667. **Esser K,** Kück U, Lang-Hinrichs C, Lemke P, Osiewacz HD, Stahl U, Tudzynski P (1986) Plasmids of eukaryotes, fundamentals and application. Springer, Berlin Heidelberg New York.

**Feng DF,** Doolittle RF (1987) J Mol Evol 25:351–360. **Fox TD** (1987) Annu Rev Genet 21:67–91. **Francou F** (1981) Mol Gen Genet 184:440–444. **Fujimura H,** Hishinuma F, Gunge N (1987) Curr Genet 12:99–102.

**Garcia E,** Gomez A, Ronda C, Escarmis C, Lopez R (1983) Virology 128:92–104. **Gessner-Ulrich K,** Tudzynski P (1992) Curr Genet 21:249–254. **Giasson L,** Lalonde M (1987) Curr Genet 11:131–134. **Giese H,** Christiansen SK, Jensen HP (1990) Theor Appl Genet 79:56–64. **Gingeras TR,** Sciaky D, Gelinas RE, Bing-Dong J, Yen CE, Kelly MM, Bullock PA, Parsons BL, O'Neill KE, Roberts RJ (1982) J Biol Chem 257:13475–13491. **Griffith AJF,** Kraus SR, Barton R, Court DA, Myers CJ, Bertrand H (1990) Curr Genet 17:139–145. **Gunge N,** Sakaguchi K (1981) J Bacteriol 147:155–160. **Gunge N,** Tamaru A, Ozawa S, Sakaguchi K (1981) J Bacteriol 145:382–390. **Gunge N,** Murata K, Sakaguchi T (1982) J Bacteriol 151:462–464. **Gunge N,** Kitada K (1988) Eur J Epidemiol 4:409–414. **Günther E** (1984) Lehrbuch der Genetik. 4. Aufl. Gustav Fischer, Stuttgart. **Gutierrez J,** Vinos J, Prieto I, Mendez E, Hermoso JM, Salas M (1986) Virology 155:474–483.

**Hagen EW,** Reilly BE, Tosi ME, Anderson DL (1976) J Virol 19:501–517. **Harding NE,** Ito J, David GS (1978) Virology 84:279–292. **Hashiba T,** Homma Y, Hyakumachi M, Matsuda I (1984) J Gen Microbiol 130:2067–207. **Hay RT** (1985a) EMBO J 4:421–426. **Hay RT** (1985b) J Mol Biol 186:129–136. **Hayakawa T,** Tanaka T, Sakaguchi K, Otake N, Yonehar H (1979) J Gen Appl Microbiol 25:255–260. **Hayman CT,** Bolen PL (1991) Curr Genet 19:389–393. **Hermanns J,**

Osiewacz HD (1990) Molecular analysis of an extrachromosomal long-lived mutant of *Podospora anserina*: characterization of a linear plasmid. Dechema Biotechnol Conf Vol. 4 A. VCH Verlagsgesellschaft, Weinheim, pp 349–352. **Hermoso JM,** Salas M (1980) Proc Natl Acad Sci USA 77:6425–6428. **Hirochika H,** Sakaguchi K (1982) Plasmid 7:59–65. **Hirochika H,** Nakamura K, Sakaguchi K (1984) EMBO J 3:761–766. **Hishinuma F,** Hirai K (1991) Mol Gen Genet 226:97–106. **Hishinuma F,** Nakamura K, Hirai K, Nishizawa R, Gunge N, Maeda T (1984) Nucleic Acids Res 12:7581–7597. **Honeyman AL,** Currier TC (1983) The islation and characterization of two linear DNA elements from *Gaeumannomyces graminis* var. *tritici*, the causative agent of "take-all-disease" of wheat. 83th Annu Meeting Am Soc Microbiol New Orleans Abstr, pp 45. **Honeyman AL,** Currier TC (1986) Appl Environ Microbiol 52:924–929.

**Johnston SA,** Anziano PQ, Shark K, Sanford JC, Butow RA (1988) Science 240:1538–1541. **Jung G,** Leavitt MC, Ito J (1987a) Nucleic Acids Res 15:9088. **Jung G,** Leavitt MC, Hsieh JC, Ito J (1987b) Proc Natl Acad Sci USA 84:8287–8291.

**Kämper J,** Meinhardt F, Gunge N, Esser K (1989a) Mol Cell Biol 9:3931–3937. **Kämper J,** Meinhardt F, Gunge N, Esser K (1989b) Nucleic Acids Res 17:1781. **Kämper J,** Esser K, Gunge N, Meinhardt F (1991) Curr Genet 19:109–118. **Katayose Y,** Kajiwara S, Shishido K (1990) Nucleic Acids Res 18:1395–1400. **Kaudewitz F** (1983) Genetik 1. Aufl. Ulmer, Stuttgart. **Keen CL,** Mendelovitz S, Cohen G, Aharonowitz Y, Roy KL (1988) Mol Gen Genet 212:172–176. **Kemble RJ,** Thompson DJ (1982) Nucleic Acids Res 10:8181–8190. **Kemble RJ,** Gunn RE, Flavell RB (1980) Genetics 95:451–458. **Kemble RJ,** Carlson JE, Erickson LR, Sernyk JL, Thompson DJ (1986) Mol Gen Genet 205:183–185. **Kempken F,** Meinhardt F, Esser K (1989) Mol Gen Genet 215:46–52. **Kim WK,** Withmore E, Klassen GR (1990) Curr Genet 17:229–233. **Kistler HC,** Leong SA (1986) J Bacteriol 167:587–593. **Kitada K,** Hishinuma F (1987) Mol Gen Genet 206:377–381. **Kuzmin EV,** Levchenko IV (1987) Nucleic Acids Res 15:6758. **Kuzmin EV,** Levchenko IV, Zaitseva GN (1988) Nucleic Acids Res 16:4177.

**Lechner RL,** Kelly TJ Jr (1977) Cell 12:1007–1020. **Lederberg J** (1952) Physiol Rev 32:403–430. **Leon P,** Walbot V, Bedinger P (1989) Nucleic Acids Res 17:4089–4099. **Levings III CS,** Sederoff RR (1983) Proc Natl Acad Sci USA 80:4055–4059. **Levings III CS,** Sederoff RR, Hu WWL, Timothy DH (1984) In: Ciferri O, Owe L III (eds) Structure and function of plant genomes. Life Sciences 63. Plenum, New York, pp 363–371. **Lewin B** (1988) Gene. VCH Verlagsgesellschaft, Weinheim. **Ligon JM,** Bolen PL, Hills DJ, Bothast RJ, Kurtzman CP (1989) Plasmid 21:185–194. **Lopez R,** Ronda C, Garcia P, Escarmis C, Garcia E (1984) Mol Gen Genet 197:67–74.

**Manson JC,** Liddell AD, Leaver CJ, Murray K (1986) EMBO J 5:2775–2780. **McGraw T,** Yang HL, Mindich L (1983) Mol Gen Genet 190:237–244. **Meinhardt F,** Esser K (1984) Curr Genet 8:15–18. **Meinhardt F,** Esser K (1987) Appl Microbiol Biotechnol 27:276–282. **Meinhardt F,** Kempken F, Esser K (1986) Curr Genet 11:243–246. **Meinhardt F,** Kempken F, Kämper J, Esser K (1990) Curr Genet 17:89–95. **Mogen KL,** Siegel MR, Schardl CL (1991) Curr Genet 20:519–526. **Mohan M,** Meyer RJ, Anderson JB, Horgen PA (1984) Curr Genet 8:615–619. **Mohr S,** Esser K (1990) Mobile genetische Elemente bei Eucaryoten – Grundlage und Anwendung für die Biotechnologie. In: Präve P, Schlingmann M, Crueger W, Esser K, Thaner R, Wagner F (eds) Jahrbuch Biotechnologie Bd 3. Hanser, München, pp 5–25.

**Normand P,** Simone P, Giasson L, Ravel-Chapius P, Fortin J, Lalonde M (1987) Curr Genet 11:335–338. **Newton KJ** (1988) Annu Rev Plant Physiol Plant Mol Biol 39:503–532.

**Oeser B** (1988) Nucleic Acids Res 16:8729. **Oeser B,** Tudzynski P (1989) Mol Gen Genet 217:132–140. **Olsen RH,** Siak JS, Shipley PL (1974) J Virol 14:689–699. **Osiewacz HD,** Hermanns J, Marcou D, Triffi M, Esser K (1989) Mutat Res 219:9–15.

**Paillard M,** Sederoff RR, Levings III CS (1985) EMBO J 4:1125–1128. **Palmer JD,** Shields CR, Chen DB, Orton TJ (1983) Nature 301:725–728. **Petit CS,** Horwitz MS, Engler JA (1988) J Virol 62:496–500. **Plastkerk RHA,** Simon MI, Barbour AG (1985) Nature 318:257–263. **Prieto I,** Lazaro JM, Garcia JA, Hermoso JM, Salas M (1984) Proc Natl Acad Sci USA 81:1639–1643. **Pring DR,** Lewings III CS, Hu WWL, Timothy DH (1977) Proc Natl Acad Sci USA 72:2904–2908. **Pring DR,** Conde MF, Schertz KF, Levings III CS (1982) Mol Gen Genet 186:180–184.

**Robison MM,** Royer JC, Horgen PA (1991) Curr Genet 19:495–502. **Rohe M,** Schrage K, Meinhardt F (1991) Curr Genet 20:527–533. **Rohe M,** Schründer J, Tudzynski P, Meinhardt F (1992) Curr Genet 21:173–176. **Ronda C,** Lopez R, Garcia E (1981) J Virol 40:551–559. **Rubidge T** (1986) Trans Br Mycol Soc 87:463–466.

**Sakaguchi K** (1990) Microbiol Rev 54:66–74. **Salas M** (1988) Curr Top Microbiol Immunol 136:71–88. **Samac DA,** Leong SA (1986) J Cell Biochem 10c:27. **Samac DA,** Leong SA (1988) Plasmid 19:56–67. **Samac DA,** Leong SA (1989a) Mitochondrial plasmids of filamentous fungi: characteristics and use in transformation vectors. Molecular Plant-Microbe Interactions, vol 2, no 4, pp 155–159. **Samac DA,** Leong SA (1989b) Curr Genet 16:187–194. **Samitou-Laprade P,** Pannenbecker G, Magguta F, Jean R, Michaelis G (1989) Curr Genet 16:181–186. **Savilahti H,** Bamford DH (1986) Gene 49:199–205. **Savilahti H,** Bamford DH (1987) Gene 57:121–130. **Schaffrath R,** Stark MJR, Gunge N, Meinhardt F (1992) Curr Genet 21:357–363. **Schardl CL,** Pring DR, Lonsdale DM (1985) Cell 43:361–368. **Schründer J,** Debaud JC, Meinhardt F (1991) Adenoviral-like genetic elements in *Hebeloma circinans*. Abstr. 3rd Europ Symp Mycorrhizas, Sheffield. **Schründer J,** Rohe M, Debaud JC, Meinhardt F (1992) Characterization of a linear DNA-molecule from the ectomycorrhizal fungus *Hebeloma circinans* and its relationship to other protein primed replication genomes. Abstr. VAAM Meeting 1992, Düsseldorf. **Schuster W,** Hiesel R, Schobel W, Wissinger B, Brennicke A (1987) Biol Unserer Zeit 5:144–147. **Sheperd HS,** Ligon JM, Bolen PL, Kurtzman CP (1987) Curr Genet 12:297–304. **Shinagawa M,** Yoichi I, Matsuda A, Tsukiyama T, Sato G (1987) Gene 55:85–93. **Shu L,** Hong JS, Wei YF, Engeler JA (1986) Gene 46:187–195. **Sinclair JH,** Stevens BJ, Sanghavi P, Rabinovitz M (1967) Science 156:1234–1237. **Sor F,** Fukuhara H (1985) Curr Genet 9:147–156. **Sor F,** Wesolowski M, Fukuhara H (1983) Nucleic Acids Res 11:5037–5044. **Stahl U,** Lemke PA, Tudzynski P, Kück U, Esser K (1978) Mol Gen Genet 162:341–344. **Stahl U,** Tudzynski P, Kück U, Esser K (1982) Proc Natl Acad Sci USA 79:3641–3645. **Stam JC,** Kwakman J, Meijer M, Stuitje AR (1986) Nucleic Acids Res 14:6871–6884. **Stark MJR,** Boyd A (1986) EMBO J 5:1995–2002. **Stark MJR,** Mileham AJ, Romanos MA, Boyd A (1984) Nucleic Acids Res 12:6011–6030. **Stark MJR,** Mileham AJ, Romanos MA, Boyd A (1990) Yeast 6:1–29. **Sugisaki Y,** Gunge N, Sakaguchi K, Yamasaki M, Tamura G (1985) J Bacteriol 164:1373–1375.

**Takano H,** Kawano S, Kuroiwa T (1991) Curr Genet 20:315–317. **Talavera A,** Jimenez F, Salas M, Vinuela E (1971) Eur J Biochem 31:367–371. **Tanguy-Rougenau C,** Chen XJ, Wesolowski-Louvel M, Fukuhara H (1990) Gene 91:43–50. **Timothy DH,** Levings III CS, Hu WWL, Goodman MM (1982) Maize Genet Coop News Lett 56:133–138. **Tipper D,** Bostian KA (1984) Microbiol Rev 48:125–150. **Tokunaga M,** Wada N, Hishinuma F (1987) Nucleic Acids Res 15:1031–1046. **Tommasino M** (1991) Yeast 7:145–252. **Tommasino M,** Ricci S, Galeotti CL (1988) Nucleic Acids Res 16:5863–5878. **Tudzynski P,** Düvell A (1985) Molecular aspects of mitochondrial plasmids in *Claviceps purpurea*. In: Quaglianello E, Slater EC, Palmieri F, Saccone C, Kroon AN (eds) Achievements and perspectives of mitochondrial research. Vol III. Biogenesis. Elsevier, Amsterdam, pp 249–256. **Tudzynski P,** Esser K (1986) Curr Genet 10:463–467. **Tudzynski P,** Düvell A, Esser K (1983) Curr Genet 7:145–150. **Turmel M,** Bellemare G, Lee RW, Lemieux C (1986) Plant Mol Biol 6:313–319.

**Vierula PJ,** Cheng CK, Court DA, Humphrey RW, Thomas DY, Bertrand H (1990) Curr Genet 17:195–201.

**Watabe K,** Leusch M, Ito J (1984) Proc Natl Acad Sci USA 81:5374–5378. **Weissinger AK,** Timothy DH, Levings II CS, Hu WWL, Goodman MM (1982) Proc Natl Acad Sci USA 79:1–5. **Wilber KA,** Profitt JH (1987) Detection of linear plasmids in Phaffia rhodozyma. UCLA Yeast Symp, pp 106. **Wilson DW,** Meacock PA (1988) Nucleic Acids Res 16:8097–8112. **Worsham H,** Bolen PL (1990) Curr Genet 18:77–80.

**Yoshikawa H,** Ito J (1981) Proc Natl Acad Sci USA 78:2596–2600. **Yoshikawa H,** Ito J (1982) Gene 17:32–35. **Yui Y,** Katayose Y, Shishido K (1988) Biochem Biophys Acta 951:53–60.

**Zabala G,** Walbot V (1988) Mol Gen Genet 211:386–392.

Friedhelm Meinhardt
Matthias Rohe
Institut für Mikrobiologie
Westfälische Wilhems-Universität Münster
Correnssstraße 3
W-4400 Münster, Germany

# VII. Molecular Genetics of Phytopathogenic Fungi

By Paul Tudzynski and Klaus-Michael Weltring

## 1. Introduction

Plant diseases caused by phytopathogenic fungi represent an enormous economical hazard. Nevertheless, most interactions between these parasites and their hosts are not well understood mainly due to considerable experimental restrictions. In the past few years, however, considerable progress has been made in this field, mainly due to the availability of a wide range of molecular techniques, which allow new and promising approaches to the understanding of host-parasite interactions. This review cannot represent a comprehensive update of the exponentially expanding literature on this topic, but intends to present an overview and report on trends and highlights, continuing the short update of Meinhardt (1991) in this series.

We will focus on the molecular methods available for analyses of phytopathogenic fungi, and we will describe a few systems where these techniques have been applied successfully.

Due to the actuality of this field, no monograph exists dealing exclusively with this topic. There are, however, some excellent recent reviews (e.g., Kerr 1987; Leong 1988; Leong and Holden 1989; Keen 1990; Bruns et al. 1991; van den Hondel and Punt 1991) and conference proceedings (e.g., Hennecke and Verma 1992; Stahl and Tudzynski 1992) dealing with certain aspects of the subjects discussed below.

## 2. Molecular Genetic Methods: New Tools for the Study of Fungus-Plant Interactions

A variety of molecular methods have been developed or optimized for fungal systems. With respect to their potential application, they fall into two main categories: first, they can be used for identification or characterization of fungal isolates, i.e., for the detection of intra- and interspecies polymorphisms, which are important for taxonomic and diagnostic purposes, and for the establishment of genetic maps, which reveal the precise location of genes of interest and hence allow their cloning via chromosome walking. Second, they can help to perform functional analyses of genes possibly involved in host-parasite interaction, i.e., to identify and characterize fungal (a)virulence and pathogenicity genes.

Progress in Botany, Vol. 54

One of the major problems of the genetic analysis of phytopathogenic fungi has been the lack of suitable genetic markers. In addition, many of these fungi do not possess clearly defined virulence markers that would allow genetic analyses, and even in those cases where such virulence markers exist, they only assess variability of virulence and need laborious in planta tests. Therefore, markers provided by molecular techniques are rapidly becoming the tools of choice for studying fungal genetics, because they are easy to assay, precise, selectively neutral, and almost unlimited in numbers.

The best-established technique so far is the restriction fragment lenght polymorphism (RFLP) analysis. In several cases, RFLPs of mitochondrial DNA have been used to measure genetic distance between closely related species and to study intraspecies variation.

Results obtained with different systems varied considerably: in *Claviceps purpurea*, for example, variability in mtDNA restriction pattern is extremly high. All field isolates tested so far exhibited unique restriction patterns of mtDNA (Düvell et al. 1988; Tudzynski, unpubl.). On the other hand, in *Phytophthora infestans*, only very few RFLPs could be detected in a worldwide analysis involving 24 isolates from 11 countries (Carter et al. 1990); in *Phytophthora megasperma* RFLPs of mtDNA were used to differentiate the host specific isolates f.sp. *glycinea* and f.sp. *medicagines* from each other and from non host specific isolates. In *Fusarium oxysporum* f.sp. *melonis*, mtDNA restriction patterns of 78 strains were found to correspond to vegetative compatibility groupings (Jacobson and Gordon 1990).

RFLPs in nuclear DNA have been reported for a variety of pathogenic fungi. They usually involve use of heterologous probes like the ribosomal DNA of *Saccharomyces cerevisiae*, used for RFLP analysis in *Rhynchosporium secalis* (McDermott et al. 1989), or random lambda or plasmid clones of genomic DNA. The degree of variability detectable with such random clones turned out to be generally high, e.g., 97% of such genomic clones of *Septoria tritici* detected RFLPs between at least two out of six isolates (McDonald and Martinez 1990a). Nuclear RFLPs have been used to study pathogen population genetics, e.g., in *Bremia lactuca* (Hulbert and Michelmore 1988) and in *R. secalis* (McDermott et al. 1989).

In *S. tritici*, McDonald and Martinez (1990b) analyzed variation within a population from a single field. They found differences at all levels tested, among locations within a field, among plants within a location, among lesions within a leaf, and among pycnidia within a lesion. An extremely sensitive tool for the study of variation within fungal populations is the use of DNA probes recognizing dispersed repetitive sequences (e.g., in *Magnaportha grisea*: Levy et al. 1991).

In several species, RFLP markers were used to establish genetic maps, e.g., in *Erysiphe graminis* f.sp. *hordei* (Christiansen and Giese 1990), *Phanerochaete chrysosporium* (Raeder et al. 1989), *Magnaporthe grisea* (Valent and Chumley 1991), and *Cochliobolus heterostrophus* (Tzeng et al. 1992).

In the latter case, 128 RFLPs and 4 phenotypic markers were used. In combination with pulsed field gel electrophoresis (see below) 125 markers could be assigned to specific chromosomes confirming the genetic linkage data of 120 markers. The overall genetic map length was estimated by this data to be about 1500 CM.

A new and promising approach for the detection of molecular variability is the use of the polymerase chain reaction (PCR) technique, i.e., the amplification of genomic DNA fragments defined by specific primers. This highly sensitive technique was first used to define variability of ribosomal intergenic sequences in *Rhizoctonia* (Cubeta et al. 1991) and *Verticillium* (Nazar et al. 1991). Recently, a variation of this technique turned out to be very efficient in variability analyses: the random amplified polymorphic DNA (RAPD) assay (review: Williams et al. 1991a, b). This technique is based on the use of random synthetic 10-mer oligo-nucleotides as primers for DNA amplification reactions. Only single primers are used, therefore the occurrence of distinct PCR products depends on the presence of two priming sites close to each other in an inverted orientation; several primers have to be checked for (1) their ability to generate distinct amplified DNA fragments, and (2) for their use to detect variability between different fungal isolates.

In *Leptosphaeria maculans*, RAPD analysis allowed differentiation of aggressive and nonaggressive isolates (Goodwin and Annis 1991; Schäfer and Wöstemeyer 1992). In *F. solani* f.sp. *cucurbitae*, four isolates, which could not be assigned to a specific mating population by traditional means, could be distinguished by their RAPD pattern (Crowhurst et al. 1991). In several other fungi, RAPD assays were successfully used to differentiate various isolates for example in *Ascochyta rabiei* (D. Klein-Bölting and W. Barz, pers. comm.), *Gibberella fujikuroi* (B. Brückner, pers. comm.), *C. purpurea* (U. Jungehülsing and P. Tudzynski, unpubl.).

Apart from the high sensitivity, this method has many advantages in comparison to RFLP analyses: there is no need for establishment of genomic libraries in order to obtain specific probes, for use of restriction enzymes, blotting, radioactive labeling; and it is considerably faster. In addition, RAPD products may be used as probes in RFLP analyses; in *F. solani* f.sp. *cucurbitae* it turned out that five out of seven polymorphic RAPD products were specific for one of two mating populations when used as probes in Southern blots (Crowhurst et al. 1991). RAPD products may also be sequenced directly or be used as probes in chromosome blots (see below), opening the possibility of assigning the observed variability to specific chromosomes, i.e., using them as genetic markers.

An important new tool for the study of genetic variability in phytopathogenic fungi is the recently developed pulsed field gel electrophoresis (PFGE), which allows separation of fungal chromosomes up to 10 megabases (Mb) and establishment of so-called electrophoretic karyotypes (reviews: Eby 1990; Mills and McCluskey 1990; Skinner et al. 1991). Application of this technique to a variety of phytopathogenic fungi led to astonishing results (see Table 1). Different isolates of fungal species and even populations showed considerable differences in chromosome length (chromosome length polymorphism, CLP) and number, revealing an unexpectedly high degree of variability at this level. Karyotype analyses were used, e.g., to distinguish highly virulent from weakly virulent isolates of *L. maculans* (Taylor et al. 1991) and different pathogen types of *Colletotrichum gloeosporioides* occurring in Australia (Masel et al. 1990). PFGE allows detection of translocations and duplications, which occur at an unexpectedly high rate in these fungi, e.g., in *S. tritici* (McDonald and Martinez 1991), in *C. heterostrophus*, and *Nectria haematococca* (see below). Especially interesting is the detection of very small chromosomes, which may vary in number between different isolates. In some cases (e.g., *N. haematococca*, Miao et al. 1991a, b), their inheritance in sexual crosses does not follow the Mendelian rules, and they are obviously dispensable. Therefore, they are referred to as B chromosomes (Jones and Rees 1982). They may carry important functions in host-parasite interaction, e.g., in *N. haematococca* (see below).

**Table 1.** Electrophoretic karyotypes of phytopathogenic fungi. Updated from Cooley (1992) and Mills and McCluskey (1990).

| Species | No. of chromosomes A | No. of chromosomes B[a] | Size range chromosomes (Mb) A | Size range chromosomes (Mb) B[a] | Reference |
|---|---|---|---|---|---|
| *Cladosoprium fulvum* | 9–11 | | 1.9–5.4 | | Talbot et al. 1991 |
| *Colletotrichum gloeosporioides* | 3–5 | 3–10 | 2.0–6.0 | 0.27–1.2 | Masel et al. (1990) |
| *Chochliobolus heterostrophus* | 15 | 1 | 1.3–3.7 | 1.3 | Tzeng et al. (1992) |
| *Fusarium oxysporum* | | | | | |
| f.sp. *conglutinans* | 8 | | 2.2–6.2 | | Mills and |
| f.sp *raphani* | 11 | | 0.63–6.4 | | McCluskey (1990) |
| *Leptosphaeria maculans* | 6–14 | | 0.39–2.150 | | Taylor et al. (1991) |
| *Magnaporthe grisea* | 5–6 | 1–4 | 3.0–12.0 | 0.5–2.0 | Mills and McCluskey (1990) |
| *Nectria haematococca* | 10–20 | | 0.4–4.0 | | Miao et al. (1991a) |
| *Phoma tracheiphila* | 12 | | 0.7–1.6 | | Rollo et al. (1989) |
| *Phytophthora megasperma* | 9–14 | | 1.4–4.0 | | Howlett (1989) |
| *Schizophyllum commune* | 6–11 | | 1.2–5.1 | | Horton and Raper (1991) |
| *Septoria nodorum* | 14–19 | | 0.5–3.5 | | Cooley (1992) |
| *S. tritici* | 17–18 | | 0.33–3.5 | | McDonald and Martinez (1991) |
| *Tilletia caries* | 12–15 | | 0.88–4.0 | | McCluskey et al. (1990) |
| *T. controversa* | 12–16 | | 0.88–4.0 | | McCluskey et al. (1990) |
| *Ustilago hordei* | 16–21 | | 0.17–3.15 | | McCluskey and Mills (1990) |
| *U. maydis* | 20 | | 0.3–2.0 | | Kinscherf and Leong (1988) |
| *U. tritici* | 13 | | 0.25–3.0 | | Cooley (1992) |

[a] Several authors do not differentiate between A and B chromosomes; especially the character "dispensability" is often questionable.

Apart from revealing variability within fungal populations, PFGE has immense practical importance for gene mapping and cloning in fungi. Cloned genes, RFLP probes, and RAPD products can be assigned to linkage groups by hybridization following Southern blotting of PFGE gels. This combination of molecular methods is a highly efficient and rapid way for the establishment of genetic maps, as shown for *C. heterostrophus* (Tzeng et al. 1992). PFGE can also efficiently be used to produce chromosome-specific libraries or to help order genomic libraries by using labelled DNA of a single chromosome as probe. Such ordered libraries will be very helpful in chromosome walking and complementation experiments.

For a functional analysis and characterization of fungal pathogenicity-related genes, molecular genetic methods provide the means to induce specific mutations in such genes leading to specific zero-mutants without affecting any other character, which is always the case when using conventional mutagenesis. In addition, they provide the tools to study gene expression in axenic and parasitic culture. One prerequisite for such studies is the development of efficient transformation systems. Such systems are now available for a wide range of phytopathogenic fungi; since there are several excellent reviews on this topic (e.g., van den Hondel and Punt 1991), no list will be presented here.

In general, in these systems dominant selective markers predominate, due to the lack of specific (complemetable) mutant strains as recipients. In most cases, bacterial antibiotic resistance genes are used, controlled by fungal promotors. Since promotors are obviously more or less exchangeable between fungi at least within the ascomycetes and basidiomycetes, often heterologous promotors are used.

An example for a widely used vector is the plasmid pAN7-1, carrying the bacterial hygromycin resistance gene under the control of the strong constitutive *Gpd* promoter of *Aspergillus nidulans* (van den Hondel and Punt 1991).

In some cases, also homologous promotors were used, which usually allow higher transformation rates, e.g., in *G. pulicaris* (Weltring, unpubl.) and *Chryphonectria parasitica* (Choi and Nuss 1992). For some fungi also transformation systems were developed based on complementation of positively screened mutants. Pyrimidine-auxotrophic mutants, for example, can easily be screened via their resistance against the toxic analog 5-fluor-orotate. For *C. purpurea* a transformation system was established based on such pyr-mutants and the homologous orotidine monophosphate decarboxylase gene as a selectable marker (Smit and Tudzynski 1992). In only very few cases, where defined auxtrophic mutants were available, which cannot be positively screened, they were used as recipients in transformation experiments, e.g., an *arg B* mutant of *G. fujikuroi* was complemented by the heterologous *A. nidulans* gene (Brückner et al. 1992).

Compared to the ascomycetes and, more recently, to the basidiomycetes, where genetic transformation can now be considered a standard procedure, the oomycetes have proven recalcitrant to transformation. A breakthrough was reached by Judelson et al. (1991), who report transformation of *P. infestans* using the bacterial hygromycin and G418 resistance genes and expression signals from the oomycete *Bremia lactuca*.

Transforming DNA in filamentous fungi normally integrates into the chromosomal DNA and is stabely maintained, even without selective pressure. In most cases it was shown that the transformed phenotype is relatively stable also in parasitic culture of the fungus (e.g., Keller et al. 1991; Smit and Tudzynski 1992), which is an important prerequisite for application of these methods for the study of host-pathogen interactions. Integration may occur at homologous sites within the chromosome, opening the possibility of introducing in vitro-deleted copies of a gene of interest, thereby disrupting the target gene. This approach has been successfully used for the functional analysis of pathogenicity genes (see below); such analyses are hampered by the possibility of random (ectopic) integration of vector DNA in those fungi. Therefore, transformants must be carefully analyzed by Southern hybridization. Since multi-copy integration of transforming DNA is also possible in most cases, the study of overexpression of pathogenicity genes is possible (see, e.g., the detailed study on the correlation of expression level of the cutinase gene in *N. haematococca* and aggressiveness, see below).

Taken together, the wide range of molecular techniques now available for filamentous fungi allows new and fascinating approaches for the analysis of the interaction between phytopathogenic fungi and their hosts. Some of the most interesting data and promising developments resulting from application of these methods will be described in the following section.

## 3. Applications and Achievements

### a) Identification and Characterization of an Avirulence Gene

In *Cladosporium fulvum*, for the first time a classical avirulence gene (*avr9*) was cloned and characterized (review: van Kan et al. 1992). According to the gene-for-gene hypothesis the *avr9* gene is responsible for the avirulent phenotype of *C. fulvum* races on resistant tomato genotypes. The product of the *avr9* gene is a race-specific elicitor, a small polypeptide (28 aminoacids), which induces necrosis (hypersensitive response, HR) on tomato cultivars with the *Cf9* resistance gene. DNA probes derived from the amino acid sequence of the elicitor were used to clone the *avr9* gene. By transformation experiments it was proven that an avirulence gene had indeed been cloned. The prospective *avr9* gene was introduced into a fungal race virulent on *Cf9* tomato cultivars. The resulting transformants became avirulent on these *Cf9* plants, i.e., they showed the *avr9* phenotype.

Interestingly, Southern hybridization experiments using the cloned *avr9* gene as probe showed that two races of *C. fulvum*, which are virulent on *Cf9* plants, did not contain *avr9* homologous sequences. Virulence in this case seems to be correlated with the loss of the avirulence gene.

### b) Cutinase and Penetration

The importance of cutinase as a pathogenicity factor for phytopathogenic fungi has been investigated for almost 20 years (Purdy and Kolattukudy 1973). The best-studied cutinase of a phytopathogenic fungus is the enzyme produced by *Nectria haematococca* MPVI (anamorph *Fusarium solani* f.sp. *pisi*) (Kolattukudy 1985; Kolattukudy et al. 1989a; Köller 1991). Kolattukudy and co-workers have purified the protein, produced antibodies against it, and have isolated the gene from a cDNA library. Using the antibodies, they were able to show that cutinase is produced during penetration, and that these antibodies and other inhibitors of cutinase prevent penetration, when applied together with the spores to the plant surface (Kolattukudy et al. 1989a; Köller 1991). A promotor analysis and nuclear run-off experiments have shown that cutinase is inducible by cutin monomers (Podila et al. 1988; Soliday et al. 1989; Bajar et al. 1991). A field isolate, which produced low levels of cutinase not enough to penetrate intact pea stems, was transformed with the isolated gene. Transformants showing a high level of cutinase activity also showed a high degree of infection of intact pea stems, while transformants with low activities were not virulent (Kolattukudy et al. 1989b). All these data strongly suggested that cutinase of *N. haematococca* is necessary for the fungus to penetrate the intact surface of its host pea.

These data were recently contradicted by results of a gene disruption experiment by Stahl and Schäfer (1992). They isolated and in vitro mutagenized the same cutinase gene from a different isolate of *N. haematococca* containing only one copy of the cutinase gene, and retransformed the mutated gene back into the isolate. By Southern analysis, transformants were identified, which had integrated the mutated copy of the gene into the resident wild-type copy. The resulting Cut⁻ transformants did not produce any cutinase, but were as virulent as the untransformed wild-type in different bioassays. These results prove that at least this isolate of *N. haematococca* does not need cutinase to penetrate the intact surface of its host. The question remains if the same gene disruption experiment in the isolate used by the group of Kolattukudy, which contains two copies of the gene, would lead to the same result.

*Magnaporthe grisea* is another example where the importance of cutinase for penetration was analyzed by gene disruption. Sweigard et al. (1992a, b) have cloned the cutinase gene *cut1* of this fungus and have produced disruption mutants in two different strains in the same way as in *N. haematococca*. These mutants make no *cut1* mRNA and no *cut1* protein, but they can penetrate. This means that this cutinase is not required for penetration. The situation is complicated by the fact that they can measure another weak cutinase activity in the standard enzyme assay. The fact that the isolated cutinase gene is present in only one copy means that *M. grisea* produces an esterase with cutinase activity, which must have a very different nucleotide sequence compared to the cutinase gene. It cannot be ruled out that this enzyme plays a role in penetration.

Certainly more important than cutinase for penetration of *M. grisea* is the melanization of the appressorium. Howard and Ferrari (1989) have shown direct correlation between melanization and ability to penetrate. Chumley and Valent (1990) have described melanin biosynthesis mutants like *buf1, rsy1,* and *alb1*, which prevent penetration. Subsequent genetic analysis of these mutants demonstrated the critical role in penetration of the *alb1*, *rsy1*, and *buf1* genes.

Comparable results have been obtained with *Colletotrichum lagenarium*. Bonnen and Hammerschmidt (1989) have shown by using cutinase inhibitors and mutants that only a poor correlation exists between cutinase production and penetration and disease expression, but a good correlation between penetration and disease expression. This is supported by Kubo et al. (1991), who cloned a melanin biosynthesis gene, which complemented a melanin-deficient albino mutant. The melanin-restored transformants were able to penetrate and to cause disease. These two cases indicate that cutinase may not be essential in penetration for fungi, which produce an appressorium before penetration.

The best evidence to date for the importance of cutinase in a plant-pathogen interaction comes from the wound pathogen *Mycosphaerella* spp. This pathogen of papaya was transformed by Dickman et al. with the cutinase gene of *N. haematococca* (Dickman et al. 1989). The resulting transformants were able to penetrate the cuticula of intact papaya fruits, proving the importance of cutinase for penetration in this case. Although this experiment gave a very clear cut result, its value for the assessment of the possible role of cutinases in a host-parasite interaction is of limited value, because *Mycosphaerella* is a wound pathogen and therefore does not need a cutinase to infect its natural host. In addition, such experiments are certainly questionable under safety aspects.

### c) Toxins

Toxins produced by phytopathogenic fungi are another pathogenicity factor, which has been studied for a long time. Most of the research has focused on structure elucidation, mode of action, and biochemical studies of the biosynthesis of toxins. This area has been reviewed for example by Walton (1990), Ballio (1991), Bronson (1991), Kohmoto and Otani (1991), and Hohn et al. (1992).

Recently, a gene involved in the biosynthesis of a trichothecene toxin was isolated and characterized (Hohn and Beremand 1989). Trichothecenes are potent phytotoxic sesquiterpenoids produced by certain species of *Fusarium* and related genera. The first step in sesquiterpenoid biosynthesis is catalyzed by a cyclase, leading in the case of the trichothecenes to the first product trichodiene. The gene (*Tox5*) for the so-called trichodiene synthase (TS) of *Fusarium sporotrichioides* has been cloned and sequenced (Hohn and Beremand 1989). It was isolated from a lambda-gt11 library using antiserum against TS. Southern analysis indicated that the *Tox5* gene is present as a single copy.

Using the *Tox5* gene of *F. sporotrichioides* as a probe, the same gene was isolated from *Gibberella pulicaris* (anamorph *Fusarium sambucinum*), the causal agent of potato dry rot (Hohn and Desjardins 1992). A survey of over 70 field isolates of *G. pulicaris* revealed a strong correlation between high levels of trichothecene production and virulence (Beremand et al. 1991). In order to prove or disprove the involvement of trichothecenes in pathogenicity of *G. pulicaris*, an attempt was made to disrupt the *Tox5* gene, which is present in one copy in the genome of this fungus, too (Hohn and Desjardins 1992). For this purpose, a plasmid containing a doubly truncated *Tox5* gene was constructed and used to transform *G. pulicaris*. Five Tox$^-$ transformants were identified out of a total of 82 transformants analyzed (Hohn and Desjardins 1992). These transformants were examined for their ability to rot potato slices and parsnip slices. Parsnip (*Pastinaca sativa*) is another host of *G. pulicaris* and it was previously shown that trichothecene production is important for virulence of *F. sporotrichioides* on this host (Desjardins et al. 1989). The Tox$^-$ transformants of *G. pulicaris* showed greatly reduced virulence on parsnip slices, but no reduction in virulence on potato tubers (Desjardins et al. 1992b). This result proves that trichothecene production is important for virulence of *G. pulicaris* on parsnip, but not on potato, where a correlation between virulence and toxin production was previously suggested by analysis of field isolates. This result also confirms that only gene disruption experiments can provide conclusive evidence for the importance of virulence factors in a given host-pathogen interaction, and that caution should be used in generalizing a result from one interaction to another one.

*Cochliobolus carbonum* race 1 is a pathogen of corn and produces the cyclic tetrapeptide HC-toxin, which is the primary disease determinant. A 22 kb DNA fragment was isolated containing a 15.7 kb open reading frame (ORF) encoding a four-partite cyclic peptide synthetase of predicted MW 570 000. This MW corresponds to the MW of HTS, the enzyme involved in the biosynthesis of the toxin. The 22 kb region is duplicated in all Tox$^+$ strains, but absent in Tox$^-$ strains. When one or the other copy of the *Tox* gene was disrupted, enzyme activities dropped but pathogenicity and toxin production remained. When both copies were disrupted, toxin was no longer detectable and the strain was no longer pathogenic (Panaccione et al. 1991; B. Walten, pers. comm.). This proves that toxin production is essential for pathogenicity of *C. carbonum*.

An even more complicated situation is found in *Cochliobolus heterostrophus*, another pathogen of corn. This fungus also produces a toxin called T-toxin, a family of linear polyketols, which specifically inhibit mitochondrial function in cms-T maize (Holden and Sze 1987). A RFLP map and electrophoretic karyotype analysis of the genome of this fungus has shown that the *Tox1* locus controlling the ability or inability to produce the toxin is tightly linked to the breakpoint of a reciprocal translocation (Tzeng et al. 1992). This means that production of toxin is not a matter of mutation or absence or presence of a gene, but a matter of distinct chromosomal states, which separate or combine DNA regions, thereby controlling the biosynthesis of the toxin. It will certainly take some time and effort to isolate and analyze this translocation breakpoint in order to understand the molecular basis of T-toxin production.

d) Phytoalexin Metabolism

Phytoalexins are toxic plant defence compounds synthesized by plants in response to microbial infection. Since 1964, when Uehara (1964) suggested that pathogenicity of a fungus may depend on its ability to detoxify its host phytoalexin, this theory has been discussed controversially. A lot of data have been accumulated in favor of this hypothesis, but almost all of them have been circumstantial (VanEtten et al. 1982a, 1989; Weltring 1992). The strongest arguments for a linkage of phytoalexin metabolism and virulence have been obtained by genetic analyses of this correlation in *Nectria haematococca* (anamorph *Fusarium solani* f.sp. *pisi*) and *Gibberella pulicaris* (anamorph *Fusarium sambucinum*).

The latter fungus is the causal agent of potato dry rot, and virulent isolates are able to metabolize rishitin and lubimin, two phytoalexins produced by potato tubers in response to microbial infection (Desjardins et al. 1992a). Genetic analysis of the correlation between rishitin metabolism and virulence has shown that the metabolism phenotype is controlled by more that one gene (Desjardins and Gardner 1989). One of these genes, designated *Rim1*, is always connected to virulence (Desjardins and Gardner 1991). Attempts to isolate this and other metabolism genes have been initiated (Desjardins et al. 1992b).

*N. haematococca*, a pathogen of garden pea (*Pisum sativum*), detoxifies the phytoalexin pisatin of its host plant by demethylation. An extensive genetic analysis of the importance of pisatin demethylating ability (pda) for virulence has resulted in the formal identification of six different *Pda* loci each conferring a distinct *Pda* phenotype characterized by the lag time for induction of the enzyme, a cytochrome P-450 monooxygenase, and by the maximal level of enzyme activity obtained (VanEtten et al. 1989b). *Pda* genes representing each phenotype have been cloned and sequenced.

The first gene of isolate T-9 was isolated from a genomic library by expression in *Aspergillus nidulans* (Weltring et al. 1988). The sequence of this gene revealed five exons interrupted by four small introns (VanEtten et al. 1989b). At the 3' end it contained a region with homology to known cytochrome P-450 genes and a heme-binding site proving that the structural *Pda* gene was actually cloned.

Southern analysis of $Pda^+$ and $Pda^-$ field isolates using PdaT9 as a probe revealed that all but one $Pda^-$ isolate did not hybridize to the Pda probe (Miao et al. 1991a). This suggested that none of the 6 *Pda* genes is present in $Pda^-$ isolates. A possible explanation for this phenomenon was obtained from crosses, where *Pda* genes failed to be transmitted during meiosis in some of the tetrads, resulting in an excess of Pda⁻ progeny (Miao and VanEtten 1992). When the DNA of such tetrads was probed with PdaT9, no hybridizing band was found in any of the DNAs, suggesting that the *Pda* genes can be lost completely during meiosis.

The loss of the *Pda* gene could be correlated with changes involving a small 1.6 Mb chromosome by karyotype analysis of several tetrads of a cross between two strains each carrying the same *Pda* gene (Miao et al. 1991a). $Pda^-$ progeny either had lost the entire chromosome or carried a deleted version of 0.9 or 1 Mb, which no longer hybridized to the *Pda* gene. The complete loss of the chromosome or the deletion of parts of it was verified by using the 1.6 Mb chromosome as a probe. It either hybridized not at all or only to the deletion derivatives, but not to other chromosomes of the $Pda^-$ strains (Miao et al. 1991b).

A cross between a $Pda^+$ strain with a 1.5 Mb derivative of the 1.6 Mb chromosome and a $Pda^-$ strain carrying a 1 Mb derivative resulted in some progeny with a reconstructed 1.6 Mb chromosome; a presumptive reciprocal product of 0.9 Mb was also generated (Miao et al. 1991b). This suggests that this chromosome may contain sites that are favored for breakage and reunion.

The 1.6 Mb chromosome, which can be regarded as a B chromosome because of its dispensability, plasticity, and meiotic instability, may contain an additional phytoalexin metabolism gene, the so-called *Mak* gene. This gene codes for the detoxification of the phytoalexin maackiain, which is produced by chickpea, another host plant of *N. haematococca*. The *Mak* gene is lost concurrently with the *Pda* gene in the crosses described above, suggesting that it is carried on the same chromosome (Miao and VanEtten 1992). Circumstantial evidence suggests that most of the *Pda* genes, and maybe also the *Mak* genes, are located on chromosomes less than 2 Mb in size (Miao et al. 1991a). Chromosomes in this size range have been found in other phytopathogenic fungi, too (see Table 1).

### e) DsRNA Involved in Hypovirulence

Virulence of phytopathogenic fungi is not only determined by different alleles of virulence- or pathogenicity genes. DsRNA genetic elements also seem to influence the virulence of at least some phytopathogenic fungi. They are found in encapsidated "virus-like" forms or unencapsidated (Nuss and Koltin 1990), and they are reported for all major classes of fungi (Buck 1986). The genomes of dsRNA elements are mostly segmented and sequence data have shown that most of the elements are informational and contain open reading frames (Rae et al. 1989; Choi et al. 1991; Shapira et al. 1991).

The presence of dsRNAs in phytopathogenic fungi has been reported, for example, for *Phytophthora infestans* (Tooley et al. 1989), *Ustilago maydis* (Wood and Bozarth 1973), *Septoria nodorum* (Newton 1987), *Pyricularia oryzae* (Hunst et al. 1986), *Mycospherella pinodes* (Yamada et al. 1991), *Rhizoctonia solani* (Castanho et al. 1978; Zanzinger et al. 1984), *Ophiostoma ulmi* (Brasier 1983), *Helminthosporium victoriae* (Ghabrial 1986), and *Cryphonectria parasitica* (Day et al. 1977). Investigations of the last four fungi have created evidence for the involvement of dsRNAs in hypovirulence or "diseased states" of some naturally occurring isolates of these fungi (Nuss and Koltin 1990). Controversial results have been obtained with *R. solani* (Koltin et al. 1987; Nuss and Koltin 1990). Early studies of Castanho et al. (1978) indicated a negative effect of dsRNA on virulence of *R. solani*, while studies by Finkler et al. (1985) established a clear correlation between the presence of dsRNA viruses and virulence. It appears that resolution of this problem will require the development of a transformation system for *R. solani* and the cloning and expression of the dsRNA genes in isogenic strains under controlled environmental conditions.

Currently, *C. parasitica* is certainly the most intensely studied example of the abovementioned fungi. The attention stems from the consistent correlation between the presence of dsRNAs and reduced virulence (hypovirulence), transmissibility of the hypovirulent phenotype, and the successful use of this transmissibility to biocontrol chestnut blight (Van Alfen et al. 1975; Fulbright et al. 1988). The dsRNAs are transmitted by anastomosis from hypovirulent strains to virulent strains, converting the latter into hypovirulent strains (Anagnostakis and Day 1979; Fulbright 1984). The time course of movement of dsRNAs between colonies has been determined by in situ hybridization techniques (Martin and VanAlfen 1991). According to these results the presence of dsRNAs can be detected in virulent colonies 1 to 2 days after the colonies touched.

Hypovirulent strains of *C. parasitica* usually produce superficial cancers that are contained and eventually heal as a result of host defence mechanisms, whereas virulent strains kill bark and cambium layers, causing wilting and death (Nuss and Koltin 1990). The hypovirulent phenotype can be associated with symptoms like suppressed conidation, reduced oxalate accumulation, reduced laccase production, and reduced pigment production (Nuss and Koltin 1990). The reduction in these and other traits is due to a down regulation of mRNA and polypeptide accumulation (Powell and Van Alfen 1987a, b; Kazmierczak et al. 1991). Examples are the reduction of laccase mRNA in a hypovirulent strain by 85% (Rigling and Van Alfen 1991) and of the mRNA coding for cryparin, a cell-surface, tissue-specific lectin-like protein (Carpenter et al. 1992).

The availability of transformation systems for *C. parasitica* (Churchill et al. 1990) and of cDNA clones of open reading frames (ORF) of hypovirulence-associated dsRNAs (Nuss and Koltin 1990; Choi et al. 1991; Shapira et al. 1991) made it possible to transform a cDNA of ORF A present in the largest dsRNA of the hypovirulent strain EP713 into the virulent isogenic strain EP155 (Choi and Nuss 1992). Transformants showed traits similar to those exhibited by hypovirulent strains like reduced conidiation, reduced pigmentation, and reduced laccase accumulation, but they were still as virulent as the untransformed strain, indicating an apparent uncoupling of hypovirulence from associated traits. These results for the first time established a direct cause-and-effect relationship between a viral dsRNA genetic element present in a hypovirulent strain and specific phenotypic traits. In addition, they prove that these traits are caused by a specific viral coding domain. Experiments like these demonstrate that the tools are available now for elucidating the molecular basis of transmissible hypovirulence.

## 4. Conclusion

The achievements in applying molecular genetic techniques to phytopathogenic fungi described in this chapter clearly demonstrate the rapid progress which has been made in understanding the molecular basis of pathogenicity of phytopathogenic fungi. The genome analyses by RFLP mapping and PFGE, for example, have provided for the first time the means to analyse and explain the variability and unusual genetic behavior of these fungi. The gene disruption experiments have proven to be the method of choice for evaluation of the importance of single traits in different host-parasite interactions, and transformation and expression of genes are the only way to identify and characterize genetically defined genes involved in pathogenicity, especially those with no known function, like avirulence genes. The application of the available techniques and the described experimental approaches to other phytopathogenic fungi will rapidly increase our knowledge about the molecular basis of fungus-plant interactions.

## References

**Anagnostakis SL,** Day PR (1979) Phytopathology 69:1226–1229.

**Bajar A,** Podila GK, Kolattukudy PE (1991) Proc Natl Acad Sci 88:8208–8212. **Ballio A** (1991) Experientia 47:783–790. **Beremand MN,** Desjardins AE, Hohn TM, VanMiddlesworth E (1991) Phytopathology 81:1452–1458. **Bonnen AM,** Hammerschmidt R (1989) Physiol Mol Plant Pathol 35:475–481. **Brasier CM** (1983) Nature 305:220–222. **Bronson CR** (1991) Experientia 47:771–776. **Brückner B,** Uncles S, Kinghorn J, Weltring KM (1992) Curr Genet 22:313–316. **Bruns TD,** White TJ, Taylor JW (1991) Annu Rev Ecol Syst 22:525–564. **Buck KW** (1986) Fungal virology – an overview. In: Buck KW (ed) Fungal virology. CRC, Boca Raton, pp 1–84.

**Carpenter CE,** Mueller RJ, Kazmierczak P, Zhang L, Villalon DK, Van Alfen NK (1992) Mol Plant-Microbe Interact 4:55–61. **Carter DA,** Archer SA, Buck KW, Shaw DS, Shattock RC (1990) Mycol Res 94:1123–1128. **Castanho B,** Butler EE, Shepherd RJ (1978) Phytopathology 68:1515–1518. **Choi GH,** Nuss DL (1992) EMBO J 11:473–477. **Choi GH,** Shapira R, Nuss DL (1991) Proc Natl Acad Sci 88:1167–1171. **Christiansen SK,** Giese H (1990) Theor Appl Genet 79:705–712. **Chumley FG,** Valent B (1990) Mol Plant-Microbe Interact 3:135–143. **Churchill ACL,** Ciuffetti LM, Hansen DR, VanEtten HD, Van Alfen NK (1990) Curr Genet 17:25–31. **Cooley RN** (1992) The use of RFLP analysis, electrophoretic karyotyping and PCR in studies of plant pathogenic fungi. In: Stahl U, Tudzynski P (eds) Molecular biology of filamentous fungi. Verlag Chemie, Weinheim, pp 13–26. **Crowhurst RN,** Hawthorne BT, Rikkerink EHA, Templeton MD (1991) Curr Genet 20:391–396. **Cubeta MA,** Echandi E, Abernethy T, Vilgalys R (1991) Phytopathology 81:1385–1400.

**Day PR,** Dodds JA, Elliston JE, Jaynes RA (1977) Phytopathology 67:1393–1396. **Desjardins AE,** Gardner HW (1989) Mol Plant-Microbe Interact 2:26–34. **Desjardins AE,** Gardner HW (1991) Phytopathology 81:429–435. **Desjardins AE,** Spencer GF, Plattner RD, Beremand MN (1989) Phythopathology 79:170–175. **Desjardins AE,** Gardner HW, Weltring KM (1992a) J Ind Microbiol 9:201–211. **Desjardins AE,** Hohn TM, McCormick SP (1992b) Mol Plant-Micorbe Interact 5:(in press). **Dickman MB,** Podila GK, Kolattukudy PE (1989) Nature 342:446–448. **Düvell A,** Hessberg-Stutzke H, Oeser B, Rogmann-Backwinkel P, Tudzynski P (1988) Mol Gen Genet 214:128–134.

**Eby MJ** (1990) Bio/Technology 8:243–245.

**Finkler A,** Koltin Y, Barash I, Sneh B, Poszniak D (1985) J Gen Virol 66:1221–1232. **Förster H,** Kinscherf TG, Leong SA, Maxwell DP (1989) Can J Bot 67:529–537. **Fulbright DW** (1984) Phytopathology 74:722–724. **Fulbright DW,** Paul CP, Garrod SW (1988) Hyporvirulence: a natural control of chestnut blight. In: Mukerji KG, Garg KL (eds) Biocontrol of plant diseases. CRC, Boca Raton, II.121–138.

**Ghabrial SA** (1986) A transmissible disease of *Helminthosporium victoriae* – evidence for a viral etiology. In: Buck KW (ed) Fungal virology. CRC, Boca Raton, pp 163–176. **Goodwin PH,** Annis SL (1991) Appl Environ Microbiol 57:2482–2486.

**Hennecke H,** Verma DPS (eds) (1992) Advances in molecular genetics of plant-microbe interactions. Kluwer, Dordrecht. **Hohn TM,** Beremand PD (1989) Gene 79:131–138. **Hohn TM,** Desjardins AE (1992) Mol Plant-Microbe Interact 5:249–256. **Hohn TM,** Proctor RH, Desjardins AE (1992) Biosynthesis of sesquiterpenoid toxins by fungal pathogens. In: Stahl U, Tudzynski P (eds) Molecular biology of filamentous fungi. Verlag Chemie, Weinheim, pp 99–110. **Holden MJ,** Sze H (1987) Plant Physiol 84:670–676. **Hondel van den CAMJJ,** Punt PJ (1991) Gene-transfer systems and vector development for filamentous fungi. In: Peberdy JF (ed) BSM Symposium Vol 18, Applied molecular genetics of fungi. Br Mycol Soc, Cambridge Univ Press, Cambridge, pp 1–29. **Horton JS,** Raper CA (1991) Curr Genet 19:77–80. **Howard RJ,** Ferrari MA (1989) Exp Mycol 13:403–418. **Howlett BJ** (1989) Exp Mycol 13:199–202. **Hulbert SH,** Michelmore RW (1988) Mol Plant-Microbe Interact 1:17–24. **Hunst PL,** Latterell FM, Rossi AE (1986) Phytopathology 76:674–678.

**Jacobson DJ,** Gordon TR (1990) Mycol Res 94:734–744. **Jones RN,** Rees H (eds) (1982) B chromosomes. Academic Press, London. **Judelson HS,** Tyler BM, Michelmore RW (1991) Mol Plant-Microbe Interact 4:602–607.

**Kan van JAL,** van den Acherveken GFJM, de Witt PJGM (1992) Molecular aspects of virulence of the tomato pathogen *Cladosporium fulvum*. In: Stahl U, Tudzynski P (eds) Molecular biology of filamentous fungi. Verlag Chemie, Weinheim, pp 87–98. **Kazmierczak P,** Zhang L, Pfeiffer P, Van Alfen NK (1991) Phytopathology 81:1156. **Keen NT** (1990) Annu Rev Genet 24:447–462. **Keller NP,** Bergstrom GC, Yoder OC (1991) Curr Genet 19:227–233. **Kerr A** (1987) Annu Rev Phytopathol 25:87–110. **Kinscherf TG,** Leong SA (1988) Chromosoma 96:427–433. **Kohmoto K,** Otani H (1991) Experientia 47:755–764. **Kolattukudy PE** (1985) Annu Rev Phytopathol 22:223–250. **Kolattukudy PE,** Podila GK, Mohan R (1989a) Genome 31:342–349. **Kolattukudy PE,** Podila GK, Roberts E, Dickman MD (1989b) Gene expression resulting from the early signals in plant-fungus interaction. In: Staskawicz BJ, Ahlquist A, Yoder O (eds) Molecular biology of plant-pathogen interactions. Liss, New York, pp 87–102. **Köller W** (1991) The plant cuticle: a barrier to be overcome by fungal plant pathogens. In: Cole GT, Hoch HC (eds) The fungal spore and disease initiation in plants and animals. Plenum, New York, pp 219–246. **Koltin Y,** Finkler A, Ben-Zvi BS (1987) Double-stranded RNA viruses of pathogenic fungi: virulence and plant protection. In: Pegg GF, Ayres PG (eds) Fungal infection of plants. Cambridge Univ Press, pp 334–348. **Kubo Y,** Nakamura H, Kobayashi K, Okuno T, Furusawa I (1991) Mol Plant-Microbe Interact 4:440–445.

**Leong SA** (1988) Adv Plant Pathol 6:1–26. **Leong SA,** Holden DW (1989) Annu Rev Phytopathol 27:463–481. **Levy M,** Romao J, Marchetti MA (1991) Plant Cell 3:95–102.

**Martin RM,** Van Alfen NK (1991) Mol Plant-Microbe Interact 4:507–511. **Masel A,** Braithwaite K, Irwin J, Manners J (1990) Curr Genet 18:81–86. **McCluskey K,** Mills D (1990) Mol Plant-Microbe Interact 3:366–373. **McCluskey K,** Russels BW, Mills D (1990) Curr Genet 18:385–386. **McDermott JM,** McDonald BA, Allard RW, Webster RK (1989) Genetics 122:561–565. **McDon-**

**ald BA,** Martinez JP (1990a) Curr Genet 17:133–138. **McDonald BA,** Martinez JP (1990b) Phytopathology 80:1368–1373. **McDonald BA,** Martinez JP (1991) Curr Genet 19:265–271. **Meinhardt F** (1991) Molecular genetics of phytopathogenic fungi. In: Behnke HD, Esser K, Kubitzki K, Runge M, Ziegler H (eds) Progess in Botany. Springer, Berlin Heidelberg New York 52:264–270. **Miao VP,** Matthews DE, VanEtten HD (1991a) Mol Gen Genet 226:214–223. **Miao VP,** Covert SF, VanEtten HD (1991b) Science 254:1773–1776. **Miao VP,** VanEtten HD (1992) Appl Environ Microbiol 58:801–808. **Mills D,** McCluskey K (1990) Mol Plant-Microbe Interact 3:351–357.

**Nazar RN,** Hu X, Schmidt J, Culham D, Robb J (1991) Physiol Mol Plant Pathol 39:1–11. **Newton AC** (1987) Trans Br Mycol Soc 88:113–116. **Nuss DL,** Koltin Y (1990) Annu Rev Phytopathol 28:37–58.

**Pannacione DG,** Scott-Craig JS, Walton JD (1991) Phytopathology 81:1156. **Podila GK,** Dickman MB, Kolattukudy PE (1988) Science 242:922–925. **Powell WA,** Van Alfen NK (1987a) Mol Cell Biol 7:3688–3693. **Powell WA,** Van Alfen NK (1987b) J Bacteriol 169:5324–5326. **Purdy RE,** Kolattukudy PE (1973) Arch Biochem Biophys 159:61–69.

**Rae BP,** Hillman BI, Tartaglia J, Nuss DL (1989) EMBO J 8:657–663. **Raeder U,** Thompson W, Broda P (1989) Mol Microbiol 3:911–918. **Rigling D,** Van Alfen NK (1991) J Bacteriol 173:8000–8003. **Rollo F,** Ferractui T, Pacilli A (1989) Curr Genet 16:477–479.

**Schäfer C,** Wöstemeyer J (1992) J Phytopathol 136:124–136. **Shaphira R,** Choi GH, Nuss DL (1991) EMBO J 10:731–739. **Skinner DZ,** Budde AD, Leong SA (1991) Molecular karyotype analysis of fungi. In: Bennett JW, Lasure LL (eds) More gene manipulations in fungi. Academic Press, New York, pp 86–103. **Smit R,** Tudzynski P (1992) Mol Gen Genet 234:297–305. **Soliday CL,** Dickman MB, Kolattukudy PE (1989) J Bacteriol 171:1942–1951. **Stahl DJ,** Schäfer W (1992) The Plant Cell 4:621–629. **Stahl U,** Tudzynski P (1992) Molecular biology of filamentous fungi. Verlag Chemie, Weinheim. **Sweigard JA,** Chumley FG, Valent B (1992a) Mol Gen Genet 232:174–182. **Sweigard JA,** Chumley FG, Valent B (1992b) Mol Gen Genet (in press)

**Talbot NJ,** Oliver RP, Coddington A (1991) Mol Gen Genet 229:267–272. **Taylor JL,** Borgmann I, Seguin-Swartz G (1991) Curr Genet 19:273–277. **Tooley PW,** Hewings AD, Falkenstein HF (1989) Phytopathology 79:470–474. **Tzeng T-H,** Lyngholm LK, Ford CF, Bronson CR (1992) Genetics 130:81–96.

**Uehara K** (1964) Ann Phytopathol Soc Jpn 29:103–110.

**Valent B,** Chumley FG (1991) Annu Rev Phytopathol 29:443–467. **Van Alfen NK,** Jaynes RA, Anagnostakis SL, Day PR (1975) Science 189:890–891. **VanEtten HD,** Matthews DE, Matthews PS (1989a) Annu Rev Phytopathol 27:143–164. **VanEtten HD,** Matthews DE, Matthews PS, Miao V, Maloney A, Staney D (1989b) A family of genes for phytoalexin detoxification in the plant pathogen *Nectria haematococca*. In: Lugtenberg B (ed) Signal molecules in plants and plant-microbe interactions. NATO ASI Series, Springer, Berlin Heidelberg New York, H36:219–228. **VanEtten HD,** Matthews DE, Smith DA (1982) Metabolism of phytoalexins. In: Bailey JA, Mansfield JW (eds) Phytoalexins. Blackie, Glasgow, pp 181–217.

**Walton JD** (1990) Peptide phytotoxins from plant pathogenic fungi. In: Kleinkauf H, von Döhren H (eds) Biochemistry of peptide antibiotics. De Gruyter, Berlin, pp 179–203. **Weltring KM** (1992) Phytoalexins in the relation between plants and their fungal pathogens. In: Stahl U, Tudzynski P (eds) Molecular biology of filamentous fungi. Verlag Chemie, Weinheim, pp 111–124. **Weltring KM,** Turgeon BG, Yoder O, VanEtten HD (1988) Gene 68:335–344. **Williams JGK,** Kubelik AR, Rafalski JA, Tingey SV (1991a) Genetic analysis with RAPD markers. In: Bennett JW, Lasure LL (eds) More gene manipulations in fungi. Academic Press, New York, pp 431–439. **Williams JGK,**

Kubelik AR, Livak KJ, Rafalski JA, Tingey SV (1991b) Nucleic Acids Res 18:6531–6535. **Wood HA,** Bozarth RF (1973) Phytopatholgy 63:1018–1023.

**Yamada T,** Seno M, Shiraishi T, Kato H, An C, Yoshida T, Oku H (1991) Plant Cell Physiol 32:65–71.

**Zanzinger DH,** Bandy BP, Tavantzis SM (1984) J Gen Virol 65:1601–1605.

Professor Dr. Paul Tudzynski
Dr. Klaus-Michael Weltring
Westfälische Wilhelms-Universität
Lehrstuhl f. Allg. Botanik/Mikrobiologie
Institut für Botanik und Botanischer Garten
Schloßgarten 3
W-4400 Münster, Germany

# D. Taxonomy

## I. Systematics of the Bryophytes

By Jan-Peter Frahm

### 1. General Aspects

Modern trends in bryophyte systematics are emphasized in the recent volume of the series *Advances in Bryology* (Miller 1991). This volume contains an overview of significant research topics, e.g., application of enzyme electrophoresis to bryophyte systematics and population biology, ecological genetics, laboratory and greenhouse cultures and experimental taxonomy, and the use of ontogenetic data in phylogenetic analysis. It should, however, not be overlooked that these contributions concern studies in single laboratories in the world and do not represent worldwide activities. The papers may, however, show trends for the future. The main activities in taxonomic bryology concern basic work such as treatments of single taxa, monographs, and revisions. Taxonomic work is, however, no longer merely descriptive but more and more combined with phytogeographical evaluations and interpretations illustrating possible evolutionary pathways (see Sect. 6).

A multilingual glossary for bryology was edited by Magill (1990). With the help of several colleagues, 1155 bryological terms are given in seven languages, English, French, German, Japanese, Latin, Russian, and Spanish. In the English, French, and Spanish lists, definitions for the terms are given. All entries are numbered allowing cross-references to other languages. In this way this glossary can also be used as a dictionary.

### 2. Morphology, Anatomy

Pseudoparaphyllia have become an important character in the taxonomy of pleurocarpous mosses. Recently, the morphological features of pseudoparaphyllia of 70 species were illustrated with SEM photographs (Nishimura and Matsui 1990a, b).

During the last 10 years, rhizoid tubers have been observed in an increasing number of moss species. Now rhizoid tubers have also been found in Japanese species of the genus *Fissidens* (Imura and Iwatsuki 1988).

A morphogenetic re-evaluation of *Haplomitrium* was published by Bartholomew-Began (1991). In a series of papers, morphological and anatomical adaptations of bryophyte species of the Judean desert have been studied. Life forms, water conducting and water storing structures, and structures serving as protection against light and temperature have been investigated along a transect between 780 m above to 400 m below sea level (Frey and Kürschner 1991a). In the same region, life strategies were determined and interpreted

Progress in Botany, Vol. 54

with respect to climatic conditions (Frey and Kürschner 1991b). A similar attempt to interpret climatic and edaphic conditions of the Judean desert with respect to bryophyte communities has been made by Frey et al. (1990b). Similar studies have been made with tropical rainforest bryophytes along a transect on Mt. Kinabalu, Borneo (Frey et al. 1990a).

In *Sphagnum*, the cell wall ornamentation of the hyalocysts has been studied with SEM by Anderson and Ammann (1991) with special regard to papillae and ridges in certain species.

## 3. Chemistry

Bryophytes are known to possess compounds with antitumor activities. In a large study, 184 species of mosses, 23 species of liverworts and 1 species of hornwort have been screened in a program of the National Cancer Institute in the United States for antitumor agents (Spjut et al. 1986). Extracts of 43 species were active against lymphocytic leukemia and therefore bryophytes are regarded as a promising source for biologically active compounds.

Beside numerous papers on special chemical components of many taxonomic groups of bryophytes (mostly hepatics), chemical constituents of bryophytes and their biological activities are summarized by Zinsmeister et al. (1991). Bryophytes are regarded as a remarkable source for rare and new compounds, of which numerous have biological activities. In-vitro culture produces large quantities of compounds.

The results of the international symposium on bryophyte chemistry held at Saarbrücken in 1988 are included in a special volume published by Zinsmeister and Mues (1990). It is the most extensive treatment of research on the chemical composition, biochemistry, chemotaxonomy, and pharmacological relevance of bryophytes.

Taxonomic conclusions regarding the separation of the hepatic genera *Eopleurozia* and *Pleurozia* were made by Mues et al. (1991), based on flavonoid chemistry.

The species of *Plagiothecium* in Europe are notoriously difficult to separate. Several attempts have been made in the past to recognize the species based on morphological studies. Now a new attempt has been made using electrophoresis (Hofman 1991). Based on this method, three species pairs, *P. laetum* and *P. curvifolium*, *P. denticulatum* and *P. ruthei*, and also *P. nemorale* and *P. succulentum* are inseparable by their isozymes. Therefore the species pairs previously differentiated by unsatisfactory morphological characters and without distinct delimitations can probably be regarded as conspecific.

## 4. Systematics and Evolution

### a) General

For the hepatics, the *Index Hepaticarum* has been completed with the 12th volume (Bischler 1990). This important work now contains 40 000 checked references for hepatic taxa. In contrast to the *Index Muscorum*, which was completed within 15 years after the proposal for both projects at the International Botanical Congress in Paris in 1954, the preparation of the Index to hepatics took longer, but finally this gigantic work has been finished.

Most taxonomic changes concern pleurocarpous mosses. Pleurocarps are probably a phylogenetically young branch of the mosses and therefore the differences between genera and families are usually smaller than in acrocarps. As shown by Buck and Crum (1990), Hedwig in 1801 recognized four genera of pleurocarpous mosses, Schimper in 1855 distinguished 37, Brotherus in 1924 enumerated 367, and today about 485 genera are known. The authors demonstrate that generic concepts were based mainly on European species and many species described from the tropics were placed into these existent genera. Therefore Buck and Crum evaluated the definitions of genera traditionally placed into the Thuidiaceae and Leskeaceae and drew numerous nomenclatural conclusions. The most important changes concern the re-establishment of the Pterigynandraceae, comprising *Pterigynandrum, Myurella, Habrodon*, and *Heterocladium*, and the establishment of the new family Helodiaceae for the genus *Helodium*, which was separated from the Thuidiaceae. For the first time, the theoretical background for the familial classification of pleurocarpous mosses is discussed by Buck (1991), illustrating the basis for delimitations in bryophyte taxonomy.

The trend to split genera into smaller ones is continuing, e.g., *Aerolindigia* from *Lindigia* (Menzel 1991a). Also, the description of new families and genera of pleurocarpous mosses is being continued, mainly by the same author, who names predominantly after bryologists, with the new family *Cochlearidium* (Ochyra and Mahu 1990) and (all monotypic!) genera *Miehea* (Ochyra 1989), *Crassiphyllum* (Ochyra 1991b) for the endemic species *Thamnobryum fernandezii* Sergio from Madeira, *Streimannia* (Brachytheciaceae) from Australia (Ochyra 1990b), and *Gradsteinia* (Donrichardiaceae) from Colombia (Ochyra 1990c). The subgenus *Parathamnium* is raised to generic rank (Ochyra 1991a) and the necessary new combinations in the species are provided. Another new genus described is *Corynotheca* (Funariaceae) from the Himalayas (Ochyra 1990a).

Cultivation studies have been performed with European species of the genus *Polytrichum* to clarify the taxanomic value of infraspecific categories. In *P. alpinum, P. juniperinum*, and *P. piliferum*, all varieties proved to be modifications (Schriebl 1991). In *P. commune* the varieties *perigoniale, jensenii, uliginosum*, and *commune* remained unchanged in culture. It remains doubtful, however, to give these taxa specific rank, as proposed by the author.

In Tasmania, a strange new species of *Sphagnum* has been found and described as new within a new section (Yamaguchi et al. 1990). The remarkable characters include stems with few branches growing upright in cushions, similar to a species of *Leucobryum*, large imbricate stem leaves and ellipsoid antheridia borne laterally along the stem. In all other species of the genus, antheridia are globose and produced on branches. This and the unusual habit may indicate a very ancient type of a *Sphagnum*.

Usually taxonomic work is confined to monographs and revisions. Only rarely evolutionary trends are discussed. A noteworthy exception is the study by Robinson (1990) of the functional evolution of Leucobryaceae. The unusual leaf anatomy in the different genera of this family is used to explain the evolution of this group.

The morphology of axillary hairs is used for the first time for the taxonomy of bryophytes, for separation of species in Bartramiaceae (Griffin 1990), and for pleurocarpous mosses by Hedenäs (1990).

*Marchantia polymorpha*, the probably best-known hepatic, has been lectotypified because of the problematic nature of the type material (Bischler-Causse and Boisselier-Dubayle 1991). Based on electrophoretic, ecological and morphological characters, three subspecies are recognized: Subspecies *polymorpha*, a taxon which corresponds with the taxon formerly called *aquatica* at various taxonomic ranks. The illustration of the type of *Marchantia polymorpha* must be referred to what later was called *aquatica*. Subspecies *ruderalis* ssp. nov. is the taxon which was formerly called *polymorpha* s. str. Subspecies *montivagans* ssp. nov. corresponds to the taxon formerly called *alpestris*. A new name was chosen, since the application of this name is questionable and older names exist, for which, however, type specimens do not exist, are in poor condition or difficult to locate.

Neotenic species in the hepatic genus *Radula* are treated and illustrated by Schuster (1991). In these species, leafy shoots are produced from ramified thalli as highly specialized adaptation to tropical rain forest habitats.

Based on the study of flavonoid chemistry, Mues et al. (1991) proposed a modified infrageneric classification of the hepatic family Pleuroziaceae.

### b) Monographs, Revisions

Major monographs and revisions have been published for the Asiatic and Oceanic taxa of *Marchantia* (Bischler 1989), Indian Hornworts (Asthana and Srivastava 1991), the genus *Octoblepharum* (now placed into Calymperaceae and no more in Leucobryaceae) from Central America (Salazar Allen 1991), the genus *Crossomitrium* (worldwide) by Allen (1990), and Latin American species of *Riccia* (Jovet-Ast 1991). The first comprehensive volume in the Flora Neotropica series has been published dealing with bryophytes. The subfamilies Campylopodioideae and Paraleucobryoideae with 87 species were monographed by Frahm (1991a).

### c) Cytology

A new edition of the Index to bryophyte chromosome count was published by Fritsch (1991) only 9 years after the first edition. Data are given based on 6700 chromosome counts in 900 publications of 830 taxa of liverworts and 1550 taxa of mosses. The first edition comprised 6000 chromosome counts in 700 publications of 700 taxa of liverworts and 1300 taxa of mosses. An index to chromosome numbers of Polish Bryophytes has been compiled by Przywara et al. (1991). Chromosome numbers are known for one species of hornworts, 54 species of liverworts and 126 species of mosses. These are only 20% of the 909 species of bryophytes known from Poland. For 12 species, new chromosome races have been detected in material from Poland. New counts have been published for seven species from Chile (Deguchi and Oginuma 1990b) and seven species from

Nepal (Inoue and Higuchi 1990). All four Japanese species of *Ptychomitrium* (mosses) have n = 13, one of the commonest number (Deguchi and Oginuma 1990a).

Chromosome counts are now known for approximately 700 species of hepatics and 1300 species of mosses, which is about one fifth of the bryophyte species worldwide.

There is evidence that polyploidy in bryophytes increases with latitude in polar areas. This was recently supported by karyological studies in the genus *Bryum* from Antarctica (Kanda and Okada 1990).

Instructions for microscopical examination of bryophyte chromosomes are given by Newton (1989). Based on the extensive knowledge of the author, detailed information is given on preparation and staining, including Giemsa C-banding techniques.

### d) Fossil Bryophytes

From Paleogene deposits in the Eifel, Germany, two fossil mosses have been provisionally described as *Dicranites* sp. and *Muscites* sp. (Wilde 1990). The photograph of *Dicranites* sp. with curved setae shows clearly that this fossil belongs to the recent genus *Campylopus*. This is the first record of a fossil *Campylopus* and insofar interesting as the species of *Campylopus* in Europe are regarded as relicts from the Tertiary by phytogeographical reasons (predominantly tropical montane species occurring in Europe only under atlantic conditions).

Seventeen species in 12 genera of mosses are described from Russian Permian deposits (Ignatov 1990). Eight new genera are described. The photographs give an excellent and important impression of Paleozoic fossil bryophytes, and illustrate the close affinities between the fossils and extant mosses.

## 5. Floristics

The second volume of the *Handbook of Malesian Mosses* was published (Eddy 1991), comprising Leucobryaceae to Buxbaumiaceae.

The floristic and taxonomic results of the German Bryotrop Expedition to Borneo in 1986 are reported by numerous specialists. So far, 15 contributions have been published in the journal Willdenowia. As a byproduct of this expedition, a guide to the mosses and liverworts of the Kinabalu Park has been published (Frahm et al. 1990), which is the first illustrated guide for tropical bryophytes.

Amongst numerous new country records and extension of ranges, the new record of the heavy metal moss *Scopelophila cataractae* from Spain is worth mentioning. This species was originally not known from Europe, has been reported for the first time from England in 1985, and subsequently from Belgium, the Netherlands, Germany, and France. It was found in Spain over hundreds of square meters around an abandoned mine in Navarra. As in all other European sites, only male plants were found. Whereas all other localities of *Scopelophila cataractae* are at zink-melting furnaces, where the species could have been introduced with foreign minerals from overseas, which are washed and being added in the melting process, the Spanish locality is considered as possibly indigenous.

A completely new Flora for the British liverworts was presented (Smith 1990), replacing the outdated Flora by McVicar of 1926 and complementing the moss Flora for Britain and Ireland published earlier by the same author. For The Netherlands, a new Flora for the mosses was published, including distribution maps of all species based on a revision of 65 000 specimens in Dutch herbaria (Touw and Rubers 1989).

For students or amateurs intending to collect bryophytes in the tropics, a guide has been produced by O'Shea (1989), including suggestions for collecting, labeling, drying, research topics, and identification, as well as an extensive bibliography.

*Checklists*. Conspicuously many checklists have been published during the last years, including inventories especially of tropical regions. A first checklist for Australia is given by Streimann and Curnow (1989). From El Salvador, only 233 species of mosses are known (Menzel 1991b), which illustrates the poor knowledge of tropical bryofloras. Two or three times as many species can be expected. In contrast, the hepatic flora of the better investigated Costa Rica comprises 426 taxa (Morales 1991). A catalog of the mosses of the Guianas comprises 238 species (Florschütz-de Waard 1990). In an annotatated catalog of the bryophytes of SW-Asia ("Middle-East", Turkey, Iran. Afghanistan, Near East and Arabian Peninsula), four species of hornworts, 210 liverworts and 922 mosses are enumerated (Frey and Kürschner 1991c). From the Maldive Islands, only seven species of mosses are known, and these comprise pantropical lowland elements, predominantly weedy acrocarpous species (Menzel and Passow-Schindhelm 1990). For the moss flora of Europe, additions and amendments are given to the checklist published 10 years ago (Corley and Crundwell 1991). The catalog of the Chinese Hepaticae and Anthocerotae contains 884 species (Piippo 1990). A few areas in China are relatively well known (e.g., Taiwan with 489 species, Yunnan with 251 species), while most other provinces are largely undercollected (under 100 species), hardly explored (under 10 species known) or even not known (no species recorded yet). For Israel, 232 taxa of mosses are reported (Herrnstadt et al. 1991), including 45 new records for this country. For North America, already the third version of a checklist for the mosses was presented (Anderson et al. 1990). For the United States and Canada, 1248 species of mosses are reported. Seventeen years ago, only 1170 species were listed. The increase depends partly on a narrower species concept. The checklist includes extensive lists of synonyms but no annotations. Therefore it is not apparent how some species got into this checklist. For the same region, 72 species of peat mosses are enumerated (Anderson 1990), which is distinctly more than in Europe, with about 45 species (depending on taxonomic concepts).

For Germany, the first part of an annotated list of bryophytes was published, comprising hepatics and acrocarpous mosses to Pottiales (Düll and Meinunger 1989). The book contains 98 distribution maps, which do not reflect the distribution but the poor state of mapping, covering only one third of the many grid squares with only a few records. In contrast to Germany, the hepatic flora of Britain and Ireland is completely mapped (Hill et al. 1991). Based on 30 years of work of the British Bryological Society, all literature records, herbarium records, and field records have been incorporated in a database. The distribution is shown on a 10 x 10 km grid.

*Bibliographies.* A bibliography of the West Indian mosses has been presented by Buck (1989). The literature on the ecology of tropical bryophytes has been compiled by Frahm and Gradstein (1990). For tropical Africa, a reference list of basic literature to identify bryophytes is given by Pócs and O'Shea (1991). This is a very useful help for everybody trying to identify tropical African bryophytes, since no standard bryofloras are available for this region. The compilation includes a list of checklists, an alphabetic reference list of taxa and a large bibliography. A so-called "secondary bibliography" of bryophytes has been compiled by Rückert and Korth (1990). It lists publications which contain at least 50 references. By this way, 1468 titles of important publications are included in one volume, giving access to 50–100 000 titles, completed by an extensive subject index.

## 6. Geography

In the first half of this century, disjunctions of tropical bryophytes were known only on the genus level. Many genera were known to occur in tropical America, Africa, and SE Asia, but with different species. This view was influenced by bryologists who were specialized in the bryofloras of one of these regions and had hardly on overview over other parts of the world, or worked on a worldwide basis but with accordingly less special knowledge. Since the 1970s, worldwide revisions of genera revealed the identity of many species in the tropics, which are either pantropical in distribution or are disjunct between tropical America and Africa or Africa and SE Asia. More examples are still being found: *Entodon macropodus*, described under several names from India, Ceylon, S. Africa, Malawi, and Haiti (Townsend 1991), provides another example of a Gondwana range. It shows that many bryophyte species are 200 Mio years old and older. *Caduciella mariei* (Besch.) Enroth (*Pinnatella mariei* (Besch.) Broth.) was revealed to be disjunct between SE Asia and E Africa.

Another difference between flowering plants and bryophytes concerns the percentage of endemic species. Especially island floras are known to be rich in endemic phanerogams. This is not the case in most bryophytes. The rate of endemism in the phanerogamic flora of Cuba is 51.4%, whereas it is only 11.4% in Cuban liverworts and 12.2% in Cuban mosses (Pócs 1988). The rate of endemism will certainly decrease if revisionary work continues. Twenty years ago, the rate of endemism for the whole Caribbean bryoflora was given with 30% (Crosby 1969). A noteworthy exception to the above is the genus *Dicnemon*. A revision of the New Caledonian species revealed seven endemic species (Allen 1989). The reason is that the spores of *Dicnemon* are not dispersed but are retained in the capsule. In addition, they germinate within the capsule to structures of over 1000 cells, which totally lack any ability even for short distance wind dispersal. The rate of endemism (54%) in *Dicnemon* in New Caledonia indicates a high speciation rate there. The rate of endemism found depends on the intensity of taxonomic work, as shown by Iwatsuki (1990), again for New Caledonia. According to Iwatsuki, the percentage of endemic Fissidentaceae (Musci) decreased from 80.6% in a checklist from 1982 to 56.5% in a preliminary treatment of the Fissidentaceae in 1982 and at last to 10.7% in a final revision in 1989. The rate of endemism in New Caledonia is therefore only slightly higher than in New Guinea (8%), the Malay Peninsula (4.1%), and Japan (4.7%). In the *Campylopus* flora of New Caledonia, the total number of species was reduced to 43% and the rate of endemism from 50% to zero (Frahm 1990b). The geographical elements comprise subantarctic species (which inhabit high elevations and probably drifted on these islands northwards) and SE Asian species, which probably reached the islands from the North when New Caledonia reached its present position. For the same genus, the rate of endemism in Hawaii decreased from 70 to 33% (Frahm 1991b), whereas it is 94.5% in flowering plants. All nonendemic species of this genus from Hawaii are of Indo-Pacific origin, which is in strong contrast with the phanerogamic flora, in which only 40% of the anchestral forms are of paleotropic origin. The rate of endemism depends also on the monographer of a taxonomic group, as shown by Iwatsuki (1990). In the Frullaniaceae (Hepaticae), the percentage of endemic species in New Caledonia is 66.6%, in New Guinea 62.3%, and in Japan even 28.6%. The discrepancy between the two families is also distinctly correlated with spore size, which is usually about 10 µm in the Fissidentaceae but 25–80 µm in the Frullaniaceae. It is estimated that aerial transport over long distances is only possible with a spore size of less than 25 µm. Interestingly, the only species of *Fissidens* endemic to New Caledonia has a spore size of 16–26 µm.

For the Antarctic (south of 60 S lat.), an annotated checklist of the hepatics is given by Ochyra and Vana (1989). Only 22 species are known. Only one species is known from Antarctica, all others from the antarctic islands. The number of mosses is also low; there are approximately 70–80 species of mosses in the same region. Compared with the arctic, this is a strikingly low number of species. In contrast, arctic Alaska has about 415 species of mosses and 135 species of hepatics; there are 280 species of mosses in Spitsbergen and 136 species of hepatics in West Greenland. As reason for the paucity of bryophyte species in the Antarctic, the glaciations during the Pleistocene are considered. Because of the absence of refugia during Pleistocene glaciations, all species must be regarded as the result of recent dispersal. Half of the hepatic species of the Antarctic belong to the southern temperate element, five are of subantarctic distribution, five belong to the bipolar floristic element and one species (*Marchantia polymorpha*) is cosmopolitan.

For the Falkland Islands, a taxonomic and phytogeographic study of all 131 species of liverworts known from these islands is provided (Engel 1990). The Falkland liverwort flora has an austral character and shows many phytogeographic relations with South Africa, Australia, and New Zealand.

Data on pre-angiospermous distribution patterns of bryophytes are presented by Frey (1990). Based on the interpretation of present distribution patterns, three geno-elements are distinguished, the Pangaean, Circum Tethyan, and Gondwanian element. An interpretation of morphological and anatomical structures of these species makes it possible to define the time when such characters have been established and also an interpretation of the paleoclimate under which they have evolved.

Several small floristic publications indicate range extensions of species with an Atlantic type of distribution in Europe towards the east. The moss *Ulota phyllantha* distributed along the coasts of NW Europe has been found in Poland (Ochyra 1991b). The moss *Oxystegus hibernicus*, so far known only from the British Isles, has been found in the Vosges Mountains, France (Frahm 1991c).

The origin of the bryophyte flora of the Columbian Andes is discussed by Gradstein et al. (1989) based on the data of the ECOANDES project. It is shown that the number of temperate and Andean species increases with elevation, whereas wide tropical and neotropical species decrease with elevation. Most endemic species are found in the upper montane forest.

For about 20 years, a Gondwanan origin of many bryophyte taxa has been postulated based on interpretations of the present ranges in comparison with the geological data on plate tectonics. Still more examples are provided. The revision of the genus *Desmotheca* (Musci, Orthotrichaceae) led to new assumptions concerning the evolution and dispersal of this genus (Vitt 1990). Of the seven species of *Desmotheca*, two are accepted which occur from New Caledonia to Burma. The genus belongs to the Macromitrioideae, for which a Gondwanan origin has been postulated. Migration of such taxa to Laurasia so far has been explained by drifting on the Indian plate northwards. However, there are no representatives of Desmotheca on the Indian subcontinent today. Therefore another pathway is proposed corresponding with new geological results. According to these geological data, Burma, western Thailand, and the Malay Peninsula comprise continental fragments that were rifted from northern Australia 100 to 44 Mio years ago. As a consequence, a direct pathway for tropical bryophytes to Laurasia was possible rather than rafting northwards on the Indian Plate and subsequent migration into SE Asia.

Studies on the origin of neotropical bryophytes have been performed by Schuster (1990) and Frahm (1990a). Schuster deals in general with neotropical Hepaticae. The amount of endemism is shown to be higher than in any other comparable region of the globe due to tectonic activities and creation of many biotic islands. For the creation of endemics, a time span of 40–60 Mio years is postulated. This cannot explain endemics in alpine habitats of the Andes, which have existed for only a few million years. The endemics are explained as derived from cool-temperate Gonwanan suborders and belong nearly exclusively to Jungermanniales but not to Metzgeriales. A phytogeographical analysis of the 65 species of *Campylopus* in the neotropics has been performed by Frahm (1990a). For the first time, bryophyte distribution is compared with Pleistocene forest refugia, as proposed by zoologists and phanerogamists. The distribution of several rainforest species of *Campylopus* coincides with the major part of the montane refugia. Based on an interpretation of present day ranges, a hypothetical survey is given of the origin and evolution of the genus in South and Central America. For most species, an origin on the south coast of the Gondwana continent can be postulated, with subsequent northward migration and origin of new species. Several species pairs and species chains illustrate this theory. The Andean species are interpreted as having originated from subantarctic ancestors. A similar attempt has been undertaken for the species of *Campylopus* from Africa (Frahm 1990e), which has strong floristic affinities to tropical America. Disjunct intercontinental distributions and sister species in both continents are indicators of continental drift. Different ranges of afro-montane species are explained by presence or absence of long-distance spore dispersal.

## 7. Ecology

### a) General

Edaphic factors for a whole family, the Encalyptaceae, were determined by an analysis of soil samples (Horton 1988). Calcium and pH influence most taxa more than any other factor, but in the subspecies of *Encalypta affinis*, organic matter appears to be more significant. Although the Encalyptaceae traditionally have been reported as calcicoles, the result of this study shows that 12 of 15 taxa also occur on acidic soil and four are most frequently found in slightly to moderately acidic conditions.

The altitudinal zonation of bryophytes in tropical rainforest is subject of several studies. Five altitudinal zones are distinguished in New Guinea (Enroth 1990) on the basis of floristic data, which show a striking similarity with the zonations worked out previously by other authors for other tropical regions (e.g., Colombia, Peru) and recently by Frahm (1990d) for Mt. Kinabalu, Borneo, based on floristic or ecological data. This leads to the conclusion that the same zonation is present worldwide in equatorial latitudes. This zonation, based on bryophytes, can be a useful tool for the zonation of rainforest types, since bryophytes are the best indicators of microclimate, are fewer in number than flowering plants, and have wider (in part even pantropical) ranges. Bryophytes are therefore proposed as useful tools for the construction of a general scheme of an altitudinal zonation of tropical rainforests (Frahm and Gradstein 1991).

The ecology of epiphytic bryophytes on Mt. Kinabalu, Borneo, was studied by Frahm (1990c). For the first time, pH measurements of the bark of tropical host trees were made, ranging between 4.6 and 6.2. These were not correlated with the structure of the bark or altitude. Measurements of the phytomass of epiphytic bryophytes revealed an increase from hardly recordable quantities in the lowlands to a maximum of 440 g/m$^2$ at the timber line. Accordingly, the weigth of epiphytic bryophytes increases from approximately 1 kg/ha to 1320 kg/ha. At the timber line, up to 2760 l of water can be stored by epiphytic bryophytes on a hectare, which is relatively low compared with the daily rainfall of an estimated 110 000 l/ha. Therefore the main function of the epiphytic bryophyte cover is not storage of water but the incorporation of nutrients in the rainfall into the ecosystem.

For the first time, indicator values of bryophytes for ecological parameters such as temperature, light, humidity, or pH have been proposed (Düll 1991). This attempt is based solely on the personal field experience of the author and therefore partly differs from ecological measurements previously published.

In contrast to a common point of view and publications based on studies carried out in parts of Western Europe, host specificity of epiphytic bryophytes has been found in only a few species in a comparative study of the Adirondack Mts and the Blue Ridge Mts in Eastern North America (Schmitt and Slack 1990). The reason is probably the high humidity in these areas, whereas host specificity is more pronounced in drier regions.

It is known that many desert bryophytes live in close relation with Cyanolichens from which the bryophytes obtain nitrogen. This connection between the moss species *Gigaspermum mourettii* and the lichens has been illustrated in SEM pictures for the first time (Frey and Kürschner 1991d). *Gigaspermum mourettii* is a species with low chlorophyll contents and a long subterraneous stolon system. It is assumed that these stolons with their rhizoids function in nitrogen uptake from *Nostoc* species associated with the moss.

b) Pollution

It is known that aquatic bryophytes can store large amounts of heavy metals. In Japan, crystals of mercury sulfide could be observed in the cell walls of the liverworts *Jungermannia vulcanicola* and *Scapania undulata* by electron microscopy with X-ray microanalysis (Satake et al. 1990). Single particles have a size of 5–15 nm, polycrystalline particles are 10–200 nm. The same authors tested the rate of accumulation by transplantation experiments, which was as high as 300–1000 μg per month and 0.6 μg/l of water. By the same technique, lead was found in the shoots of *Scapania undulata* from England. The lead concentration of the water was 0.02 mg/l but in the shoots 0.7–2.4% of dry weight (Satake et al. 1989). Lead and cadmium contents of aquatic bryophytes in the Lower Rhine, Germany, have been determined by Frost (1990). After transplantations of specimens into more polluted parts of the river, an increase of the cadmium contents of the plants of up to 500% and the lead contents of up to 200% was measured within only 4 weeks.

Heavy metal tolerance was measured in several populations of *Bryum argenteum* from various different habitats including industrial sites in the United States, but no adaptation was found in tolerance to any of the metals (Shaw and Albright 1990). These results are in contrast to studies on flowering plants, in which heavy metal tolerance is almost always combined with ecotypic variation.

The same conclusion was supported by tests with *Bryum bicolor*, in which no evidence was found that plants from mine site populations were physiologically adapted to contaminated soils (Shaw 1990).

## References

**Allen BH** (1989) Hikobia 10:245–267. **Allen BH** (1990) Trop Bryol 2:3–34. **Anderson LE** (1990) Bryologist 93:500–501. **Anderson LE,** Ammann K (1991) J Hattori Bot Lab 69:49–64. **Anderson LE,** Crum HA, Buck WR (1990) Bryologist 93:448–499. **Asthana AK,** Srivastava SC (1991) Bryophytorum Bibl 42.

**Bartholomew-Began SE** (1991) Bryophytorum Bibl 41. **Bischler H** (1989) Bryophytorum Bibl 38. **Bischler H** (1990) Bryol Times 57/58:1–2. **Bischler H,** Boisselier-Dubayle MC (1991) J Bryol 16:361–366. **Buck WR** (1989) Moscosoa 5:324–339. **Buck WR** (1991) In: Miller NG (ed) Advances in Bryology vol 4. Cramer, Stuttgart, pp 169–186. **Buck WR,** Crum H (1990) Contrib Univ Mich Herb 17:55–69.

**Casas C** (1991) Orsis 6:1–26. **Corley MFV,** Crundwell AG (1991) J Bryol 16:337–356. **Crosby MR** (1969) Ann Mo Bot Gard 56:409–416.

**Deguchi H,** Oginuma K (1990a) Hikobia 10:423–427. **Deguchi H,** Oginuma K (1990b) Bull Nat Sci Mus Tokyo Ser B 16:105–108. **Düll R** (1991) Zeigerwerte von Laub- und Lebermoosen. In: Ellenberg H et al. (eds) Zeigerwerte von Pflanzen in Mitteleuropa. Scr Geobot XVIII:175–214. **Düll R,** Meinunger L (1989) Deutschlands Moose. Ohlerath, 368 pp.

**Eddy A** (1991) A handbook of Malesian mosses vol 2. Natural History Publ, London. **Engel JJ** (1990) Fieldiana Bot New Ser 25:1–209. **Enroth J** (1990) Trop Bryol 2:61–90. **Enroth J** (1991) J Bryol 16:611–618.

**Florschütz-de Waard J** (1990) Trop Bryol 3:89–104. **Frahm JP** (1990a) Trop Bryol 3:1–18. **Frahm JP** (1990b) Cryptogam Bryol Lichenol 11:369–376. **Frahm JP** (1990c) Nova Hedwigia 51:121–132. **Frahm JP** (1990d) Nova Hedwigia 51:87–119. **Frahm JP** (1990e) J Bryol 16:289–304. **Frahm JP** (1991a) Flora Neotropica Monograph vol 54. N Y Bot Gard, New York. **Frahm JP** (1991b) Bryologist 94:60–66. **Frahm JP** (1991c) Cryptogam Bryol Lichenol 12:165–67. **Frahm JP,** Gradstein SR (1990) Trop Bryol 3:75–77. **Frahm JP,** Gradstein SR (1991) J Biogeogr 18:669–678. **Frahm JP,** Frey W, Kürschner H, Menzel M (1990) Mosses and liverworts of Mt. Kinabalu, Sabah Parks Publ 12, Kota Kinabalu, 91 pp. **Frey W** (1990) Bot Jahrb Syst 111:433–456. **Frey W,** Kürschner H (1991a) Bot Jahrb Syst 112:529–552. **Frey W,** Kürschner H (1991b) Bot Acta 104:172–182. **Frey W,** Kürschner H (1991c) Bryophytorum Bibl 39. **Frey W,** Kürschner H(1991d) Cryptogam Bot 2/3:73–84. **Frey W,** Goßow R, Kürschner H (1990a) Nova Hedwigia 51:87–119. **Frey W,** Herrnstadt I, Kürschner H (1990b) Phytocoenologia 19:233–265. **Fritsch R** (1991) Bryophytorum Bibl 40. **Frost U** (1990) Cryptogam Bryol Lichenol 11:339–352.

**Gradstein SR,** van Reenen GBA, Griffin III D (1989) Acta Bot Neerl 38:439–448. **Griffin DG** (1990) J Bryol 16:61–66.

**Hedenäs L** (1990) Lindbergia 15:166–180. **Herrnstadt I,** Heyn CC, Crosby MR (1991) Bryologist 94:168–178. **Hill MO,** Preston CD, Smith AJE (1991) Atlas of the bryophytes of Britain and Ireland vol 1. Liverworts, Cardiff, 351 pp. **Hofman A** (1991) Phylogeny and population genetics of the genus *Plagiothecium*. Thesis Univ Groningen, 132 pp. **Horton DG** (1988) Beih Nova Hedwigia 90:261–282.

**Ignatov MS** (1990) Palaeontogr Abt B 217:147–189. **Imura S,** Iwatsuki Z (1988) Hikobia 10:225–229. **Inoue S,** Higuchi M (1990) Cryptogams of the Himalayas vol 2. National Science Museum, Tsukuba, pp 117–119. **Iwatsuki Z** (1990) Trop Bryol 2:139–148.

**Jovet-Ast S** (1991) Cryptogam Bryol Lichenol 12:189–370.

**Kanda H,** Okada H (1990) Antarct Rec 34:1–7

**Magill RE** (ed) (1990) Glossarium Polyglottum Bryologiae. Missouri Botanical Garden St. Louis, 297 pp. **Menzel M** (1991a) Nova Hedwigia 52:319–335. **Menzel M** (1991b) Cuscatlania 1:1–24. **Menzel M,** Passow-Schindhelm R (1990) Cryptogam Bryol Lichenol 11:363–367. **Miller NG** (ed) (1991) Adv Bryol 4. **Morales MI** (1991) Trop Bryol 4:25–58. **Mues R,** Klein R, Gradstein SR (1991) J Hattori Bot Lab 70:70–90

**Newton ME** (1989) A practical guide to bryophyte chromosomes. Br Bryol Soc Spec Vol 20:22. **Nishimura N,** Matsui S (1990a) Hikobia 10:429–434. **Nishimura N,** Matsui S (1990b) Bull Hiruzen Res Inst Okayama Univ Sci 16:117–138.

**Ochyra R** (1989) Nova Hedwigia 49:323–332. **Ochyra R** (1990a) Polish Bot Stud 1:59–65. **Ochyra R** (1990b) J Bryol 16:223–229. **Ochyra R** (1990c) Trop Bryol 3:19–28. **Ochyra R** (1991a) Fragm Flor Geobot 36(1):71–79. **Ochyra R** (1991b) Fragm Florist Geobot 36:57–70.

**Ochyra R,** Mahu M (1990) J Bryol 16:215–222. **Ochyra R,** Vana J (1989) Pol Polar Res 10:211–229. **O'Shea BJ** (1989) A guide to collectioning bryophytes in the tropics. Br Bryol Soc Spec Vol 3:28.

**Piippo S** (1990) J Hattori Bot Lab 86:1–192. **Pócs T** (1988) Taxon 37:615–621. **Pócs T,** O'Shea (1991) Trop Bryol 4:69–84. **Przywara L,** Kuta L, Ochyra R (1991) Fragm Florist Geobot 36(1):89–104.

**Robinson H** (1990) Trop Bryol 2:223–238. **Rückert D,** Korth W (1990) Bibliographie bryologischer Bibliographien. Courier Forschungsinst Senckenberg 132, Frankfurt, 144 pp.

**Salazar Allen N** (1991) Trop Bryol 4:85–98. **Satake K,** Tkamatsu T, Soma M, Shibata K, Nishikawa M, Say PJ, Whitton BA (1989) Aquat Bot 33:111–122. **Satake K,** Shibata K, Bando Y (1990) Aquat Bot 36:325–341. **Schmitt CK,** Slack NG (1990) Bryologist 93:257–274. **Schriebl A** (1991) Carinthia 101:461–506. **Schumacker R,** Brugués M (1991) J Bryol 16:486–487. **Schuster RM** (1990) Trop Bryol 2:239–264. **Schuster RM** (1991) J Hattori Bot Lab 70:51–62. **Shaw AJ** (1990) Bryologist 93:1–6. **Shaw AJ,** Albright DL (1990) Bryologist 93:187–192. **Smith AJE** (1990) The liverworts of Britain and Ireland. Cambridge University Press, Cambridge, 362 pp. **Spjut RW,** Suffness M, Cragg GM, Norris DH (1986) Econ Bot 40:310–338. **Streimann H,** Curnow J (1989) Catalogue of mosses of Australia and its external territories. Australia Government Publishing Service, Canberra, 479 pp.

**Touw A,** Rubers WV (1989) De Nederlandse Bladmossen. Stichting Uitgeverij Koninklijke Nederlande Naturhistorische Vereniging, Zutphen, 532 pp. **Townsend CC** (1991) J Bryol 16:601–606.

**Vitt DH** (1990) Trop Bryol 3:78–88.

**Wilde V** (1990) Mainzer Naturwiss Arch 28:1–6.

**Yamaguchi T,** Seppelt RD, Iwatsuki Z, Buchanan AM (1990) J Bryol 16:45–54.

**Zinsmeister HD,** Mues R (1990) Bryophytes, their chemistry and chemical taxonomy. Clarendon Press, Oxford, 470 pp. **Zinsmeister HD,** Becker H, Eicher T (1991) Angew Chem 103:134–151.

Professor Dr. Jan-Peter Frahm
Universität Duisburg
Gesamthochschule
Fachbereich 6, Botanik
Postfach 10 15 03
W-4100 Duisburg, Germany

# II. Systematics of the Pteridophytes

By Karl U. Kramer

## 1. Systematics

Detailed investigations of a limited alliance of plants often lead to the discovery of new differentiating characters. These are then not rarely strongly emphasized in the process of evaluation, the points of similarity being underrated. This usually leads to the establishment of new genera and even families, the former resulting in many new combinations or names. A number of fern taxonomists thought that the time for a re-evaluation and for some measure of consolidation had come. They set themselves the task of critically re-examining all characters employed for splitting off families and especially genera of ferns, of which there was a great amount in recent times. All currently available evidence from biosystematics, karyology, chemotaxonomy, SEM study of spores and other organs, etc., was incorporated as far as possible in an attempt to arrive at a balanced system of classification above the level of species. The results are to be found in the contributions of some 25 specialists in the pteridophyte treatment in Vol. 1 of the newly started series *The Families and Genera of Vascular Plants*, edited by K. Kubitzki (Kramer and Green 1990). In many cases, there is an appreciable reduction in the number of families and especially genera upheld. Numerous genera recognized solely on the basis of local studies, based largely or entirely on characters like type of venation, shape of sorus, presence or absence of an indusium, all criteria that had gradually become rather obsolete, have been relegated to the synonymy. Special literature pertaining to (part of) a family or a genus is cited as fully as possible. Artificial, i.e., practical keys are provided to families and genera in general as well to the genera of each family. Illustrations are provided for many of the genera, partly drawings (often taken from other publications), partly original photographs. Descriptions are provided for all families and genera recognized, where necessary with critical notes on their circumscription and delimitation. Concepts about affinities between families are not discussed at great length, but under the (alphabetically arranged) families the genera are arranged in what is thought to represent a natural sequence.

Few major monographs have been published in the period covered by this review. One is Bosman's study of *Microsorum* (Polypodiaceae) (1991). The genus is redefined against *Phymatosorus*, *Neocheiropteris*, *Colysis*, and *Leptochilus*, mainly on the basis of correlations of venation type, paraphyses, and anatomical features, much less on soral characters. Parts of *Neocheiropteris*, *Leptochilus*, and *Colysis* had to be included in the study. Twenty species remain in this exclusively paleotropical genus.

Progress in Botany, Vol. 54

Hensen's (1990) revision of the *Polypodium loriceum* complex unfortunately leaves much to be desired. Hennipman (in Baas et al. 1990, pp. 105–120) discussed the methods employed for and the problems met with in current taxonomic work on ferns, exemplified by the Polypodiaceae. Hoshizaki (1991) reported the occurrence of a hybrid between *Aglaomorpha* and *Drynaria* in cultivation: the distinctness of the two genera should be re-examined. Hoshizaki and Price (1990) challenged the soundness of the cladistics underlying the monograph of *Platycerium* (Hennipman and Roos 1982) and arrived at a strongly revised infrageneric classification; critical notes also on species. The morphology of the gametophytes of the genus confirms its placement in the Polypodiaceae, which has been doubted by some authors (Bhattachayya and Sen, pp. 137–143 in Bhardwaja and Gena 1991).

An entire issue of Hooker's *Icones Plantarum* is devoted to S. E. Asiatic Grammitidaceae (Parris 1990) with detailed descriptions and high-standard illustrations of 31 less well-known species, including several novelties. The traditional genera are upheld. This contrasts with another newly established segregate of *Grammitis* sens. lat., *Lellingeria*, containing species formerly placed in *Xiphopteris*, *Ctenopteris*, or *Grammitis* sen. str. (Smith et al. 1991).

Holttum and Lin (1990) re-examined the poorly known genus *Pseudotectaria* from Madagascar, showing that it is distinct and stands closer to *Ctenitis* than to *Tectaria*; four species newly transferred to it make six in all. The imperfectly understood neotropical genus *Stigmatopteris*, last revised in 1913, was monographed by Moran (1991); this is a substantial contribution to current fern taxonomy. The genus is critically delimited; it now has 24 species. Stolze (1990) gave critical notes on *Ctenitis* and two New World genera related to it, with which it was often confused; see also Pichi Sermolli (1991). Fraser-Jenkins (in Bhardwaja and Gena 1991, pp. 249–287) published a critical overview of *Polystichum* in the Indian Subcontinent where it has an important distributional center. The 45 species are classed in 7 sections. Many names disappear in the synonymy, others are re-interpreted; some are hybrids. The Chinese authors who described so many new species are criticized; of the 38 species recently described from Tibet not a single one can be maintained. The species of the N. W. Himalayas were revised by Pande and Pande (in Bhardwaja and Gena 1991, pp. 305–324). Darnaedi et al. (1989a) revised the E.-S. E. Asiatic complex of *Dryopteris sparsa*, with three new species; see also Darnaedi et al. (1989b). Kato et al. (1991b) re-examined the E. Asiatic *Matteuccia intermedia*, concluding that it is not a hybrid but a valid species that is transferred to *Onoclea*.

A detailed study of *Cheilanthes* in Australia (Chambers and Farrant 1991) revealed the presence of 15 species, many very similar but all distinguished by a series of correlated, often newly evaluated characters, including features of spores. *Pityrogramma triangularis*, of somewhat uncertain affinity, was shown by Yatskievych et al. (1990) to belong to an as yet undescribed genus, named here *Pentagramma*, with two species in western North America. Salvo et al. (1990) argued on the basis of spore characters that *Notholaena vellea* represents the monotypic genus *Cosentinia*, related to *Anogramma* and *Pityrogramma* rather than to *Cheilanthes* and *Notholaena*. Morphological and biochemical evidence supports the treatment of *Hemionitis* and *Gymnopteris* as one, and of *Bommeria* as a related but distinct genus; both are strictly neogean (Ranker 1990). Zimmer (1991) summed up the differences between *Cheilanthes*, *Notholaena*, and *Cosentinia*, further discussing their taxonomy and distribution in Greece. Four species and two hybrids of the former genus are present there.

Paris (1991) revised the *Adiantum pedatum* complex in North America. *A. pedatum* var. *aleuticum* is raised to species rank. An allotetraploid hybrid of this and true *A. pedatum* is the new *A. viridimontanum* of Vermont. Lellinger (1991b) published some critical notes on confused species of the *A. tetraphyllum* group. Kato (1988a) revised *Taenitis* and *Syngramma* for the southern Moluccas, with critical notes on generic delimitation and affinity. Kato and Kramer (1990) gave an account of *Acrostichum* and *Pteris* for the same area, with some new taxa.

Study of the gemmae of the independently growing *Vittaria* gametophytes in the Appalachians led Farrar and Mickel (1991) to the conclusion that it belongs to an as yet undescribed species, *V. appalachiana*, the sporophyte of which is not known. A similar situation was encountered in the case of the Appalachian *"Hymenophyllum tunbrigense"* gametophytes which were described as the new *H. tayloriae* (Raine et al. 1991). Lellinger (1991a) unraveled several neotropical species complexes in the Hymenophyllaceae. Roux (1990a, b) raised the number of species of *Mohria* (Schizaeaceae) in South Africa from one to five. The chromosome number, rhizome anatomy, spore morphology, etc., show that *Asplenium* (*"Boniniella"*) *cardiophyllum* is a member of *A.* sect. *Hymenasplenium* (Kato et al. 1990). Page (1989) reported two new subspecies of *Pteridium aquilinum* from Britain, ssp. *latiusculum* (raised from variety to subspecies) and the previously undescribed ssp. *atlanticum*, very locally occurring on limestone in Scotland; ssp. *aquilinum* is supposed to be a hybrid between these two.

Rolleri et al. (1991) investigated the epidermal morphology of *Angiopteris, Marattia,* and *Danaea* in great detail. Their placement in a single family was once more confirmed. The former two are not quite as close as concluded by other workers. Wagner and Wagner (1991) revised the group of species centered on *Botrychium lunaria* for North America, with descriptions of two new species.

## 2. Bibliography, Collections, Nomenclature

The history of pteridology in Britain was outlined in a booklet edited by Camus (1991). Special emphasis is on the activities of the British Pteridological Society, the centennial of which was celebrated in 1991, and on changing methods in pteridology. Labatut and Boudrie (1991) gave a brief overview of the history of pteridology in France.

The volume of the Annual Review of Pteridological Research covering 1990 was compiled by Moran and Moran (1992). It includes a list of active research workers and of projects in progress. There are over 600 titles of papers dealing with or touching upon research on ferns.

Bibliographic lists of pteridological papers recently published in India or contributed by Indian pteridologists were compiled by Khullar et al. (1989) and Vasudeva and Bir (1990). Vasudeva (in Bhardwaja and Gena 1991, pp. xxi–xxxviii) listed S. S. Bir's contributions to botany up to 1990, mostly pteridological.

According to Pichi Sermolli (1991), the type specimen of *Triplophyllum protensum*, the type species of the genus, is a hybridogenous plant, perhaps *T. heudelotii* x *T. jense-*

*niae*. The former is described as a new species for what has usually been called *T. protensum*. Some important nomenclatural notes on the pteridophytic flora of North America are by Kartesz and Gandhi (1991).

## 3. Floristics

In contrast to the few monographic publications, a considerable number of floristic works have been published.

Asia, Australia, Pacific. A second instalment of the fern flora of China covers Pteridaceae, Adiantaceae, and related families (Shing et al. 1990). Families, genera, and species are very narrowly circumscribed. After a considerable interruption, another fern group was revised for Flora Malesiana: the *Tectaria* group of genera in the Dryopteridaceae (Holttum 1991b). *Tectaria* has 105 Malesian species, *Ctenitis* 28, *Pleocnemia* 17; a few more are in satellite genera. Genera like *Hemigramma*, *Stenosemia*, *Quercifilix*, and *Ctenitopsis* are included in *Tectaria*. Critical notes on *Tectaria* were previously given by the same author (Holttum 1991a). Salgado (1990) gave a checklist of the ferns of the Philippines, with modern taxonomy and nomenclature. Critical notes on Philippine ferns, with new taxa, were given by Price (1990).

A very fine pteridophytic flora of Japan was produced by Iwatsuki (1992). It is a full treatment of all ferns and fern-allies of the country, with keys and descriptions, notes on ecology and distribution, variability, hybrids, etc., and an extensive general introduction, all in Japanese. A large number of excellent color photographs, mostly taken in situ.

The first modern fern Flora of Queensland is by Andrews (1990). It is a critical, largely up-to-date, well-illustrated Flora of the ferns and fern allies, except for somewhat exaggerated splitting of families and genera, and a most adequate treatment for the part of Australia that is richest in ferns. The pteridophyte treatment in the Flora of New South Wales (Harden 1990) is by Wilson: profusely illustrated, partly in color. Equally well produced is the new fern Flora of New Zealand by Brownsey and Smith-Dodsworth (1989), again with excellent illustrations.

Kato and collaborators began a series of papers on the pteridophytes of Ambon and Seram in the southern Moluccas: Kato 1989a, b; Kato and Price 1990; see also Kato (1988a), Kato and Price (1990) and Kato and Kramer (1990). Several regional fern Floras, or instalments of them, were published for India, e.g., for N. E. India by Bir and Vasudeva (1989), Kachroo et al. (1989), Vasudeva et al. (1990); for part of Orissa by Das et al. (1989), for Kumaun by Pande (1990); Thelypteridaceae of the Western Ghats by Manickam and Irudayaraj (1990).

Africa. A very useful checklist of the pteridophytes of tropical East Africa, with notes on distribution, synonymy, etc. and with some illustrations, is by Johns (1991). The last instalment of the pteridophytic flora of Bioko (Fernando Po) has now been issued (Benl 1991). New records for the fern flora of Zaire were given by Kornas et al. (1990). A new pteridophyte flora of southern Africa, the third to be published since 1986, was written by Burrows (1990); outstanding illustrations and particularly much ecological information.

America. Several new regional fern floras or treatments for the US should be mentioned: Snyder and Bruce for Georgia (1986), Evans for Tennessee (1989), Brooks (in Anonymous 1986) for the Great Plains. A field guide to hybrid ferns of the North-East was edited by Thorne and Thorne (1989). See also Dunbar (1989), reviewed in Section 8.

Palacios-Rios continued her series of fern treatments for the flora of Veracruz, Mexico, with accounts of the Marattiaceae (1990a) and Osmundaceae (1990b). Proctor's (1989) fern flora of Puerto Rico and the Virgin Islands includes the fern allies: it is a worthy successor to its predecessor of 1926. The number of taxa known from the island has increased since that time from 285 to 376! Some are here newly described. The treatment is completely updated, except for the somewhat antiquated taxonomy at genus and family level; a checklist is included. The first fascicle of the fern flora of the Guianas, issued as fascicle 4, deals with Dennstaedtiaceae and Hymenophyllopsidaceae (Cremers et al. in Görts-van Rijn 1991). Cremers and Hoff (1990) published a checklist of the pteridophytes of French Guiana, with distribution maps within the country. Smith (1990) described a number of new taxa from Venezuelan Guayana. The fern flora of Peru was continued with a treatment of the Dryopteridaceae sens. lat. (Tryon and Stolze 1991). Ortega (1991) gave a checklist, with keys and ecological data, of the ferns of the Venezuelan state of Portuguesa.

Europe. One of the few fern floras for a European country to be published within the last 2 years is the field guide to the pteridophytes of the British Isles (including Ireland) by Jermy and Camus (1991). It is a handy identification guide, strongly relying on the excellent illustrations by P. J. Edwards, with concise descriptions and emphasis on diagnostic characters. Another is the guide to the ferns and fern-allies of the Iberian Peninsula and the Balearic Islands by Salvo (1990), again with many very good illustrations; also many distribution maps and an important general introduction. The taxonomy suffers, however, from undue "splitting".

## 4. Geography and Ecology

The importance of pteridophytes as a basis for the phytogeographical subdivision of a given area, exemplified by the Iberian Peninsula, was demonstrated by Pichi Sermolli et al. (1988): similar work by Salvo and García-Verdugo (in Rita 1990, pp. 115–149) includes the Macaronesian and other North Atlantic islands. Whereas the fern floras of tropical America and of tropical Asia-Australasia are comparable in terms of species numbers, the (sub)generic diversity is much higher in the Eastern than in the Western Hemisphere; this is as yet unexplained (Kramer 1990). An analysis of the Venezuelan pteridophyte flora led Ortega and Rivero (1989) to the conclusion that the higher parts of the Andes and the sandstone table mountains (tepuís) are phytogeographically like islands, not only for seed plants but also for spore plants. Iversen's (1991) analysis of the vascular flora of the Usambara Mountains in Tanzania includes pteridophytes. The number of species is quite large, especially for Africa, but in contrast to the seed plants, there are very few endemics.

Young and León (1991), working in the Peruvian Andes, found the highest species diversity in the montane rainforest, through overlap of the conterminous zones. Cremers

(1991) inventorized the pteridophytes in the valleys of a mountain group in French Guiana. The flora is very diversified, according to the various available microhabitats. The importance of edaphic factors governing the occurrence of pteridophytes in Amazonian Perú was pointed out by Young and León (1989); the importance of soil types for the diversity of the flora was formerly underestimated. Kato (in Baas et al. 1990, pp. 225–234) discussed the rich fern flora of Seram in the southern Moluccas – nearly 700 species –, ascribing it to the great ecologial diversity of the island as well as to the proximity of New Guinea, which with about 2000 species is an important source area. Kato et al. (1991a) summed up the knowledge of rheophytic ferns of Borneo, perhaps their major center of concentration; a considerable number of new records. Cronk's (1989) account of the vegetation of St. Helena (mainly what remains of it) contains many data on the ecology of its pteridophytes. Godoy (1990) analyzed the composition of the Chilean pteridophytic flora as to correlations between vegetation zones and frequency of life forms. The correlation is indeed strong, but Raunkiaer's system of life forms is perhaps not too well applicable to pteridophytes. On Santa Cruz Is., Galapágos, pteridophytes proved to be very good indicators of vegetation types and zones, notwithstanding the small number of species present. Low humidity reduces the size of the plants (van der Werff 1990). Bidin (in Baas et al. 1990, pp. 323–328) described the destructive effects of "development" on the fern flora of some parts of Malaya. Equally threatened is the rich fern flora of the West Indian island Guadeloupe, as emphasized by Lazare et al. (1991) who also gave a survey of the distribution of its ferns over the various vegetation zones.

Size and spore production of fertile leaves decrease with increasing altitude in the Alps (Sato et al. 1989). In the tropical American genus *Danaea*, spore release is in the wet season; only about 10% of the plants are fertile in a given year. Sporophylls develop fast but remain on the plant for only half a year, whereas the slowly developing trophophylls last about six times as long (Sharpe and Jernstedt 1990). *Blechnum* species with longer axes have a distinctly higher water conductivity than short-stemmed species; the twining *Salpichlaena* has particularly numerous, long, and especially wide tracheids (Veres 1990).

Genetic homogeneity within populations and heterogeneity between populations of *Asplenium ruta-muraria* show that colonization of new habitats must as a rule be a rare single-spore event, and that expansion within such a habitat proceeds from a single parent plant (Schneller in Schmid and Stöcklin 1991, pp. 53–61). Further evidence for the existence of a spore bank in the soil containing viable spores the year round was found for four fern species by Lindsay and Dyer (in Rita 1990, pp. 243–253). The marsh fern *Thelypteris palustris* regularly produces enormous numbers of spores, but, like many other plants from swampy habitats, reproduces mostly by vegetative growth (Marti 1990). Further studies by Odland (1991) on the autoecology of *Thelypteris limbosperma* in Atlantic Norway confirmed that climate is the main factor governing its occurrence, with edaphic and anthropogenic factors in the second rank of importance.

## 5. Morphology and Anatomy

Viane (in Rita 1990, pp. 69–89) coined the new term "epidermology" for a new tool in the identification of ferns by micromorphological characters: size, distribution, and shape of epidermis cells; size of stoma guard-cells; fine structure of the indusium and of the paraphyses; etc. The technique is described, and the importance for establishing the identity of incomplete or sterile specimens, taxa in biosystematic aggregates, and such is emphasized. Computer programs can be applied for the evaluation of the character complex. As an example the method is successfully applied to the discrimination of the taxa in the *Dryopteris pallida* group. Oteo et al. (in Rita 1990, pp. 171–190) studied the structure of the receptacle, the annulus, and other details of the characters related to the spore-producing organs in some Spanish *Asplenium* species. Some of the characters are taxonomically relevant. Davies (1991) published a first attempt at a comprehensive study of aerophores and pneumatophores in ferns. Anatomical and morphological observations are promising, but taxonomical and ecological implications, as stated, are sometimes weakly founded. Hovenkamp (1990) examined the morphology of the rhizome, especially its branching patterns, in the Polypodiaceae, and demonstrated the importance of this character complex for classification within the family. The peculiar morphology of the "proliferous" filmy fern *Trichomanes proliferum* was re-examined by Hébant-Mauri (1990); he concluded that the peculiarity had been rather over-emphasized and that the only curious features are nonphyllome-conjunct dichotomies of the rhizome and epiphyllous budding. The rhizome structure of two species of the ant-fern genus *Lecanopteris* was studied by Gay (1991). The space for the ants is provided in two different ways: in *L. sarcopus* by differentiation into regular rhizomes and hollow, leafless ramifications, in *L. darnaedii* by the presence of two superimposed gallery and chamber systems in all parts of the rhizome. The ontogeny of the vascular system in the shoot-apex of *Blechnum orientale* was described by Gopalakrishnan and Nayar (1990).

Philipson (1990), reviewing the structure of apical meristems in cormophytes, pointed out that the presence of a pyramidal apical cell, without periclinal divisions, unites Equisetatae, Psilotatae, Selaginellales, and bryophytes, whereas an apical meristem with superposed initials, with periclinal divisions, is common to seed plants, Lycopodiales, and Isoëtales. The conclusion that Selaginellaceae therefore stand apart from the other Lycopodiatae is, however, not quite convincing. In *Helminthostachys* there is a kind of anomalous secondary tissue called accessory xylem, anomalous because it does not stem from a regualar cambium. This fits in well with the idea that Ophioglossaceae are a reduced, secondarily herbaceous offshoot of (woody) progymnosperms (Takahashi and Kato 1990). Similar conclusions were reached by the same authors' (Takahashi and Kato 1988) studies of the vascular cambium and periderm of *Botrychium (Botrypus) virginianum*. Stipular budding, well known in *Angiopteris*, was demonstrated for *Danaea*, a related genus, by Sharpe and Jernstedt (1991); it was enhanced by decapitation.

Spores. Spores are currently popular as objects of study, as shown by the considerable number of books and articles devoted to them.

The principal – and uniquely comprehensive – work is the book *Spores of the Pteridophyta* by Tryon and Lugardon (1991). Representatives of virtually all genera of pteridophytes have their spores illustrated by SEM and TEM pictures of very high standard, at different magnifications, and usually for several to many species per genus. Their features are described in detail, and there are remarks on similarity with those of other taxa, homogeneity vs. heterogeneity within a group, etc., but taxonomic implications are carefully worded and not particularly numerous. Examples are: homogeneity of the Ophioglossaceae with the three traditional genera, and their place among the true ferns; the heterogeneous nature of *Ctenitis* even in its current, restricted sense; Psilotaceae are most closely related to ferns. An important general introduction explains details of spore wall structure, terminology, and methodology. Valuable bibliography for every family.

New local spore floras have been produced for Barro Colorado Is., Panama, by Roubik and Moreno (1991) within the framework of a general treatment of local pollen and spores as observed by light microscopy; and by Large and Braggins (1991) for New Zealand. The latter work also provides keys and descriptions and gives interesting juxtapositions of SEM and light microscopy photographs. See also Mehra and Gupta (1986), reviewed below.

Sporogenesis in pteridophytes was reviewed by Lugardon (in Blackmore and Knox 1990, pp. 95–120). Heterosporous ferns diverge from homosporous taxa by the possession of an extra-exospore (epispore), formed between the tapetum and the regular exospore; Equisetaceae have a comparable spore-wall layer. Microspores of *Isoëtes* and *Selaginella* share another special layer, the para-exospore. Sporogenesis in Psilotaceae is similar to that observed in Gleicheniaceae. Equisetaceae have some points of similarity in their spore-wall structure, but Lycopsida are widely divergent. A. Tryon (1990) gave an overview of the three main spore types met with in ferns: primitive, usually trilete, with thick exospore and thin, conform perispore; derived, monolete, with thin exospore and elaborated, nonconform perispore; and specialized, trilete or monolete, with the exo- and/or the perispore having a particularly complicated structure. Some correlations of surface structure with the ecology of the species are emerging. Hennipman (in Claugher 1990, pp. 23–44) discussed the details of spore-wall sculpture in the Polypodiaceae and emphasized that they vary so much from one species group to another, with many instances of parallel development, that their importance for indicating generic relationships and boundaries seems quite limited in the family. Braggins and Large (1990) studied the spores of the New Zealand species of *Cyathea* and *Asplenium*. Subdividing the former genus is not advocated. In *Asplenium*, natural groups can be discerned but their delimitation is very difficult. All Iberian *Asplenium* species can be reliably recognized by their spore structure (Pangua and Prada 1988). Dettman and Clifford (1991) examined the spores of *Anemia*, *Mohria*, and *Ceratopteris*. They have certain features in common, but every one is unique. Morbelli and Michelena (1989) studied the spores of *Cheilanthes* in N. W. Argentina, Michelena (1989) those of the Pteridaceae (Adiantaceae) in the province of Buenos Aires (with key).

The Gametophyte. Gametophytes of 111 species of Himalayan ferns were grown from spores from nature and described and extensively illustrated, in various stages of development, by Mehra and Gupta (1986). The spores, and also – and this makes the work particularly valuable – the sporelings (juvenile sporophytes) are included in the study, which also describes and illustrates sex organs, appendages of the surface, anatomical details, and aspects of regeneration. Few, if any, works contain such complete life-histories of the gametophytic generation from the spore to the appearance of the young sporophyte. The drawings are good, the photomicrographs much less so. The taxonomy is here and there open to debate. See also Bhattacharyya and Sen (in Bhardwaja and Gena 1991, pp. 137–143).

Farrar and Johnson-Groh (1990) discovered the presence of subterranean gemmae on the sporophytes of four out of ten species of *Botrychium*. Their development into complete sporophytes resembles that of sporelings.

A method for studying gametophytes by SEM without damaging these delicate structures was described by Sheffield et al. (1991): impressions made in dental impression material yield casts that show all essential details.

## 6. Chemotaxonomy

A useful overview of the application of the phytochemistry of flavonoids and xanthones to questions of pteridophyte systematics is by Wallace (in Bhardwaja and Gena 1991, pp. 117–135). The Psilotales stand apart; they have nothing to do with ferns. The two genera of this family are chemically very close. Recently proposed segregates from *Lycopodium* are also chemically well characterized. Somewhat surprisingly, Isoëtales are closer to Lycopodiales than to Selaginellales. Relatively little is as yet known about Equisetales. *Marattia* stands somewhat apart in the Marattiales, with *Danaea* providing a link with *Angiopteris* and relatives. The classification of the Gleicheniaceae in more than two genera is confirmed by their chemistry. *Marsilea* is chemically quite diverse. The distinction of a *Trichomanes* and a *Hymenophyllum* group of genera is borne out chemically, as is the heterogenetity of these two genera in their traditional, comprehensive circumscription. Extensive bibliography.

A number of *Dryopteris* species, especially from the *D. villarii* complex, were examined for the constitution of their phloroglucinols by Widén et al. (1991). The chemistry of 17 taxa, all from subgenus *Dryopteris*, by and large bears out the infrageneric classification of the genus as proposed by Fraser-Jenkins (1986). Phloroglucinols are particularly suitable for chemotaxonomic research because of their great constancy; besides, they can be extracted from not-so-fresh herbarium material. Related (sub)species are in some cases indistinguishable by their phloroglucinols. On the other hand, subspecies *balearica* of *D. pallida* is chemically so distinct that it probably merits specific rank.

## 7. Cytotaxonomy, Biosystematics, Hybridization

The principal aspects of the life-cycle of pteridophytes (and other plants) were summarized by Bell (in Callow 1989, pp. 55–93), with discussion of many comparative aspects; see also Bell (in Rita 1990, pp. 29–37). Prelli (1991) reviewed the importance of hybridization in the evolution of ferns, without which their taxonomy cannot be understood.

In biosystematic research on ferns, the genera *Asplenium*, *Polypodium*, and *Polystichum* were again in the focus.

The derivation of *Polypodium virginianum*, previously shown to be tetraploid, has at last been elucidated. One of its ancestors is the recently described *P. sibiricum*, now known to be fairly widespread also in North America; the other is the newly described *P. appalachianum* (Haufler and Windham 1991). Triploid backcrosses are also known. The western North American *P. amorphum* is also related, but more distantly (Haufler and Wang 1991). In western North America the complex of *P. californicum* and *P. glycyrrhiza*, both diploid, proved to contain an undescribed allotetraploid, *P. calirhiza*, probably derived from these two; it may resemble *P. californicum* quite closely. A key is given (Whitmore and Smith 1991).

Allozyme analysis provided evidence for the derivation of the allotetraploid *Polystichum californicum* from *P. dudleyi* and *P. imbricans*; another species that was previously regarded as a possible parent, *P. munitum*, is definitely not involved (Soltis and Soltis pp. 103–104 in Kawano 1990; see also Soltis and Soltis 1991 and Soltis et al.

1991). The genetic variation of two narrowly distributed, endemic *Polystichum* species of western North America proved to be much less than that of widespread species of the genus in the same area. This is of importance if conservation measures need to be taken. *P. potteri*, a North American hybrid of *P. acrostichoides* and *P. braunii*, has chloroplast DNA of either of the parents, depending on the individual, showing that the crossing must have occurred repeatedly and reciprocally (Stein and Barrington 1990). Barrington (1990) studied four *Polystichum* species and an equal number of hybrids in Costa Rica by means of cytological and isozyme methods. There was a reticulate pattern of relationships, two of the four species proving to be allotetraploids.

Haufler et al. (1990) investigated the North American *Cystopteris tennesseensis* complex which consists of allotetraploid *C. tennesseensis* and its two diploid parent species, *C. bulbifera* and *C. protrusa*. Triploid backcrosses were found, but no diploid plants of primary hybrid origin. The allozyme pattern of *C. tennesseensis* is purely additive, suggesting that it is a relatively recent hybrid species. Some of the triploids turned out to be autotriploid *C. protrusa*. The pattern of variation thus becomes extremely complex and difficult to interpret except by karyological and biochemical methods. Another species was added to this complex by Haufler and Windham (1991), *C. utahensis*, an allotetraploid derived from *C. reevesiana* and *C. bulbifera*. *C. fragilis* is probably an ancient tetraploid that has begun to diversify as such, not as a consequence of its having a double genome. Hexaploids in the same complex are of multiple derivation and their taxonomy has as yet not been unraveled. *C. dickieana* is thought not to be separable from *C. fragilis*.

*Asplenium trichomanes* nothosubsp. *melzeri* is a newly found hybrid of ssp. *inexpectans* and ssp. *quadrivalens* from Styria. In contrast to the calcifuge notosubsp. *lusaticum* (ssp. *trichomanes* x ssp. *quadrivalens*), it is calciphile (Lovis et al. 1989). The same hybrid was described as nothosubsp. *lucanum* from Mallorca (Cubas et al. 1989). Another new subspecies is subsp. *coriaceifolium* (Rasbach et al. 1990a, 1991a) from Spain; its hybrids with ssp. *inexpectans* and ssp. *quadrivalens* are also described. The autotetraploid *A. billotii* has probably arisen from the diploid *A. obovatum* var. *protobillotii*, newly described from Anatolia. It is better regarded as a subsp. of *A. obovatum*, *A. o.* subsp. *lanceolatum* (Demiriz et al. 1990). These two taxa were subsequently discovered in Spain where they produce a hybrid, *A. o.* nothosubsp. *cyrnosardoum* var. *ibericum* (Rasbach et al. 1990b). The allotetraploid *A. balearicum* originated from a cross between *A. onopteris* and *A. obovatum*. The western Mediterranean *A.* x *tyrrhenicum* is very probably a backcross of *A. balearicum* with *A. onopteris* (Cubas et al. 1988). The diploid form of *A. petrarchae*, ssp. *bivalens*, was found as a new record on Mallorca, Balearic Is., where it forms a triploid hybrid with *A. trichomanes* ssp. *quadrivalens*, newly described as *A.* x *helii* nothosubsp. *calobraense*. *A. majoricum*, endemic to Mallorca, is derived from *A. petrarchae* subsp. *bivalens* and *A. fontanum* (Bennert et al. 1990). The $F_1$ hybrid was recently found in Spain and described as *A.* x *protomajoricum* (Pangua et al. 1992). *A.* x *artanense* is a remarkable new sterile hybrid from Mallorca, very probably *A. sagittatum* x *A. trichomanes* ssp. *inexpectans*, two most dissimilar species (Rosselló et al. 1990).

The hybrid between *Athyrium filix-femina* and *A. distentifolium* in Europe which are both diploid occurs in a diploid and two triploid forms: it is called *A.* x *reichsteinii*. One of the triploid hybrids was recently described as nothosubsp. *microderris*. It has originated by somatic polyploidization of the diploid hybrid and subsequent backcrossing with *A. distentifolium*. Another triploid hybrid, nothosubsp. *praetermissum*, was formed in the same way but by backcrossing with *Athyrium filix-femina* (Rasbach et al. 1991b). Kramer (1991) discussed the hybrids so far known in *Lindsaea* (none, however, karyologically checked) and described a new one between two very disparate species from different sections. Sporogenesis in Asiatic lindsaeoid ferns was investigated by Lin et al. (1990). The presence of only 32 spores in the sporangia of 20 sexually reproducing, i.e., non-apogamous species, is the first report of such a low number. Conant (1990) studied the reproductive biology of two species and two hybrids of tree-fern in Puerto Rico. The germination rate of the spores of the hybrids was lower than that of the species, yet considerable; their gametophytes produced sporophytes, apparently non-apogamously.

Rasbach and Reichstein (1990) and Queiros et al. (in Rita 1990, pp. 235–242) established the chromosome number of *Anogramma leptophylla* as unequivocally n = 26. As *A. chaerophylla* has n = 29, there is a suspicion that the genus may be taxonomically heterogeneous. Walker (1990) confirmed the base numbers known for the (sub)genera of Gleicheniaceae. Out of ten species of *Pteris* occurring in China, only five proved to be sexual diploids; *Pt. vittata* is a complex of different ploidy levels, some sterile; four species are apogamous. Ploidy levels up to 4 x were found (Wang 1989). Loyal (in Bhardwaja and Gena 1991, pp. 171–248) looked for correlations between karyological and morphological features in Himalayan Thelypteridaceae, but these proved to be slight. A lineage based on x = 36 seems to be distinguishable from another with x numbers from 30 to 35. Darnaedi et al. (1989b) investigated the karyology and breeding system of the species of the *Dryopteris sparsa* complex (see also Darnaedi et al. 1989a). Most species are diploid and sexual, but some contain tetraploid and also aneuploid apogamous forms; three hybrids were also found.

Chromosome counts of Nepalese ferns by Matsumoto and Nakaike (1990) largely confirm earlier counts for the same taxa from elsewhere, especially from Japan. Pichi Sermolli (1990) studied the distribution of allopolyploid species and their progenitors in Europe (especially in the South), arriving at the conclusion that most of them probably arose polytopically. Stergianou and Fowler (1990) examined the karyology of *Azolla*. *A. caroliniana*, *A. filiculoides*, and three other species are 2 n = 44; some triploids and a tetraploid were also found, which is interesting in view of the relative rarity of polyploidy in heterosporous pteridophytes. *Azolla nilotica* has 2 n = 52, which tallies with the isolated position of the species within the genus.

Chromosome counts made for, i.a., pteridophytes in 1988 and 1989 are cited in Goldblatt and Johnson's Index (1991).

## 8. Folklore, Uses

Moran (1992) once more told the story of the *"Vegetable lamb of Tartary"*, arguing that the plant behind this story is not, as generally stated, *Cibotium barometz* (or other species of this tree-fern genus) but cotton (*Gossypium*). Dunbar's (1989) well illustrated, semi-popular fern flora of the coastal plain of the eastern USA lays special emphasis on the (now mostly obsolete) use of ferns; their suitability for cultivation is also reported. Notes on such and on many more aspects of ferns, especially their ecology, are assembled in the semi-popular, lavishly illustrated book *A World of Ferns* by Camus et al. (1991), published on the occasion of the centennial of the British Pteridological Society.

Jacquat's (1990) attractive book on plants from the markets of Thailand contains numerous entries for pteridophytes used as food and medicine, for packing, as ornamentals, etc.; Thai names are cited.

## 9. Fern Allies

Recent data on *Equisetum* hybrids, especially their morphology, ecology, and distribution, were assembled by Page (in Rita 1990, pp. 151–156). Paeger and Bennert (1990) investigated the spore production and germination of three European *Equisetum* species. They found between ca. 1000 and 3000 spores per sporangium, and between half a million and over seven million spores per cone. Germination rates were between 63 and 88% (but 0% in the hybrid *E.* x *litorale*). Germinability lasts from 1 to 9 days, depending on humidity. Dark germination proved impossible.

Ibars et al. (in Rita 1990, pp. 225–234) examined the distribution of stomata in five European Lycopodiaceae; growth habit and ecology proved to influence this. Stronger differentiation between sporophylls and trophophylls was associated with greater differences in stomatal distribution. Holub (1991) re-examined the genus *Huperzia* and made some emendations in its subdivision. *Lycopodium* sect. *Magellanica* was raised by him to generic rank as *Austrolycopodium*. Many new combinations were made under four generic names. No new evidence is adduced to support this splitting, which will not be acceptable to all.

Dahlen (1990), after examining the surface sculpture of the *Selaginella* species of Hong Kong in great detail by SEM, concluded that the differences in this character between microspores and megaspores are functional: they facilitate adhesion upon contact (synaptospory), enhancing the chance for fertilization. Fresh megaspores have to be used for maximum discrimination of surface features. Adhesion of microspores to megaspores was also observed by Sen et al. (1989), who made a detailed study of the sporoderm of Indian *Selaginella* species. The spores of five European *Selaginella* species are figured and described in the account of the genus for the Northwest European Pollen Flora (Stafford in Punt and Blackmore 1991, pp. 1–22). Imaichi and Kato (1991) found further evidence for the sui generis nature of the *Selaginella* rhizophore. When it branches dichotomously, the original apical cell is lost and two new ones are formed. A root cap is never present. The roots are endogenously formed.

A new species of *Selaginella* from Madagascar is the first species that produces tubers on its rhizophores (Rauh and Hagemann 1991). *Selaginella* was revised for part of the State of Mexico by Abundiz-Bonilla and Tejero-Diaz (1990).

The very involved taxonomy of *Isoëtes* is gradually becoming clarified by application of modern methods. Since these have become available, the genus is receiving rather much attention. Hickey (1990) pointed out that three South American species form a distinct subgenus, *Euphyllum*, characterized by the presence of completely laminate leaves lacking a distal subula. The fossil species classed under *Isoëtites* are very probably closely related to this group. *Isoëtes harveyi*, fairly widespread in North America, is very probably a hybrid between *I. macrospora* and *I. tuckermanii* (Britton 1991). Another natural hybrid from eastern North America is the newly described *I. brittonii* (Brunton and Taylor 1991). An unusual new species from eastern Canada is *I. prototypus*, a diploid with reduced megaspore sculpture (Britton and Goltz 1991). *I. taiwanensis* is confined to a single pond in northern Taiwan. Britton and Brunton (1991) re-examined its megaspores and concluded that its closest relative is *I. kirkii* from New Zealand (!). It is a diploid, sexual species that may be of some importance as an ancestor of polyploids in the western Pacific. The ecology of the two Scandinavian species of *I.* was reviewed by

Rørslett and Brettum (1989). *I. sectacea* is less susceptible to physical or chemical disturbance of its habitat than is *I. lacustris*. *Isoëtes* was revised for the Indian state of Rajasthan by Sharma et al. (1987).

## References

**Abundiz-Bonilla L,** Tejero-Diaz JD (1990) Acta Bot Mex 11:23–47. **Andrews SB** (1990) Ferns of Queensland. Queensld Dep Primary Ind, Brisbane. **Anonymous** (1986) Flora of the Great Plains. Univ Kansas Press, Lawrence, Kansas.

**Baas P,** Kalkman K, Geesink R (eds) (1990) The plant diversity of Malesia. Proc F1 Males Symp comm C G G J van Steenis. Kluwer, Dordrecht. **Barrington DS** (1990) Ann Mo Bot Gard 77:297–305. **Benl G** (1991) The Pteridophyta of Bioko (Fernando Po) V. Acta Bot Barc 40:1–106. **Bennert HW,** Rasbach H, Rasbach K (1990) Farnblätter 21:15–26. **Bhardwaja TN,** Gena CB (eds) (1991) Perspectives in pteridology. Present and future. Aspects Plant Sci 13:xxxviii + 341 pp. **Bir SS,** Vasudeva SM (1989) Indian Fern J 6:30–55. **Blackmore S,** Knox RB (1990) Microspores. Evolution and ontogeny. Academic Press, London. **Bosman MTM** (1991) A monograph of the fern genus *Microsorum* (Polypodiaceae). Leiden Bot Ser 14:ix + 161pp. **Braggins JE,** Large MF (1990) Rev Palaeobot Palynol 64:149–158. **Britton DM** (1991) Can J Bot 69:634–640. **Britton DM,** Brunton DF (1991) Fern Gaz 14(2):73–81. **Britton DM,** Goltz JP (1991) Can J Bot 69:277–281. **Brownsey PJ,** Smith-Dodsworth JC (1989) New Zealand ferns and allied plants. Bateman, Auckland, N Z. **Brunton DF,** Taylor WC (1991) Am Fern J 81:82–89. **Burrows JE** (1990) Southern African ferns and fern allies. Frandsen, Fourways, S Afr.

**Callow JA** (ed) (1989) Adv Bot Res 16:xii + 273 pp. **Camus JM** (1991) The history of British pteridology 1891–1991. Br Pteridol Soc Spec Publ 4, 127 pp. **Camus JM,** Jermy AC, Thomas BA (1991) A world of ferns. Nat Hist Mus Publ, London. **Chambers TC,** Farrant PA (1991) Telopea 4:509–557. **Claugher D** (ed) (1990) Scanning electron microscopy in taxonomy and plant morphology. Clarendon, Oxford. **Conant DS** (1990) Ann Mo Bot Gard 77:290–296. **Cremers G** (1991) Bot Helv 101 (1991):69–76. **Cremers G,** Hoff M (1990) Inventaire taxonomique des plantes de la Guyane française. I – Les ptéridophytes. Mus Natl Hist Nat, Paris. **Cronk QCB** (1989) J Biogeogr 16:47–64. **Cubas P,** Rosselló JA, Pangua E (1988) Anal Jard Bot Madrid 45:75–92. **Cubas P,** Rosseló JA, Pangua E (1989) Candollea 44:181–190.

**Dahlen MA** (1990) Farnblätter 22:20–27. **Darnaedi D,** Kato M, Iwatsuki K (1989a) J Jpn Bot 64:330–340. **Darnaedi D,** Kato M, Iwatsuki K (1989b) J Jpn Bot 64:299–310. **Das PK,** Mishra MK, Panigrahi G (1989) Plant Sci Res 11:7–44. **Davies KL** (1991) Bot J Linn Soc 107:115–137. **Demiriz H,** Viane R, Reichstein T (1990) Candollea 45:241–259. **Dettman ME,** Clifford HT (1991) Am J Bot 78:303–321. **Dunbar L** (1989) Ferns of the Coastal Plain. Univ S Carol Press, Columbia, SC.

**Evans AM** (1989) J Tenn Acad Sci 64:103–105.

**Farrar DR,** Johnson-Groh CL (1990) Am J Bot 77:1168–1175. **Farrar DR,** Mickel JT (1991) Am Fern J 81:69–75. **Fraser-Jenkins CR** (1986) Bull Br Mus Nat Hist (Bot) 14(3):183–218.

**Gay H** (1991) Bot J Linn Soc 106:199–208. **Godoy R** (1990) Anal Jard Bot Madrid 46:599–603. **Goldblatt P,** Johnson DE (eds) (1991) Index to plant chromosome numbers 1988–1989. Monogr Syst Bot Mo Bot G 40:1–238. **Gopalakrishnan V,** Nayar BK (1990) Indian Fern J 7(1/2):24–34.

**Görts-van Rijn ARA** (ed) (1991) Flora of the Guianas Ser B, 4: Dennstaedtiaceae (Cremers G, Kramer KU), Hymenophyllopsidaceae (Lellinger DB). Koeltz, Koenigstein.

**Harden GJ** (1990) Flora of New South Wales 1. N S. Wales Univ Press, Kensington, N S W. **Haufler CH,** Wang Z (1991) Am J Bot 78:624–629. **Haufler CH,** Windham MD (1991) Am Fern J 81:7–23. **Haufler CH,** Windham MD, Ranker TA (1990) Ann Mo Bot Gard 77:314–329. **Hébant-Mauri R** (1990) Can J Bot 68:1091–1097. **Hennipman E,** Roos MC (1982) Verh K Ned Akad Wet Afd Natuurkd II 80:1–126. **Hensen RV** (1990) Nova Hedwigia 50:279–336. **Hickey RJ** (1990) Ann Mo Bot Gard 77:239–245. **Holttum RE** (1991a) Blumea 35(1991):547–557. **Holttum RE** (1991b) Flora Malesiana Series II – Pteridophyta. Ferns and fern allies. Vol 2 pt 1: Tectaria Group. Rijksherbarium, Leiden. **Holttum RE,** Lin YX (1990) Kew Bull 45:257–263. **Holub J** (1991) Folia Geobot Phytotaxon 26:81–94. **Hoshizaki BJ** (1991) Am Fern J 81:37–43. Hoshizaki BJ, Price MG (1990) Am Fern J 80:53–69. **Hovenkamp P** (1990) Am Fern J 80:33–43.

**Imaichi R,** Kato M (1991) Am J Bot 78:1694–1703. **Iversen ST** (1991) Symb Bot Ups 29:1–234. **Iwatsuki K** (ed) (1992) Ferns and fern allies of Japan. Heibonsha, Tokyo (Jap).

**Jacquat C** (1990) Plants from the markets of Thailand. Duang Kamol, Bangkok. **Jermy AC,** Camus J (1991) The illustrated field guide to ferns and allied plants of the British Isles. Nat Hist Mus Publ, London. **Johns RJ** (1991) Pteridophytes of tropical East Africa. A preliminary checklist of the species. R Bot Gard, Kew, Richmond.

**Kachroo P,** Bir SS, Vasudeva SM (1989) Indian Fern J 6:78–99. **Kartesz JT,** Gandhi KN (1991) Phytologia 70:194–208. **Kato M** (1988a) Acta Phytotaxon Geobot 39:133–146. **Kato M** (1988b) J Fac Sci Univ Tokyo III. 14:161–182. **Kato M** (1989a) J Fac Sci Univ Tokyo III. 14:219–241. **Kato M** (1989b) Acta Phytotaxon Geobot 40:77–92. **Kato M,** Kramer KU (1990) Acta Phytotaxon Geobot 41:155–168. **Kato M,** Price MG (1990) Acta Phytotaxon Geobot 41:61–76. **Kato M,** Nakato N, Akiyama S, Iwatsuki K (1990) Bot Mag (Tokyo) 103:461–468. **Kato M,** Darnaedi D, Iwatsuki K (1991a) J Fac Sci Univ Tokyo III. 15:91–110. **Kato M,** Suzuki T, Nakato N (1991b) Fern Gaz 14(2):65–71. **Kawano C** (ed) (1990) Biological approaches and evolutionary trends in plants. Academic Press, London. **Khullar SP,** Sharma A, Verma S (1989) Indian Fern J 6:258–266. **Kornas J,** Bodenghien A, Malaisse F (1990) Mitt Inst Allg Bot Hamb 236:791–801. **Kramer KU** (1990) Ann Mo Bot Gard 77:330–333. **Kramer KU** (1991) Bot Helv 101:227–233. **Kramer KU,** Green PS (eds) (1990) The families and genera of vascular plants. In: Kubitzki K (ed) I: Pteridophytes and gymnosperms. Springer, Berlin Heidelberg New York.

**Labatut AJ,** Boudrie M (1991) Pteridologist 2,2:68–72. **Large MF,** Braggins JE (1991) Spore atlas of New Zealand ferns and fern allies. Suppl N Z J Bot 1991:167 pp. **Lazare J-J,** Vivant J, Sastre C (1991) Bull Soc Bot Fr 138, Actual Bot 2:197–214. **Lellinger DB** (1991a) Am Fern J 81:24–37. **Lellinger DB** (1991b) Am Fern J 81:99–102. **Lin S-J,** Kato M, Iwatsuki K (1990) Am Fern J 80:97–109. **Lovis JD,** Rasbach H, Reichstein T (1989) Cadollea 44:543–553.

**Manickam VS,** Irudayaraj V (1990) Indian Fern J 7(1/2)100–117. **Marti K** (1990) Farnblätter 22:1–19. **Matsumoto S,** Nakaike T (1990) In: Watanabe M, Malla SB (eds) Cryptogams of the Himalayas 2. Central and Eastern Nepal. Dep Bot Nat Sci Mus, Tsukuba, Japan, pp 163–172. **Mehra PN,** Gupta A (1986) Gametophytes of Himalayan ferns. Dep Bot, Punjabi Univ, Chandigarh. **Michelena IG** (1989) Bol Asoc Latinoam Palaeobot Palinol 12:25–31. **Moran RC** (1991) Ann Mo Bot Gard 78:857–914. **Moran RC** (1992) Fiddlehead Forum 19:2–8. **Moran CKR,** Moran RC (1992) Annual review of pteridological research. Mo Bot Gard, St. Louis. **Morbelli MA,** Michelena IG (1989) Grana 28:295–304.

**Odland A** (1991) Nord J Bot 10:637–659. **Ortega F** (1991) Helechos del Estado Portuguesa. Biollania Ed Espec 2, Guanare/Caracas. **Ortega F,** Rivero R (1989) Pittiera 18:20–43.

**Paeger J,** Bennert HW (1990) Flor Rundbr (Bochum) 24:46–56. **Page CN** (1989) Watsonia 17:429–434. **Palacios-Rios M** (1990a) Flora de Veracruz 60: Marattiaceae. Xalapa, México. **Palacios-Rios M** (1990b) Flora de Veracruz 61: Osmundaceae. Xalapa, México. **Pande PC** (1990) Indian Fern J 7(1/2):58–64. **Pangua E,** Prada C (1988) Lagascalia 15 (Extra):157–167. **Pangua E,** Prada C, Pajarón S, Salvo E (1992) Bot J Linn Soc 108:1–13. **Paris CA** (1991) Rhodora 93:105–121. **Parris BS** (1990) Hooker's Icones Plantarum XL (IV) i–iv, 1–127. **Philipson WR** (1990) Plant Syst Evol 173:17–38. **Pichi Sermolli REG** (1990) Anal Jard Bot Madrid 46:489–518. **Pichi Sermolli REG** (1991) Webbia 45:117–135. **Pichi Sermolli REG,** España L, Salvo AE (1988) Lazaroa 10:187–205. **Prelli R** (1991) Bull Soc Bot Fr 138, Actual Bot 2:117–125. **Price MG** (1990) Contrib Mich Herb 17:267–278. **Proctor GR** (1989) Ferns of Puerto Rico and the Virgin Islands. Mem N Y Bot Gard 53:389 pp. **Punt W,** Blackmore S (eds) (1991) The Northwest European Pollen Flora VI. Rev Palaeobot Palynol 69:1–275.

**Raine CA,** Farrar DR, Sheffield E (1991) Am Fern J 81:109–118. **Ranker TA** (1990) Syst Bot 15:442–453. **Rasbach H,** Reichstein T (1990) Fern Gaz 13(6):341–348. **Rasbach H,** Rasbach K, Reichstein T, Bennert HW (1990a) Willdenowia 19:471–474. **Rasbach H,** Rasbach K, Reichstein T, Viane RLL, Bennert HW (1990b) Bot Helv 100:3–16. **Rasbach H,** Rasbach K, Reichstein T, Bennert HW (1991a) Willdenowia 21:239–261. **Rasbach H,** Reichstein T, Schneller JJ (1991b) Bot Helv 101:227–233. **Rauh W,** Hagemann W (1991) Plant Syst Evol 176:205–219. **Rita J** (ed) (1990) Taxonomia, biogeografía y conservación de pteridófitos. Inst Menorquí d'Estudios, Maó, Menorca. **Rolleri C,** Deferrari A, Lavalle MC (1991) Am Fern J 81:44–62. **Rørslett B,** Brettum P (1989) Aquat Bot 35:223–261. **Rosselló JA,** Cubas P, Gradaille JL, Sastre B (1990) Fern Gaz 13(6):349–355. **Roubik DW,** Moreno PJE (1991) Monogr Syst Bot Mo Bot Gard 36:1–268. **Roux JP** (1990a) S Afr J Bot 56:266–270. **Roux JP** (1990b) S Afr J Bot 56:399–402.

**Salgado EA** (1990) Philipp J Sci 119:107–148. **Salvo TE** (1990) Guía de los helechos de la Península Ibérica y Baleares. Pirámide, Madrid. **Salvo AE,** Trigo M, Pangua E, Prada C (1990) Webbia 44:233–247. **Sato T,** Grabherr G, Washio K (1989) J Biogeogr 16:449–455. **Schmid B,** Stöcklin J (eds) (1991) Populationsbiologie der Pflanzen. Birkhäuser, Basel. **Sen T,** Mukhopadhyay R, Sen U (1989) Indian Fern J 6:184–194. **Sharma BD,** Singh R, Vyas MS (1987) Bull Bot Surv India 27 ("1985"):58–62. **Sharpe JM,** Jernstedt JA (1990) Am J Bot 77:1040–1049. **Sharpe JM,** Jernstedt JA (1991) Am Fern J 81:119–127. **Sheffield E,** Cutter EG, Douglas GE, Lindsay S (1991) Am Fern J 81:128–133. **Shing K-H,** Lin Y-X, Wu S-H, Wu S-K (1990) Flora Reipublicae Popularis Sinicae 3(1):Pteridophyta (seq). Science Press, Beijing. **Smith AR** (1990) Ann Mo Bot Gard 77:249–273. **Smith AR,** Moran RC, Bishop LE (1991) Am Fern J 81:76–88. **Snyder LH,** Bruce JG (1986) Field guide to the ferns and other pteridophytes of Georgia. Univ Georgia Press, Athens. **Soltis PS,** Soltis PE (1991) Aliso 13:215–223. **Soltis PS,** Soltis DE, Wolf DG (1991) Syst Bot 16:245–256. **Stein DB,** Barrington DS (1990) Ann Mo Bot Gard 77:334–339. **Stergianou KK,** Fowler K (1990) Plant Syst Evol 173:223–239. **Stolze RG** (1990) Ann Mo Bot Gard 77:274–280.

**Takahashi A,** Kato M (1988) Bot Mag Tokyo 101:373–385. **Takahashi A,** Kato M (1990) Plant Syst Evol 173:119–127. **Thorne F,** Thorne L (1989) Henry Potter's field guide to the hybrid ferns of the North-East. Vermont Inst Nat Sci, Woodstock, Vt. **Tryon AF** (1990) Plant Syst Evol Suppl 5:71–79. **Tryon AF,** Lugardon B (1991) Spores of the Pteridophyta. Springer, Berlin Heidelberg New York. **Tryon RM,** Stolze RG (1991) Pteridophyta of Peru Part IV. 17. Dryopteridaceae. Fieldiana Bot New Ser 27:1–176.

**Vasudeva SM,** Bir SS (1990) Indian Fern J 7(1/2):196–202. **Vasudeva SM,** Bir SS, Kachroo P (1990) Indian Fern J 7(1/2):66–85. **Veres JS** (1990) Am J Bot 77:1610–1625.

**Wagner WH Jr,** Wagner FS (1991) Am Fern J 81:73–81. **Walker TG** (1990) Fern Gaz 13(7):385–390. **Wang Z-R** (1989) Acta Phytotaxon Sin 27:421–438 (Chinese, Engl summ). **Werff H van der** (1990) Monogr Syst Mo Bot Gard 32:79–92. **Whitmore SA,** Smith AR (1991) Madroño 38:233–248. **Widén C-J,** Widén K, Vida G, Reichstein T (1991) Bot Helv 101:77–120.

**Yatskievych G,** Turner J (1990) Catalogue of the flora of Missouri. Monogr Syst Bot Mo Bot Gard 37:1–345. **Yatskievych G,** Windham MD, Wollenweber E (1990) Am Fern J 80:9–17. **Young KR,** León B (1989) Brittonia 41:388–395. **Young KR,** León B (1991) Fern Gaz 14(1):25–30.

**Zimmer B** (1991) Bot Chron 10:221–237.

Professor Dr. Karl U. Kramer
Institut für Systematische Botanik
der Universität Zürich
Zollikerstraße 107
CH-8008 Zürich, Switzerland

# E. Geobotany

## I. The History of Flora and Vegetation During the Quaternary

By Burkhard Frenzel

During the past decades, Soviet scientists have greatly contributed to progress in Quaternary vegetation history and paleoecology. Due to the economic difficulties now existing in the states of the former Soviet Union, the intensity of publishing scientific results may well diminish for some tine. So it seems worth summarizing the most important results published during the past years in Russian (additional information in Frenzel 1982, 1984, 1987, 1989, 1991a). In this report only those non-Russian publications will be mentioned which enable a better understanding of the problems discussed in the Russian literature.

### 1. Paleoecology of the Younger Part of the Last Glaciation

It is widely held that during the younger part of the last glaciation, the maximum ice advances of the Upper Pleistocene occurred. Whether this is correct or not has been discussed repeatedly (for Central and Northern Europe: de Beaulieu et al. 1991; Dricot et al. 1991; Frenzel 1991b; Haeberli 1991; Mangerud 1991; Menke 1991; Petersen and Kronborg 1991; Schlüchter 1991; for Siberia see literature in Frenzel 1991a). This will also be dealt with here to some extent. However, regarding the upper pleniglacial of the last glaciation, several major problems have been investigated intensively: what were the size and extent of the North Eurasian full glacial inland ice? Was the Barents Sea covered by grounded ice or by drift-ice only? When and how did the preceding and the ensuing warm-climate phases end or begin, respectively? What was the paleoecological situation during the last full glacial in Northern Eurasia, particularly in Siberia?

According to Grosswald (1988) the New Siberian Islands were overridden during the last glaciation by an inland ice coming from the northeast (further information: Grosswald 1983). Several geomorphological arguments had been published, which stimulated the author to reconstruct a continuous ice dome all over the Arctic Ocean during that time. Grosswald (1988) relies strongly on the reconstruction of the former climatic snowline, which is claimed to have lain during that time at the present-day sea level in the lowlands of the rivers Yana and Indigirka. The broad geomorphological and paleoglaciological literature on northeastern Siberia does not confirm this view. Kagan et al. (1987) discuss the problem of the existence of sea ice during the pleniglacial of the last glaciation in the Arctic Ocean from a theoretical point of view, and Velichko and Faustova (1989) reject the assumption of a strong glacierization at that time in the Arctic Basin (for more information see Frenzel et al. 1992). The northern part of the West

Progress in Botany, Vol. 54

Siberian lowlands seems to have been invaded during the last three glaciations by ice coming from the northern surroundings of the Jamal and Gydan Peninsulas, damming up the rivers Ob, Taz, and Jenisseij (Astakhov 1976). Yet it is difficult to find convincing evidence for this when traversing this region by plane. Repeatedly, massive ice was found in some exposures in the northern part of the West Siberian lowlands. Kaplyanskaya and Tarnogradskiy (1976) interpret this as indicating former glacier ice, though convincing proofs were not given. On the other hand, Vasilchuk et al. (1986) suggest that this ice is fossil drift ice only, which may even be included into last interglacial (Kazantsevo) sediments. This is concluded from the heterogenous salt content within parts of these ice blocks and by the pollenflora of the surrounding sediments, which seems to be governed by long-distance transport or by rebedded older material. The general conclusion drawn by the authors is that the last interglacial was in those regions an appreciably cool one. Yet this is in contrast to other paleoclimatological reconstructions (see Frenzel et al. 1992).

A comparably complicated problem is that of the former ice cover of the Barents Sea. Mangerud et al. (1987) suggest that the Barents Sea was covered during full glacial times by a continuous mass of glaciers, and Blazhchishin and Linkova (1977) report on Late Pliocene and Early Pleistocene glacial sediments on the shelf area of the Barents Sea. On the other hand, Starovoytov et al. (1982) found no indications for a former glaciation in the southern part of the Barents Sea. Yet several modern observations favor the view of a formerly glaciated Barents Sea shelf at least (Epshtein et al. 1983; Lavrushin and Chistyakova 1988; Dunaev et al. 1989). It is suggested from foraminiferal evidence that at the same time the center of the Barents Sea was covered by long lasting pack ice (Khusid 1989). Recently, more facts in support of a continuous ice cover of the whole Barents Sea were reported (Gataullin and Polyak 1990; Lavrushin et al. 1990). This would help in better understanding the pronounced influence of ice pressure coming from the north, the northeast, and the west on sediments of the coastal area of northernmost northeast Europe (Balashova 1976; Gornostai 1990; Lavrov 1981; Prokopchuk 1985). This ice cover seems already to have disintegrated very early during late glacial times, due to the eustatically rising sea level and the influx of warmer Atlantic waters (Khusid 1989). This rising sea level, on the other hand, caused a stepwise isolation of the present-day islands within the northern Pacific Ocean. This is held to have had considerable influence on the endemism of these islands (Velizhanin 1976).

Thus it seems that during full glacial times of the last glaciation, a continuous cover of glaciers and of very thick sea ice existed in the Barents Sea and on Novaya Zemlya. The coverage of this ice strongly diminished farther to the east, with no glacier ice in the eastern part of the Arctic Ocean. On the other hand, this means, in connection with the then eustatically lowered sea level, that vast areas of the Arctic shelf had become dry land at that time. This directly pertains to the problem of vegetation and animal life during full glacial times in these regions.

A widely distributed general pattern of sediment structure in Northern and Northeastern Siberia and on the islands farther to the north is the sequence of gravel or medium-grained sediments at the base repeatedly containing peat layers, stumps and trunks of various tree species, and fossils of a rich fauna. These sediments are overlain by fine grained to silty sediments, rich in ice wedges and ice veins. They are covered in their turn by lacustrine and peaty sediments, again with layers rich in plant remains. The fine grained silty sediments are called the Yedom Series. The uppermost lacustrine or peaty sediments prove to be of thermokarst origin. In various parts of the regions mentioned, $^{14}C$-datations were done on the whole of this sequence. The results are summarized in Fig. 1, according to the data provided by Adamenko et al. (1989), Alekseeva and Tikhomirov (1987), Arkhangelov et al. (1986), Astakhov and Isaeva (1985), Danilov and Parunin (1982), Dobrodeev and Parunin (1973), Dodonov (1986), Furrer (1991), Goncharov (1986), Grekhova (1990), Ivanova et al. (1989), Kaplina and Lozhkin (1977), Katasonov et al. (1979), Laukhin et al. (1986), Lazarev and Tomskaya (1987), Lozhkin (1977, 1984, 1989, 1990), Lozhkin and Postolenko (1989), Lozhkin et al. (1975, 1988), Makeev et al. (1989), Markova (1975), Melekestsev et al. (1988), Morozova (1981), Nikolaeva et al. (1989), Nikonov et al. (1988), Pshenin et al. (1984), Serebryanniy et al. (1984), Shevyrev et al. (1987), Shilo et al. (1977a, b, c), Sinitsyn et al. (1974), Stanko and Svezhentsev (1988), Stefanovich et al. (1986), Tarakanov et al. (1974), Tomirdiaro (1972), Tomirdiaro et al. (1984), Ukraintseva (1990), Vasilchuk (1988, 1989, 1990), Vasilchuk and Trofimov (1984), Vasilchuk et al. (1985), Vasilev et al. (1986). In this figure not only observations made in northernmost Siberia are incorporated but also those of some other regions. The infinite data will not be discussed here. They only indicate that the lowermost layer rich in organic matter dates not only from the middle of the last glaciation but also from some older phases of the Pleistocene, the exact ages of which are unknown. This is nevertheless important, since some of these infinite data were repeatedly used in the Russian literature to prove the age of the so-called Karga interglacial or interstadial, which is said to have occurred during an earlier part of the last glaciation, from about 55 000 to approximately 25 000 B.P. [For further information see Frenzel (1991a); about a possible confusion between Karga interstadial and Kazantsevo interglacial sediments in the northern part of the Taimyr Peninsula: Krupnik et al. (1990); Laukhin and Rybakova (1982) report on a remarkable warming period in the western part of the Chukchi Peninsula, which might be correlated with the Karga interstadial, here called the Topolevo warm period. The vegetation history of that time is described at some length. It is held that is was the time of warmest climates during the Upper Pleistocene, from which Danilov and Parunin (1982) reported the paleoecological situation regarding the widespread formation of carbonate concretions within the then existing vegetation. Comparable concretions were formed during the early Holocene in the Kolyma lowlands at about $9190 \pm 40$ B.P., when a dry and appreciably warm climate favored the spreading of a tundra-steppe vegetation, instead of the present-day tundra: Gubin and Kartashova (1989)].

From Fig. 1, it will be learnt that in the regions mentioned various phases of an intensive bioproduction had occured during Upper Quaternary times: from about 35 000 to approximately 28 000 B.P., from about 25 000 to ca. 20 000 B.P., and from 13 500 to ca. 12 000, finally from 10 500 to 7 000 B.P. All these phases mentioned were characteristic by climatic conditions which enabled thermokarst features to develop, in which later lacustrine and peaty sediments were formed, while at the same time an appreciably rich vegetation of various types surrounded lakes and mires. This vegetation is described at some length in the papers mentioned, which were used for the construction of Fig. 1. The two oldest warm climate phases are repeatedly suggested to be equivalent of the younger parts of the Karga interglacial or interstadial (for further information see Frenzel 1991a). On the other hand, it is worth mentioning that the younger warm climate phases corrobo-

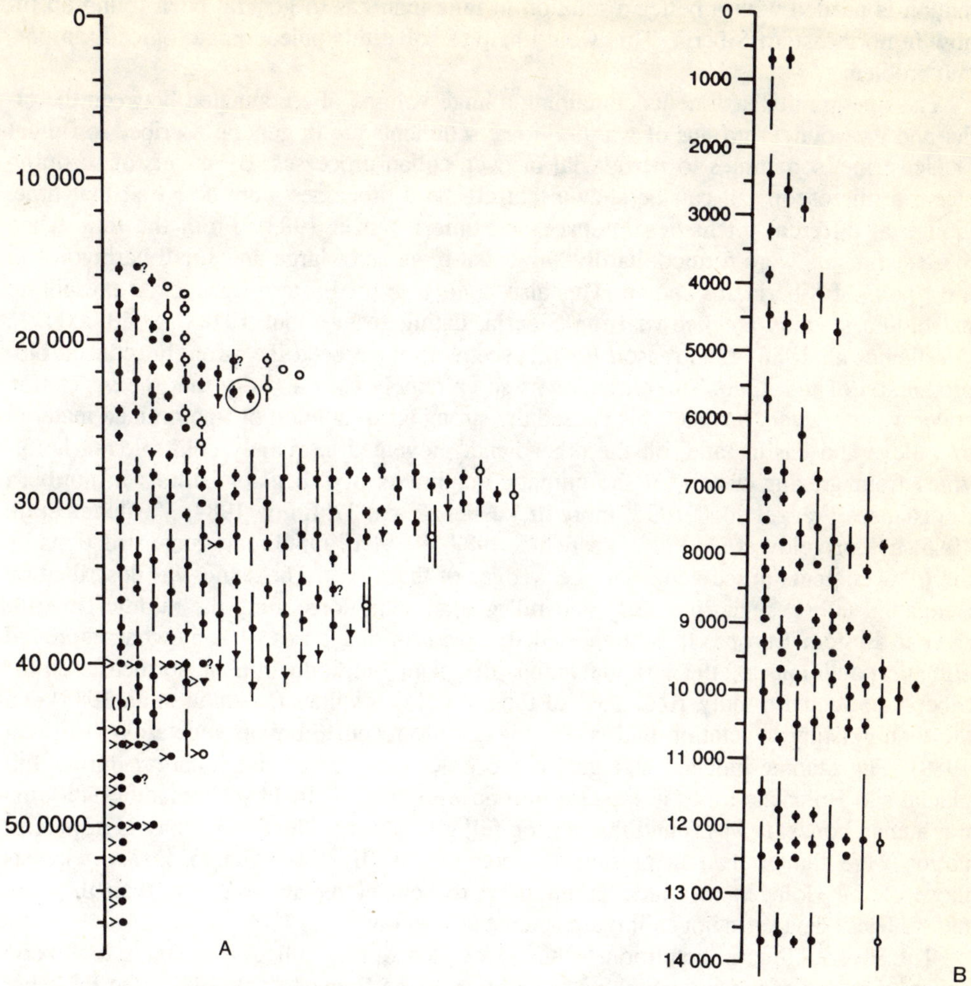


**Fig. 1.** [14]C-ages (*vertical axis*) of North Eurasian Upper Pleistocene and Holocene organic matter, mainly from thermokarst features. **A** Older part. **B** Younger part. *Black dots* Siberian material; *open circles* East European material; *triangles* organic matter from the Kamchatka Peninsula and the Kuriles Isles, dating volcanic eruptions. *Big circle* various datations of the same sample. *Vertical bars* standard deviation.

rate observations made repeatedly in Alaska. They point to either the absence or a faint development of the Younger Dryas oscillation at the end of the last glaciation only, if the warm climate phase of about 13 500 to 12 000 or 11 700 (Fig. 1) does not reflect the Bølling-Allerød oscillation.

Like in Alaska, it is widely held that the strongest late glacial or Holocene warming had occurred in northern Siberia at the transition from late glacial to early Holocene times, with an increase in temperatures against present-day conditions by about 8 to 9 °C at ca. 9200 B.P. (Gubin and Kartashova 1989). Here, much more pollen-analytic information is needed with a better resolution in time than has in general been found up till now in northeastern Siberia. This would help to solve this paleoclimatologically important problem.

The fine grained sediments containing a huge volume of ice situated between the older and the younger groups of warm-climate sediments are in general ascribed to fluvial or lacustrine, sometimes to periglacial or even eolian processes. By means of scanning electron microscopy, it can be shown that all these processes were active at that time, though at different intensities in space and time (Frenzel 1983). From the time when these sediments were formed, hardly any fossil remains of large and small herbivores in most parts of Siberia are known. This also holds true for Eastern Europe. At present no paleolithic stations are known from Siberia, dating from about 19 000 to 18 000 B.P. (Vasilev et al. 1986). The reason for this seems to have been the strongly reduced bioproduction of that time. The climate was so extremely cold and dry that the vegetation cover was only very sparse. This caused the strong accumulation of fine grained material in valleys and basins, and, on the other hand, prevented extremely cold-tolerant herbivores from grazing there (for the climatic conditions of full glacial times in northern Eurasia, see Frenzel et al. 1992; more in Vasilchuk and Trofimov 1984; Vasilchuk et al. 1985; Arkhangelov et al. 1986; Vasilchuk 1988, 1989, 1990: climate reconstructions by the $\delta^{18}O$ contents in ice veins and ice wedges of that time). The same was described by Rengarten (1981) regarding early and full glacial conditions along the middle Dniester river in Eastern Europe. It is suggested that even during times of somewhat improved climatic conditions of the last glaciation, the plant bioproduction in northeasternmost Siberia amounted to only 10 to 25% of the present-day values (Lozhkin et al. 1988) and the then existing vegetation had no analogs in the recent Siberian vegetation (Lozhkin 1989). The climate-induced changes in the $\delta^{18}O$ contents of precipitation during full glacial and Holocene times in Eastern Europe were investigated by Nikolaev (1988), using human bones. It was found that during full glacial times the $\delta^{18}O$ contents of precipitation were lighter than in present-day precipitation by about 3.5 to 4.0‰, whereas those of the Holocene climatic optimum were heavier by about 2.5‰ than they are today. These data were not calibrated against temperature.

The severe climatic conditions which prevailed during full glacial times in Siberia caused a strong expansion of the permafrost area (see Frenzel et al. 1992). On the other hand, the formation of permafrost seems to date back to early Pleistocene times, between 2.4 and 1.8 million years (Laukhin et al. 1992) and the first mountain glaciers are held to have been formed in northeastern Siberia during the first part of the Tertiary (Laukhin 1986). This must have influenced the development of terrestrial biota tremendously. Zamolotchikova (1980) established the history of permafrost on the Kamchatka Peninsula since the Middle Pleistocene, yet regrettably, only via reconstructions of vegetation and climate, not by frost-geological facts. In view of the problem of former glacier ice incorporated into the perennially frozen ground of Northern Siberia, it is worth mentioning that Danilov (1989) points to the fact that under these severe conditions fresh water from springs can easily be transformed into thick masses of compact ice which later may be distorted and thrusted by periglacial processes, thus creating a false impression of former glacier ice. Regarding the extent, thickness, and Holocene history of the active layer and the permafrost area: Bobov (1977); Kachurin (1961); Nekrasov (1970); Tomirdiaro (1974).

## 2. Last Interglacial and the Beginning of the Last Glaciation

In view of the possibility that the future climate will become warmer, due to an increase in the contents of the greenhouse gases in the atmosphere, the ecological situation of the last interglacial attracts general interest, since it is well known that the climate during the phases of optimal conditions was much warmer than it is now. Several attempts were made to elucidate the last interglacial paleoecological setting, using various disciplines. Morozova (1981) mapped the distribution pattern of the most important types of interglacial paleosoils all over Europe. It could be shown that soils of the lessivé type had spread much farther south than can be observed today under natural conditions (more than 350 km in Central Russia, more than 75 to 100 km in the Oka-Don-plain, with no southward displacement farther to the west).

Yet the problem is complicated due to the fact that this interglacial was followed by strong interstadial phases, the paleoecological conditions of which can easily be erroneously taken as indicating interglacial ones, if the dating of the fossil samples and features was incorrect. Thus in the Central Russian hilly region the interglacial podzolic soils were soon followed by early last glacial tschernozem soils (Agadzhanyan and Glushankova 1990). Rich faunas of small mammals seem to indicate an interglacial mosaic of various vegetation types, from forest to steppe in the northern Ukraina and in the lower Danube area, dating from the last interglacial (Markova 1985; Markova and Mikhailescu 1990). On the other hand, sedimentological facts and the mollusk fauna of presumably last interglacial sediments on the Kertch Peninsula are held to indicate arid conditions during that time (Shekhotkin 1990; first finds of coccoliths in these Karangat sediments: Shumenko 1987). In this respect, observations by Shevyrev et al. (1987) on the geological history of the middle Don area are interesting, since they indicate that during the time of formation of the last interglacial Shkurlatov layers, initially prevailing oak-linden-elm forests were replaced during the optimum of the last interglacial by a steppe vegetation, which in its turn was later followed again by these broadleaved forests. Thus it may be that the strange distribution pattern of last interglacial forest, forest-steppe and steppe communities, evidenced by the fossil mollusk fauna of various sites in the Ukraine, may be explained not only by regional differences but also by ecological changes in time (Kunitsa 1969). Repeatedly, the fossil rodent faunas of the ensuing interstadial phases may falsify the interpretation of the interglacial phase proper (Markova 1975). Valueva et al. (1974) report on a very interesting last interglacial flora (Sosnovki, 75 km to the north of Kalinin) which is extremely rich in pollen and macrofossils. Climate during its optimal phase seems to have resembled that of present-day central Czeckoslovakia. At the time the vegetation in the foothill region of Transcarpathia was governed by oak-, oak-hornbeam- and beech forests. The herb layer in these forests was rich in various herb species and under the warm and moist climate strongly developed forest soils were formed (Adamenko et al. 1989). The ensuing interstadial phases brought the prevalence of coniferous forests, with much fir and *Pinus cembra*, or with oak, linden, ash, and elm, respectively. From this it is suggested that climate should have been warm and moist. This also seems to be indicated by the fossil soils (Adamenko et al. 1989), yet does not fit into the picture of the paleoecological situation on the Dniester river, Moustérian site Ketrosy, Chernovsk Oblast, where only faint fossil steppe soils and their paleobiological situation could be analyzed (archeology: Anisyutkin 1981; geological setting: Ivanova 1981; fossil soils: Gubin 1981; sedimentology and soil chemistry; Rengarten 1981; vegetation history: Bolikhovskaya 1981; Levkovskaya 1981; mammal fauna: David 1981). Regrettably, no absolute datings of this most interesting site are available.

Nevertheless, it must be suggested that the phases of soil formation at Ketrosy date from much younger phases of climate history than the interstadials investigated by Adamenko et al. (1989) in Transcarpathia. Evidently the climatic and paleoecological development during the earlier phases of the last glaciation need much more investigation in order to better understand this most interesting period.

Devyatova et al. (1981) report on a remarkably complete last interglacial pollen sequence, from the preceding late glacial to a strong interstadial warming after the interglacial in the Arkhangelsk region. There the strong poleward interglacial and interstadial migration of exacting tree species is worth mentioning. Laukhin et al. (1989) investigated the last interglacial diatom flora in the Vankarem lowlands of the Chukchi Peninsula (77 taxa), which points to formerly much warmer climates there than at present. On the other hand, Alekseev et al. (1986), when investigating the Pleistocene vegetation and climate history of Central Yakoutia, described a boreal forest, characterized most of all by spruce, pine *Pinus sibirica, Betula* Sect. *albae*, which is held to have occurred at the same time in the lower Aldan river region (Chuya outcrop) and which is said to date from the time of the paleomagnetic Blake event, which in reality dates from a much later time. So I feel that the correlation of the interglacial phase mentioned with the last interglacial may be wrong. Perhaps it is much older, dating from Holsteinian (Likhvin) time? This might by corroborated by the fact that at the time permafrost did not influence the soils there (the paper mentioned contains a wealth of most interesting reports on the Pleistocene vegetation history of Central Yakoutia).

## 3. Pleistocene Vegetation History of Siberia, the Far East, and Central Asia

In the Polar Urals and in the Pechora Lowlands, over deepened valleys and tectonic basins are filled with thick Pleistocene sediment sequences. They frequently also contain interesting pollenfloras. Amongst others, Simonov and Stepanov (1988) described from the lower Pechora valley a vegetation sequence, which led from open pine-birch forests (type redkolesye) to pine-spruce forests with some broadleafed thermophilus trees, to fir-spruce forests and finally back again to open pine-birch forests. This sequence is held to date from the Likhvin interglacial (equivalent to the Holsteinian interglacial of Central Europe), yet the correlation is based only on the characteristics of the vegetation history, far away from other well comparable sites. The same holds true for an analysis of the Pleistocene vegetation history in various parts of the Polar Urals (Rindzyunskaya et al. 1987, 1990). A sound correlation is, moreover, hampered by the fact that pollen diagrams were not published.

Northeasternmost Siberia is dealt with intensively in several papers. Lazarev and Tomskaya (1987) describe in great detail the Pleistocene vegetation history of Northern Yakoutia, combining pollenanalytical investigation with research in plant macrofossils, on a wealth of animal remains, on sedimentology, and on absolute datings. They cover period from Mid-Tertiary times to the end of the Pleistocene. In this paper, as well as in the following ones, the remarkably strong migratious of various ecosystems and even biomes are well demonstrated. As already stated, permafrost seems to have developed there at the beginning of the Pleistocene. Laukhin et al. (1988) report on the hitherto oldest ice-wedges, dating from about 2.5 million years ago. These authors stress that the oldest pollen assemblages of modern type date there from approximately 0.9 to 0.7 million years B.P. In northeastern Siberia, the older Pleistocene pollen floras are repeatedly strongly influenced by exotic taxa, like *Juglans, Tsuga, Pinus* Sect. *strobus*, etc. It is extremely difficult to discriminate there between autochthonous and allochthonous pollen floras of the older Pleistocene. Romanovskiy et al. (1970) regard these thermophilous taxa as having grown there at that time, whereas Kaplina and Lozhkin (1977), Lozhkin (1977), Alekseev et al. (1986), and Laukhin et al. (1988) discuss these pollen floras as probably reworked older material. From my personal experience of the masses of older clay bolders and gravels in the huge Central Yakoutian outcrops, I would follow the latter authors. It will be remembered that Grosswald (1983) suggested that a huge ice-dome covered the Arctic Basin during the last glaciation. It was most of all reconstructed by the climatic snow-line, which was held to have been situated at that time at the level of the present-day coastline. Thus observations by Zhezhel et al. (1975) within the northeast Siberian Cherskiy Mts. are welcome, indicating that even there within the intramontane basins proofs for glacigenous sediments are nearly completely lacking. This is also stressed by other authors. Thus within the mountainous regions of Eastern Siberia there might have existed even during full glacial times small refuge areas for the tree vegetation, as was suggested by Lozhkin (1989) concerning *Picea obovata, Alnus hirsuta*, etc. on the river Yama, Sea of Okhotsk. On the other hand, the intramontane basins of northeastern Siberia were filled during full glacial times with thick fine grained clastic sediments, as was already stated. Svitoch (1975) suggests that these sediments were accumulated by a damming-up of rivers, caused by marine transgressions. Yet during the glaciations from which these sediments date, regressions of the Polar Sea should have taken place, not transgressions. It is thus much more likely that this sedimentation was caused by strong periglacial processes under an extreme continental climate.

The history of the Bering Sea, so important as a possible connection between Asia and the Americas, proves to have been relatively complicated: it existed between 4.5 and 3.8 million years, between 3.6 and 3.25 million years, and again between 2.2 and 0.3 million years. Thereafter its existence seems to have been governed by the glacial to interglacial cycles (Gladenkov et al. 1991). The end-Tertiary vegetation history for this region is dealt with by Baranova and Biske (1990). Svitoch et al. (1990) describe at some length the vegetation history of the southern part of the isle of Sakhalin during the European Cromer complex (dated by paleomagnetics), and Kuzmina and Eremeeva (1990) dealt with some aspects of the allegedly Middle Pleistocene vegetation history on the west coast of the Kamchatka Peninsula, together with geochemical results. Here the formerly strong expansion of the dark spruce-taiga is worth of mention, though it is unknown whether this forest type occurred there only once during the Middle Pleistocene or several times. It is thought that approximately simultaneously marine sediments were formed in the vicinity of Vladivostok, which at first contained the pollenflora of a steppe-like vegetation with much *Artemisia*, Chenopodiaceae, and *Ephedra*, and which later was replaced by a larch-*Alnaster-Betula nanae* vegetation. The first phase is interpreted as belonging to a warm-climate oak-birch forest-steppe, though no proofs for this suggestion are evident. The sediments are held to date from about 180 000 ± 45 000 years B.P. (obtained from a gravel horizon), yet it must be questioned whether this TL-date on marine gravel is correct (Kuzmina et al. 1984).

Kazachikhina (1972) and Boyarskaya et al. (1977) report on the Mid-Pleistocene to Holocene vegetation history of the middle and lower parts of the Amur valley and its surroundings. A wealth of information is found, yet regrettably the pollen diagrams are in general very poor. Interesting observations on Holocene phases of intensified permafrost or peat formation, respectively, on the northern coasts of the Sea of Okhotsk are reported by Lozhkin et al. (1991), together with a discussion of the main trends in vegetation history of that time. It is striking to learn how deeply the Holocene tundra vegetation differed even there from the already mentioned full glacial steppe-tundra vegetation. Of course, these Holocene changes in the area influenced by permafrost were connected with the poleward migration of warm water during Mid-Holocene times in the Japan Sea (Lutaenko 1991). It seems that the ensuing phase of a colder climate also influenced the position of the northern oak limit in the Amur region (Dobrynin et al. 1990).

The Pamir Mts. are held to have experienced the first glaciation at about 660 000 to 400 000 years B.P., later again between 320 000 to 150 000 years B.P., with the strongest glaciation at about 230 000 to 180 000 ± 50 000 years B.P. This is much earlier than was suggested previously (Vlasov et al. 1984). On the other hand, strong landslides and debris flows are able to dam up rivers there to high relative elevations, thus causing the impression of climate-controlled river terraces. They may even date from Mid-Holocene times only (Nesmeyanov 1983). Tadzhikistan is rich in paleolithic sites. Some of them are very old, as for instance Kuldara (approximately 800 000 years B.P.) or Karamaidan (about 200 000 years). Sometimes these sites contain a rich mammal fauna (Karamaidan: Dodonov et al. 1989), which may be compared with other faunas of this region, though they were not associated with remains of primitive man, e.g., Obigarn and Zilfi (Dodonov et al. 1991; more than 1.65 million years) or Esekartkan (Tleuberdina 1982; Middle Pliocene). Together with these faunas, pollenfloras were repeatedly encountered, which I feel are somewhat problematic. It is stressed that they are characterized by *Cedrus, Juglans, Tilia, Rhus, Carya, Pistacia, Platanus*, etc, and the exotic tree species are said to amount to nearly 45%. From this it is concluded that the mountains were much lower, thus enabling the summer monsoon to penetrate there easily. Yet when the data are published in some detail, it can easily be seen that the percentages of tree pollen are in general only very low, so that the taxa mentioned may point much more to long-distant transport than to plant geographical conditions strongly differing from those of today (Pakhomov 1983; Pakhomov and Baiguzina 1986; Ranov et al. 1987; additional references see in Frenzel 1991a). The long sequence of fossil soils in the thick loess layers of southern Tadzhikistan is relatively well dated now, with ten soil complexes dating from 12 000 to 13 000 B.P. up to 735 000 to 750 000 B.P. (Dodonov 1986). They are interpreted as indicating interstadial or interglacial conditions. This is corroborated by their ages and sometimes by their faunal assemblages. Yet Pakhomov (1983) suggests, according to his pollenanalytical work, that the loesses date from interglacial or interstadial times, whereas the fossil soils are said to have been formed during the glacial periods. The main reason for this is that the author interprets the steppe-pollenflora as indicating warm-arid conditions, yet convincing temperatures indicating plants are lacking. In the

Pamir-Alai Mts. climate between 7500 and 5500 B.P. seems to have been moister than before or afterwards. This is evidenced by more arboreal pollen at that time, most of all of *Betula* and *Salix*. A much moister period is held to have occurred at about 18 000 to 23 000 B.P., yet this age could only be estimated, and it must be stressed that most of the pollen spectra from lower elevations were formed in a herb-Chenopodiaceae-steppe vegetation only (Nikonov et al. 1988).

According to Ivanova and Glukhovskaya (1987), the southeastern part of the Mongolian Altai experienced during the Middle Pleistocene a remarkably strong glaciation, the end moraines of which are situated at about 1500 to 2000 m a.s.l. It is suggested that the glaciers moved into a vegetation rich in bushes of birch and *Alnaster*, with possibly some larch. Later, yet still during this glaciation, a desert steppe is held to have spread, the flora of which impoverished strongly. It is difficult to check the reliability of these data since the pollen floras mentioned were obtained from the till itself. Syrnev et al. (1986) dealt with the Middle Pleistocene vegetation history of eastern Mongolia (at about 260 000 to 180 000 B.P.). It is suggested that the vegetation changed from a forest steppe with some *Corylus*, *Quercus*, *Tilia*, *Ulmus*, and *Acer* to birch forests. Yet it is possible that this development is much more governed by a sediment change than by a change in the prevailing vegetation (sand against clay).

Dodonov (1986) published a pollen diagram of the Siwalik series, Siwalik Mts., dating from the beginning of the Matuyama (Early Pleistocene) period. It is governed by arboreal pollen, with prevailing *Picea* and *Betula* and some demanding tree species. The exact age of the monotonous pollen sequence is unknown. Regarding the interglacial and interstadial soil formation since approximately 350 000 B.P. in the Kashmir Valley: Bronger et al. (1987).

## 4. Problems of the European Middle Pleistocene Vegetation History

Ten years ago, the problem of climate and vegetation history in the time between the Dneprovsk and the Moskva ice advances to Middle Russia were already dealt with intensively (Frenzel 1982). Is was shown that the published geological and pollen evidence was not strong enough to prove convincingly that an interglacial period had existed between both these ice advances, though this hypothetical interglacial was at that time held to have consisted of three warm climate phases and two phases of a deteriorating climate. This complex interglacial was called Odintsovo interglacial, which should have been older than the Mikulino-(Eemian) interglacial, but younger than the Likhvin (Holsteinian) interglacial. Meanwhile, more observations have been published, which exhibit still more complications (Table 1). Motuz (1970) investigated the mollusk fauna in sediments of this age in the Tambov area; Adamenko et al. (1989) studied the geological setting and the paleobotanical and paleopedological situation of the Paleolithic site Korolevo, rayon Vinogradovo, Russian Transkarpathia; Agadzhanyan and Glushankova (1990) focused on the geology and paleopedology of the Central-Russian hilly area; Pisareva and Chebotareva (1986) reported on the vegetation history in approximately the same region; Maudina et al. (1985) investigated geology and vegetation history of the classical Odintsovo site near Moscow, and Agadzhanyan et al. (1988) studied primarily

**Table 1.** The problem of Middle Pleistocene interglacials

| Motuz (1970)<br>Tambov Oblast | Agadzhanyan and Glushankova (1990)<br>Central Russia | | Adamenko et al. (1989)<br>Russian Transkarpathia |
|---|---|---|---|
| Mollusk fauna of the 60 000 to 70 000 years long Roslavl interglacial. The facies of the sediments changes repeatedly | Mikulino interglacial: chernozem, podozolic soil | | Mikulino interglacial: yellow brown spotted loam; oak, oak-hornbeam and oak-beech forests. Climate warmer and moister than today. Yellow-brown loam, thin humus layers, loess. TL: 150 000 ± 20 000 |
| | Moscow glaciation | | |
| | Brown soil, xerotic oak-hornbeam forests | | |
| | Steppe vegetation, loess accumulation | | |
| | Likhvin interglacial, chernozem, prairie soil<br>Forest steppe with loess formation<br>Likhvin interglacial, brown earth; broad-leaved and mixed deciduous-coniferous forests | Fourth soil | Upper part, yellow-brown loam; forests, mostly conifers; meadow steppe. Pseudo-gleyification<br>Lower part, presumably separate soil. Gleyification. TL: 220 000 ± 35 000 (upper strata). Lower strata: gleyification. Coniferous forests: pine stone pine, spruce, larch, fir. High precipitation |
| | Oka glaciation: periglacial steppe vegetation, steppe tundra | | Loess-like loam, continuously developing into lower part of the fifth soil. TL: 260 000 ± 40 000 |
| | Chernozem of the Muchkap warm phase | | |
| | | Sixth soil, Likhvin interglacial | Warm-climate period XII. Climate appreciable moister than in all the other warm-climate periods, except warm-period X. Forests; much *Pterocarya*, together with *Juglans*, and *Ulmus*.<br>Cold-climate period XI. Herb vegetation with some *Picea* and *Ulmus*. **Pterocarya** and *Abies* possibly redeposited.<br>Warm-climate period X. Comparable to XII. *Pterocarya* always present |
| | | | Bright yellow loam; steppe vegetation, some subarctic birches. Permafrost features. TL: 360 000 ± 50 000 |
| | | Seventh soil | Upper part: Forest steppe with birches, salty soil. Lower part with gleyification: warm-climate period VIII. Open vegetation common. Within the forests much elm, at the beginning much hornbeam, too. *Juglans*, *Zelkova*, *Pterocarya* rare |

| Pisareva and Chebotareva (1986) Central Russia | Maudina et al. (1985) Odintsovo | Agadzhanyan et al. (1988) Roslavl |
|---|---|---|
| Mikulino interglacial | | |
| Moscow glaciation { herbs-artemisia steppe / meadow steppe | Moscow glaciation. Till, Scandinavian erratics = Upper Till | Till |
| Galich climatic optimum | Galich climatic optimum: ombrogenous peat bogs. Dark coniferous forests | Roslavl climatic optimum |
| Colder climate, several hiatusses | | Cold climate period |
| Roslavl climatic optimum. Abundant spruce, oak, elm, pine | Cold-climate period, no indicators of periglacial plant communities | Interglacial: more than 70% broad leaved taxa, more than 30% *Carpinus*: Konakhorsk climatic optimum; sometimes held to be the Likhvin interglacial. |
| ?? | Roslavl climatic optimum. Mixed oak forests without *Tilia tomentosa* and *Zelkova*. | Cooler climate. Loam with rodents of the faunistic Tiraspol complex: *Pitymys*, *Mimomys* |
| Distorted older warm-climate material | Cold-climatic period | |
| Till | Glazovo climatic optimum. Mixed oak forest, very rich in oak, elm, *Zelkova*, *Celtis*, *Tilia platyphyllos*, *T. tomentosa* | |
| Hiatus | | Glazovo climatic optimum |
| Till | Middle Till, erratics from the Timan Mts. Rebedded material from a warm-climate period. | Till |
| | Lower, Till, Scandinavian erratics. | |
| Old Pleistocene or Pliocene sediments | | |
| | Akulovo interglacial, reversed magnetism. High amount of local or even regional exotics, most of all North American-European and North American-Mediterranean taxa. *Azolla interglacialica* and *Salvinia natans* abundant | |
| | Cold climate | |
| | Oldest Till: Likov glacination; normal magnetism | |

the small rodent fauna in the Roslavl region. It will be seen from Table 1 that even today the opinions on the vegetation and climate history of the period between the classical Likhvin and the Mikulino interglacials strongly differ from each other, which seems to be mainly caused by several hiatuses, at least in the Roslavl and Odintsovo sites. Moreover, if the rodent fauna or some of its parts investigated by Agadzhanyan et al. (1988) was not redeposited, even higher ages for the interglacial sequence mentioned are possible, since the faunistic Tiraspol complex, which was found on the top of the Glazovo climatic optimum (i.e., oldest warm-climate phase of this tripartite interglacial) dates from the time before the Oka glaciation, i.e., before the formation of the till layers of the former glacial Don lobe, which is definitely older than the Dneprovsk glaciation (Gerasimov and Velichko 1980; Valueva et al. 1983; Agadzhanyan and Glushankova 1988). Faustov et al. (1974) and Sudakova (1975) try to elucidate these problems by reporting on thermoluminescence datings of several steps in the vegetation history of those Middle Pleistocene times. Here ages of 186 000 ± 21 000 years are mentioned for the loess which was accumulated before the beginning of the Mikulino interglacial, of about 280 300 ± 30 600 for the Dneprovsk moraine, of 336 000 ± 41 000 for the loess and a fossil soil on which the Dneprovsk moraine rests, and of 459 000 ± 56 000 or 453 000 ± 52 000 B.P., respectively, for the Likhvin interglacial (Faustov et al. 1974) or of 459 000 ± 56 000 to 400 600 ± 45 000 B.P. for the same interglacial sediments (Sudakova 1975). Yet it might be doubted that the TL-data obtained then approach reality. Moreover, the complexity of the Middle Pleistocene vegetation history in Central Russia is also evident by two other papers. On the middle Oka river, near the town of Tarusa, a 40-m-high exposure is found, which contains below two loess layers and a fossil soil a tripartite till, the uppermost layer of which, according to its sedimentpetrography, belongs to the Moscow glaciation. The three till layers are underlain by interglacial lacustrine sediments, which rest on a still older till. The interglacial sediments are characterized from bottom to top by (1) 70 to 73% of broadleaved trees, with 40 to 45% *Carpinus* and 9 to 10% *Fagus*. Elm, hazel, oak, and chestnut are much rarer. (2) The percentage of the broadleaved trees amounts to 80%, with 29% *Carpinus* and up to 40% *Fagus*. Only some *Juglans*, *Pterocarya*, *Tilia*, and *Corylus* are found. (3) Beech and hornbeam decline (in total 1 to 6% only). They are replaced by the coniferous trees: *Abies* 19%, *Picea* 16%, pine up to 50%. (4) The broadleaved trees are represented by 6% only, whereas pine (62%), *Abies*, *Picea* increase. It is thought (Sudakova et al. 1982) that this pollen sequence dates from the Likhvin interglacial, yet the high amounts of *Fagus* are surprising, so it might be questioned whether they could belong to an older interglacial. From approximatley the same region, Valueva et al. (1983) report on another interesting pollen sequence. Here at Karamyshevo on the river Oka below an end morain, thick loam and sand layers were encountered, which in their lower parts contain abundantly *Azolla interglacialica*, *Selaginella* cf. *helvetica*, *Potamogeton manshuriensis*, *Najas* cf. *macrosperma*, etc., together with the pollen flora of *Pinus silvestris*, *P. cembrae*, mixed broadleaved forests during the climatic optimum. This pollenflora does not resemble the typical Likhvin flora. It is held that it dates from some warm-climate phases of the Cromerian complex. The sequence of loesses, fossil soils, and of geochemistry of these sediments and soils in the northern foothill region on the Kaukasus Mts. are described by Galai (1983, 1984, 1986). Absolute or paleontological datings are lacking in these long profiles. On more regional problems in the Pleistocene vegetation history in the former USSR: Boyarskaya

(1976). It is held that there was a metachronous development of the vegetation in this vast area during the Pleistocene.

## 5. Correlation of Old and Middle Pleistocene Sediments

The various sediment traps in the southern part of Eastern Europe, caused by differential tectonic movements, necessitated different stratigraphical terminologies in various regions. Thus it becomes important to exactly correlate these local or regional sequences. According to paleomagnetic measurements the famous Kuyalnik sediments of the Black Sea region date from about 3.4 to 1.8 million years B.P. (Pevzner 1989). The approximately contemporaneous Akchagyl transgression of the Caspian Sea was divided into two parts. The older one began at about 3.32 million years B.P., the younger one lasted from 2.42 to about 1.61 or 1.79 million years B.P. (Trubikhin 1977). Timoshina (1980) reports on the biocoenoses existing in the surroundings of the Caspian Sea. The datation is corroborated by Vangengejm et al. (1990) as middle Akchagylian: younger than 2.5 yet older than 1.97 million years. This roughly coincides with the faunistic Moldavian complex (3.3 to 2.7 million years) and the early Khaprovsk complex (2.7 to 1.4 million years: Alekseeva 1977). The same authors suggest that the faunistic Taman complex dated from about 1.0 to 0.7 million years. It was preceded by the faunistic later Khaprovsk complex (1.4 to 1.0 million years and it was thought that the ensuing Tiraspol complex just mentioned lasted to about 0.4 million years. The older part of this faunistic sequence is evidently the eqivalent of the classical Villafranchian (3.15 to 2.48 million years; Aleksandrova 1989), whereas according to Velichko et al. (1983), the Brunhes/Matuyama boundary at approximately 700 000 years B.P. is situated between the early and the middle parts of the faunistic Tiraspol complex. It is held that the Karangat transgression of the Black Sea, which is in general looked upon as being the equivalent of the last interglacial, in reality had lasted from approximately 400 000 or 420 000 to about 100 000 years. The preceding Uzunlar transgression of the Black Sea is said to date from about 600 000 to 470 000 years B.P. (Zubakov et al. 1982), yet it might be questioned whether the stratigraphical definition of the Karangat sediments dated was fine enough. Chepalyga et al. (1986) contribute to a better paleontological definition of the preceding Uzunlar transgression and its correlative terrestrial sediments. For paleomagnetic datings of Upper Pleistocene fossil soils in Eastern Europe: Velichko and Svetlitskaya (1988).

## 6. Paleoecology of Southeasternmost Europe and Middle Asia

Figure 2 summarizes the age data on phases of a moister climate or of locally or regionally moister conditions on the northern rim of the Black Sea, in the Lake Aral depression, and in the Pamir-Alai. It is based on: Dobrodeev and Parunin (1973); Pshenin et al. (1984); Vasilev et al. (1986); Nikonov et al. (1988); Stanko and Svezhentsev (1988).

Comparing Fig. 2 with Fig. 1, it will be seen that in both cases the periods of relatively moister conditions coincided quite well. At any rate, even in Middle Asia the last

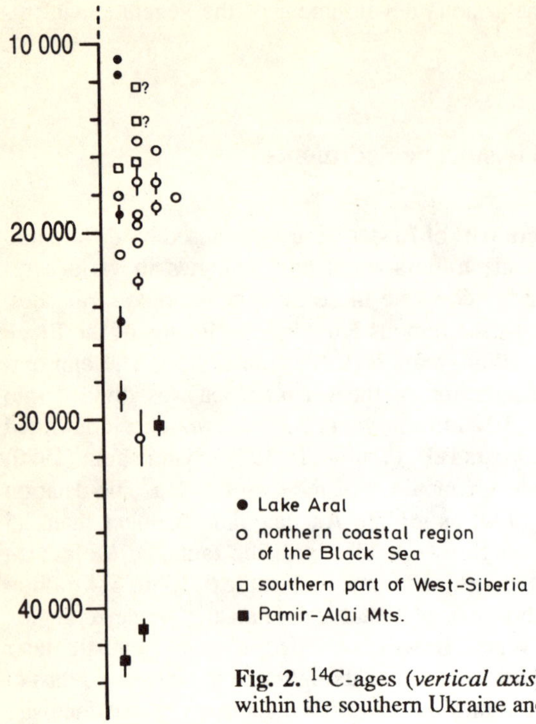


**Fig. 2.** $^{14}$C-ages (*vertical axis*) of sediments indicating moister conditions within the southern Ukraine and Middle Asia during the Upper Pleistocene

glacial phases of severest cold in the Northern Hemisphere was the time of severest drought in the Middle Asian desert belt. A somewhat different picture is obtained from recent U/Th datings on sediments of the Caspian region. Here it could be demonstrated that the Early Khazar or Gyurgyan transgression is older than approximately 300 000 years. The Late Khazar transgression dates from about 78 400 ± 2600 to 100 000 ± 3500 years B.P. (Arslanov et al. 1988). It is suggested that this transgression was the equivalent of the Mikulino-Eemian interglacial, the younger ages of the transgression mentioned being presumably caused by a later incorporation of uranium into the shells of the mollusks dated. On the other hand, I feel that these younger ages in reality might point to the relatively moist climate of the first steps in the development of the last glaciation. According to Arslanov et al. (1988), the Early Khvalyn transgression took place between about 24 000 ± 700 and approximately 16 510 ± 710 or 14 400 ± 400 years B.P. Thus it would date from the time during which other parts of Middle Asia experienced a very dry climate. This will be explained by the influence the river Volga exerted on the Caspian Lake. The Late Khvalyn transgression was repeatedly dated to about 16 500 to 12 900 years B.P., so both these Khvalynian transgressions seem to have followed each other very rapidly. The Holocene New Caspian transgression reached its maximum levels at about 8000 B.P. and between 2000 and 1500 B.P. The reasons for some of these oscillations of the Caspian lake level are discussed very controversially. Golitsyn et al. (1990), by analysis of modern meteorological observations, detect strong influence of a decrease in the wind speed on an increase of the water level via a reduction of the evaporation rate. Shilo (1989) and Shilo and Krivoshey (1991) find evidence for neotectonic movements, having squeezed deep-lying groundwater into the basin, whereas Vdovykin (1990) considers a combination of neotectonics and solar activity. Measuring the $\delta$ D and $\delta$ $^{18}$O contents in rivers and in meltwater of glaciers, Voytov et al. (1986) present good evidence for a strong contribution of deep groundwater storeys on the influx to the Caspian Sea. Yet, of course, the changes in the lake level of the Caspian Sea occurred much more slowly and were more retarded than those of the endorheic lakes in the basins of lakes Aral and Balkhash. Measuring the $^{234}$U/$^{238}$U ratios in bottom sediments and water of Lake Balkhash, it is concluded that this lake was formed for the first time at about 37 000 ± 7000 years B.P. Various aspects of the water budget and the chemistry of waters and sediments of this lake were analyzed by Turovskiy and Sheko (1974), Turovskiy et al. (1973), and Verzilin and Utsal (1990). From these data it is evident that results obtained in one part of the lake cannot be directly extrapolated to another, since strong hydrogeological differences exist between various parts of the lake, strongly influencing the limnology. According to Khrustalev and Chernousov (1983), the Holocene history of Lake Balkhash was characterized by several trans- and regressions. The oldest transgression had occurred before 10 300 ± 200 years B.P. The next strong transgression took place between 8300 ± 100 B.P. and 5600 ± 200 B.P. The strong New Balkhash transgression occurred at about 4380 ± 150 to 3580 ± 120 B.P., when Lake Balkhash, the Sassyk Kul, and the Ala Kul had joined. The last transgression was observed after 2690 ± 120 B.P., culminating at about 1920 ± 100 B.P. These changes in the lake levels are explained by paleoclimatological conditions only. The same holds true for the Holocene history of Lake Lyavlyakan in the southern Kyzyl Kum (Mamedov and Trofimov 1986) and for the history of the Usboi (Trofimov 1986). On the other hand, it is well known that the human influence in these steppe and desert regions has been very strong since the beginning of neolithic times (Mamedov and Trofimov 1986; Dolukhanov 1988), and it may be questioned whether human impact can be found in the history of these lakes and the middle Asian rivers also. In this respect, much more research work is needed in order to differentiate between natural and man-made changes in the hydrology of this sensitive region. This would include an investigation of the changing sedimentation rate of the Ili river (Khrustalev and Chernousov 1983). For a widespread and thick layer of plant remains of unknown age in the central Lake Aral, see Rubanov (1982). It is thought that these organic sediments were formed in reed swamps and algal mats during a pronounced regression.

## 7. Human Influences on Natural Ecosystems and the History of Soils

Madeyska (1986) summarized the available data about timing and distribution pattern of Paleolithic man in Northern Africa and Eurasia, including the hunting activities. It becomes evident again that full glacial times of the last glaciation in northern Eurasia were in general characterized by the absence of or by faint traces of human activities only (for more details, see Frenzel et al. 1992). According to the rich mollusk fauna of the old Paleolithic sites Ketrosy (Moustérian, middle Dniester area), it seems quite reasonable to suggest even for the beginning of the last glaciation the establishment of a negative feedback system caused by human settlements: the intensive destruction of the natural vege-

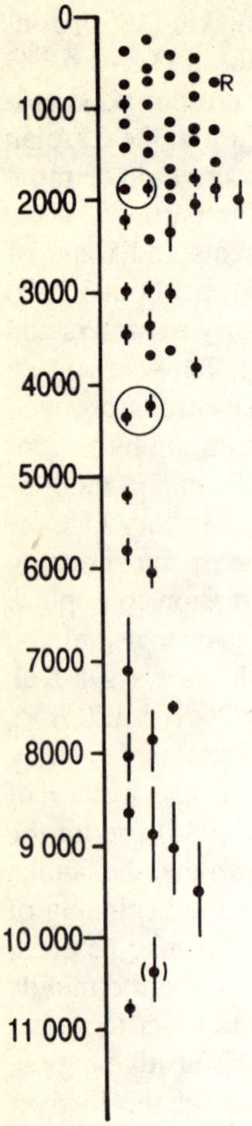


Fig. 3. $^{14}$C-ages (*vertical axis*) of traces indicating human activity during the Holocene in northeast Siberia. *Vertical bars* standard deviation; *big circles* $^{14}$C-ages of the same sample; *R* domesticated reindeer

tation by cutting bushes to obtain firewood, by trampling, etc. caused loess slopes to become instable, and man had to move to new sites for settlement (Motuz 1981). On the other hand, it is very often held in the Russian literature that human populations were much too sparse in those remote times to endanger the ecosystem (Blagoveshchenskaya 1987; upper Paleolithic times in the middle Volga area not more than 1000 persons; during Mesolithic times about 2000 to 3000 persons only). Thus it is interesting to summarize the hitherto known data on traces of human activities within a climatically extremely unfavorable region, i.e., in northeasternmost Siberia (Fig. 3). Figure 3 is based on the data in Shilo et al. (1976; 1977a, b). Here the oldest traces of domesticated reindeer date from 710 ± 40 B.P. (Shilo et al. 1976). Yet it is unknown how intensively man may have influenced vegetation at that time, even though he had been present there nearly continuously since about late glacial times.

Dolukhanov (1971, 1973, 1975, 1979) described regional aspects of neolithization in Europe and the Near East, together with the periodization of various East-European Holocene cultures (Dolukhanov 1971). The small number of factual data about cultivated plants during various phases of neolithic, bronze age, and iron age times astonishes in these papers (more information: Frenzel 1992). This causes Sejbutis (1987) to deal with the problem of the introduction of cultivating rye in Eastern Europe theoretically only. Of course there existed strong regional differences in the timing and intensity of early human impact on vegetation. It seems to have begun intensively already during Neolithic times in the middle Volga region, most of all during its second part, together with a strong expansion of weeds, such as *Urtica dioica*, Brassicaceae, *Galium aparine*, *Polygonum convolvulus*, *Taraxacum officinale*, *Pteridium aquilinum*, *Ranunculus repens*, *Chenopodium album*, *Atriplex* sp., and *Artemisia* ssp. They most of all point to cattle grazing within the forests. Here *Triticum spelta* was grown at about 5730 ± 60 B.P. Forests were opened at that time by grazing and steppe communities were able to spread into the former forested area, in which the clearing was done most of all by burning (charcoal layers! Blagoveshchenskaya 1987). On the other hand, stronger activities of man seem to have begun in some parts of Estonia at the beginning of bronze age times only (Dolukhanov 1987; Raukas and Róuk 1987). Sometimes the prehistoric activity of man is ascribed to various ethnic or lingiustic groups (Dolukhanov 1987). Yet I feel that these connections are impossible to give by the factual archeological or paleontological/paleobotanical material. Agriculture has dominated over animal husbandry in Estonia since about 600 B.P. and in between the 1st and the 5th century A.D. rye cultures are said to have dominated all other plant-cultivating practices in Lithuania (Sejbutis 1987). In the middle Volga region, natural forests were already intensively damaged during bronze age times. They had been replaced by secondary pine forests since about the iron ages (Blagoveshchenskaya 1987). For bronze age commercial connections between the Krasnoyarsk or the western Baikal area and the surroundings of Vladivostok, by which lead was transported from the west to the southern part of the Far East: Fefelov et al. (1989). Though the influence of man had been faintly felt in the upper Don area since approximately late Neolithic to early bronze age times, a strong impact on the different ecosystems can only be proven there by about the early Middle Ages (Aleksandrovskiy 1987; Khotinskiy 1987). Here a strong depopulation caused by the Mongolian raids initiated a reestablishment of natural soils (Aleksandrovskiy 1987) and of spontaneous vegetation types (Khothinskiy 1987, 1988). In the Central Russian Kursk area, impact of man has been felt since approximately early bronze age times, though interruptions occurred there repeatedly. At that time animal husbandry and agriculture were already practised, and quite a lot of plants were cultivated on plowed fields. This led during the pre-Roman iron ages to a strong soil erosion (Sycheva and Uzyanov 1987). This might also have influenced river activity, just as it did in Central and Western Europe. Vasilev (1988) investigated fossil peat bogs, buried in the riverine terraces of the river Ob, Tyumen Oblast. They date from about 7330 to ca. 2020 B.P., with a local interruption in peat growth by river aggradation between 3500 and 3000 B.P. Some of these fossil riverine bogs were formerly interpreted as having been buried under man-made fine grained sediments, due to forest clearances and agricultural activities. Vasilev neglects this possibility, because the strong human impact of today only causes high floods. Yet are these floods ineffective in accumulation of sand and loam? The river Oka, Central Russia, contains in its riverine terraces at least four fossil soils, amongst which the second fossil soil is the best developed (1800 to 800 B.P.). Other phases of soil formation are from 500 to 300 B.P., 3200 to 2500 B.P., 6000 to 4500 B.P. In general they contain remnants of human activities, whereas the fluvial sediments in between in general do not. It is held that the human influence on river activity has been most felt since 1000 or at the latest since 300 B.P. (Aleksandrovskiy et al. 1987). Khotinskiy (1987, 1988) interprets comparable soils along other river systems of Eastern Europe as indicating climate history only. In this respect much more research work should be done, taking the variable consequences of human activities into consideration. One of these consequences might have been the Late Pleistocene/Early Holocene extinction of some herbivorous species: Alekseeva (1990: further information: Frenzel 1989).

Mednikova et al. (1988) reported on changes in the geochemistry of soils in sites of prehistory settlements of the Kola Peninsula, due to fishing activities etc. Here the contents of P, Pb, Ni, Zn, Cr, Cd, Co, and Cu are markedly elevated in comparison to noninhabited nearby sites and the humus content is increased. The problem of changes in the humus content was investigated by Zharikov (1987) regarding dernopodsols in the northwestern part of the Russian Plain during about the last 300 years, where the humus content has also increased, and the pH has risen to 5.0 to 6.0, with a remarkable increase in the base status. These alterations touch the problem of natural versus man-made pedological changes during the Holocene. Of course, soil formation had already begun during the late glacial. This is felt in Byelorussia in the ratios of resistant to nonresistant minerals, or in the amounts of autigenous minerals, like carbonates, pyrite, and phosphates (Shimanovich 1987). Studying these processes, the problem of reliable $^{14}C$-datations becomes crucial. In Russia, the datation of soil profiles is done primarily on well-adsorbed humic acids. Yet even they can easily be dislocated by burrowing animals. Thus an exact pedological investigation of the soil to be dated is an inevitable prerequisite for analyses of this type (Chichagova et al. 1984). It is in general held that soils of alpine meadows are very young, due to reworking by mud flows, frost action, various erosional processes, etc. Yet lower parts of the humic layers of alpine meadow soils under *Avenastrum* repeatedly gave ages of about 3600 years B.P. This holds true most of all under moist conditions (Grishina et al. 1987). Model calculations show that humic substances of chernozem soils show increasing ages up to approximately 2000 years. Thereafter the ages should remain constant. In humic acids the increase in ages is calculated to occur up to about 4500 years. Then they should remain constant (Cherkinskiy and Brovkin 1989). Yet Arslanov et al. (1970) demonstrated by a comparison between $^{14}C$ ages of well-adsorbed humic acids in lowland and mountain steppe soils that, at least in the lowland, the ages of the humic horizons may increase up to more than 6000 years, and that the $^{14}C$ ages of the upper humic horizons are the same under both the conditions compared. Thus the C turnover seems to be the same in lowland and mountain steppe soils. The increasing ages of the humic acid with increasing depth are said to prove that in this soil type sediment accumulation is more or less uninterruptedly taking place. I wonder whether a downward movement of humic acids can be excluded within these steppe soils? The age problem of humic horizons of dernopodzols attracted some interest. Here two horizons rich in humic substances are repeatedly encountered. The lowermost may either be the $B_h$-horizon or an older A-horizon. The ages of the lower humus horizon strongly depend on the vegetation cover and on the moisture content within the soil profile: under fields the mean residence time of the $C_{org}$ is much less than under forests, and it is remarkably less under drier than under moister conditions, the differences amounting to more than 3000 $^{14}C$-years (Cherkinskiy et al. 1989). This was corroborated by

Tolchelnikov and Kostarev (1982), who pointed out that the mean residence time of $B_h$-horizons of recent moister soils in the northwestern part of the former RSFSR can amount to 4500 ± 130 years, whereas it is only some 1000 to 2000 years in better-drained soils. It is concluded that most of the humic substances of the $B_h$-horizon are relic, dating from other climatic conditions. They were analyzed for about the same region by Chernavskaya and Fogel (1989), using a principal component analysis on pollen diagrams. Yet to me the results are not too convincing. Aleksandrovskiy et al. (1990), during their investigations of the age problem of Central Russian dernopodsols, combined $^{14}C$-datings and pollen analyses. Concerning the age problem, the same results were obtained as were already described. Moreover, it could be clearly demonstrated that clefts in the soils act as transport ways for pollen and $^{14}C$, thus strongly falsifying the data. According to Dobrovolskiy et al. (1970), dernopodsols of the second terrace of the river Ob, western Siberia, have $B_h$-horizons, the C-content of which is much higher than that of the uppermost humus horizon. Moreover, the organic material of the upper humus horizon is typical for meadow podzols there, whereas that of the lower humus horizon strongly resembles that of grass chernozems. This tendency is corroborated by $^{14}C$ datations: the upper humus horizon gave ages of about 1230 ± 80 years B.P. whereas the lower had ages of between 5020 ± 130 and 7000 ± 160 years B.P. Thus it is concluded that the lower humus horizon dates from an Early Holocene steppe vegetation, whereas that of the upper was formed under present-day conditions. It can be shown indeed palynologically and by analyses of plant macrofossils that in the nearby Yenisseiy Kryazh a forest steppe vegetation with vast steppe areas had prevailed up to approximately 6100 years B.P., when it was replaced by mountain coniferous forests, which since then have not changed remarkably (Savina et al. 1988). Comparable changes had happened approximately simultaneously in the surroundings of the northern part of Lake Baikal at about 8000 B.P., when the open steppe or forest-steppe vegetation was replaced by the spreading of subalpine forest and bush communitites. This was interpreted by Bezrukova et al. (1991) as indicating the transition form strong meltwater influxes before 8000 to 8200 B.P., when nearly no diatoms existed in the lake, to a period without larger glaciers, when the turbidity of the water had definitely decreased, thus enabling the diatoms to develop and spread. Yet it may be that a superimposed climatic change had influenced the development of the lake and its surrounding terrestrial ecosystems just as it did in the Yenisseiy Kryazh and in the West Siberian lowlands. On ages and types of Upper Pleistocene fossil soils in the Russian Plain: Dobrodeev and Parunin (1973).

## References

**Adamenko OM,** Adamenko RS, Gladilin VN, Grodetskaya GD, Levkovskaya GM, Pashkevich GA, Pospelova GA, Soldatenko LV (1989) Byull Kom Po Izuch Chetvert Per 58:5–25. **Agadzhanyan AK,** Glushankova NI (1988) Byull Kom Po Izuch Chetvert Per 57:62–77. **Agadzhanyan AK,** Glushankova NI (1990) Byull Kom Po Izuch Chetvert Per 59:66–80. **Agadzhanyan AK,** Biryukov IP, Shik SM (1988) Dokl Akad Nauk SSSR 299:1191–1195. **Aleksandrova LP** (1989) Byull Kom Po Izuch Chetvert Per 58:97–104. **Aleksandrovskiy AL** (1987) Antropogennaya evolyuciya pochv Kulikova Polya. In: Khotinskiy NA, Sycheva SA (eds) Antropogennaya evolyuciya geosistem i ikh komponentov. Inst Geogr Akad Nauk SSSR, Moskva, pp 88–104. **Aleksandrovskiy AL,** Glasko MP, Folomeev BA (1987) Byull Kom Po Izuch Chetvert

Per 56:123–128. **Aleksandrovskiy AL,** Targulyan VO, Cherkinskiy AE, Chichagova OA (1990) Dokl Adad Nauk SSSR 310:454–457. **Alekseev MN,** Golubeva LV, Liskun IG, Sokolovskaya VT (1986) Byull Kom Po Izuch Chetvert Per 55:14–26. **Alekseeva LI** (1977) Teriofauna rannego antropogena Vostochnoy Evropy. Tr Geol Inst Akad Nauk SSSR 300. Nauka, Moskva, 214 pp. **Alekseeva LI** (1990) Teriofauna verkhnego pleistotsena Vostochnoy Evropy (krupnye mlekopitayushchie). Tr Geol Inst Akad Nauk SSSR 455. Nauka, Moskva, 109 pp. **Alekseeva LI,** Tikhomirov SV (1987) Byull Kom Po Izuch Chetvert Per 56:109–111. **Anisyutkin NK** (1981) Arkheologicheskoe izuchenie musterskoy stoyanki Ketrosy. In: Praslov ND (ed) Ketrosy. Musterskaya stoyanka na Srednem Dnestre. Nauka, Moskva, pp 7–53. **Arkhangelov AA,** Vaykmyae RA, Mikhalev DV, Punning YAMK, Solomatin VI (1986) Dokl Akad Nauk SSSR 290:415–417. **Arslanov KHA,** Gerasimov IP, Zubkov AI, Kozyreva MG, Rubilin EV, Chichagova OA (1970) Dokl Akad Nauk SSSR 192:1141–1144. **Arslanov KHA,** Lokshin NV, Mamedov AV, Aleskerov BD, Gerasimova SA, Tertychniy NI, Tertynaya TV, Chernov SB (1988) Byull Kom Po Izuch Chetvert Per 57:28–38. **Astakhov VI** (1976) Dokl Akad Nauk SSSR 231:1178–1181. **Astakhov VI,** Isaeva LL (1985) Dokl Akad Nauk SSSR 283:438–440.

**Balashova TA** (1976) Dokl Akad Nauk SSSR 231:426–429. **Baranova YUP,** Biske SF (1990) Izv Akad Nauk SSSR Ser Geol Vyp 12:57–67. **Bezrukova EV,** Bogdanov YUA, Vilyams DF, Granina LZ, Grachev MA, Igazntova NV, Karabanov EB, Kuptsov VM, Kurylev AV, Letunova PP, Likhoshvay EV, Chernyaeva GP, Shimaraeva MK, Yakushin AO (1991) Dokl Akad Nauk SSSR 312:1032–1037. **Blagoveshchenskaya NB** (1987) Antropogennye izmeneniya landshaftov tsentralnoy chasti Privolzhskoy vozvyshennosti v golotsene. In: Khotinskiy NA, Sycheva SA (eds) Antropogennaya evolyuciya geosistem i ikh komponentov. Inst Geogr Akad Nauk SSSR, Moskva, pp 143–159. **Blazhchishin AN,** Linkova TI (1977) Dokl Akad Nauk SSSR 236:696–699. **Bobov NG** (1977) Dokl Akad Nauk SSSR 232:1410–1413. **Bolikhovskaya NS** (1981) Rastitelnost i klimat srednego Pridnestrovya v pozdnem pleystotsene. Rezultaty palinologicheskogo analiza otlozheniy Kishlyanskogo Yara. In: Praslov ND (ed) Ketrosy. Musterskaya stoyanka na Srednem Dnestre. Nauka, Moskva, pp 103–124. **Boyarskaya TD** (1976) O metakhronnom razvitii rastitelnosti v pleistotsene na territorii SSSR. In: Agadzhanyan AK, Dobrodeev OP (eds) Problemy obshchei fizicheskoy geografii i paleogeografii. Mosk Univ, Moskva, pp 245–256. **Boyarskaya TD,** Sokhina EN, Chernyuk AV (1977) Vopr Geogr Dalnego Vostoka 18:41–48. **Bronger A,** Pant RK, Singhvi AK (1987) Quat Res 27:167–181.

**Chepalyga AL,** Markova AK, Mikhailescu KD (1986) Dokl Akad Nauk SSSR 290:433–437. **Cherkinskiy AE,** Brovkin VA (1989) Dokl Akad Nauk SSSR 307:1227–1231. **Cherkinskiy AE,** Zaidelman FR, Danilova GA (1989) Dokl Akad Nauk SSSR 305:965–969. **Chernavskaya MM,** Fogel GA (1989) Dokl Akad Nauk SSSR 307:1474–1477. **Chichagova OA,** Cherkinskiy AE, Tolchelnikov YUS (1984) Dokl Akad Nauk SSSR 275:1176–1178.

**Danilov ID** (1989) Dokl Akad Nauk SSSR 306:1201–1203. **Danilov ID,** Parunin OB (1982) Dokl Akad Nauk SSSR 262:402–404. **David AI** (1981) Ostatki mlekopitayushchikh iz raskopok paleoliticheskoi stoyanki Ketrosy. In: Praslov ND (ed) Ketrosy. Musterskaya stoyanka na Srednem Dnestre. Nauka, Moskva, pp 135–142. **De Beaulieu J-L,** Monjuvent G, Nicaud G (1991) Chronology of the Wurmian glaciation in the French Alps. A survey and new hypotheses. In: Frenzel B (ed) Paläoklimaforschung 1. Fischer, Stuttgart, pp 435–448. **Devyatova EI,** Raukas AV, Rayamyae RA, Khyutt GI (1981) Byull Kom Po Izuch Chetvert Per 51:38–50. **Dobrodeev OP,** Parunin OB (1973) Dokl Akad Nauk SSSR 209:426–427. **Dobrovolskiy GV,** Afanasyeva TV, Vasilenko VI, Devirts AL, Markova NG (1970) Dokl Akad Nauk SSSR 192:651–653. **Dobrynin AP,** Nedoluzhko VA, Saboldashev SA, Starodubtsev VN (1990) Bot Zh 75:75–78. **Dodonov AE** (1986) Antropogen yuzhnogo Tadzhikistana. Trudy Geol Inst Akad Nauk SSSR 409, Nauka, Moskva, 165 pp. **Dodonov AE,** Ranov VA, Sharapov SHSH (1989) Byull Kom Po Izuch Chetvert Per 58:26–36. **Dodonov AE,** Zhegallo VI, Pankov AV, Sotnikova MB (1991) Izv Akad Nauk SSSR Ser Geol Vyp 5:12–21. **Dolukhanov PM** (1971) EAZ Ethnogr Archäol Z 12:161–170. **Dolukhanov PM** (1973) The neolithization of Europe: a chronological and ecological approach. In: Renfrew C (ed) The explanation of culture change models in prehistory. Duckworth, London, pp 329–343. **Dolukhanov PM** (1975) Die ökologischen und ökonomischen Grundlagen der Neolithisierung Mittel- und Osteuropas. In: Moderne Probleme der Archäologie. Berlin, pp 73–82. **Dolukhanov PM** (1979) Ecology and economy in Neolithic eastern Europe. Duckworth, London, 212 pp. **Dolukhanov PM** (1987) Istoriya osvoenia landshaftov severo-zapada Russkoy Ravniny v golotsene. In: Khotinskiy NA, Sycheva SA (eds) Antropogennaya evolyuciya geosistem i ikh komponentov. Inst Geogr Akad Nauk SSSR, Moskva, pp 19–29. **Dolukhanov PM** (1988) Paléoécologie de l'Asie Centrale aux âges de la pierre et du bronze. In: L'Asia Centrale et ses rapports avec les civilisations orientales des origines à l'âge du fer. Colloque Franco-Soviétique sur l'archéologie de L'Asie Centrale. Boccard, Paris, pp 215–221. **Dricot E,** Pétillon M, Seret G (1991) When and why did glaciers grow or melt in the Vosges Mountains (France)? In: Frenzel B (ed) Paläoklimaforschung 1. Fischer, Stuttgart, pp 363–376. **Dunaev NN,** Levchenko OV, Merklin LR, Pavlidis YUA (1989) Dokl Akad Nauk SSSR 306:1437–1441.

**Epshtein OG,** Lavrushin YUA, Valpeter AL, Yukhnevich KB, Stephanov EF (1983) Doklady Akad Nauk SSSR 272:180–183.

**Faustov SS,** Ilichev VA, Bolshakov VA (1974) Dokl Akad Nauk SSSR 214:1160–1162. **Fefelov NN,** Konkova LV, Zarudneva NV (1989) Dokl Akad Nauk SSSR 308:172–175. **Frenzel B** (1982) In: Ellenberg H, Esser K, Kubitzki K, Schnepf E, Ziegler H (eds) Progress in Botany 44. Springer, Berlin Heidelberg New York, pp 406–417. **Frenzel B** (1983) Eiszeitalter und Gegenwart 33:163–168. **Frenzel B** (1984) In: Esser K, Kubitzki K, Runge M, Schnepf E, Ziegler H (eds) Progress in Botany 46. Springer, Berlin Heidelberg New York, pp 344–359. **Frenzel B** (1987) In: Behnke H-D, Esser K, Kubitzki K, Runge M, Ziegler H (eds) Progress in Botany 49. Springer, Berlin Heidelberg New York, pp 364–380. **Frenzel B** (1989) In: Behnke H-D, Esser K, Kubitzki K, Runge M, Ziegler H (eds) Progress in Botany 50. Springer, Berlin Heidelberg New York, pp 327–340. **Frenzel B** (1991a) In: Behnke H-D, Esser K, Kubitzki K, Runge M, Ziegler H (eds) Progress in Botany 52. Springer, Berlin Heidelberg New York, pp 359–380. **Frenzel B** (1991b) Über einen frühen letzteiszeitlichen Vorstoß des Rheingletschers in das deutsche Alpenvorland. In: Frenzel B (ed) Paläoklimaforschung 1. Fischer, Stuttgart, pp 377–400. **Frenzel B** (1992) In: Behnke H-D, Esser K, Kubitzki K, Runge M, Ziegler H (eds) Progress in Botany 53. Springer, Berlin Heidelberg New York, pp 361–400. **Frenzel B,** Pécsi M, Velichko AA (1992) Atlas of paleoclimates and paleoenvironments of the Northern Hemisphere; Late Pleistocene-Holocene. Geographical Research Institute, Hungarian Academy of Sciences, Budapest. Fischer, Stuttgart, 153 pp. **Furrer G** (1991) Zum Alter vonWallmoränen um das Vorfeld des Erikbreen und auf der westlichen Lernerinsel – ein Beitrag zur Gletschergeschichte des Liefdefjordes (erste Ergebnisse der geowissenschaftlichen Spitzbergenexpedition 1990 des Teilprojektes Gletscher und Moränen als Klima- und Datierungsphänomene). Sonderveröff Geol Inst Univ Köln 82, Festschrift Karl Brunnacker. Geol Inst Univ Köln, pp 113–121.

**Galai BF** (1983) Dokl Akad Nauk SSSR 272:146–149. **Galai BF** (1984) Dokl Akad Nauk SSSR 279:160–163. **Galai BF** (1986) Dokl Akad Nauk SSSR 288:437–440. **Gataullin VN,** Polyak LV (1990) Dokl Akad Nauk SSSR 314:1463–1467. **Gerasimov IP,** Velichko AA (1980) (eds) Vozrast i rasprostranenie maksimalnogo oledeneniya Vostochnoy Evropy. Nauka, Moskva, 212 pp. **Gladenkov YUB,** Barinov KB, Basilyan AE (1991) Izv Akad Nauk SSSR Ser Geol Vyp 9:79–86. **Golitsyn GS,** Dzyuba AV, Osipov AG, Panin GN (1990) Dokl Akad Nauk SSSR 313:1224–1227. **Goncharov SV** (1986) Dokl Akad Nauk SSSR 290:1436–1439. **Gornostai BA** (1990) Byull Kom Po Izuch Chetvert Per 59:152–155. **Grekhova LV** (1990) Byull Kom Po Izuch Chetvert Per 59:111–116. **Grishina LA,** Cherkinskiy AE, Zhakova OE, Onipchenko VG (1987) Dokl Akad Nauk SSSR 296:218–220. **Grosswald MG** (1983) Ice sheets of the continental shelves. Nauka, Moskva, 216 pp. **Grosswald MG** (1988) Dokl Akad Nauk SSSR 302:654–659. **Gubin SV** (1981) Pogrebennye pleistotsenovye pochvy raiona paleoliticheskoi stoyanki Ketrosy. In: Praslov ND (ed) Ketrosy, Musterskay stoyanka na Srednem Dnestre. Nauka, Moskva, pp 93–103. **Gubin SV,** Kartashova GG (1989) Byull Kom Po Izuch Chetvert Per 58:79–85.

**Haeberli W** (1991) Zur Glaziologie der letzteiszeitlichen Alpenvergletscherung. In: Frenzel B (ed) Paläoklimaforschung 1. Gustav Fischer, Stuttgart, pp 409–420.

**Ivanova IK** (1981) Geologiya i geomorfologiya okrestnostei stoyanki Ketrosy. In: Praslov ND (ed) Ketrosy, Musterskaya stoyanka na Srednem Dnestre. Nauka, Moskva, pp 59–80. **Ivanova IK,** Bolikhovskaya NS, Rengarten NV (1981) Geologicheskiy vozrast i prirodnaya obstanovka musterskoi stoyanki Ketrosy. In: Praslov ND (ed) Ketrosy, Musterskaya stoyanka na Srednem Dnestre. Nauka, Moskva, pp 152–161. **Ivanova IK,** Lyubin VP, Praslov ND (1989) Byull Kom Po Izuch Chetvert Per 58:49–57. **Ivanova NG,** Glukhovskaya NB (1987) Byull Kom Po Izuch Chetvert Per 56:86–94.

**Kachurin SP** (1961) Termokarst na territorii SSSR. Akad Nauk SSSR, Moskva, 291 pp. **Kagan BA,** Ryabchenkov VA, Safray AS (1987) Dokl Akad Nauk SSSR 296:1469–1472. **Kaplina TN,** Lozhkin AV (1977) Dokl Akad Nauk SSSR 233:1162–1164. **Kaplyanskaya FA,** Tarnogradskiy VD (1976) Dokl Akad Nauk SSSR 231:1185–1187. **Katasonov EM,** Ivanov MS, Pudov GG, Siegert CH (1979) Stroenie i absolyutnaya geokhronologiya alasnykh otlozhenij tsentralnoi Yakutii. Akad Nauk SSSR, Sibirsk Otdel, Novosibirsk, 95 pp. **Kazachikhina LL** (1972) Vopr Geogr Dalnego Vostoka 10:259–268. **Khotinskiy NA** (1987) Antropogennaya transformaciya landshaftov Kulikova Polya i perspektivy ikh okhrany i vosstanovleniya. In: Khotinskiy NA, Sycheva SA (eds) Antropogennaya evolyuciya geosistem i ikh komponentov. Inst Geogr Akad Nauk SSSR, Moskva, pp 69–87. **Khotinskiy NA** (1988) Kovyl-trava na Kulikovom Pole. Mysl, Moskva, 175 pp. **Khrustalev YUP,** Chernousov SYA (1983) Dokl Akad Nauk SSSR 271:1468–1471. **Khusid TA** (1989) Byull Kom Po Izuch Chetvert Per 58:105–116. **Krupnik VM,** Kalnichenko SS, Kostyaev AG (1990) Byull Kom Po Izuch Chetvert Per 59:48–57. **Kunitsa NA** (1969) Dokl Akad Nauk SSSR 188:1352–1354. **Kuzmina NN,** Eremeeva GP (1990) Dokl Akad Nauk SSSR 310:1425–1428. **Kuzmina NN,** Polyakova EI, Shumova GM, Voskresenskaya TN, Nedesheva GN (1984) Dokl Akad Nauk SSSR 282:679–683.

**Laukhin SA** (1986) Dokl Akad Nauk SSSR 291:1445–1449. **Laukhin SA,** Rybakova NO (1982) Dokl Akad Nauk SSSR 267:895–900. **Laukhin SA,** Panychev VA, Rybakova NO, Orlova LA, Kolpakov VV, Laukhina NI (1986) Byull Kom Po Izuch Chetvert Per 55:117–122. **Laukhin SA,** Rybakova NO, Velichko SV, Laukhina NI (1988) Dokl Akad Nauk SSSR 303:696–700. **Laukhin SA,** Pushkar VS, Velichko SV (1989) Dokl Akad Nauk SSSR 309:947–950. **Laukhin SA,** Grinenko OV, Fradkina AF (1992) Izv Akad Nauk SSSR Ser Geol Vyp 1:131–139. **Lavrov AS** (1981) Byull Kom Po Izuch Chetvert Per 51:51–60. **Lavrushin YUA,** Chistyakova IA (1988) Dokl Akad Nauk SSSR 303:173–177. **Lavrushin YUA,** Alekseev VV, Chistyakova IA, Khasankaev VB (1990) Izv Akad Nauk SSSR Ser Geol Vyp 2:82–90. **Lazarev PA,** Tomskaya AI (1987) Mlekopitayushchie i biostratigrafiya pozdnego kajnozoya Severnoy Yakutii. Akad Nauk SSSR, Yakutskiy Filial SO, Yakutsk, 170 pp. **Levkovskaya GM** (1981) Palinologicheskaya kharakteristika musterskogo kulturnogo sloya stoyanki Ketrosy. In: Praslov ND (ed) Ketrosy, Musterskaya stoyanka na Srednem Dnestre. Nauka, Moskva, pp 125–133. **Lozhkin AV** (1977) Dokl Akad Nauk SSSR 235:435–437. **Lozhkin AV** (1984) Dokl Akad Nauk SSSR 275:1143–1146. **Lozhkin AV** (1989) Dokl Akad Nauk SSSR 308:1194–1196. **Lozhkin AV** (1990) Byull Kom Po Izuch Chetvert Per 59:156–159. **Lozhkin AV,** Postolenko GA (1989) Dokl Akad Nauk SSSR 307:1184–1188. **Lozhkin AV,** Prokhorova TP, Pariy VP (1975) Dokl Akad Nauk SSSR 224:1395–1398. **Lozhkin AV,** Pavlov GF, Ryabchun VK, Gorbachev AL, Zadalskiy SV, Shubert EI (1988) Dokl Akad Nauk SSSR 302:1440–1443. **Lozhkin AV,** Skorodumov IM, Meshkov AP, Robako LG (1991) Dokl Akad Nauk SSSR 316:184–188. **Lutaenko KA** (1991) Dokl Akad Nauk SSSR 321:596–598.

**Madeyska T** (1986) Remarks on human settlement distribution in the Old World during the period 24 000 to 15 000 BP. In: Pleistocene perspective 2. Allen and Unwin, Southampton, pp 1–19. **Makeev VM,** Arslanov KHA, Baranovskaya OF, Kosmodamianskiy AV, Ponomareva DF, Tertychnaya TV (1989) Byull Kom Po Izuch Chetvert Per 58:58–69. **Mamedov ED,** Trofimov GN (1986) Byull Kom Po Izuch Chetvert Per 55:102–106. **Mangerud J** (1991) The Scandinavian Ice Sheet Through the Last Interglacial/Glacial Cycle. In: Frenzel B (ed) Paläoklimaforschung 1. Fischer, Stuttgart, pp 307–330. **Mangerud J,** Bolstad M, Elgersma A, Helliksen D, Landvik JK, Lycke AK, Lonne J, Salvigsen O, Sandahl T, Sejrup HP (1987) Polar Res 5:275–278. **Markova AK** (1975) Dokl Akad Nauk SSSR 222:913–916. **Markova AK** (1985) Dokl Akad Nauk SSSR 284:718–721. **Markova AK,** Mikhailescu KD (1990) Byull kom Po Izuch Chetvert Per 59:94–101. **Maudina MJ,** Pisareva VN, Velichkevich FYU (1985) Dokl Akad Nauk SSSR 284:1195–1199. **Mednikova EYU,** Zagryanskaya MF, Kotova EYU, Tsarkova SI (1988) Byull Kom Po Izuch Chetvert Per 57:120–124. **Melekestsev IV,** Braitseva OA, Sulerzhitskiy LD (1988) Dokl Akad Nauk SSSR 300:175–181. **Menke B** (1991) Zur stratigraphischen Stellung der ältesten Weichsel-Möranen in Schleswig-Holstein. In: Frenzel B (ed) Paläoklimaforschung 1. Fischer, Stuttgart, pp 343–352. **Morozova TD** (1981) Razvitie pochvennogo pokrova Evropy v pozdnem pleistotsene. Nauka, Moskva, 282 pp. **Motuz VM** (1970) Dokl Akad Nauk SSSR 194:657–659. **Motuz VM** (1981) Fauna mollyuskov chetvertichnykh otlozheniy Kishlyanskogo Yara i ego ekologiya. In: Praslov ND (ed) Ketrosy, Musterskaya stoyanka na Srednem Dnestre. Nauka, Moskva, pp 143–151.

**Nekrasov IA** (1970) Dokl Akad Nauk SSSR 194:643–645. **Nesmeyanov SA** (1983) Dokl Akad Nauk SSSR 274:395–397. **Nikolaev SD** (1988) Dokl Akad Nauk SSSR 298:450–453. **Nikolaeva IV,** Panychev VA, Orlova LA (1989) Byull Kom Po Izuch Chetvert Per 58:125–132. **Nikonov AA,** Pakhomov MM, Sulterzhitskiy LD (1988) Dokl Akad Nauk SSSR 298:944–948.

**Pakhomov MM** (1983) Dokl Akad Nauk SSSR 270:967–972. **Pakhomov MM,** Baiguzina LL (1986) Dokl Akad Nauk SSSR 293: 945–947. **Petersen KST,** Kronborg CHR (1991) Late Pleistocene history of the inland glaciation in Denmark. In: Frenzel B (ed) Paläoklimaforschung 1. Fischer, Stuttgart, pp 331–342. **Pevzner MA** (1989) Byull Kom Po Izuch Chetvert Per 58:117–124. **Pisareva VV,** Chebotareva NS (1986) Byull Kom Po Izuch Chetvert Per 55:45–51. **Prokopchuk BI** (1985) Dokl Akad Nauk SSSR 283:446–447. **Pshenin GN,** Steklenkov AP, Cherkinskiy AE (1984) Dokl Akad Nauk SSSR 276:675–677.

**Ranov VA,** Dodonov AE, Lomov SP, Pakhomov MM, Penkov AV (1987) Byull Kom Po Izuch Chetvert Per 56:65–75. **Raukas A,** Róuk A-M (1987) On the effect of human economic activities on Estonian lake ecosystems. In: Raukas A, Saarse L (eds) Palaeohydrology of the temperate zone II, Lakes, Valgus, Tallinn, pp 147–153. **Rengarten NV** (1981) Sostav i usloviya obrazovaniya chetvertichnykh otlozheniy Kishlyanskogo Yara. In: Praslov ND (ed) Ketrosy, Musterskaya stoyanka na Srednem Dnestre. Nauka, Moskva, pp 80–93. **Rindzyunskaya NM,** Reverdatto MV, Ivanov NM, Nedashkovskaya NM, Zelikzon EM (1987) Byull Kom Po Izuch Chetvert Per 56:111–118. **Rindzyunskaya NM,** Reverdatto MV, Nedashkovskaya NN (1990) Byull Kom Po Izuch Chetvert Per 59:29–38. **Romanovskiy NM,** Kaplina TN, Kondratyeva KA, Rybakova NO, Pirumova LG (1970) Merzlotnye issledov 10:80–95. **Rubanov IV** (1982) Dokl Akad Nauk SSSR 264:927–929.

**Savina LN,** Zubarev AP, Koshkarova VL, Karpenko LV, Starikov EV, Zhidovlenko VA (1988) Byull Kom Po Izuch Chetvert Per 57:124–131. **Schlüchter CHR** (1991) Fazies und Chronologie des letzteiszeitlichen Eisaufbaus im Alpenvorland der Schweiz. In: Frenzel B (ed) Paläoklimaforschung 1. Fischer, Stuttgart, pp 401–408. **Sejbutis AA** (1987) Arkheologicheskie izmeneniya landshaftov Litvy. Geograficheskie, paleobotanicheskie, istoricheskie i linguisticheskie dannye. In: Khotinskiy NA, Sycheva SA (eds) Antropogennaya evolyuciya geosistem i ikh komponentov. Inst Geogr Akad Nauk SSSR, Moskva, pp 30–43. **Serebryanniy LR,** Tishkov AA, Malyasova ES, Solomina ON (1984) Izv Akad Nauk SSSR Ser Geogr Vyp 6:75–84. **Shekhotkin VV** (1990) Dokl Akad Nauk SSSR 311:716–719. **Shevyrev LT,** Alekseeva LI, Spiridonova EA, Tikhomirov SV, Arslanov CHA, Gej NA, Antsiferova GA (1987) Byull Kom Po Izuch Chetvert Per 56:45–64. **Shilo NA** (1989) Dokl Akad Nauk SSSR 305:412–416. **Shilo NA,** Krivoshey MJ (1991) Dokl Akad Nauk SSSR 317:177–180. **Shilo NA,** Dikov NN, Lozhkin AV (1976) Dokl Akad Nauk SSSR 231:1204–1205. **Shilo NA,** Dikova TM, Lozhkin AV (1977a) Dokl Akad Nauk SSSR 237:180–181. **Shilo NA,** Dikov NN, Lozhkin AV, Starikov AV (1977b) Dokl Akad Nauk SSSR 237:688–689. **Shilo NA,** Lozhkin AV, Titov EE, Shumilov YUV (1977c) Dokl Akad Nauk SSSR

236:1449–1452. **Shimanovich S** (1987) Minerals in alluvial deposits of Byelorussia as indicators of late- and postglacial palaeogeographical conditions. In: Raukas A, Saarse L (eds) Palaeohydrology of the temperate zone II, rivers and lakes. Valgus, Tallinn, pp 120–131. **Shumenko SJ** (1987) Dokl Akad Nauk SSSR 459–461. **Simonov AN,** Stephanov AN (1988) Byull Kom Po Izuch Chetvert Per 57:39–49. **Sinitsyn VV,** Prokopchuk BI, Pakhomov MM (1974) Dokl Akad Nauk SSSR 217:657–658. **Stanko VN,** Svezhentsev YUS (1988) Byull Kom Po Izuch Chetvert Per 57:116–120. **Starovoytov AV,** Kalinin AV, Spiridonov MA, Rybalko AE, Stuchkov VA (1982) Dokl Akad Nauk SSSR 270:1179–1181. **Stefanovich EN,** Klimanov VA, Borisova ZK, Vinogradova SN (1986) Byull Kom Po Izuch Chetvert Per 55:97–102. **Sudakova NG** (1975) Dokl Akad Nauk SSSR 221:168–171. **Sudakova NG,** Bolshakov VA, Boyarskaya TD, Bazilevskaya LI, Nemtsova GM, Glushankova NI (1982) Dokl Akad Nauk SSSR 267:909–912. **Svitoch AA** (1975) Dokl Akad Nauk SSSR 224:665–668. **Svitoch AA,** Bolikhovskaya IS, Bolshakov VA, Gunova VS, Denisenko OV (1990) Dokl Akad Nauk SSSR 314:928–931. **Sycheva SA,** Uzyanov AA (1987) Istoriya antropogennogo vliyaniya na prirodu Kurskogo Posejamya. In: Khotinskiy NA, Sycheva SA (eds) Antropogennaya evolyuciya geosistem i ikh komponentov. Inst Geogr Akad Nauk SSSR, Moskva, pp 105–120. **Syrnev IP,** Malaeva EM, Kramarenko GS, Leflat ON (1986) Byull Kom Po Izuch Chetvert Per 55:35–44.

**Tarakanov LV,** Kaplin PA, Kursalova VI (1974) Dokl Akad Nauk SSSR 216:1128–1130. **Timoshina NA** (1980) In: Stepanov DI, Khozatskiy LI (eds) Ekostratigrafiya i ekologicheskie sistemy geologicheskogo proshlogo. Trudy XXII sessii vsesoyuznogo paleontologicheskogo obshchestva. Nauka, Leningrad, pp 144–150. **Tleuberdina PA** (1982) Mater Po Ist Fauny Flory Kazachstana 8:65–71. **Tolchelnikov YUO,** Kostarev AS (1982) Dokl Akad Nauk SSSR 264:1236–1238. **Tomirdiaro SV** (1972) Vechnaya merzlota i osvoenie gornykh stran i nizmennostei, na primere Magadanskoi Oblasti i Yakutskoi ASSR. Magadanskoe knizhnoe izdat, Magadan, 174 pp. **Tomirdiaro SV** (1974) Dokl Akad Nauk SSSR 219:179–182. **Tomirdiaro SV,** Arslanov KHA, Chernenkin BI, Tertychyana TV, Prokhorova TN (1984) Dokl Akad Nauk SSSR 278:1446–1449. **Trofimov GN** (1986) Byull Kom Po Izuch Chetvert Per 55:107–110. **Trubikhin VM** (1977) Paleomagnetizm i stratigrafiya akchagylskikh otlozheniy Zapadnoy Turkmenii. Trudy Geol Inst Akad Nauk SSSR 301. Nauka, Moskva, 79 pp. **Turovskiy DS,** Sheko AB (1974) Dokl Akad Nauk SSSR 217:931–934. **Turovskiy DS,** Lubchenko IYU, Cherkasova EV (1973) Dokl Akad Nauk SSSR 212:475–478.

**Ukraintseva VN** (1990) Bot Zh 75:70–74.

**Valueva MN,** Kozlov VB, Kolesnikova TD, Semenenko LT (1974) Dokl Akad Nauk SSSR 219:1196–1199. **Valueva MN,** Tsukurova AM, Krasnenkov RV (1983) Dokl Akad Nauk SSSR 273:166–170. **Vangengejm EA,** Pevzner MA, Tesakov AS (1990) Byull Kom Po Izuch Chetvert Per 59:81–93. **Vasilchuk YUK** (1988) Dokl Akad Nauk SSSR 298:425–429. **Vasilchuk YUK** (1989) Dokl Akad Nauk SSSR 309:920–924. **Vasilchuk YUK** (1990) Dokl Akad Nauk SSSR 310:154–157. **Vasilchuk YUK,** Trofimov VT (1984) Dokl Akad Nauk SSSR 275:425–428. **Vasilchuk YUK,** Esikov AD, Oprunenko YUF, Petrova EA, Serova AK, Sulerzhitskiy LD (1985) Dokl Akad Nauk SSSR 281:904–907. **Vasilchuk YUK,** Petrov OM, Serova AK (1986) Byull Kom Po Izuch Chetvert Per 55:111–117. **Vasilev SA,** Tseitlin SM, Svezhentsev YUA (1986) Byull Kom Po Izuch Chetvert Per 55:88–93. **Vasilev SV** (1988) Byull Kom Po Izuch Chetvert Per 57:131–136. **Vdovykin GP** (1990) Dokl Akad Nauk SSSR 310:673–675. **Velichko AA,** Faustova MA (1989) Dokl Akad Nauk SSSR 309:1465–1468. **Velichko AA,** Svetlitskaya TV (1988) Dokl Akad Nauk SSSR 300:444–448. **Velichko AA,** Markova AK, Pevzner MA, Udartsev VP (1983) Dokl Akad Nauk SSSR 269:1147–1150. **Velizhanin AG** (1976) Dokl Akad Nauk SSSR 231:205–207. **Verzilin NN,** Utsal KR (1990) Dokl Akad Nauk SSSR 314:686–691. **Vlasov VK,** Kulikov OA, Nikonov AA (1984) Dokl Akad Nauk SSSR 279:1429–1433. **Voytov GI,** Esikov AD, Osika LD, Osika DG, Orlova TG (1986) Dokl Akad Nauk SSSR 292:702–705.

**Zamolotchikova SA** (1980) Merzlotnye issledov 19:102–110. **Zharikov SN** (1987) Istoriya ispolzovaniya dernovo-podzolistykh pochv severozapada Russkoy Ravniny i izmenenie ikh svojstv. In: Khotinskiy NA, Sycheva SA (eds) Antropogennaya evolyuciya geosistem i ikh komponentov. Inst Geogr Akad Nauk SSSR, Moskva, pp. 44–60. **Zhezhel ON,** Zamoruev VV, Zdanskaya GG, Klimov GI (1975) Dokl Akad Nauk SSSR 224:658–660. **Zubakov VA,** Bogatina NV, Pisarevskiy SA (1982) Dokl Akad Nauk SSSR 267:426–429.

Professor Dr. Dr. h.c. Burkhard Frenzel
Botanisches Institut der Universität Hohenheim
W-7000 Stuttgart 70, Germany

## II. Plant Geography

By Eckehart J. Jäger

### 1. Trends of Investigation

The most urgent problems confronting plant geography with during the past years resulted from man's activities: the extinction of plant species, and the effect of expected climatic changes.

Expanded investigations on the human impact upon plant distribution and on the response of plants to former and recent catastrophic environmental changes are needed if plant sciences are to make an effective contribution to the further survival of mankind. The knowledge of migration rates and of the ecological reasons of the range limits is crucial to predict future changes of plant distribution.

In connection with the influence of man's activities on the distribution of plants, the rate and the geographical bases of evolution are discussed. Trees obviously responded to Holocene climatic change by migration rather than by evolution, but some weeds expanded their climatic tolerance limits within the short timespan of some decades. A lot of man-made species point to the high evolution rate under extraordinary pressure of disruptive selection. The relative role of different kinds of speciation (allopatric versus sympatric) is discussed controversially.

Real progress is to be seen in many distribution histories based on fossil evidence, which show that reconstructions of old (pre-Miocene) migrations and centers of origin are purely guesswork, if they are only based on modern distribution patterns.

Phytogeographical foundations of species conservation include iterative mapping, but also the search for important centers of endemism, that are at the same time the base of phytogeographical subdivisions, which have been revised especially in Eurasia and Africa.

It must be stressed that the first precondition for all phytogeographic investigations are taxonomic monographs, floras, and distribution maps, that will therefore be considered here first. Especially in the regions with the highest diversity of species, the basic data are still very incomplete and need accelerated research.

Progress in Botany, Vol. 54

## 2. Terminology

In a dictionary of the floristic terms and their synonymy, Jurcev and Kamelin (1991) arrange the terms on flora and elements of a flora in a logical system. The term "dealpine" was misapplied to many distribution patterns, as Skalicky (1990) states for Europe and Czechoslovakia.

## 3. Taxonomic and Floristic Foundations of Plant Geography

a) New Findings and Range Extensions

There are always many publications on "new and interesting plants" for special regions, that include also surprising discoveries. To mention only two of them: the "king's flower" *Daphne blagayana*, that had been the object of an expedition of Friedrich August II, King of Saxony, to northwestern Illyria, was found to grow also in Italy (Martini and Poldini 1990). On the island of Chios, that has been visited by botanists for at least 150 years, Boratynski et al. (1987) found among a total of 108 species of woody plants 32 new for the island.

b) Distribution of Marine Algae

Especially the distribution of lower plants is incompletely known. In an account of the Rhodophyta of eastern Australia, Millar (1990) mentions 45% of the species new to the region and 13% new to the Southern Hemisphere.

Nevertheless, the biogeography of the seaweeds is often treated. The richness of the algal flora of Macaronesia is explained by vicariance, migration, the rather complex coasts, and the relatively high temperature during the Pleistocene (Prud'Homme and Hoek 1990). Temperature conditions are considered to control most distribution limits of seaweeds, e.g., the northern limit of *Ceramium shuttleworthianum* at the Norwegian coast (Rueness et al. 1990) or the sharp boundary of the temperate algal flora of Namibia at the political border of Angola (Lawson et al. 1990), whereas the influence of Quaternary glaciations was stressed in a symposium on the biogeography of the algae in the North Atlantic (Garbary and South 1990).

### c) New Floras

#### α) Large Monographic Floras

A valuable review of the floristic inventory of all tropical countries is given by Campbell and Hammond (1989). The number of herbarium specimens/100 $km^2$ varies between 1 (Gabon, Central African Republic) and over 500 (Taiwan). At the recent rate of botanic inventory, it will take hundreds of years to obtain 100 herbarium specimens/100 $km^2$ in most countries, a density that is considered to be minimal to describe a flora. Priority areas for a floristic inventory are listed by the authors. At the present rate of publishing, none of the large monographic floras of the tropics will be completed before the middle of the next century. *Flora Malesiana* (1989: vol. 10/4, Chrysobalanaceae, Sabiaceae, biography of Blume; 1991: ser. II vol. 2/1, *Tectaria* group; good dot maps and discussion of fossil and recent distribution; in all, hitherto 5000 species treated on 7000 pages) will take another 160 years to complete.

To write a flora of the whole Republic of Mexico would take as long as 300 years, provided the staff and effectivity of work is the same as in the production of *Flora fanerogamica del valle de Mexico* (vol. 3, finishing the work, Rzedowski and Calderon de Rzedowski 1990a). The number of species of vascular plants of Mexico (22 000) equals almost that of Malesia (25 000 plus 3000 ferns). The flora of many parts of Mexico is known insufficiently. Twenty one percent of the 2071 flowering plants were new to the capital area, certainly one of the better-known parts of the country. The number of *Carex* species discovered in Mexico increased by the same percent within 15 years (Rezniczek 1990). In the alpine zone of northeastern Mexico, McDonald (1990) found 170 species, twice as many as were known formerly.

More than 200 years will go by before the last volume of *Flore du Cambodge, du Laos et du Viêt-Nam* will be published (1989: vol. 24, Caryophyllales; 1990: vol. 25, Dipterocarpaceae, 46 species). The same is true for the monographic *Flora of India* series (1988: fasc. 19, Malvaceae and some small families, 167 species). The *Flora of South Africa*, one of the large floras of temperate regions, comprised by 1989 only 13% of a total number of 22 000 species to be treated. (The third approximation of a computer-aided checklist is a forerunner of this flora, Leistner 1990.) One of the worlds most underexplored regions is Central Asia, that is treated in the monographic flora *Rastenija Central'noj Azii* (1988: vol. 8a, Fabaceae excl. *Oxytropis* and *Astragalus*, 181 species, 32 dot maps; most of the species are western elements, others like *Sophora, Thermopsis, Lespedeza, Caragana* have East Asiatic relationship; 1989: vol. 9, 75 Salicaceae, 106 Polygonaceae, 13 dot maps; included are also the data of the Chinese floras, e.g., those of Tibet).

Encouraging progress was made in the *Flora of Australia* (1990: vol. 18, Podostemales, Haloragales, Myrtales excl. Myrtaceae; 1988: vol. 19, *Eucalyptus*, 564 dot maps), but the planned publishing period of 20 years cannot be kept – only 10 of 50 scheduled volumes were published, 10 years after the first one. The *Flora Reipublicae Popularis Sinicae* was continued in 1990 with four volumes describing over 1000 species, among them the Gesneriaceae and Panicoideae. A project to translate this flora into English has been initiated in collaboration with the Missouri Botanical Garden. It is expected to deal with some 30 000 species (Wu 1990).

In consideration of the slow progress of most monographic floras, few-volume flora manuals are to be favored, that reflect but the status quo, provide at least keys, point to open questions, and may encourage further investigations. Also the *Flora Malesiana* board is now aiming at a more condensed version for the following volumes in order to reduce the time span needed for completion to "only" 50 years.

### β) Regional Flora Handbooks

Such floras with keys, short descriptions, selected synonymy and concise site and distribution data (some also with illustrations) are now on the way for the following large regions or countries: Europe (revised edition of *Flora Europaea*, vol. 1 in press, vol. 2 in preparation), Scandinavia and Iceland (*Flora Nordica*, 5 vols. scheduled, see Svensk Bot. Tidskr. 81, 1987: 332), the European part of the former URSS (*Flora evrop. časti*, 1989: vol. 8, Cichorioideae, 465 *Hieracium*, 151 *Taraxacum*, not yet exhausting the richness of these genera in the region, most of the Choripetalae still missing), Siberia (*Flora Sibiri*, 1990: vol. 2, Poaceae, dot maps for 306 of the 440 species, 10% of the species are new descriptions or new findings for the region), the Russian Far East from Chukotia to Manchuria (*Sosud. rast.*, 1989: vol. 4 of 10 scheduled, in sequence of completion, all species mapped), Middle Asia (*Opredelitel rastenij Srednej Azii* 1987: vol. 9 Lamiaceae-Campanulaceae), North America (first vol. in press, 12 planned, to deal with 17 000 species) and Ethiopia (Hedberg and Edwards 1989: first of 8 volumes, Pittosporaceae through Araliaceae, 1100 species, among them 620 Fabales, sequence according to Hutchinson's system; stored on computer discs enabling the production of a revised edition in a few years time, richly illustrated, distribution given on the base of a subdivision into 14 districts, always herbarium specimens in the herbarium in Adis Abeba cited, that will help to determine material in the country). A publication time of another 10 or 15 years is assumed to be the shortest possible time. The latter flora is a good example for other projects now initiated, like the flora of the Greater Antilles (see Brittonia 43, 1991:209) and the flora of Chile (see Prog Bot 51:316).

Parallel to the multi-state floras of southeastern USA (Isely 1990: vol. 3/2, the second published of the set, ca. 400 species of the Fabales) the North-East (revised edition: Gleason and Cronquist 1991, complete in 1 vol.) and the Intermountain West (Barneby in Cronquist et al. 1989: vol. 3 Fabales, illustrated, vertical and horizontal distribution given), a new Rocky Mountain flora is planned to be published in the short time of 12 years (6 vols., Hartmann 1990).

### γ) Local and State Floras (Eurasia and Australia)

Europe and Mediterranean Countries. Important contributions to the knowledge of the flora of Central Europe that have been published during the last 3 years are two floras of Czechoslovakia. In the flora of Bohemia and Moravia (Hejny and Slaviik 1990: vol. 1 and 2 of 6 planned vols., Magnoliidae, Caryophyllidae, Dilleniidae excl. Brassicaceae), the authors provide excellent phytogeographic information, mention trends of range expansion or reduction, and describe distribution types on the base of grid maps. The Slovakian flora (*Flora Slovenska*, 1992: vol. 4/3, Rosaceae) contains many dot maps. Additionally, the two-volume flora of Czechoslovakia was completely rewritten by Dostál (1989).

A special advantage of the new, well-illustrated flora of Württemberg is the citation of the oldest fossils and literature documents, that enables judgement on the status of the species in the country (Sebald et al. 1990, vols. 1 and 2 of 6 scheduled). Maps on the base of a 5 x 6 km grid are provided for every species. The new fascicles of the illustrated flora of central Europe (Hegi 1989 vol. I 3/5 Poaceae, 1990 Vol. IV/2B Rosaceae) contain maps of general distribution, distribution diagnoses on the base of a three dimensional subdivision of the earth for the grasses and the Central European Alchemillas.

For the new project of a flora of the whole Alpine chain, much exploration work is needed, especially in the western part of this mountain range. This is to be seen from the new edition of the flora of Vaucluse (Girerd 1991), that comprises 109 species not mentioned in the first edition published 13 years before. Another step forward here is Charpin and Salanon's conspectus (final part: 1988) complementing Burnat's unfinished *Flore des Alpes Maritimes*.

Also Briquet's unfinished flora of Corsica was now completed (Jeanmonod and Burdet 1987–1990). For the production of Mediterranean floras the Med-checklist is an important help (Greuter et al. 1989 vol. 4, Lauraceae-Rhamnaceae). It provides a profound terminology and synonymy of all plants of the countries that adjoin the Mediterranean sea. Remarkable progress for the floristics in Spain and Portugal is the publication of the second volumes of the excellent *Flora Iberica* (1990) and of the *Flora dels paisos Catalans* (Bolós and Vigo 1990). The latter, now half way to completion, is fully illustrated and contains maps of the local and European distribution, an advantage that it shares only with Hulten's flora of Alaska. At the same time a one-volume Catalonian flora manual appeared (Bolós et al. 1990), omitting the maps, but with site description in terms of phytosociology and information on the horizontal and vertical distribution in the Catalonian countries (including parts of France and Valencia) for the 3580 species that have been found over the area of 70 000 km$^2$. With the second volume, the mountain flora of Greece is completed (Strid and Kit Tan 1991), setting a high standard also with respect to the clearly arranged distribution data. About one third of the 1980 species either had not been recorded for Greece in *Flora Europaea* or were recognized under a different name or rank.

The flora of Libya (Jafri and El Gadi 1989) is now also complete, the most comprehensive complete flora of Northern Africa. A voluminous richly illustrated excursion flora of Israel contains small computer-generated maps for all 2800 species treated (Feinbrun-Dothan and Danin 1991). Thus, most parts of the Mediterranean region received good, modern floras during the last two decades. Exceptions are Sardinia, Sicilia, Albania, Morocco, some parts of Yugoslavia and Southern France.

Asia. The last and most comprehensive of all republic floras of the former URSS is now complete: *Flora Tadžikskoj SSR* (1991: vol. 10, Cynareae and Lactuceae, e.g., 121 species of *Cousinia*, in all now 4513 species). Turcmenia is three times larger than Tadshikistan, but as a result of the more arid climate only 2800 species are recorded from there (excursion flora: Nikitin and Gel'dichanov 1988, illustrated, distribution data based on 18 phytogeographic provinces). For the adjoining Iran, a new national flora project has been initiated (Assadi et al. 1989, in Persian) which is based on more material than the well-known, almost complete *Flora Iranica*.

The flora of Saudi Arabia was almost unknown up to the 1950's. Studies of plant distribution were military espionage, because the supply of arms depended on camels and their food. The first flora could only recently be written, such as Mandaville's (1990) *Flora of Eastern Saudi Arabia* (565 species of higher plants, 30% not native, dot maps of 41 species) and a list of Dhofar (Miller and Morris 1988, 734 species hitherto known).

Among the 1068 higher plants of the Mongolian Changaj, a mountain range as large as France, are steppe plants like *Stipa krylovii* and *Cleistogenes*, taiga trees, and semidesert grasses like *Stipa gobica* (Bjazrov et al. 1989). In China, floras of various provinces and districts are coming up "like bamboo shoots after a spring rain" (review: Wu 1990), e.g., *Flora Yunnanica* (1991: vol. 5), *Flora Sichuanica* (1989: vol. 6, Pteridophyta; 1991: vol. 5/2, Gramineae). The richness of species is especially high in south-west China. The Huping Mts. in the western part of Hunan, where evergreen vegetation ascends to 1100 m, harbor 1961 species on only 420 km$^2$ (Li et al. 1991).

In contrast, the floras of the Deccan monsoon forest regions are astonishingly poor and uninteresting, despite of yearly precipitation sums of about 1000 mm: that of the Godavari district (Andhra Pradesh) includes only 809 species over 7795 km$^2$ (Sesagiriri et al. 1986); that of Bilaspur district

(Madhya Pradesh) 852 species over 20 000 $km^2$ (Panigrahi and Murti 1989). (Malyshew 1991 assumes there are an average of 2000–3000 species/100 000 $km^2$.) Perhaps one should wait for additions. This is shown for the flora of the Indian desert by Bhandari (1990). The author now enumerates 682 species, 15% more than in the first edition.

Whereas the *Flora of Bangladesh* is still in its initial phase, the *Flora of Pakistan* (Ali and Nasir 1990: Boraginaceae, 135 species, Labiatae, ca. 200 species) is near to completion. The floras of Ceylon (Dassanayake and Fosberg 1991: vol. 7, Fabales) and of Thailand (Yamazaki 1990: Scrophulariaceae) were continued. The first volume of an annotated conspectus of the Vietnamese flora (with distribution data) was published in Leningrad (Cvelev 1990, e.g., 815 Orchidaceae).

Australasia and the Pacific Islands. With respect to the much treated biogeography of Hawaii, a new, very thorough, illustrated two-volume flora handbook of these islands is especially welcomed. The 1094 native species are believed to have developed after about 280 long-distance dispersal events within a time of 4 million years. By now, 10% of the native species are extinct and 861 species of newcomers have become naturalized (Wagner et al. 1990). The *New Flora of Fiji* (Smith 1991, vol. 5: Solanales, Asterales, Orchidales) is now complete; a checklist of the ferns includes 958 species (Salgado 1990) and a checklist of the Solomon Islands (Hancock and Henderson 1988) 3210 species.

In Australia the publication of a first comprehensive flora handbook of New South Wales has begun (Harden 1990, 1991, 2 vols. of 4, illustrated, ferns included, many naturalized plants from Europe, e.g., most of the Caryophyllaceae, Brassicaceae, and Rosaceae). A new checklist for South Australia, including data on distribution and status, reveals also the high percent of aliens (Jessop 1989). The flora of southeastern Queensland includes on 180 600 $km^2$ (10% of the state) 3700 species (Stanley and Ross 1989, complete with the third volume). The ferns of the state have been treated in a checklist by Andrews (1990). The success of invaders is to be seen also in a local flora of the semiarid western coast, the Perth region (1510 native and 547 naturalized species/10 500 $km^2$, Marchant et al. 1987) and in a voluminous book on the naturalized dicotyledons of New Zealand (Webb et al. 1988), where the aliens almost equal the native species in number.

## 4. Plant Distribution Mapping

The world index of distribution maps (*Index Holmiensis*, 1992: vol. 7, to letter H) is half way to completion, but the amount of yearly published maps (about 30 000) would fill another volume every year. Alone the books and papers cited in this review contain about 20 000 maps.

### a) Grid Mapping Projects

The different sources and aims of distribution mapping result in different kinds of maps. In national and regional projects, hundreds of professional and amateur botanists collect millions of data. To handle these data and to print the maps with the aid of computers is today a matter of course (e.g., programs for Belgium: Fabri 1989). The detailed range limits in such maps reveal the influence not only of the climate, but also of edaphic factors and human activities on the distribution. A very good example is the atlas of the Bavarian flora (Schönfelder and Bresinsky 1990), the 2496 maps of which are based on a grid map of 5 x 6 km squares. They include 300 taxa not treated in the atlas of the FRG. The signs are differentiated as to the time of registration and status. In the case of 340 threatened plants of Slovenia, also the sources of the dots are given (Wraber and Skoberne 1989, 5 x 6 km grid squares). The same grid is used by Zimmermann et al. (1989) in an atlas of threatened higher plants of Steiermark, another example for many atlasses of plants in danger of extinction, being but a first step to save them.

The 10 x 10 km UTM grid use Bolós and Romo (1991) in mapping 203 Catalonian plants of different distribution type, Villar and Lazare (1991, first results of mapping the flora of the Pyrenees), Fernandez Casas and Gamarra (1990: 475 maps of Iberian plants), Molina (1990: maps of Iberian plants) and Slavík (1990: 616 maps of plants of Bohemia and Moravia). In the cited Spanish projects, all or at least some of the sources are given. Further useful information would be information on the vertical distribution, site characteristics, and general distribution.

The first 640 species (about 20% of the French flora) have been mapped for France on a 20 x 20 km grid (Dupont 1990). Even more generalized (50 x 50 km squares) are the maps of the Flora Europaea Mapping Committee (Jalas and Suominen 1989, 1991). Twenty years after the publication of the first volume, this series comprises 12% of the whole flora of Europe (2109 maps up to the Capparaceae in the Englerian system). Many new country records and taxonomic changes are added to the knowledge of the European flora. The density of record is now very scarce only in northwestern and southeastern Russia (e.g., the Lower Don region), whereas the distribution in northeastern and central Russia, as well as in Spain and Greece, is now well documented. The grid of Donner's (1990) small maps of Turkey has the largest squares (150 x 230 km). All additions to Davis' flora have been included there. County maps of all 2010 higher plants of Minnesota (Ownbey and Morley 1991) are especially useful for the exact delimitation of the ranges of boreal and temperate plants. With over 3000 county maps of the plants of Georgia, Jones and Coile (1988) aim to contribute to the "maintenance of quality of life".

Grid mapping is also proceeding for the mosses. An atlas of all species of Musci of the Netherlands with 5 x 5 km grid maps (Touw and Rubers 1989) and some instalments of maps of Iberian Bryophyta (Casas et al. 1989, map 51–100, with site characteristics, general distribution, data on fertility) have only few parallels in European countries.

### b) Dot Maps

Dot mapping is applied today mainly in taxonomic work. Cody and Britton's (1989) treatment of the ferns and fern allies of Canada is illustrated with 140 dot maps. The thorough dot maps of the woody plants of the Near East are based on proved herbarium specimens (Browicz and Zielinski 1988, 1990, 1991). These authors take into consideration also subshrubs like *Phlomis* and *Teucrium* and difficult taxa like *Tamarix* and *Ephedra*. Ciaciura (1989) illustrates the different degree of descent of mountain plants into the Silesian lowland with 204 dot maps.

c) Discussion of the Distribution Patterns

In regional atlasses of plants, the importance and use of the maps for conservation, ecology of species, and landscape planning is usually stressed in the foreword, but then the reader is left with the maps and sometimes with some transparent overlays of physicogeographic factors. A shining exception is the *Atlas van de Nederlandse Flora.* Its maps are compared with the results of a first mapping 40 to 90 years before, and the reason for the changes are discussed. In the final third part (1989) besides maps of 583 common species, 34 synthetic maps of ecological groups are included, and the reason for changes in the collective distribution patterns are discussed.

The final two volumes of the *Comparative Chorology of the Central European Flora* (Meusel and Jäger 1992) for the first time contain maps of the general distribution of almost all Central European species and genera of Rubiaceae through Asteraceae (Englerian system). The distribution patterns are described in short formulas on the base of a system of phytogeographic zones, altitudinal belts and oceanity sectors, thus making it possible to compare them easily. The ecological and historical factors of the distribution patterns are discussed, and on the basis of a broad comparison, a system of distribution types of all Mediterranean-Central European plants is proposed.

## 5. Chorology and Taxonomy

A reliable taxonomic treatment is a prerequisite for plant mapping and for all conclusions on distribution history. A "disjunction" in the genus *Commelina* proved to be only an old introduction, just as Faden (1989) had showed *C. caroliniana* (North America) merely to be a synonym of *C. hasskarlii* from India. On the other hand, a distribution pattern that fits into the chorological experience is a test of a sound circumscription of a taxon. Most monographs therefore include maps of distribution. The close relationship of 70 alpine species of the Caucasus to plants of European mountains had been obscured as a result of a narrow species concept (Korotkov 1989). Thorne (1989) emphasized the high value of distribution patterns for the estimation of relationship in a reinvestigation of the system of Pittosporales.

Distribution patterns provide valuable ecological characters for the taxonomist. Meusel and Kästner (1990) use the distribution spectra of the associated vegetation to characterize the species of *Carlina* in terms of ecogeography; but monographers make use of the distribution patterns primarily for evolutionary deductions. The ancestral types of Carlina survived as relicts on the Canary Islands and on an island near Crete. Low, single-headed mountain plants are supposed to have evolved in connection with the alpidic orogenesis; short lived species were dispersed broadly over the Mediterranean countries when the climate became dry during the Miocene, and finally, a few biennials expanded the range of *Carlina* into the temperate lowlands and as far as central Siberia during the late Quaternary.

## 6. Reconstruction of Geographic Origin and Migration Routes

Modern centers of high species diversity and of ancestral types do not convincingly indicate the geographical origin of a taxon. That has already been emphasized by many authors, e.g., by Taylor (1989) in his excellent monograph of *Utricularia*. The fact that most primitive Utricularias grow in Australia and that most advanced species are cosmopolitan is not considered to be significant for the geographic origin.

A recent concentration of species should rather be interpreted as indicating the ecological environment during the time of evolution than as evidence of the geographic origin, as is done for *Gentiana* by Ho and Liu (1990: concentration of sections in eastern Tibet), for *Cyananthus*, a primitive genus of Campanulaceae closely related to *Codonopsis*, by Hong and Ma (1991, centered in southwest China), or for the Ranunculaceae by Ziman and Keener (1989: speciation connected with alpidic orogenesis, centers in China, southwestern Eurasia, and America). Similar reconstructions of distribution history are in common use also for mosses. Frey (1990) considers some disjunctions of mosses to be of pre-angiosperm, pangean origin, although they are paralleled by similar distribution types of angiosperms, e.g., *Tortula atrovirens* by *Frankenia*.

Reconstructions of time and place of origin and migration history are difficult to test if no fossils are available. "An ounce of historical fact is worth a ton of the historical speculation" (Godwin). The oldest fossils of *Keteleeria*, *Pinus*, and other Pinaceae have not been found in south-west China, where the modern centers of diversity lie, but up to the middle of the Tertiary only in the north (Ying 1989). Another informative example of the danger of deductions of the early distribution history that are based on the modern distribution is given by *Rhododendron*. With respect to the high diversity of the modern sections in south-west China, Ming and Fang (1990) consider this genus to be traced back to a late Cretaceous cradle in this region. On the other hand, Krutzsch (1991), on the basis of fossil evidence, reconstructed quite another distribution history: the oldest range obviously lies in Europe. Only by the end of the Oligocene did *Rhododendron* reach East Asia, that had been covered by savanna vegetation before. Far from the modern distribution limits are also the finding places of the oldest fossils of *Nypa*, *Ctenolophon*, Olacaceae, Bombacaceae, *Symplocos*, *Alangium*, and Restionaceae. Krutzsch (1989) presents distribution maps for these taxa from the upper Cretaceous, Paleocene, Eocene, Oligocene, Miocene up to the present.

On the basis of such distribution histories that are substantiated by fossil evidence, also the so-called molecular clocks have to be calibrated, a method to estimate the time of the segregation of taxa using the degree of difference in the sequence of DNA bases. In *Liriodendron*, allozyme data and divergence in the plastid genomes suggest a minimal time of 13 million years for the segregation of Asiatic and American species (Parks and Wende 1990). The analysis of chloroplast DNA reveals that at least six groups of *Poa* independently invaded North America from Eurasia. Two of them later colonized South America. Australia was reached from there as well as directly from East Asia (Soreng 1990). Paton (1990) dates a similar migration of *Scutellaria* from Asia to North America back to the Miocene and its range expansion to South America to 3 Ma B.P.

Karyological progressions and parallelism with the history of continents and islands can provide additional arguments for the reconstruction of distribution histories. Karyological progressions corroborate the picture of ecogeographical differentiation designed by Schönbeck-Temesy and Ehrendorfer (1989) for the perennial species of *Crucianella*. Plesiomorphic, stenochore taxa overlap in the warm temperate forests of Hyrkan, whereas apomorphic, eurychore species radiated into various Irano-Turanian habitats, a pattern similar to those in the *Veronica persica* group and *Centaurea* subgen. *Psephellus*.

## 7. Vicariance Biogeography and Geographic Speciation

A matter of much discussion is vicariance biogeography. Its aims are to find historical relationships between the different distribution areas and to test, on the basis of distribution patterns, the generic cladogram by comparing it with cladograms of independent groups of plants and animals.

This is exemplified by Welzen (1989) in an analysis of *Guioa* (Sapindaceae). According to the author, the ancestral species of this genus grew in Gondwana. The orogenesis in New Guinea resulted in a high degree of speciation and endemism. After the collision of the Australian plate with the Pacific one, the genus could invade the Pacific Islands. The collision of the Eurasian and Australian continent is thought to have taken place approximately 15–10 Ma B.P. Then and possibly later, when the sea level started lowering by the end of the Tertiary, the distribution of *Fordia* (Fabaceae) could expand, possibly in waves (Schott 1991). The distribution of ancestral species of *Senna* in Australia leads Randell (1989) to consider this genus to have originated there and not to have invaded from Africa via Asia, as was assumed by Raven. (Other examples of vicariance biogeography in Australasia: Barlow in Baas et al. 1990 on Loranthaceae, Rödl-Linder 1990 on the fern genus *Goniophlebium*).

Vicariance biogeography is based on the assumption of allopatric speciation and of a splitting of an ancestral species into two sister species after a more or less continuous range expansion. Difficulties arise therefore in the cases of long-distance dispersal and of polytopic origin of species. A polytopic allotetraploid origin of some fern species is to be presumed (e.g., *Cheilanthes tinaei–Ch. guanchica*) or accepted (*Asplenium balearicum*, Pichi-Sermolli 1990). With regard to the hybridogenous origin of many species of *Symphytum*, Sandbrink et al. (1990) consider cladistics not to be applicable in that genus. Evidence in favor of long-distance dispersal is increasing, e.g., for Afromontane plants (White 1990). From late Miocene pollen grains that are very similar to modern *Trochetiopsis* (Sterculiaceae) pollen found on St. Helena, Cronk (1990) concludes that this genus arrived by long-distance dispersal from Africa. Under the heading *Allopatric speciation – deduced but not detected*, Sauer (1990) challenges the supporters of the predominance of allopatric speciation. He puts forward arguments that speak for sympatric speciation, like the adaptive radiation on islands after the arrival of a single species and genetic barriers in wind-pollinated genera (e.g., two dozen pine species native to Mexico).

The cause of speciation patterns in neotropical regions have elicited much speculation in recent years. Hypotheses about fundamental processes of ecogeographical differentiation in which the tropics play a key role are proposed, often without the systematic data needed to test them (Simpson 1988). A well-balanced review of the Quaternary history in that region, taking into account not only the refugia concept but also alternative explanations, is provided by Whitmore and Prance (1987). Simpson (1988) argues that most distribution patterns and diversity could be attributed to climatic changes that occurred during the last 5 million years. The forest refugia are supposed to have been the places of speciation in *Himatanthus* during the late Quaternary dry periods (Plumel 1990) and in *Tillandsia* (Winkler 1990, ecogeographical evolution of the genus).

## 8. Sernander's Nunatak Hypothesis

There has also been much discussion about the nunatak hypothesis (survival on rocks that have not been covered by the surrounding Quaternary ice sheet). With regard to the site conditions, Dahl (1989a, 1991) claims that the plants immigrated from refugia along the Atlantic coast, and rejects the hypothesis of a survival of the northern boreal and alpine plants south of the ice sheets in Europe. In contrast, the analysis of flower substances in *Oxytropis deflexa* led to suppose a migration from Northern Asia westward to Norway, and a subsequent extinction in the region between (Hoiland and Laane 1989). Weak arguments to support the nunatak hypothesis are the Scandinavian endemics, that are mainly young and apomictic (Dahl 1989b).

The discussion on "nunatak survival" contra "tabula rasa" is on principle finished now, due to macrofossil data from an sediment column from northwestern Norway spanning 20 000 to 11 000 B.P. The seeds found there support neither of the theories – the truth lies somewhere between. Some arctic plants such as *Papaver radicatum* s. l., *Puccinellia, Chrysosplenium tetrandrum, Polygonum viviparum,* and *Braya* survived the entire Weichselian glaciation on the island Andoya. For most other phanerogams, a history of immigration and subsequent in situ survival, migration, or disappearance is suggested (Alm and Birks 1991). In Greenland, *Betula nana* arrived in 8000 B.P., *B. glandulosa* in 5700 B.P.; *B. pubescens* did not arrive until 3500 B.P. The total distribution has changed since then owing to climatic changes (Fredskild 1990).

That also fossil data have to be proved critically is to be seen from a surprising disparity between the pollen (recent arrival) and macrofossil ($^{14}C$ -dating: 10 000 B.P.) results in Alaskan lodgepole pine (Peteet 1991). The uncertainty in the use of pollen percentages to indicate the presence of a species has important implications for paleomigration research.

For the late Quaternary history of plant distribution present-day distribution patterns and investigations of their factors are no less important than direct historical evidence. Thus, by mapping collectively the associates (*Podocarpus, Nuxia*, and others) of the disjunctly distributed *Ptaeroxylon* and *Ternstroemia*, White (1990) detected a southern migratory track connecting the mountains of East and West Africa during more humid periods.

## 9. Ecological Factors of Plant Distribution

### a) Climate

Species ranges are dynamic structures varying with changing environmental conditions. A search for correlation between distribution limits and climatic or edaphic factors may contribute to the knowledge of the ecological constitution of the plant and should therefore be attempted for every distribution pattern (Jäger 1990).

Many investigations are concerned with temperature conditions of distribution. Kozarinov (1989) compares the distribution limits of forest trees, shrubs, and herbs in Belorussia with 37 different climatic factors, and determines by correlation calculation a formula of probability of occurrence for these plants. Salvesen (1989) infers from cultivation experiments that failing to remain frost-tolerant may preclude the south-west-coast-avoiding species from establishing and persisting in the oceanic climate in western Scandinavia.

Ecophysiological experiments on limits of temperature tolerance are comparatively easy to make, although complicated enough when regarding the influence of preceding conditions and the different response of individuals and developmental stages. Woodward (1988) believes that one can confine oneself to a few stages of the life cycle which, when diminished by extremes of temperature, may have the greatest impact on plant survival; but the climatic extremes usually do not reach the limits of physiological tolerance; the climate acts rather on the balance of competition (Grace 1988, 1989; Jäger 1990).

A major determinant of the competition and thus of the distribution is the productivity. An increase of 1 °C increases the productivity by about 10% in an northern temperate climate, provided the other factors do not become limiting (Grace 1988). Therefore, a range limit directed to decreasing global productivity will be less pushed back by competition than a range limit at which the plant is confronted with higher productivity of competitors. Competition, however, is almost ubiquitous, and Woodward (1988, 1990) certainly goes too far in saying that northern limits in England are controlled by climate, whereas southern ones are caused by competition.

Especially well investigated are temperature conditions of boreal limits of tree species. The width of annual rings and the flower initiation are clearly correlated with temperature in Scandinavia (Treter 1984; Grace 1989; Kullmann 1991). The correlation between temperature and the boreal limits of zonal forests in Eastern North America has been investigated by Greller (1989).

Trees respond rather quickly to secular cooling or warming of climate. The realized maximum migration rates of most trees are in the range of 150–500 m/year, in some cases also to 2000 m. The taxa respond individualistically, not as closed plant communities (Huntley 1991). During the climatic optimum in Scandinavia from 1930 to 1950 (a rather cool time in Spain, Schweingruber et al. 1991) pines invaded the birch region without a lag-phase and reached the forest limit in many places (Treter 1984). On the other hand, some cold and dry winters, especially that of 1987, resulted in a cataclysmic response of Scots pine in northern Sweden (Kullmann 1991).

The response to climatic change is of special interest in connection with the climatic change forecast for the next century, the rate of which will be 10–100 times faster than the rate of deglacial warming (Huntley 1991). All organisms will face demands to migrate across landscapes that may have been severaly altered by human activities and in which the habitats available to them will be fragmented. Without artificial dispersal, the prospect for many organisms will be extinction (Lyr and Wersenger 1990 for taiga trees in Central Europe; Huntley 1991).

Huntley argues that fast climatic change is hardly followed by evolutionary adaptions. This may be right in the case of the slowly evolving trees, but some observations on agricultural weeds reveal expansions of tolerance limits within some decades. The northern limits of *Abutilon theophrasti* and of some Paniceae had been controlled by temperature already in the 1950s, but have moved distinctly north since then (Warwick 1990).

The geographical differentiation of closely related taxa is also very often controlled by climate, e.g., in *Chaerophyllum hirsutum*/*Ch. villarsii* in the Alps (Wörz 1989), *Ecballium elaterium* subsp. *elaterium* and the southern subsp. *dioicum* in Spain (controlled by humidity, Costich and Galan 1988) or in the "ecophyletic pair of taxa" *Lophira alata*/*L. lanceolata* in Cameroun (Satabie 1991).

### b) Edaphic Factors

Edaphic conditions of distribution emerge from detailed large-scale maps. A numerical approach (twinspan) to explain the distributional patterns of vascular plants in Fennoscandia does not demonstrate any urgent need for nontraditional thinking or new hypotheses on edaphic or climatic range factors in the area studied (Pedersen 1989). A cluster analysis revealed the content of Ca, Cd, Pb, and Zn in the soil to be of importance for the occurrence of *Minuartia verna* and *Thlaspi alpestre* in the British Isles (Garcia Gonzalez and Clark 1989). The distribution of water plants is closely connected with the chemistry of water, in both the local and the general distribution (Arts and Den Hartog 1990). A similar correlation between edaphic demands in Scandinavia and general distribution has been stated by Dahl (1991): calcicolous species usually have a wider range in America, silicolous plants in Europe.

### c) Dispersal

The connection between dispersal mechanisms and rate of migration is evident in the case of long-distance dispersal (e.g., amphitropical ranges occupied by means of ornithochory), but it is almost irrelevant to the outline of coherent distribution patterns. For the dispersal to the islands in the Great Lakes mainly water (63%), birds (61%), and wind (14%) have been responsible (Morton and Hogg 1989). In a symposium on seed dispersal, Murray (1986) states that the range of dispersal pathways open to any diaspore is much wider than usually considered. Nakanishi (1988) lists for Japanese herbaceous plants the widest observed distance of mechanical seed dispersal. The fly distance of seeds with pappus-like structures results from wind speed and from the relation seed weight/pappus weight (Athey and Pippen 1989). The average distance is astonishingly small: in *Cacalia* 168 cm at a wind speed of 20 km/h.

## 10. Man's Impact on Plant Distribution

The most important factor of plant distribution, man's activities, was the theme of various symposia (e.g., Di Castri et al. 1990; Drake et al. 1990). In an interesting review of anthropogene migration of plants, Trepl (1990) states that there nationalistic attitudes have often been displayed to the "invaders". The urban floras are investigated as to the species diversity in city zones, the change in relative abundance of species (Kowarik 1990b), the correlation between species number and area or number of inhabitants (Klotz 1990; Pyšek 1989; Wittig 1991).

In many cases, the origin of aliens is obscure. Zajac (1988) tries to obtain evidence for the origin of the archeophytes in Poland from phytosociological criteria. Kadereit (1990) deduces the origin of *Papaver rhoeas, P. dubium* and *P. hybridum* from the distribution of the closest relatives. Karyological data and reduced interfertility suggest that the Mediterranean+Atlantic disjunction of *Brassica* sect. *Brassica* is a result of range disruption, not of introduction (Snogerup et al. 1990).

Case studies on the range expansion of single aliens, e.g., of *Robinia* in Europe (Kowarik 1990a) or *Prunus serotina* in Central Europe (Startinger 1990) will be of more use to control the spread of plant invasions than the modeling of the range extension on the base of area and number of primary foci (Moody and Mack 1988).

The success of European weeds is especially high in the Southern Hemisphere. To the province of Buenos Aires came 337 species from Mediterranean countries, i.e., 83.5% of all aliens, among them all 29 common weeds (Soyrinki 1991). In Mexico, a notable portion of invaders came from Africa. Rzedowski and Calderon de Rzedowski (1990b) deal with the 33 most important species, mainly grasses. Exceedingly high is the percent of aliens on islands. In the Cape Verde Islands, only 27.6% of 621 species are supposed to be native and 42.2% introduced. For 30.2% the status is doubtful (Lobin and Zizka 1990). For this classification, the authors had to rely on indirect evidence (ecology, phytosociology, general distribution). The number of established aliens decreases rapidly towards the north in Finland, whereas the number of native species remains fairly constant throughout the country (Lahti et al. 1988). On the Czukotsk peninsula only native species grow on man-made habitats (Sumina and Jackevic 1990). The wide natural distribution of mosses may be one of the reasons that only three alien moss species (less than 0.5%) have been found in East Germany, whereas naturalized angiosperms comprise 19% of the flora there (Berg and Meinunger 1988).

During the next century, about 34 000 species of higher plants will become extinct (Soulé 1990). There are many consequences of the extinction of species. The detrimental ones are especially evident in the case of useful plants. Palm oil production, although a multi-billion dollar industry, had been founded on a very narrow genetic base, deriving mainly from four individual plants. Resampling of wild material from Africa has resulted in the development of new cultivars with increased yield potential and quality (Rajanady 1990). A thorough investigation of the ecology and life history, as offered for *Adenophora liliifolia* in Bavaria by Gaggermeier (1991), is the best precondition for conservative measures. In central Europe, the rapid change of floras will continue until a species assemblage is selected that is adapted to a higher trophic level of soil and until most of the dangered species are protected in nature reserves (Westhus and Zündorf 1990).

## 11. Phytogeographic Subdivision of the Continents

The increasing knowledge of plant distribution makes it possible to scrutinize the phytogeographic subdivision, especially in Eurasia and Africa. The subdivisions proposed in recent years are based either on endemism alone or in connection with the overall floristic similarities. The principles of comparative floristics are treated in a new Russian book (*Teoreticeskie problemy sravnitel'noj floristiki* 1987). Two problems remain unsolved: the different weighting of common and rare species and the circumscription of the floras which are to be compared in the subdivision research. The quantitative assessment of floristic richness (floristic heterogeneity and species abundance per unit area), according to Malyshew (1991), reveals the most important characters of a flora. Temporarily lowered richness and low competition pressure result in the success of apomictic taxa and of man-made, homeless species, which may develop rapidly (Scholz 1991).

The hierarchic concept of phytogeographical subdivision is applied by Chochrjakov (1989) in a subdivision of the Circumboreal subkingdom into seven regions: Far East, East Asiatic, Siberian-NE-European, European-West-Siberian, Balcano-Caucasian, Atlantic-American, and Canado-Alaskan. To the first region belong the Kolyma Highlands, a floristic province characterized by three endemic genera and some dozen endemic species. For the rank of a province, the Russian authors demand at least one genus and some species to be endemic, but exceptions are admitted.

For a new subdivision of Middle and Central Asia, Kamelin (1990) assessed the floristic similarity by a coefficient modified after Jaccard (the latter is applicable only when comparing floras of similar richness). Long lists of endemics are given in order to characterize the six provinces of the Turano-Gobian subregion: the Tibetian, Dsungarian, Central Iranian, and Syrio-Mesopotamian provinces. This subdivision overlaps that of southwestern Eurasia and northern Africa by Léonard (1988, 1989). This author follows White's concept of regional centers of endemism and transitional zones, i.e., plant geographic regions relatively poor in endemism. He discusses 39 formerly proposed subdivisions (from Boissier to Danin and Plitman) and returns to a wide Saharo-Sindian zone with Saharian, Arabian, and Nubo-Sindian subzones. Freitag (1991) confirms that this chorotype classification fits the facts of distribution of Chenopodiaceae better than the older one by Zohary.

A subdivision on the basis of endemism alone is criticized by Linder (1990), because it may obscure the real similarities in composition, structure, and function in the floras, especially the relationship between the floras of the Afromontane and the Cape regions of Africa. To White's centers of endemism, Linder adds the South Drakensberg center. Considering the nonendemic elements, the taxonomic relationships, and the overall similarity of the floras leads inevitably to a hierarchy of the phytochoria (that had been refused by White). An investigation of the distribution of the Lamiaceae of Africa resulted therefore in regions instead of centers of endemism (Ayobangira 1990). A survey of all formerly proposed subdivisions for Africa from Schouw through Wickens and White is given by Iversen (1991) in a study of the flora of the Usambara Mountains, a meeting place for four main African phytochoria. The Cape Verde Islands may be assigned to different phytogeographical regions: the lowland to the Saharian, the mountains to the Macaronesian region (Beyhl et al. 1990). The affinities of the South-West Malagasy flora are first with Arabia, Somaliland, and Socotra (Thomasson and Thomasson 1991).

A numerical approach based on the moss flora shows floristic change in the Gulf of St. Lawrence region to be gradual and to follow a north-south gradient and an east-west gradient. The lack of distinct boundaries is attributed to the high diversity of species distribution possible in a region with complex environmental gradients (Belland 1989). This will also apply to many other regions.

Subdivisions for some other countries have been proposed in recent years, e.g., for The Netherlands by Weeda (1989), for the Northern Apennine Mts. by Foggi (1990, relation with the Alps), for the Crimea by Diduch et al. (1990), for the czernosem region of Russia by Buchalo (1989), for nonczernosem Russia by Aleksandrova and Jurkovskaja (1989), for Cuba by Borhidi and Muñiz (1990), for Sumatra by Laumonier (1990), and for the Australian Northern Territory by Bowman et al. (1988).

## References

**Aleksandrova VD,** Jurkovskaja TK (1989) Geobotaničeskoe rajonirovanie nečernozem'ja evropejskoj časti RSFR. Nauka, Leningrad, 61 pp. **Ali SI,** Nasir YJ (1990) Flora of Pakistan 191, 192. Stewart Herbarium, Rawalpindi, 200, 310 pp. **Alm T,** Birks HH (1991) Nord J Bot 11:465–476. **Andrews SB** (1990) Ferns of Queensland. Queensl Dept Prim Ind, Brisbane, 427 pp. **Arts GHP,** Den Hartog C (1990) Acta Bot Neerl 39:369–378. **Assadi M,** Khatamsaz M, Ramak Maassoumi AA, Mozaffarian V (1989) Flora of Iran. Rep. of Iran, Ministry Agric, 212 pp. **Athey JT,** Pippen RW (1989) Mich Bot 28:89–95. Atlas van de Nederlandse Flora 3 (1989) Rijksherbarium, Leiden, 264 pp. **Ayobangira FX** (1990) Mitt Inst Allg Bot Hamb 23b:643–656.

**Baas P,** Kalkman C, Geesink R (1990) The plant diversity of Malesia. Kluwer, Dordrecht, 420 pp. **Belland RJ** (1989) Can J Bot 67:1633–1649. **Berg C,** Meinunger L (1988) Proc 6th CEBWG Meeting Liblice, pp 103–107. **Beyhl FE,** Lösch R, Mies B, Schweihofen B (1990) Cour Forschungsinst 129:47–53. **Bhandari MM** (1990) Flora of the Indian Desert. MPS Repros, Jodhpur, 435 pp. **Bjazrov LG,** Gan'bold E, Gubanov IA, Ülzijchutag N (1989) Flora Changaja. Nauka, Leningrad, 192 pp. **Bolós O de,** Romo AM (1991) Atlas Corológic de la Flora Vascular dels Països Catalans. Inst. Estudis Catal., Barcelona, map 104–306. **Bolós O de,** Vigo J (1990) Flora dels Països Catalans, vol. 2. Barcino, Barcelona, 921 pp. **Bolós O de,** Vigo J, Masalles RM, Ninot JM (1990) Flora Manual dels Paisos Catalans. Pòrtic, Barcelona, 1233 pp. **Boratynski A,** Browicz K, Zielinski J (1987) Arboretum Kornickie 32:37–92. **Borhidi A,** Muñiz O (1990) Acta Bot Acad Sci Hung 32:3–48. **Bowman DMJS,** Wilson BA, Dunlop CR (1988) Austr J Bot 36:503–517. **Browicz K,** Zielinski J (1988, 1990, 1991) Chorology of trees and shrubs in South-West Asia and adjacent regions. Vols 6–8. Polish Sci Publ, Warszawa, 86 pp. **Buchalo MA** (1989) Bot Zurn 74:638–650. **Burnat E** (1892–1931) Flore des Alpes Maritimes. Vol 1–7. Georg, Genève.

**Casas C,** Montserrat B, Cros RM, Sérgio C (1989) Bryophyte cartography. Iberian Peninsula, Balearic and Canary Islands, Azores and Madeira, Fasc II. Univ Auton, Barcelona, 158 pp. **Campbell DG,** Hammond HD (1989) Floristic inventory of tropical countries. N Y Bot Gard, Bronx/NY, 545 pp. **Charpin A,** Salanon R (1988) Boissiera 41:1–339. **Chochrjakov AP** (1989) Analiz flory Kolymskogo Nagor'ja. Nauka, Moskva, 152 pp. **Ciaciura M** (1988) Characterystyka rozmiesczenia gorskich gatunkov naczyniowych na Slasku. Akad Med, Wroclaw, 157+204 pp. **Cody WJ,** Britton DM (1989) Ferns and fern allies of Canada. Can Gvt Publ. Ottawa, 430 pp. **Costich DA,** Galan F (1988) Lagascalia 15 extra 697–710. **Cronk QCB** (1990) New Phytol 114:159–165. **Cronquist A,** Holmgren A, Holmgren NH, Reveal JL, Holmgren PK (1989) Intermountain flora. Vol 3B. N Y Bot Gard, Bronx/NY, 279 pp. **Cvelev NN** Konspekt sosudistych rastenij flory V'etnama. Vol 1. Nauka, Leningrad, 200 pp.

**Dahl E** (1989a) Blyttia 47:124–133. **Dahl E** (1989b) Blyttia 47:163–172. **Dahl E** (1991) Blyttia 49:17–33. **Dassanayake MD,** Fosberg FR (1991) A revised handbook of the flora of Ceylon. Vol 7. Amerind, New Delhi, 439 pp. **Di Castri F,** Hansen AJ, Debusche M (1990) Biological invasions in Europe and the Mediterranean Basin, Kluwer, Dordrecht, 463 pp. **Diduch JAP,** Litvinskaja SA, Novosad VV (1990) Bot Žurn 75:494–507. **Donner J** (1990) Linzer Biol Beitr 22:381–515. **Dostál J** (1989) Nová květena ČSSR. Academia, Praha, 1548 pp. **Drake JA,** Mooney HA, Di Castri F, Groves RH, Kruger FJ, Rejmanek M, Williamson M (1990) Biological invasions. SCOPE, Wiley, 525 pp. **Dupont P** (1990) Atlas partiel de la flore de France. Mus Nat Hist Nat, Paris, 442 pp.

**Fabri R** (1989) Bull Soc R Bot Belg 122:135–141. **Faden RB** (1989) Taxon 38:43–53. **Feinbrun-Dothan N,** Danin A (1991) Analytical flora of Eretz – Israel. CANA, Jerusalem, 1040 pp. **Fernandez Casas J,** Gamarra R (1990) Fontqueria 30:226–234. **Flora Europaea** (1992) Vol 1, revised edn. Univ Press, Cambridge (in press). **Flora evropejskoj časti SSSR.** Vol 8 (1989) Leningrad, Nauka, 412 pp. **Flora Iberica.** Vol 2 (1990) Real Jard Bot, Madrid, 897 pp. **Flora Malesiana.** Ser 1 vol 10/4 (1989), ser 2 vol 2/1 (1991) Kluwer Dordrecht, 157, 132 pp. **Flora of Australia.** Vol 18 (1990), vol 19 (1988) Aust Government Canberra, 349, 542 pp. **Flora of India** (1988) Monographs Fasc 19. Bot Surv India, Calcutta, 235 pp. **Flora of North America** (1992) Vol 1. Introductory essays. Vol 2. Ferns and gymnosperms. Univ Press, Oxford (in press). **Flora Reipublicae Popularis Sinicae** (1990) Vols 10/1, 56, 59/2, 69. Sci Press, Beijing, 445, 132, 581, 321 pp. **Flora Sibiri** (1990) Vol 2. Nauka, Novosibirsk, 361 pp. **Flora Sichuanica.** Vol 6 (1989) Pteridophyta, vol 5/2 (1991) Gramineae. Sci Press, Chengdu, 410, 410 pp. **Flora Slovenska** (1992) Vol 4/3 Rosales. Slov Akad Vied, Bratislava, 564 pp. **Flora Tadžikskoj SSR** (1991) Vol 10 Asteraceae-Cichorieae. Nauka, Leningrad, 619 pp. **Flora Yunnanica** (1991) Vol 5 Spermatophyta. Sci Press, Yunnan, 809 pp. **Flore du Cambodge,** du Laos et du Viêt-Nam. Vol 24 (1989), vol 25 (1990). Mus Nat Hist Nat, Paris, 44, 123 pp. **Foggi B** (1990) Webbia 44:169–196. **Fredskild B** (1990) Nord J Bot 11:393–412. **Freitag H** (1991) Flora et Vegetatio Mundi 9:281–292. **Frey W** (1990) Bot Syst 111:433–456.

**Gaggermeier H** (1991) Hoppea 50:287–322. **Garbary DJ,** South GR (1990) Evolutionary biogeography of the marine algae of the North Atlantic. Springer, Berlin Heidelberg New York, 429 pp. **Garcia-Gonzales A,** Clark SC (1989) Vegetatio 84:87–98. **Girerd B** (1991) La flore du département de Vaucluse. Aphéleny, Le Pontet, 391 pp. **Gleason HA,** Cronquist A (1991) Manual of vascular plants of Northeastern United States and adjacent Canada. N Y Bot Gard, Bronx/NY, 910 pp. **Grace J** (1988) In: Long SF, Woodward FI (eds) Plants and temperature. Comp of Biologists, Cambridge, pp 91–107. **Grace J** (1989) Philos Trans R Soc Lond B 324:233–245. **Greller AM** (1989) Bull Torrey Bot Club 116:145–163. **Greuter W,** Burdet HM, Long G (1989) Med-checklist. Vol 4 Lauraceae-Rhamnaceae. Cons Jard Bot, Genève; Bot Garten, Berlin-Dahlem, 458+129 pp.

**Hancock IR,** Henderson CP (1988) Flora of the Solomon Islands. Res Bull 7:1–203. **Harden GJ** (1990, 1991) Flora of New South Wales. Vols 1 + 2. NSW Univ Press, Kensington, 601, 620 pp. **Hartmann RL** (1990) Am J Bot 77(Suppl):134–135. **Hedberg I,** Edwards S (1989) Flora of Ethiopia. Vol 3, Biol Dept, Addis Abeba; Dept Syst Bot Uppsala, 660 pp. **Hegi G** (1990) Illustrierte Flora von Mitteleuropa. Vol IV 2B/1–3 *Alchemilla*, (1989) Vol I 3/5 Gramineae. Parey, Berlin, 248, 80 pp. **Hejny S,** Slavík B (1990) Květena Česke Republiky. Vol 2. Praha, Academia, 540 pp. **Ho TN,** Liu SW (1990) Bull Br Mus Nat Hist Bot 20:169–192. **Hoiland K,** Laane MM (1989) Can J Bot 67:218–224. **Hong DY,** Ma LM (1991) Acta PhytotaxonSin 29:25–51. **Huntley B** (1991) Ann Bot 67 (Suppl 1):15–22.

**Index Holmiensis** (1992) Vol 7 G–H. Lundquist J, Nordenstam B (eds) Swed Mus Nat Hist, Stockholm, 216 pp. **Isely D** (1990) Vascular flora of the Southeastern United States. Vol 3/2. Univ North Carolina, Chapel Hill, 258 pp. **Iversen ST** (1991) Symb Bot Ups 29:1–234.

**Jafri SMH,** El-Gadi A (1989) Flora of Libya. Parts 145–147. Al Faateh Univ, Tripoli, 422 pp. **Jäger EJ** (1990) Biol Rundsch 28:295–306. **Jalas J,** Suominen J (1989, 1991) Atlas florae Europaeae. Vols 8 + 9. Committee Mapping Flora Europe, Helsinki, 261, 110 pp. **Jeanmonod D,** Burdet HM (1987–1990) Compléments au Prodrôme de la flore Corse. Fasc 1–6. Cons Jard Bot, Genève, 18, 40, 56, 14, 19 pp. **Jessop JP** (1989) J Adelaide Bot Gard 12:1–163. **Jones SB,** Coile NC (1988) The distribution of the vascular flora of Georgia. Univ of Georgia, Athens, 230 pp. **Jurcev BA,** Kamelin RV (1991) Osnovnye ponjatija i terminy floristiki. Permsk Univ, Perm, 80 pp.

**Kadereit JW** (1990) Bot J Linnean Soc 103:221–231. **Kamelin RV** (1990) Flora Syrdar'inskogo Karatau. Nauka, Leningrad, 146 pp. **Klotz S** (1990) In: Sukopp H, Hejny S (eds) Urban ecology. SPB Acad, The Hague, pp 99–103. **Korotkov KO** (1989) Bjull Mosk Obšč Ispyt Prir Otd Biol 94:119–128. **Kowarik I** (1990a) Verh Berl Bot Ver 8:33–67. **Kowarik I** (1990b) In: Sukopp H, Hejny S (eds) Urban ecology.SPB Acad, The Hague, pp 45–75. **Kožarinov AV** (1989) Klimatochorologičeskij analiz populjacij lesnych rastenij Belorussii. Nauka Technika, Minsk, 176 pp. **Krutzsch W** (1989) Plant Syst Evol 162:5–61. **Krutzsch W** (1991) Feddes Repert 102:11–28. **Kullman L** (1991) New Phytol 117:351–360.

**Lahti T,** Kurtto A, Väisänen RA (1988) Ann Bot Fenn 25:281–291. **Laumonier Y** (1990) In: Baas P, Kalkman C, Geesink R (eds) The plant diversity of Malesia. Kluwer, Dordrecht, pp 193–211. **Lawson GW,** Simons RH, Isaac WE (1990) Bull Br Mus Nat Hist Bot 20:153–168. **Leistner OA** (1990) S Afr J Bot 56:706–707. **Léonard J** (1988, 1989) Contribution a l'étude de la flore et da la végétation des deserts d'Iran. Fasc 8+9. Jard Bot Nation Belg, Meise, 190, 123 pp. **Li LQ,** Zhang CF, Song SY, Chen Y (1991) Acta Phytotax Sin 29:113–130. **Linder HP** (1990) Mitt Inst Allg Bot Hamb 23b:777–790. **Lobin W,** Zizka G (1990) Mitt Inst Allg Bot Hamb 23b:739–743. **Lyr H,** Wersenger H (1990) Biol Rundsch 28:185–194.

**Malyshew LI** (1991) In: Nimis PL, Crovello TJ (eds) Quantitative approaches to phytogeography. Kluwer, Dordrecht, pp 15–33. **Mandaville JP** (1990) Flora of Eastern Saudi Arabia. Kegan Paul, London, 482 pp. **Marchant NG,** Wheeler JR, Rye BL, Bennet EM, Lander NS, Macfarlane TD (1987) Flora of the Perth region. Western Austr Herbarium, Baron-Hay Court, 1080 pp. **Martini F,** Poldini L (1990) Webbia 44:295–306. **Mc Donald JA** (1990) Sida 14:21–28. **Meusel H,** Jäger EJ (1992) Vergleichende Chorologie der zentraleuropäischen Flora. Vol 3 (2 parts). Fischer, Jena, 333+264 pp. **Meusel H,** Kästner A (1990) Denkschr Österr Akad Wien 127:1–294. **Millar AJK** (1990) Austr Syst Bot 3:293–593. **Miller A,** Morris M (1988) Plants of Dhofar. Diwan of Royal Court, Sultanate of Oman, 361 pp. **Ming T,** Fang R (1990) Acta Bot Yunn 12:353–365. **Molina A,** (1990) Bot Complutensis 16:129–172. **Moody ME,** Mack RN (1988) J Appl Ecol 25:1009–1021. **Morton JK,** Hogg EH (1989) Can J Bot 67:1803–1820. **Murray DR** (1986) Seed dispersal. Academic Press, Sydney, 322 pp.

**Nakanishi H** (1988) Hikobia 10:129–133. **Nikitin VV,** Gel'dichanov AM (1988) Opredelitel' rastenij Turkmenistana. Nauka, Leningrad, 680 pp.

**Ownbey GB,** Morley T (1991) Vascular plants of Minnesota: a checklist and atlas. Univ Minnesota, Minneapolis, 307 pp. **Opredelitel' rastenij Srednej Azii** (1987) Vol 9 Lamiaceae-Campanulaceae. FAN, Taškent, 400 pp.

**Panigrahi G,** Murti SK (1989) Flora of Bilaspur district (Madhya Pradesh) Vol 1. Bot Surv India, Calcutta, 396 pp. **Parks CR,** Wende JF (1990) Am J Bot 77:1243–1256. **Paton A** (1990) Notes R Bot Gard Edinb 46:345–359. **Pedersen B** (1989) Nord J Bot 10:163–189. **Peteet DM** (1991) Can J Bot 69:786–796. **Pichi Sermolli REG** (1990) An Jard Bot Madr 46:489–518. **Plumel MM** (1990) C R Seances Soc Biogeogr 66/3:103–127. **Prud'Homme WF,** Hoek C (1990) Cour Forschungsinst Senckenb 129:55–73. **Pyšek P** (1989) Preslia 61:329–334.

**Rajanady N** (1990) Mitt Inst Allg Bot Hamb 23a:39–52. **Randell BR** (1989) J Adelaide Bot Gard 12:165–272. **Rastenija Central'noj Azii** (1988, 1989) Vol 8a + 9. Nauka, Leningrad, 125, 148 pp. **Rezniczek AA** (1990) Contrib Univ Mich Herb 17:279–293. **Rödl-Linder G** (1990) Blumea 34:277–423. **Rueness J,** Jacobsen T, Asen PA (1990) Blyttia 48:21–26. **Rzedowski J,** Calderon de Rzedowski G (1990a) Flora Fanerogamica del Valle de Mexico. Vol 3. Inst de Ecologia, Patzcuaro/Mich, 494 pp. **Rzedowski J,** Calderon de Rzedowski G (1990b) Acta Bot Mexicana 12:21–24.

**Salgado AE** (1990) Philipp J Sci 119:107–148. **Salvesen PH** (1989) Blyttia 47:143–153. **Sandbrink JM,** Brederode J, Gadella TWJ (1990) Proc Kon Nederl Acad Wetensch 93/3:295–334. **Satabie B** (1991) Candollea 46:85–94. **Sauer JD** (1990) J Biogeogr 17:1–3. **Scholz H** (1991) Flora et Vegetatio Mundi 9:105–112. **Schönbeck-Temesy E,** Ehrendorfer F (1989) Plant Syst Evol 165:101–136. **Schönfelder P,** Bresinsky A (1990) Verbreitungsatlas der Farn- und Blütenpflanzen Bayerns. Ulmer, Stuttgart, 752 pp. **Schott AM** (1991) Blumea 36:205–234. **Schweingruber FH,** Briffa KR, Jones PD (1991) Vegetatio 92:5–71. **Sebald O,** Seybold S, Philippi G (1990) Die Farn- und Blütenpflanzen Baden-Württembergs. Vols 1+2. Ulmer, Stuttgart, 613, 442 pp. **Sesagiriri R,** Rolla PV, Reddy TA (1986) Flora of West Godavari district, Andhra Pradesh, India. Meerut Univ, Meerut, 520 pp. **Simpson B** (1988) Symb Bot Ups 28:150–158. **Skalicky V** (1990) Preslia 92:97–102. **Slavík B** (1990) Fytokartografické syntézy ČR. Vol 2. Bot Ustav ČSAV, Pruhonice, 179 pp. **Smith AC** (1991) Flora Vitiensis Nova. A new flora of Fiji. Vol 5. Pacif Trop Bot Gard, Lawaii/Hawaii, 626 pp. **Snogerup S,** Gustafsson M, Bothmer R (1990) Willdenowia 19:271–365. **Soreng RJ** (1990) Am J Bot 77:1383–1400. **Sosudistye rastenija Sovetskogo Dal'nego Vostoka** (1989) Vol 4. Nauka, Leningrad, 380 pp. **Soulé ME** (1990) Ann Mo Bot Gard 77:4–12. **Soyrinki N** (1991) Ann Bot Fenn 28:59–79. **Stanley TD,** Ross EM (1989) Flora of South-East Queensland. Vol 3. Queensland Dept Prim Industries, Brisbane, 532 pp. **Startinger U** (1990) Die Einbürgerung der Spätblühenden Traubenkirsche (*Prunus serotina* Ehrh.) in Mitteleuropa. Techn Univ, Berlin, 119 pp. **Strid A,** Kit Tan (1991) Mountain Flora of Greece. Vol 2. Univ Press, Edinburgh, 974 pp. **Sumina OI,** Jackevic VV (1990) Vestnik Leningr Univ Ser 3 Biol 3:41-45.

**Taylor P** (1989) The genus *Utricularia* – a taxonomic monograph. Her Majestys Stat Off, London, 724 pp. **Teoretičeskie problemy sravnitel'noj floristiki** (1987) Nauka, Leningrad, 292 pp. **Thomasson M,** Thomasson G (1991) Adansonia Bot 13:71–89. **Thorne RF** (1989) Rhodora 91:10–24. **Touw A,** Rubers VW (1989) De Nederlandse bladmosen – Flora en verspreidingsatlas van de Nederlandse Musci (*Sphagnum* uitgesondert). Stichting, Utrecht, 532 pp. **Trepl L** (1990) In: Sukopp H, Hejny S (eds) Urban ecology. SPB Acad, The Hague, pp 75–97. **Treter U** (1984) Die Baumgrenzen Skandinaviens. Steiner, Wiesbaden, 111 pp.

**Villar L,** Lazare J (1991) Itinera Geobot 5:481–504. **Wagner WL,** Herbst DR, Sohmer SH (1990) Manual of the flowering plants of Hawaii. Vol 1+2. Univ Hawaii, Honolulu, 1854 pp. **Warwick SJ** (1990) Plant Syst Evol 169:41–54. **Webb CJ,** Sykes WR, Garnock-Jones PV (1988) Flora of New Zealand. Vol 4 Naturalized pteridophytes, gymnosperms, dicotyledons. DSIR Bot Div, Christchurch, 1365 pp. **Weeda EJ** (1989) Gorteria 15:119–124. **Welzen PC** (1989) *Guioa* Cav. (Sapindaceae): taxonomy, phylogeny, and historical biogeography. Leiden Bot Ser vol 12, 315 pp. **Westhus W,** Zündorf HJ (1990) Landschaftspflege Naturschutz Thüringen 27:6–21. **White F** (1990) Adansonia Bot 12:139–185. **Whitmore TC,** Prance GT (1987) Biogeography and Quaternary history in tropical Latin America. Oxford Univ, Oxford, 564 pp. **Winkler S** (1990) Bot Jahrb Syst 112:43–77. **Wittig R** (1991) Ökologie der Großstadtflora. Fischer, Stuttgart, 261 pp. **Woodward FI** (1988) In: Long SF, Woodward FI (eds) Plants and temperature. Company of Biologists, Cambridge, pp 59–75. **Woodward FI** (1990) Philos Trans R Soc London B 326:585–593. **Wörz A** (1989) Bot Jahrb Syst 110:493–510. **Wraber T,** Skoberne P (1989) Rdeci seznam ogroženih praprotnic in semenk SR Slovenije. Varstvo Narave, Ljubljana, 14–15 428 pp. **Wu ZY** (1990) Ann Mo Bot Gard 77:427–429.

**Yamazaki T** (1990) Flora of Thailand. Vol 5/2 Scrophulariaceae. Thai Nation Doc Centre, Bangkok, pp 139–238. **Ying TS** (1989) Acta Phytotaxon Sin 27:17–26.

**Zajac A** (1988) Zesz Nauk Univ Jagiell Prace Bot 17:23–51. **Ziman SN,** Keener CS (1989) Ann Mo Bot Gard 76:1012–1049. **Zimmermann A,** Kniely G, Melzer H, Maurer W, Hollriegl R (1989) Mitt Abt Bot Landesmuseum Joanneum Graz 18/19:1–302.

Professor Dr. Eckehart J. Jäger
Institut für Geobotanik
und Botanischer Garten
Martin-Luther-Universität Halle
Neuwerk 21
O-4020 Halle, Germany

# III. Ecological Aspects of Nitrogen Nutrition

By Marianne Popp

## 1. Introduction

Nitrogen availability is one of the most decisive factors for plant performance in agricultural systems as well as in natural habitats. Not solely the amount of nitrogen, but also the presence of its different forms – nitrate or ammonium – has been considered important for plant distribution (cf. Lee et al. 1983; Runge 1983). A number of reviews and proceedings have dealt with these ecological aspects of nitrogen assimilation (Smirnoff and Stewart 1985; Lambers et al. 1986; Sprent 1987). However, in view of some recent developments, it seems justified to dedicate a further review to this subject.

More ample use of $^{15}N$ measurements has substantially increased our knowledge about nitrogen fixation in natural environments (Shearer and Kohl 1989). The use of this stable isotope is also of great help in determining preferential nitrate or ammonium uptake by plant roots, as well as the extent of deposition of these ions on aerial plant parts (Winner et al. 1989; Garten and Hanson 1990).

Another point of concern are the effects of nitrogen-containing air pollutants (e.g., $NO_x$) on the vegetation (Schulze 1989). In the latter paper, it is hypothesized that the imbalance between increasing nitrogen availability (via roots and canopy) from air pollutants and decreasing cation supply caused by soil acidification lead to the observed decline symptoms in spruce forests. In other words: the growth stimulus caused by the abundance of nitrogen is not matched by sufficient amounts of other nutrients.

The interrelation between nitrogen allocation and photosynthetic carbon gain of whole canopies are a further field of recent interest and research (Field and Mooney 1986; Werger and Hirose 1991). Facing the enormous body of literature dealing with ecological aspects of nitrogen nutrition makes clear that only some of it can be covered in this contribution. The present chapter will focus on the following topics:

1. measurement of $\delta^{15}N$ and its application on nitrogen-fixing, insectivorous and parasitic plants,
2. nitrogen assimilation and nitrogen allocation in relation to photosynthesis,
3. effects of nitrogen-containing air-pollutants.

Progress in Botany, Vol. 54

## 2. The Use of the Stable $^{15}N$ Isotope in Ecophysiological Research

### a) Methodology

Studies in nitrogen uptake have for long been hampered by the nonavailability of an "easy-to-handle" radioactive isotope. Only a few scientists had access to cyclotron-generated $^{13}N$, which has an half-life of 9.97 min (Glass et al. 1985; Lee and Clarkson 1986; Hole et al. 1990) and is therefore only suited for short-term experiments. However, like many other elements, also nitrogen exists in form of different stable isotopes. The $^{14}N$ isotope has an average abundance of 99.63 atom % and the $^{15}N$ isotope makes up for the missing 0.37 atom %.

Stable isotope ratios are measured by mass spectrometry after samples and elements respectively have been converted to gases (e.g., $N_2$, $CO_2$, $SO_2$). Values are generally expressed as a ratio between the sample and a standard.

$$\delta^{15}N = \frac{\text{atom \% }^{15}N\text{ (sample)} - \text{atom }^{15}N\text{ (standard)}}{\text{atom \% }^{15}N\text{ (standard)}} \cdot 1000‰\ ^{15}N$$

Shearer and Kohl (1989):

This definition of $\delta^{15}N$ is slightly different form the one used in geochemistry, namely

$$\delta^{15}N = \frac{(^{15}N/^{14}N)\text{ sample} - (^{15}N/^{14}N)\text{ standard}}{(^{15}N/^{14}N)\text{ standard}} \cdot 1000‰\ ^{15}N$$

However, at the level of natural $^{15}N$ abundance, the values for $\delta^{15}N$ calculated from the two equations are virtually identical, the difference being well within experimental error.

In the case of nitrogen, air is used as the standard, since the abundance of $^{15}N$ in air (in contrast to other stable isotopes like $^{13}C$) is constant (Ehleringer and Rundel 1989). The natural $^{15}N$ abundance expressed as $\delta^{15}N$ for biological materials ranges between –5 and +10‰. This means, compared to atmospheric $N_2$ with its 0.3663% $^{15}N$ abundance, biological samples may be depleted in $^{15}N$ content down to 0.3645 atom % or enriched up to 0.3699 atom % (Shearer and Kohl 1989). The reason for depletion or enrichment of biological material in $^{15}N$ lies in the different reactivity of molecules bearing the $^{14}N$ or $^{15}N$ isotope. Molecules containing the lighter $^{14}N$ isotope react somewhat faster than those with $^{15}N$. This kinetic isotope effect leads to an $^{15}N$ enrichment of substrates and depletion of products (Melander and Saunders 1980, cited in Shearer and Kohl 1989). Thus principially nearly all soil nitrogen is enriched in $^{15}N$ compared to atmospheric $N_2$, since nitrification, denitrification, as well as volatilization of $NH_3$ discriminate against $^{15}N$ (Högberg 1990). Values for $\delta^{15}N$ of total soil nitrogen are therefore in most cases positive and can reach levels up to +15 $\delta^{15}N$. As several investigations show (cf. Shearer and Kohl 1989), there is great variation in $^{15}N$ abundance of soils in respect to soil depth, time of the year, humidity, and so on. However, the difference in $^{15}N$ abundance in the soil and the constant $^{15}N$ abundance in air is the basis for estimating $N_2$ fixation, especially in natural ecosystems. If a plant derives its nitrogen solely from soil nitrogen, the $\delta^{15}N$ in the plant material will be close to that of the soil and thus more or less positive. If a nitrogen-fixing plant is grown hydroponically with no other nitrogen source than $N_2$ from the air, the $\delta^{15}N$ value of this plant might be even lower than that in the air, since there is also some slight discrimination against $^{15}N$ during the nitrogen-fixing process. The picture is obscured under natural conditions, where nitrogen-fixing plants use both sources (soil and air) for their nitrogen nutrition. Therefore, great care has to be taken when applying the $\delta^{15}N$ method for estimating $N_2$ fixation (for details see Shearer and Kohl 1989). For a qualitative statement of whether or not a plant is nitrogen-fixing, the plant under question and several nonfixing reference plants, which should if possible be of the same growth form and have similar rooting depth, have to be analyzed for their $^{15}N$ abundance. The basic assumption is that the soil-derived nitrogen would lead to the same $\delta^{15}N$ values in nitrogen-fixing and nonfixing plants, but the $\delta^{15}N$ of the nitrogen-fixing plant is lowered by the input of atmospheric nitrogen. This comparison of nitrogen-fixing plants to nonfixing reference plants has proven to be more useful than relating $\delta^{15}N$ plant material to that of the soil, because of the huge heterogenity of $\delta^{15}N$ in soils mentioned above.

To obtain quantitative estimate of the contribution of nitrogen fixation in a certain plant, one needs the $\delta^{15}N$ of this plant when grown in a N-free medium and depending solely on atmospheric $N_2$. The formula proposed by Shearer and Kohl (1989) runs as follows:

$$N_f = \frac{\delta^{15}N_o - \delta^{15}N_t}{\delta^{15}N_o - \delta^{15}N_a} \cdot 100,$$

where $N_f$ proportion of nitrogen derived from nitrogen fixation,

$\delta^{15}N_a$ $\delta^{15}N$ of the $N_2$-fixing plant growing on N-free medium, having only atmospheric $N_2$ as nitrogen source,

$\delta^{15}N_t$ $\delta^{15}N$ of the $N_2$-fixing plant under natural conditions, having excess to atmospheric and soil nitrogen,

$\delta^{15}N_o$ $\delta^{15}N$ neighboring nonfixing reference plants, depending solely on soil nitrogen.

Taking into account that the difference between $N_2$-fixing and nonfixing plants is usually less than 10 ‰, and that on the other hand the difference between parts of one and the same plant may account for 2 ‰, makes clear that great care has to be taken when using this method. However, there are only few and not much better alternatives (Sprent and Sprent 1990). The final sentences in the paper of Shearer and Kohl (1989) makes the point: "The $\delta^{15}N$ method is particularly well suited to ecosystem studies in which it is desirable to avoid experimental manipulation. In combination with N productivity data, the $\delta^{15}N$ method is capable of providing estimates of the amount of N introduced to an ecosystem by symbiotic $N_2$ fixation".

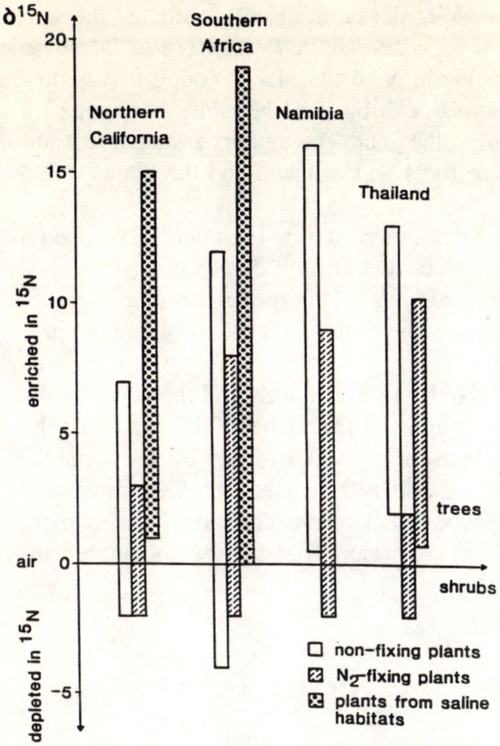


**Fig. 1.** Range of $\delta^{15}N$ values from plants of various ecosystems. Northern California (Virginia and Delwiche 1982); 115 samples, only one nitrogen-fixing tree included, all other $N_2$-fixing plants herbs or shrubs. Southern Africa (Heaton 1987); 140 samples, only 15 species of $N_2$-fixers included; $N_2$-fixers mainly trees, nonfixers include a high number of grasses and herbs. Thailand (Yoneyama et al. 1990); 68 species, $N_2$-fixing shrub and tree legumes and *Casuarina spp*; nonfixers: all kinds of growth forms. Namibia (Schulze et al. 1991a); 108 samples, $N_2$-fixing and nonfixing trees

## b) Nitrogen Fixation in Natural Ecosystems

From Fig. 1, where $\delta^{15}N$ values from plants of various ecosystems are assembled, it becomes evident that distinguishing between nitrogen-fixing and nonfixing plants is not as easy as differentiating between the $C_3$ and $C_4$ pathways of photosynthesis by $\delta^{13}C$ values. Only in the study of Thailand (Yoneyama et al. 1990) were the shrub legumes different in their $\delta^{15}N$ compared to the reference plants. In all other cases there was considerable overlap between $\delta^{15}N$ of fixing and nonfixing plants, when all data were pooled together. However, with a few exceptions, when focussing on a single site or a single growth form, fixing and nonfixing plants could be separated by their $\delta^{15}N$ values (Virginia and Delwiche 1982; Yoneyama et al. 1990; Schulze et al. 1991a). Comparing different growth forms, Virginia and Delwiche (1982) found the biggest differences in $\delta^{15}N$ between $N_2$-fixing and other plants in annual herbs, followed by perennial herbs and shrubs. Since they included only one nitrogen-fixing tree, they could not make a statement on this growth form. Differences in $\delta^{15}N$ between growth forms might be due to their different rooting depth and therefore varying access to soil layers with different $^{15}N$ abundance (Virginia et al. 1989). This also makes clear how delicate the choice of the "right" reference plant is. The same $\delta^{15}N$ value of a nitrogen-fixing plant can be brought about by having either a high proportion of atmospheric $N_2$ fixed and excess to soil nitrogen, which is very enriched in $^{15}N$ or having excess to soil nitrogen depleted in

$^{15}N$ and rather little uptake of $N_2$ from the air. Virginia et al. (1989) attributed the seasonal pattern of foliar $^{15}N$ abundance that they observed in *Prosopis glandulosa* to changes in the contribution of soil and atmospheric N to the plant. Young leaves produced in the early season, when the soil was moist, exhibited $\delta^{15}N$ values of about +1.5 decreasing to –1.5, when the soil dried out during the year. The authors assumed that under the latter conditions, the activity of surface roots was reduced and the plants relied more on $N_2$ fixation.

In his study on mainly not $N_2$-fixing plants in Southern Africa Heaton (1987) stated a slight positive correlation between plant $\delta^{15}N$ values and aridity: $\delta^{15}N$ values decreased with increasing rainfall (–0.39‰ per 100 mm rain $yr^{-1}$). This mirrors former observations that $\delta^{15}N$ values of total nitrogen in soils are correlated with the mean annual precipitation (Shearer et al. 1978, cited in Heaton 1987).

Another factor which seems to influence $^{15}N/^{14}N$ ratios of soils and therewith in the plants growing on them seems to be salinity. Virginia and Delwiche (1982) explained the relatively high $\delta^{15}N$ values they found in plants of coastal habitats by the input of oceanic nitrate, which has $\delta^{15}N$ values between +5 and +10‰. Heaton (1987) confirmed these findings for coastal plants, but moreover reported also considerable $^{15}N$ enrichment for plants from an inland salt pan (Fig. 1). The reason for this awaits further investigation.

### c) Insectivorous Plants

Dixon et al. (1980) were the first to use the stable isotope technique to elucidate the proportion of prey-derived nitrogen in the rosette sundew *Drosera erythrorhiza*. For this plant they estimated that 11–17% of the plant's nitrogen uptake during one growing season came from their insect prey. For the same plant species, Schulze et al. (1991c) calculated between 14.3 and 32.2% $N_{dI}$ ($N_{dI}$ = proportional dependence on insect N) when glandless plants of *D. erythrorhiza* were used as reference plants. The values were much lower (10.4%) or even zero when calculations were based on other reference plants, although the difference between $\delta^{15}N$ of insects (+2.56 to 3.58) and soils (–0.81 to +0.47) seemed to be sufficiently high. In the same study, a much higher percentage (up to 50%) of insect derived nitrogen was described for climbing and erect species of *Drosera*. This agrees with findings by Thum (1989) on *D. rotundifolia* where the catching success increased with higher plant position. However, the reverse was true for *D. intermedia* and a number of other factors, like trapping area and intraspecific neighborhood, are also important for the acquisition of prey (Thum 1989).

The "fertilizing effect" of animal food supply for carnivorous plants is out of the question (Thum 1988; Schulze and Schulze 1990), nevertheless it has been hypothesized that the response of carnivorous plants to prey depends on the soil nutrient levels (Givnish et al. 1984, cited in Karlsson et al. 1991).

Karlsson et al. (1991) proved this hypothesis by growing *D. rotundifolia, Pinguicula alpina, P. villosa*, and *P. vulgaris* at two different nutrient levels (0.005 and 5 mM nitrogen in the nutrient solution, respectively) and three levels of prey supply. Five cases out of 28 confirmed the hypothesis; in one case it was disproved, the response to prey being higher in the plants of the high nutrient treatment. In the majority of cases (78%) there was no significant difference in response to prey between the two different soil nutrient treatments. The authors concluded that the response to prey can be considered to be independent of soil nutrient status, although the two bog species (*D. rotundifolia, P. villosa*) were more responsive to prey supply than the two species (*P. alpina, P. vulgaris*) from mineral soil.

### d) Parasitic Plants

Parasitic angiosperms derive their nitrogen solely from their hosts, either in inorganic or already reduced form (Stewart and Press 1990). Sometimes appreciable activities of *n*itrate *r*eductase (up to 238 nmol $NO_2$ $g^{-1}$ *f*resh *w*eight $h^{-1}$ or 66 pkat $g^{-1}$ f.w.) were demonstrated for stem parasites, while many root parasites did not exhibit measurable NR activities (Hunter and Visser 1986; McNally and Stewart 1987). However, in nutrient-poor habitats, when host xylem saps are low in nitrate, the bulk of nitrogen for the parasites comes from organic forms, mainly amino acids, which are transported via either apoplastic or symplastic routes (Coetzee and Fineran 1989; Pate et al. 1991a,b). Feeding $^{15}N$-labeled aspartic acid or $^{15}N$-(amide) labeled aspargine to cut shoots or whole haustoria-bearing plants of the misteltoe *Amyema preissii* led to similar amounts of $^{15}N$ in the ethanol-soluble amino compounds of the shoot, thus demonstrating that the haustorium had relatively little effect on processing these incoming solutes.

In agreement with this observation, Schulze et al. (1991b) found only little or no difference in foliar $\delta^{15}N$ values in different host-parasitic associations from Namibia. Mistletoe (e.g., *Odontella welwitschii*) could be depleted compared to the corresponding host by up to 2.3‰. This tendency fits the general view that nitrogen sources become enriched in $^{15}N$ compared to their sinks (see above) and vice versa.

Total nitrogen content in mistletoes and hosts is also of interest in regard to the hypothesis put forward on the occurrence of mimicry (Ehleringer et al. 1986; Barlow and Wiens 1977, cited in Bannister 1989). Barlow and Wiens (1977, cited in Bannister 1989) assumed that those mistletoes which closely resemble their host avoid browsing by vertebrate predators. Ehleringer et al. (1986) extended this hypothesis on the basis of total nitrogen measurements on Australian mistletoes which give an estimate of their leaf protein. According to these authors, mistletoes which mimicked their hosts had higher leaf protein than their hosts, whereas nonmimics have lower values. The latter group was supposed to gain some selective advantage by "advertising" their lower nutrient values to the consumers by looking different from their host. Bannister (1989) collected a similar data set for mistletoes from New Zealand confirming the findings from Australian mistletoes: cryptic mistletoes having nitrogen concentration equal to or higher than their hosts while noncryptic species are lower in total N compared to their hosts. However, the hypothesis of selective grazing cannot be applied to the New Zealand mistletoes, since they almost certainly evolved in the absence of herbivorous mammals (Bannister 1989). No other explanation is offered at the moment, why leaf mimicry and nitrogen accumulation coincide in so many cases. Hall et al. (1987) showed that in mimicking situations (*Amyema miquelii* on *Eucalyptus polyanthemos*) types and concentrations of cytokinins in the xylem saps of mistletoe and host were more similar than in a nonmimicking association (*Amyema pendulum* on *Eucalyptus cinerea*). Although we mentioned above that the haustorium of *Amyema preissii* did not alter the nitrogen intake from the xylem sap very much, one interpretation might be that in haustoria of nonmimicking species changes of phytohormones of the host xylem sap take place while this is not the case in mimicking species. Also, a different proportion of apoplastic and symplastic transport could lead to the observed effects.

## 3. Nitrogen Assimilation and Nitrogen Allocation in Relation to Photosynthesis and Plant Performance

### a) Nitrate or Ammonium?

One of the "everlasting" questions for ecophysiologists is why certain plants occur in certain environments and others not. Decisive soil factors like pH and aeration have also a great bearing on the availability of nitrate and ammonium. Although the calculated cost of growth per unit N with $NO_3^-$ as N source is from 5 to 12% greater than that with $NH_4^+$ as N source (Raven 1985), most plants prefer $NO_3^-$ as N source. In several cases, a close relationship between *n*itrate *r*eductase (NR) activity and nitrogen content of plants was shown (Gebauer et al. 1988; Högbom and Ohlson 1991). In their study on 48 herbaceous plant species from 14 natural habitats in Central Europe, Gebauer et al. (1988) demonstrated good correlations between nitrate, NR activity as well as organic nitrogen of plant species and the "nitrogen figure" (Stickstoffzahl) introduced by Ellenberg (1979) for vascular plants of Central Europe.

However, no such relationship was observed by Stewart et al. (1988) working on plants of Australian rainforest communities. Twenty species were investigated in both disturbed (open) and undisturbed (closed) sites. Total nitrogen between plants from the open and closed sites differed by less than 5%, while NR activity was three times greater in plants growing in the gap compared with corresponding values of closed forest. Thus species at the closed-forest seemingly utilized sources of nitrogen other than $NO_3^-$. There are several examples for a preferential $NH_4^+$ uptake in a variety of plants. Marschner et al. (1991) showed that uptake rates of $NH_4^+$ were three to four times higher than those of $NO_3^-$ in nonmycorrhizal roots of *Picea abies*. Using $^{15}N$-labeled ammonium and nitrate revealed also for *Carex bigelowii* a higher uptake of ammonium-N, but nitrate-N was translocated at a higher rate (Iónsdóttir and Callaghan 1990). Preference for $NH_4^+$ uptake is what one expects from a plant of a wet, nutrient-poor environment, such as *C. bigelowii* from Swedish lapland. Koch et al. (1991) compared two Alaskan sedges *Eriophorum scheuchzeri* and *Eriophorum vaginatum*, which are native to high- and low-fertility sites, respectively, although from contrasting edaphic habitats kinetics of $NH_4^+$ and $NO_3^-$ absorption, as well as the capability to assimilate $NO_3^-$, were similar in both species. Growth of *E. vaginatum* under $NO_3^-$-nutrition was only slightly less than under $NH_4^+$. Rode and Runge (1991) showed for a number of plants (*Avenella flexuosa*, *Calluna vulgaris*, and others) from acid soils that they grew equally well with $NH_4NO_3^-$ and with only $NO_3$-supply. In contrast, plants from alkaline to moderately acid soils (*Carex remota, Origanum vulgare*, and others) exhibited a marked preference for $NO_3$-supply. On the other hand, coexisting plants may vary appreciable in their capacity to use $NO_3^-$ as N source, as was described by Högbom and Ohlson (1991) for 14 species from a central Swedish mire.

More ample use of $^{15}N$-labeling methods will help to elucidate the preference of either $NO_3^-$ or $NH_4^+$ uptake in a wider range of species.

b) $NO_3^-$ Assimilation Where?

It is widely believed that assimilation of nitrate in leaves carries a lower energy cost than root assimilation as long as photosynthesis is light-saturated (cf. Smirnoff and Stewart 1985). Since nitrate reductase, glutamine synthase (GS), and glutamate oxo-glutarate amino transferase (GOGAT) are located in the chloroplasts, they can make direct use of excess reductant and ATP generated by the photochemical reactions. In the case of root assimilation, reductant and ATP are provided by the mitochondria which have to be supplied with sucrose from the shoots. Smirnoff and Stewart (1985) argued that in shade environments, where photosynthesis is light-limited, root assimilation may allow a greater control between carbon and nitrogen assimilation. However, in a study on six species of the genus *Piper* growing under a range from high to low light conditions, root NR activities were not influenced by the available photosynthetically active photon flux density (PFD) (Fredeen et al. 1991). Root NR activities were an order of magnitude lower than leaf NR activities. The latter correlated highly with (PFD) and showed pronounced differences in their reaction to $NO_3^-$ addition to the soil. Leaf NR was highly inducible in the gap, but not in the shade species (Fredeen et al. 1991).

The importance of root nitrate assimilation has for long been overestimated in woody plants. Based on only a few measurements of nitrate and organic nitrogen compounds in xylem saps it was generally assumed that woody plants reduce nitrate mostly in their roots (cf. Smirnoff et al. 1984). This is evidently not the case, since Smirnoff et al. (1984) showed for 41% of 555 woody species investigated NR activities greater than 0.2 μmol $h^{-}$ $g^{-1}$ fresh weight (55 pkat $g^{-1}$ f.w.), excluding Gymnosperms, Ericaceae, and Proteaceae (which had consistently low activities) the number went even up to 73%. This indicated that leaf nitrate assimilation is of common occurrence in woody plants. Comparing 39 species of tropical and subtropical trees, Stewart et al. (1989) reported only for 30% shoot/root ratios in NR activity below 1, in 40% of the species this ratio ranged between 1 and 5, in 30% the ratio was greater than 5. The group with higher root than shoot NR activities included several mangrove species as well as some $N_2$-fixing trees, thus suggesting that the preponderance of root nitrate reduction may be confined to trees which utilize mainly $NH_4^+$ (mangroves) or exhibit symbiotic dinitrogen fixation (Stewart et al. 1989).

A preference for leaf nitrate assimilation was shown for pioneer species from subtropical-tropical (Stewart et al. 1988, 1989) and temperate forest sites (Clough et al. 1989; Pearson et al. 1989). In investigation from several English woodland communities, climax species had on average NR activity of 117 pkat $g^{-1}$ f.w. compared with 340 pkat $g^{-1}$ f.w. for pioneer species (Clough et al. 1989). In pioneer (e.g., *Sambucus niger*) as well as climax (e.g., *Betula pendula*) species a marked seasonal variation in leaf NR activities was observed (Clough et al. 1989; Pearson et a. 1989). This might be the consequence of different $NO_3$ availability or different "sharing" of $NO_3^-$ assimilation between roots and shoots during the seasons. Pronounced seasonal fluctuations of root NR activities were reported for the weedy perennial *Cichorium intybus* by Sechley et al. (1991).

Thus further studies are needed to elucidate if there is a shift between root and shoot nitrate assimilation during the course of seasons in one and the same plant.

### c) Nitrogen Allocation

Once the nitrogen is taken up and assimilated, the question of distribution occurs. From a vast number of investigations, it is evident that photosynthetic capacity of single leaves to the whole canopy is strongly geared to nitrogen content (cf. Field and Mooney 1986). The ratio between photosynthetic capacity (A max) (which is the photosynthetic rate measured under saturating light intensity, optimum temperature, relatively high humidity and $CO_2$ concentration of normal air) and total leaf nitrogen is termed *p*hotosynthetic *ni*trogen-*use* *e*fficiency (PNUE). Values given for plants of different growth forms and various habitats ranged between 142 (Death Valley annuals) to 46 (California evergreens) µmol $CO_2$ [mol N]$^{-1}$ s$^{-1}$ (Field and Mooney 1986). In a comparison of 24 wild species differing in relative growth rate, Poorter et al. (1990) observed that fast-growing species had a higher PNUE than slow-growing ones. Fast-growing species had also a higher total organic nitrogen concentration per unit plant weight and allocated more nitrogen to their leaves.

Nitrogen distribution within a canopy is not uniform, but affected by the light climate in the canopy (Werger and Hirose 1991) and leaf age (Field 1983; Hirose et al. 1989). For *Solidago altissima*, Hirose and Werger (1987) described that with the actual distribution of leaf nitrogen in the canopy 20% more photosynthesis was achieved than with an uniform distribution of nitrogen throughout the canopy. For *S. altissima*, as well as for *Lysimachia vulgaris* (Hirose et al. 1988), it was shown that in open stands nitrogen distribution was more uniform than in dense stands, which was only partly explained by the different light climates of the stands. As calculations revealed, in some dense stands whole carbon gain of the canopy could have been higher if the available total leaf nitrogen had been distributed on a smaller leaf area. However, the development of a higher leaf area index might have been of advantage in shading out competitors (Werger and Hirose 1991).

As pointed out by Küppers et al. (1988), extrapolations from photosynthetic rates to carbon gains can be only made under defined environmental conditions. In this study *Raphanus sativus* x *raphanistrum* was grown at high and low N-nutrition, high and low light intensities, which, in combination with long and short photoperiods, gave the same daily light dose. High N, high light and long photoperiod increased A max and therewith growth. Low N, low light and short photoperiod reversed these responses. At other combinations, such as low light and long photoperiod or high light and short photoperiod, A max did not correlate with carbon gain of the canopy or whole plant growth. In respect to plant growth, low N-nutrition could be compensated by a long photoperiod.

The close interrelationship between nitrogen and photosynthesis observed on the leaf canopy level is backed up by a number of biochemical investigations, which show how closely nitrogen and carbon assimilation are connected.

Exposing spinach leaves to low $CO_2$ in the light resulted in a rapid decline of extractable NR activity, which was quickly reversed when leaves were supplied with air again (Kaiser and Brendle-Behnisch 1991). It seems likely that NR activity is modulated by the cytoplasmic ATP/AMP ratio, which might either regulate phosphorylation/dephosphorylation of the NR protein itself or of a NR-regulating protein factor (Kaiser and Spill 1991). For cotyledons of *Cucurbita maxima*, it was postulated that NR gen expression is regulated not only by $NO_3^-$, but also by light, the phytochrome system and a plastidic factor being involved (Redingbaugh and Campbell 1991).

On the other hand, phosphoenolpyruvate carboxylase and pyruvate orthophosphate dikinase were reduced in their synthesis by nitrogen deficiency in maize leaves (Sugiharto et al. 1991). This was not only because of the lack of nitrogen for protein synthesis, but regulation took place already at the level of gen expression.

Following up nitrogen allocation on the subcellular level revealed that additional nitrogen will be mainly located in the stroma (Makino and Osmond 1991) where in sun plants already the main proportion of total leaf nitrogen is stored (Evans 1989). In shade plants like *Alocasia* sp. under low light conditions, 71% of total leaf nitrogen was located in the thylakoids, which changed under high light conditions to 29% comparable to 24% thylakoid-N of total leaf-N found in spinach leaves.

Thus nitrogen allocation seems to be tuned by light availability on the subcellular as well as on the canopy level.

## 4. Nitrogen Deposition

Eutrophication by nitrogen-containing air pollutants may lead to species decline (Schulze 1989) or changes in the competitive relations among plant species (Heil et al. 1988).

In several parts of Europe, atmospheric $NH_x$ depositions originating from intensive farming systems are causing severe problems. In the Netherlands, *Calluna* heaths are turned into grass heaths, since *Deschampsia flexuosa* benefits much more from foliar uptake of $NH_3$ than *Calluna vulgaris*, as was shown by $^{15}NH_3$ uptake experiments (Dueck et al. 1991). In Finland, winter dieback of Scots pine (*Pinus sylvestris*) was observed in the vicinity of fur farms. The $NH_3$ emitted increased free amino acids and proteins in the needles of close-by trees, but the formation of certain polypeptides, which were possibly decisive for frost hardening, was inhibited.

Not only $NH_3$, but also $HNO_3$, $NO_2$, and NO are taken up by the aerial part of plants, although to different degrees. When leaves of *Acer rubrum* and *Quercus alba* were exposed to either $^{15}N$-nitrate of $^{15}N$-ammonium, foliar retention of $NO_3^-$ was with 13% in red maple and 27% in white oak much lower compared to the $^{15}NH_4^+$ treatment, where retentions accounted for 58 and 84% respectively (Garten and Hanson 1990).

Following the routes of $HNO_3$ dry deposition on coniferous seedlings, Cadle et al. (1991) found that 80% of the nitric acid deposited via the transcuticular and stomatal routes was assimilated by the plants. They also pointed out that deposition to branches and stems can account for as much as 40% of the total recovered $NO_3^-$. Absorption of atmospheric $NO_2$ by spruce (*Picea abies*) showed a linear increase with increasing transpiration, but also an uptake component, which might be due to $NO_2$ flux through the cuticle and bark (Thoene et al. 1991). Uptake of $NO_2$ and NO was demonstrated for both isolated cuticles and intact needles from *Picea abies* (Kisser-Priesack et al. 1987).

Fumigation with $NO_2$ led to a marked increase of NR activity in the needles of *Picea abies* (Thoene et al. 1991). According to Raven (1988), who assumed that 2 mol $NO_2$ react with 1 mol $H_2O$ to give 1 mol $HNO_2$ and 1 mol $HNO_3$, $NO_2$ dry deposition yields 0.22 $H^+$ mol excess per mol N assimilated. A number of other examples in this paper (Raven 1988) make clear that acquisition of nitrogen by the aerial parts of plants are of important consequence for acid-base regulation. For $NH_4$ wet deposition (for instance from $(NH_4)_2SO_4$) even 1.22 $H^+$ mol are produced per N assimilated.

Since the aerial parts do not have the possibility of extruding $H^+$ as the roots do, acid-base perturbations at the cellular level are likely to be a possible reason for decline symptoms.

As mentioned in the introduction, an imbalance between the nitrogen status and the lowered availability of other nutrients was also assumed to be responsible for the different performance of *Picea abies* in certain stands (Oren et al. 1988)

## 5. Outlook

The impact of air-borne nitrogen on ecosystems will certainly be a further field of research, where future investigations will also have to take into consideration the possible increases in $CO_2$. Looking at the numerous experiments on the consequences of elevated $CO_2$ on plant performance raises the question of how different levels of nitrogen availability would influence the outcome of these investigations.

The use of stable isotope techniques would be suited for several tasks in nitrogen research, the high costs of appropriate mass spectrometers being the only hindrance for their broader application.

## References

**Bannister P** (1989) Oecologia 79:128–132.

**Cadle SH,** Marshall JD, Mulawa PA (1991) Environ Pollu 72:287–305. **Clough ECM,** Pearson J, Stewart GR (1989) Ann Sci For 46:669s–672s. **Coetzee J,** Fineran BA (1989) New Phytol 112:377–381.

**Dixon KW,** Pate JS, Bailey WJ (1980) Aust J Bot 28:283–297. **Dueck THA,** van der Eerden LJ, Beemsterboer B, Elderson J (1991) Acta Bot Neerl 40:257–267.

**Ehleringer JR,** Rundel PW (1989) In: Rundel PW, Ehleringer JR, Nagy KA (eds) Stable isotopes in ecological research. Ecological Studies 68. Springer, New York Berlin Heidelberg. **Ehleringer JR,** Ullmann I, Lange OL, Farquhar GD, Cowan IR, Schulze ED, Ziegler H (1986) Oecologia 70:243–247. **Ellenberg H** (1979) Scr Geobot 9:1–122. **Evans JR** (1989) Oecologia 78:9–19.

**Field C** (1983) Oecologia 56:341–347. **Field C,** Mooney HA (1986) In: Givnish TJ (ed) On the economy of plant form and function. Cambridge Univ Press, Cambridge. **Fredeen AL,** Griffin K, Fiew CB (1991) Oecologia 86:441–446.

**Garten CT Jr,** Hanson PJ (1990) Environ Exp Bot 30:333–342. **Gebauer G,** Rehder H, Wollenweber B (1988) Oecologia 75:371–385. **Glass ADM,** Thompson RG, Bordeleau L (1985) Plant Physiol 77:379–381.

**Hall PJ,** Badenoch-Jones J, Parker CW, Letham DL, Barlow BA (1987) Aust J Plant Physiol 14:429–438. **Heaton THE** (1987) Oecologia 74:236–246. **Heil GW,** Werger MJA, De Mol W, van Dam D, Heijne B (1988) Science 239:764–765. **Hirose T,** Werger MJA (1987) Oecologia 72:520–526. **Hirose T,** Werger MJA, Pons TL, Van Rheenen JWA (1988) Oecologia 77:145–150. **Hirose T,** Werger MJA, Van Rheenen JWA (1989) Ecology 70:1610–1618. **Högberg P** (1990) Oecologia 84:229–231. **Högbom L,** Ohlson M (1991) Oecologia 87:495–499. **Hole DJ,** Emran AM, Fares Y, Drew MC (1990) Plant Physiol 93:642–647. **Hunter JJ,** Visser JH (1986) S Afr Tydskr Plantk 52:81–84.

**Ionsdottir IS,** Callaghan TV (1990) New Phytol 114:419–428.

**Kaiser WM,** Brendle-Behnisch E (1991) Plant Physiol 96:363–367. **Kaiser WM,** Spill D (1991) Plant Physiol 96:368–375. **Karlsson PS,** Nordell KO, Carlsson BA, Svensson BM (1991) Oecologia 86:1–7. **Kisser-Priesack GM,** Scheunert I, Gnatz G, Ziegler H (1987) Naturwissenschaften 74:550–551. **Koch GW,** Bloom AJ, Chapin FS III (1991) Oecologia 88:570–573. **Küppers M,** Koch G, Mooney HA (1988) Aust J Plant Physiol 15:289–298.

**Lambers H,** Neeteson JJ, Stulen I (1986) Fundamental ecological and agricultural aspects of nitrogen metabolism in higher plants. Nijhoff, Dordrecht. **Lee JA,** McNeill S, Rorison IH (1983) Nitrogen as an ecological factor. Blackwell, Oxford. **Lee RB,** Clarkson DT (1986) J Exp Bot 37:1753–1767.

**Makino A,** Osmond CB (1991) Plant Physiol 96:355–362. **Marscher H,** Häussling M, George E (1991) Trees 5:14–21. **McNally SF,** Stewart GR (1987) In: Weber HC, Forstreuter W (eds) Parasitic flowering plants. Proc 4th Int symp parasitic flowering plant. Philipps Univ, Marburg.

**Oren R,** Schulze ED, Werk KS, Meyer J (1988) Oecologia 77:163–173.

**Pate JS,** True KC, Kuo J (1991a) J Exp Bot 42:429–439. **Pate JS,** True KC, Rasins E (1991b) J Exp Bot 42:441–451. **Pearson J,** Clough ECM, Kershaw JL (1989) Ann Sci For 46:663s–665s. **Pietilä M,** Lähdesmäki P, Pietiläinen P, Ferm A, Hytönen J, Pätila A (1991) Environ Pollut 72:103–115. **Poorter H,** Remkes C, Lambers H (1990) Plant Physiol 94:621–627.

**Raven JA** (1985) New Phytol 101:25–77. **Raven JA** (1988) New Phytol 109:1–20. **Redinbaugh MG,** Campbell WH (1991) Physiol Plant 82:640–650. **Rode MW,** Runge M (1991) In: McMichael BL, Persson H (eds) Plant roots and their environment. Elsevier, Amsterdam, pp 263–272. **Runge M** (1983) In: Pirson A, Zimmermann MH (eds) Encyclopedia of Plant Physiology New Series, vol 12C. Springer, Berlin Heidelberg New York, pp 163–200.

**Schulze ED** (1989) Science 244:776–783. **Schulze ED,** Gebauer G, Ziegler H, Lange OL (1991a) Oecologia 88:451–455. **Schulze ED,** Lange OL, Ziegler H, Gebauer G (1991b) Oecologia 88:457–462. **Schulze ED,** Gebauer G, Schulze W, Pate J.S. (1991c) Oecologia 87:240–246. **Schulze W,** Schulze ED (1990) Oecologia 82:427–429. **Sechley KA,** Oaks A, Bewley JD (1991) Plant Physiol 97:322–329. **Shearer G,** Kohl DH (1989) In: Rundel PW, Ehleringer JR, Nagy KA (eds) Stable isotopes in ecological research. Springer, Berlin Heidelberg New York, pp 342–374. **Smirnoff N,** Stewart GR (1985) Physiol Plant 64:133–140. **Smirnhoff N,** Tood P, Stewart GR (1984) Ann Bot 54:363–374. **Sprent JI** (1987) The ecology of the nitrogen cycle. Cambridge Univ Press, Cambridge. **Sprent JI,** Sprent P (1990) Nitrogen fixing organisms. Chapman and Hall, London. **Stewart GR,** Press MC (1990) Annu Rev Plant Physiol Mol Biol 41:127–151. **Stewart GR,** Hegarty EE, Specht RL (1988) Physiol Plant 74:26–33. **Stewart GR,** Pearson J, Kershaw JL, Clough ECM (1989) Ann Sci For 46:648s–653s. **Stewart GR,** Gracia CA, Hegarty EE, Specht RL (1990) Oecologia 82:544–551. **Sugiharto B,** Miyata K, Nakamoto H, Sasakawa H, Sugiyama T (1991) Plant Physiol 92:963–969.

**Thoene B,** Schröder P, Papen H, Egger A, Renenberg H (1991) New Phytol 117:575–585. **Thum M** (1988) Oeclogia 75:472–480. **Thum M** (1989) Oecologia 81:401–411.

**Virginia RA,** Delwiche CC (1982) Oecologia 54:j317–325. **Virginia RA,** Jarrell WM, Rundel PW, Shearer G, Kohl DH (1989) In: Rundel PW, Ehleringer JR, Nagy KA (eds) Stable isotopes in ecological research. Springer, Berlin Heidelberg New York, pp 375–394.

**Werger MJA,** Hirose T (1991) Vegetatio 97:11–20. **Winner WE,** Berg VS, Langston-Unkefer PJ (1989) In: Rundel PW, Ehleringer JR, Nagy KA (eds) Stable isotopes in ecological research. Springer, Berlin Heidelberg New York, pp 454–470.

**Yoneyama T,** Murakami T, Boonkerd N, Wadisirisuk P, Siripin S, Kouno K (1990) Plant Soil 128:287–292.

Professor Marianne Popp
Institut für Angewandte Botanik
Hindenburgplatz 55
W-4400 Münster, Germany

# F. Special Topics

## I. Floral Ecology: Report on the Years 1988 (1987) to 1991 (1992)

By Gerhard Gottsberger

### 1. Introduction

Studies on floral ecology and flower evolutionary aspects have never been as broad as at present. Research in biotic and abiotic pollination and nearly every facet of flowers, their biology and function has grown enormously. The reason for this was outlined in my last report (Prog Bot 50). The most fascinating perspectives are that floral ecologists, studying and measuring variation and morphological and behavioral modifications in the interactions they observe, are more and more in a position to evaluate directly selectional pressure, its dynamics in space and time, which then allows the understanding of the basic forces of adaptive evolution. In other words, many recent studies utilize animal-flower interactions to make quantitative tests of general hypotheses in evolutionary ecology (Feinsinger 1987; Armbruster 1992). Feinsinger (1987) proposes two major directions for future research, namely its application to agriculture and conservation biology. Barrett and Eckert (1990) expect that future studies will be even more comprehensive, as they will not concentrate on single life cycle stages only, but consider also the traits expressed at different stages of the life history; selection does not act on isolated traits but "evaluates" the whole phenotype during its entire life.

Meetings: Sixth International Symposium of Pollination, Tilburg, the Netherlands (August 1990); Plant Reproductive Ecology: Progress and Perspectives, Uppsala, Sweden (September 1991). Books, some of them proceedings of previous workshops and meetings: Rebelo (1987), Lovett Doust and Lovett Doust (1988a), Sedgley and Griffin (1989), Bawa and Hadley (1990), Dafni and Eisikowitch (1990), Gottsberger and Prance (1990), Heß (1990).

### 2. Functional Aspects of Flowers and Inflorescences, Flower Longevity, Metabolism, and Movement

Sometimes it seems necessary to consider the whole population of a plant species or its total geographical area in order to fully understand the functional significance of certain organs or mechanisms. While such structures or events may be not functional at one place, they may be important at another. In this context, Heyn (1988) suggested that investing in adaptation to rare reproductive events may be a rather widespread reproductive strategy and may have far-reaching consequences. Asclepiadaceae have one

Progress in Botany, Vol. 54

of the most complicated and complex floral structures with a variety of adaptations to highly specialized functions. Pollinaria removal and insertion by the insects is aided by various structural devices such as hairs or corona elements guiding the proboscis towards the entrance of the anther slit (Kunze 1991; Kunze and Liede 1991). By slow motion pictures, the explosive style movements in the Marantaceae *Thalia geniculata* is shown (Classen-Bockhoff 1991). When a visiting bee touches the style, the style becomes erect in a very short time (0.03 s), scrapes the pollen from the insect's proboscis into its stigmatic cave and deposits its own pollen onto the proboscis in the form of a sticky paste. The not very attractive flower has just one chance to be pollinated, which is compensated by a high synchronization and a complicated and efficient movement. For correlations between structural features and phenological patterns in *Conophytum* (Mesembryanthemaceae) see Liede and Hammer (1990), and for supposed functional implications of stigma morphology in *Cassia, Chamaecrista*, and *Senna* see Owens and Lewis (1989). In order to relate variation in floral form to variation in function, Armbruster (1988, 1990) used form-function relationships in *Dalechampia* (Euphorbiaceae) to predict the rates of pollen receipt and pollen dispersion (pollination function) resulting from different combinations of morphological characters. Field observations did not exactly correlate with this form-function relationships, which may have several reasons. Ågren and Schemske (1991) tested and proved the hypothesis that the rewardless female flowers of *Begonia* mimic male flowers and are pollinated by deceit by *Trigona* bees. For false anthers or anther dummies in Scrophulariaceae see Magin et al. (1989).

Usually flowers with larger corollas receive more pollinator visits (for wild *Raphanus sativus* see Stanton and Preston 1988). By measuring variation of corolla length of *Calathea ovadensis* (Marantaceae) over a period of 3 years, it was verified that selection on corolla length varies significantly in time due to temporal variation in the pollinator assemblage and is generally weaker than that created by artificial selection (Schemske and Horvitz 1989).

The evolution of inflorescence size and number in relation to resource allocation is discussed by Schoen and Dubuc (1990). In Ericaceae and Fabaceae, nectar production was either unaffected by inflorescence size, or flowers in large inflorescences produced more nectrar (Harder and Cruzan 1990). Classen-Bockhoff (1990) surveyed patterns in pseudanthia, which apparently have developed along independent evolutionary lines, and are not bound to a particular inflorescence structure and to no special pollinator group. Usually the showy attractive organs are surrounding the functional small flowers, but in flag-blossom type inflorescences from the Mediterranean region, like *Lavandula stoechas* and *Muscari comosum*, a prominent show-apparatus is at the top of the inflorescence. The longevity of the functional flowers of an inflorescence, which still continue to produce nectar and to attract insects while withering and ceasing to invest in turgescent floral tissue, is remarkable (Gottsberger and Hartmann 1988). The size of the inflorescence should influence pollinator visition rate. Schmid-Hempel and Speiser (1988) tested this for *Epilobium angustifolium* and found that on larger inflorescences the rate of bee arrival was larger than on small ones. The number of bumblebee approaches also increased with increasing number of flowers per plant in *Cynoglossum officinale* (Klinkhamer et al. 1989) and *Echium vulgare* (Klinkhamer and de Jong 1990), or with number of umbels per inflorescence in *Aralia hispida* (Thomson 1988).

Corbet (1990) reviewed climatic effects on flowers, flower movement, pollen release, and pollinator behavior, while Kevan (1989) related what is known about thermoregulation in Arctic insects and flowers. Irregular flowering patterns to Inghe (1990) may be partly explained by chaos, which to this author often might be a neutral trait in reproductive dynamics. Longevity of individual flowers of 110 species in Costa Rica was determined by Stratton (1989). He found that taxonomic constraints were the most important determinants. In *Streptanthus tortuosus* (Brassicaceae), the length of the pistillate phase depends on pollinator activity and the staminate phase of the flower is three times longer than the pistillate phase (Preston 1991). This is also the case in *Campanula rapunculoides* flowers, which under conditions of natural pollinator visitation lasted significantly longer than when pollen was removed fast (Richardson and Stephenson 1989).

A study on petal senescence (Woltering and van Doorn 1988) showed that the initial symptoms were either wilting or abscission. Ethylene was involved in the natural senescence of only a minority of the wilting type of flowers and in the majority of the abscising type. In *Cymbidium* flowers senescence (with production of ethylene and a precursor) is regulated by the column (Woltering 1990). In *Petunia hybrida*, pollination resulted in the production of short-chain saturated fatty acids, which were transported to the corolla and induced an increase in ethylene sensitivity (Whitehead and Halevy 1989; see also Han et al. 1991). Stead (1992) in a review gave examples of the various pollination-induced strategies shown by plants and the role of ethylene and other potential mediators of senescence in these processes. In *Mimulus guttatus* corolla abscission 1–9 days after anthesis is a mechanism for delayed self-fertilization when there were no visits by pollinators (Dole 1990).

## 3. Means of Attraction, Rewards

### a) Color Vision, Visual Patterns, and Pigmentation

The chemical mechanisms to display an extraordinary diversity of flower colors using only a limited number of anthocyanins include loss of color of the anthocyanin and participation of a copigment (Brouillard et al. 1990). For floral color polymorphism in *Trillium sessile* see Les et al. (1989). Variation in floral pigmentation patterns seems genetically relatively simple at least in *Clarkia gracilis* (Gottlieb and Ford 1988; Gottlieb 1989). For *Petunia hybrida*, Weiss and Halevy (1989) found that a removal of stamens or even anthers, at an early stage of corolla development before the start of main anthocyanin production, inhibited both growth and pigmentation of attached corollas; application of gibberellic acid could replace the effect of the anthers. The studies of Stead and Reid (1990) showed that the color change in the banner spot of *Lupinus albifrons* with an increase in anthocyanins is regulated by ethylene. In a very broad study, Rosen and Barthlott (1991) examined the UV reflection of flowers of 832 entomophilous species and showed that they often contrast with the natural background in UV-light. Sympatrically occurring taxa of similar flower color and shape are often differentiated by different amounts of ultraviolet reflection. In reviewing present knowledge of color vision in flower-visiting insects, Menzel (1990) outlined that nearly all Hymenoptera studied have similar color vision systems with three receptor types (UV, blue, and green), and that only a few species have tetrachromatic color vision with a receptor absorbing red. The studies of Lunau (1991) convincingly prove that bumblebees possess an inate mechanism that responds to optical signals of anthers and triggers the landing reaction. This confirms the conclusion of previous studies by Osche and Gack, that stamens and stamen mimics are a highly standardized signal for the pollinating insects.

Age-dependent change in flower color usually is used by the pollinators as a signal for visiting the "best" flowers (review: Weiss 1991; birds and insects in *Banksia ilicifolia*: Lamont and Collens 1988; birds in *Fuchsia exortica*: Delph and Lively 1989; insects in *Oenothera drummondii*: Eisikowitch and Lazar 1987). However, the red, pink, and white color morphs of *Ipomopsis aggregata* within a given population were not preferentially visited by the hummingbirds and hawkmoths present as could have been expected (Elam and Linhart 1988). There was neither ethological isolation nor any evidence for differences between red, white, and pink flowers in any aspect of reproductive biology. Dukas and Shmida (1989), who analyzed 89 crucifers in Israel, detected a close connection between flower color and structural traits. Large violet flowers forming a tube were visited by large long-tongued bees, while smaller white and yellow flowers without tube were visited by small short-tongued bees and flies. This correlation of color with structural adaptations to different pollinators may have various reasons, one implying that selection pressure does not always work on flower color divergence.

### b) Production of Scent and Heat

The work on the role of scent glands (osmophores) in pollination by Vogel, originally published in 1963 in German, was translated into English (Vogel 1990c), which will make this important contribution accessible to non-German-speaking scientists. Kaiser (1991) introduces the methods used by different investigators to trap, investigate, and reconstruct flower scents.

Male euglossine bees not only collect fragrances from flowers, but they also mix these fragrances with their labial gland secretions; the biological importance of this behavior is unknown (Whitten et al. 1989). Why *Euplusia violacea* changes its preferences for scent baits from vanillin to cineole during the season is not at all clear yet (Wittmann et al. 1989). Gerlach and Schill (1991), who broadly investigated and analyzed fragrance composition of flowers from 100 plant species, especially orchids, found out that plants pollinated by euglossine bees usually produce a large amount of a few fragrant substances while other melittophilous species often produce many rare compounds in small amounts. Further studies on orchids: structure of the osmophores of *Stanhopea* (Curry et al. 1991): floral odors of *Platanthera* species (Patt et al. 1988; Tollsten and Bergström 1989).

In *Moneses, Pyrola* and *Rosa*, the petals and the anthers have different fragrance profiles (Dobson et al. 1990; Knudsen and Tollsten 1991) or, as in many Solanaceae, the anthers are the only scent-producing organs (D'Arcy et al. 1990), with implications for the visiting insects, their behavior, and pollination.

Fragrance emission often follows an endogenous circadian rhythm (for *Hoya carnosa* see Altenburger and Matile 1988). In species which have odor morphs, the differences in odor compounds may be too small to attract different pollinators (for *Bartsia alpina* see Bergström and Bergström 1989), or the odor quality is so different that different pollinator groups are attracted which affect seed set. Galen and Newport (1988) found such a situation in *Polemonium viscosum*, which has individuals with large sweet and small skunky-scented flowers. In populations frequented by a fly-dominated assemblage of pollinators, skunky-scented flowers, tended to receive more outcross pollen than sweet-scented flowers, regardless of their size. In contrast, in predominantly bumblebee-pollinated populations, floral scent and size were both correlated with pollination quality and had complentary effects on seed set. Young (1989) found that floral fragrance oils distilled from *Theobroma* species function as attractants for cacao-pollinating Ceratopogonidae and Cecidomyiidae.

Since it was discovered that the hormone calorigen, which triggers the metabolic explosion and heating of the *Sauromatum* inflorescence, is salicylic acid, the effect of this substance for thermogenesis was tested. Since salicylic acid was not found in the thermogenic *Victoria amazonica* and *Bactris major*, but on the other hand occurs in comparable concentration in other both thermogenic and nonthermogenic species, it is still open what its regulatory role is (Raskin et al. 1990). In several thermogenic species of *Victoria, Nelumbo, Sauromatum,* and *Arum*, the cyanide-insensitive respiratory pathway was verified, with ultrastructural changes in the mitochondria, and a strong oxidation of proline and glutamate (Skubatz et al. 1989; 1990a; Leport and Larher 1991). By an elegant method, the pattern of surface temperature of Araceae inflorescence via infrared thermography was shown (Skubatz et al. 1990b, 1991).

There are new accounts of heat production, promoting odor volatilization and attracting insects for pollination in Cycadaceae (Tang 1987), *Annona* (Gottberger 1989a, b), and Cyclanthaceae (Gottsberger 1990, 1991a).

### c) Nectaries and Nectar

The question whether rate of nectar production and flowering phenology influences pollinator behavior is addressed by Zimmerman (1988). In answering this question positively, he shows what has to be done in order to reveal if pollinators can be manipulated to the plant's advantage. Possingham (1988) developed a model of resource dynamics in a patchy environment, which can be applied to animals exploiting nectar. The way in which the mean and standard deviation of nectar content varies with foraging intensity can be predicted. That nectar production is a costly process for a plant, was shown by Pyke (1991), who reported that removal of nectar from flowers of *Blandfordia nobilis* increased the plant's net nectar production but reduced its ability to produce seeds, while addition of artificial nectar only partly resulted in increased seed set (Pyke et al. 1988). That nectar removal increases the total yield of nectar in flowers was shown also for *Heliconia imbricata* (Gill 1988a). Defoliation experiments in *Polemonium foliosissim* (Zimmerman and Pyke 1988c) and *Epilobium angustifolium* (Michaud 1989) had only a limited influence on nectar production rate or bee visitation, and seed production was not notably affected.

Nectar may have several functions besides being a reward to pollinating animals. The nectar which flows from the stigmatic chamber to the cuculli of *Asclepias syriaca* flowers is a germination medium for the pollen grains of the pollinia. Pollen germination was inhibited by the presence of the yeast *Metschnikowia reukaufii* in the nectar, which is widespread in nature and transported from flower to flower by pollinating insects (Kevan et al. 1989; Eiskikowitch et al. 1990). In *Croton suberosus*, a primarily wind-pollinated species, the floral nectar attracts mainly the wasp *Polistes instabilis*, which seems to defend the plant against predators (Domínguez et al. 1989). Still another important function of floral and extrafloral nectar was observed especially in dry tropical lowlands, where it attracts ants, which reduce herbivore damage and enhance fruit production (Rico-Gray 1989). Not only floral but also extrafloral nectaries can attract pollinating insects (for *Acacia longiflora* see Thorp and Sugden 1990).

Smets and Cresens (1988) characterize the floral nectary types of the angiosperms and arrange them in a hierarchical order (see also Smets 1988). A survey of floral nectaries in the Melastomataceae revealed them to be of the rare nonstructural androecial type. Additionally, there is secretion from the petal tips in *Medinilla* and probably from the stigma in *Miconia* (Stein and Tobe 1989). The Limnanthaceae have their nectaries at the base of each of the five stamens (Link 1992), *Polygonum* and related genera have axial nectaries (Decraene and Smets 1991a), and the nectar-secreting appendages of *Harungana madagascariensis* (Clusiaceae) seem to represent a whorl of nectariferous staminodes (Decraene and Smets 1991b). In the Myrtaceae *Tryptomene calycine* from Australia, the nectaries of the floral tube secrete a nectar rich in sugars, while glands on the anther connectives apparently produce lipids (Beardsell et al. 1989).

New methods of approaches to assess nectar amount are described by Armstrong and Paton (1990), who extracted nectar by centrifuging whole inflorescences of *Banksia*, by McKenna and Thomson (1988), who sampled and measured small amounts of floral nectar by using small strips of filter paper as wicks to collect nectar samples for subsequent colorimetric analysis, and by Pahlke (1989), who used flower density and average production rate of nectar sugar in *Vaccinium myrtillus* flowers for calculating the production rate per square meters.

Gilbert et al. (1991), using *Cerinthe major*, tested the hypothesis of Bell (1986), who proposed that plants should be able to produce some flowers which do not secrete any nectar, thus having no costs of nectar production while being visited. The number of such "cheater" and secretor flowers did not significantly differ from Bell's prediction. The nectar standing crop (at least nectar volume) was found to be positively correlated in neighboring flowers on the same plant of *Anchusa strigosa* (see Shmida and Kadmon 1991), among neighboring flowers in inflorescences of *Blandfordia nobilis* (see Zimmerman and Pyke 1988b), while Waser and Mitchell (1990) were not able to detect a regular pattern in nectar standing volumes among neighboring plants of *Delphinium nelsonii*, as documented for this species by previous authors. It is expected that patchiness in nectar reward influences pollinator foraging behavior. The magnitude of variability of nectar production within individuals is shown for the ericaceous shrub *Kalmia latifolia* (Real and Rathcke 1988), and its daily variation for the Vitaceae *Cayratia japonica* (Kakutani et al. 1989).

Studies on nectar chemicals and their qualitative and quantitative influence on pollinators brought many new data. The cation spectra of the nectar of 20 species showed that potassium is the predominating ion (Heinrich 1989). Samples of nectar of *Rhododendron* species had relatively low sugar contents, but showed antibiotic activity by toxic acetylandromedol, which is thought to prevent bacterial growth in the weak sugar solution (Martini et al. 1990). Floral nectar composition (sugar and partly also amino acids) was investigated in the Loranthaceae *Ligaria cuneifolia* (Galetto et al. 1990), and compared among the New World genera of the Antirrhineae (Elisens and Freeman 1988), diverse Malaysian species (Freeman et al. 1991), Lamiaceae species in Jerusalem (Dafni et al. 1988), Argentinean Bromeliaceae (Bernardello et al. 1991), and several *Nicotiana* species (Galetto 1991). In most cases, nectar sugar or amino acid ratios lie more or less within certain limits corresponding with pollinator types. The value of nectar chemistry for the recognition of pollinator types, as elaborated by Baker and Baker in the past years, may indeed work in most cases, but one certainly should not forget the interesting exceptions. *Erythrina crista-galli*, which because of its hexose-dominated nectar is suggested to be a passarine-bird-pollinated species (Baker and Baker 1990), was found by us to be frequently visited by hummingbirds only in its natural habitat in Brazil (Gottsberger and Silberbauer-Gottsberger, unpubl.). That sugar composition might not be stable, but can change during the nectar-producing period or by stress situations, was demonstrated by the studies of Kronestedt-Robards et al. (1989) for *Strelitzia reginae*, and by Villarreal and Freeman (1990) for *Ipomopsis longiflora*.

Since the significance of amino acids in floral nectar is far from being clear, feeding preferences of several flower-visiting insect groups were tested. In the laboratory, flesh flies preferred solutions containing amino acids over artificial nectar with sucrose only (Potter and Bertin 1988; Rathman et al. 1990). Such a preference for nectar substitutes containing amino acids was found also in the cabbage white butterfly (*Pieris rapae*) and *Apis mellifera* (Alm et al. 1990), in ants which feed on the extrafloral nectaries of *Passiflora* species (Lanza 1988), but not in the butterfly *Battus philenor*, for which this aversion is explained as a posssible indication of a disadvantageous taste effect of the amino acids (Erhardt 1991). Erhardt and Baker (1990), who studied the increase of amino acid concentration due to pollen contamination in artifical nectar, concluded that a significant increase requires contamination with a fair amount of pollen. However, that as few as five or ten pollen grains can distinctly increase the amino acid content of 20 µl genuine floral nectar, was shown in *Hibiscus rosa-sinensis* (Gottsberger et al. 1990). The study of J. Herrera (1989a) on amino acid concentration in the nectar of 61 plant species in the Iberian Peninsula did not show the tendency predicted by Baker and Baker, namely that higher amino acid concentrations occur in herbaceous or bilaterally flowered species versus woody or radiate-flowered species. We could show that amino acid composition and concentration in floral nectar of several species varied strongly from sample to sample within each species (Gottsberger et al. 1989a), that amino acid number and concentration can increase with aging of flowers, and that it increased also drastically after pollen addition and piercing of flowers (Gottsberger et al. 1990), and that there is no significant positive or negative correlation between the concentration of amino acids and sugars in floral nectar of the investigated species (Gottsberger et al. 1989b).

### d) Oil Secretion and Oil-Flowers

Vogel (1988) gave a very thorough overview about the oil-flowers and the oil-collecting bees with biogeographical, historical, and phylogenetical aspects of this remarkable relationship, showing the parallel development in different plant families (see also Vogel 1989). Finally, the third part of his monograph of oil-flowers and oil-collecting bees has appeared (Vogel 1990b), which deals with the peculiar relationship between the cucurbitaceous genera *Thladiantha* and *Momordica* and fatty oil-collecting bees of the genus *Ctenoplectra* (Ctenoplectridae), mainly in Africa and subtropical and temperate China. On more than 180 pages, this relationship is treated in detail, including its morphological, anatomical, phytochemical, taxonomical, phylogenetic, and biogeographical implications. Considering Malpighiaceae, Vogel (1990a) presents convincing data to show that the New World genera with their calyx-borne elaiophores and with their adaptation to oil-collecting bees are not the basic group, but are derived from the Old World genera. He informs that the calyx glands found in some Old World genera are not vestigial elaiophores, but are extranuptial nectaries of the vegetative type widespread in the family. There are even American genera which lack elaiophores, which is a reduction, a situation which occurs also in eglandular morphs of malpighiaceous *Banisteriopsis* and *Heteropterys* (Sazima and Sazima 1989). In this case the eglandular morphs attract oil-gathering bees by deceit and probably act as mimics of glandular flowers of the same species. How broad the spectrum of oil- and pollen-collecting bees can be, is shown by the studies of de Albuquerque and Rêgo 1989; and Rêgo and de Albuquerque (1989) on *Byrsonima crassifolia* (see also Lobreau-Callen 1989). The only two genera of Malpighiaceae in Chile, *Dinemandra* and *Dinemagonum*, have calyx glands borne on stalks, which, according to Simpson (1989), provide an intergland distance and surface area needed for the robust *Centris* bees that serve as pollinators.

Flowers of *Diascia* (Scrophulariaceae) section Racemosae have a transclucent window-like structure at the base of the upper lip, which is strongly UV-absorptive and probably is used as an orientation cue by the oil-collecting *Rediviva* bees (Steiner 1990). The evolution of spurs is probably analogous to the situation outlined for the nectar spurs of moth-pollinated orchids (see below). Spurs longer than the legs of bees were more effective in pollen dispersal and receipt. Longer spurs in the population probably had then selected for longer-legged bees that could remove residual oil that other bees were unable to reach (Steiner and Whitehead 1991). Regardless of this, Steiner (1990) believes that the reciprocal influence or coevolution of *Diascia* flowers and *Rediviva* bees has been diffuse and somewhat asymmetrical in most cases. The relationship of the oil-producing Brazilian *Angelonia* species with female oil-collecting and pollinating *Centris* bees is reported for the first time (Vogel and Machado 1991). *Centris hyptidis*, with elongated front legs, and *Angelonia pubescens*, with deep pouches, can be regarded as a neotropical parallel to the South African/*Diascia*/*Rediviva* complex. Another genus of the Scrophulariaceae, *Monttea*, unexpectedly was shown to be an oil-plant with trichome elaiophores on the inside of the lower lip, which finally also explained why oil-collecting bees occur in regions of Argentine where oil-secreting flowers were previously thought to be absent (Simpson et al. 1990). For the solanaceous *Nierembergia*, Cocucci (1991) recognized different pollination modes involving oil-collecting bees. Especially interesting is the anthophorid bee genus *Tapinotaspis*, which has its oil collectors on the middle legs, from where the oil is transferred to the scopae while in flight.

e) Pollen and Pollen Presentation

The ability of a plant to restrict pollen removal, involving inflorescence size, flower morphology, anthesis patterns, nectar production, or dichogamy, is an important evolutionary trend (Harder and Thomson 1989). Pollen dispersal and gene flow in natural populations is sometimes surprisingly restricted when compared with gene flow inferred from pollinator behavior (for *Lotus corniculatus* see Rasmussen and Brodsgaard 1992), and is often difficult to assess (for *Ipomopsis aggregata* see Campbell 1991). Using analytical and simulation models, Galen and Rotenberry (1988) show that natural selection should favor floral characters that stabilize pollen carry-over rates; estimating the different levels of gene flow in predominantly self-, animal-, and wind-pollinated species, Govindaraju (1988) concluded that pollination mechanisms play an important role in determining the levels of gene flow and patterns of micro-evolution in plants.

The acetolysis method, which destroys all pollen material with the exception of sporopollenin, is not suitable to give a balanced view of pollen structure and function. This is instructively shown with some examples by Hesse and Waha (1989). A formation of pollen aggregates has occurred repeatedly in angiosperms (Ali 1988) and is not limited to any special pollination system. The function of such aggregates is not always clear and may vary. The pollen connecting threads in *Porcelia*, the first described in the Annonaceae, are interpreted to function as stabilizers for the polyad, and as an aid in better and more economic transport (Morawetz and Waha 1991). Two new methods were developed for general labeling of pollen and marking of individual pollen grains. For one technique, the pollen grains of *Solanum rostratum*, covered with metal powders, were analyzed by the use of backscatter scanning electron microscopy and X-ray microanalysis (Wolfe et al. 1991). Another technique was radioactive labeling of the pollinia of Asclepiadaceae (Pleasants et al. 1990). Both techniques provided an efficient and convenient method for tracking pollen movement within and between flowers. Dobson (1989) reviews the importance of pollenkitt in plant reproduction, considering ultraviolet radiation, desiccation, microbial attack, visual and olfactory attraction, animal nutrition, adhesion, and pollen-pistil interactions. In this context it is also interesting that some species have pollen with extremely high (1.0% of dry matter) proline content (Pálfi et al. 1987). Since such high concentration of this amino acid occurs in entomophilous, anemophilous, and autogamous species, it is not clear if it has any function in pollination or for pollinators or is just necessary for the growth of the pollen tubes. Viability of pollinia collected from *Asclepias syriaca* allows a transport by an insect for at least 4 days. Inserted in a stigmatic chamber of a recipient flower, the sometimes high glucose concentration of the nectar stores the pollinia for another 48 h before viability decreases (Eisikowitch et al. 1987). The exine ultrastructure and the consistency, distribution, and total amount of pollenkitt present were found to be well correlated with anemophilous (*Brasenia schreberi*) and entomophilous (*Cabomba caroliniana*) Cabombaceae (Osborn et al. 1991). For pollen analysis as a means to discover host specifity in pollen-collecting bees see Westrich and Schmidt (1987) and for pollen samples in honeys from Brazil and Spain see Barth (1990) and Ortiz (1990). Galil (1990) found that pollen of *Arisarum vulgare* needs aeration for germination, which does not occur within the inflorescence chamber but on the body of a flying insect.

Studies of the structure and mechanisms of secondary pollen presentation in Brunoniaceae (Erbar and Leins 1988), Goodeniaceae (Leins and Erbar 1989), Campanulaceae, and Lobeliaceae (Erbar and Leins 1989), showed the variety of special mechanisms for secondary pollen presentation for the Campanulales-Asterales complex, with pump, brushing, deposition, cup, and combined cup-brushing mechanisms (Leins and Erbar 1990). The different mechanisms are based on three successive events: elongation of the filaments, opening of the anthers, and elongation of the style. It is proposed that secondary pollen presentation has the advantage of prolongation of the male phase of anthesis through portioned pollen release. For variation of pollen brush types with trichomes on the distal end of the style of Fabaceae and functioning in secondary pollen presentation see Lavin and Delgado (1990).

## 4. Pollinator Behavior, Pollinator Efficiency, Flower Constancy, Foraging Strategies

The first comprehensive treatment of the foraging theory in book form is by Stephens and Krebs (1986). Parker and Maynard Smith (1990), in reviewing the optimality theory, state that optimization models help us to test our insight into the biological constraint that influences the outcome of evolution. It is important for researchers to keep in mind that the models serve to improve our understanding about adaptations, rather than to demonstrate that natural selection produces optimal solutions. The marginal-value theorem, the rule that a nectarivore animal has to make a decision whether to visit another flower on the same inflorescence or move to a new inflorescence was tested and compared with computer simulations (Pleasants 1989). The behavior of the simulated bees using the predicted multiple thresholds was indistinguishable from actual behavior of bumblebes foraging on *Delphinium nelsonii*. While investigating the abundance and flower visitation rate of the pollinators of *Lavandula latifolia*, C.M. Herrera (1989) found that regardless of proboscis length, hymenopterans had intrinsically shorter handling times than lepidopterans, and that within each group, handling time decreased exponentially with increasing proboscis length. Judging from pollen loads of insects visiting the pistillate flowers of three dioecious rain forest tree species, House (1989) detected that more than 90% of individual insects carried more than one pollen type. Flower constancy of two *Bombylius* (Diptera) species on blue-flowering *Hedyotis caerulea* (Rubiaceae) was detected at several flower-rich sites in eastern North America (Grimaldi 1988).

Bees and bumblebees: flight directionality is not influenced by energy gain but by the amount of turning within inflorescences (Kipp et al. 1989); the normal upward movement within an inflorescence is often modified into a downward movement when bees leave the inflorescence to visit another one at a certain distance (Kevan 1990); departure decisions after visiting an inflorescence are based on the amount of nectar in the last flower probed (Cresswell 1990); pollen-collecting bees remove more pollen but deposit less of it than nectar-collecting bees while visiting *Impatiens capensis* (Wilson and Thomson 1991); patterns in bumblebee nectar-foraging profits mirrored temporal patterns in nectar availability (Wright 1988); risk-sensitive foraging decisions (based on both reward mean and variance) are affected and changed by colony energy requirements (Cartar and Dill 1990; Cartar 1991); temperature was the most important factor influencing foraging activity in a subalpine meadow (Young and Owen 1989). The size of patches in which plants grow influences the pollinator community and their foraging decisions. Short-tongued bumblebees which often rob nectar by biting holes in the flowers are dominant in large patches, while long-tongued species are forced to small patches, where nectar-robbing occurred only exceptionally (Sowig 1989).

Honeyeaters (Meliphagidae) in Australia become more aggressive towards other intruding birds when nectar availability in their territory decreases during the season (Armstrong 1991). The smaller honeyeater species seem to be more effective in cross-pollination of *Banksia menziesii*, since they were frequently chased from inflorescences by the larger species (Ramsey 1989), and the optimum of nectar sugar concentration they prefer seems to be around 20% (Mitchell and Paton 1990). The black-chinned hummingbird preferred a sucrose concentration of 40% (Stromberg and Johnsen 1990). While migrating, the travel distance influenced both amount and type of food loaded by male Calliope hummingbird (Tamm 1989). Because of competition, the traplining *Phaethornis superciliosus* had to visit a feeder frequently in order not to lose accumulated nectar to competitors (Gill 1988b). Movement patterns of two species of hummingbirds between inflorescences of *Ipomopsis aggregata* indicated that they visited nonrandomly tall inflorescences with many flowers in the neighborhood, unless the inflorescence was of low quality, which led them to fly further away (Wolf and Hainsworth 1990). Two Acanthaceae species from Costa Rica, *Razisia spicata* and *Hansteinia blepharorachis*, were seen to be visited by long-distance and short-distance pollinators (hummingbirds). It was found that lower pollen-ovule ratios and greater levels of self-compatibility in *Razisia* were correlated with the long-distance pollinators in comparison to the other species pollinated by nearest-neighbor foragers (Linhart et al. 1987).

## 5. Flower Classes and Their Evolution

### a) Zoophily

#### α) Evolution and Pollination of Early Angiosperms; Cantharophily, Flies and Thrips as Flower Visitors

If the synthesis of recent molecular and paleobotanical findings is correct, then we have to look to the end of the Carboniferous or the Permo-Triassic time, during which the primary angiosperm radiation supposedly occurred, an idea proposed by Axelrod and others many years ago. Molecular evidence suggests that angiosperm ancestors diversified more than 300 million years ago (Martin et al. 1989), at about the Carboniferous-Permian boundary. The interesting fossil *Sanmiguelia lewisii* from the late Triassic is interpreted as a very primitive pre-magnoliid angiosperm that combined both monocot and dicot characters. Although its flowers were unisexual, direct and indirect evidence is presented for pollination by phytophagous insects (Cornet 1989). If further studies should confirm the status of *Sanmiguelia* as the oldest-known angiosperm fossil, then many prevailing ideas about early angiosperm diversification must necessarily be modified. Under these circumstances, it is not surprising that studies of the extinct Bennettitales and the extant Gnetales (for double fertilization in *Ephedra* see Friedman 1990a, b) the supposed sister groups of the angiosperms, are being revived (see, e.g., Crepet et al. 1991). A partial dependence on entomophily for seed set in *Ephedra* species with certain features like adhesive pollen (but without pollenkitt) is shown by Meeuse et al. (1990a, b). For a review of wind and insect pollination in *Ephedra*, *Gnetum*, and *Welwitschia*, and their ambophilous characteristics see Kubitzki (1990). Such ambophilous situation was documented in the last few years also for cycads, in which, besides wind pollination, especially beetles (Curculionidae) are prominent and sometimes exclusively associated and possibly pollen-transporting insects (Norstog 1987; Vovides 1991). For a comparison of cycad cone volatiles with those from magnoliid flowers see Pellmyr et al. (1991).

Endress (1990) gave a broad overview of morphological and functional aspects of the Magnoliidae, considering patterns and plasticity of floral structures, their pollination biology (with prevalence of flies and beetles), and breeding system, concluding that in some families still almost nothing is known about their reproductive biology. The Chloranthaceae, which, when considering the fossil pollen of *Clavatipollenites*, belong to the oldest angiosperms found in the lower Cretaceous, apparently show insect- and wind-pollination syndromes in their early Cretaceous members (Crane et al. 1989). The Hamamelidaceae and extant lower Hamamelididae also show such diversification into insect- and wind-pollinated taxa (Endress 1989). The Winteraceae, as was outlined already by previous authors (compare Gottsberger 1988), have unspecific interactions with many insect groups, which was confirmed by studies of New Caledonian Winteraceae (Pellmyr et al. 1990): moths, weevils, and thrips were the pollinators of *Zygogynum, Exospermum,* and *Belliolum* species, and the authors correctly conclude that the observed associations do not seem to be extensively coadapted or mutually dependent. The New Zealand *Pseudowintera colorata*, pollinated by chironomid flies and pollen-eating halodid beetles, with increasing ambient temperature had a higher pollination frequency and a shortening of the female phase of the protogynous flowers (Wells and Lloyd 1991).

Although it has been documented several times that exclusive beetle pollination, even in the Magnoliales and Laurales, is a secondary, derived, and specialized situation paralleled by significant flower adaptations (e.g., Gottsberger 1988, 1989b, 1991b), there are still authors (e.g., Meeuse 1990) who continue to repeat the old arguments against cantharophily being the original condition in angiosperms. That this idea loses more and more support is shown by several studies considering the morphology and function of flowers and inflorescences of archaic and modern plant groups.

Two Annonaceae were studied in detail by Deroin (1988a, b, 1989, 1991). In *Annona senegalensis*, because of strong dichogamy and only one open flower per tree, the plant first functions as a dioecious one, passing gradually to a functional monoecious stage when more flowers are opening. On the Ivory Coast, this species is pollinated by *Endaeus* (Curculionidae). *Cananga odorata*, also from this country, has Nitidulidae as early and Chrysomelidae as late pollinators during the flowering season. For phenology, breeding system, temperature elevation, and flower visitors (mostly beetles) of several Annonaceae species in the neo-and paleotropics see Nagel et al. (1989), Okada (1990), and van Tol and Meijdam (1991). Synchronous dichogamy, promoting interplant movement of beetles and therefore increasing the probability of cross-pollination, was described for *Rollinia jimenezii* (Murray and Johnson 1987). In large flowers of *Annona* species, measurable temperature elevation (up to 15 °C above the ambient air) and concomitant odor production attract night-active *Cyclocephala* species, which stay, species-specific, for few hours to 3 days within the floral chamber (Gottsberger 1989a, b). Since similar phenomena can be found in Araceae (for *Xanthosoma* species see Valerio 1988; Thompson and Rawlins 1990, for *Philodendron* see review by Mayo 1991), Cyclanthaceae, *Victoria*, and palms, dynastid scarab beetle pollination apparently has caused a convergent evolution with the result of similar functional adaptations in flowers and inflorescences of very different families. Such scarab beetles may have varying effect as pollinators and on genetic interchange in a population, as was shown by Young (1988a, b) for *Dieffenbachia longispatha*. Experiments and observations have shown that one of these pollinating beetles, *Erioscelis emarginata*, the sole pollinator of *Philodendron selloum*, is guided at dusk to the inflorescences in their female stage by a combination of olfactory and visual stimuli (Gottsberger and Silberbauer-Gottsberger 1991). The fragrances, which serve as attractants for the pollinating beetles, were identified for two *Eupomatia* species and revealed to be short chain, fatty esters (Bergström et al. 1991). The volatile constituents of *Chimonanthus praecox* (Calycanthaceae) consist of 161 compounds with prevailing aliphatic, monoterpene, and sesquiterpene hydrocarbons (Ueyama et al. 1990). A *Victoria* hybrid contained mainly aliphatic or benzyl esters (Kite et al. 1991), and the New Caledonian Winteraceae flower volatiles (see above) were found to be simple in composition, being usually dominated by short esters; experiments showed that ethyl acetate elicited alighting of insects (Pellmyr et al. 1990). In a review of pollination in magnolias, including newly collected insects, the predominance of beetles, especially Nitidulidae as flower visitors, is emphasized (Peigler 1988). A new and interesting association between dioecious *Siparuna* (Monimiaceae) and gall midges (Cecidomyiidae) was shown by Feil and Renner (1991). At night, female cecidomyiids visit male and female flowers of *Siparuna* for oviposition (with only the male flowers being suitable egg-laying sites), with floral odors acting as cues. This is the first record of cecidomyiid pollination in Magnoliidae and widens the pollinator spectrum of this group.

For several families outside the Magnoliidae predominant or exclusive beetle pollination was verified as well. One family in which cantharophily or at least beetle visitation of flowers is detected quite commonly are the palms, with new interesting observations for *Acrocomia aculeata* (Scariot et al. 1991), and *Orbignya phalerata* (Anderson et al. 1988); especially prominent are certain curculionid and nitidulid groups. However, many palms have a mixed pollination syndrome with a large array of different insect groups visiting the flowers and inflorescences, with specialization and exclusiveness of beetles as visitors apparently being a derived event (Silberbauer-Gottsberger 1990). Another group, which even seems to be exclusively cantharophilous, are the Cyclanthaceae-Carludovicoideae, with the conspicuous staminodes being scent-emitting osmophores, attracting weevils in the early morning hours (Gottsberger 1990, 1991a).

The first thorough study of a Balanophoraceae is by Borchsenius and Olesen (1990), who showed beetles (Chrysomelidae, Curculionidae, and Nitidulidae) to be the pollinators of the Amazonian rain forest root holoparasite *Lophophytum mirabile*. Further data on beetles being exclusive flower visitors are from Australian Myrtaceae (Hawkeswood 1987a, b, c) for which more information is badly needed. Beetles as additional pollinators were found on Australian Asteraceae (Hawkeswood 1987d), the Araceae *Arum dioscorides, A. orientale* (Drummond and Hammond 1991), *Anchomanes difformis* (Thompson and Rawlins 1986), and Pandanaceae (Crowson 1988). The previous idea that *Hibbertia* (Dilleniaceae) is cantharophilous was discarded by Hawkeswood (1989). An interesting association exists between red bowl-shaped flowers of *Anemone, Papaver, Ranunculus*, and *Tulipa* species and scarabaeid beetles (*Amphicoma*) in the East Mediterranean region. Experiments showed that these beetles were attracted primarily by the red color, which is a new facet in cantharophilous syndromes (Dafni et al. 1990). Irvine and Armstrong (1988) discuss beetle pollination and its importance in Australian tropical rain forests, and demonstrate with examples different stages of interactions from quite lose to highly specialized ones.

Considering the Diptera, the rediscovered *Permotipula patricia* seems to be the oldest known fly and is the first one found in the Upper Permian (Willmann 1989). This is about 50 million years earlier than any previously known fossil fly. This will obviously have implications for the discussion of spore (pollen) dispersal by this insect group. Fossils of Bibionidae and Tipulidae (of an extant genus) were detected in the Upper Cretaceous, and indicate stasis of more than 90 million years (Rayner 1987; Rayner and Waters 1990). An interaction with an ascomycete, in which a fly transfers conidia, is described by Bultman and White (1988), and the phenomenon of flies transporting sticky spores of the moss *Splachnum*, attracted by color and volatile compounds emanating from the hypophyses of the sporogon, is reviewed by Koponen (1990).

Pollen-feeding by Bombyliidae (Deyrup 1988) and Syrphidae and their selectivity in diets (Haslett 1989), as well as an inborne reaction to small, yellow floral guides of *Eristalis pertinax*, were described (Lunau 1988). Another syrphid fly, *Scaeva pyrastri*, seems to be able to collect pollen of Apiaceae and Asteraceae flowers by buzzing (Nickol 1991). Du Toit (1987) showed that nectar flies (*Drosophila hirta*) can be detrimental for beekeeping and honey production, since the larvae of this fly species live in the flower cup and utilize nectar as food source. On the other hand, Drosophilidae, together with Mycetophilidae (fungus gnats), are pollinators of *Arisarum simorrhinum* (J. Herrera 1988b). A strong, fetid smell along with nectar and pollen attract mainly Diptera as pollinators of the Rhamnaceae *Discaria americana* (Medan 1991). The sapromyophilous pollination syndrome of *Rafflesia pricei*, working on optical and olfactory cues, involves carrion flies which have to travel between the unisexual flowers (Beaman et al. 1988).

Thysanoptera were finally found also on Cycadaceae, namely within the male cones of *Macrozamia communis* in Australia (Mound 1991), and thrips species were detected to be important for within-flower pollen transfer in self-compatible *Ranunculus scleratus* and *Potentilla rivalis* in Iowa (Baker and Cruden 1991).

### β) Melittophily and Ant Pollination

Two important and nicely illustrated books on bee biology, including interaction with flowers have appeared in the last few years. One is by Westrich (1989) on the bees of the German state Baden-Württemberg, and the other by Roubik (1989) on the ecology and natural history of tropical bees. The dating and description of a *Trigona* bee from Late Cretaceous amber (Michener and Grimaldi 1988a, b) was in itself a small revolution. The fossil is 96-74 million years old, which is about twice the age of the oldest previously known fossil bee. Since the fossil is a worker and closely similar to modern neotropical species, this means that not only social organization had arisen by then, but that bee differentiation and morphological evolution was quite advanced at that time. This also implies that bee/flower relationships are much older than previously thought, perhaps going back to the Early Cretaceous or even earlier. Bees of the *Trigona hypogea* group are obligate necrophages foraging upon carrion in tropical America. This unusual use of food is correlated with a lack of corbicula on the hind legs, lack of stored pollen in nests, and complete absence of pollen grains in larval provisions (Camargo and Roubik 1991). Since honey is abundant in all nests, the use of floral nectar is quite likely. However, until now there exists no observation of any of these obligate necrophages on flowers.

New data were published on physiological, morphological, and behavioral correlates of bees with flowers. An increase in metabolic expenditure as a function of the load carried (nectar and pollen) is demonstrated for the first time for the honeybee (Wolf et al. 1989). During a foraging cycle the energetic costs of resource collection increase as the load increases. Different bumblebees have different pollen-collecting preferences (Kratochwil and Kohl 1988), and there are also specialists and generalists, handling the preferred food type more or less efficiently (for *Aconitum* see Laverty and Plowright 1988). Bumblebees can detect differences between flowers with respect to pollen availability, alter their behavior to obtain the highest reward (Harder 1990a), and also make use of age-related morphological differences in the flowers (*Anemopsis macrophylla*) to discriminate in favor of young, pollen-rich flowers (Pellmyr 1988). Harder (1990b) studied the pollen removal pattern on six species in Canada and found that removal increases with visit duration, but did not depend on the number of visits involved. Observations on bumblebees of different body size and proboscis length visiting the flowers of *Symphytum officinale* (Sowig and Paulus 1988) allowed the conclusion that nectar-robbing can be considered a strategy for short-tongued species to explore new resources and to restrict long-tongued species to smaller patches. Apparently, bumblebees also use chemical cues deposited by other foraging individuals to increase efficiency of visiting and nectar-gathering (for *Impatiens* see Kato 1988). Their pollination behavior was found to be influenced also by the presence of endoparasitic larvae of parasitoid flies (Conopidae). Schmid-Hempel and Schmid-Hempel (1990) showed that nonparasitized *Bombus pascuorum* individuals preferentially visited the flowers of *Prunella grandiflora*, while parasitized individuals visited the less deep and less difficult flowers of *Betonica officinalis*. Bumblebee flight distances while visiting *Phyteuma*, the variation in pollen flow when bumblebees are disturbed by honeybees, and competition and seed set in *Malampyrum pratense* and *Viscaria vulgaris* with healthy and *Ustilago*-infected flowers were studied by Kwak (1987), Kwak et al. (1991), Kwak and Jennersten (1991), and Jennersten and Kwak (1991).

In the colletid bee *Niltonia virgilii*, the enormous labial palpi can form a tube through which nectar, especially of the long-tubed *Jacaranda puberula*-flowers, is sucked (Laroca et al. 1989). Odor-marking of vegetation and patroling of male bees is known for many species, and is reported for the Malagasy anthophorid bee *Pachymelus limbatus*, which marks its food-resource area (Nilsson and Rabakonandrianina 1988b). Out of the enormous literature on bee/flower relationships, a few studies are selected: *Lathyrus grandiflorus* floral construction with a twisted wing-keel-complex is highly specialized, and only carpenter bees, which tilt to the right side of the flower when sucking nectar, are able to work this complicated mechanism and to transmit pollen which is deposited on the bee's back (Teppner 1988). *Schrankia nutallii*, a Mimosaceae from the tallgrass prairie is visited and pollinated by polylectic, female bees of five families (Bernhardt 1990a). An interesting case of oligolecty is found in the southern Brazilian andrenid bee *Callonychium petuniae*, which forages only on purple *Petunia* flowers and copulates on the flowers (Wittmann et al. 1990). For co-flowering shrubs in Rhode Island, competitive avoidance was achieved not by specialization and divergence as is commonly hypothesized, but by having diverse pollinator assemblages (small bees and bumblebees), and by characteristics compensating for rare pollinator visits like long-lived flowers or self-pollination (Rathcke 1988a). Remarkable size correlation of bees and corresponding flowes in the Lamiaceae were shown by Shmida and Dukas (1990). Solitary bees that are active from March to August are progressively smaller with the seasonal increase in air temperature. This decrease in body size is correlated with a progressive decrease in flower size of the different bee-pollinated species of the mint family flowering each month.

Dukas and Dafni (1990), after having observed buzzing on nectariferous Boraginaceae flowers, speculate about the evolutionary pathway of typical buzz-pollinated flowers without nectar. A new method of pollen collecting from poricidal *Solanum* anthers, the buzz-milking by a *Protandrena* species, is described by Cane and Buchmann (1989), and Buchmann and Cane (1989) demonstrate experimentally that while buzzing *Solanum* flowers, solitary and social bees directly assess pollen harvest on a flower-by-flower basis. Armbruster and McCormick (1990) reveal interesting correlations between foraging patterns of male euglossine bees and evolutionary responses of plants. Presentation of floral rewards and receptiveness for pollination in the morning in a *Dalechampia* species and the early morning hour activity of *Euglossa* species seem to be correlates of an intimate bee/flower interaction.

There is much discussion on floral resource utilization of Meliponinae (Ramalho et al. 1989; Ramalho 1990) and their generalistic collecting on very many different plant species. Studies of interactions on floral resources of the Africanized honeybee *Apis mellifera* and native bee communities (Cortopassi-Laurino and Ramalho 1988; Ramalho et al. 1990; Roubik and Boreham 1990; de Menezes and Camargo 1991; Roubik 1991) are not yet conclusive, and it appears that many groups of bees, including the Meliponinae, have little overlap with *Apis* in their food sources. However, as Roubik (1991) states, we do not yet know enough about the influence of competition of *Apis* with native bees, their domination of whole forest tree crowns, and their influence on a plant population or its outcrossing rate. For the solitary bee fauna of Arctic Alaska and their flower associations see Armbruster and Guinn (1989). Meanwhile, in the Old World, perhaps stimulated by the agressiveness of the African race of *Apis*, heavy discussion and polemics have started, involving the honeybee and its importance as pollinator in the light of nature preservation and pollination (Dreher 1988; Dustmann 1988; Lindauer 1990; Westerkamp 1990b). In Westerkamp's (1991) opinion, honeybees are not only poor pollinators but, because of their unpredictable behavior, there are not even flowers which are adapted to them.

Vespid wasps were found to eat and digest pollen at levels equal to those of some bees (Hunt et al. 1991). In *Hyptis capitata* in Panama, eumenid wasps regularly tripped and pollinated the flowers (Keller and Armbruster 1989), and in South Africa masarid wasps were found to be most commonly associated with Mesembryanthemaceae and Asteraceae, obtaining both nectar and pollen rather specifically (Gess and Gess 1989).

The idea that antibiotics from the metapleural glands in ants disrupt pollen viability was advanced further by exposing pollen to *Atta texana* and other North American ants (Hull and Beattie 1988), with the result that a high number of grains showed membrane dysfunction and reduced viability. Nevertheless, there exist plant species pollinated by ants. The orchid *Microtis parviflora* from Australia is the first case known in which a wingless worker ant (*Iridomyrmex gracilis*) is the exclusive pollinator. Pollen carried by this ant is unaffected by antibiotic secretions from the metapleural glands, because pollinia are separated from the ant integument by a short stalk (Peakall and Beattie 1989). Since ant foraging is restricted, the populations are highly inbred, however, with some outcrossing occurring (Peakall and Beattie 1991). Another orchid, *Leporella fimbriata*, also from Australia, is pollinated by male ants of *Myrmecia urens*, which have no metapleural glands. This orchid shows a unique pollination by sexually attracted male winged ants, which pseudocopulate with the flower (Peakall 1989b).

### γ) Psychophily, Sphingophily, and Phalenophily

The study of Boggs (1988) of a nymphalid species confirmed the previous results that butterflies may ingest concentrated nectar with peak sucrose rates between 30 and 40%. May (1988), who investigated flower selection and foraging energetics of a nymphalid and a pierid species in Florida, found that the flowers visited showed a nectar concentration which varied between 1 and 40% sucrose-equivalents. The cabbage butterfly *Pieris rapae*, when foraging in a highly diverse, natural environment, exhibits a significant consistency in visits to several species, which was quantified by Lewis (1989) and might be a kind of flower constancy comparable with the learning abilities of the better-known hymenopterans. This species discriminated against white flowers and preferentially visited yellow ones of *Raphanus raphanistrum* (Stanton et al. 1989).

Haber and Frankie (1989) describe the Costa Rica dry forest hawkmoth community, including floral and hawkmoth biology, their foraging behavior and flower specificity. Among others they found that long-tongued moths did not restrict their visits to long-tubed flowers. A somewhat comparable situation was found by Nilsson and Rabakonandrianina (1988a) in the orchid *Aerangis ellisii* in Madagascar, which, besides their pollinating moths, was visited also by species with proboscises too long to act as good pollinators. Nilsson (1988), after having investigated the long-tongued hawkmoth guild of Madagascar and their relationship with the native orchids, which exhibit extremely long nectariferous spurs, tested Darwin's original proposal considering *Angraecum sesquipedale*. He shows convincingly for *Platanthera bifolia* and *P. chlorantha* that experimental spur shortening gives a consistent reduction in fitness, since pollinia removal and pollen receipt through moths dicreased. Thus, long-tongued visitors exert a direct selectional pressure in natural populations for longer-than-average-tubed individuals. Such pollinators favor the evolution of longer floral tubes, which in turn favors the evolution of still longer tongues in a reciprocal and escalating process. Nilsson confirmed experimentally that Darwin's ingenious conclusion from the Madagascar Star Orchid and its predicted phantom sphinx provided the basic idea for the general principle about interactional evolution between animals and flowering plants. *Viola cazorlensis*, an endemic of Spain pollinated by day-active hawkmoths, shows broad intraspecific variation in size and proportions of floral parts, with no obvious influence on probability of fruit set. The differences among populations in spite of spatial constancy in the identity of pollinators suggest low adaptedness of the floral phenotype to its pollinators (C.M. Herrera 1990a, b).

Pollination by nocturnal moths, phalenophily, was shown for the Malagasy treelet *Ixora platythyrsa*, which shows secondary pollen presentation on the stylar heads (Nilsson et al. 1990), and for *Silene vulgaris* (Pettersson 1991). The pollinating moth of the genus *Hadena* visiting *Silene* also acts as predator and synchronous flowering of the population is probably a way to minimize seed predation.

### δ) Ornithophily

The fact that birds can see and discriminate between UV guides is further strengthened by the report that a passerine bird even has its highest spectral sensitivity in the UV (Burkhardt and Maier 1989). From a review by Waldvogel (1989), we learn that probably most, if not all, birds possess a functional olfactory system. The level of sensitivity is very low, but might be sufficient to permit the detection and use of environmental odors as orientation cues. Since such olfactory orientation has already been shown to work in a hummingbird species, this opens unexpected possibilities for future research into the bird/flower/environment relationship.

Westerkamp (1990a) correctly reminds us that not all ornithophilous New World flowers are used by hovering hummingbirds (for e.g., *Combretum fruticosum* in Mexico see Gryj et al. 1990) and not all Old World species are adapted to perching birds. However, the basic question remains why most hummingbirds have evolved hovering? Apparently, perching does not pay enough (calculated in net gain savings) for small hummingbirds, but seems to increase with body size (Hainsworth 1991).

In parallel to Nilsson (1988), who tested selective pressure on long corolla tubes of some orchids by hawkmoths, Fenster (1991) used the great diversity of hummingbirds and hummingbird-pollinated taxa to test and confirm the prediction that variance in corolla tube length is less in long-tubed flowers than in shorter ones; apparently, pollinator specialization and selective pressure increase with flower depth.

A comparison of morphology and function of bills and tongues of nectar-feeding birds is given by Paton and Collins (1989). Anthocyanidin pigments of bird-visited flowers have a typical pigment syndrome across wide geographical distances, with the exception of Australia and New Zealand (Scogin 1988). Interestingly enough, the pigment syndromes of perching and hovering birds are distinct. Hummingbirds, even when they take up pollen, derive little nutritional benefit from it, and pollen passes the bird's body more than 90% undigested.

For the New Zealand flora as a whole, bird pollination seems to be relatively unimportant as compared with other regions, and there is no evidence of tight coevolution between particular bird species and particular flowers (Clout and Hay 1989). The New Zealand flax (*Phormium tenax*), according to Craig (1989) is pollinated mainly by Meliphagidae, which preferentially visit male phase flowers. That in many cases ornithophily has evolved from bee-pollinated groups is demonstrated by ambiguous characters or by a mixed pollination system (for *Vellozia* in Brazil see Sazima and Sazima 1990; for *Banksia menziesii* in Australia see Ramsey 1988; for *Aloe ferox* in South Africa see Hoffman 1988; and for *Echium wildpretii* in the Canary Isles see Olesen 1988). However, pollination efficiency and effectiveness of bees and birds may vary greatly. Waser and Price (1990) showed this for *Delphinium nelsonii*. Bumble bees deposited about ten times as much pollen while visiting a flower and also caused much better seed set. At the level of entire populations, however, hummingbird visitation rates may be over ten times as great as those of bumblebees, since birds visit flowers more quickly.

*Fritillaria imperialis*, a common European garden plant, seems to be a species pollinated by passarine birds in its native range. Flower morphology, absence of sucrose and a low amino-acid concentration in the nectar point to this syndrome (Búrquez 1989). In Britain this species is visited by blue tits, *Parus caeruleus*, an observation confirmed also for Central Germany (M. Pirl, pers. comm.). Vanstone and Paton (1988) studied the function of the extrafloral nectaries of *Acacia pycnantha* in Australia and found that a variety of birds, including silvereyes and several species of honeyeaters and thornbills, consumed extrafloral nectar and in the process brushed against inflorescences. Since pod production was significantly reduced after bird exclusion from flowering branches, the authors confirm that the extrafloral secretions aid in pollination and not, as was previously thought, in reduction of herbivore damage. For corolla curvature in Andean *Centropogon* species in relation to billshape of hummingbirds see Stein (1992).

### ε) Mammal Pollination and Visits of Lizards

Methods and problems of diet analysis for nectariverous bats are outlined in a paper by Thomas (1988). Since the Old World Pteropodidae (Megachiroptera) are endangered by human activity throughout Southeast Asia, Kitchener et al. (1990) had to investigate the nectariverous habits and the pollen dietary niche overlap of the flower-visiting bat community on a island with still enough primary forest. There also exist observations of an extension of the range of nectar-feeding bats (for Arizona see Sidner and Davis 1988). In *Freycinetia baueriana* from New Zealand, an introduced possum becomes the pollinator in some areas where its main pollinator, the bat *Mystacina tuberculata*, is extinct, and the survival of the plant species is guaranteed (Lord 1991). Another well-documented case of a mixed pollination system, which might even be the ancient condition, is the African tree *Parkia bicolor*, in which the rather archaic "bat-flowers" in Cameroon are pollinated mainly by three megachiropteran bat species and additionally also by a dormouse and two prosimian species (Grünmeier 1990). In the New World, Gribel (1988) found the Brazilian cerrado tree *Pseudobombax tomentosum* with large and strong flowers visited more frequently by the marsupial *Caluromys lanatus* than by a large phyllostomid bat; the characteristics of the flowers make an adaptation to both nonflying mammals *and* large bats likely. Visitation rate and pollinating efficiency of nonflying mammals in *Banksia integrifolia* and *B. spinuosa* in Australia have a significant influence on fruit-set of at least the former species. *Petaurus australis*, which is the predominant nectar-feeder on several *Eucalyptus* species, may be an important cross-pollinator especially of tree individuals with significantly more flowers than others (Goldingay 1990). Judging from pollen loads, nonflying mammals may be as important as pollinators of proteaceous plants as are nectar-feeding birds (Goldingay et al. 1991). Chiropterophily is documented for *Helicteres ovata*, supposedly a relatively recent derivate from the ornithophilous stock of *Helicteres* (M. Sazima and Sazima 1988), and the Bromeliaceae *Encholirium glaziovii* (I. Sazima et al. 1989), both from Brazil. From the Old World, chiropterophily is reported for the wild banana *Musa acuminata* subsp. *halabanensis* in Sumatra (Itino et al. 1991), and for the neotropical *Ochroma lagopus* by several Megachiroptera when planted in Cameroon (Grünmeier 1989). A laser illuminator for pollination studies with a night vision device was designed by Farnsworth and Cox (1988).

Two apparently curious cases deserve attention. Geckos of the genus *Hoplodactylus* take nectar from flowers of several species in New Zealand and transport and transmit pollen especially of brush-structured flowers (Whitaker 1987). Du Toit (1990) monitored for 2 years giraffe feeding on flowers of *Acacia* in the southern Africa savanna and concluded that several factors, like inflorescence structure and color, pollen morphology, thorn structure, and flowering phenology suggest that at least in *A. nigrescens* giraffe could well be a regular pollen vector.

### b) Anemophily and Hydrophily

Cox (1991b), in his discussion of abiotic pollination, mentioned recent studies which show highly sophisticated cases of wind and water pollination in angiosperms. He also emphasized the superiority of abiotic pollination in areas of low species diversity, newly colonized habitats, and places with adverse climatic conditions. Usually wind pollination is not a feature especially associated with the life in the aquatic habitat (Cook 1988). That exceptionally one species can be pollinated by wind, but also at the water surface, and even underwater, is instructively shown for the Potamogetonaceae *Groenlandia densa* (Guo and Cook 1990).

There exist new approaches to a better understanding of anemophily and pollen dispersion by different models (Di-Giovanni et al. 1989; Okubo and Levin 1989; Smith et al. 1990). Taking into account the relationship between the size of pollen grains and female receptors, Paw U and Hotton (1989) identified anemophily with an accuracy of approximately 75%, which might be a help in the determination of anemophily of extant and especially fossil plant taxa. Niklas (1988) elaborated a computer program which can predict the motion and behavior of any pollen for which physical properties (size and density) are specified or empirically known. It is interesting that a comparison of the composition of pollen in the atmosphere with the pollen captured by the flower (for *Pistacia vera* see Niklas and Buchmann 1988) or by the pollination droplet in *Ephedra trifurca* (Buchmann et al. 1989) showed that there is a much higher number of conspecific pollen on the seed or flower structure than could be expected from the distribution of pollen grains in the air. This indicates that these structures operate as aerodynamic filtration devices which selectively capture conspecific pollen.

The protogynous *Brasenia schreberi*, although visited by flies, shows clear adaptations to wind pollination, which is the first case known in the Nymphaeaceae sensu lato (Osborn and Schneider 1988). In the genus *Espeletia*, a few species at very high elevation, where lack of pollinators is typical, have switched from entomophily to anemophily (Berry and Calvo 1989). Coppola (1989) tested and verified that *Arceuthobium americanum* pollen indeed can become airborne and dispersed several hundred meters from its source. Very often the total stigmatic surface (*Celtis iguanae*, de Arruda and Sazima 1988) or at least the papillae of the stigmas (*Acer*, Peck and Lersten 1991) are augmented and contribute to more efficient pollen capture in wind pollination. *Salix* is remarkable in having transfer of pollen by insects and air currents. However, in this ambophilous genus, different species behave differently in this respect, with slight tendencies to adapt to environmental conditions (Sacchi and Price 1988; Vroege and Stelleman 1990).

Hydrophily, which is found in one dicot and seven monocot families, seems to be of polyphyletic origin (Philbrick 1991), and probably has arisen from aerial pollination systems passing through an intermediate system involving air bubble pollination as in Potamogetonaceae (Philbrick 1988, see also Guo and Cook 1990). Although hydrophyles show a prevalence of dicliny, many water-pollinated species often seem to fail to meet all conditions necessary for outcrossing, resulting in very low levels of variability (Les 1988). When pollen is dispersed on the water surface there is two-dimensional pollination, which is made more efficient by aggregation of many pollen grains in rafts or search vehicles. These floating search vehicles have larger diameters, thus greatly increasing the probability of encountering a stigma (for *Halodule, Halophila, Lepilaena, Ruppia, Amphibolis*, and *Thalassodendron* see Cox and Knox 1988, 1989; Cox 1991a). Even in underwater pollen dispersion, and strictly submarine seagrasses, pollen movement, which is also in large search vehicles, is not tri- but bidirectional because of wave motion. It appears that there has been an evolutionary convergence towards efficient search strategies in underwater plants (Cox and Tomlinson 1988; Cox 1988a; Cox et al. 1991). For both submarine and surface pollination in *Syringodium filiforme* see Cox et al. (1990). That pollination is less efficient underwater than at the surface was tested on *Potamogeton pectinatus* (Guo and Cock 1989). Remarkable was the low (2.5–7.5%) but constant seed set underwater. In *Zannichellia palustris* fruit-set efficiency through underwater pollination is as high as 55–91% (Guo et al. 1990).

## 6. Pollination of Particular Groups; Flower Biological Radiation

### a) Differentiation of Various Angiosperms at the Species, Genus, and Family Level

Alismataceae: there is a correlation of self-incompatibility in *Baldellia* with fewer, larger, more showy flowers and higher pollen production (Vuille 1988); Apiaceae: spiders have no effect on insect visitation and female reproductive success in *Daucus carota* (Wilkinson et al. 1991); Araliaceae: Gradual presentation of pollen and nectar favors traplining behavior of pollinating bumblebees (Thomson et al. 1989); Asclepiadaceae: Quantification of pollen dispersal within a population of *Asclepias exaltata* in Vermont (Broyles and Wyatt 1991); reproductive limitations and scarcity of *A. quadrifolia* (Cabin et al. 1991); short-distance dispersal of pollinia of *A. syriaca* detected by labeling with $^{14}C$ (Pleasants 1991); fewer flowers and lower reproductive success of West Slovakian *A. syriaca* populations as compared to North American ones (Valachovic 1989); contribution of diurnal and nocturnal insects to pollination and maternal success after cross- and self-pollination of *A. syriaca* (Jennersten and Morse 1991; Kahn and Morse 1991); pollination and reproductive strategy of *Sarcostemma viminale* in adaptation to unpredictable weather conditions (Liede and Whitehead 1991); unattractive *Tylophora hirsuta* shows autogamy and self-compatibility (Chaturvedi 1989a); pollination biology of *Calotropis procera* in Pakistan (Ali and Ali 1989); a lycid beetle as pollinator of *Marsdenia fraseri* in Queensland (Forster 1989); Arecaceae: pollination biology and evolution in palms, with unspecialized general entomophily as supposed original pollination mode and with cantharophily being derived (Silberbauer-Gottsberger 1990); the Mediterranean dwarf palm *Chamaerops humilis* presumably is wind-pollinated although bearing traits (nectar production or attraction of weevils) relating to an earlier insect-pollinated condition (J. Herrera 1989b); *Geonoma macrostachys* inflorescences are visited in Ecuador by 22 species of insects and arachnids during the staminate phase and 10 species during the pistilate phase, of which trigonids and drosophilids seem the most important pollinators (Olesen and Balslev 1990); the north Brazilian babassu palm, *Orbignya phalerata*, although being visited by a large array of different insects, is pollinated mainly by the nitidulid *Mystrops mexicana* (Anderson et al. 1988); Asteraceae: there is no correlation between the different types of styles in Asteraceae, the breeding system, and the various pollinators visiting their capitula (Meiri 1988); the mosquito *Anopheles arabiensis* (Culicidae) was found to respond with flight and landing to the odor (monoterpenes) of *Achillea millefolium* inflorescences (Healy and Jepson 1988); many-headed and large inflorescences of *A. ptarmica* were better visited by pollinating flies than smaller ones (Andersson 1991); lack of insect pollinators was not the reason for rarity of *Aster curtus* (Clampitt 1987); sequential flowering of *Erigeron glabellus* and *Aster sibiricus* seems not to be maintained by natural selection generated by reproductive interference (Armbruster and McGuire 1991); for origin and adaptation of the arthropod fauna, including their feeding habits, and importance as pollinators of Espeletiinae see Sturm (1989, 1990), Klimaszewski and Sturm (1991); pollinator attractivity and outcrossing rate in radiate and nonradiate morphs of *Senecio vulgaris* (Abbott and Irwin 1988; Warren 1988); the results of a 5 year study strongly support natural selection for flowering time in *S. integrifolius* (Widén 1991); the showiness of capitula, their attractiveness to flower-visiting insects, life form, biomass allocation, and pollen-ovule ratio of several investigated Central European *Senecio* species are clearly associated with the evolutionary shift from outbreeding to inbreeding (Graumann and Gottsberger 1988); Avicenniaceae: flowering, pollination, and reproductive phenology of *Avicennia marina* (Clarke and Myerscough 1991); Bignoniaceae: phenology, pollination, and reproductive success of several cerrado species in Brazil (Stevens 1991); Bombacaceae: anthophorid bees as pollinators of two *Eriotheca* species with contrasting breeding systems (late-acting self-compatibility and apomixis) (Oliveira et al. 1992); Boraginaceae: in *Anchusa officinalis* the calculated effective visitation rate per flower was highest in plants of intermediate size (Andersson 1988); the Canarian *Echium wildpretii* has traits related to both bird and bee pollination (Olesen 1988); in the gynodioecious *Echium vulgare*, bumblebees visited more the nectar-rich flowers on hermaphrodite plants, with the result, however, of a lower seed set as compared with female plants (Klinkhamer et al. 1991); Brassicaceae: a genetic basis of effects of pollen competition on subsequent generations in *Raphanus raphanistrum* could not be demonstrated (Snow and Mazer 1988); pollen of this species produced under low nutrient conditions sired fewer seeds than pollen produced under better conditions (Young and Stanton 1990); pollination ecology of the rare Central Texas endemic *Streptanthus bracteatus* (Dieringer 1991); Cactaceae: phenology, breeding system, and pollinators of several *Opuntia* species in southern Colorado (Osborn et al. 1988; McFarland et al. 1989); Caesalpiniaceae: in several *Bauhinia* species the flower morphology, floral biology, pollinator species, nectar composition, and timing and amount of secretion were associated with life form and subgeneric designation (Hokche and Ramirez 1990); flowering phenology, flower differentiation, and buzz pollination in *Cassia*, *Senna*, and *Chamaecrista* (Gottsberger and Silberbauer-Gottsberger 1988);

for pollination of *Ch. ramosa* see Pinheiro et al. (1988); Caprifoliaceae: dichotomy of pollination systems within the genus *Lonicera* in Spain, with long-tubed sphingophilous, and short-tubed melittophilous species (Jordano 1990); insect visitors to *Viburnum opulus* in Ontario and pollen loads (Krannitz and Maun 1991); Caryophyllaceae: pollen-tube growth after cross-pollination was faster than after self-pollination in *Dianthus chinensis* (Aizen et al. 1990); habitat fragmentation had negative effects on visitation rates of pollinators and seed set of *D. deltoides* (Jennersten 1988a); pollination and nectar composition in *Dianthus gratianopolis* and *D. silvester*, with the second species being intermediate in the transition of butterfly to moth pollination (Erhardt 1988, 1990); stamen number and stamen fertility is influenced by crossing distance in *Scleranthus annuus* (Svensson 1988); *Viscaria vulgaris* produced nectar and was visited and pollinated by insects for 24 h a day with 90% of all visitors being diurnal (Jennersten 1988b); nectar standing crop in this species and seasonal differences in pollinator visitation (Jennersten et al. 1988); Cistaceae: flowering phenology, pollinating insects, and self-incompatible breeding system in *Cistus, Halimium*, and *Tuberaria* species in Portugal and Spain, with sensitive stamina as devices to facilitate cross-pollination (Brandt and Gottsberger 1988, J. Herrera 1992); Clusiaceae: nectar is restricted to the large discoid stigma or pistillode in the facultatively agamospermous *Garcinia hombroniana*, which is pollinated by *Trigona* bees (Richards 1990b); buzz pollination by carpenter bees in the self-incompatible cerrado species *Kielmeyera coriacea* and *K. speciosa* (de Oliveira and Sazima 1990); Convolvulaceae: *Ipomoea wolcottiana* has nocturnal anthesis and constant nectar secretion until after midday, attracting sphingid moths and bees (Bullock et al. 1987); Euphorbiaceae: preliminary studies of floral biology in European *Euphorbia* species showed lack of specialization and extensive range of visitors (Meeuse et al. 1989); Fabaceae: inbreeding depression in geographically restricted *Astragalus linifolius* but not in widespread *A. lonchocarpus* (Karron 1989); self-incompatibility and carpenter bees as pollinators of *Crotalaria juncea* in Brazil (Couto et al. 1988); outcrossing rates in *Robinia pseudoacacia* (Surles et al. 1990); self-compatibility of cultivated *Vigna unguiculata* in Nigeria (Ugborogho and Agomo 1989); Flacourtiaceae: number of visits per flower by bees did not increase with flower number of individual trees of *Prockia crucis* in Mexico (Bullock et al. 1989); Gentianaceae: bumblebee pollination in self-compatible *Gentiana andrewsii* (Costelloe 1988); Geraniaceae: inbreeding depression in geitonogamously pollinated *Geranium caespitosum* (Hessing 1988); crossing experiments and breeding barriers between 11 *Geranium*-species (Widler-Kiefer and Yeo 1987); Gesneriaceae: studies on the reproductive biology of the monotypic *Jankaea heldreichii*, a Tertiary relict species at Mt. Olympus in Greece, showed that it is not endangered (Vokou et al. 1990); Weber (1989) described the new *Didymocarpus geitleri* as having deceptive pollen flowers; Hydrophyllaceae: variation of pollinator visitation among populations of *Phacelia linearis* affecting reproductive success (Eckhart 1991); Iridaceae: bee pollination in *Moraea* species (Goldblatt et al. 1989), and entomophily in *Witsenia maura* (Wright et al. 1989), both from South Africa; Juglandaceae: pollination, breeding system, and variability in flower and fruit production of *Carya ovata* and *C. tomentosa* (McCarthy and Quinn 1989, 1990); Lecythidaceae: pollination modes in actinomorphic and zygomorphic species from French Guiana (Mori and Boeke 1987); high outcrossing rate of *Bertholletia excelsa* determined by isozyme analysis (O'Malley et al. 1988); Lamiaceae: the Mediterranean shrub *Lavandula latifolia* and its insect pollinators in their daily cycles represent independent responses to daily rhythmicity of the physical environment (C.M. Herrera 1990c); in *Glechoma hederacea* fruit-set decreased with increasing pollination distance (Widén and Widén 1990); Liliaceae: inflorescence size is important to fecundity and pollen donation in *Amianthium muscaetoxicum*, which has great percentage of long-distance pollination (Palmer et al. 1988); reciprocal incompatibility between two varieties of honeyeater seems sufficient to induce maximal seed set in *Blandfordia nobilis* (Zimmerman and Pyke 1988); self-incompatibility and pollen-ovule ratios of *Muscari* species from southern Spain (Valdés and Mejías 1988); Linaceae: honeybee visitation did not increase capsule or seed production in self-compatible, autogamous *Linum usitatissium* cultivar (Williams et al. 1991); Lobeliaceae: natural selection on floral traits in hummingbird-pollinated *Lobelia cardinalis* and bumblebee-pollinated *L. siphilitica*, and reproductive success (Johnston 1991a, b); Lythraceae: outcrossing rate and population density in two *Cuphea* species (Krueger and Knapp 1991); Malvaceae: self-compatibility and bee-pollination in *Hibiscus moscheutos* (Spira 1989); Melastomataceae: for principal pollinators, pollination and breeding systems in the Melastomataceae without and with nectar-bearing flowers see Renner (1987a, 1989a, 1990); Mimosaceae: high level of outcrossing of two *Acacia* species in Australia and Papua New Guinea detected with isozyme markers (Moran et al. 1989); intraspecific variation in *A. terminalis* indicates two different pollination syndromes (bees and birds) within this species (Kenrick et al. 1987); bee and bird pollination in Australian *Acacia* species with corresponding flower and inflorescence adaptations (Bernhardt 1989); *A. nilotica*, flowering during the rainy season in Kenya, has about 1.1 million pollen grains per inflorescence and is pollinated by solitary bees (Tybirk 1989); Myrtaceae: flower activity and nectar flow in *Syzygium cormiflorum* allows bats, birds, and insects to be pollinators (Crome and Irvine 1986); Nyctaginaceae: when insect visitors are lacking, the flowers of *Boerhaavia diffusa* self-pollinate successfully (Chaturvedi 1989b); Nymphaeaceae: Schneider et al. (1990) review the pollination of thermogenic *Victoria* and *Nelumbo* species; the self-compatible *Euryale ferox* in Japan has chasmogamous and cleistogamous flowers (Kadono and Schneider 1987); reproductive biology of *Nymphea* species, including flowering, floral odors, color, and the form and function of various flower parts were evaluated by Wiersema (1988); Ochnaceae: considerations about functional morphology, including pollination (Amaral 1991), indicate that possession of poricidal anthers is most likely to be the ancestral condition in the family, and that several genera by a transference of function developed a novel structure, also with a porus, from staminodia (Kubitzki and Amaral 1991); Onagraceae: for pollinating insects of several *Circaea* species see Boufford (1987); Pandanaceae: vertebrate pollination and dioecy appear to be ancestral in the family and anemophily and entomophily derived (Cox 1990); Phormiaceae: the importance of pollen source as mediated by pollinator movements appears important in explaining seed set in *Phormium tenax* (Craig 1989); Poaceae: perennial grasses of the tribe Triticeae contain many more self-fertile species than self-sterile ones (Jensen et al. 1990); Polemoniaceae: divergent pollination systems in the cleistogamous *Collomia grandiflora* (Sanders et al. 1990); for flexible life history traits in response to pollinator abundance see Paige and Whitham (1987); measurement of phenotypic selection of floral traits in *Ipomopsis aggregata* indicated that several traits are selected through male or female functions of the hermaphroditic flowers (Campbell 1989; Campbell et al. 1991); Campbell and Waser (1989) found for this species that pollen dispersal varied strongly among populations and years: bumblebee behavior and flower visitation of *Polemonium viscosum* (Galen and Blau 1988), and determination of a positive relationship between morphometric floral traits and components of male function (Galen and Stanton 1989; Galen 1989); deposition of foreign pollen influenced germination rate of proper pollen and also reduced fertilization success (Galen and Gregory 1989). Self-pollination in the self-incompatible *P. viscosum*, which is impeded by spatial and temporal separation of pollen and stigma presentation, negatively influenced the normal functioning of outcross pollen (Galen et al. 1989); Pontederiaceae: reproductive biology of *Eichhornia* species and other aquatic plants, with a review of the prevailing pollination modes (Haynes 1988); Portulacaceae: autogamy, cleistogamy, and biotic pollination of *Portulaca* species in Hawaii (Kim and Carr 1990); Proteaceae: intra-specific variation in the breeding system, with self-compatibility found in New England National Park (Vaughton 1988); general floral biology (Vaughton and Ramsey 1991), and effect of inflorescence damage on fruit-set in *Banksia spinulosa* (Wallace and O'Dowd 1989); there is limitation of fruit set by ineffective pollination in co-occurring *Banksia* species (Copland and Wheland 1989), and an improved fruit set and seed to flower ratio after cross-pollination as compared to selfing and open pollination (Fuss and Sedgley 1991); beetles as main pollinators of South African *Leucadendron* species (Hattingh and Giliomee 1989); Pyrolaceae: differences in pollinator preferences may prevent interbreeding of sympatric *Chimaphila* species in Massachusetts (Standley et al. 1988); Ranunculaceae: the spring-flowering *Pulsatilla vulgaris* requires at least 12 °C for flower opening, which is also the temperature by which bees become active as flower visitors (37 bee species, with *Lasioglossum lineare* being the main pollinator) (Kratochwil 1988b); the flowers of the alpine *Ranunculus adoneus* track the sun's movement from the early morning until mid-afternoon with temperature in the flower rising several °C above ambient temperature and better attraction of the pollinating fly species (Stanton and Galen 1989); *Trollius europaeus* is exclusively pollinated by four *Chiastocheta* (Diptera) species that mate in the flowers, eat pollen and nectar, and complete larval development in the seeds. Pellmyr (1989) analyzed the effects of pollinator density and behavior on seed production and seed loss. This interesting study of a mutualism based on seed parasites as pollinators provides an extensive cost/benefit analysis; Rhizophoraceae: studies on the carpenter bee-pollinated *Kandelia candel* and the bird-pollinated *Bruguira gymnorrhiza* (Tanaka 1989); Salicaceae: sexual dimorphism and considerable variation in flowering and pollinator behavior in *Salix myrsinifolia-phylicifolia* (Elmqvist et al. 1988), and reproductive effort is greater in females than in males of *Salix rigida* (Elmqvist et al. 1991); Sarraceniaceae: common buzz-pollination in *Heliamphora tatei* and other plants from Cerro de la Neblina in Venezuela (Renner 1989b); Saxifragaceae: experiments showed that forced synchronous flowering of two sympatrically occurring *Saxifraga* species had no negative reproductive effect (McGuire and Armbruster 1991); pollinators, flight distance, and pollen flow in *Saxifraga hirculus* (Olesen and Warncke 1989a, b); Scrophulariaceae: the hemiparasitic *Melampyrum arvense* was pollinated and robbed by different bumblebees and artificial pollination increased fruiting success (Kwak 1988); syrphid flies but no birds or bumblebees were found to be pallinators of *Mimulus cardinalis* and *M. lewisii* in Utah (Vickery 1990); for multiple paternity and self-fertilization in relation to floral age in *M. guttatus* see Dudash and Ritland (1991), and for resource allocation in *Mimulus* see Sutherland and Vickery (1988); studies on pollination ecology in Japanese *Pedicularis* species indicate close parallels with characteristics of North American species (Macior 1988); Solanaceae: flower visitors and buzz pollination by bees in 18 *Solanum* species in Australia (Anderson and Symon 1988), in *S. lycocarpum* in Brazil (de Oliveira-Filho and de Andrade Oliveira 1988), and *S. sysimbrifolium* in India (Babu et al. 1987); Styracaceae: the self-incompatible Brazilian cerrado species *Styrax camporum* and *St. ferrugineus* were found to be pollinated by *Bombus* bees and Scoliidae wasps (Saraiva et al. 1988); Tecophilaeaceae: enantiostylous *Cyanella* should be buzz-pollinated, but nothing is known about flower visitation in its natural habitat (Dulberger and Ornduff 1987); Tiliaceae: reproductive biology of *Grewia occidentalis* in South Africa (Zietsman 1991); Urticales: Berg (1990b) gives an overview about differentiation of flowers and inflorescences of Urticales in relation to their protection against breeding insects and to pollination; Velloziaceae: the self-incompatible Brazilian cerrado species *Vellozia squamata* has a pulsed flowering phenology and is pollinated by generalistic bee species (Oliveira et al. 1991); Verbenaceae: *Verbena* species in Iowa and Michigan are facultatively xenogamous, being pollinated by several Hymenoptera, Lepidoptera, and Diptera (Cruden et al. 1990).

### b) Ficus

An International Symposium in Coral Gables in 1988 was dedicated to the biology of figs and fig/pollinator mutualism in seasonal environments (Bronstein and McKey 1989). The mutualism is mainly a tropical event, but there are certain adaptations allowing it to occur at the edge of the range of the genus. There seems to be more plasticity of fruit and wasp developmental time and also a selective advantage of dioecy in seasonal habitats (Bronstein 1989). Verkerke (1989) showed the different strategies in monoecious and gynodioecious species, with the former producing wasps and seeds in a single fig type and showing imperfect heterostyly, and the latter having specialized figs for wasps and for seed production and showing perfect heterostyly. Frank (1989) proposed several hypotheses about the ecological interactions among figs, pollinator wasps, and the many arthropod species associated with each fig species (for specificity between African figs and their Chalcidoidea wasps see Wiebes 1989a, b; Wiebes and Compton 1990). Likewise, Addicott et al. (1990) focus on aspects of life-histories important for understanding the origin, maintenance, and breakdown of mutualistic interactions for *Ficus* and its wasps (also for *Yucca* and its moths), and Berg (1990a) reviews aspects of differentiation, like heterostyly, protogyny, the number of staminate flowers, evolution of the syconium, phenology, and dioecy, which are of importance in the rearing of the pollinators. The pollinator specificity may have its reason in specific attractant volatiles, as is indicated by the study of *F. burtt-davyi* (van Noort et al. 1989). Structural changes of gall and seed flowers are related to wasp development and behavior (for *F. ingens* see Baijnath and Naicker 1989) and pollen pockets can be understood (at least in *F. asperifolia*) as tools for the pollination in oviposited gall flowers and the prevention of seed formation in neighboring, not oviposited, gall flowers, which otherwise would strongly compete for space with the developing galls. In *Ficus* subgenus *Sycomorus*, the interior of the syconium is partly filled with liquid; the male fig wasps possess morphological adaptations in response to living in the aquatic environment. This allows them to be active and to reach the females (Compton and McLaren 1989). In *F. carica*, hermaphrodite flower primordia indicate that this species is of gynomonoecious ancestry, but presently has a complex breeding system with two tree morphs and three floral forms. The short-styled flowers, which, contrary to the long-styled ones with wet stigma, have a nearly dry stigma, seem to lose their ability to function as fully viable females (Beck and Lord 1988a, b). The flowering phenology may be responsible for pollen limitation and fruit abortion (Bronstein 1988b), and also the within-tree reproductive synchrony seems critical for the reproductive success (Bronstein et al. 1990). Additionally to the Chalcidoidea: Agaonidae-wasps there are nonpollinating wasps and predators which may or may not have an impact on successful pollen transfer (Compton and Robertson 1988; Bronstein 1988a, c, 1991).

### c) Orchids

Benzing (1987) discussed the phenomenon of abrupt cladogenesis effected by pollinators, supposedly the main reason for orchid speciation. Similarities and differences in pollination of terrestrial orchids of southern Australia and the Mediterranean regions are compared by Dafni and Bernhardt (1990), who show the broad convergences in the orchid pollination systems in the two regions. On the other hand, pseudocopulatory syndromes are much more diverse in southern Australia than in the Mediterranean region, while shelter flowers have not been described in Australia.

The detailed study and quantification of pseudocopulation of *Drakaea glyptododon* by a male thynnine wasp showed that of the male wasps present at a site, 93% landed on the flowers, 21% attempted to grasp the female decoy, and only 6% attempted to copulate with the flower (Peakall 1990). *Leporella fimbriata*, pollinated through pseudocopulation by winged male ants (see above), forms large clones. Some outcrossing, however, is mediated by the ant pollinator (Peakall and James 1989). Borg-Karlson (1990), while reviewing chemical and ethological studies of pollination in the genus *Ophrys*, concluded that insect-flower relationships in the various pollination systems are complex and that the degree of selectivity in the attraction of insects is still unclear. Paulus and Gack (1990a, b) gave several examples of speciation in this genus which, to them, is always a consequence of a change in pollinator. Several examples of closely related species pairs are given, which are the result of genetic isolation due to varying pollinators. Also nonsexual deception is a common phenomenon in orchids. New examples are *Cleistes divaricata*, a rewardless bee-food-flower mimic (Gregg 1989), the nectarless *Orchis spitzelii* (Fritz 1990), the generalized food-flower mimic *Cymbidiella flabellata* (Nilsson et al. 1986), and the rewardless *Thelymitra epipactoides* (Cropper and Calder 1990). For evolutionary trends and pollination of the tribe Thelymitreae see Burns-Balogh and Berhardt (1988), and for different fragrance profiles in interfertile *Coryanthes* as a primary reproductive isolating mechanism see Gerlach and Schill (1989).

There are two pollination ecotypes in *Platanthera ciliaris* as the result of supposed selection pressure by butterflies with different proboscis length in the mountains and the coastal plain in the southeastern United States (Robertson and Wyatt 1990). In *Cleistes divaricata* there are no pollinia but loosely aggregated pollen tetrad masses, which are deposited onto the dorsal thorax of the pollinating bees; this system showed pollination limitation and reduced seed set (Gregg 1991). Such a situation of limited pollination seems to be very common in orchids. Examples for tropical and temperate species are (some of them also resource or otherwise limited): *Epidendrum ciliare* (Ackerman and Montalvo 1990); *Aspasia principissa* (Zimmerman and Aide 1989); *Dactylorhiza purpurella* and *D. fuchsii* (Hazeldon et al. 1991); *Dendrobium speciosum* (Slater and Calder 1988); *Encyclia krugii* (Ackerman 1989); *Cyclopogon cranichoides* (Calvo 1990a); *Platanthera stricta* (Patt et al. 1989); *Myrosmodes cochleare* (Berry and Calvo 1991); *Epidendrum exasperatum, Lepanthes wendlandii, Oeceoclades maculata* (Calvo 1990b). Only rarely is there a high visitation rate, which might be a consequence of conspicuousness or continuous and/or copious nectar production: for *Spiranthes romanzoffiana* see Larson and Larson (1987, 1990).

A new technique for monitoring pollen flow in orchids was described by Peakall (1989a), who labeled pollinia with colored histochemical stains without interfering with pollinarium removal. Such methods are especially promising in view of the fact that pollinium export of an inflorescence normally exceed pollinium import (for *Epipactis helleborine* see Piper and Waite 1988), or also in cases of long-distance pollinators (e.g., male ithomiine butterflies pollinating *Epidendrum paniculatum*, DeVries and Stiles 1990). Sex ratio in *Catasetum viridiflavum* is largely determined by light intensity (Zimmerman 1991), and its effect on flowering phenology, which is asynchronous with the activity of its pollinator, indicates that it is not pollinator-determined but influenced by other phenomena (Zimmerman et al. 1989).

### d) Economy of Pollination and Crop Plants

Applied pollination ecology is an active field with an enormous output of information (not to be reviewed here). For sexual reproduction of tree crops see Sedgley and Griffin (1989). The Sixth International Symposium on Pollination, held at Tilburg, The Netherlands, in 1990, brought encouraging results and also showed that the effective management of pollination systems is a goal within reach (Corbet 1991). Since the European race of honeybees is still the most commonly used pollinator for crops, and in certain geographical regions may even be the main or the only pollen vector (e.g., for *Vicia faba* in Southern Australia, Stoddard 1991), the value attributed to its pollination was calculated for US crops to be several billion dollars (Robinson et al. 1989). This value, however, is in dispute, since it seems difficult to measure the contribution of *Apis* to the pollination of various crops (Southwick and Southwick 1989). The benefits from honeybee pollination certainly are very impressive, but there is a plea for modification and diversification of pollination strategies for crops. For North America this has several reasons. The effects of the invasion of Africanized bees certainly will have a negative impact on beekeeping, as also have diseases of the honeybees, the tracheal mite, and *Varroa jacobsoni*. Torchio (1990a) and Kevan et al. (1990a, b) show alternatives and suggest that much can be done to avert a potential crisis in agricultural production brought about by pollinator problems. A large number of bee species are now being tested as primary or alternative pollinators of crop species, with the establishment of a profile of each commercialized nonhoneybee pollinator (Torchio 1987). Examples of such alternative bee species are *Osmia lignaria* as a mobile pollinator of orchard crops (Torchio 1991). *Osmia ribifloris*, bumblebees, and the anthophorid *Habropoda laboriosa* for blueberry, canberry, and lignonberry (Mohr and Kevan 1987; Cane and Payne 1988; Torchio 1990b), *Peponapis pruinosa* and bumble bees for *Curcurbita* and *Cucumis* (Fisher and Pomeroy 1988; Kevan et al. 1988), or *Megachile rotundata* for red clover (Fairey et al. 1989). In some cases, e.g., in orchards in Ontario, *Apis mellifera* may be the most frequent pollinator, but other bee species may be more effective (Boyle-Makowski 1987).

Interference of nectar thieves can be quite deleterious, as was shown by I. Sazima and Sazima (1989), who observed *Trigona spinipes* at a site in Brazil, which not only diminished nectar supply and attractivity of passionflower (*Passiflora edulis*), but also aggressively defended the flowers against the pollinating carpenter bees, with a consequent reduction in fruit production. Other problems show up when a species like *Acca* (*Feijoa*) *sellowiana*, which is a native of South America pollinated by birds (also in New Zealand, Patterson 1989), is cultivated in Italy (Intoppa and Piazza 1989). It is not surprising that there serious production problems and lack of fruit formation occur. The correct timing of artificial pollination is important. Since the normal time of anthesis in *Hevea brasiliensis* is in the afternoon, hand pollination in the afternoon causes higher fruit set (Sedgley and Attanayake 1988). Double pollination in apple and pear enhances seed quality. There is stronger pollen competition after double pollination, which means that after repeated pollination – as occurs in nature by pollinators – the survival of the fittest pollen is promoted (Visser and Verhaegh 1988).

## 7. Breeding Systems and Gender Distribution

### a) Pollen-Pistil Interaction

A new pollen-stigma interaction was described for the submarine-pollinated seagrass *Posidonia australis* (McConchie and Knox 1989). The pollen shows precocious germination before contact with the stigma in seawater; the stigma seems to release a germination substance. Lu et al. (1990) found that different *Arachis* species had different stigma forms, with perennial species having very small stigmas and consequent difficulties in pollination and limitation in seed set; annual species have larger stigmas and are more easily pollinated. The Onagraceae show a unique heterogeneity of stigma form and physiology, with wet-stigma species showing self-incompatibility, and species of the dry-stigma type being self-compatible but not always inbreeding (Heslop-Harrison 1990). There exist structural differences in the dimorphic flowers of *Collomia grandiflora*, with chasmogamous flowers having a solid style and cleistogamous styles being partly hollow (Sanders et al. 1990). Differences in the cuticular thickness of the two S (incompatibility) genotypes of *Brassica oleracea* are responsible for differences observed in the rate of pollen hydration on the stigma surface (Elleman et al. 1988). Stigmatic inhibition of pollen grains also occurs in *Leucaena leucocephala*, with germination of pollen grains only occurring when a certain minimum of pollen reaches the stigma. This number-dependent inhibition of pollen grain germination is effected by a pH-dependent proteinaceous inhibitor active at the stigmatic pH (Ganeshaiah and Shaanker 1988). Hybridization between *Rhododendron* species depends, among other things, on comparative style length (Williams and Rouse 1990). Pollen competition in the style was discussed by Charlesworth (1988), who warns that only few papers contain real evidence for it (see Snow and Spira 1991a, for *Hibiscus moscheutos*).

Stigmas and phenomena of secretion were reexamined by Dumas et al. (1988) in the light of new data. In distylous *Linum* species, Dulberger (1987) found dissimilar properties of the papillae, i.e., the occurrence of wet thrum stigmas and dry pin stigmas, with the well-known inhibition in the two intramorph incompatible combinations. Differences in stigma types also occur among different species of a genus (for *Populus* see Villar et al. 1987). Stylar exudates after cross- or self-pollination differ in *Lilium longiflorum* (Amaki and Higuchi 1991), and stylar sap collected from cross-pollinated styles significantly promoted growth of incompatible pollen tubes (Amaki et al. 1989). In *Dendrobium speciosum*, after deposition onto the wet stigma, the pollen grains in the pollinia progressively hydrated and germinated from the outside of the pollinium to the inside (Slater 1991).

### b) Incompatibility Versus Compatibility, Cleistogamy; Apomixis

Within a population of plants there is an optimal outcrossing distance (for *Delphinium nelsonii* it was found to be 3–10 m, Waser et al. 1987), and this kind of mate choice, although difficult to study, certainly will be an important opportunity for future research (Marshall and Folsom 1991). The "Zavada hypothesis", again defended (Zavada 1990), was again rebutted by Gibbs (1991), who repeated that it is unlikely that there is any general correlation between the different types of self-incompatibility mechanisms in flowering plants and pollen exine sculpturing or pore size, and that it would be inadvisable to attempt to deduce the self-incompatibility system (SI) of the early angiosperms on the basis of fossil pollen data. The traditional view of three relatively clear-cut self-incompatibility mechanisms (GSI, SSI, Het SI) is now very difficult to sustain and the evidence shows that SI systems are much more diverse and complex than previously thought (Gibbs 1988). For example, there is a multifactorial SI system in *Borago officinalis* (Leach et al. 1990), a unique system in *Dendrobium*, expressed by flower abscission and not by inhibition of pollen germination or pollen tube growth (Johansen 1990), or a system like that in *Crocus vernus*, in which the SI barrier can be overcome by mixed pollination, which allows selfing according due to a mentor effect (Chichiricco 1990). In a SI species like *Espeletia schultzii*, seed abortion at the highest elevations in the Andes is not a genetic effect but is simply correlated with altitude (Sobrevila 1989). Tropical trees not only show much variation within their populations, but also have sufficient outcrossing and show a high index of SI or dioecy (Bawa and O'Malley 1987; Hamrick and Loveless

1989). In the Cruciferae of Israel, the 60% xenogamous species are more common in the Mediterranean part of the country, while the remainder are self-compatible and largely autogamous, and occur more frequently in the desert (Boaz et al. 1990). For a screening of breeding systems and pollen-ovule ratios in more than 160 taxa of the Polemoniaceae see Plitmann and Levin (1990), and for quantitative variation and population structure in outcrossing and selfing *Phlox* and *Clarkia* species see Clay and Levin (1989), and Vasek and Weng (1988). The major pathway of breeding system evolution in the genus *Eichhornia* involves the repeated breakdown of tristyly to give rise to selfing semi-homostylous populations (Barrett 1988a).

Facultative xenogamy is a mixed mating system, with self-compatibility, delayed autogamy, and adaptations for cross-pollination. The balance between self- and cross-pollination is important in stable communities where pollinator activity may be low or unreliable (Cruden and Lyon 1989). Such mixed mating systems were found for *Cavanillesia platanifolia* (Murawski et al. 1990), in *Amphicarpaea bracteata*, with a tendency in inbred individuals to form cleistogamous flowers (Trapp and Hendrix 1988), in all Hawaiian *Bidens* species (Sun and Ganders 1988), in *Bartsia alpina* (Molau et al. 1989), or in *Senecio vulgaris* (J.M. Warren et al. 1988). Such mixed mating systems form more easily in sparse than in dense populations and seem evolutionarily stable (Holsinger 1991).

Cox (1989), while reviewing Baker's Law, which says that self-compatible taxa have the advantage over other taxa in establishment after long-distance dispersal, especially on islands, confirmed this principle as well justified and proven, with the only exception that dioecious taxa seen to be far better in long-distance dispersal than predicted. Self-compatibility is also the predominating breeding system on the Galapagos Islands (McMullen 1990).

The evolution of plant mating systems, including self-fertilization, is discussed in relation to its advantages and disadvantages (Barrett 1988b; Holsinger 1988, 1991b; Ziehe and Gregorius 1988). Inbreeding depression can vary in its effect in a population (Charlesworth et al. 1990). It often results in reduced seed production or lower seed weight (Bertin et al. 1989; Hessing 1989a; Kalisz 1989; Weller and Ornduff 1991). Inbreeding reduces the number of fertile stamens in *Scleranthus annuus* (Svensson 1990). High temperature can weaken SI in *Lycopersicum peruvianum* (Webb and Williams 1988). Self-fertility did not vary along an elevation gradient in *Geranium caespitosum* (Hessing 1989b), or in different populations of *Crepis tectorum* on the Baltic island Öland (Andersson 1989), and was found to be stimulated by cross pollen in *Campsis radicans*, which previously was considered self-sterile. This is a warning to be cautious when using pure self-pollen for assessing compatibility reactions (Bertin and Sullivan 1988). Self-fertility is higher in *Arrhenaterum elatius* individuals growing in toxic as compared with normal habitats (Cuguen et al. 1989). Different compatibility reactions in two subspecies of *Stephanomeria exigua* are governed by a single gene (Brauner and Gottlieb 1987).

Apomictic species or populations often have greater geographical ranges and occur at higher latitudes or elevations than their sexual relatives. Their plasticity make apomicts well suited for colonization (Bennett and Elgar 1987; Bara 1989; Bierzychudek 1989; Michaels and Bazzaz 1989). Apomixis is common in Maloideae (Rosaceae) (Campbell et al. 1991), in *Rubus* (Nybom 1988), in triploid and tetraploid *Tripsacum dactyloides* (Burson et al. 1990), and in triploid *Eupatorium tanacetifolium* (Rozenblum et al. 1988). First report of a predominantly agamospermous dioecious genus – *Garcinia* (Richards 1990a). For apomictic nucellar embryos in triploid and tetraploid *Gymnigritella* and *Nigritella* species see Teppner (1991) and Teppner and Klein (1989). In *Triodanis perfoliata* (Campanulaceae) the seeds produced by cleistogamous flowers showed inbreeding depression (as seen from seed development) in comparison to the seeds from chasmogamous flowers (Gara and Muenchow 1990).

### c) Heterostyly

Barrett and Richards (1990) gave a comprehensive review of heterostyly in tropical plants with many woody genera as examples. The few existing data already indicate that tropical heterostylous groups display large variation in sex expression. This seems to be related to a greater range of potential pollinators in tropical communities, as compared with temperate ones. One of these phenomena is the replacement of distyly by various gender polymorphisms in *Cordia* and several Rubiaceae. Another pathway involves the breakdown of the polymorphism toward increased fertilization and floral monomorphism. Homostylous species or populations derived from distylous ancestors include *Turnera ulmifolia* (Shore and Barrett 1990), *Villarsia albiflora* (Ornduff 1988a), the apomictic *Erythroxylum undulatum* (Berry et al. 1991), and *Primula vulgaris* (Boyd et al. 1990). *Villarsia* is self-incompatible, whereas *Turnera* and the homostylous *Primula* have changed to self-compatibility. Barrett (1990), referring to the evolution and adaptive significance of heterostyly, describes experiments in *Jepsonia heterandra* and *Pontederia cordata*, which provide support for the Darwinian hypothesis that the style-stamen polymorphism promotes disassortative (intermorph) pollination and mating.

In most cases, floral dimorphism is accompanied by a very effective self-incompatibility system. Further examples are *Luculia gratissima* (Rubiaceae) (with interactions of the compatible and incompatible pollen tubes in the style, Murray 1990), *Aconogonon campanulatum* (Polygonaceae) (Hong 1991), *Pulmonaria affinis* (Richards and Mitchell 1990), *Menyanthes trifoliata* (Nic Lughadha and Parnell 1989), *Villarsia congestiflora* (Ornduff 1988b), *Oxalis violacea* (Bernhardt 1990b), or *Pontederia cordata* (Richards and Barrett 1987). There is partial or total breakdown of self-incompatibility in *Eichhornia paniculata* (Morgan and Barrett 1989; Seburn et al. 1990), *Oxalis physocalyx* (de Barros Pâcheco and Coleman 1989), and in the genus *Nivenia* (Iridaceae) (Goldblatt and Bernhardt 1990). *Epacris impressa* in southeastern Australia with bimodality in style length, but without heteromorphic incompatibility, can at most be interpreted as the earliest stage in the evolution of heterostyly (O'Brien and Calder 1989). The variation in the diverse and unspecialized pollinator fauna of the tristylous *Pontederia cordata* in terms of collecting preferences for the three morphs is shown by Wolfe and Barrett (1987, 1988). Boyd et al. (1990), when comparing heterostylous and homostylous populations of *Primula vulgaris* in Great Britain, have found that the self-fertile homostylous populations have not spread since about 50 years. Since nocturnal moths are the main pollinators of *P. vulgaris*, the self-compatible homostylous population should have advantages at least in years of poor pollination when heterostyle seed set is low. However, in such pollinator-limited years, the few developing heterostyle seeds in a capsule were significantly larger than homostyle seeds, which might have given them a competitive advantage.

### d) Sex Distribution; Monoecism Versus Dioecism, and the Costs and Benefits of Reproduction

Since selection in plants might act on male and female traits separately, statistical and experimental difficulties arise in the interpretation of results with regard to sexual selection (Lyons et al. 1989). The sexuality of plants as compared with animals is more diffuse, although this modular nature of sexuality in plants facilitates greater control of resource allocations to maleness and femaleness and maternity (Lovett Doust 1990). Certainly, reproductive traits are not free to evolve independent of plant form (Waller 1988), and competition also affects the reproductive behavior of individuals and populations (Weiner 1988).

Sex determination refers to the complicated balance between expression of female versus male sexuality within plants. Correlated with these processes is the differential allocation to the sexually different functions. Meagher (1988) in a review emphasizes that there is growing evidence that even within hermaphroditic populations there is variation in maternal versus paternal success among plants, which indicates that, e.g., dioecy is only one end point of many possible developments. When discussing sexual strategies in plants, it becomes necessary to take the modular construction of plants into consideration and to estimate both "phenotypic gender", which is the plant's total investment in male and female functions, and the "functional gender", which is the relative success of a plant as a maternal and paternal parent at the level of populations (see review in Cox 1988b). Lovett Doust and Lovett Doust (1988b) speak about an emerging synthesis called "sociobotany", which brings together ideas and evidence concerning sex allocation, sex habits and dimorphisms, sexual selection, paternity, female choice, and aspects of sexual incompatibility in plants. The literature on all these interesting aspects and on this synthesis is very vast and cannot be evaluated in all its details. There is a plea for more empirical and quantitative data, since the subject of gender modification and sex allocation has a direct influence on our understanding of the pressures of natural selection on reproductive function (Schlessman 1988; Charlesworth and Morgan 1991). Also, when evaluating paternity in plants, Bertin (1988) emphasizes a considerable theoretical literature exists with which empirical developments have not kept pace. Willson (1991) explored the phenomenon of sexual dimorphism and concluded that, parallel to animals, also in plants frequently the males (of dioecious taxa) or the male portions of the flowers (in co-sexual flowers) have diversified structurally.

Sex determination in monoecious and dioecious plants has a physiological and genetic basis, but it seems difficult to distinguish the factors that influence sex determination from those coding for floral differentiation programs (Irish and Nelson 1989). The theories and empirical data on the evolution of dioecy are manifold (see Prog Bot 50). A further interesting case was shown for *Cucurbita foetidissima*, which has monoecious and dioecious individuals. The purely female plants produce seeds which survive their first year in nature 2.8 times more frequently than seeds from monoecious plants, apparently because seeds from monoecious plants are mostly self-fertilized and selfing reduces seedling survival (Kohn 1988). Thomson and Brunet (1990), while discussing the manifold hypothesis for the evolution of dioecy, express the opinion that since few of the old or new hypotheses have been rejected, this could well indicate that most of them are correct. Since it seems indisputable that the separation of sexes has arisen by various pathways, it is better not to speak of one dioecy but of several dioecies. There is a continuous range between gynodioecy and dioecy, with less constant sex expression in males than females (Charlesworth 1989; Delph and Lloyd 1991). In *Daphne kamtchatica*, clogging of flowers by self pollen is assumed to be the main factor driving the evolution towards gynodioecy (Kikuzawa 1989). In *Silene acaulis* gynodioecy is maintained through higher fecundity of female than hermaphrodite plants (Shykoff 1988). Dioecy in *Rosa setigera* is cryptic and difficult to see without careful examination (Kevan et al. 1990c); in *Bauhinia malabarica* and *B. divaricata* it is a result of organ suppression (Tucker 1989), and certainly derived from hermaphroditism in *Wikstroemia* and *Schiedea* in the Hawaiian Islands (Mayer 1990; Weller et al. 1990). The conditions which promote andromonoecy are discussed by Spalik (1991). In the genus *Solanum* andromonoecy and dioecy seem to be polyphyletic, and dioecy in this genus is likely to have arisen from hermaphroditic ancestors (Anderson and Symon 1989). In contrast, androdioecy – a very rare condition in angiosperms – in *Datisca glomerata* (Datiscaceae) supposedly evolved from a dioecious precursor (Liston et al. 1990).

Reproductive success is defined as the appearance of individuals forming a new generation in a demographic sense. Urbanska (1989) proposes a revised approach to reproductive strategies based on performance of various reproductive units. Reproductive ecology and reproductive strategies, considering resource allocation and seed size, were discussed and reviewed by Lovett Doust (1989) and Lloyd (1989). There is wide variation in reproductive success among individual plants related to physiological, structural, and phenological characteristics (Lechowicz and Blais 1988), and there are also innate constraints, like, e.g., the production of large seeds by large plants (Thompson and Rabinowitz 1989). To Lloyd (1988), the expenditure on reproduction may be divided into the "direct" costs of structures (pollen and seeds) that contribute genes to the next generation and "accessory" costs of structure (flower stalks, sepals, petals, nectar, and parts of the sporophylls) that contribute to maternal and/or paternal fitness indirectly. If reproduction is resource-limited, the pattern of fruit and seed abortion is related to time of fruit initiation, proximity of the fruit to resources, number of developing seeds in fruits, damages, number of pollen donors, etc. (Lee 1988). The common surplus of flowers on a plant is a cost and reduces potential seed production, but surplus flowers have another function, namely to serve as reserve ovaries (Ehrlén 1991). Very interesting and unexpected are the results from *Ipomopsis aggregata* and *Delphinium nelsonii*, where not only crosses between nearby plants cause low fruit set (inbreeding depression), but also crosses between more widely separated plants (outbreeding depression), which mostly is a result of disruption of local adaptation (Waser and Price 1989, 1991). In *Cypripedium acaule* the high cost of reproduction after fruit set could be demonstrated only after 3 or 4 years. Only then did plants show lower growth and flowering rates (Primack and Hall 1990).

## 8. Pollination Studies at the Community Level, Distribution of Flower Types, Flowering Phenology, and Nature Preservation

It is important to recognize and to keep in mind that flower/animal interactions are constant and fixed not even at the level of communities. They rather have to be considered as being part of a dynamic system with strong fluctuations and modifications. This was demonstrated by Hartmann (1988) for several Asteraceae species in a dry meadow in Central Germany, and by C.M. Herrera (1988) for the pollinator assemblage of *Lavandula latifolia* in southeastern Spain. These authors have found that diversity, composition, and abundance of pollinators varied markedly during the season or between years. This might have an influence on the selective forces and on the evolution of flower/pollinator interrelationships. The effect of competition between co-flowering conspecific and heterospecific plants visited by a common hummingbird in a tropical forest in Costa Rica was shown during laboratory and field experiments (Feinsinger and Tiebout III 1991; Feinsinger et al. 1991). These fascinating studies demonstrated instructively how slight modifications in floral composition and density influence pollination levels and female success. "The collective results demonstrate that studying the role of particular interactions in community structure must involve examinations of mechanisms at the level of individuals, processes at the population level, and patterns of coexistence rather than just one level of this hierarchy" (Feinsinger and Thiebout III 1991).

Pollination and reproductive biology, including female success and seed dispersal, is not only an integrating factor of biocenoses (Vogel and Westerkamp 1991), but also provides basic information concerning function, maintenance, and regeneration of biotopes with a broad range of application in forestry and nature and landscape preservation (Gottsberger et al. 1991). Kratochwil (1988a, 1991) emphasizes the importance of co-phenological studies of plants and anthophilous insects (for syrphid flies and forests in southwest Germany see Ssymank 1991).

Relevant community studies and reviews refer to the tropical rain forest (Bawa 1990), the arid winter rainfall region of South Africa (Struck 1990), locations at different elevations in Utah (S.D. Warren et al. 1988), in Michigan (Rotenberry 1990), and Rhode Island (Rathcke 1988b), the Snowy Mountains of Australia (Inouye and Pyke 1988), Mediterranean shrublands (Arroyo 1988; J. Herrera 1988a; Petanidou and Vokou 1990), shrublands in the Venezuelan highlands (Ramirez 1989; Ramirez et al. 1990), the Brazilian cerrado vegetation (Silberbauer-Gottsberger and Gottsberger 1988), tropical dry forests in Mexico (Arizmendi and Ornelas 1990), boreal forests in Canada (Barrett and Helenurm 1987; Helenurm und Barrett 1987), and warm and cool temperature forests in Japan (Yumoto 1987a, b, 1988). Different flowering times can act as a reproductive barrier, e.g., between sympatric *Salix* species around Toronto (Mosseler and Papadopol 1989). Phenological data have now become available for spring-blooming plants of the Washington, D.C., area (Shetler and Wiser 1987), for the Rio Grande Valley in Texas (Vora 1990), for mesotrophic grassland in northern England as influenced by hay cutting (Smith and Jones 1991), for tree layer species from the timber line in Central Himalaya (Rawal et al. 1991), for South Indian hill ranges (Sivaraj and Krishnamurthy 1989), for subtropical heathland in Queensland (McFarland 1990), for Mediterranean shrubland and other formations in Spain (Arroyo 1990a, b; Guitián and Guitián 1990; Guitián et al. 1990), for tropical deciduous forest in Mexico (Bullock and Solís-Magellanes 1990), for tropical cloud forest in Costa Rica (Koptur et al. 1988), for rain forests in the Central Philippines (Heitsman 1989), in Singapore (Corlett 1990), and Australia (Hopkins and Graham 1989), for semideciduous forests (Morellato et al. 1989; Morellato and Leitão-Filho 1990), for cerrado vegetation (Mantovani and Martins 1988), and coastal dune plant communities (Cordazzo and Seeliger 1988) in Brazil, for subtropical humid forests of India (Baruah and Ramakrishnan 1989), for desert annuals and geophytes (Fox 1990a, b; Boeken and Gutterman 1991) in Arizona and Israel, for species in arid Patagonia (Bertiller et al. 1991), in the swampy "morichal" community in Venezuela (Ramirez and Brito 1987), and in alpine regions of southern British Columbia (Ratcliffe and Turkington 1989). In several of these above-mentioned studies the distribution of floral attributes, like color, morphological type, size, sex, and reward for pollinators is given, and the relationships with the visiting and pollinating animals are shown. This kind of study obviously will have to be intensified in the future.

*Acknowledgments*. The author is very thankful to Ilse Silberbauer-Gottsberger, Elke Hochgesand, Ullrich Hartmann, and Christian Westerkamp for assistance, especially concerning the tedious work of locating, requesting, and handling of literature. Angelika Hofmann kindly typed the manuscript.

## References

**Abbott RJ,** Irwin JA (1988) Heredity 60:295–298. **Ackerman JD** (1989) Syst Bot 14:101–109. **Ackerman JD,** Montalvo AM (1990) Ecology 71:263–272. **Addicott JF,** Bronstein J, Kjellberg F (1990) In: Gilbert F (ed) Insect life cycles: genetics, evolution and coordination. Springer, Berlin, Heidelberg, New York. **Ågren J,** Schemske DW (1991) Biotropica 23:235–241. **Aizen MA,** Searcy KB, Mulcahy DL (1990) Am J Bot 77:671–676. **Ali SI** (1988) Pak J Bot 20:21–44. **Ali T,** Ali SI (1989) Phyton (Austria) 29:175–188. **Alm J,** Ohnmeiss TE, Lanza J, Vriesenga L (1990) Oecologia 84:53–57. **Altenburger R,** Matile P (1988) Planta 174:248–252. **Amaki W,** Higuchi H (1991) Sci Hortic 46:147–154. **Amaki W,** Tezuka T, Yamamoto Y (1989) J Jpn Soc Hortic Sci 57:648–654. **Amaral MCE** (1991) Bot Jahrb Syst 113:105–196. **Anderson AB,** Overal WL, Henderson A (1988) Biotropica 20:192–205. **Anderson GJ,** Symon DE (1988) Ann Mo Bot Gard 75:842–852. **Anderson GJ,** Symon DE (1989) Evolution 43:204–219. **Andersson S** (1988) Oecologia 76:125–130. **Andersson S** (1989) Plant Syst Evol 168:227–236. **Andersson S** (1991) Holarct Ecol 14:186–191. **Arizmendi M del C,** Ornelas JF (1990) Biotropica 22:172–180. **Armbruster WS** (1988) Ecology 69:1746–1761. **Armbruster WS** (1990) Am Nat 135:14–31. **Armbruster WS** (1992) BioScience 42:12–20. **Armbruster WS,** Guinn DA (1989) J Kans Entomol Soc 62:468–483. **Armbruster WS,** McCormick KD (1990) Biotropica 22:160–171. **Armbruster WS,** McGuire AD (1991) Am J Bot 78:1449–1457. **Armstrong DP** (1991) Behav Ecol Sociobiol 29:103–111. **Armstrong DP,** Paton DC (1990) Aust J Ecol 15:291–297. **Arroyo J** (1988) Lagascalia 15:43–78. **Arroyo J** (1990a) Flora 184:43–49. **Arroyo J** (1990b) Isr J Bot 39:249–262.

**Babu GVS,** Bhatt SC, Pandeya SC (1987) Proc Indian Acad Sci 97:295–300. **Baijnath H,** Naicker S (1989) S Afr J Bot 55:409–421. **Baker HG,** Baker I (1990) Isr J Bot 39:157–166. **Baker JD,** Cruden RW (1991) Am J Bot 78:959–963. **Bara II** (1989) Acta Soc Bot Pol 58:127–139. **Barrett SCH** (1988a) Ann Mo Bot Gard 75:741–760. **Barrett SCH** (1988b) In: Lovett Doust J, Lovett Doust L (eds) Plant reproductive ecology: patterns and strategies. Oxford Univ, New York, pp 98–124. **Barrett SCH** (1990) TREE 5:144–148. **Barrett SCH** Eckert CG (1990) Isr J Bot 39:5–12. **Barrett SCH,** Helenurm K (1987) Can J Bot 65:2036–2046. **Barrett SCH,** Richards JH (1990) Mem N Y Bot Gard 55:35–61. **Barth OM** (1990) J Apic Res 29:89–94. **Baruah U,** Ramakrishnan PS (1989) Vegetatio 80:63–67. **Bawa KS** (1990) Annu Rev Ecol Syst 21:399–422. **Bawa KS,** Hadley M (eds) (1990) Reproductive ecology of tropical forest plants. Unesco Parthenon, Paris. **Bawa KS,** O'Malley DM (1987) Rev Biol Trop 35 Suppl 1:177–188. **Beaman RS,** Decker PJ, Beaman JH (1988) Am J Bot 75:1148–1162. **Beardsell DV,** Williams EG, Knox RB (1989) Aust J Bot 37:65–80. **Beck NG,** Lord EM (1988a) Am J Bot 75:1904–1912. **Beck NG,** Lord EM (1988b) Am J Bot 75:1913–1922. **Bell G** (1986) J Theor Biol 118:253–258. **Bennett ATD,** Elgar MA (1987) TREE 2:353–354. **Benzing DH** (1987) In: Arditti J (ed) Orchid biology, reviews and perspectives IV. Cornell, Ithaca, pp 33–77. **Berg CC** (1990a) Mem N Y Bot Gard 55:169–185. **Berg CC** (1990b) Sommerfeltia 11:13–34. **Bergström G,** Groth I, Pellmyr O, Endress PK, Thien LB, Hübener A, Francke W (1991) Phytochemistry 30:3221–3225. **Bergström J,** Bergström G (1989) Nord J Bot 9:363–365. **Bernardello LM,** Galetto L, Juliani HR (1991) Ann Bot 67:401–411. **Bernhardt P** (1989) In: Stirton CH, Zarucchi JL (eds) Advances in legume biology. Monogr Syst Bot Mo Bot Gard 29:263–281. **Bernhardt P** (1990a) Plant Syst Evol 170:247–255.

**Bernhardt P** (1990b) Plant Syst Evol 171:147–155. **Berry PE,** Calvo RN (1989) Am J Bot 76:1602–1614. **Berry PE,** Calvo RN (1991) Plant Syst Evol 174:93–101. **Berry PE,** Tobe H, Gómez JA (1991) Am J Bot 78:595–600. **Bertiller MB,** Beeskow AM, Coronato F (1991) J Arid Environ 21:1–11. **Bertin RI** (1988) In: Lovett Doust J, Lovett Doust L (eds) Plant reproductive ecology: patterns and strategies. Oxford Univ, New York, pp 30–59. **Bertin RI,** Sullivan M (1988) Am J Bot 75:1140–1147. **Bertin RI,** Barnes C, Guttman SI (1989) Bot Gaz 150:397–403. **Bierzychudek P** (1989) Evolution 43:1456–1466. **Boaz M,** Plitmann U, Heyn CC (1990) Isr J Bot 39:31–42. **Boeken B,** Gutterman Y (1991) Isr J Bot 40:17–31. **Boggs CL** (1988) Funct Ecol 2:289–295. **Borchsenius F,** Olesen JM (1990) J Trop Ecol 6:501–505. **Borg-Karlson A-K** (1990) Phytochemistry 29:1359–1387. **Boufford DE** (1987) Acta Phytotax Geobot 38:133–154. **Boyd M,** Silvertown J, Tucker C (1990) J Ecol 78:799–813. **Boyle-Makowski RMD** (1987) Proc Entomol Soc Ont 118:125–141. **Brandt U,** Gottsberger G (1988) Lagascalia 15:625–634. **Brauner S,** Gottlieb LD (1987) Syst Bot 12:299–304. **Brice AT,** Dahl KH, Grau CR (1989) Condor 91:681–688. **Bronstein JL** (1988a) Biotropica 20:215–219. **Bronstein JL** (1988b) Ecology 69:207–214. **Bronstein JL** (1988c) Ecology 69:1298–1302. **Bronstein JL** (1989) Experientia 45:622–637. **Bronstein JL** (1991) Oikos 61:175–186. **Bronstein JL,** McKey D (1989) Experientia 45:601–604. **Bronstein JL,** Gouyon P-H, Gliddon C, Kjellberg F, Michaloud G (1990) Ecology 71:2145–2156. **Brouillard R,** Wigand M-C, Cheminat A (1990) Phytochemistry 29:3457–3460. **Broyles SB,** Wyatt R (1991) Am Nat 138:1239–1249. **Buchmann SL,** Cane JH (1989) Oecologia 81:289–294. **Buchmann SL,** O'Rourke MK, Niklas KJ (1989) Bot Gaz 150:122–131. **Bullock SH,** Solis-Magallanes JA (1990) Biotropica 22:22–35. **Bullock SH,** Ayala R, Baker I, Baker HG (1987) Madroño 34:304–314. **Bullock SH,** Martínez del Rio C, Ayala R (1989) Oecologia 78:389–393. **Bultman TL,** White JF (1988) Oecologia 75:317–319. **Burkhardt D,** Maier E (1989) Naturwissenschaften 76:82–83. **Burns-Balogh P,** Bernhardt P (1988) Plant Syst Evol 159:19–47. **Búrquez A** (1989) Oikos 55:335–340. Burson BL, Voigt PW, Sherman RA, Dewald CL (1990) Crop Sci 30:86–89.

**Cabin RJ,** Ramstetter J, Engel RE (1991) Rhodora 93:1–10. **Calvo RN** (1990a) Am J Bot 77:736–741. **Calvo RN** (1990b) Am J Bot 77:1378–1381. **Camargo JMF,** Roubik DW (1991) Biol J Linn Soc 44:13–39. **Campbell CS,** Greene CW, Dickinson TA (1991) Syst Bot 16:333–349. **Campbell DR** (1989) Evolution 43:318–334. **Campbell DR** (1991) Evolution 45:1965–1968. **Campbell DR,** Waser NM (1989) Evolution 43:1444–1455. **Campbell DR,** Waser NM, Price MV, Lynch EA, Mitchell RJ (1991) Evolution 45:1458–1467. **Cane JH,** Buchmann SL (1989) J Insect Behav 2:431–436. **Cane JH,** Payne JA (1988) Ann Entomol Soc Am 81:419–427. **Cartar RV** (1991) Ecology 72:888–895. **Cartar RV,** Dill LM (1990) Behav Ecol Sociobiol 26:121–127. **Charlesworth D** (1988) Am Nat 132:298–302. **Charlesworth D** (1989) J Theor Biol 139:327–342. **Charlesworth D,** Morgan MT (1991) Philos Trans R Soc Lond B 332:91–102. **Charlesworth D,** Morgan MT, Charlesworth B (1990) Evolution 44:1469–1489. **Chaturvedi SK** (1989a) Asklepios 1989:58–62. **Chaturvedi SK** (1989b) Beitr Biol Pflanz 64:55–58. **Chichiriccò G** (1990) Plant Syst Evol 172:77–82. **Clampitt CA** (1987) Am J Bot 74:941–946. **Clarke PJ,** Myerscough PJ (1991) Aust J Bot 39:283–293. **Classen-Bockhoff R** (1990) Plant Syst Evol 171:57–88. **Classen-Bockhoff R** (1991) Bot Acta 104:183–193. **Clay K,** Levin DA (1989) Am J Bot 76:577–588. Clout MN, Hay JR (1989) N Z J Ecol 12:27–32. **Cocucci AA** (1991) Plant Syst Evol 174:17–35. **Compton SG,** McLaren FAC (1989) Proc K Ned Akad Wet C 92:57–71. **Compton SG,** Robertson HG (1988) Ecology 69:1302–1305. **Cook CDK** (1988) Ann Mo Bot Gard 75:768–777. **Copland BJ,** Whelan RJ (1989) J Ecol 77:509–523. **Coppola J** (1989) Southwest Nat 34:291–293. **Corbet SA** (1990) Isr J Bot 39:13–30. **Corbet SA** (1991) TREE 6:3–4. **Cordazzo CV,** Seeliger U (1988) Vegetatio 75:169–173. **Corlett RT** (1990) J Trop Ecol 6:55–63. **Cornet B** (1989) Evol Trends Plants 3:25–51. **Cortopassi-Laurino M,** Ramalho M (1988) Apidologie 19:1–24. **Costelloe BH** (1988) Ohio J Sci 88:132–138. **Couto RHN,** Costa JA, da Silveira RCM (1988) Científica 16:85–88. **Cox PA** (1988a) Ann Rev Ecol Syst 19:261–280. **Cox PA** (1988b) In: Lovett Doust J, Lovett Doust L (eds) Plant reproductive ecology: patterns and strategies. Oxford Univ, New York, pp 80–97. **Cox PA** (1989) In: Bock JH, Linhart YB (eds) The evolutionary ecology of plants. Westview, Boulder, pp 209–224. **Cox PA** (1990) Ann Mo Bot Gard 77:816–840. **Cox PA** (1991a) Biotropica 23:159–165. **Cox PA** (1991b) Philos Trans R Soc Lond B 333:217–224. **Cox PA,** Knox RB (1988) Ann Mo Bot Gard 75:811–818. **Cox PA,** Knox RB (1989)

Am J Bot 76:164–175. **Cox PA,** Tomlinson PB (1988) Am J Bot 75:958–965. **Cox PA,** Elmqvist T, Tomlinson PB (1990) Biotropica 22:259–265. **Cox PA,** Cromar S, Jarvis T (1991) In: Blackmore S, Barnes SH, (eds) Pollen and spores. Syst Assoc Spec Vol 44. Clarendon, Oxford, pp 363–375. **Craig JL** (1989) Oecologia 81:1–5. **Crane PR,** Friis EM, Pedersen KR (1989) Plant Syst Evol 165:211–226. **Crepet WL,** Friis EM, Nixon KC (1991) Philos Trans R Soc Lond B 333:187–195. **Cresswell JE** (1990) Oecologia 82:450–460. **Crome FHJ,** Irvine AK (1986) Biotropica 18:115–125. **Cropper SC,** Calder DM (1990) Plant Syst Evol 170:11–27. **Crowson RA** (1988) Entomol Gen 14:61–62. **Cruden RW,** Lyon DL (1989) In: Bock JH, Linhart YB (eds) The evolutionary ecology of plants. Westview, Boulder, pp 171–207. **Cruden RW,** Baker KK, Cullinan TE, Disbrow KA, Douglas KL, Erb JD, Kirsten KJ, Malik ML, Turner EA, Weier JA, Wilmot SR (1990) J Iowa Acad Sci 97:178–183. **Cuguen J,** Acheroy M, Loutfi AL, Petit D, Vernet P (1989) Evol Trends Plants 3:17–24. **Cunningham SA** (1991) Oecologia 87:86–90. **Curry KJ,** McDowell LM, Judd WS, Stern WL (1991) Am J Bot 78:610–623.

**Dafni A,** Bernhardt P (1990) In: Hecht MK, Wallace B, Macintryre RJ (eds) Evolutionary biology, vol 24. Plenum, New York, pp 193–252. **Dafni A,** Eisikowitch (eds) (1990) Adv Poll Ecol. Isr J Bot 39. **Dafni H,** Lensky Y, Fahn A (1988) J Apicult Res 27:103–114. **Dafni A,** Bernhardt P, Shmida A, Ivri Y, Greenbaum S, O'Toole C, Losito L (1990) Isr J Bot 39:81–92. **D'Arcy WG,** D'Arcy NS, Keating RC (1990) Rhodora 92:50–53. **de Albuquerque PMC,** Rêgo MMC (1989) Bol Mus Pará Emílio Goeldi Sér Zool 5:163–178. **de Arruda VLV,** Sazima M (1988) Rev Bras Bot 11:113–122. **de Barros Pâcheco RP,** Coleman JR (1989) Rev Bras Genet 12:347–359. **Decraene LPR,** Smets E (1991a) Flora 185:165–185. **Decraene LPR,** Smets E (1991b) Plant Syst Evol 178:179–194. **Delph LF,** Lively CM (1989) Evolution 43:1252–1262. **Delph LF,** Lloyd DG (1991) Evolution 45:1957–1964. **de Menezes PSR,** de Camargo JMF (1991) Apidologie 22:397–415. **de Oliveira PEAM,** Sazima M (1990) Plant Syst Evol 172:35–49. **de Oliveira-Filho AT,** de Andrade Oliveira LC (1988) Rev Bras Bot 11:23–32. **Deroin T** (1988a) Aspects anatomiques et biologiques de la fleur des annonacées. Doct thesis, Univ Paris-Sud. **Deroin T** (1988b) Bull Mus Natn Hist Nat, Paris 10:377–393. **Deroin T** (1989) CR Acad Sci Paris 308:307–311. **Deroin T** (1991) CR Acad Sci Paris 312:561–566. **DeVries PJ,** Stiles FG (1990) Biotropica 22:290–297. **Deyrup MA** (1988) Fla Entomol 71:597–605. **Dieringer G** (1991) Southwest Nat 36:341–343. **Di-Giovanni F,** Beckett PM, Flenley JR (1989) Grana 28:129–139. **Dobson HEM** (1989) In: Bock JH, Linhart YB (eds) The evolutionary ecology of plants. Westview, Boulder, pp 227–246. **Dobson HEM,** Berström G, Groth I (1990) Isr J Bot 39:143–156. **Dole JF** (1990) Am J Bot 77:1505–1507. **Domínguez CA,** Dirzo R, Bullock SH (1989) Oikos 56:109–114. **Dreher K** (1988) Biene 8:412–415. **Drummond DC,** Hammond PM (1991) Entomol Mon Mag 127:151–156. **Dudash MR,** Ritland K (1991) Am J Bot 78:1746–1753. **Dukas R,** Dafni A (1990) Plant Syst Evol 169:65–68. **Dukas R,** Shmida A (1989) Oikos 54:281–286. **Dulberger R** (1987) Ann Bot 59:203–217. **Dulberger R,** Ornduff R (1987) In: Halevy AH (ed) CRC Handbook of flowering. CRC, Boca Raton, pp 275–280. **Dumas C,** Bowman RB, Gaude T, Guilly CM, Heizmann P, Roeckel P, Rougier M (1988) Phyton (Austria) 28:193–200. **Dustmann JH** (1988) Naturschutz Naturparke 128:8–14. **Du Toit AP** (1987) S Afr For J 143:54–54. **Du Toit JT** (1990) Afr J Ecol 28:63–68.

**Eckhart VM** (1991) Evol Ecol 5:370–384. **Ehrlén J** (1991) Am Nat 138:918–933. **Eisikowitch D,** Lazar Z (1987) Bot J Linn Soc 95:101–111. **Eisikowitch D,** Kevan PG, Fowle S, Thomas K (1987) Pollen Spores 29:121–128. **Eisikowitch D,** Kevan PG, Lachance M-A (1990) Isr J Bot 39:217–225. **Elam DR,** Linhart YB (1988) Am J Bot 75:1262–1274. **Elisens WJ,** Freeman CE (1988) Am J Bot 75:971–978. **Elleman CJ,** Willson CE, Sarker RH, Dickinson HG (1988) New Phytol 109:111–117. **Elmqvist T,** Ågren J, Tunlid A (1988) Oikos 53:58–66. **Elmqvist T,** Cates RG, Harper JK, Gardfjell H (1991) Oikos 61:65–72. **Endress PK** (1989) Plant Syst Evol 162:193–211. **Endress PK** (1990) Mem N Y Bot Gard 55:5–34. **Erbar C,** Leins P (1988) Bot Jahrb Syst 110:263–282. **Erbar C,** Leins P (1989) Bot Jahrb Syst 111:29–55. **Erhardt A** (1988) In: Cresti M, Gori P, Pacini E (eds) Sexual reproduction in higher plants. Springer, Berlin Heidelberg New York, pp 351–356. **Erhardt A** (1990) Plant Syst Evol 170:125–132. **Erhardt A** (1991) Ecol Entomol 16:425–434. **Erhardt A,** Baker I (1990) Plant Syst Evol 169:111–121.

**Fairey DT,** Lefkovitch LP, Lieverse JAC (1989) J Appl Entomol 107:52–57. **Farnsworth PB,** Cox PA (1988) Biotropica 20:334–335. **Feil JP,** Renner SS (1991) Am J Bot 78:186. **Feinsinger P** (1987) Rev Chil Hist Nat 60:285–319. **Feinsinger P,** Tiebout III HM (1991) Ecology 72:1946–1952. **Feinsinger P,** Tiebout III HM, Young BE (1991) Ecology 72:1953–1963. **Fenster CB** (1991) Biotropica 23:98–101. **Fisher RM,** Pomeroy N (1989) J Econ Entomol 82:1061–1066. **Forster PI** (1989) Coleopt Bull 43:311–312..**Fox GA** (1990a) Am J Bot 77:1508–1518. **Fox GA** (1990b) Evolution 44:1404–1423. **Frank SA** (1989) Experientia 45:674–680. **Freeman CE,** Worthington RD, Jackson MS (1991) Biotropica 23:568–574. **Friedman WE** (1990a) Am J Bot 77:1582–1598. **Friedman WE** (1990b) Science 247:951–954. **Fritz A-L** (1990) Nord J Bot 9:577–587. **Fritz-Sheridan JK** (1988) Am J Bot 75:1–14. **Fuss AM,** Sedgley M (1991) Ann Bot 68:377–384.

**Galen C** (1989) Evolution 43:882–890. **Galen C,** Blau S (1988) Ecol Entomol 13:11–17. **Galen C,** Gregory T (1989) Oecologia 81:120–123. **Galen C,** Newport MEA (1988) Am J Bot 75:900–905. **Galen C,** Rothenberry JT (1988) J Theor Biol 135:419–429. **Galen C,** Stanton ML (1989) Am J Bot 76:419–426. **Galen C,** Gregory T, Galloway LF (1989) Am J Bot 76:1675–1680. **Galetto L** (1991) Kurtziana 21:165–176. **Galetto L,** Bernardello LM, Juliani HR (1990) Darwiniana 30:155–161. **Galil J** (1990) Flora 184:51–61. **Ganeshaiah KN,** Shaanker RU (1988) Oecologia 75:110–113. **Gara B,** Muenchow G (1990) Am J Bot 77:1–6. **Gerlach G,** Schill R (1989) Plant Syst Evol 168:159–165. **Gerlach G,** Schill R (1991) Bot Acta 104:379–391. **Gess SK,** Gess FW (1989) Ann Cape Prov Mus 18:95–134. **Gibbs PE** (1988) Lagascalia 15:17–28. **Gibbs PE** (1991) Taxon 40:583–595. **Gilbert FS,** Haines N, Dickson K (1991) Funct Ecol 5:29–39. **Gill FB** (1988a) Biotropica 20:169–171. **Gill FB** (1988b) Ecology 69:1933–1942. **Goldblatt P,** Bernhardt P (1990) Isr J Bot 39:93–111. **Goldblatt P,** Bernhardt P, Manning J (1989) Plant Syst Evol 163:201–209. **Goldingay RL** (1990) Oecologia 85:191–199. **Goldingay RL,** Carthew SM, Whelan RJ (1991) Oikos 61:79–87. **Gottlieb LD** (1989) Madroño 36:1–8. **Gottlieb LD,** Ford VS (1988) Heredity 60:237–246. **Gottsberger G** (1988) Taxon 37:630–643. **Gottsberger G** (1989a) Plant Syst Evol 167:165–187. **Gottsberger G** (1989b) Plant Syst Evol 167:189–194. **Gottsberger G** (1990) Bot Acta 103:360–365. **Gottsberger G** (1991a) Bot Jahrb Syst 113:221–235. **Gottsberger G** (1991b) Rep DFG German Res 1/91:10–12. **Gottsberger G,** Hartmann U (1988) Lagascalia 15:635–641. **Gottsberger G,** Prance T (eds) (1990) Reproductive biology and evolution of tropical woody angiosperms. Mem N Y Bot Gard 55. **Gottsberger G,** Silberbauer-Gottsberger I (1988) Phyton (Austria) 28:293–320. **Gottsberger G,** Silberbauer-Gottsberger I (1991) Biotropica 23:23–28. **Gottsberger G,** Arnold T, Linskens HF (1989a) Bot Acta 102:141–144. **Gottsberger G,** Arnold T, Linskens HF (1989b) Proc K Ned Akad Wet C 92:461–464. **Gottsberger G,** Arnold T, Linskens HF (1990) Isr J Bot 39:167–176. **Gottsberger G,** Hartmann U, Pirl M (1991) Phytocoenologia 20:95–110. **Govindaraju DR** (1988) Oikos 52:31–35. **Graumann S,** Gottsberger G (1988) Lagascalia 15:673–679. **Gregg KB** (1989) Castanea 54:57–78. **Gregg KB** (1991) Am J Bot 78:350–360. **Gribel R** (1988) Biotropica 20:344–347. **Grimaldi D** (1988) J Nat Hist 22:1–10. **Grünmeier R** (1989) 9. Symp Morphol Anat Syst Wien, Sept 1989, Abstr: 15. **Grünmeier R** (1990) Mem N Y Bot Gard 55:83–104. **Gryj E,** Martínez del Rio C, Baker I (1990) Biotropica 22:266–271. **Guitián J,** Guitián P (1990) An Jard Bot Madr 48:53–61. **Guitián J,** Oubiña JR, Fernández JMS (1990) An Jard Bot Madr 47:401–409. **Guo Y-H,** Cook CDK (1989) Aquat Bot 34:381–384. **Guo Y-H,** Cook CDK (1990) Aquat Bot 38:283–288. **Guo Y-H,** Sperry R, Cook CDK, Cox PA (1990) Aquat Bot 38:341–356.

**Haber WA,** Frankie GW (1989) Biotropica 21:155–172. **Hainsworth FR** (1991) Anim Behav 41:902–903. **Hamrick JL,** Loveless MD (1989) Bock JH, Linhart YB (eds) The evolutionary ecology of plants. Westview, Boulder, pp 129–146. **Han SS,** Halevy AH, Reid MS (1991) J Am Soc Hortic Sci 116:68–72. **Harder LD** (1990a) Oecologia 85:41–47. **Harder LD** (1990b) Ecology 71:1110–1125. **Harder LD,** Cruzan MB (1990) Funct Ecol 4:559–572. **Harder LD,** Thomson JD (1989) Am Nat 133:323–344. **Hartmann U** (1988) Lagascalia 15:651–659. **Haslett JR** (1989) Oecologia 78:433–442. **Hattingh V,** Giliomee JH (1989) S Afr J Bot 55:387–393. **Hawkeswood TJ** (1987a) Entomol Z 97:257–272. **Hawkeswood TJ** (1987b) G Ital Entomol 3:261–269. **Hawkeswood TJ** (1987c) G Ital Entomol 3:285–290. **Hawkeswood TJ** (1987d) Victorian Nat 104:76–80.

**Hawkeswood TJ** (1989) Plant Syst Evol 168:1–5. **Haynes RR** (1988) Ann Mo Bot Gard 75:805–810. **Hazeldon E,** Naisbitt T, Richards AJ (1991) Watsonia 18:391–393. **Healy TP,** Jepson PC (1988) Bull Entomol Res 78:651–657. **Heideman PD** (1989) J Ecol 77:1059–1079. **Heinrich G** (1989) Beitr Biol Pflanz 64:293–308. **Helenurm K,** Barrett SCH (1987) Can J Bot 65:2047–2056. **Herrera CM** (1988) Biol J Linn Soc 35:95–125. **Herrera CM** (1989) Oecologia 80:241–248. **Herrera CM** (1990a) Biol J Linn Soc 40:263–274. **Herrera CM** (1990b) Biol J Linn Soc 40:275–291. **Herrera CM** (1990c) Oikos 58:277–288. **Herrera J** (1988a) J Ecol 76:274–287. **Herrera J** (1988b) Lagascalia 15:25–41. **Herrera J** (1989a) An Jard Bot Madr 45:475–482. **Herrera J** (1989b) Principes 33:27–32. **Herrera J** (1992) Plant Syst Evol 179:245–255. **Heslop-Harrison Y** (1990) Nord J Bot 10:1–19. **Heß D** (1990) Die Blüte. 2nd edn. Ulmer, Stuttgart. **Hesse M,** Waha M (1989) Plant Syst Evol 163:147–152. **Hessing MB** (1988) Am J Bot 75:1324–1333. **Hessing MB** (1989a) Bot Gaz 150:404–410. **Hessing MB** (1989b) Plant Syst Evol 166:225–241. **Heyn CC** (1988) Lagascalia 15:29–36. **Hoffman MT** (1988) S Afr J Bot 54:345–350. **Hokche O,** Ramirez N (1990) Ann Mo Bot Gard 77:559–572. **Holsinger KE** (1988) Acta Oecol/Oecol Plant 9:95–102. **Holsinger KE** (1991a) Am Nat 138:606–622. **Holsinger KE** (1991b) TREE 6:307–308. **Hong S-P** (1991) Plant Syst Evol 176:125–131. **Hopkins MS,** Graham AW (1989) Aust J Ecol 14:399–413. **House SM** (1989) Aust J Ecol 14:77–94. **Hull DA,** Beattie AJ (1988) Oecologia 75:153–155. **Hunt JH,** Brown PA, Sago KM, Kerker JA (1991) J Kans Entomol Soc 64:127–130.

**Inghe O** (1990) J Theor Biol 147:449–469. **Inouye DW,** Pyke GH (1988) Aust J Ecol 13:191–210. **Intoppa F,** Piazza MG (1989) Riv Frutticolt 7:73–75. **Irish EE,** Nelson T (1989) Plant Cell 1:737–744. **Irvine AK,** Armstrong J (1988) Proc Ecol Soc Aust 15:107–113. **Itino T,** Kato M, Hotta M (1991) Biotropica 23:151–158.

**Jennersten O** (1988a) Conserv Biol 2:359–366. **Jennersten O** (1988b) Oikos 52:319–327. **Jennersten O,** Kwak MM (1991) Oecologia 86:88–98. **Jennersten O,** Morse DH (1991) Am Midl Nat 125:18–28. **Jennersten O,** Berg L, Lehman C (1988) J Ecol 76:1111–1132. **Jensen KB,** Zhang YF, Dewey DR (1990) Can J Plant Sci 70:215–225. **Johansen B** (1990) Bot J Linn Soc 103:165–196. **Johnston MO** (1991a) Ecology 72:1500–1503. **Johnston MO** (1991b) Evolution 45:1468–1479. **Jordano P** (1990) An Jard Bot Madr 48:31–52.

**Kadono Y,** Schneider EL (1987) Plant Spec Biol 2:109–115. **Kahn AP,** Morse DH (1991) Am Midl Nat 126:61–67. **Kaiser R** (1991) In: Muller PM, Lamparsky D (eds) Perfumes: art, science and technology. Elsevier, Barking Essex, pp 213–250. **Kakutani T,** Inoue T, Kato M (1989) Res Popul Ecol 31:381–400. **Kalisz S** (1989) Evolution 43:1263–1272. **Karron JD** (1989) Am J Bot 76:331–340. **Kato M** (1988) Oecologia 76:364–370. **Keller S,** Armbruster S (1989) Biotropica 21:190–192. **Kenrick J,** Bernhardt P, Marginson R, Beresford G, Knox RB, Baker I, Baker HG (1987) Plant Syst Evol 157:49–62. **Kevan PG** (1989) In: Mercer JB (ed) Thermal physiology. Elsevier, Barking Essex, pp 747–753. **Kevan PG** (1990) Ethol Ecol Evol 2:233–242. **Kevan PG,** Mohr NA, Offer MD, Kemp JR (1988) Proc Entomol Soc Ont 119:9–15. **Kevan PG,** Eisikowitch D, Rathwell B (1989) Bot Gaz 150:266–270. **Kevan PG,** Clark EA, Thomas VG (1990a) Am J Altern Agric 5:13–22. **Kevan PG,** Clark EA, Thomas VG (1990b) Proc Entomol Soc Ont 121:43–48. **Kevan PG,** Eisikowitch D, Ambrose JD, Kemp JR (1990c) Biol J Linn Soc 40:229–243. **Kikuzawa K** (1989) Oikos 56:196–202. Kim I, Carr GD (1990) Pac Sci 44:123–129. **Kipp LR,** Knight W, Kipp ER (1989) J Insect Behav 2:453–472. **Kitchener DJ,** Gunnell A, Maharadatunkamsi I (1990) Mammalia 54:561–578. **Kite G,** Reynolds T, Prance GT (1991) Biochem Syst Ecol 19:535–539. **Klimaszewski J,** Sturm H (1991) Coleopt Bull 45:1–13. **Klinkhamer PGL,** de Jong TJ (1990) Oikos 57:399–405. **Klinkhamer PGL,** de Jong TJ, de Bruyn G-J (1989) Oikos 54:201–204. **Klinkhamer PGL,** de Jong TJ, Wesselingh RA (1991) Neth J Zool 41:130–143. **Knudsen JT,** Tollsten L (1991) Plant Syst Evol 177:81–91. **Kohn JR** (1988) Nature 335:431–433. **Koponen A** (1990) Bot J Linn Soc 104:115–127. **Koptur S,** Haber WA, Frankie GW, Baker HG (1988) J Trop Ecol 4:323–346. **Krannitz PG,** Maun MA (1991) Can Field-Nat 104:13–17. **Kratochwil A** (1988a) Entomol Gen 13:67–80. **Kratochwil A** (1988b) Flora 181:261–324. **Kratochwil A** (1991) Beih Verhand Ges Ökol 2:9–44. **Kratochwil A,** Kohl A (1988) Mitt Bad Landesver Naturkd Naturschutz NF 14:697–715. **Kronestedt-Robards EC,** Greger M, Robards

AW (1989) Physiol Plant 77:341–346. **Krueger SK,** Knapp SJ (1991) Theor Appl Genet 82:221–226. **Kubitzki K** (1990) In: Kramer KU, Green PS (eds) Vol I: Pteridophytes and gymnosperms. In: Kubitzki K (ed) The families and genera of vascular plants. Springer, Berlin Heidelberg New York, pp 378–391. **Kubitzki K,** Amaral MCE (1991) Plant Syst Evol 177:77–80. **Kunze H** (1991) Plant Syst Evol 176:227–253. **Kunze H,** Liede S (1991) Plant Syst Evol 178:95–105. **Kwak MM** (1987) In: van Antel J et al. (eds) Disturbance in grasslands. Junk, Dordrecht, pp 273–283. **Kwak MM** (1988) Acta Bot Neerl 37:153–163. **Kwak MM,** Jennersten O (1991) Oecologia 86:99–104. **Kwak MM,** Kremer P, Boerrichter E, van den Brand C (1991) Proc Exp Appl Entomol Amst 2:131–136.

**Lamont BB,** Collins BG (1988) Austr J Ecol 13:129–135. **Lanza J** (1988) Biotropica 20:341–344. **Laroca S,** Michener CD, Hofmeister RM (1989) J Kans Entomol Soc 62:400–410. **Larson KS,** Larson RJ (1990) Can Field-Nat 104:519–525. **Larson RJ,** Larson KS (1987) Lindleyana 2:176–179. **Laverty TM,** Plowright RC (1988) Anim Behav 36:733–740. **Lavin M,** Delgado SA (1990) Am J Bot 77:1294–1312. **Leach CR,** Mayo O, Bürger R (1990) Theor Appl Genet 79:427–430. **Lechowicz MJ,** Blais PA (1988) J Evol Biol 1:255–273. **Lee TD** (1988) In: Lovett Doust J, Lovett Doust L (eds) Plant reproductive ecology: patterns and strategies. Oxford Univ, New York, pp 179–202. **Leins P,** Erbar C (1989) Flora 182:43–56. **Leins P,** Erbar C (1990) Bot Acta 103:87–92. **Leport L,** Larher F (1991) C R Acad Sci Paris 312:567–572. **Les DH** (1988) Ann Mo Bot Gard 75:819–835. **Les DH,** Whitkus R, Bryan FA, Tyrell LE (1989) Am J Bot 76:23–29. **Lewis AC** (1989) J Anim Ecol 58:1–13. **Liede S,** Hammer S (1990) Plant Syst Evol 172:229–240. **Liede S,** Whitehead V (1991) S Afr Bot 57:115–122. **Lindauer M** (1990) Spectrum Wiss 1990:10. **Linhart YB,** Busby WH, Beach JH, Feinsinger P (1987) Evolution 41:679–682. **Link DA** (1992) Plant Syst Evol 179:235–243. **Liston A,** Rieseberg LH, Elias TS (1990) Nature 343:641–642. **Lloyd DG** (1988) In: Michod RE, Levin BR (eds) The Evolution of Sex. Sinauer, Sunderland, pp 233–252. **Lloyd DG** (1989) In: Grubb PJ, Whittaker JB (eds) Toward a more exact ecology. Blackwell, Oxford. pp 185–208. **Lobreau-Callen D** (1989) Bull Mus Natn Hist Nat Paris 11:79–94. **Lord JM** (1991) N Z J Bot 29:83–86. **Lovett Doust J** (1989) TREE 4:230–234. **Lovett Doust J** (1990) Evol Trends Plants 4:121–133. **Lovett Doust J,** Lovett Doust L (eds) (1988a) Plant reproductive ecology: patterns and strategies. Oxford Univ, New York. **Lovett Doust J,** Lovett Doust L (1988b) In: Lovett Doust J, Lovett Doust L (eds) Plant reproductive ecology: patterns and strategies. Oxford Univ, New York, pp 5–29. **Lu J,** Mayer A, Pickersgill B (1990) Ann Bot 66:73–82. **Lunau K** (1988) Zool Jahrb Physiol 92:487–499. **Lunau K** (1991) Ethology 88:203–214. **Lyons EE,** Waser NM, Price MV, Antonovics J, Motten AF (1989) Am Nat 134:409–433.

**Macior LW** (1988) Plant Spec Biol 3:61–66. **Magin N,** Classen R, Gack C (1989) Can J Bot 67:1931–1937. **Mantovani W,** Martins FR (1988) Rev Bras Bot 11:101–112. **Marshall DL,** Folsom MW (1991) Annu Rev Ecol Syst 22:37–63. **Martin W,** Gierl A, Saedler H (1989) Nature 339:46–48. **Martini M,** Schmid A, Hess D (1990) Bot Acta 103:343–348. **May PG** (1988) Ecol Entomol 13:171–184. **Mayer SS** (1990) Mem N Y Bot Gard 55:76–82. **Mayo SJ** (1991) Kew Bull 46:601–681. **McCarthy BC,** Quinn JA (1989) Am J Bot 76:1015–1023. **McCarthy BC,** Quinn JA (1990) Am J Bot 77:261–273. **McConchie CA,** Knox RB (1989) Ann Bot 63:235–248. **McFarland DC** (1990) Aust J Bot 38:501–509. **McFarland JD,** Kevan PG, Lane MA (1989) Can J Bot 67:24–28. **McGuire AD,** Armbruster WS (1991) Am J Bot 78:214–219. **McKenna MA,** Thomson JD (1988) Ecology 69:1306–1307. **McMullen CK** (1990) Monogr Syst Bot Mo Bot Gard 32:35–45. **Meagher TR** (1988) In: Lovett Doust J, Lovett Doust L (eds) Plant reproductive ecology: patterns and strategies. Oxford Univ, New York, pp 125–138. **Medan D** (1991) N Z J Bot 29:31–42. **Meeuse ADJ** (1990) In: Malik CP, Bhatia DS, Setia RC, Singh P (eds) Advances in frontier areas of plant sciences. Narenda, Delhi, pp 155–166. **Meeuse ADJ,** Vinkenoog S, Vroege PW (1989) Acta Bot Neerl 38:493–502. **Meeuse ADJ,** de Meijer AH, Mohr OWP, Wellinga SM (1990a) In: Meeuse ADJ (ed) Flowers and fossils supplement. Eburon, Delft, pp 1–9. **Meeuse ADJ,** de Meijer AH, Mohr OWP, Wellinga SM (1990b) Isr J Bot 39:113–123. **Meiri L** (1988) Styles and stigmas, pollination mechanisms and breeding systems in Asteraceae. Doct Thesis, Tel-Aviv Univ. **Menzel R** (1990) Color vision in flower-visiting insects. KFA Jülich 1990:1–16.

**Michaels HJ,** Bazzaz FA (1989) Am Nat 134:190–207. **Michaud JP** (1989) J Apicult Res 28:181–186. **Michener CD,** Grimaldi DA (1988a) Am Mus Novit 2917:1–10. **Michener CD,** Grimaldi DA (1988b) Proc Natl Acad Sci USA 85:6424–6426. **Mitchell RJ,** Paton DC (1990) Oecologia 83:238–246. **Mohr NA,** Kevan PG (1987) Proc Entomol Soc Ont 118:149–154. **Molau U,** Carlsson M, Dahlberg A, Hill Ö (1989) Oikos 55:409–419. **Moran GF,** Muona O, Bell JC (1989) Biotropica 21:250–256. **Morawetz W,** Waha M (1991) Beitr Biol Pflanz 66:145–154. **Morellato LPC,** Leitão-Filho HF (1990) Rev Bras Biol 50:163–173. **Morellato LPC,** Rodrigues RR, de Freitas Leitão Filho H, Joly CA (1989) Rev Bras Bot 12:85–98. **Morgan MT,** Barrett SCH (1989) J Evol Biol 2:183–203. **Mori SA,** Boeke JD (1987) Mem N Y Bot Gard 44:137–155. **Mosseler A,** Papadopol CS (1989) Can J Bot 67:2563–2570. **Mound L** (1991) J Nat Hist 25:647–652. **Murawski DA,** Hamrick JL, Hubbell SP, Foster RB (1990) Oecologia 82:501–506. **Murray BG** (1990) Ann Bot 65:691–698. **Murray NA,** Johnson DM (1987) Contrib Univ Mich Herb 16:173–178.

**Nagel J,** Peña JE, Habeck D (1989) Fla Entomol 72:207–211. **Nic Lughadha EM,** Parnell JAN (1989) Bot J Linn Soc 100:337–354. **Nickol M** (1991) Mainz Naturwiss Arch 29:205–213. **Niklas KJ** (1988) Am J Bot 75:433–444. **Nicklas KJ,** Buchmann SL (1988) Am J Bot 75:1813–1829. **Nilsson LA** (1988) Nature 334:147–149. **Nilsson LA,** Rabakonandrianina E (1988a) Bot J Linn Soc 97:49–61. **Nilsson LA,** Rabakonandrianina E (1988b) J Zool Lond 215:475–489. **Nilsson LA,** Jonsson L, Rason L, Randrianjohany E (1986) Nord J Bot 6:411–422. **Nilsson LA,** Rabakonandrianina E, Pettersson B, Ranaivo J (1990) Plant Syst Evol 170:161–175. **Norstog K** (1987) Am Sci 75:270–279. **Nybom H** (1988) Plant Syst Evol 160:207–279.

**O'Brien SP,** Calder DM (1989) Aust J Bot 37:43–54. **Okada H** (1990) Plant Syst Evol 170:237–245. **Okuba A,** Levin SA (1989) Ecology 70:329–338. **Olesen JM** (1988) Acta Bot Neerl 37:509–513. **Olesen JM,** Balslev H (1990) Principes 34:181–190. **Olesen JM,** Warncke E (1989a) Holarct Ecol 12:21–30. **Olesen JM,** Warncke E (1989b) Oecologia 79:205–211. **Oliveira PE,** Gibbs PE, Bianchi M (1991) Bot Acta 104:392–398. **Oliveira PE,** Gibbs PE, Barbosa AA, Talavera S (1992) Plant Syst Evol 179:207–219. **O'Malley DM,** Buckley DP, Prance GT, Bawa KS (1988) Theor Appl Genet 76:929–932. **Ornduff R** (1988a) Ann Mo Bot Gard 75:761–767. **Ornduff R** (1988b) Plant Syst Evol 159:81–83. **Ortiz PL** (1990) Lagascalia 15:165–177. **Osborn JM,** Schneider EL (1988) Ann Mo Bot Gard 75:778–794. **Osborn JM,** Taylor TN, Schneider EL (1991) Am J Bot 78:1367–1378. **Osborn MM,** Kevan PG, Lane MA (1988) Plant Syst Evol 159:85–94. **Owens SJ,** Lewis GP (1989) Plant Syst Evol 163:93–105.

**Pahlke KM** (1989) Beitr Biol Pflanz 64:329–334. **Paige KN,** Whitham TG (1987) Ecology 68:1691–1695. **Pálfi G,** Gulyás S, Szöllösi I (1987) Acta Biol Szeged 33:25–34. **Palmer M,** Travis J, Antonovics J (1988) Oecologia 77:19–24. **Parker GA,** Maynard Smith J (1990) Nature 348:27–33. **Paton DC,** Collins BG (1989) Austr J Ecol 14:473–506. **Patt JM,** Rhoades DF, Corkill JA (1988) Phytochemistry 27:91–95. **Patt JM,** Merchant MW, Williams DRE, Meeuse BJD (1989) Am J Bot 76:1097–1106. **Patterson KJ** (1989) Acta Hortic 240:197–200. **Paulus HF,** Gack C (1990a) Isr J Bot 39:43–79. **Paulus HF,** Gack C (1990b) Plant Syst Evol 169:177–207. **Paw U KT,** Hotton C (1989) Am J Bot 76:445–453. **Peakall R** (1989a) Oecologia 79:361–365. **Peakall R** (1989b) Plant Syst Evol 167:137–148. **Peakall R** (1990) Funct Ecol 4:159–167. **Peakall R,** Beattie AJ (1989) Funct Ecol 3:515–522. **Peakall R,** Beattie AJ (1991) Evolution 45:1837–1848. **Peakall R,** James SH (1989) Heredity 62:161–167. **Peck CJ,** Lersten NR (1991) Bull Torrey Bot Club 118:20–23. **Peigler RS** (1988) Magnolia 1988 (45):1–8. **Pellmyr O** (1988) Ann Entomol Soc Am 81:792–797. **Pellmyr O** (1989) Oecologia 78:53–59. **Pellmyr O,** Thien LB, Bergström G, Groth I (1990) Plant Syst Evol 173:143–157. **Pellmyr O,** Tang W, Groth I, Bergström G, Thien LB (1991) Biochem Syst Ecol 19:623–627. **Petanidou T,** Vokou D (1990) Am J Bot 77:986–992. **Pettersson MW** (1991) Holarct Ecol 14:45–50. **Philbrick CT** (1988) Ann Mo Bot Gard 75:836–841. **Philbrick CT** (1991) Rhodora 93:36–50. **Pinheiro MCB,** Ormond WT, de O Leite C, de Lima HA (1988) Rev Bras Biol 48:665–672. **Piper JG,** Waite S (1988) Funct Ecol 2:35–40. **Pleasants JM** (1989) Am Nat 134:51–71. **Pleasants JM** (1991) Funct Ecol 5:75–82. **Pleasants JM,** Horner HT, Ng G (1990) Funct Ecol 4:823–827. **Plitmann U,** Levin DA (1990) Plant Syst Evol 170:205–214.

**Possingham HP** (1988) Theor Popul Biol 33:138–160. **Potter CF,** Bertin RI (1988) Am Midl Nat 120:156–162. **Preston RE** (1991) Am J Bot 78:1044–1053. **Primack RB,** Hall P (1990) Am Nat 136:638–656. **Pyke GH** (1991) Nature 350:58–59. **Pyke GH,** Day LP, Wale KA (1988) Aust J Ecol 13:279–284.

**Ramalho M** (1990) J Apicult Res 29:61–67. **Ramalho M,** Kleinert-Giovannini A, Imperatriz-Fonseca VL (1989) Apidologie 20:185–195. **Ramalho M,** Kleinert-Giovannini A, Imperatriz-Fonseca VL (1990) Apidologie 21:469–488. **Ramirez N** (1989) Biotropica 21:319–330. **Ramirez N,** Brito Y (1987) Acta Bot Venez 38:376–381. **Ramirez N,** Gil C, Hokche O, Seres A, Brito Y (1990) Ann Mo Bot Gard 77:383–397. **Ramsey MW** (1988) Oecologia 76:119–124. **Ramsey MW** (1989) Aust J Ecol 14:33–40. **Raskin I,** Skubatz H, Tang W, Meeuse BJD (1990) Ann Bot 66:369–373. **Rasmussen IR,** Brødsgaard B (1992) Oecologia 89:277–283. **Ratcliffe MJ,** Turkington R (1989) Can Field-Nat 103:348–352. **Rathcke B** (1988a) Ecology 69:446–457. **Rathcke B** (1988b) J Ecol 76:975–994. **Rathman ES,** Lanza J, Wilson J (1990) Am Midl Nat 124:379–389. **Rawal RS,** Bankoti NS, Samant SS, Pangtey YPS (1991) Vegetation 93:109–119. **Rayner RJ** (1987) Lethaia 20:123–127. **Rayner RJ,** Waters SB (1990) Zool J Linn Soc 99:309–318. **Real L,** Rathcke BJ (1988) Ecology 69:728–735. **Rebelo AG** (ed) (1987) A preliminary synthesis of pollination biology in the Cape flora. Found Res Dev, Pretoria. **Rêgo MMC,** de Albuquerque PHC (1989) Bol Mus Pará Emílio Goeldi Sér Zool 5:179–193. **Renner SS** (1986/87a) Acta Amazonica 16/17:197–208. **Renner SS** (1987b) Brittonia 39:441–446. **Renner SS** (1989a) Ann Mo Bot Gard 76:496–518. **Renner SS** (1989b) Plant Syst Evol 163:21–29. **Renner SS** (1990) Mem N Y Bot Gard 55:143–152. **Richards AJ** (1990a) Bot J Linn Soc 103:233–250. **Richards AJ** (1990b) Bot J Linn Soc 103:251–261. **Richards AJ,** Mitchell J (1990) Bot J Linn Soc 104:369–380. **Richards JH,** Barrett SCH (1987) Am J Bot 74:1831–1841. **Richardson TE,** Stephenson AG (1989) AM J Bot 76:532–538. **Rico-Gray V** (1989) Biol J Linn Soc 38:173–181. **Robertson JL,** Wyatt R (1990) Evolution 44:121–133. **Robinson WS,** Nowogrodzki R, Morse RA (1989) Am Bee J 129:411–423, 477–486. **Rosen D,** Barthlott W (1991) Decheniana (Bonn) 144:72–112. **Rotenberry JT** (1990) Holarct Ecol 13:1–10. **Roubik DW** (1989) Ecology and natural history of tropical bees. Cambridge Univ, Cambridge. **Roubik DW** (1991) In: Spivak M, Fletcher DJC, Breed MD (eds) The "African" honey bee. Westview, Boulder, pp 259–281. **Roubik DW,** Boreham MM (1990) Interciencia 15:146–153. **Rozenblum E,** Maldonado S, Waisman CE (1988) Am J Bot 75:311–322.

**Sacchi CF,** Price PW (1988) Am J Bot 75:1387–1393. **Sanders LC,** Eckard KJ, Lord EM (1990) Protoplasma 159:26–34. **Saraiva SC,** Cesar O, Monteiro R (1988) Rev Bras Bot 11:71–80. **Sazima I,** Sazima M (1989) Rev Bras Entomol 33:109–118. **Sazima I,** Vogel S, Sazima M (1989) Plant Syst Evol 168:167–179. **Sazima M,** Sazima I (1988) Bot Acta 101:269–271. **Sazima M,** Sazima I (1989) Bot Acta 102:106–111. **Sazima M,** Sazima I (1990) Bot Acta 103:83–86. **Scariot AO,** Lleras E, Hay JD (1991) Biotropica 23:12–22. **Schemske DW,** Horvitz CC (1989) Evolution 43:461–465. **Schlessman MA** (1988) In: Lovett Doust J, Lovett Doust L (eds) Plant reproductive ecology: patterns and strategies. Oxford Univ, New York, pp 139–153. **Schmid-Hempel P,** Schmid-Hempel R (1990) Naturwissenschaften 77:450–452. **Schmid-Hempel P,** Speiser B (1988) Oikos 53:98–104. **Schneider EL,** Williamson PS, Whitenberg DC (1990) Water Gard J 6:41–51. **Schoen DJ,** Dubuc M (1990) Am Nat 135:841–857. **Scogin R** (1988) Bot Gaz 149:437–442. **Seburn CNL,** Dickinson TA, Barrett SCH (1990) J Evol Biol 3:103–123. **Sedgley M,** Attanayake DPSTG (1988) Euphytica 39:83–91. **Sedgley M,** Griffin A R (1989) Sexual reproduction of tree crops. Academic Press, London. **Shetler SG,** Wiser SK (1987) Proc Biol Soc Wash 100:993–1017. **Shmida A,** Dukas R (1990) Isr J Bot 39:133–141. **Shmida A,** Kadmon R (1991) Vegetatio 94:95–99. **Shore JS,** Barrett SCH (1990) Heredity 64:105–112. **Shykoff JA** (1988) Am J Bot 75:844–850. **Sidner R,** Davis R (1988) Southwest Nat 33:493–495. **Silberbauer-Gottsberger I** (1990) Phyton (Austria) 30:213–233. **Silberbauer-Gottsberger I,** Gottsberger G (1988) Rev Bras Biol 48:651–663. **Simpson BB** (1989) Syst Bot 14:408–426. **Simpson BB,** Neff JL, Dieringer G (1990) Plant Syst Evol 173:209–222. **Sivaraj N,** Krishnamurthy KV (1989) Vegetatio 79:85–88. **Skubatz H,** Meeuse BJD, Bendich AJ (1989) Plant Physiol 91:530–535. **Skubatz H,** Nelson TA, Dong AM, Meeuse BJD, Bendich AJ (1990a) Planta 182:432–436. **Skubatz H,** Williamson PS,

Schneider EL, Meeuse BJD (1990b) J Exp Bot 41:1335–1339. **Skubatz H,** Nelson TA, Meeuse BJD, Bendich AJ (1991) Plant Physiol 95:1084–1088. **Slater AT** (1991) Aust J Bot 39:273–282. **Slater AT,** Calder DM (1988) Aust J Bot 36:145–158. **Smets E** (1988) Candollea 43:709–716. **Smets EF,** Cresens EM (1988) Acta Bot Neerl 37:121–128. **Smith CC,** Hamrick JL, Kramer CL (1990) Am Nat 136:154–166. **Smith RS,** Jones L (1991) J Appl Ecol 28:42–59. **Snow AA,** Mazer SJ (1988) Evolution 42:1065–1075. **Snow AA,** Spira TP (1991a) Am J Bot 78:1419–1426. **Snow AA,** Spira TP (1991b) Nature 352:796–797. **Sobrevila C** (1989) Plant Syst Evol 166:45–67. **Southwick L,** Southwick EE (1989) Am Bee J 1989:805–807. **Sowig P** (1989) Oecologia 78:550–558. **Sowig P,** Paulus HF (1988) Mitt Dtsch Ges Allg Angew Entomol 6:109–115. **Spalik K** (1991) Biol J Linn Soc 42:325–336. **Spira T** (1989) In: Bock JH, Linhart YB (eds) The evolutionary ecology of plants. Westview, Boulder, pp 247–255. **Ssymank A** (1991) Phytocoenologia 19:307–390. **Standley LA,** Kim SS-H, Hjersted IM (1988) Rhodora 90:233–244. **Stanton ML,** Galen C (1989) Oecologia 78:477–485. **Stanton ML,** Preston RE (1988) Am J Bot 75:528–539. **Stanton ML,** Snow AA, Handel SN, Bereczky J (1989) Evolution 43:335–346. **Stead AD** (1992) Plant Growth Regul 11:13–20. **Stead AD,** Reid MS (1990) Ann Bot 66:655–663. **Stein BA** (1992) BioScience 42:27–33. **Stein BA,** Tobe H (1989) Ann Mo Bot Gard 76:519–531. **Steiner KE** (1990) Bot J Linn Soc 102:175–195. **Steiner KE,** Whitehead VB (1990) Evolution 44:1701–1707. **Steiner KE,** Whitehead VB (1991) Evolution 45:1493–1501. **Stephens DW,** Krebs JR (1986) Foraging theory. Princeton Univ, Princeton. **Stevens A-D** (1991) Zur Reproduktionsbiologie einiger Bignoniaceen im Cerrado Brasiliens. Doct Thesis, Univ Giessen. **Stoddard FL** (1991) Aust J Agric Res 42:1173–1178. Stratton DA (1989) Biotropica 21:308–318. **Stromberg MR,** Johnsen PB (1990) Condor 92:606–612. **Struck M** (1990) Vergleichende phänologische und blütenbiologische Studien an Daueruntersuchungsflächen im ariden Winterregengebiet des südlichen Afrika. Doct Thesis, Univ Hamburg. **Sturm H** (1989) Stud Neotrop Fauna Environ 24:137–155. **Sturm H** (1990) Rev Acad Colomb Cienc 17:667–685. **Sun M,** Ganders FR (1988) Evolution 42:516–527. **Surles SE,** Hamrick JL, Bongarten BC (1990) Silvae Genet 39:35–40. **Sutherland S,** Vickery RK Jr (1988) Oecologia 76:330–335. **Svensson L** (1988) Evol Trends Plants 2:31–37. **Svensson L** (1990) Am J Bot 77:889–896.

**Tamm S** (1989) Ecology 70:195–205. **Tanaka H** (1989) J Phytogeogr Taxon 37:65–68. **Tang W** (1987) Bot Gaz 148:165–174. **Teppner H** (1988) Phyton (Austria) 28:321–336. **Teppner H** (1991) 9th Europ Orchid Congr Rome, April 1991, Proc:14–15. **Teppner H,** Klein E (1989) Phyton (Austria) 29:161–173. **Thomas DW** (1988) In: Kunz TH (ed) Ecological and behavioral methods for the study of bats. Smithsonian Inst, Washington, pp 211–220. **Thompson K,** Rabinowitz D (1989) Am Nat 133:722–728. **Thompson SA,** Rawlins JE (1986) Am J Bot 73:791. **Thompson SA,** Rawlins JE (1990) V Congr Latinoam Bot, Havana, Cuba, Abstr: 269. **Thomson JD** (1988) Evol Ecol 2:65–76. **Thomson JD,** Brunet J (1990) TREE 5:11–16. **Thomson JD,** McKenna MA, Cruzan MB (1989) Ecology 70:1061–1068. **Thorp RW,** Sugden EA (1990) Isr J Bot 39:177–186. **Tollsten L,** Bergström J (1989) Nord J Bot 9:359–362. **Torchio PF** (1987) Proc Entomol Soc Ont 118:111–124. **Torchio PF** (1990a) Environ Entomol 19:1649–1656. **Torchio PF** (1990b) J Kans Entomol Soc 63:427–436. **Torchio PF** (1991) Environ Entomol 20:590–596. **Trapp EJ,** Hendrix SD (1988) Oecologia 75:285–290. **Tucker SC** (1988) Am J Bot 75:1584–1597. **Tybirk K** (1989) Nord J Bot 9:375–381.

**Ueyama Y,** Hashimoto S, Nii H, Furukawa K (1990) Flavour Frag J 5:85–88. **Ugborogho RE,** Agomo CO (1989) Feddes Repert 100:137–145. **Urbanska KM** (1989) Bot Helv 99:49–63.

**Valachovič M** (1989) Biológia (Bratisl) 44:37–42. **Valdés B,** Mejías JA (1988) Lagascalia 15:95–103. **Valerio CE** (1988) Rev Biol Trop 36:55–61. **van Noort S,** Ware AB, Compton SG (1989) S Afr Tydskr Wet 85:323–324. **Vanstone VA,** Paton DC (1988) Aust J Bot 36:519–531. **van Tol IAV,** Meijdam NAJ (1991) Field research on pollination and seed dispersal of Annonaceae (soursop familiy). Doct Thesis, Univ Utrecht. **Vasek FC,** Weng V (1988) Syst Bot 13:336–350. **Vaughton G** (1988) Aust J Bot 36:633–642. **Vaughton G,** Ramsey M (1991) Aust J Bot 39:167–177. **Verkerke W** (1987) Proc K Ned Akad Wet C 90:461–492. **Verkerke W** (1989) Experientia 45:612–622. **Vickery RK** (1990) Great Basin Nat 50:155–159. **Villar M,** Gaget M, Said

C, Knox RB, Dumas C (1987) J Cell Sci 87:483–490. **Villarreal AG,** Freeman CE (1990) Bot Gaz 151:5–9. **Visser T,** Verhaegh JJ (1988) In: Cresti M, Gori P, Pacini E (eds) Sexual reproduction in higher plants. Springer, Berlin Heidelberg New York, pp 369–374. **Vogel S** (1988) Z Zool Syst Evolutionsforsch 26:341–362. **Vogel S** (1989) Akad Wiss Lit Mainz, Jubiläumsband 1989:113–130. **Vogel S** (1990a) Mem N Y Bot Gard 55:130–142. **Vogel S** (1990b) Ölblumen und ölsammelnde Bienen. Dritte Folge: *Momordica, Thladiantha* und die Ctenoplectridae. Akad Wiss Lit Mainz. Steiner, Stuttgart. **Vogel S** (1990c) The role of scent glands in pollination. Renner S (ed) Smithsonian Inst Libr Nat Sci Found, Washington. **Vogel S,** Machado IC (1991) Plant Syst Evol 178:153–178. **Vogel S,** Westerkamp C (1991) In: Seitz A, Loeschckev (eds) Species conservation: a population-biological approach. Birkhäuser, Basel, pp 159–170. **Vokou D,** Petanidou T, Bellos D (1990) Biol Conserv 52:125–133. **Vora RS** (1990) Tex J Sci 42:137–142. **Vovides AP** (1991) Biotropica 23:102–104. **Vroege PW,** Stelleman P (1990) Isr J Bot 39:125–132. **Vuille F-L** (1988) Plant Syst Evol 159:173–183.

**Waldvogel JA** (1989) In: Power DM (ed) Current ornithology, Chapt 6. Plenum, New York, pp 269–321. **Wallace DD,** O'Dowd DJ (1989) Oecologia 79:482–488. **Waller DM** (1988) In: Lovett Doust J, Lovett Doust L (eds) Plant reproductive ecology: patterns and strategies. Oxford Univ, New York, pp 203–227. **Warren JM** (1988) Heredity 61:161–166. **Warren JM,** Crawford TJ, Oxford GS (1988) Heredity 60:33–38. **Warren SD,** Harper KT, Booth GM (1988) Am Midl Nat 120:325–330. **Waser NM,** Mitchell RJ (1990) Ecology 71:116–123. **Waser NM,** Price MV (1989) Evolution 43:1097–1109. **Waser NM,** Price MV (1990) Collect Bot 19:9–20. **Waser NM,** Price MV (1991) Ecology 72:171–179. **Waser NM,** Price MV, Montalvo AM, Gray RN (1987) Evol Trends Plants 1:29–33. **Webb MC,** Williams EG (1988) Ann Bot 61:395–404. **Weber A** (1989) Plant Syst Evol 165:95–100. **Weiner J** (1988) In: Lovett Doust J, Lovett Doust L (eds) Plant reproductive ecology: patterns and strategies. Oxford Univ, New York, pp 228–245. **Weiss D,** Halevy AH (1989) Planta 179:89–96. **Weiss MR** (1991) Nature 354:227–229. **Weller SG,** Ornduff R (1991) Am J Bot 78:801–804. **Weller SG,** Sakai AK, Wagner WL, Herbst DR (1990) Syst Bot 15:266–276. **Wells MS,** Lloyd DG (1991) Evol Ecol 5:310–326. **Westerkamp C** (1990a) Bot Acta 103:366–371. **Westerkamp C** (1990b) Spektrum Wiss 1990:10. **Westerkamp C** (1991) Plant Syst Evol 177:71–75. **Westrich P** (1989) Die Wildbienen Baden-Württembergs. Ulmer, Stuttgart. **Westrich P,** Schmidt K (1987) Apidologie 18:199–214. **Whitaker AH** (1987) N Z J Bot 25:315–328. **Whitehead CS,** Halevy AH (1989) Plant Growth Regul 8:41–54. **Whitten WM,** Young AM, Williams NH (1989) J Chem Ecol 15:1285–1295. **Widén B** (1991) Oikos 61:205–215. **Widén B,** Widén M (1990) Oecologia 83:191–196. **Widler-Kiefer H,** Yeo PF (1987) Plant Syst Evol 155:283–306. **Wiebes JT** (1989a) Proc K Ned Akad Wet C 92:251–266. **Wiebes JT** (1989b) Proc K Ned Akad Wet C 92:395–407. **Wiebes JT,** Compton SG (1990) Proc K Ned Akad Wet C 93:203–222. **Wiersema JH** (1988) Ann Mo Bot Gard 75:795–804. **Wilkinson K,** Westmoreland D, Westmoreland GR (1991) Am Midl Nat 125:364–367. **Williams EG,** Rouse JL (1990) Sex Plant Reprod 3:7–17. **Williams IH,** Simpkins JR, Martin AP (1991) J Agric Sci Cambridge 117:75–79. **Willmann R** (1989) Naturwissenschaften 76:375–377. **Willson MF** (1991) Evol Ecol 5:69–87. **Willson MF,** Ågren J (1989) Oikos 55:23–29. **Wilson P,** Thomson JD (1991) Ecology 72:1503–1507. **Wittmann D,** Radtke R, Hoffmann M, Blochtein B (1989) Entomol Gen 14:217–221. **Wittmann D,** Radtke R, Cure JR, Schifino-Wittmann MT (1990) Z Zool Syst Evolutionsforsch 28:157–165. **Wolf LL,** Hainsworth FR (1990) Funct Ecol 4:149–157. **Wolf TJ,** Schmid-Hempel P, Ellington CP, Stevenson RD (1989) Funct Ecol 3:417–424. **Wolfe AD,** Estes JR, Chissoe III WF (1991) Am J Bot 78:1503–1507. **Wolfe LM,** Barrett SCH (1987) Oecologia 74:347–351. **Wolfe LM,** Barrett SCH (1988) Can J Zool 66:1421–1424. **Woltering EJ** (1990) J Exp Bot 41:1021–1029. **Woltering EJ,** van Doorn WG (1988) J Exp Bot 39:1605–1616. **Wright DH** (1988) Southwest Natur 33:219–227. **Wright MG,** Wright GEP, Smith P (1989) S Afr J Bot 55:273–277.

**Young AM** (1989) Turrialba 39:454–458. **Young CG,** Owen RE (1989) Can Field-Nat 103:341–347. **Young HJ** (1988a) Ecology 69:832–844. **Young HJ** (1988b) Oecologia 76:461–466. **Young HJ,** Stanton ML (1990) Science 248:1631–1633. **Yumoto T** (1987a) Ecol Res 2:133–145. **Yumoto T** (1987b) In: Kawano S, Connell JH, Hidaka T (eds) Evolution and coadaptation in biotic communities. Univ Tokyo, Tokyo, pp 165–182. **Yumoto T** (1988) Ecol Res 3:117–129.

**Zavada MS** (1990) Taxon 39:442–447. **Ziehe M,** Gregorius H-R (1988) J Evol Biol 1:233–253. **Zietsman PC** (1991) S Afr J Bot 57:348–351. **Zimmerman JK** (1991) Ecology 72:597–608. **Zimmerman JK,** Aide TM (1989) Am J Bot 76:67–73. **Zimmerman JK,** Roubik DW, Ackerman JD (1989) Ecology 70:1192–1195. **Zimmerman M** (1988) In: Lovett Doust J, Lovett Doust L (eds) Plant reproductive ecology: patterns and strategies. Oxford Univ, New York, pp 157–178. **Zimmerman M,** Pyke GH (1988a) Aust J Ecol 13:93–99. **Zimmerman M,** Pyke GH (1988b) Aust J Ecol 13:301–309. **Zimmerman M,** Pyke GH (1988c) J Ecol 76:777–789.

Professor Dr. Gerhard Gottsberger
Botanisches Institut I
der Universität Giessen
Senckenbergstrasse 17–25
D-W-6300 Giessen, Germany

# II a. Mycorrhizae: Ectomycorrhiza and Ectendomycorrhiza

By Reinhard Agerer

Mycorrhizae can presently be divided in seven different groups (Harley 1991) which may be designated after Wilcox (1983) as classes.

The seven classes can be arranged with respect to their nutrition. In *ectotrophic* mycorrhizae, the fungal hyphae are restricted to the cortical cell walls, leading to a net-like structure (Hartig net), and they mostly possess a hyphal envelope around the root (hyphal mantle). In *endotrophic* mycorrhizae the hyphae grow mainly intracellularly, and in *ectendotrophic* mycorrhizae features of both the others are combined. Typical characteristics of the mycorrhizal classes are compiled in Fig. 1. Sometimes a single fungus can form mycorrhizae which lie in more than one class but with different hosts, e.g., ectomycorrhiza and monotropoid mycorrhiza, or ectomycorrhiza and ectendomycorrhiza.

## 1. Ectomycorrhiza

### a) Symbiotic Organisms and Morphology/Anatomy of the Symbiotic Organs

#### α) Keys, Including Detailed Descriptions of Ectomycorrhizae

Ingleby et al. (1990) described in detail and provided pictures of ectomycorrhizae mostly on young trees of *Picea sitchensis* and *Betula pendula*. Three additional deliveries of the *Colour Atlas of Ectomycorrhizae* have been published, altogether now five deliveries (Agerer 1987–1991); a key for the identification of 16 tree genera based on the anatomy of thicker roots of boreal temperature forests is now included.

Jansen and de Vries (1989) published a key for identification of ectomycorrhizae on *Pseudotsuga menziesii*, Treu (1990) for ectomycorrhizae of several fungal species of alpine stands of *Larix decidua, Pinus cembra, P. mugo*, and *Alnus viridis*, Miller et al. (1991) for ectomycorrhizae of several fungi on *Alnus rubra*, and Brand (1991) keyed out naturally grown ectomycorrhizae on *Fagus sylvatica*.

Agerer (1991b) compiled the methods for characterization of ectomycorrhizae and reviewed the most important characteristics. It was forcibly emphasized that from all described ectomycorrhizae, voucher specimens must be preserved as fixed material in official herbaria, and one reference specimen has to be selected.

Progress in Botany, Vol. 54

CLASSES OF MYCORRHIZAE

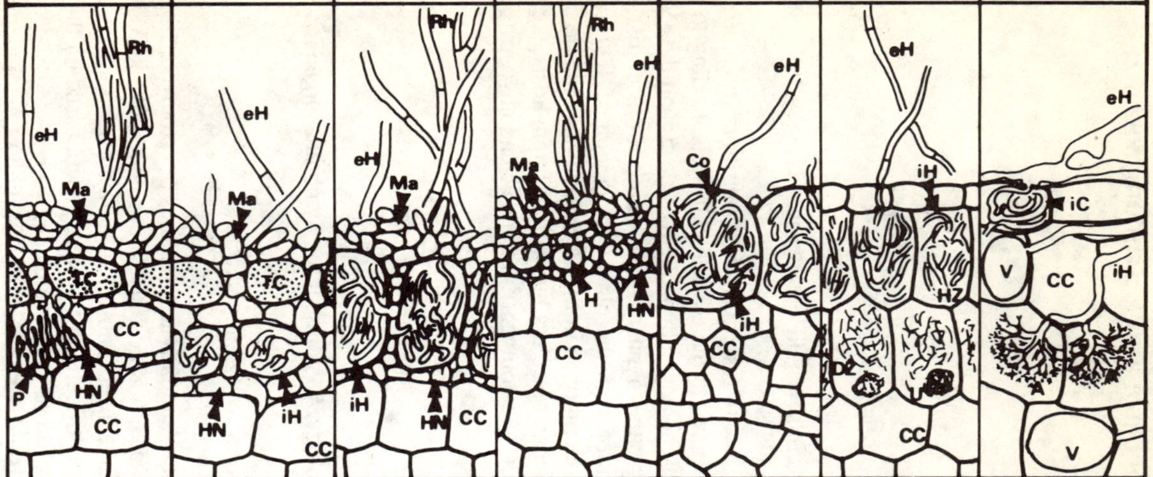


| Ecto-Mycorrhiza | Ectendo-M. | Arbutoid M. | Monotropoid M. | Ericoid M. | Orchid M. | Vesicular-Arbuscular M. |
|---|---|---|---|---|---|---|
| ectotrophic | ectendotrophic | | | endotrophic | | |
| Basidiomycetes Ascomycetes (Zygomycetes) | Ascomycetes | Basidiomycetes | Basidiomycetes | (Basidiomycetes) Ascomycetes | Basidiomycetes | Zygomycetes |
| Trees of seasonal climates, p.pte. also in moist tropics | Gymnosperms | Ericales: Arbutus Arctostaphylos Pyrolaceae | Monotropaceae | Ericaceae Epacridaceae Empetraceae | Orchidaceae | Mostly on trees in moist tropics and on herbs |

A = Arbuscule; CC = Cortex cell; Co = Collar (thickening of cell wall); DZ = Digestion Zone; eH = emanating Hypha; H = Haustorium; HN = Hartig Net; HZ = Host Zone; iC = infection Coil; iH = intracellular Hyphae; Ma = Mantle of hyphae; P = Palmetti (HN in plan view); Rh = Rhizomorph; TC = Tannin Cell (only in Gymnosperms); U = Undigested residues (hyphal walls); V = Vesicle

**Fig. 1.** Classes of Mycorrhizae

### β) Comprehensive Descriptions of Selected Ectomycorrhizae

Taylor and Alexander (1991) synthesized ectomycorrhizae between *Tylospora fibrillosa* (Corticiaceae) and *Picea sitchensis*. Weiss (1991) described ectomycorrhizae of *Amphinema byssoides, Laccaria bicolor,* and *Lactarius mitissimus,* synthesized on *Picea abies* in nonsterile culture. Agerer and Weiss (1989) characterized naturally grown as well as synthesized ectomycorrhizae of *Thelephora terrestris* on *Picea abies*. Agerer (1991) investigated ectomycorrhizae of *Sarcodon imbricatus* on *Picea abies* in detail.

Tylor and Alexander (1989b) succeeded in taking cultures of *Russula aeruginea* using a half strength modified Melin Norkrans medium (MMN) and synthesized ectomycorrhizae with *Picea sitchensis*. Cuvelier (1990) studied the ectomycorrhizae of *Amanita muscaria, Cortinarius armillatus,* and *Dermocybe phoenicea* on *Betula pendula*.

Agerer (1990b, 1991c) described naturally occurring ectomycorrhizae of some Gomphidiaceae. Brand (1989) characterized comprehensively the ectomycorrhizae of *Xerocomus chrysenteron* on *Fagus sylvatica*. Müller and Agerer (1990) compared the ectomycorrhizae of three *Leccinum* species (*L. scabrum, L. holopus, L. variicolor*) and found no differences.

Tuberculate ectomycorrhizae were described by Haug et al. (1991) on *Castanopsis borneensis* (Fagaceae) and *Engelhardtia roxburghiana* (Juglandaceae) from Taiwan, and by Dell et al. (1990b) on *Eucalyptus pilularis* from Australia.

### γ) Unidentified Ectomycorrhizae Named Binomially

Berndt et al. (1990) critizised the system of applying a binomial for ectomycorrhizae with an unknown fungal partner, with the only argument being "it is hard to abandon once it has been established", and continued to designate ectomycorrhizae with "Type" followed by a number. Agerer (1991b) substantiated again in detail the great advantage of binomials for yet unidentified ectomycorrhizae, and compiled all binomial names of unidentified ectomycorrhizae described up to 1990 (Agerer 1991a).

### δ) Verification of Ectomycorrhizal Nature of Fungi, Including Short Descriptions of Ectomycorrhizae

Warcup (1990c) synthesized ectomycorrhizae of the Zygomycetes *Sclerogone eucalypti, Endogone aggregata,* and *E. tuberculosa* with *Eucalyptus* and other Australian trees; *E. flammicorona* formed ectomycorrhizae only with *Pinus*.

Warcup and Talbot (1989) and Warcup (1990b) proved as ectomycorrhizal fungi the Ascomycetes: *Boudiera tracheia, Labyrinthomyces varius, Lachnea vinosobrunnea, Muciturbo reticulata, M. truncata, M. verrucosa, Nothojafnea cryptotricha, Peziza whitei, Pulvinula tetraspora, Ruhlandiella berolinensis,* and *Sphaerosoma mucidum*.

Tommerup et al. (1991) synthesized ectomycorrhizae of *Laccaria fraterna* and *Hydnangium carneum* each separately on *Eucalyptus* spp. and *Allocasuarina littoralis; Hydnangium carneum* ectomycorrhizae were also synthesized on *Eucalyptus pilularis* by Moore et al. (1989). Thoen and Ducousso (1989b) synthesized ectomycorrhizae of *Phlebopus sudanicus* with ***Acacia*** and ***Pinus*** species. Richter and Bruhn (1989) were able to grow ectomycorrhizae of *Xerocomus parasiticus* with *Pinus resinosa*. Bougher et al. (1991) described a new *Hebeloma* species (*H. westraliense*) and included in the description a short characterization of synthesized ectomycorrhizae. Bougher and Malajczuk (1990) synthesized ectomycorrhizae of *Descolea maculata* on *Eucalyptus diversicolor*. Dell et al. (1990b) found ectomycorrhizae of *Mesophellia trabalis, M. labyrintha,* and of *Castorium camphoratum* formed with *Eucalyptus diversicolor* within the peridium of their hypogeous fruitbodies.

Brunner (1991) examined the effect of different synthesis techniques on the ectomycorrhizae of *Hebeloma crustuliniforme* and *H. cylindrosporum* and concluded that the structure of the mantle and Hartig net within the two species are stable; however, thickness of the Hartig net between cortical cells are dependent on techniques and substrates.

### ε) Strain Variability of Fungi Regarding Ectomycorrhizae Formation

Danell and Fries (1990) synthesized ectomycorrhizae between *Cantharellus cibarius* and *Picea abies*; the only strain out of several which formed ectomycorrhizae was isolated from a fruitbody associated with *Picea abies*. Tonkin et al. (1989) pointed out that genotype, maturity, and fungal specifity of *Pisolithus tinctorius* and also of *Eucalyptus marginata* are key factors influencing successful formation of mycorrhizae in vitro. Malajczuk et al. (1990) and Lei et al. (1990) found strain-specific differences of *Pisolithus tinctorius* collected under *Eucalyptus* or *Pinus* with respect to their ability to form ectomycorrhizal with their "host or nonhost" tree.

Sib-monokaryotic and dikaryotic strains of *Laccaria bicolor* were compared with respect to their ability to form ectomycorrhizae on *Pinus banksiana* by Wong et al. (1989, 1990a, b). Strains incapable of forming ectomycorrhizae and differences in the development of the Hartig net and in the mantle thickness and density were found. Wagner et al. (1989) studied the genetic variability of monokaryotic and dikaryotic populations of the ectomycorrhizal fungus *Hebeloma cylindrosporum* with respect to nitrate reductase (NR). The results indicated that breeding with sib-monokaryons could be of use in improving NR activity. Lahamedi and Fortin (1991) found differences in the amount of extramatrical mycelium due to genetic variation of reconstituted dikaryons of *Pisolithus*. Lahamedi et al. (1990) isolated monokaryotic mycelia from one fruitbody of *Pisolithus tinctorius*, reconstituted dikaryotic mycelia, and showed differences in the ectomycorrhization ability between monokaryotic and between dikaryotic strains; the dikaryotic state was necessary for full ectomycorrhizal development.

### ζ) Verification of Ectomycorrhizal Nature of Some Selected Plants, Including Short Descriptions of Ectomycorrhizae

Bledsoe et al. (1990) found ectomycorrhizae on *Potentilla hyparctica*. Kohn and Stasovski (1990) found ectomycorrhizae on *Pedicularis capitata*, *Saxifraga oppositifolia*, and *Kobresia myosuroides*. Warcup (1990a) came to the conclusion that a restricted number of Australian Asteraceae form ectomycorrhizae; all appear to belong to the tribus Inuleae.

Thoen and Ducousso (1989a) observed ectomycorrhizae on *Cinnamomum zeylandicum* (Lauraceae). Alexander (1989a) compiled the presently known ectomycorrhizae-forming legumes. Theodorou and Reddell (1991) synthesized ectomycorrhizae between several fungi and members of the Casuarinaceae.

### η) Influence of Tree Clones on Ectomycorrhiza Formation

Catinus et al. (1990) found differences between clones of *Betula pendula* inoculated with the same *Paxillus involutus* strain with respect to growth and NaCl salt tolerance. Coutts and Nicoll (1990a) reported that *Picea sitchensis* clones significantly altered the growth rate of emanating hyphae and rhizomorphs. Adams and O'Neill (1991) found significant differences in the percentage of mycorrhization by *Pisolithus tinctorius* due to genotype of *Pinus taeda* seedlings. Thomson et al. (1990) observed on seedlings of *Picea marinana* from lowland seed source significantly more mycorrhizal second-order lateral roots than on seedlings from an upland seed source after inoculation with *Laccaria laccata.*

## b) Ontongeny and Ultrastructure

Weiss (1990) demonstrated that initial infection takes place in the subapical region of the short root, and after the formation of a primary hyphal cover, the mantle and Hartig net formed simultaneously. Ectomycorrhizae with pseudoparenchymatous mantles were organized plectenchymatically in early ontogenetical stages. Agerer (1991d) compared the ontogeny of the ectomycorrhizal rhizomorphs of *Pisolithus tinctorius* and the rhizoid strands of the moss *Polytrichum juniperinum* and pointed out that backwards ramifications of hyphae or of rhizoids are a common and apparently a basic means of construction for both complex structures. Massicotte et al. (1989) mentioned that the shape of synthesized ectomycorrhizae of *Alpova diplophloeus* on *Alnus rubra* can be influenced by the stage of root elongation at the time of colonization by the hyphae.

Hyphae can change their orientation of growth after contact with root hairs, and can partially envelope the elongating hairs (Massicotte et al. 1989; Thomson et al. 1989). Massicotte et al. (1989a) showed degradation of middle lamellae of the cortex cells at the growing hyphal front of the Hartig net. Massicotte et al. (1989b) found differences with respect to Hartig net development on comparing *Laccaria bicolor* ectomycorrhizae on *Betula alleghaniensis* and *Pinus resinosa*. Massicotte et al. (1990) studied the ontogeny of *Pisolithus tinctorius* ectomycorrhizae with *Betula alleghaniensis*. Jacobs et al. (1989) found an altered morphogenesis of hyphae in contact to the root surface prior to the formation of a full hyphal mantle as a consequence of fusions of proliferating, branching hyphae.

Kottke and Oberwinkler (1990) showed that in *Larix decidua* the Hartig net can be initiated as early as the primary endodermal stage, possessing only the casparian strip, whereas in *Picea abies* it commenced after the secondary stage with an endodermis possessing already a regular sequence of elongated, suberized, degenerating cells and short, unsuberized, active passage cells. Kottke and Oberwinkler (1989) regarded the formation of multibranched palmetti of the Hartig net as a method for amplification of root-fungus interface.

Boutekrapt and Pargney (1991) demonstrated that the outer hyphal wall layers were implicated in the formation of mantle matrix of ectomycorrhizae of *Tuber melanosporum* on *Quercus spp*. Pargney (1991) found that matrix elements of the mycorrhizal mantle include pectic material originating from medial layers of root cells. Cytochemical tests by

Pargney and Leduc (1990) on ectomycorrhizae of *Tuber melanosporum* on *Corylus avellana* showed polysaccharides associated with cystine-rich proteins in walls and matrix.

Berndt et al. (1990) recorded that ascomycetous and basidiomycetous ectomycorrhizae can best be held apart by the septal pore apparatus, whereas the structure of mantle matrix is not stable enough to be used as a distinguishing feature. Hofmann (1989) found electron-dense wall modifications in ectomycorrhizal rhizomorphs of *Xerocomus chrysenteron* between adjacent hyphae.

### c) Pigments

Keller (1989) found specific differences between seven types of ectomycorrhizae on *Pinus cembra* with respect to their pigment composition. Two tuberculate ectomycorrhizae could be identified as *Suillus plorans* and *S. placidus* by comparison with fruitbody pigments.

### d) Nucleic Acid Researches

Gardes et al. (1991) used the PCR technique to compare axenically grown fungi with ectomycorrhizae, using specific regions of the nuclear and mitochondrial genome of *Laccaria* species and *Thelephora terrestris*. Differences could be demonstrated, and identification of an accidental contamination of *Pinus banksiana* with *Thelephora terrestris* was possible.

### e) Physiology

#### α) Protoplasts

Protoplasts have been successfully produced already from some ectomycorrhizal species (Barret et al. 1989; Zel et al. 1989; Anunciacao et al. 1990; Farquhar and Peterson 1990). Barret et al. (1990) succeeded in genetic transformation of *Laccaria laccata* protoplasts.

#### β) Substances Assumed as Important for Formation of Ectomycorrhizae

##### i) Hormones

Treatment of roots with indolylacetic acid (IAA) led to increased production of *Laccaria laccata* ectomycorrhizae on pine (Unestam and Stenström 1989). Stein et al. (1990) compared the effect of indolbutyric acid (IBA) on rooting of cuttings of *Picea mariana* and *Larix decidua* with that caused by *Laccaria bicolor* and *Boletinus cavipes*. Only *Picea mariana* showed increased rooting with the fungi, but it was half as high as that of the hormone treatment. Gay (1990) could show that *Hebeloma hiemale* was able to induce adventitious roots of derooted hypocotyls of *Pinus halepensis* to a higher degree than IAA treatment alone, but only if tryptophan was added.

Wullschleger and Reid (1990) discovered a significant elevation of cytokinin levels in *Pinus taeda* seedlings, mediated by *Pisolithus tinctorius*. The mycelium, however, did not release cytokinins into the medium. *Suillus punctipes* exudated cytokinins in axenic culture (probably zeatin and zeatin ribside), but did not influence cytokinin levels in *Pinus taeda*. Rupp et al. (1989) studied the influence of ethylene on ectomycorrhizae formation and concluded that it, as well as associated changes in root morphology, may be influenced by endogenous ethylen.

Krizaj (1989) hypothesized that an additional factor guiding initiation of the ectomycorrhizal infection are ionic currents and their associated electric fields produced by both partners and that they act synergistically with chemical gradients, e.g., hormones like IAA.

ii) Phenolics

Münzenberger et al. (1990) compared the phenolics of synthesized *Lactarius deterrimus* and *Laccaria amethystina* ectomycorrhizae on *Picea abies* and found the same phenolics in ectomycorrhizae as in nonmycorrhizal roots, but the amounts of several phenolics are reduced in ectomycorrhizae.

### γ) Recognition

Horan and Chilvers (1990) showed that there is a selective chemotropic attraction of mycorrhizal fungi to substrates diffusing from compatible host root apices. Thomson et al. (1989) argued that the initial contact between root hairs and hyphae may be influenced by a fibrillar material on the root hair surface. Lei et al. (1991) found particularly intensive Concanavalin A-colloidal gold binding complexes at the hypha-root interfaces in the more aggressive root colonizer of two *Laccaria* **bicolor** strains.

### δ) Enzymes

Hilbert and Martin (1989) found protein changes and ectomycorrhizae-specific polypeptides in ectomycorrhizae of *Pisolithus tinctorius* with *Eucalyptus globulus*, the so-called ectomycorrhizins. During ectomycorrhiza formation, a massive decrease of proteins in comparison to free-living mycelium and uninfected roots could be found.

Sarjala (1991) detected higher nitrate reductase (NR) activity in ectomycorrhizae of *Cenococcum, Piloderma croceum*, and *Paxillus involutus* on *Pinus sylvestris* seedlings than in the mycelium alone, but in ectomycorrhizae it was lower than in nonmycorrhizal roots. Vezina et al. (1989) found that ectomycorrhizal infections by different fungal species on roots of *Pinus banksiana* seedlings altered the enzymatic activity. *Laccaria bicolor* ectomycorrhizae had significantly greater NR and glutamate dehydrogenase (GDCO) activities than those of *Pisolithus tinctorius*. Nonectomycorrhizal root systems had higher specific activities than ectomycorrhizal ones for NR, glutamine synthetase (GS), glutamate dehydrogenase (GDH), and GDCO, but glutamate oxoglutarate aminotrasferase (GOGAT) activities were the same. Dell et al. (1989) showed that NADP-dependent GDH was present in ectomycorrhizae of *Hebeloma sp.* on *Picea abies*, but it was strongly suppressed in *Hebeloma crustuliniforme* and *Paxillus involutus* ectomycorrhizae on *Fagus sylvatica*, usually a nonhost of these species. Chalot et al. (1989b) detected NADP-GDH both of the fungus and of nonmycorrhizal roots in the ectomycorrhizae of *Picea abies* with *Hebeloma crustuliniforme* and *Pseudotsuga menziesii* with *Laccaria laccata* respectively. By contrast, fungal NADP-GDH was suppressed in mycorrhizae of *Fagus sylvatica* with *H. crustuliniforme* and *Paxillus involutus*, respectively; aspartate aminotransferase (AAT) found in ectomycorrhizae corresponded to the host roots form, whereas the fungal isoform was strongly repressed.

Ahonen et al. (1989) reported on a 1.66 and 2.77 times higher peroxidase activity in nonmycorrhizal *Pinus sylvestris* seedling roots than in their ectomycorrhizae with *Piloderma croceum* and *Rhizopogon roseolus*.

Sauter and Hager (1989) showed that chitinase activity in *Picea abies* suspension-cultured cells was increased by *Amanita muscaria* cell wall fractions; it was also increased in roots upon incubation with the fungal elicitor. Treatment of cells with auxin and cytokinin resulted in a severalfold depression of chitinase activity; the chitinase activity of hormone-treated as well as hormone-free cells showed an elicitor-induced increase.

### ε) Carbon Nutrition of Ectomycorrhizae

Miller et al. (1989) using $^{14}CO_2$ showed that extramatrical mycelium of *Hebeloma crustuliniforme Pinus ponderosa* ectomycorrhizae was a strong sink for carbon. Bauer et al. (1991) found label in cortex, Hartig net, and the mantle hyphae, after application of $^{14}CO_2$ to *Pinus silvestris* seedlings ectomycorrhizal with *Suillus variegatus*. Cairney et al. (1989) detected more radioactivity originating from $^{14}CO_2$ in ectomycorrhizal roots of *Eucalyptus pilularis* with *Pisolithus tinctorius* than in nonmycorrhizal roots; greater $^{14}C$ accumulation occurred in young than in older ectomycorrhizae. Finlay (1989a) reported on a significantly greater accumulation of labeled assimilate in the roots of *Pinus sylvestris* seedlings when ectomycorrhizal with the *Pinus*-specific fungus *Suillus bovinus* in comparison to being infected with the *Larix*-specific *Suillus grevillei* and *Boletinus cavipes*.

Guehl et al. (1990) found a higher $CO_2$ assimilation capacity in *Pinus pinea* seedlings inoculated with *Rhizopogon roseolus* in comparison with those with a strain of *Suillus collinitus* (another strain of *S. collinitus* caused the same $CO_2$ assimilation capacity as *R. roseolus*). Dosskey et al. (1990) mentioned a significant increase in photosynthesis of *Pseudotsuga menziesii* seedlings if inoculated with *Rhizopogon vinicolor*, whereas *Hebeloma crustuliniforme* and *Laccaria laccata* had no effect. Guehl and Garbaye (1990) reported that ectomycorrhizae of *Laccaria laccata* on *Pseudotsuga menziesii* are more efficient in increasing photosynthesis and water use efficiency than those of *Thelephora terrestris*.

Niederer et al. (1989) showed that excised Norway spruce ectomycorrhizae were able to synthesize trehalose from glucose; the more vital the mycorrhizae were, the more trehalose they formed.

Abuzinadah and Read (1989b) highlighted that *Betula pendula* seedlings ectomycorrhizal with *Hebeloma crustuliniforme* may, over a 55–day period with proteins as the sole carbon source, derive up to 9% of their C content from the protein. In addition, they detected that the carbon drain from the protein consumed by the ectomycorrhizae and transported to the plant is greatest in low irradiance.

### ζ) Growth Responses of Plants

Guehl et al. (1990) observed a higher root and total plant biomass of *Pinus pinea* seedlings inoculated with *Rhizopogon roseolus* in comparison to those ectomycorrhizal with *Suillus collinitus*. Bougher et al. (1990) presented differences in growth responses of *Eucalyptus diversicolor* ectomycorrhizal with *Pisolithus tinctorius*, *Laccaria laccata*, and *Descolea maculata*; soil moisture also exerts an influence (Bougher and Malajczuk 1990). Rousseau and Read (1991) concluded that only under conditions where seedlings maintain a relatively low shoot/root ratio, a sufficient carbon cost would exist to cause a reduction in the relative growth rate of ectomycorrhizal seedlings.

Stenström (1990) noted that *Pinus sylvestris* seedlings, inoculated with *Amanita muscaria,* were 50% smaller than uninoculated seedlings at outplanting in the field; but after 18 months they were about 20% larger than the controls. Stenström et al. (1990) pointed out that nursery *Pinus sylvestris* seedlings inoculated with *Laccaria laccata*, *Hebeloma crustuliniforme*, and *Cenococcum*, had reduced seedling growth. Dosskey et al. (1990) reported that seedlings colonized by *Hebeloma* were significantly smaller than noninoculated ones. *Laccaria* had no effect, whereas seedlings ectomycorrhizal with *Rhizopogon* were smaller only in low level phosphate treatments.

Last et al. (1989) found that in an unheated glasshouse trial, the heights of *Picea sitchensis* seedlings when inoculated with *Laccaria proxima* or *Paxillus involutus*, was exponentially related to total number of ectomycorrhizae. McAfee and Fortin (1989) showed that on outplanted seedlings of *Picea mariana* the percent of mycorrhizal colonization was positively correlated with seedling dry weight, while with *Pinus banksiana* there was no correlation; a positive correlation was also found by Newton (1991).

### η) Phosphate Nutrition

Cumming and Weinstein (1990a) showed that *Pinus rigida* seedlings, ectomycorrhizal with *Pisolithus tinctorius*, were able to extract phosphate, in contrast to nonmycorrhizal seedlings, from the insoluble salt $AlPO_4$. Measurement of matrix solution pH indicated that increased proton production in association with high proliferation of mycorrhizal roots under P-limiting conditions, led to increased $AlPO_4$ solubilization.

Finlay (1989a) reported on a significantly greater uptake and translocation of $^{32}P$ orthophosphate by the *Larix*-specific fungi *Boletinus cavipes* and *Suillus grevillei* to *Larix* than the *Pinus*-specific *Suillus bovinus* translocated to *Larix*. Finlay (1989b) pointed out that in mixed microcosm communities, translocation of $^{32}P$ through ectomycorrhizae to

*Larix eurolepis* by the *Larix*-specific fungus *Boletinus cavipes* was greater than to *Pinus sylvestris*; in addition the fungus gained more carbon from *Larix* than from *Pinus*.

Jones et al. (1990) compared *Salix viminalis* inoculated with *Laccaria proxima* or *Thelephora terrestris*, and found a relationship between production of extramatrical hyphae per gram soil, unit length of mycorrhizal root, P uptake per unit root weight, and total P content; all values were greater for *L. proxima* than for *T. terrestris*. MacFall et al. (1989) presented data that the degree of ectomycorrhizal colonization and growth enhancement of *Pisolithus tinctorius* and *Hebeloma arenosa* ectomycorrhizae on *Pinus resinosa* decreased with increasing phosphorus additions. Bougher et al. (1990) found no growth responses with *Descolea maculata*, *Pisolithus tinctorius*, or *Laccaria laccata*. Rousseau and Reid (1991) observed that almost twice as much of the absorbed P in *Pinus taeda* seedlings ectomycorrhizal with *Pisolithus tinctorius* was retained in the fine root component compared to nonmycorrhizal seedlings and there seemed to be a more immediate transfer of P from the fine root fraction to the foliage in nonmycorrhizal seedlings. It was suggested that P was preferentially retained in the fungal tissue of the ectomycorrhizae, and this could explain an initial growth depression of ectomycorrhizal seedlings in comparison to nonmycorrhizal ones.

Kammerbauer et al. (1989) determined the translocation velocity and flux rate of $^{32}$P-orthophosphate in ectomycorrhizal rhizomorphs of *Thelephora terrestris* on *Picea abies*. Comparing with literature data for *Suillus bovinus* ectomycorrhizal rhizomorphs on *Pinus silvestris* they concluded that the higher organization of the latter rhizomorphs reflect the higher flux rates found for *S. bovinus*.

θ) Nitrogen Nutrition

Abuzinadah and Read (1989a) investigated the access to nitrogen from protein applied to ectomycorrhizal *Betula pendula* seedlings. *Hebeloma crustuliniforme*, an "early stage" fungus, provided significantly more nitrogen to the seedlings than the "late stage" fungus *Amanita muscaria* or *Paxillus involutus*. Plants without ectomycorrhizae were unable to use these substances for N-nutrition (Abuzinadah and Read 1989a). Finlay et al. (1989), feeding *Paxillus involutus* mycorrhizae on *Pinus sylvestris* with $^{14}$N-ammonium, found as principle sinks alanine and glutamate/glutamine: a significant proportion of label was found in the needles. Bauer et al. (1991) showed that *Pinus sylvestris* seedlings ectomycorrhizal with *Suillus variegatus* supplied the hyphae of the ectomycorrhizae with a higher amount of photosynthates if chitin was the sole nitrogen source.

Högberg (1989) synthesized *Pisolithus tinctorius* ectomycorrhizae on *Pinus sylvestris* in nonaxenic sand cultures fertilized with $NH_4^+$ and calculated the inflow of N via ectomycorrhizae as 0.48 µg per tip per day in comparison to 0.16 for ordinary roots. Scheromm et al. (1990) found that the amount of xylem transported reduced nitrogen was similar in seedlings ectomycorrhizal with *Hebeloma cylindrosporum* and nonmycorrhizal seedlings; however phloem transport was approximately twice that found in nonmycorrhizal plants.

ι) Micronutrients

Chalot et al. (1989a) found higher K, Ca, and Mg contents in needles of *Pseudotsuga menziesii* seedlings due to inoculation with *Laccaria laccata*. Mitchell et al. (1990) found evidence that boron fertilization may indirectly effect the mineral nutrition of *Pinus echinata* seedlings ectomycorrhizal with *Pisolithus tinctorius* by increasing root colonization. They observed significant negative correlations between elemental concentrations and mycorrhizal colonization, but significant positive correlation for mycorrhizal colonization, biomass accumulation and elemental content. Chakravarty and Chatarpaul (1990a) analyzed significantly higher K, Ca, and Mg concentrations in seedlings ectomycorrhizal with *Laccaria laccata* than in nonmycorrhizal seedlings.

κ) Water

Boyle and Hellenbrand (1991) found differences between ectomycorrhizal fungi in pure culture due to water stress, which correlated well to seedling drought tolerance when ectomycorrhizal with these fungi.

Lehto (1989) in a droughting experiment, found considerably lower water contents in the soil of *Picea sitchensis* seedlings lacking ectomycorrhizae than when they were inoculated with *Paxillus involutus*, but there were no differences in plant water potentials, leaf conductances, and net photosynthesis. Dosskey et al. (1991) pointed out that *Rhizopogon vinicolor* ectomycorrhizal with *Pseudotsuga menziesii* significantly enhanced both net photosynthesis rate and stomatal conductance, but leaf water potentials were lower compared to nonmycorrhizal controls over a range (–0.05 to –0.50 MPa) of soil water potential. *Hebeloma crustuliniforme* tended to enhance, while *Laccaria laccata* decreased net photosynthesis rate and stomatal conductance of host seedlings, but neither fungus affected leaf potential.

Walker et al. (1989) found that the xylem pressure potential of 4-year-old *Pinus taeda* seedlings bearing *Pisolithus tinctorius* ectomycorrhizae was significantly less negative than in those possessing *Thelephora terrestris* ectomycorrhizae. Coleman et al. (1990) found the higher hydraulic conductivity of *Pseudotsuga menziesii* seedlings ectomycorrhizal with different fungi correlated with increased tissue-P and decreased root/shoot ratio, but when adjusted for the effect of these factors, no effect on hydraulic conductivity could be detected.

f) Ecology

α) Ecological Laboratory Research

Leyval and Berthelin (1991) demonstrated that phlogopite (mica) particles closely attached to *Laccaria laccata* ectomycorrhizae lost more K than elsewhere, due to an increase of exchange surface area rather than to an increase in acidification; dual inoculation with *Agrobacterium* and *Laccaria laccata* significantly promoted phlogopite cation-exchange capacity. Cairney and Ashford (1989) demonstrated that ectomycorrhizae of *Pisolithus tinctorius* had a reducing ability.

Periodic flooding of ectomycorrhizae, as well as waterlogging, can influence some species negatively (Unestam and Stenström 1989; Coutts and Nicoll 1990b; Stenström 1991). Rhizomorphs of *Thelephora terrestris* can survive after waterlogging, and regenerate (Coutts and Nicoll 1990). Bougher and Malajczuk (1990) found prominent differences of ectomycorrhizal *Eucalyptus diversicolor* seedlings with three different fungi with respect to four levels of soil moisture.

Kasuya and Muchovej (1989) found slight differences in ectomycorrhizae formation of different fungi due to temperature. Mycorrhizal fungi isolated from tropical conditions did not always show affinity for higher temperatures.

Gibson and Deacon (1990) detected differences in the mycorrhiza formation of "early-stage fungi" and of "late-stage fungi" dependence of phosphate and glucose contents in an agar medium. Stenström (1989) reported on a higher infection level of early-stage fungi on *Pinus sylvestris* seedlings than of late-stage fungi. Brunner et al. (1990) found no ectomycorrhiza formation of *Hebeloma crustuliniforme* and *Alpova diplophloeus* with *Alnus tenuifolia*, if the seedlings were still unnodulated, but other ectomycorrhizal fungi could do so.

β) Researches in Natural Habitats

Air pockets in the soil were found to be favorable for formation of ectomycorrhizae (Unestam and Stenström 1989; Wöllmer and Kottke 1990).

Von Alten and Roßbach (1989) found a greater diversity of mycorrhizal types on *Picea abies* in soils overlaying limestone than in those over sandstone. Taylor and Alexander (1989a) found that mycorrhizal types of a *Picea sitchensis* stand were related to spatial variation in organic layer and to variation in soil moisture availability. Glenn et al. (1991) detected a reverse relationship of two different morphotypes of ectomycorrhizae on *Picea rubens* with respect to mesic and wetland sites. Väre (1989a) found a significantly higher ectomycorrhiza percentage on roots of *Pinus sylvestris* which were growing among decaying birch logs. Blasius and Oberwinkler (1989) showed that no differences in the occurrence of mycorrhizal types in relation to tree age were apparent in a natural *Picea abies* stand. Newton (1991) studied *Quercus robur* and *Betula pendula* seedlings in natural soil with respect to the occurrence of ectomycorrhizal types. Only 5 types out of 36 were found to infect both *Quercus* and *Betula*.

Termorshuizen and Schaffers (1989) found a high correlation between number of mycorrhizae and fruitbodies of mycorrhizal fungi. Similar results have been published by Agerer (1990a). Jansen (1991) calculated for *Pseudotsuga menziesi* stands that numbers of ectomycorrhizal fruitbodies summed up over three seasons were positively correlated to number of ectomycorrhizae per soil volume. In old stands fewer ectomycorrhizae per soil volume were found than in young stands.

Griffiths et al. (1990) found differences between mat and nonmat soils of ectomycorrhizal fungi, e.g., in respiration rates and levels of mineralizable nitrogen, which were generally higher in the mat soils, pH and denitrification rates were generally lower in nonmat soils. Entry et al. (1991b) found that in hyphal mats of the ectomycorrhizal fungus *Hysterangium setchellii* there was nearly twice as much release of N and P than in nonmat soils. Entry et al. (1991a) detected higher cellulose and lignin decomposition rates in mat soils of *Hysterangium setchellii* than outside these mats. Griffiths et al. (1991) observed significant ecologically important differences between hyphal mats of the ectomycorrhizal fungi *Hysterangium setchellii* and *Gautieria monticola*, and also differences between mats primarily formed in mineral soil and those formed in litter. Unestam (1991) and Griffiths et al. (1991) found that most forest-derived ectomycorrhizal species were hydrophobic with respect to their ectomycorrhizae and their hyphal mats; some were, however, hydrophilic. Light microscopical studies by Ponge (1990) revealed circumstantial evidence that in *Pinus sylvestris* forest humus, ectomycorrhizal mycelia seem to protect the partly decayed plant material and the feces deposited by soil animals from subsequent attack by soil bacteria.

γ) Coexistence with Other Organisms

i) Fungi

Summerbell (1989) isolated several fungi from mantles of *Picea mariana* ectomycorrhizae and could not find any specifity to ectomycorrhizae because most of the isolated fungi could also be found on suberized root surfaces and on roots of endomycorrhizal plants.

Fungitoxic activity of *Paxillus involutus* ectomycorrhizae against *Fusarium* has been shown in artificial media by Duchesne et al. (1989c). Duchesne et al. (1989a) found that the degree of protection activity of *Paxillus involutus* against *Fusarium* is dependent on the *Pinus* species. Duchesne et al. (1989b) could identify oxalic acid as a toxic substance against *Fusarium*. Chakravarty et al. (1991) found a significant protecting effect of *Paxillus involutus* ectomycorrhizae on *Pinus resinosa* against *Fusarium moliniforme* and *F. oxysporum* in vitro, but not against *Cylindrocarpon destructans*, if inoculated simultaneously. Farquhar and Peterson (1991) showed that *Fusarium* became vacuolized or filled with slightly electron-dense substances and only a few hyphae invaded the roots, but only when the seedlings were protected by *P. involutus*. Chakravarty and Hwang (1991) pointed out that *Laccaria laccata* protected 37.7% of *Pinus banksiana* seedlings when grown concomitantly with the damping-off causing *Fusarium oxysporum* in vitro. *Pisolithus tinctorius*, as well as *Tricholoma pessundatum*, inhibited the ectomycorrhizal fungus *Sphaerosporella brunnea* (Kope and Fortin 1989).

Agerer (1990b, 1991c) proved that ectomycorrhizae of *Rhizopogon* sp. and of several *Suillus* spp. can possess intracellular hyphae (= haustoria) in their root cortex cells originating from foreign ectomycorrhizal Gomphidiaceae. *Gomphidius glutinosus* hyphae very likely intrude into hyphae of *Rhizopogon* rhizomorphs (Agerer 1991c). Brand (1991) analyzed the tripartite relationships between the ectomycorrhizae of *Lactarius subdulcis* on *Fagus sylvatica* and the ascomycete *Leucoscypha leucotricha*, and came to the conclusion that the ascomycete is a consistent inhabitant of these ectomycorrhizae.

Haug (1989b) could show intercellular hyphae in the apical region of the unidentified ectomycorrhiza *Piceirhiza gelatinosa*. They were able to penetrate between meristematic and immature cortical cells.

ii) Bacteria

Ali and Jackson (1989) detected bacteria in fruitbodies of ectomycorrhizal fungi, ectomycorrhizae, and soil, which were used for germination studies of spores; only three out of seven tested fungal species germinated.

Fluorescent *Pseudomonas* species were found by Mamoun and Olivier (1989) as dominant on the mycorrhizoplane of *Tuber malanosporum* on *Corylus avellana*, one species was characteristic of it; the more powerful iron-chelating strains were also closely related to ectomycorrhizal surface. Schisler and Linderman (1989) showed significantly increased bacterial populations due to the influence of volatiles originating from ectomycorrhizae of *Hebeloma* sp. and *Laccaria laccata*; actinomycetes and phosphate solubilizing bacteria were not significantly influenced.

Garbaye and Bowen (1989) and Duponnois and Garbaye (1991) found stimulating effects of some bacteria on ectomycorrhization and called them Mycorrhization Helper Bacteria (MHB); but some of these bacteria inhibited the formation of *Thelephora terrestris* ectomycorrhizae. It is suggested that the MHB act directly on the plant and/or on the fungus. Amaranthus et al. (1990) isolated an *Azospirillum* species from within the mycorrhiza of inoculated *Pseudotsuga menziesii* seedlings and found that seedling growth paralleled associative $N_2$ fixation.

iii) Animals

Gehring and Whitham (1991) found that *Pinus edulis* trees susceptible to chronic insect attack have fewer ectomycorrhizae than resistant trees, and that after removal of the insects from the susceptible trees, the mycorrhizal levels increased to a level comparable to that of resistant trees.

Agerer (1991e) argued that chlamydospores formed on ectomycorrhizae could perhaps be disseminated, e.g., by invertebrates, and could act as effective propagules for the fungi, given the opportunity to germinate adjacent to roots.

Wöllmer and Kottke (1990) observed in situ slugs, collembola, and other soil organisms feeding on the hyphal mantles of ectomycorrhizae, and found an influence of soil animals on the quantity of fine roots. Unestam (1991) reported on insect larvae which were "eating hyphae like spaghetti" of ectomycorrhizae. Stumpp (1990) highlighted that protura feed by sucking on ectomycorrhizal mantles. In a *Picea abies* forest, Proturan abundance, especially of larval stages, was highly correlated with fluctuations in mycorrhiza frequency.

iv) Plants

Timbal et al. (1990) found a depressive effect of the grass *Molinia* on the growth of *Quercus rubra*, together with significant modifications in the mycorrhizal status of the roots: regression of ectomycorrhizae formed by *Laccaria* sp. and an increasing number of those formed by *Cenococcum*, of nonmycorrhizal and of dead short roots. Carleton and Read (1991) showed that shoots of the moss *Pleurozium schreberi*, applied to root chambers containing ectomycorrhizal plants, were colonized by the ectomycorrhizal hyphae, particularly intensive growth occurred in the senescent region of the moss shoots.

δ) Influences by Man

i) Fertilization with Lime or Nitrogen

Ritter et al. (1989b), using five vitality classes, discovered on limed plots of *Abies alba* and *Picea abies* a higher percentage of ectomycorrhizae with medium vitality, whereas in the unlimed plots the classes of vitality were rather evenly distributed resulting in a surprisingly high percentage of completely vital ectomycorrhizae. Erland and Söderström (1990) found in a laboratory liming (CaO) experiment of *Pinus sylvestris* seedlings in a forest humus that the number of root tips that became mycorrhizal increases from pH 4 to around pH 5 and decreased to a minimum at pH 7.5. The species of ectomycorrhizae involved differed specifically in their response to pH. Erland and Söderström (1991) recognized a replacement of an unidentified pink ectomycorrhiza and of *Cenococcum* ectomycorrhizae by *Paxillus involutus* grown in humus taken from a *Pinus sylvestris* forest which had been fertilized with lime 22 months previously.

Nitrogen fertilization reduced the percentage of ectomycorrhizae in a *Picea sitchensis* stand, and affected the relative proportions of mycorrhizal types (Taylor and Alexander 1989a). Marx (1990) observed that application of $NH_4NO_3$ resulted in soil concentrations of $NO_3$-N ranging from 60 to 120 kg/ha and $NH_4$-N ranging from 90–130 kg/ha at pH 4.8 and 5.8 were associated with the most abundant ectomycorrhizal development. Newton and Pigott (1991) found in a pot experiment with natural soil, as well as in a field experiment, a reduction of 38% of ectomycorrhizae on *Betula pendula* following N-application, the influence on *Quercus robur* was less pronounced; the relative proportion of ectomycorrhizal types was also influenced. Väre (1989b) studied the influence of various N forms on *Pinus sylvestris* seedlings ectomycorrhizal with *Suillus variegatus* and showed that $NaNO_3$ treatment resulted in the lowest amount of ectomycorrhizae; the highest value was caused by the $NH_4NO_3$ treatment; the growth of the seedlings paralleled these results.

ii) Pesticides

Unestam et al. (1989) concluded that in vitro application of fungicides on mycelia were not reliable when estimating effects on mycorrhizae.

Chakravarty and Chatarpaul (1990b) came to the conclusion that the herbicide glyphosate (at 0.54 and 3.23 kg/ha) did not influence ectomycorrhizal development of *Paxillus involutus* on *Pinus resinosa*. Sidhu and Chakravarty (1990) characterized triclopyr as most toxic on *Pinus contorta* and *Picea glauca* ectomycorrhizal seedlings under aseptic conditions; hexazinone application reduced growth and mycorrhizal infection significantly but showed recovery with time. Under field conditions overall effects were less intensive. Marks and Becker (1990) wrote that propazine applied in normal and half-normal concentration of field application and chlorthal-dimethyl in normal concentration both significantly suppressed mycorrhizal formation, but both increased shoot growth at the end of 6 months in *Pinus radiata* seedlings. The species of ectomycorrhizae were influenced differently in their number. Chakravarty et al. (1990) found a stimulating effect on ectomycorrhiza formation of *Paxillus involutus* on *Pinus resinosa* by benomyl, but an inhibitory effect was obvious in an oxine benzoate treatment; both fungicides significantly reduced damping-off of seedlings caused by *Fusarium oxysporum*. De la Bastide and Kendrick (1990) reported that addition of benomyl after simultaneous inoculation of *Pinus strobus* with the pathogen *Mycelium radicis atrovirens* and *Laccaria bicolor* improved mycorrhization by *Laccaria bicolor* and reduced pathogenicity.

iii) Pollution

*Acidification.* After acidification of forest soil (F/H horizons, and mineral soil) with elemental S in a greenhouse experiment, *Pinus contorta* seedlings failed to form ectomycorrhizae below pH 3.3 (Danielson and Visser 1989b). Application of simulated acid rain of different pH to *Pinus rubens* seedlings in natural soil by Meier et al. (1989) indicated that short-term acid deposition does not induce significant changes in frequency of ectomycorrhizae, but species composition was changed.

Metzler and Oberwinkler (1989) examined *Amanita muscaria, Cenococcum* and *Pisolithus tinctorius* ectomycorrhizae on *Pinus sylvestris*; apart from specific optimal conditions for mycorrhization, no aberrant morphological and ultrastructural growth pattern could be detected due to pH between 3 and 6.

*Effects of $SO_2$.* Termorshuizen et al. (1989) reported that ectomycorrhiza formation of *L. proxima* on *Pinus sylvestris* was not affected by different $SO_2$ concentations applied for 7 weeks, whilst that of *P. involutus* decreased significantly from 120 mg/m$^3$ onwards. Holopainen (1989) observed a decrease of mycorrhizal types and a dominance of *Cenococcum* on *Pinus sylvestris* near a pulp mill (with $SO_2$ as the most important emission); generally, accumulation of tannin in cortical and pericycle cells and an increase of electron-dense vacuolar accumulation in the fungal cells occurred. Termorshuizen and Schaffers (1989) found a close correlation between the amount of $NH_3$ and $SO_2$ pollution with the number of branched mycorrhizae on *Pinus silvestris* stands in The Netherlands, but not with the number of tips.

*Effects of Ozone (and $SO_2$).* In an experiment with *Pinus taeda* seedlings planted in nonsterile soil and treated in a combination of different levels of ozone, acid irrigation of pH 3.8 or 5.2, and fertilization with Mg, Simmons and Kelly (1989) found a greater number of ectomycorrhizae in subambient concentrations of ozone than in ambient or twice ambient concentrations. Increasing rainfall acidity and soil Mg concentration resulted in a significantly greater number of ectomycorrhizae, due primarily to an increased occurrence of one coralloid mycorrhizal type. Meier et al. (1990) observed in *Pinus taeda* seedlings a negative linear dose-response relationship between $O_3$ and the number of ectomycorrhizae.

Wöllmer and Kottke (1990) found after a long-term exposure to $SO_2$, $SO_2 + O_3$, and simulated acid precipitation an increased mycorrhizal production by *Abies alba.* However, on *Picea abies*, a decreased number of mycorrhizae was found in combination with a high proportion of dead fine roots when polluted with $SO_2$. The cation analyses showed an accumulation of $Ca^{2+}$ ad $Zn^{2+}$ in the mycorrhizae of spruce exposed to ozone. Blaschke (1990) found a significant decrease in the proportion of ectomycorrhizal root tips of 8-year-old *Picea abies* after long-term irrigation with acid rain of pH 4 and a simultaneous fumigation with $SO_2$ and ozone. The acid treatment alone resulted in a drastic reduction of the proportion of ectomycorrhizae to nonmycorrhizal root tips. Adams and O'Neill (1991) found no significant variation of mycorrhizal infection with varying ozone levels in a combined treatment with acid irrigation on *Pinus taeda* seedlings ectomycorrhizal with *Pisolithus tinctorius.* After 12 weeks, there was a tendency towards fewer mycorrhizae with higher concentrations of ozone. Blaschke and Weiss (1990) found that 4-year-old *Picea abies* cuttings grown in acid or calcareous soils did not show differences in the relative number of ectomycorrhizae (number/mg dry weight of fine roots) after a 14-months exposure to various concentrations of ozone and acid mist.

*Effects of Nitrogen.* In a greenhouse experiment, Van Dijk et al. (1990) applied different levels of ammonium-nitrogen to young coniferous trees. Trees treated with 48 kg $NH_4^+$/ha/yr showed no reactions. The highest treatment (480 kg) resulted in a strong decrease of the numbers of mycorrhizae, levels of K, Mg, and Ca of needles decreased sharply. Haug and Feger (1991) investigated plots of *Picea abies* fertilized with $MgSO_4$ or $(NH_4)SO_4$, and observed that on the N-treated plot the relative mycorrhiza frequency (number per 100 mg dry weight of fine roots) declined, whereas in the $MgSO_4$-treated plot no decline could be found. Elevated concentrations of potentially toxic Al and an extremely low Ca/Al ratio appeared in the soil of both treatments, but did not result in elevated Al concentrations in fine roots, indicating that the damage of trees (needle yellowing) may not be due to Al.

Wallander and Nylund (1989) reported on a negative correlation between mycorrhizal infection and N-concentration of the shoots at different N-concentrations in the growing media of *Pinus sylvestris* ectomycorrhizal with *Laccaria laccata*. Ammonium had a greater inhibitory effect than nitrate. Termorshuizen and Ket (1989) applied nitrate (as Na salt) and ammonium (as sulphate) to *Pinus sylvestris* seedlings ectomycorrhizal with *Suillus bovinus* or *Paxillus involutus*. Both nitrogen sources enhanced plant dry weight to a fertilization level of 50–100 kg/yr onwards. Ammonium was more inhibitory than nitrate. Haug (1989a) applied ammonium or nitrate in five different amounts on *Picea abies* ectomycorrhizal with *Pisolithus tinctorius*. Above 252 ppm ammonium N, mycorrhizae formed only a loose hyphal mantle, with no Hartig net and an intracellular growth in the dead outer cortical cells. Even at concentrations of 207 ppm nitrate N, mycorrhization was still abundant. Root and needle concentrations of N, Ca, Mg, Mn increased with increasing ammonium concentrations, K contents either decreased or remained unchanged. Increasing nitrate concentrations decreased Mg, Ca, Mn, and increased K and N contents of the needles.

Jentschke et al. (1991a) found no effect of nitrate on ectomycorrhizal development of *Lactarius rufus* on *Picea abies* seedlings, only a slight decrease with respect to ammonium treatment was apparent.

*Effects of Toxic Metals.* Jentschke et al. (1991c) reported that the mycorrhizal infection approached 100% in *Picea abies* seedlings inoculated either with *Lactarius rufus* or *Paxillus involutus*, and exposed to $PbCl_2$, but the number of root tips per dry weight decreased after a 19 week treatment. There was no accumulation of Pb in the hyphal cell walls of the mantle or of the Hartig net, but in plant cell walls of mycorrhizal as well as of nonmycorrhizal seedlings. The endodermis acted as a barrier to Pb radial transport in both mycorrhizal and non-mycorrhizal roots. Chappelka et al. (1991) grew *Pinus taeda* seedlings in natural soil with different levels of applied $PbCl_2$. The treatments showed that growth was highest in controls and in the highest treatments (240 mg or 480 mg/kg soil), and least in the intermediate levels, but ectomycorrhiza formation decreased significantly and there were changes in the mycorrhizal species composition.

Kasuya et al. (1990) tested several ectomycorrhizal fungi for their capacity to form ectomycorrhizae under different Al-regimes. They found specific differences, and the fungi appeared to ameliorate Al damage to plant roots even in treatments where no mycorrhizae are formed. Cumming and Weinstein (1990b) found that *Pinus rigida* seedlings ectomycorrhizal with *Pisolithus tinctorius* exhibit unaltered growth, physiological function, and ionic relation when exposed to 200 μM Al, in comparison to inoculated seedlings untreated with Al; seedlings grown in Al lacking ectomycorrhizae showed drastic changes. Jentschke et al. (1991a) found no major difference in physiological response to aluminium exposure of *Picea abies* seedlings, whether they were ectomycorrhizal with *Lactarius rufus* or not; in all cases, Mg uptake into roots and needles was strongly reduced. Jentschke et al. (1991b) obtained results suggesting that the endodermis of *Lactarius rufus* ectomycorrhizae on *Picea abies* seedlings is the primary barrier to radial Al transport and that the presence of the hyphal mantle did not prevent Al from reaching the root cortex and from displacing $Mg^{2+}$ and $Ca^{2+}$.

Heyser and Donner (1989) and Donner and Heyser (1989), using X-ray analysis, reported on an accumulation of Ti, Fe, and Al within the fungal mantle of a white beech ectomycorrhiza collected from an air pollution-stressed stand, indicating selective exclusion and/or specific accumulation, whereas highest amounts of Co, Pb, and to a lesser extent Zn were found in the root cortex and the endodermis.

*Damage of Trees.* Vincent (1989) found in humus-free topsoil a correlation between the absolute number and relative frequency (number of ectomycorrhizae on 100 mg fine roots) of ectomycorrhizae on *Fagus sylvatica* and its decline classes in Bavaria. Von Alten and Roßbach (1989) found no correlation between needle symptoms and development of mycorrhizae in natural stands of *Picea abies* affected by die-back. Perrin and Estivalet (1989) came to the conclusion that declining (yellowing) *Picea abies* trees have fewer mycorrhizae than healthy ones.

Estisvalet et al. (1990) concluded that low pH and fungi which cause no root necrosis (e.g., *Trichoderma, Cylindrocarpon*), rather than other microorganisms could be involved in poor mycorrhizal infection. Ritter et al. (1989a,b) found no alterations in the main anatomical features of ectomycorrhizae of *Abies alba* or *Picea abies*, irrespective of whether the trees were damaged or not. The occurrence of vital ectomycorrhizae, however, differed. Healthy trees had a maximum of medium vital ectomycorrhizae, whereas damaged trees showed a rather even distribution of all five vitality classes. In healthy-looking trees they found fewer potential pathogens, *Cylindrocarpon destructans* and *Trichoderma viride*, than on ectomycorrhizae of damaged trees, suggesting that root pathogens may play an important role with respect to decline of trees.

Jansen (1991) found that the numbers of ectomycorrhizae per soil volume were inversely related to the extent of "crown transparency" in *Pseudotsuga menziesii* stands, loss of twigs and incidence of damaged bark, but there were no significant correlations with needle-loss and discoloration.

Kropacek et al. (1989) found a correlation between degree of stand damage of *Picea abies* and the "Mycorrhizal Inoculation Potential"; the more severe the damage, the lower the MIP. Rejsek (1991) showed a significant decrease of acid phosphomonoesterase activity in ectomycorrhizal spruce roots affected by natural as well as by artificial (elemental sulfur) pollution.

### iv) Afforestation, Clear-Cut

Väre (1989c) indicated that plowing in clear-cut areas of northern Finland had caused a marked reduction in the formation of ectomycorrhizae of "injured-looking" as well as of healthy-looking *Pinus sylvestris* saplings in comparison to unplowed sites. Borchers and Perry (1990) found evidence that pioneering hardwoods strongly influence soil biological activity in clear-cuts, and impose one or more soil patterns that favor establishment and growth of conifer seedlings. In Sweden, Dahlberg and Stenström (1991) found that clear-cuts in general had only minor effects on the composition of species of ectomycorrhizae on outplanted seedlings in comparison to forest stands. However, *Piloderma croceum* infected seedlings more frequently in natural stands than in clear-cuts; *Cenococcum* had an opposite reaction.

Danielson and Visser (1989a) found that ectomycorrhizae, formed after inoculation of seedlings by nine fungal species, were replaced almost completely by indigenous fungi after 3 years in oil-sand tailings; only *Complexipes* survived in substantial quantities. Stenström (1990) detected in the field a replacement of inoculated ectomycorrhizae by indigenous fungi, which also colonized the originally fungus-free roots. Chu-Chou and Grace (1990), after outplanting of *Pinus radiata* seedlings from nurseries into natural stands, found soil type and fertility-dependent changes in the ectomycorrhiza composition as well as species-dependent changes. Villeneuve et al. (1991) studied the survival of inoculated *Laccaria bicolor* ectomycorrhizae on *Pseudotsuga menziesii* in competition with natively formed ectomycorrhizae and came to the conclusion that *Laccaria* ectomycorrhizae survived on the old roots and colonized newly formed roots, limiting the colonization by other naturally occurring fungi. Danielson (1991) studied the species composition and succession of ectomycorrhizae on *Picea glauca* and *Pinus banksiana* grown over 10 years in subalpine coal mine soil and soil sands tailings under different fertilization regimes.

## g) Methods

Norris et al. (1991) edited a compilation of several methods for the study of ectomycorrhizae, comprising characterization, synthesis, histochemistry, NMR studies, carbon metabolism, enzymology, DNA analysis, application of radioisotopes, ultrastructure and electron loss spectroscopy.

Sen (1990) used isoenzymes to distinguish *Suillus variegatus* and *S. bovinus* ectomycorrhizae. Dighton et al. (1989, 1990) applied in the field $^{32}P$ tracer to measure phosphate uptake by birch mycorrhizae. Högberg (1990) used the relation between $^{15}N/^{14}N$ to determine the availability of organic nitrogen to mycorrhizal plants.

## h) Reviews

Alexander (1989a) reviewed the ecology and biogeography of ectomycorrhizal legumes.

Vogt et al. (1991) critically reviewed the role of ectomycorrhizae in forest ecology and identified areas for future research. Harley (1989) stressed the significance of mycorrhizae for nutrition of the symbionts, for cycling of carbon compounds, and for cycling of soil-derived nutrients. Lapeyrie (1990) reviewed the role of ectomycorrhizae in calcareous soils for calcitolerant woody plants. De Oliveira and Garbaye (1989) compiled the literature on stimulatory effects of microorganisms on mycorrhizal symbiosis. Kropp and Langlois (1990) reviewed the application of ectomycorrhizae in reforestation. Dighton (1991) outlined the meaning of ectomycorrhizal autotrophic plants for acquisition of nutrients from organic resources. Alexander (1989b) pointed out the importance of ectomycorrhizae in tropical forests with respect to phosphorus and nitrogen nutrition and ecological adaptations. Kubitzki (1989) stressed the importance of ectomycorrhizae for the development of the flora in oligotrophic areas of the Amazonian region.

Gogala (1991) compiled the literature on regulation of ectomycorrhizal infection by hormonal factors. Gianinazzi-Pearson and Gianinazzi (1989) reviewed cellular and genetical aspects of root-fungus interactions.

Wilkins (1991) reviewed the literature on the influence of ectomycorrhizae on uptake and toxicity of metals and came to the conclusion that the presence of ectomycorrhizae decreases metal accumulation in shoots. Concentrations in roots are little altered, but large amounts of metal may be found in emanating hyphae and rhizomorphs. Jansen and Dighton (1990) reviewed the effects of air pollutants on ectomycorrhizae.

## 2. Ectendomycorrhiza

Ivory and Pearce (1991) synthesized ectendomycorrhizae between *Wilcoxinia mikolae* and *Pinus caribea*; these had a well-developed mantle, a scanty coarse Hartig net and some root cells with intracellular hyphae. Scales and Peterson (1991a) synthesized typical ectomycorrhizae on *Picea mariana* and *Betula alleghaniensis* with *Wilcoxinia mikolae*, which formed ectendomycorrhizae on *Pinus banksiana* (Scales and Peterson 1991b).

## 3. Some Highlights of This Report Period

Considerable progress has been made in the characterization of ectomycorrhizae. However, decidedly more research is necessary in man-made forests, as well as in natural systems, especially in the tropics. Several additional fungal species are now proved to form ectomycorrhizae; most of the new data concentrate on Ascomycetes of the order Pezizales. A new development is the inclusion of ectomycorrhizae in the characterization of new fungal taxa, also being an advance in fungal taxonomy. Examination of strain variability inevitably calls attention to host-specifity of ectomycorrhizal fungi, e.g., *Pisolithus tinctorius*, assumed to be a broad host range fungus, might be a conglomerate of minor species with narrower host ranges. From the view point of a plant systematist, the occurrence of ectomycorrhizae in the tribe Inuleae of the Compositae is of interest. First results in nucleic acid researches and isoenzyme analysis give rise to considerable hope of these being further tools for the identification of ectomycorrhizae.

Ontogenetic and ultrastructural researches advanced twofold. Early stages of development gave some hints as to how hyphal contacts to roots are established due to recognition reactions by either partner. Secondly, ontogeny and ultrastructure of rhizomorphs could provide an insight into the physiological and ecological function of ectomycorrhizae.

Additional evidence has been found that hormones, produced by the fungi, might play a decisive role in ectomycorrhiza formation and function. The detection of concanavalin A binding sites at the hypha-root interface offer a valuable insight into the recognition phenomena of ectomycorrhizal partners and of host specifity. Protoplast formation and genetic transformation, as well as monokaryotic and reconstituted dikaryotic isolates, are also of great importance for researches of recognition phenomena and of ectomycorrhizal function.

It becomes more and more evident that function of ectomycorrhizae with respect to tree nutrition and tree growth is highly dependent on the fungal species involved. This applies for a large set of enzymatic activities, for macronutrients, micronutrients, and water relations. Of special interest is the obvious competition between isoenzymes from roots and fungi. An additional facet of tree nutrition by ectomycorrhizae was contributed by the proof of possible heterotrophic growth for carbon by ectomycorrhizal seedlings.

Ecological research has advanced in several respects. The development and persistence of ectomycorrhizae have recently been shown to be strongly influenced by biological as well as by environmental and soil conditions. Great advances have been made in studying the influence and the role of hyphal mats of ectomycorrhizal fungi on soil ecology.

Important insights are presented in the last report period with respect to co-existence between ectomycorrhizae and other fungi, regardless if parasitic, saprophytic, or even ectomycorrhizal fungi are involved, bacteria, animals, or roots of other plants.

Fertilization with lime or nitrogen is shown to influence ectomycorrhizal populations, with the number of ectomycorrhizae usually appearing to decrease. Large nitrogen applications appear decidedly unfavorable for ectomycorrhizae, the original soil type, however, has a great influence. Several pesticides (herbicides, fungicides) have been tested and they differ in their efficacy with respect to mycorrhiza formation, seedling growth, and survival under pathogen stress.

The extensive literature on pollution effects on ectomycorrhizae and tree health published in the last period is chronically inconsistent, and hardly a common denominator can be found. The acidification of soils appears to influence species populations and amount of ectomycorrhizae rather than their anatomy. Sulfur dioxide, ozone, and nitrogen may negatively influence, but not consistently, the number of ectomycorrhizae in dependence on applied concentrations as well as in dependence on tree and fungal species. Ammonium appears to be more detrimental than nitrate. Currently, it perhaps seems more probable that nitrogen emission is more harmful than ozone, $SO_2$ or acid rain to disturb the symbiotic balance in ectomycorrhizae. Of the metals thought to be toxic, lead seems to reduce the number of ectomycorrhizae; not the hyphae but the endodermis appears to act as a barrier. Contradictory results are published for aluminium with regard to if hyphae or the endodermis act as a barrier for preventing its transfer to the plant. Correlations between some kind of tree damage symptoms and number or/and frequency of ectomycorrhizae are published, only exceptionally is the vitality of ectomycorrhizae included. The abundance of differently vital ectomycorrhizae, as well as the associated microfungi, are most likely more important than the actual numbers of ectomycorrhizae. Unfortunately, most analyses of mycorrhizal systems still do not include any attempt to determine the species involved, even though a considerable amount of research indicates that different species of ectomycorrhizae can behave and function differently with respect to biological and environmental changes.

The decisive overall progress of this report period probably can be detected in the realization that there are species-specific differences in ectomycorrhizal behavior and in presenting data on their undoubted ecological and physiological importance. As a consequence, research has to be focused on the recognition of individual species and on their individual capabilities.

## References

**Abuzinadah RA,** Read DJ (1989a) New Phytol 112:55–60. **Abuzinadah RA,** Read DJ (1989b) Trees 3:17–23. **Abuzinadah RA,** Read DJ (1989c) New Phytol 112:61–68. **Adams MB,** O'Neill EG (1991) For Sci 37:5–16. **Agerer R** (ed) (1987–1991) Colour atlas of Ectomycorrhizae. 1st–5th del. Einhorn, Schwäbisch Gmünd. **Agerer R** (1990a) Z Mykol 56:155–158. **Agerer R** (1990b) Nova Hedwigia 50:1–63. **Agerer R** (1991a) Mycorrhiza 1:45–46. **Agerer R** (1991b) In: Norris JR, Read DJ, Varma AK (eds): Methods Microbiol 23:25–73. **Agerer R** (1991c) Nova Hedwigia

53:127–170. **Agerer R** (1991d) Crypt Bot 2/3:85–92. **Agerer R** (1991e) Mycorrhiza 1:21–30. **Agerer R,** Weiss M (1989) Mycologia 81:444–453. **Ahonen U,** Markkola AM, Ohtonen R et al. (1989) Aquilo Ser Bot 26:25–30. **Alexander I** (1989a) In: Stirton CH, Zarucchi JL (eds) Monogr Syst Bot Mo Bot Gard 29:607–624. **Alexander I** (1989b) In: Proctor J (ed) Spec Publ Br Ecol Soc 9:169–188. **Ali NA,** Jackson RM (1989) Mycol Res 93:182–186. **Amaranthus MP,** Li CY, Perry DA (1990) Can J For Res 20:368–371. **Anunciacao CE,** Guimaraes WV, Araujo EF et al. (1990) Biotechnol Tech 4:215–220.

**Barret V,** Lemke AP, Dixon RK (1989) Appl Microbiol Biotechnol 30:381–387. **Barret V,** Dixon RK, Lemke PA (1990) Appl Microbiol Biotechnol 33:313–316. **Bauer T,** Blechschmidt-Schneider S, Eschrich W (1991) Trees 5:36–43. **Berndt R,** Kottke I, Oberwinkler F (1990) New Phytol 115:471–482. **Blaschke H** (1990) Environ Pollut 68:409–418. **Blaschke H,** Weiss M (1990) Environ Pollut 64:255–264. **Blasius D,** Oberwinkler F (1989) Ann Sci For (Paris) 46 (Supl):758s–761s. **Bledsoe C,** Klein P, Bliss LC (1990) Can J Bot 68:1848–1856. **Borchers SL,** Perry DA (1990) Can J For Res 20:712–721. **Bougher NL,** Malajczuk N (1990) New Phytol 114:87–92. **Bougher NL,** Grove TS, Malajczuk N (1990) New Phytol 114:77–86. **Bougher NL,** Tommerup IC, Malajczuk N (1991) Mycol Res 95:683–688. **Boutekrabt A,** Pargney JC (1991) Cryptogam Mycol 12:25–46. **Boyle CD,** Hellenbrand KE (1991) Can J Bot 69:1764–1771. **Brand F** (1989) Nova Hedwigia 48:469–483. **Brand F** (1991) Libri Bot 2:1–228. **Brunner I** (1991) Trees 5:90–94. **Brunner IL,** Brunner F, Miller OK (1990) Can J Bot 68:761–767.

**Cairney JWG,** Ashford AE (1989) Aust J Plant Physiol 16:99–105. **Cairney JWG,** Ashford AE, Allaway WG (1989) New Phytol 112:495–500. **Carleton TJ,** Read DJ (1991) Can J Bot 69:778–785. **Catinus B,** Guerrier G, Strullu DG (1990) Can J For Res 20:1120–1124. **Chakravarty P,** Chatarpaul L (1990a) Can J For Res 20:245–248. **Chakravarty P,** Chatarpaul L (1990b) Pestic Sci 28:243–248. **Chakravarty P,** Hwang SF (1991) Eur J For Pathol 21:97–106. **Chakravarty P,** Peterson RL, Ellis BE (1990) Can J For Res 20:1283–1288. **Chakravarty P,** Peterson RL, Ellis BE (1991) J Phytopathol 132:207–218. **Chalot M,** Botton B, Banvoy J (1989a) Agric Ecosyst Environ 28:59–64. **Chalot M,** Dell B, Botton B et al. (1989b) Agric Ecosyst Environ 28:65–70. **Chappelka AH,** Kush JS, Runion GB et al. (1991) Environ Pollut 72:307–316. **Chleyet-Marel JC,** Bousquet N, Mousain D (1989) Agric Ecosyst Environ 28:79–84. **Chu-Chou M,** Grace LJ (1990) Soil Biol Biochem 22:959–966. **Coleman MD,** Bledsoe CS, Smit Ba (1990) New Phytol 115:275–284. **Coutts MP,** Nicoll BC (1990a) Can J For Res 20:861–868. **Coutts MP,** Nicoll BC (1990b) Can J For Res 20:1894–1899. **Cumming JR,** Weinstein LH (1990a) New Phytol 116:99–106. **Cumming JR,** Weinstein LH (1990b) Plant Soil 125:7–18. **Cuvelier J-J** (1990) Belg J Bot 123:73–91.

**Dahlberg A** (1990) Scand J For Res 5:103–112. **Dahlberg A,** Stenström E (1991) Plant Soil 136:73–86. **Danell E,** Fries N (1990) Mycotaxon 38:141–148. **Danielson RM** (1991) Agric Ecosyst Environ 35:261–281. **Danielson RM,** Visser S (1989a) Can J For Res 19:1412–1421. **Danielson RM,** Visser S (1989b) New Phytol 112:41–48. **De La Bastide PY,** Kendrick B (1990) Can J Bot 68:444–448. **Dell B,** Botton B, Martin F et al. (1989) New Phytol 111:683–692. **Dell B,** Malajczuk N, Thompson GT (1990a) New Phytol 114:635–640. **Dell B,** Malajczuk N, Grove TS et al. (1990b) New Phytol 114:449–456. **De Oliveira VL,** Garbaye J (1989) Eur J For Pathol 19:54–64. **Dighton J** (1991) Experientia 47:362–369. **Dighton J,** Mason PA, Harrison AF (1989) Agric Ecosyst Environ 28:95–99. **Dighton J,** Mason PA, Poskitt JM (1990) New Phytol 116:655–661. **Donner B,** Heyser W (1989) Forstwiss Centralbl 108:150–163. **Dosskey MG,** Lindermann RG, Boersma L (1990) New Phytol 115:269–274. **Dosskey MG,** Boersma L, Linderman RG (1991) New Phytol 117:327–334. **Duchesne LC,** Campbell SE, Koehler H et al. (1989a) Symbiosis 7:139–148. **Duchesne LC,** Ellis BE, Peterson RL (1989b) Can J Bot 67:2726–2730. **Duchesne LC,** Peterson RL, Ellis BE (1989c) New Phytol 111:693–698. **Duponnois R,** Garbaye J (1991) Ann Sci For (Paris) 48:239–252.

**Entry JA,** Donnelly PK, Cromack K (1991a) Biol Fertil Soils 11:75–78. **Entry JA,** Rose CL, Cromack K (1991b) Soil Biol Biochem 23:285–290. **Erland S,** Söderström B (1990) New Phytol

115:675–682. **Erland S,** Söderström B (1991) New Phytol 117:405–411. **Estivalet D,** Perrin R, Le Tacon F et al. (1990) For Ecol Manage 37:233–248.

**Farquhar ML,** Peterson RL (1990) New Phytol 116:107–114. **Farquhar ML,** Peterson RL (1991) Can J Bot 69:1372–1383. **Finlay RD** (1989a) New Phytol 112:185–192. **Finlay RD** (1989b) Agric Ecosyst Environ 28:127–132. **Finlay RD,** Ek H, Odham G et al. (1989) Agric Ecosyst Environ 28:133–138.

**Garbaye J,** Bowen GD (1989) New Phytol 112:383–388. **Gardes M,** White TJ, Forth JA et al. (1991) Can J Bot 69:180–190. **Gay G** (1990) Can J Bot 86:1265–1270. **Gehring CA,** Whitham TG (1991) Nature (Lond) 353:556–557. **Gianinazzi-Pearson V,** Gianinazzi S (1989) Genome 31:336–341. **Gibson F,** Deacon JW (1990) Mycol Res 94:166–172. **Glenn MG,** Wagner WS, Webb SL (1991) Can J For Res 21:741–749. **Gogala N** (1991) Experentia 47:331–340. **Griffiths RP,** Caldwell BA, Cromack K et al. (1990) Can J For Res 20:211–218. **Griffiths RP,** Ingham ER, Caldwell BA et al. (1991) Biol Fertil Soil 11:196–202. **Guehl JM,** Garbaye J (1990) Ann Sci For (Paris) 47:551–564. **Guehl JM,** Mousain D, Falconnet G et al. (1990) Ann Sci For (Paris) 47:91–100.

**Harley JL** (1989) Mycol Res 92:129–139. **Harley JL** (1991) In: Norris JR, Read DJ, Varma AK (eds) Methods Microbiol 23:1–23. **Haug I** (1989a) Agric Ecosyst Environ 28:167–170. **Haug I** (1989b) New Phytol 111:203–208. **Haug I,** Feger KH (1991) Water Air Soil Pollut 54:453–467. **Haug I,** Weber R, Oberwinkler F et al. (1991) New Phytol 117:25–36. **Heyser W,** Donner B (1989) Agric Ecosyst Environ 28:175–179. **Hilbert JL,** Martin F (1989) Agric Ecosyst Environ 28:181–184. **Hofmann H-P** (1989) Nova Hedwigia 48:455–468. **Högberg P** (1989) For Ecol Manage 28:7–18. **Högberg P** (1990) New Phytol 115:483–486. **Holopainen T** (1989) Agric Ecosyst Environ 28:185–189. **Horan DP,** Chilvers GA (1990) New Phytol 116:297–302.

**Ingleby K,** Mason PA, Last FT, Fleming LV (1990) Inst Terrestr Ecol Res Publ 5, Edinburgh. **Ivory MH,** Pearce RB (1991) Mycol Res 95:250–253.

**Jacobs PF,** Peterson RL, Massicotte HB (1989) Scanning Microsc 3:249–256. **Jansen AE** (1991) Agric Ecosyst Environ 35:191–208. **Jansen AE,** de Vries FW (1989) Agric Ecosyst Environ 28:197–200. **Jansen AE,** Dighton J (1990) Air Pollut Res Rep 30 Comm Europ Comm Brussels. **Jentschke G,** Godbold DL, Hüttermann A (1991a) Physiol Plant 81:408–416. **Jentschke G,** Schlegel H, Godbold DL (1991b) Physiol Plant 82:266–270. **Jentschke G,** Fritz E, Godbold DL (1991c) Physiol Plant 81:417–422. **Jones MD,** Durall DM, Tinker PB (1990) New Phytol 115:259–267.

**Kammerbauer H,** Agerer R, Sandermann H (1989) Trees 3:78–84. **Kasuya MCM,** Muchovej RMC (1989) S Afr For J 151:32–35. **Kasuya MC,** Muchovej RMC, Muchovej JJ (1990) Plant Soil 124:73–78. **Keller G** (1989) Centralbl Gesamte Forstwes 106:129–147. **Kohn LM,** Stasovski E (1990) Mycologia 82:23–35. **Kope HH,** Fortin JA (1989) New Phytol 113:57–64. **Kottke I,** Oberwinkler F (1989) Ann Sci For (Paris) 46(Suppl):737s–740s. **Kottke I,** Oberwinkler F (1990) Trees 4:41–48. **Krizaj D** (1989) Spec Sci Techn 12:209–211. **Kropacek K,** Kristinova M, Chmelikova E et al. (1989) Agric Ecosyst Environ 28:271–277. **Kropp BR,** Langlois C-G (1990) Can J For Res 20:438–451. **Kubitzki K** (1989) Amazoniana 11:1–12.

**Lahamedi MS,** Fortin JA (1991) Can J Bot 69:1927–1934. **Lahamedi MS,** Fortin JA, Kope HH et al. (1990) New Phytol 115:689–698. **Lapeyrie F** (1990) Ann Sci For 21:579–589. **Last FT,** Wilson J, Mason P (1989) Agric Ecosyst Environ 28:293–298. **Lehto TH** (1989) Agric Ecosyst Environ 28:299–303. **Lei J,** Lapeyrie F, Malajczuk N et al. (1990) New Phytol 116:115–122. **Lei J,** Wong KKY, Piche Y (1991) Mycol Res 95:357–363. **Leyval C,** Berthelin J (1991) Soil Sci Soc Am J 55:1009–1016.

**MacFall JS,** Iyer J, Slack S et al. (1989) Agric Ecosyst Env 28:321–324. **Malajczuk N,** Lapeyrie F, Garbaye J (1990) New Phytol 114:627–632. **Mamoun M,** Olivier JM (1989) Agronomie (Paris) 9:345–352. **Marks GC,** Becker SL (1990) Aust J Bot 38:341–350. **Marx DH** (1990) For Sci 36:224–245. **Massicotte HB,** Peterson RL, Melville LH (1989a) Can J Bot 67:191–200. **Massicotte HB,** Peterson RL, Melville LH (1989b) Am J Bot 76:1654–1667. **Massicotte HB,** Ackerley CA, Peterson RL (1989c) Can J Bot 67:201–210. **Massicotte HB,** Peterson RL, Ackerley CA et al. (1990) Can J Bot 68:579–593. **McAfee BJ,** Fortin JA (1989) Plant Soil 116:9–17. **Meier S,** Robarge WP, Bruck RI et al. (1989) Environ Pollut 59:315–324. **Meier S,** Grand LF, Schoenberger MM et al. (1990) Environ Pollut 64:11–28. **Metzler B,** Oberwinkler F (1989) Agric Ecosyst Environ 28:339–342. **Miller SL,** Durall DM, Rygiewicz PT (1989) Tree Physiol 5:239–250. **Miller SL,** Koo CD, Molina R (1991) Can J Bot 69:516–531. **Mitchell RJ,** Garrett HE, Cox GS et al. (1990) J Plant Nutr 13:1555–1574. **Moore AEP,** Massicotte HB, Peterson RL (1989) New Phytol 112:193–204. **Müller W,** Agerer R (1990) Nova Hedwigia 51:381–410. **Münzenberger B,** Heilemann J, Strack D et al. (1990) Planta 182:142–148.

**Newton AC** (1991) New Phytol 117:53–60. **Newton AC,** Pigott CD (1991) New Phytol 117:45–52. **Niederer M,** Pankow W, Wiemken A (1989) Eur J For Pathol 19:14–20. **Norris JR,** Read DJ, Varma AK (eds) (1991) Methods Microbiol 23:1–480.

**Pargney J-C** (1991) Cryptogam Mycol 12:47–62. **Pargney J-C,** Leduc J-P (1990) Bull Soc Bot Fr Lett Bot 137:21–34. **Perrin R,** Estivalet D (1989) Agric Ecosyst Environ 28:381–387. **Ponge JF** (1990) Eur J For Pathol 20:290–303.

**Rejsek K** (1991) Soil Biol Biochem 23:667–671. **Richter DL,** Bruhn JN (1989) Symbiosis 7:211–228. **Ritter T,** Weber G, Kottke I et al. (1989a) BIUZ 19:9–15. **Ritter T,** Kottke I, Oberwinkler F (1989b) Agric Ecosyst Environ 28:415–420. **Rousseau JVD,** Reid CPP (1991) New Phytol 117:319–326. **Rupp LA,** Mudge KW, Negm FB (1989) Can J Bot 67:477–482.

**Sarjala T** (1991) Physiol Plant 81:89–94. **Sauter M,** Hager A (1989) Planta (Berl) 179:61–66. **Scales PF,** Peterson RL (1991a) Can J Bot 69:2149–2157. **Scales PF,** Peterson RL (1991b) Can J Bot 69:2135–2148. **Scheromm P,** Plassard C, Salsac L (1990) New Phytol 114:93–98. **Schisler DA,** Linderman RG (1989) Soil Biol Biochem 21:397–402. **Sen R** (1990) New Phytol 114:617–626. **Sidhu SS,** Chakravarty P (1990) Eur J For Path 20:77–94. **Simmons GL,** Kelly JM (1989) Water Air Soil Pollut 44:159–172. **Stein A,** Fortin A, Vallee G (1990) Can J Bot 68:468–470. **Stenström E** (1989) Agric Ecosyst Environ 28:479–482. **Stenström E** (1990) Can J For Res 20:914–918. **Stenström E** (1991) Plant Soil 131:247–250. **Stenström E,** Ek M, Unestam T (1990) Can J For Res 20:1796–1803. **Stumpp J** (1990) Zool Beitr N F 33:345–432. **Summerbell RC** (1989) Can J Bot 67:1085–1095.

**Taylor AFS,** Alexander IJ (1989a) Agric Ecosyst Environ 28:493–496. **Taylor AFS,** Alexander IJ (1989b) Mycol Res 92:103–107. **Taylor AFS,** Alexander IJ (1991) Mycol Res 95:381–384. **Termorshuizen AJ,** Ket PC (1989) Agric Ecosyst Environ 28:497–501. **Termorshuizen AJ,** Schaffers AP (1989) Agric Ecosyst Environ 28:509–512. **Termorshuizen AJ,** van der Eerden LJ, Dueck TA (1989) Agric Ecosyst Environ 28:513–518. **Theodorou C,** Reddell P (1991) New Phytol 118:279–288. **Thoen D,** Ducousso M (1989a) For Trop 221:45–63. **Thoen D,** Ducousso M (1989b) Agric Ecosyst Environ 28:519–523. **Thomson J,** Melville LH, Peterson RL (1989) Am J Bot 76:632–636. **Thomson J,** Matthes-Sears I, Peterson RL (1990) Can J For Res 20:1739–1745. **Timbal J,** Gelpe J, Garbaye J (1990) Ann Sci For (Paris) 47:643–649. **Tommerup IC,** Bougher NL Malajczuk N (1991) Mycol Res 95:689–698. **Tonkin CM,** Malajcuk N, McComb JA (1989) New Phytol 111:209–214. **Treu R** (1990) Bibl Mycol 134:1–196.

**Unestam T** (1991) Mycorrhiza 1:13–20. **Unestam T,** Stenström E (1989) Scand J For Res 4:51–58. **Unestam T,** Chakravarty P, Damm E (1989) Agric Ecosyst Environ 28:535–538.

**Van Dijk HFG,** De Louw MHJ, Roelofs JGM et al. (1990) Environ Pollut 63:41–60. **Väre H** (1989a) Agric Ecosyst Environ 28:539–545. **Väre H** (1989b) Aquilo Ser Bot 26:19–24. **Väre H** (1989c) Can J For Res 19:341–346. **Vezina LP,** Margolis HA, McAfee BJ et al. (1989) Physiol Plant 75:55–62. **Villeneuve N,** Le Tacon F, Bourchard D (1991) Plant Soil 135:95–108. **Vincent J-M** (1989) Eur J For Pathol 19:167–177. **Vogt KA,** Publicover DA, Vogt DJ (1991) Agric Ecosyst Environ 35:171–190. **Von Alten H,** Roßbach B (1989) Agric Ecosyst Environ 28:13–19.

**Wagner F,** Gay G, Debaud JC (1989) New Phytol 113:259–264. **Walker RF,** West DC, McLaughlin SB et al. (1989) For Sci 35:569–581. **Walander H,** Nylund J-E (1989) Agric Ecosyst Environ 28:547–552. **Warcup JH** (1990a) Muelleria 7:179–187. **Warcup JH** (1990b) Mycol Res 94:1065–1069. **Warcup JH** (1990c) Mycol Res 94:173–178. **Warcup JH,** Talbot PHB (1989) Mycol Res 92:95–100. **Weiss M** (1990) Nova Hedwigia 50:361–393. **Weiss M** (1991) Mycotaxon 40:53–78. **Wilcox HE** (1983) Annu Rev Phytopathol 21:221–242. **Wilkins DA** (1991) Agric Ecosyst Environ 35:245–260. **Wöllmer H,** Kottke I (1990) Environ Pollut 68:383–408. **Wong KKY,** Piche Y, Montpetit D et al. (1989) Can J Bot 67:1717–1726. **Wong KKY,** Montpetit D, Piche Y et al. (1990a) New Phytol 116:669–680. **Wong KKY,** Piche Y, Fortin JA (1990b) Mycol Res 94:876–884. **Wullschleger SD,** Reid CPP (1990) New Phytol 116:681–688.

**Zel J,** Dermastia M, Gogala N (1989) Biol Vestn 37:93–100.

Professor Dr. Reinhard Agerer
Institut für Systematische Botanik
der Universität München
Menzingerstraße 67
W-8000 München, Germany

## II b. Symbioses: Mycorrhizae

By Fritz Schönbeck and Ingo Raschen

### 1. VA-Mycorrhizae

a) Taxonomy of Endophytes

All criteria currently used to identify and classify VA-mycorrhizal fungi were critically evaluated by Morton (1988), revising the classification of the VAM fungi, proposing one new order, Glomales, two new suborders, Glomineae and Gigasporineae, and two new families, Acaulosporaceae and Gigasporaceae, with emendation of Glomaceae, and assessing the roles of species and clones in macro- and microevolutionary processes (Morton 1990; Morton and Benny 1990). Jabaji-Hare (1988) suggested the lipid and fatty acid profiles of extra and intramatrical structures of the endophytes as one means for identification, and that VA-mycorrhizal fungi had closer links with the protoctistan fungi than with Zygomycetes. The use of isozyme banding patterns after Polyacrylamide Gel Electrophoresis, in addition to morphological characters of chlamydospores, was suggested for identification, classification, and quantification of spore-cluster forming VAM fungi (Rosendahl 1989; Rosendahl et al. 1989) and Stubblefield et al. (1987) stated that they might have found the first *Sclerocystis*-like fungus in Paleozoic or Mesozoic sediments ever reported.

b) Physiology and Effects on Plants

General studies with particular reference to effects on host plants were reviewed by Amjee (1989). The so-called P-effects have been the subject of investigation to many authors, describing general positive effects of VA-mycorrhiza on host plants, such as enhanced growth on degraded alkaline soils in India (Behl 1990) or improved re-establishment after VA mycorrhizal inoculation in soils stockpiled during mineral and sand mining (Jasper et al. 1989). However, investigations on the influence of VAM and P-fertilization on cotton revealed that although the plants responded positively to both fertilization and VAM inoculation, the total length of roots of mycorrhizal fertilized plants was decreased when compared to those of nonmycorrhizal fertilized plants (Price et al. 1989). Also, Graham and Syvertsen (1989) found that VA-mycorrhiza did not alleviate salinity stress of citrus plants, but increased Cl contents of the plant material, leading to enhanced

Progress in Botany, Vol. 54

sensitivities of mycorrhizal plants. Positive effects observed in apple seedlings after inoculation with *Glomus macrocarpum* were shown to be due not only to an improved P-status of the plants, but also to enhanced levels of Zn and Cu (Gnekow and Marschner 1989). Comparing the effects of *G. macrocarpum* and *G. mosseae* on apple seedlings, both species initially reduced root and shoot fresh weight, leading to an increase only after 28 days. *G. mossae* promoted growth more effectively, spread through root-free soil faster than the contrahent used, and produced more extramatrical mycelium (Miller et al. 1989). In vitro-propagated *Pistacia integerrima* plants and grapevine responded well to inoculation with *G. caledonium* and *G. occultum* in nonamended soils only, although mycorrhiza developed to the same extend in the substrates amended with N, P, K, and appropriate amounts of micronutrients or sand and soil mixtures, respectively (Mazzitelli and Schubert 1990). Several distinct physiological and morphological effects of *G. fasciculatum* on *Plantago major* which were separable from those mediated by the increase of P concentrations were described by Baas and Kupier (1989), Baas et al. (1989) and Baas (1990). Mayer et al. (1989) reported on the general importance of VA mycorrhiza for the cultivation of vegetable seedlings presenting some facts on the effects of VAM under practical conditions. The dependence of *Apium graveolens, Allium cepa* and *Capsicum annuum* on mycorrhizal symbiosis was demonstrated, suppressing VA mycorrhizal infections by previous soil fumigation with methyl bromide. In contrast to the species mentioned above, growth of *Cucumis melo* was improved after fumigation, leading to the conclusion that crop VAM dependence and efficacy of VAM biotypes must be considered as essential components for P-fertilization recommendations on fumigated P-sorbing soils (Krikun et al. 1990). Light intensity had significant effects on the development and vitality of mycorrhiza and on the efficiency of phosphorus uptake of mycorrhizal onions inoculated with *G. mosseae* under low light and phosphorus conditions. Rates of infections were determined microscopically after staining with trypan blue for total infection rates and with nitroblue tetrazolium chloride for succinate dehydrogenase activity to determine the active fraction (Smith and Gianinazzi-Pearson 1990). The P concentration in soil and thus the rate of P uptake, rather than the total concentration of P within the roots, was considered the main factor responsible for the lateral branching and the morphology of roots of *Allium porrum* L. inoculated with *G. mosseae*. The mycorrhizal fungus increased the total root length of the plants compared to nonmycorrhizal controls (Amjee et al. 1989a). Aubergines exhibited a pronounced genotype-specific response to inoculation with VAM fungi, but were generally more responsive to *G. fasciculatum* than other *Glomus* spp. (Indi et al. 1990). Murmotsev et al. (1990) found that *G. mosseae* strains 7 and 8, isolated from strawberries, improved yields and P contents of several different plants in pot experiments and the P and N contents of clover when the plants were also inoculated with an effective strain of *Rhizobium*. Similar studies were also presented for mung bean (Mathew and Johri 1989), chickpea, pigeon pea, and various other plants. Hung et al. (1990) reported that there were interactive effects between biomass crops, VA-mycorrhizal fungi, and phosphorus fertilization, the effects of the indigenous VAM population ranging from growth enhancement to growth depression and therefore providing an opportunity for inoculation. Differences of VA mycorrhizal effects observed in *Sorghum bicolor* were reported to be temperature-dependent (Raju et al. 1990), cultivar-, (Mohan and Tilak 1990), and strain-specific for several strains of *Gigaspora calospora*. Inoculation of several cereal crops with the VA mycorrhizal fungi *Glomus macrocarpum* and *G. manihotis* resulted in more efficient utilization of rock phosphate at different soil temperatures and pH (Fabig et al. 1989a) and applying the latter endophyte mixed with N, P, K fertilizer resulted in a more balanced distribution of these elements within cassava plants, whereas the endophyte *Entrophospora colombiana* was found not to tolerate mixing with fertilizer salts at all (Sieverding and Toro 1990). Knight et al. (1989) found that $CO_2$ level and P availability in the vicinity of VA mycorrhizal root systems were significantly higher than in nonmycorrhizal systems. Furthermore, the P status of nonmycorrhizal plants grown under imposed $CO_2$ regimes could be raised to that of the mycorrhizal controls, indicating that $CO_2$ accumulation improves the availability of P. Studies on the mediation of P uptake by VA-mycorrhiza using Fe-chelators led to the conclusion that siderophore production was a feasible mechanism involved (Jayachandran et al. 1989).

VA mycorrhizal lettuce accumulated applied heavy metals in the root system only, suggesting that VA mycorrhizal infection may enhance a root/shoot barrier and thus play a major role in resistance towards these metals (Dehn and Schüepp 1990). Studies on the effects of heavy metals on VA-mycorrhizal maize and soybeans inoculated with *Bradyrhizobium japonicum* revealed that biomass in mycorrhizal plants was reduced to a lesser extent than in nonmycorrhizal controls, even though the application of heavy metals resulted in reduced infection rates. However, after VAM inoculation of *Agrostis capillaris* L., the shoot/root ratio of heavy metals increased significantly more in nontolerant cultivars than in tolerant ones (Griffioen and Ernst 1990). VA-mycorrhizal plants suffered less from Al toxicity than the nonmycorrhizal controls and Al levels above 90 μM in soil were shown to inhibit chlamydospore germination, VAM infection and development and thus plant growth (Lambais and Cardoso 1989). Manganese uptake of red clover was depressed by VA mycorrhiza, leading to a reduction of the Mn content by 55 and 80%, in shoots and roots respectively (Arines and Vilariño 1989; Arines et al. 1990).

While the application of stable manure (Harinikumar and Bagyaraj 1989; Brechelt 1990), green manure, N fertilizer (Baltruschat and Dehne 1989), and the long-term application of nonindustrial sewage had no significant effects on the development of VA mycorrhiza. Heavy metals accumulated in soils after sewage sludge application had pronounced negative effects on the development of VAM (Koomen and McGrath 1990). High concentrations of Mn, Fe, or total salt in fertilizer solutions inhibited the in vitro germination of chlamydospores of *Glomus mosseae*. The effects of various pesticides on crop and VAM development were discussed by Jalali (1990), giving a general overview. While the effects of field application rates of fungicides on three *Glomus* spp. associated with winter wheat varied with the fungicides used, herbicides showed no significant effects on VAM fungi, except for Harrier, which had detrimental effects on *G. geosporum* (Dodd and Jeffries 1989a, b). Mycorrhizal development in plants emerging from commercial fungicide-treated seeds of several crops was found not to differ from the nontreated. However, when the same fungicides were applied as soil drenches VA mycorrhizal development was severely inhibited (Spokes et al. 1989). While the fungicidal properties of Paclobutrazol (Michelini et al. 1989) and Dichlone, at the rates applied to control the alga *Nostoc*, were shown to have no significant effects on VAM development (Gupta 1990) and certain xenobiotics may be used to modify the physiological state of root systems of soybean and thus the plant-VAM interaction and development (Morandi 1989), Bavistin [carbendazim], Basalin, Captafol, and Thiram reduced the rate of colonization of *Glomus* spp. associated with pigeonpea (Singh et al. 1990a), and Captan those associated with in vitro-propagated strawberries. Furthermore, the development of VAM associated with citrus seedlings was significantly reduced by the fungicides Manozeb, Sulfex, and Vitavax [carboxin] (Singh et al. 1990a,b).

Soil disturbance was found to reduce VAM development and correspondingly the phosphate and Zn uptake of mycorrhizal plants (Evans and Miller 1990). The effects of soil temperature on VAM development and plant responses were monitored by Borges and Chaney (1989), finding that the temperature optimum for VAM and root development was at 25 °C and that the sugar contents of roots increases with soil temperature, whereas the starch content was adversely affected. Mycorrhizal infection of barley inoculated with indigenous VAM from a Canadian prairie soil was not affected by temperatures between 7 and 20 °C, although development beyond the site of infection was greatest at 12 °C (Volkmar and Woodbury 1989).

Within roots of crop plants between rows of trees, colonization and sporulation of VAM fungi increases with decreasing distance to the trees, except for *Eucalyptus tereticornis*, where VAM fungi were inhibited. Characteristics of inhibitory compounds from nonhost plants were discussed by El-Atrach et al. (1989). The highest mycorrhizal development was supported by *Leucaena leucocephala* (Rajashekhara et al. 1989). Spore production and growth enhancement per unit infected root length depended on the host species, and the development of mycorrhiza was governed and regulated by host roots (Koide and Li 1990). In vitro studies led to the conclusion that suspension-cultured plant cells stimulate hyphal growth of VAM fungi by accumulating and leaking active metabolites into the medium, which nonhost plants were found to be lacking (Paula and Siqueira 1990).

Applying the $^{14}CO_2$ method, Jakobsen and Rosendahl (1990) studied the carbon flow into soil and hyphae on the cucumber/*Glomus fasciculatum* mycorrhiza and $^{11}CO_2$ was used for demonstrating that in the VA mycorrhizal C-4 grass *Panicum coloratum* storage of assimilates in leafs was increased by 45% and doubled in the phloem (Wang et al. 1989). Mohan and Tilak (1990) reported a change in the distribution of carbohydrates mediated by VAM development to the benefit of the plant roots in sorghum. It was also concluded that reduced mycorrhizal infection upon increased P fertilization was due to induced morphological changes rendering roots resistant to VAM fungi, rather than decreased rates of exsudation (Amjee et al. 1990). The appearance of a number of new soluble proteins in plant roots after infection with VA mycorrhizal fungi was reported by Dumas et al. (1990) and the role of hydrolytic enzymes (e.g., chitinases) during the process of VA mycorrhizal infection was discussed by Garcia-Romera et al. (1990). On endophyte infection, chitinase activity was initially found to be enhanced, but decreased with the physiological age of VAM and eventually reached a level below that of the nonmycorrhizal controls. In these and during later phases of development, chitinase activity was restricted to vacuoles and extracellular spaces, suggesting that endophytes initially trigger a defence reaction in the host (Spanu et al. 1989, 1990). Succinate dehydrogenase (SDH) activity was used to detect the metabolic activity of VAM fungi in plant roots, but the specific activity per unit of infected root was not indicative of the efficiency of the fungus in terms of promoting plant growth (Vierheilig and Ocampo 1990). Digestion of host roots was always correlated with plasmolysis of vesicles and resulted in the loss of SDH activity (McGee and Smith 1990). Independently of root exsudations, cytokinin export from VA mycorrhizal roots was found to be higher than that of nonmycorrhizal controls, leading to elevated levels of cytokinins in the shoot and correlating with enhanced P uptake and growth (Dixon 1990).

A cascade of events is triggered in plant roots upon contact with appropriate symbionts, the underlying genetic mechanisms of which were investigated by Gianinazzi-Pearson and Gianinazzi (1989), using interspecific grafting and mycorrhiza-deficient plants and fungi. Chromosomal inheritance of VAM efficiency was observed in spring wheat, and differences in the behavior of the $F_1$ after reciprocal crossing indicated that there was also a cytoplasmic effect (Manske 1990). Having studied 101 diverse genotypes of *Vigna unguiculata* for the ability to form mycorrhiza, high genotypic and phenotypic coefficients of variation were estimated for the trait and heritability was moderate (Mercy et al. 1990). Nuclei of VAM roots showed hypotrophy, and chromatin was less condensed than that in nuclei of nonmycorrhizal controls. These changes could not be attributed to increasing ploidy (Berta et al. 1990). Chlamydospores of several genera were shown to comprise up to several thousand nuclei and during in vitro development there was shown to be no nuclear DNA synthesis (Burggraaf and Beringer 1989). Bécard and Piché (1989) identified two distinct mechanisms of hyphal growth of *Gigaspora magarita* in such a system, one requiring the presence of the root but being spore-dependent, the other being exclusively root-dependent. In further studies, $CO_2$ was shown to play a major role in the induction of hyphal growth, as well as other root exsudates (Bécard et al. 1990).

### c) Ecology

The physiology of VA mycorrhiza and relevance of the ecological importance were reviewed by Harley (1989). Plants in temperate ecosystems probably receive relatively little benefit from VAM and most of them derive benefits only at times when their demand for P is much greater than the capacity of their root systems to supply it (Fitter 1990). Allen and Allen (1990) enlarged on the underlying mechanisms possibly involved in the mediation of such advantages. The VA mycorrhizal influence on hosts in an early successional plant community was demonstrated by Gange et al. (1990). The relationship between mycorrhizal dependence and competitive ability of two prairie grasses was shown by Hetrick et al. (1989) and Lin and Hao (1989) reported differences in mycorrhizal dependencies for 15 plant species. Differential responses to VA mycorrhiza were observed for $C_4$ (warm season) and $C_3$ (cold season) grasses, displaying profoundly different strategies for nutrient acquisition (Hetrick et al. 1990).

Azygospores of *Gigaspora magarita* were shown not to be required for the colonization of *Paspalum notatum* (Ryan et al. 1987), although these structures are of importance for the propagation of VAM fungi, as shown by Maser et al. (1988), who identified viable chlamydospores from 21% of a variety of mammals and by Lakshman and Raghavendra (1990), who outlined the importance of invertebrates, especially earthworms, for spore dispersal. Long-range dispersal of VAM fungi through trophic linkages between primary and secondary consumers were reported to be most likely, and it was suggested that direct interactions between major groups of soil biota had influenced the feeding preferences and activity of animal consumers and/or vectors in both ecological and evolutionary time. Soil disturbance, high chemical input, and the reduction of the diversity of plant species were shown to be negatively correlated with the diversity and density of VAM fungi in soil. Consequently, it was suggested to integrate the maintenance of VAM fungi into management strategies for the development of sustainable agroecosystems (Rabatin and Stinner 1990).

Among several aspects, the effect of VA mycorrhiza on soil structure has been investigated, and Miller and Jastrow (1990) found that fungal hyphae played a role more important than fine roots in soil aggregation and that the geometric mean diameter of water-stable aggregates depended on the composition of the plant, and thus the fungal community. VA and ericoid mycorrhiza were suggested to be considered in sand dune management for better sand binding (Logan et al. 1989), and the nutritional status of eroded soils was found to hinder the establishment of mycorrhiza (Aziz and Habte 1989). Stone and Buttery (1989), however, could not find any such correlations between VAM development and aggregate stability in heavy clay soils.

Soils of salt marshes were generally found to be of low inoculum potential as determined by MPN counts. Plants with VAM were those adapted to survive in saline soils, such as *Acacia nilotica*, *Prosopis juliflora*, and *Dalbergia sissoo*. VAM colonization decreased with increasing probability of flooding with salt water and in those areas most frequently flooded, the majority of plants was nonmycorrhizal (Sengupta and Chaudhuri 1990; van Duin et al. 1990).

VAM fungi have been shown to be the major factor facilitating soil fertility of arid and desert soils. On neem trees (*Azadirachta indica*), VAM were shown to extend to a depth of 2.5 m (Kiran et al. 1989). Sieverding (1991) described the diversity of VAM fungi and plant interactions in arid soils and the need to introduce VAM fungi to disturbed tropical soils and agroecosystems. *Cucumis sativus* was found to be an appropriate bait plant for a wide variety of different VAM fungi. Associations of vegetation with a high degree of cover showed higher propagule densities than degraded bush land and cultivated areas. Differences in isoenzyme patterns from root systems of the bait plants indicated that different plant species, although growing together, may not share VAM endophytes (Michelsen and Rosendahl 1990; Rosendahl et al. 1990). That VAM fungi in the field may exhibit a high degree of ecological specificity was also found by McGonigle and Fitter (1990) for the *Holcus lanatus* – *Glomus tenue* association on meadows in N. Yorkshire, England. The infectivities of several soils in Spain and France were estimated using lucerne and leek, respectively, in a bioassay (Diaz and Honrubia 1990; Duvert et al. 1990). Seasonal variations in VAM populations were studied in sand dunes, leading to the conclusion that there were species-specific seasons for sporulation and antagonistic interactions between species (Gemma et al. 1989). Land et al. (1990) found that the species compositions of VAM populations were soil type-specific and that there were seasonal changes in VAM populations in cereal crops of north Germany. The same was reported for autumn-sown cereals in England by Dodd and Jeffries (1989c), explaining some of the variation observed as results of accumulated thermal time and for VAM in a melic-beech forest (Mayr and Godoy 1990), discussing the results in relation to climate and life cycles of the species. Newman and Eason (1989) reported on experiments suggesting that mycorrhizal links between living and dying roots may contribute to nutrient cycling and that there was major difference in the nutrient cycling process between perennial and annual crops.

When nonhosts were inoculated with VAM fungi, the result was either reduced vigor or death of the host. Mapping the infection process on the roots of *Salsola kali*, it was shown that VAM fungi initially invaded the roots, forming arbuscules and peletons, before the root segments became lignified, brown, and eventually died off and the fungus disappeared. In *Arbutus unedo* L. VAM infection resulted in plant death after 2 months. These results could form the basis for initiating studies on the genetics of mutualistic versus parasitic symbiosis (Allen et al. 1989; Giovannetti et al. 1990). *Podotheca angustifolia*, an annual composite from Australia, was found to host both ecto- and endomycorrhiza, the two competing for root space. The same findings were also reported for cotton plants in USA, where the ectomycorrhiza seemed to dominate with age (Lodge and Wentworth 1990) and several *Lobelia* spp. were dependent not only on ecto- and endomycorrhiza, but also on the functioning of the ectomycorrhiza with an accompanying plant. Seedlings were shown not to survive unless infection by an ectomycorrhizal symbiont commenced, already actively involved in another mycorrhiza, preferably with *Malaleuca uncinata* (Warcup 1988). Close associations were also shown to exist between VAM fungi and saprophytes. *Glomus macrocarpum* spores isolated from a field soil were reported to be associated with a diversity of chitin decomposing actinomycetes (Ames et al. 1989) and *Azospirillum* spp. were isolated from surface-sterilized spores of *Glomus fasciculatum, G. intraradices, G. scientillans, G. mosseae, Gigaspora gilmorei*, and *Endogone dusii* (Tilak et al. 1989). Studying the competition of different VAM fungi in the same root system, different saprophytic competitive abilities, also related to the soil type, always led to the exclusion of at least one of the endophytes introduced (Sainz et al. 1990; Sen et al. 1990).

Rhizosphere microorganisms may influence VAM development, activity, and sporulation. While the development and mycorrhizal effects of *Glomus etunicatum* on big bluestem were significantly depressed in nonsterile soil, *Gigaspora rosea* was not affected. The mechanism of the suppression observed is still unclear (Wilson et al. 1989).

## 2. Ericales-Mycorrhizae

A general account, particularly on the ericoid mycorrhiza, was given by Straker (1989) and by Leake et al. (1990), discussing the ecological implications of rhizosphere de-toxification and conferred competitive advantages. Releasing an extracellular protease the endophyte *Hymenoscyphus ericae* was shown to be capable of using proteins as sole N-source and to release amino acids into the medium when grown on substrates amended with glucose, even in the presence of phenolic compounds (Leake and Read 1990a). The results, as well as chitin activities found in several ectomycorrhizal fungi, were discussed in relation to the physiological ecology of these fungi and appropriate host plants (Leake and Read 1990b). The importance of acid phosphatase for the nutrition of ericaceous plants was discussed by Shaw and Read (1989), which were active at concentrations of $Al^{3+}$ and $Fe^{2+}$ above those normally found in soil solutions and associated almost exclusively with fungal walls and septa in living hyphae, whilst present throughout the disorganized cytoplasm in senescent hyphae.

## 3. Orchidaceae-Mycorrhizae

Symbiotic capacities and effects of mycorrhiza on seed germination and early development were investigated by Smreciu and Currah (1989) and Masuhara and Katsuya (1989), who found a new mycorrhizal fungus for germinating *Dactylorhiza* spp. In compatible host/fungus combinations in vitro growth of juvenile orchids was enhanced at low nitrogen concentrations only (Dijk 1990). Salmia (1989) investigated the features of endomycorrhizal infections of chlorophyll-free and green forms of *Epipactis helleborine. Neottia nidus-avis* mycorrhiza were ultrastructurally investigated by Barmicheva (1990) and Snetselaar and Whitney (1990) detected fungal calcium oxalate in mycorrhiza of *Monotropa uniflora.* Seasonal changes in mycorrhiza were reported for *Bletilla striata* (Masuhara et al. (1988), *Platanthera* and *Coeloglossum* spp. (Currah et al. 1990) and the radial growth rate of mycorrhizal fungi isolated from 43 species of northern orchids were compared by Lindén (1988), while Wilkinson et al. (1989) studied the interactions of soil bacteria, mycorrhiza and orchid seeds in relation to germination.

## References

**Allen EB,** Allen MF (1990) In: Gace JB, Tilman D (eds) Perspectives on Plant Competition. Academic Press, San Diego, CA. **Allen MF,** Allen EB, Friese CF (1989) New Phytol 111(1):45–49. **Ames RN,** Mihra KL, Bayne HG (1989) New Phytol 111(1):67–71. **Amjee F** (1989) Mycologist 3(4):176–180. **Amjee F,** Tinker PB, Stribley TB (1989a) Plant Soil 119(2):334–336. **Amjee F,** Tinker PB, Stribley TB (1989b) New Phytol 111(3):435–446. **Amjee F,** Stribley TB, Tinker PB (1990) Plant Soil 124(2):195–198. **Arines J,** Vilariño A (1989) Biol Fertil Soils 8(4):293–297. **Arines J,** Vilariño A, Sainz M (1990) Agric Ecosyst Environ 29(1–4):173–177. **Aziz T,** Habte M (1989) J Plant Nutr 12(7):859–869.

**Baas R** (1990) Agric Ecosyst Environ 29(1–4):17–21. **Baas R,** Kupier D (1989) Physiol Plant 76(2):211–215. **Baas R,** van der Werf A, Lambers H (1989) Plant Physiol 91(1):227–232. **Baltruschat H,** Dehne HW (1989) Plant Soil 113(2):151–156. **Barmicheva KM** (1990) Agric Ecosyst Environ 29(1–4):23–27. **Bécard G,** Piché Y (1989) New Phytol 112(1):77–83. **Bécard G,** Piché Y, Fortin JA (1990) Agric Ecosyst Environ 29(1–4):29–33. **Behl HM** (1990) Trends in mycorrhizal research. Proc Natl Conf Mycorrhiza. Haryana Agric Univ, Hisar, India, Feb 14–16. **Berta G,** Sgorbati S, Soler V, Fusconi A, Trotta A, Citterio A, Bottone MG, Sparvoli E, Scannerini S (1990) New Phytol 114(2):199–205. **Blal B,** Morel C, Gianinazzi-Pearson V, Fardeau JC, Gianinazzi S (1990) Biol Fertil Soils 9(1):43–48. **Bonfante-Fasolo P,** Faccio A, Perotto S, Schubert A (1990) Mycol Res 94(2):157–165. **Borges RG,** Chaney WR (1989) New Phytol 112(3):411–417. **Brechelt A** (1990) Agric Ecosyst Environ 29(1–4):55–58. **Burggraaf AJP,** Beringer JE (1989) New Phytol 111(1):25–33.

**Currah RS,** Smreciu EA, Hambleton S (1990) Can J Bot 68(6):1171–1181.

**Dehn B,** Schüepp H (1990) Agric Ecosyst Environ 29(1–4):79–83. **de Miranda JCC,** Harris PJ, Wild A (1989) New Phytol 112(3):405–410. **Diaz G,** Honrubia M (1990) Agric Ecosyst Environ 29(1–4):85–89. **Dijk E** (1990) Agric Ecosyst Environ 29(1–4):91–97. **Dixon RK** (1990) Agric Ecosyst Environ 29(1–4):103–106. **Dodd JC,** Jeffries P (1989a) Biol Fertil Soils 7(2):120–128. **Dodd JC,** Jeffries P (1989b) 7(2):113–119. **Dodd JC,** Jeffries P (1989c) Soil Biol Biochem 21(3):453–455. **Douglas GC,** Heslin MC, Reid C (1989) Can J Bot 67(7):2206–2212. **Dumas E,** Gianinazzi-Pearson V, Gianinazzi S (1990) Agric Ecosyst Environ 29(1–4):111–114. **Duvert P,** Perrin R, Plenchette C (1990) Plant Soil 124(1):1–6.

**El-Atrach F,** Vierheilig H, Ocampo JA (1989) Soil Biol Biochem 21(1):161–163. **Evans DG,** Miller MH (1990) New Phytol 114(1):65–71.

**Fabig B,** Moawad A, Achtnich W (1989) Z Pflanzenernähr Bodenkd 152(3):255–259. **Fabig B,** Vielhauer K, Moawad A, Achtnich W (1989b) Z Planzenernähr Bodenkd 152(3):261–265. **Fiske ML,** Thompson JP, Clewlett TG (1989) Australas Plant Pathol 18(2): 25–25. **Fitter AH** (1990) Agric Ecosyst Environ 29(1–4):137–151. **Francis R,** Hameed KM (1988) ZANCO 1(1):19–30.

**Gange AC,** Brown VK, Farmer LM (1990) New Phytol 115(1):85–91. **Garcia-Romera I,** Garcia-Garrido JM, Martinez-Molina E, Ocampo JA (1990) Soil Biol Biochem 22(2):149–152. **Gemma JN,** Koske RE, Carreiro M (1989) Mycol Res 92(3):317–321. **Giovannetti M,** Lioi L, Picci L (1990) Agric Ecosyst Environ 29(1–4):169–172. **Gnekow MA,** Marschner H (1989) Plant Soil 119(2):285–293. **Graham JH,** Syvertsen JP (1989) New Phytol 113(1):29–36. **Griffioen WAJ,** Ernst WHO (1990) Ecosyst Environ 29(1–4):173–177. **Gupta SL** (1990) Trends in mycorrhizal research. Proc Natl Conf Mycorrhiza . Haryana Agric Univ, Hisar, India, Feb 14–16.

**Harinikumar KM,** Bagyaraj DJ (1989) Biol Fertil Soils 7(2):173–175. **Harley JL** (1989) Mycol Res 92(2):129–139. **Heijne B,** Heil GW, van Dam D (1990) Agric Ecosyst Environ 29(1–4):187–192. **Hetrick BAD,** Wilson GWT, Todd TC (1990) Can J Bot 68(3):461–467. **Hung LL,** Sylvia DM, O'Keefe DM (1990) Soil Sci Soc Am J 54(3):762–768.

**Indi DV,** Konde BK, Sonar KR (1990) Trends in mycological research. Proc Natl Conf Mycorrhiza. Haryana Agric Univ, Hisar, India, Feb 14–16.

**Jabji-Hare SH** (1988) Mycologia 80(5):622–629. **Jabji-Hare SH,** Therien J, Charest PM (1990) New Phytol 114(3):481–496. **Jakobsen I,** Rosendahl L (1990) New Phytol 115(1):77–83. **Jalali BL** (1990) Trends of mycorrhizal research. Proc Natl Conf Mycorrhiza. Haryana Agric Univ, Hisar, India, Feb 14–16. **Jasper DA,** Abbott LK, Robson AD (1989) Plant Soil 115(1):99–108. **Jayachandran K,** Schwab AP, Hetrick BAD (1989) Soil Sci Soc Am J 53(6):1701–1706.

**Kamana K,** Mishra DP, Johri BN (1990) Current trends in mycorrhizal research. Proc Natl Conf Mycorrhiza. Haryana Agric Univ, Hisar, India, Feb 14–16. **Khaliel AS** (1989) Trans Mycol Soc Jpn 30(2):231–237. **Kiran B,** Rao AV, Tarafdar JC (1989) Arid Soil Res Rehabil 3(3):391–396. **Knight WG,** Allen MF, Jurinak JJ, Dudley LM (1989) Soil Sci Am J 53(4):1075–1082. **Koide R,** Elliott G (1989) Funct Ecol 3(2):252–255. **Koide RT,** Li MG (1990) New Phytol 114(1):59–64. **Koomen I,** McGrath SP (1990) Soil Biochem 22(6):871–873. **Koske RE,** Halvorson WL (1989) Mycologia 81(6):927–933. **Kough JL,** Lindermann RG (1986) Soil Biol Biochem 18(3):309–313. **Krikun J,** Haas JH, Dodd J (1990) Plant Soil 122(2):213–217.

**Lakshman HC,** Raghavendra S (1990) Current trends in mycorrhizal research. Proc Natl Conf Mycorrhiza. Haryana Agric Univ, Hilsar, India, Feb 14–16. **Lambais MR,** Cardoso EJBN (1989) Rev Bras Ciência do Solo 13(2):151–154. **Land S,** Dauck H, von Alten H (1990) Agric Ecosyst Environ 29(1–4):217–224. **Leake JR,** Read DJ (1990a) Agric Ecosyst Environ 29(1–4):225–236. **Leake JR,** Read DJ (1990b) Mycol Res 94(7):993–995. **Leake JR,** Shaw G, Read DJ (1990) Agric Ecosyst Environ 29(1–4):237–250. **Lin XG,** Hao WY (1989) Acta Bot Sin 31(9):721–725. **Lindén**

**BR** (1988) Karstenia 28(1):19–25. **Lodge DJ,** Wentworth TR (1990) Oikos 57(3):347–356. **Logan VS,** Clarke PJ, Allaway WG (1989) Aust J Plant Physiol 16(1):141–146. **Longdale AR,** Read DJ (1990) Agric Ecosyst Environ 28(1–4):285–291.

**Manske GGB** (1990) Agric Ecosyst Environ 29(1–4):273–280. **Maser C,** Maser Z, Molina R (1988) J Range Manage 41(4):309–312. **Masuhara G,** Katsuya K (1989) Sci Hortic 37(4):331–337. **Masuhara G,** Kimura S, Katsuya K (1988) Trans Mycol Soc Jpn 29(1):25–31. **Mathew J,** Johri BN (1989) Mycol Res 92(4):491–493. **Mayer C,** Venter F, Dehne HW (1989) Bodenkultur 40(1):37–45. **Mayr R,** Godoy R (1990) Agric Ecosyst Environ 29(1–4):281–288. **Mazzitelli M,** Schubert A (1990) 29(1–4):289–293. **MgGee PA,** Smith SE (1990) Mycol Res 94(3):305–308. **McGonigle TP,** Fitter AH (1990) Mycol Res 94(1):120–122. **Mercy MA,** Shrivashankar G, Bagyaraj DJ (1990) Plant Soil 121(2):292–294. **Michelini S,** Chinnery LE, Thomas JP (1989) Acta Hortic 239:427–430. **Michelsen A,** Rosendahl S (1990) Agric Ecosyst Environ 29(1–4):295–301. **Miller DD,** Bodmer M, Schüepp H (1989) New Phytol 111(1):51–59. **Miller RM,** Jastrow JD (1990) Soil Biol Biochem 22(5):579–584. **Mohan S,** Tilak KVBR (1990) Current trends in mycorrhizal research. Proc Natl Conf Mycorrhiza. Haryana Agric Univ, Hilsar, India, Feb 14–16. **Morandi D** (1989) Plant Physiol Biochem 27(5):697–701. **Mortia A,** Konishi S (1989) Soil Sci Plant Nutr 35(1):139–143. **Morton JB** (1988) Mycotaxon 32:267–324. **Morton JB** (1990) 37:493–515. **Morton JB,** Benny GL (1990) Mycotaxon 37:471–491. **Muromtsev GS,** Marshunova GN, Yakobi L (1990) Agric Ecosyst Environ 29(1–4):307–310.

**Narayanan R,** Rangarajan M, Kandasamy D (1990) Current trends in mycorrhizal research. Proc Natl Conf Mycorrhiza. Haryana Agric Univ, Hilsar, India, Feb 14–16. **Newman EI** (1988) Adv Ecol Res 18:243–270. **Newman EI,** Eason WR (1989) Plant Soil 115(2):211–215.

**Paula MA,** Siqueira JO (1990) New Phytol 115(1):69–75. **Plenchette C,** Perrin R, Duvert P (1989) Can J Bot 67(1):112–115. **Price NS,** Roncadori RW, Hussey RS (1989) New Phytol 111(1):61–66.

**Rabatin SC,** Stinner BR (1990) Agric Ecosyst Environ 27(1–4):195–204. **Rajashekhara E,** Sreenivasa MN, Rhat RS (1989) Karnataka J Agric Sci 2(3):231–233. **Raju PS,** Clark RB, Ellis JR, Maranville JW (1990) Plant Soil 121(2):165–170. **Rosendahl S** (1989) Opera Bot 100:215–223. **Rosendahl S,** Sen R, Hepper C, Azcon-Aguilar C (1989) Soil Biol Biochem 21(4):519–522. **Rosendahl S,** Rosendahl CN, Søchting U (1990) Agric Ecosyst Environ 29(1–4):329–335. **Ryan CS,** Vail WJ, Riley RK (1987) Proc W V Acad Sci 59(2–3):80–85.

**Sainz MJ,** Vilariño A, Arines J (1990) Agric Ecosyst Environ 29(1–4):337–340. **Salmia A** (1989) Ann Bot Fenn 26(1):15–26. **Schubert A,** Martinelli A (1990) Agric Ecosyst Environ 29(1–4):349–353. **Sen R,** Hepper CM, Azcon-Guilar C (1990) Agric Ecosyst Environ 29(1–4):355–395. **Sengupta A,** Chaudhuri S (1990) Plant Soil 122(1):111–113. **Shankar A,** Mathew J, Neeraj, Haur R, Mehrotra RS, Verma A (1990) Current trends in mycorrhizal research. Proc Natl Conf Mycorrhiza. Haryana Agric Univ, Hilsar, India, Feb 14–16. **Shaw G,** Read DJ (1989) New Phytol 113(4):529–533. **Sieverding E** (1991) Vesicular-Arbuscular Mycorrhiza Management in Tropical Agrosystems. Deutsche Gesellschaft für Technische Zusammenarbeit (GTZ) GmbH. Boemer Verlag, Friedland and TZ-Verlagsgesellschaft, Rossdorf. **Sieverding E,** Toro TS (1990) Agric Ecosyst Environ 29(1–4):397–401. **Singh RP,** Chhabra ML, Jalali BL (1990a) Trends in mycorrhizal research. Proc Natl Conf Mycorrhiza. Haryana Agric Univ, Hilsar, India, Feb 14–16. **Singh RS,** Hardip S, Singh OS (1990b) Trends in mycorrhizal research. Proc Natl Conf Mycorrhiza. Haryana Agric Univ, Hilsar, India, Feb 14–16. **Smith SE,** Gianinazzi-Pearson V (1990) Aust J Plant Physiol 17(2):177–188. **Smreciu EA,** Currah RS (1989) Lindleyana 4(1):6–15. **Snetselaar KM,** Whitney KD (1990) Can J Bot 68(3):533–543. **Spanu P,** Boller T, Ludwig A, Wiemken A, Faccio A, Bonfante-Fasolo P (1989) Planta 177(4):447–455. **Spanu P,** Bonfante-Fasolo P, Boller T (1990) Agric Ecosyst Environ 29(1–4):409–413. **Spokes JR,** Hayman DS, Kandasamy D (1989) Ann Appl Biol 115(2):237–241. **Stone JA,** Buttery BR (1989) Can J Soil Sci 69(1):165–169. **Straker CJ** (1989) Veld Flora 75(1):22–24. **Straker CJ,** Gianinazzi-Pearson V,

Gianinazzi S, Cleyet-Marcel JC, Bousquet N (1989) New Phytol 111(2):215–221. **Stubblefield SP,** Taylor TN, Seymour RL (1987) Mycologia 79(6):905–906. **Sukhada M** (1990) Current trends in mycorrhizal research. Proc Natl Conf Mycorrhiza. Haryana Agric Univ, Hilsar, India, Feb 14–16.

**Tang EQ** et al. (1989) Acta Agric Univ Jilinensis 11(1):39–40. **Tarafdar JC,** Rao AV (1990); **Thapar HS,** Kamla U (1990); **Tilak KVBR,** Dwivedi (1990) Current trends in mycorrhizal research. Proc Natl Conf Mycorrhiza. Haryana Agric Univ, Hilsar, India, Feb 14–16. **Tilak KVBR,** Li CY, Ho I (1989) Plant Soil 116(2):286–288.

**van Duin WE,** Rozema J, Ernst WHO (1990) Agric Ecosyst Environ 29(1–4):107–110. **Vierheilig H,** Ocampo JA (1990) Agric Ecosyst Environ 29(1–4):439–442. **Volkmar KM,** Woodbury W (1989) Can J Bot 67(6):1702–1707.

**Wang GM,** Coleman DC, Freckman DW (1989) New Phytol 112(4):489–493. **Warcup JH** (1988) Aust J Bot 36(4):461–472. **Wilkinson KG,** Dixon KW, Sivasithamparam K (1989) New Phytol 112(3):429–435. **Wilson GWT,** Hetrick BAD, Kitt DG (1989) Can J Bot 67(1):18–23.

**Yaha AI,** Abdul-Rhiba HA, Munam BH, Al-Nassiri AS (1987) J Agric Water Resour Res Soil Water Resour 6(3):43–56.

Professor Dr. Fritz Schönbeck
Ingo Raschen
Institut für Pflanzenkrankheiten und Pflanzenschutz
der Universität Hannover
Herrenhäuser Straße 2
W-3000 Hannover 21, Germany

# Subject Index

Printing: Druckhaus Beltz, Hemsbach
Binding: Buchbinderei Kränkl, Heppenheim